T0329320

Translating Gene Therapy to the Clinic
Techniques and Approaches

Translating Gene Therapy to the Clinic
Techniques and Approaches

Edited by

Jeffrey Laurence, M.D.
Weill Cornell Medical College
Payson Pavilion, NY, USA

Michael Franklin, M.S.
University of Minnesota
Minneapolis, MN, USA

AMSTERDAM • BOSTON • HEIDELBERG • LONDON • NEW YORK • OXFORD • PARIS
SAN DIEGO • SAN FRANCISCO • SINGAPORE • SYDNEY • TOKYO

Academic Press is an imprint of Elsevier

Academic Press is an imprint of Elsevier
32 Jamestown Road, London NW1 7BY, UK
525 B Street, Suite 1800, San Diego, CA 92101-4495, USA
225 Wyman Street, Waltham, MA 02451, USA
The Boulevard, Langford Lane, Kidlington, Oxford OX5 1GB, UK

British Library Cataloguing-in-Publication Data
A catalogue record for this book is available from the British Library

Library of Congress Cataloging-in-Publication Data
A catalog record for this book is available from the Library of Congress

ISBN: 978-0-12-800563-7

For information on all Academic Press publications
visit our website at http://store.elsevier.com/

Typeset by TNQ Books and Journals
www.tnq.co.in

Printed and bound in United States of America

Working together
to grow libraries in
developing countries

www.elsevier.com • www.bookaid.org

Contents

12. Gene Therapy for Hemoglobinopathies: Progress and Challenges

Alisa Dong, Laura Breda and Stefano Rivella

13. Hemophilia Gene Therapy

Christopher E. Walsh

14. Gene Transfer for Clinical Congestive Heart Failure

Tong Tang and H. Kirk Hammond

15. Gene Therapy for the Prevention of Vein Graft Disease

Sarah B. Mueller and Christopher D. Kontos

16. Gene Therapy in Cystic Fibrosis

Michelle Prickett and Manu Jain

17. Genetic Engineering of Oncolytic Viruses for Cancer Therapy

Manish R. Patel and Robert A. Kratzke

18. T Cell-Based Gene Therapy of Cancer

Saar Gill and Michael Kalos

19. Current Status of Gene Therapy for Brain Tumors

Jianfang Ning and Samuel D. Rabkin

Preface

In the summer of 1968 two staff scientists at Oak Ridge National Laboratory, Stanfield Rogers and Peter Pfuderer, suggested in a letter to *Nature* that "viral RNA or DNA information" could be used in the control of genetic deficiency diseases as well as nonheritable disorders such as cancer.[1] Their proposition was based, like many scientific breakthroughs, on an experiment of nature: the observation that circulating arginine levels are elevated in humans following infection with Shope papilloma virus, which was thought to induce a virus-specific arginase. Theirs was a prescient thought, borne on the eve of the birth of recombinant DNA technology. But it took some four decades to begin realizing its promise.

The first approved use of gene therapy occurred in 1991 under the direction of W. French Anderson. Ashanti DeSilva was a four year old girl with the enzyme-based immune deficiency disorder ADA-SCID. Retrovirus-mediated transfer of an adenosine deaminase gene into her autologous T cells led to a clinical response, albeit incomplete and temporary.[2] This was followed by a "loss of innocence" attendant on the treatment, in 1999, of Jesse Gelsinger, an 18 year old man with ornithine transcarbamylase deficiency.[3] He died as a consequence of an adenovirus vector-associated inflammatory process. Shortly there after five infants with SCID-X1 developed acute leukemia after receiving a murine retrovirus-based gene therapy to replace a defective interleukin 2 receptor H chain.[3] But then only one of those five patients died from their leukemia. And with a final enrollment of 20 SCID-X1 infants, and correction of severe immune deficiency in 17 of them over a median follow up of 9 years, gene therapy was finally established as a realistic therapeutic for those without alternatives.[4]

Since that time over 1800 gene therapy trials in 31 countries have been initiated or completed.[4] And the field's promise is not restricted to "simple" replacement or excision of a defective gene. For example, genetic engineering techniques have been used to inculcate tumor recognition or virus resistance in autologous lymphocytes of patients with metastatic cancer[5] and advanced AIDS.[6] Although there are currently no U.S. FDA licensed gene therapy products, in 2012 Glybera (alipogene tiparvovec) became the first example of this technology to be approved for clinical use, in Europe, after its endorsement by the European Medicines Agency.[7] Based on an adeno-associated virus type 2 (AAV2) vector, Glybera corrected a defect in the lipoprotein lipase gene that otherwise leads to severe pancreatitis. Like most new technologies Glybera is expensive—about $1.6 million per patient—partially related to the ultra-orphan nature of the target disorder. (There are only a few hundred cases annually in the resource rich world.[7,8]) But its clinical success, as well as preliminary data from phase 1/2 and phase 3 clinical trials for more common conditions, as outlined in our text, has led to an explosion in commercial interest; between 2013 and early 2014 US companies have invested some $600 million in gene therapy research.[9]

The text you are about to explore is an introduction for the translational and basic researcher as well as the clinician to the vast field of gene therapy technology. It is the first book in a new series, *Advances in Translational Medicine*. The project is a direct outgrowth of our editing of an illustrious journal, *Translational Research, The Journal of Laboratory and Clinical Medicine*. It is coincident with the journal's celebration of a legacy of 100 years in the promotion of excellence in clinical and translational research. This first volume is also a perfect opportunity to congratulate the Central Society for Clinical and Translational Research (CSCTR), a key partner with the journal. Albeit technically only in its 87th year, CSCTR traces its heritage to the Central Interurban Clinical Club, the establishment of which, in 1919, places it not far off the 100-year mark. Its fundamental goals, shared by our journal and this series, are critical and constant. Above all, champion the young investigator, bring in new ideas, establish diverse collaborations, and limit inbreeding. The special topics issues published annually in *Translational Research* are highly quoted. They achieved sufficient notice that the book division of Elsevier, publisher of our journal, began this series based upon expanded versions of our special issues and invited reviews.

Early on, the national importance of our society was well recognized. It also had an unanticipated impact on gene therapy related issues. The policies of genetic modification in clinical trials are regulated by the Declaration of Helsinki. And in 1966, only four societies were requested to endorse this declaration relating to "ethical principles for medical research involving human subjects"; the American Medical Association, the American Society for Clinical Investigation, and the American Federation for Clinical Research joined us. This declaration, along with the 2001 HUGO (Human Genome

Organization) consensus, covers the types of somatic gene therapies discussed in our text. Germline gene approaches by which gametes (sperm or ova) are modified, permitting a therapeutic manipulation to be passed on to future generations, are proscribed for ethical reasons in many countries, and are not covered here.

The authors of the following chapters are leaders in the field of gene therapy. They cover a range of topics and technologies with a depth and clarity to be commended, providing helpful illustrations and comprehensive citations to the literature. Several chapters focus on specific diseases, while others cover new technologies or barriers to progress. It strives to cover, in depth, disease-specific areas of particular promise. Its initial focus is on mechanisms of introducing a gene, generally via a viral vector, that either: (1) causes a protein to be expressed in a patient with a defective protein product, or two little of the normal one; or (2) introduces editing genes, "molecular scissors" that excise or disrupt genes causing a disease. As the field has evolved to encompass non-DNA-based technologies, utilizing antisense oligonucleotides, small interfering RNAs, etc. that do not alter a gene, but directly interact with RNAs or proteins, are also presented here.

These chapters also provide roadmaps to the ontogeny of current gene therapy trials and methods by which a group might design their own. I have borrowed a recently published patient-centered approach to designing a gene therapy for epilepsy[10] as an example of how the introductory chapters of this text set the stage for strategies to tackle your own areas of therapeutic interest.

1. Choose an animal model that accurately reflects the clinical problem in which to conduct preclinical studies.
2. Decide on a therapeutic approach. This is simpler when a single-gene defect is involved, limiting a functional protein product correctable by a relatively small increase in that product, as in hemophilia B. In a complex phenomenon such as epilepsy, one needs to decide if the target might best be decreasing neuronal excitation or increasing neuronal inhibitory pathways. Targeting of an entire cohort of genes could be contemplated.[10]
3. Choose a safe, effective vector. At the moment this usually means AAV, in which case limited payload size is a major impediment, or a lentivirus. But retrovirus, adenovirus, herpes simplex virus, plasmid, and other transport systems are also in various stages of clinical testing, and are outlined herein.
4. Consider all potential obstacles and explore them. Our text considers issues of payload toxicity, vector targeting, sufficiency of gene product expression, and the limits of in vitro and animal models. It also touches upon potential regulatory issues and good manufacturing-practice costs, but related details are left to other sources. For example, the American Society of Gene & Cell Therapy offers Web sites with information on issues related to the conduct of clinical gene therapy trials and the regulatory issues they raise.

This book provides coverage of the full spectrum of scientific and clinical progress, emphasizing new approaches in the field that currently have the greatest therapeutic application or potential and those areas most in need of future research. Serving both as an introduction to the field of gene therapy and as a general reference, it should prove an invaluable resource for both the expert and new investigator entering the field, as well as the clinician considering enrolling patients in clinical trials.

Jeffrey Laurence, M.D.
August 2014

REFERENCES

1. Rogers S, Pfuderer P. Use of viruses as carriers of added genetic information. *Nature* 1968;**219**:749–51.
2. Blaese RM, Culver KW, Miller AD, et al. T lymphocyte-directed gene therapy for ADA-SCID: initial trial results after 4 years. *Science* 1995;**270**:475–80.
3. Sheridan C. Gene therapy finds its niche. *Nat Biotech* 2011;**29**:121–8.
4. Ginn SL, Alexander IE, Edelstein ML, Abedi MR, Wixon J. Gene therapy clinical trials worldwide to 2012—an update. *J Gene Med* 2013;**15**:65–77.
5. Robbins PF, Morgan RA, Feldman SA, et al. Tumor regression in patients with metastatic synovial cell sarcoma and melanoma using genetically engineered lymphocytes reactive with NY-ESO-1. *J Clin Oncol* 2011;**29**:917–24.
6. Tebas P, Stein D, Tang WW, et al. Gene editing of CCR5 in autologous CD4 T cells of persons infected with HIV. *N Engl J Med* 2014;**370**:901–10.
7. Moran N. First gene therapy Glybera (finally) gets EMA approval. www.bioworld.com [accessed 14.08.14].
8. Wilson JM. Glybera's story mirrors that of gene therapy. *GENews* 2013;**33**(1).
9. Herper M. Gene therapy's big comeback. *Forbes* March 26, 2014.
10. Kullmann DM, Schorge S, Walker MC, Wykes RC. Gene therapy in epilepsy—is it time for clinical trials? *Nat Rev Neurol* 2014;**10**:300–4.

About the Editors

Dr. Jeffrey Laurence is Professor of Medicine at Weill Cornell Medical College, Attending Physician at New York Presbyterian Hospital, and Director of the Laboratory for AIDS Virus Research at those institutions. He is also Senior Scientist for Programs at amfAR, The Foundation for AIDS Research, co-founded by Dr. Mathilde Krim and Dame Elizabeth Taylor, and Editor-in-Chief of AIDS Patient Care and STDs and Translational Research (formerly the Journal of Laboratory and Clinical Medicine).

Dr. Laurence received his B.A. Phi Beta Kappa, summa cum laude from Columbia University in 1972, and his M.D. with honors from the University of Chicago Pritzker School of Medicine in 1976. He was elected a Rhodes Scholar to Oxford University in 1973. Deferring this honor, he accepted a Henry Luce Fellowship to Japan, where he worked at the Institute for Cancer Research in Osaka from 1974–1975. Dr. Laurence returned to New York to complete a residency in internal medicine and fellowship in hematology-oncology at The New York Hospital, followed by a research fellowship in immunology at The Rockefeller University.

His work focuses on the mechanisms by which HIV and antiretroviral drugs used in its treatment affect endothelial cells and osteoclasts, in models for thrombosis, cardiovascular disease and osteoporosis linked to HIV. As an outgrowth of this research he is has a long-standing interest in exploring thrombotic microangiopathies associated with complement activating disorders.

Dr. Laurence is a member of several national and international AIDS organizations, and recently received an "Award of Vision" from the Red Ribbon AIDS Foundation. He is also a recipient of the Clinician-Scientist Award of the American Heart Association and the William S. Paley Fellowship in Academic Medicine, and is an elected Fellow of the New York Academy of Sciences and a member of the American Society for Clinical Investigation. He has 3 children and lives in Greenwich, CT.

Michael Franklin, MS is a medical editor in the Division of Hematology, Oncology, and Transplantation (HOT) at the University of Minnesota. He earned his M.S. in science journalism with honors from Boston University in 2000. While in Boston, Mr. Franklin interned at the *Harvard Health Letter* at Harvard Medical School and *Boston Review* at Massachusetts Institute of Technology. He was also a staff writer for *The Daily Free Press*, the Boston University school newspaper, and contributing editor for *Stellwagen Soundings*, a newsletter covering the Stellwagen Bank National Marine Sanctuary. Shortly after graduation, Mr. Franklin began his editorial career as the Managing Editor of *The Journal of Laboratory and Clinical Medicine*, subsequently retitled *Translational Research* in 2006. During his tenure as Managing Editor, Mr. Franklin developed an interest in publication ethics while mediating breaches of scientific misconduct involving authors of the journal. He has written about publication ethics for *Minnesota Medicine* and teaches a seminar on the topic to HOT trainees. Michael has a long-standing interest in the history of science, specifically the history of experimental discoveries in chemistry and medicine, and how scientific reasoning works as an engine of human knowledge. Since becoming a medical editor in 2006, Michael has written or edited hundreds of research reports, grant proposals, book chapters, reviews, educational curricula, and other science-related material for clinicians and scientists. He regularly teaches seminars on writing in the sciences, designing visual displays of data, and how to read a journal article to HOT faculty and/or trainees, as well as other groups, including the *Association of Multicultural Scientists* and the *North Central Chapter* of the *American Medical Writers Association* (AMWA). He served as President of the *North Central Chapter* from 2010–2011 and has been a member of AMWA since 2007. When not writing or editing, Michael spends time with his partner, two daughters, and dog in Minnetonka, Minnesota.

Contributors

Mathew G. Angelos Department of Medicine, University of Minnesota, Minneapolis, MN, USA; Stem Cell Institute, University of Minnesota, Minneapolis, MN, USA

Jacopo Baglieri Department of Neurology, David Geffen School of Medicine, University of California Los Angeles, CA, USA

Carmen Bertoni Department of Neurology, David Geffen School of Medicine, University of California Los Angeles, CA, USA

Laura Breda Department of Pediatrics, Division of Hematology-Oncology, Weill Cornell Medical College, New York, NY, USA

Hildegard Büning Center for Molecular Medicine Cologne (CMMC), University of Cologne, Cologne, Germany; German Center for Infection Research (DZIF), Partner Site Bonn-Cologne, Germany; Department I of Internal Medicine, University Hospital of Cologne, Cologne, Germany

Lawrence Chan Diabetes Research Center, Houston, TX, USA; Division of Diabetes, Endocrinology and Metabolism, Department of Medicine, Baylor College of Medicine, Houston, TX, USA; Department of Molecular and Cellular Biology, Baylor College of Medicine, Houston, TX, USA

Wenhao Chen Diabetes Research Center, Division of Diabetes, Endocrinology, and Metabolism, Departments of Medicine and Molecular & Cellular Biology, Baylor College of Medicine, Houston, TX, USA

Laurence J.N. Cooper Division of Pediatrics and Department of Immunology, University of Texas MD Anderson Cancer Center, Houston, TX, USA

Alisa Dong Department of Pediatrics, Division of Hematology-Oncology, Weill Cornell Medical College, New York, NY, USA

Christopher H. Evans Mayo Clinic, Rehabilitation Medicine Research Center, Rochester, MN, USA

Charles A. Gersbach Department of Biomedical Engineering, Duke University, Durham, North Carolina, USA; Institute for Genome Sciences and Policy, Duke University, Durham, North Carolina, USA; Department of Orthopedic Surgery, Duke University Medical Center, Durham, North Carolina, USA

Steven C. Ghivizzani Department of Orthopedics and Rehabilitation, University of Florida College of Medicine, Gainesville, FL, USA

Saar Gill Division of Hematology-Oncology, Department of Medicine, University of Pennsylvania, Perelman School of Medicine, Philadelphia, PA, USA

Joseph C. Glorioso Department of Microbiology and Molecular Genetics, University of Pittsburgh School of Medicine, Pittsburgh, PA, USA

William F. Goins Department of Microbiology and Molecular Genetics, University of Pittsburgh School of Medicine, Pittsburgh, PA, USA

Perry B. Hackett Department of Genetics, Cell Biology and Development, Center for Genome Engineering and Masonic Cancer Center, University of Minnesota, Minneapolis, MN, USA

H. Kirk Hammond Department of Medicine, University of California, San Diego, La Jolla, CA, USA; VA San Diego Healthcare System, San Diego, CA, USA

Manu Jain Division of Pulmonary and Critical Care Medicine, Department of Medicine, Feinberg School of Medicine, Northwestern University, Chicago, IL, USA

Michael Kalos Department of Pathology and Laboratory Medicine, University of Pennsylvania, Perelman School of Medicine, Philadelphia, PA, USA

Dan S. Kaufman Department of Medicine, University of Minnesota, Minneapolis, MN, USA; Stem Cell Institute, University of Minnesota, Minneapolis, MN, USA

Fahad Kidwai Stem Cell Institute, University of Minnesota, Minneapolis, MN, USA; Department of Diagnostic and Biological Sciences, School of Dentistry, University of Minnesota, Minneapolis, MN, USA

Christopher D. Kontos Medical Scientist Training Program, Duke University School of Medicine; Department of Pharmacology and Cancer Biology; Department of Medicine, Duke University Medical Center, Durham, NC, USA

Robert A. Kratzke Department of Medicine, Division of Hematology, Oncology, and Transplantation, University of Minnesota Medical School, Minneapolis, USA

Robert E. MacLaren Nuffield Laboratory of Ophthalmology, Department of Clinical Neurosciences, University of Oxford, Oxford, UK; NIHR Biomedical Research Centre, Oxford Eye Hospital, Oxford, UK; Moorfields Eye Hospital & NIHR Biomedical Research Centre for Opthalmology, London, UK

Michelle E. McClements Nuffield Laboratory of Ophthalmology, Department of Clinical Neurosciences, University of Oxford, Oxford, UK; NIHR Biomedical Research Centre, Oxford Eye Hospital, Oxford, UK; Moorfields Eye Hospital & NIHR Biomedical Research Centre for Opthalmology, London, UK

Federico Mingozzi University Pierre and Marie Curie Paris, Institute of Myology, Paris, France; Genethon, Evry, France

Sarah B. Mueller Medical Scientist Training Program, Duke University School of Medicine; Department of Pharmacology and Cancer Biology; Department of Medicine, Duke University Medical Center, Durham, NC, USA

Jianfang Ning Department of Neurosurgery, Molecular Neurosurgery Laboratory, Brain Tumor Research Center, Massachusetts General Hospital and Harvard Medical School, Boston, MA, USA

Manish R. Patel Department of Medicine, Division of Hematology, Oncology, and Transplantation, University of Minnesota Medical School, Minneapolis, USA

Michelle Prickett Division of Pulmonary and Critical Care Medicine, Department of Medicine, Feinberg School of Medicine, Northwestern University, Chicago, IL, USA

Samuel D. Rabkin Department of Neurosurgery, Molecular Neurosurgery Laboratory, Brain Tumor Research Center, Massachusetts General Hospital and Harvard Medical School, Boston, MA, USA

Stefano Rivella Department of Pediatrics, Division of Hematology-Oncology, Weill Cornell Medical College, New York, NY, USA; Department of Cell and Development Biology, Weill Cornell Medical College, New York, NY, USA

Paul D. Robbins Department of Metabolism and Aging, The Scripps Research Institute, Jupiter, FL, USA

Michele Simonato Section of Pharmacology, Department of Medical Sciences, University of Ferrara, Ferrara, Italy

Timothy K. Starr Department of Genetics, Cell Biology and Development, Center for Genome Engineering and Masonic Cancer Center, University of Minnesota, Minneapolis, MN, USA; Department of Obstetrics, Gynecology and Women's Health, Center for Genome Engineering and Masonic Cancer Center, University of Minnesota Medical School, Duluth, MN, USA

Tong Tang Zensun USA Inc., San Diego, CA, USA

Pratiksha I. Thakore Department of Biomedical Engineering, Duke University, Durham, North Carolina, USA

Jakub Tolar Stem Cell Institute, Department of Pediatrics, Division of Blood and Marrow Transplantation, University of Minnesota, Minneapolis, MN, USA

Lars U. Wahlberg NsGene A/S, Ballerup, Denmark

Christopher E. Walsh Icahn School of Medicine at Mount Sinai, New York City, NY, USA

Jie Wu Diabetes Research Center, Division of Diabetes, Endocrinology, and Metabolism, Departments of Medicine and Molecular & Cellular Biology, Baylor College of Medicine, Houston, TX, USA

Aini Xie Diabetes Research Center, Division of Diabetes, Endocrinology, and Metabolism, Departments of Medicine and Molecular & Cellular Biology, Baylor College of Medicine, Houston, TX, USA

Yisheng Yang Diabetes Research Center, Houston, TX, USA; Division of Diabetes, Endocrinology and Metabolism, Department of Medicine, Baylor College of Medicine, Houston, TX, USA

Chapter 1

Translating Genome Engineering to Survival

Jakub Tolar

Stem Cell Institute, Department of Pediatrics, Division of Blood and Marrow Transplantation, University of Minnesota, Minneapolis, MN, USA

"Daring ideas are like chessmen moved forward; they may be beaten, but they may start a winning game."

Goethe

LIST OF ABBREVIATIONS

DNA Deoxyribonucleic acid
HCT Hematopoietic cell transplantation
iPSCs Induced pluripotent stem cells
mRNA Messenger ribonucleic acid
SCID X-linked severe combined immunodeficiency
CRISPR Clustered regularly interspaced short palindromic repeats (CRISPR)/Cas9 system

1.1 ORIGINS

The process of research and its clinical translation is not just driven by doctors and scientists; they are driven by the patients themselves. In the boundaries between science and society, there is a subset of individuals with genetic disorders for whom simply living life is inherently unsafe. These patients and their families frequently view risks of therapy very differently than healthy people do. Probabilities of success or failure that would be unthinkable for a healthy person can be acceptable to those with debilitating or fatal diseases. These extreme situations, in which the possibility of benefit outweighs the significant risk, are fortunately rare but critical to clinical translation. Human suffering is the same whether it is caused by a rare or a widespread disease, but the attention of experts and commitment of society may differ between the two conditions. Because of this disparity, it is important to stress that gene therapy approaches will apply equally to the orphan diseases in which they are pioneered and to the diseases and injuries that cause much of the morbidity and mortality in the world.

Gene therapy has been on the horizon for over 30 years. Encouraged by its early promise, patients and their despairing families have had their hopes raised again and again only to be repeatedly disappointed. However, in 2014, evidence of clear success in selected diseases and the application of safer technologies have added to the intellectual mass needed to forward the process of bringing gene therapy to the clinic, thus rendering countless grim disorders treatable and ultimately curable.

The concept of gene therapy for genetic disorders has its origins in Gregor Mendel's theory that cells contain small units that are messengers of inherited characteristics[1] and Erwin Schrödinger's insight that genes are blueprints of cellular function.[2] Gene therapy is one of the most appealing theories in biomedicine because it is aimed at the cause rather than the symptoms of the disease. However, the ability to harness the potential of gene transfer or DNA repair to correct a specific genomic lesion has simultaneously been one of the most daunting concepts to put into practice, although such "corrective" mutations can occur spontaneously in the setting of somatic cell mosaicism.

Mosaicism, the naturally occurring, spontaneous restoration of function in revertant cells, has been reported in patients with several genetic diseases: the skin disease epidermolysis bullosa; the metabolic disorder hereditary tyrosinemia type I; the bone marrow failure syndromes Fanconi anemia and dyskeratosis congenita; and the primary immuno deficiencies, Wiskott–Aldrich syndrome, Bloom syndrome, X-linked severe combined immunodeficiency (SCID), and adenosine deaminase (ADA)-deficient SCID.[3–14] Therefore, in theory any genetic disease can be spontaneously corrected by a gene conversion, compensatory mutation in *cis*, or by intragenic crossover between maternal and paternal alleles with two different mutations in the same gene.[15]

Despite this knowledge, there clinically are several blocks to applying revertant mosaicism. For example, it remains unclear why in some conditions the self-correction occurs rarely and why in others it may occur repeatedly even in a single individual.[16–18] Furthermore, when self-correction does occur, it may not be clinically meaningful because the revertant event is unpredictable in the molecular mechanism of reversion (and may restore the gene function only partially) and in the location in the gene. Also relevant is the unpredictability of the type of cell in which the reversion occurs, ranging from cells and tissues relevant to the disease phenotype to those that are functionally neutral. It is also currently impossible to predict the type of cell where revertant mosaicism will occur: in stem cells with significant repopulation potential, in progenitor cells—more differentiated but still possessing a variable degree of "stemness"—or in fully differentiated postmitotic cells with limited functional effect beyond the corrected cell.

If revertant mosaicism does not supply the clinical answers we seek, then where do we look next? To see the future, it is useful to first examine the past.

1.2 SYNCHRONICITY OF DISCOVERIES

The foundations for the practice of stem cell gene therapy today and for optimism about its future effect are three unrelated discoveries made 60 years ago in the 1950s. They are the following:

1. Defining the structure of DNA.[19]
2. Inducing immunological tolerance by tissue transplantation.[20]
3. Creating induced pluripotency by removing the nucleus of an egg and putting in the nucleus of a differentiated cell.[21]

Together, these three discoveries established the science needed for gene therapy to become a reality: from understanding the arrangement of the molecules in DNA, which allowed them to be rearranged; to understanding the mechanisms of immune tolerance, which allowed for cells and tissue to be transplanted; to being able to create a pluripotent cell from a gene-corrected one, which allows these cells to be multiplied and differentiated into different lineages to work in physiologically meaningful ways throughout the body.

1.3 GENE ADDITION

The discovery of the double-helix structure of DNA by James Watson and Francis Crick is obviously important to gene therapy because it showed that the nucleotides are organized in specific pairs. The understanding of this structure, and of the mechanism by which it divided, opened the new field of genome engineering, in which genes could be cut out of the genome of one cell and spliced into the genome of another as described in this volume by Gersbach. Gene therapy without this recombinant DNA technology is unimaginable.

1.3.1 Genes as Medicine

Of the more than 1800 genetic disorders described in humans, only a small fraction can be treated and even fewer cured. At this time, we are seeing possible cures using gene therapy for sickle cell disease, thalassemia, and ADA deficiency. In a field with so many challenges, these imminent steps into clinical application remind us how important and beneficial successful methods of gene therapy will be. Even with this powerful motivation, there are formidable technological barriers to accomplishing successful gene transfer, such as crossing the cellular membrane, escaping from the endosome, moving through the nuclear membrane, and integrating into the host genome.

The first technological problem is crossing the cellular membrane, the divider between the cell and its environment, a protective barrier that is quite selective about what it will let into the cell. Large molecules such as DNA, which might compromise the cell's existing DNA, are kept outside. Should the DNA cross the membrane, it must then encounter another cellular defense, the endosome, where material from outside the cell is either broken down into harmless components to be expelled or encapsulated and sent forward for use in the cell. Should the first two obstacles be surmounted, the third and most difficult remains: how to get the cell's genome, neatly and strictly paired and sequenced, to break open and accept foreign DNA.

1.3.2 Viral Gene Therapy and the "Selfish" Transgene[22]

Many of these biological hurdles are surmounted by adapting viral vectors, which are already evolutionarily perfected to penetrate a cell and transfer DNA, for use in gene therapy.[23] This method has its own risks: reactivation of the original virus, formation of tumors or other cancers, and immune reaction to the viral components. The issues of immunity are discussed by Mingozzi in another chapter in this book.

In terms of delivering a wild-type gene into a genome with a pathogenic mutation in the same gene, gene therapy as we know it today has been most advanced by its remarkably successful application to the therapy of primary immune deficiency states, namely SCID (X-linked and ADA) and Wiscott–Aldrich syndrome. Another triumph of gene therapy has been improvement of vision in individuals with Leber congenital amaurosis.[24–27] In this volume, McClements and MacLaren review gene transfer approaches to retinal disease with special consideration of the viral serotype and regulation elements of the therapeutic transgene.

In addition to the many advantages, significant questions remain about the use of viral vectors. In the brief course of this century, close to 100 individuals with (mostly) fatal genetic disorders have been treated using autologous transplantation with gene-corrected cells that were transduced with retroviral vectors carrying the therapeutic gene. This strategy relied on delivering a functional gene along with exogenous regulatory elements that promoted sustained, high-level gene expression. However, the consequences of integrating the retroviral vector and its "selfish transgene" into the host genome included loss of physiological regulation of the treated gene and the disruption and possible dysregulation of other endogenous genes.[28–30] For example, in the French-British SCID trial, 5 of 20 individuals experienced integration of the viral vector and cargo in close proximity to an oncogene, activation of which led to clonal expansion and leukemia. Four of the five children with leukemia were successfully treated, but one died. Despite these serious complications, the overall outcome of the trial provided evidence that gene therapy is superior to the previous standard of care (i.e., bone marrow transplantation) by providing better immune function with polyclonal T-cell repertoire, increased disease-free survival (>80%), and a better quality of life.

1.3.3 Safety and Bioinformatics

The biological challenges to gene therapy have not been the only obstacles. In 1974, approximately a century after Mendel was so brilliantly right about genes as "atoms of heredity," worries about the dangers of recombinant DNA technology led research scientists to create a voluntary moratorium of all research involving genome engineering.[31] In response, the 1975 Asilomar Conference on Recombinant DNA in Pacific Grove, CA, developed and released safety recommendations under which research could proceed.[32] This attempt by scientists to address public concerns through self-regulation focused attention on the research and resulted in nearly worldwide governmental regulation. In retrospect, the constraints of these recommendations may have delayed research efforts by focusing on "dangers" that, as one of the conference's intellectual leaders commented decades later, "…were strictly conjectural (that is, you *thought* they were dangers)."[33] This emphasis on safety resulted in enormous efforts being redirected to the evaluation of off-target effects using comprehensive large-scale integration site analysis to identify common integration sites and to associate clonal dominance with malignant lymphoid or myeloid proliferation.

One of the advantages of gene transfer is its spectacular modularity: with various building blocks of the vector-cargo complex adapted to deliver to specific organ locations, to express in a tissue-specific manner, or to transduce cell types committed to cell-specific differentiation programs. Gene therapy can be designed for injuries, for systemic disorders, or for diseases limited to individual organs. In this book, Tang and Hammond provide an update on gene therapy for congestive heart disease. Frazier and Kontos cover the exciting possibilities of ex vivo gene modification of a venous graft used in revascularization bypass surgery for ischemic cardiovascular disease. Prickett and Jain summarize advances in gene transfer and gene repair of the cystic fibrosis transmembrane regulator gene, and a new perspective on the gene therapy of rheumatoid arthritis and osteoarthritis is given by Evans, Ghivizzani, and Robbins.

Looking forward, this information will be amplified by the application of bioinformatics, in which huge amounts of information can be analyzed for minute connections and more complex relationships revealed (see Figure 1.1). The current knowledge of common integration sites in clonally expanded hematopoietic cells may evolve into a "genome instability index" that takes into consideration the alteration of signal cascade maps, gene regulatory pathways, and the cellular state after transduction or transfection. Thus, this future directory would better map the neighborhoods of influence on the specific transgene and quantify the risks of gene transfer in a systematic, comprehensive way. The collective intelligence on genotoxicity has already made real the concept of large-scale, computer-automated amplification of knowledge[34] and will further advance with this kind of integrative model.[35] We are now in the early stages of computational science, which is based on modeling tremendously complex interactions among genes, cells, tissues, individuals, and societies (epidemiology), and moving steadily into international cooperation.

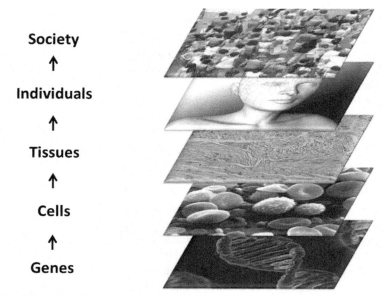

Society

↑

Individuals

↑

Tissues

↑

Cells

↑

Genes

FIGURE 1.1   Genes as medicine. When we look at gene therapy from the viewpoint of a gene itself, we may not grasp the gene in the context of a cell, a tissue, or an organism that together give the gene its physiological meaning. Bioinformatics models the complex interactions among genes, cells, tissues, individuals, and society to help pinpoint the most productive areas of research. Although experiments and clinical encounters are typically confined to one method of observation, biological systems are complex, with several interconnected organizational and functional layers. A systems approach prevents treating experimental and clinical data in a fractured and intermittent way and offers a path from mechanistic understanding of disease to a rational algorithm of effective treatment of genetic disorders.

1.3.4 Nonviral Gene Therapy

The expense associated with clinical-grade, viral-based gene therapy trials is rarely given much attention in academia, but it remains a major consideration for those who would conduct clinical trials. DNA-transposable elements are an elegant answer for the financial concerns, but their position effect after genome integration needs to be assessed with the same diligence with which viral vectors were evaluated. Hackett, Starr, and Cooper herein present a cautious, balanced view of insertional mutagenesis of transposons as a tool of discovery and as a challenge for clinical gene therapy.

1.4 FROM GENE ADDITION TO GENE EDITING

In efforts to improve the safety and efficacy of potential clinical therapies, the center of gravity for gene therapy may be shifting from gene addition (in which a whole new gene is pasted into the genome with the use of viruses or transposons) to genome editing (see Figure 1.2) (where the pathogenic mutation is corrected in its natural gene location with zinc-finger nucleases, transcription activator-like effector nucleases (TALENs), *clustered* regularly interspaced short palindromic repeats (CRISPR)/Cas9 system, or homing endonucleases).[36,37] Hybrid molecules are now being engineered to target a specific location in the genome and to introduce a break in the DNA proximal to the targeted mutation. The cleavage in the DNA is then resolved by homologous recombination between the endogenous genes and an exogenously introduced donor fragment containing the normal sequence. In this way, the pathogenic mutation is permanently changed back to the normal sequence. This has the added benefit of preserving the architecture of the genome and maintaining gene control under the normal cellular regulatory elements. Although further studies are needed, recent research gives evidence that genome editing may be highly effective and safe.[38–41]

1.5 THERAPY FOR GENETIC DISORDERS

Critical to allowing the transfer of material from one individual to another was a discovery from Peter Medawar's investigation of skin grafting. He found that tissue from one individual will be rejected by another individual because of an immune response to the donor transplant. Medawar, Billingham, and Baruch Brent[20] further showed how to manipulate the immune system to develop acquired immune tolerance. Their experiments, visually represented by a white mouse sporting a healthy spot of fur from a black mouse, demonstrated that specificity of immune response develops before birth and that a subject injected with a donor's cells in the prenatal period can later accept tissue from that donor's body. From a physiological standpoint, because

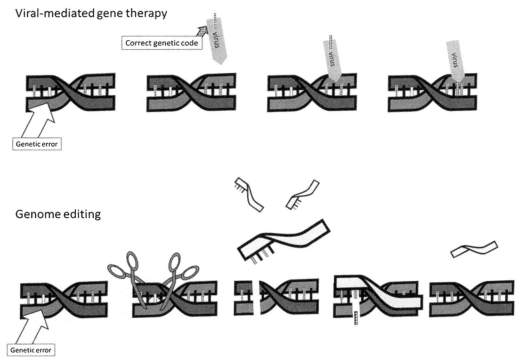

FIGURE 1.2   Gene therapy. Viral-mediated gene transfer results in the transfer of the correct genetic material, but usually in an almost random location. This added transgene may disrupt functions of other genes, potentially even causing cancer. By contrast, genome editing uses a donor template to correct the specific mutation in the genome and typically leaves nothing else behind. Thus, not only is the gene correction precisely located after gene editing but also nothing remains to disrupt normal genome function.

genes exist only in the context of a cell, the donor's cell only makes clinical sense when transplanted. Thus, immune tolerance of allogeneic wild-type or autologous gene-corrected grafts is the key to successful cell gene therapy.

1.5.1 Blood and Marrow Transplantation for Genetic Disease

Perhaps the first applied gene therapy (in a broad sense, i.e., delivery of a wild-type gene into a patient with a loss-of-function mutation in that gene) was accomplished by means of bone marrow transplantation. Pioneering work by Medawar, E. Donnell Thomas, and others[20,42–45] made it clear that transfer of allogeneic bone marrow cells can regenerate the lymphohematopoietic system. Equally impressive was the way their advances in allogeneic transplantation for malignant disease were taken in an entirely new clinical direction by Elizabeth Neufeld's discovery of enzyme cross-correction between cells derived from individuals with mucopolysaccharidosis type I (Hurler syndrome) and mucopolysaccharidosis type II (Hunter syndrome).[46] By pure coincidence, her findings were published the same month, November 1968, that Robert Good and his team performed the first successful in-human bone marrow transplant.[47,48] With this step, a new field of medicine originated—hematopoietic cell transplantation (HCT).

The concept of allogeneic transplantation as a therapy for genetic disorders has been applied to several conditions:

- Selected **enzymopathies** (e.g., mucopolysaccharidosis type I or α-mannosidosis),
- **Bone marrow failure syndromes** (e.g., osteopetrosis, Fanconi anemia, and dyskeratosis congenita),
- **Neuronopathic disorders** (e.g., adrenoleukodystrophy and metachromatic leukodystrophy),
- **Hemoglobinopathies** (e.g., sickle cell anemia and thalassemia, and those discussed by Rivella),
- **Extracellular matrix disorders** (e.g., the genodermatosis epidermolysis bullosa), and
- **Severe immune deficiencies** (e.g., X-linked SCID, ADA-deficient SCID, Wiskott–Aldrich syndrome, and chronic granulomatous disease).[49–58]

Although a life-saving measure for patients with some diseases, allogeneic HCT can result in life-threatening side effects and must be seen and understood as a radical treatment used only for otherwise deadly disorders. These complications can occur either as a result of the physical injury associated with the chemotherapy and radiation used in the preparative regimen or from the immune injury caused by the allogeneic nature of the hematopoietic graft. Injuries from the preparative regimen can present as systemic endothelial injury (sinusoid obstruction syndrome or veno-occlusive disease), renal toxicity, or pulmonary toxicity. Post-transplant immune injury leads to profound immunosuppression and, in a significant minority of HCT recipients, to an attack by the donor immune cells on the tissues of the recipient (termed graft-vs-host disease). Each of these serious and potentially fatal complications could be diminished or avoided by using the host's own cells (i.e., gene-corrected, autologous HCT).

The use of viral and adenoviral vectors, as noted in *The Selfish Transgene*, has potential complications, such as random insertional mutagenesis and triggering oncogenic processes. Another method of gene therapy involves editing the genome in a "cut and paste" manner. Various technologies (e.g., zinc finger nucleases, TALENs and CRISPRs) are all showing potential in animal models. Additional considerations include choosing the appropriate cellular vectors and creating a clinically significant number of gene-edited cells for transplant.

1.5.2 Cellular Vectors

Clearly, for the any meaningful gene therapy strategy to work, it must target the cell type relevant to the specific disease. Although tissue-specific cell types such as keratinocytes and retinal epithelial cells have been successfully gene corrected in most cases, the effects of gene correction are amplified in proportion to the proliferative, repopulating capacity of the cellular vector. For this reason, stem cells have been the target cells of choice for several gene therapy applications. The most prominent example is using hematopoietic stem cell grafts in gene therapy trials for blood disorders.[59–61] In this volume, two chapters cover the evolving knowledge of the gene therapy of nonmalignant blood disorders. Dong, Rivera, and Breda describe lentiviral gene therapy trials for thalassemia and sickle cell disease, and Walsh and Batt focus on adenoviral gene transfer and spliceosome-mediated pre-mRNA trans-splicing in hemophilia. Gene therapy for neurological disorders has not, until recently, been seen as a promising application. Simonato, Wahlberg, Goins, and Glorioso evaluate current viral gene therapies and their potential in treating pain, epilepsy, and Parkinson's disease. Bertoni outlines current gene therapy approaches to Duchenne muscular dystrophy.

1.5.3 Induced Pluripotent Stem Cells

The ability to take one kind of cell, edit the genome, and then produce another type of cell that is more suitable for therapy was made possible by the work of Sir John Gurdon.[21,62] His work, based on pioneering work by Robert Briggs and Thomas

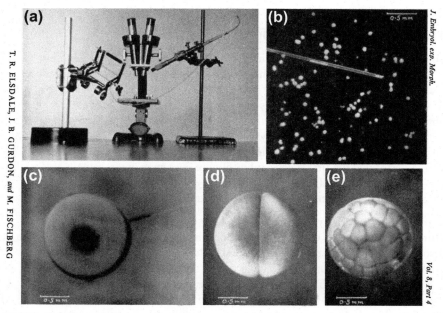

FIGURE 1.3 Simple equipment for a complex procedure. Photographs of the laboratory equipment used to successfully perform transfer of a nucleus from a mature intestinal cell into the egg cell of a frog, replacing the immature cell nucleus. The egg cell then developed into a viable tadpole, demonstrating that the DNA in all cells in the body contained the information needed to create the entire frog. (a) Low-power binocular miicroscope with Singer micromanipulator/microdissector. (b) Cells from donor embryo with micropipette. (c) Introduction of mature donor cell into frog egg. (d-e) Normal division of egg with new DNA.[64]

King,[63] established that DNA in a mature cell's nucleus contains the complete blueprint to reproduce the organism from which the cell came. This was tested by removing the nuclei from a mature cell and an immature cell. The nucleus from the mature cell was then inserted into the immature cell, and from this a complete organism (a frog) was then produced. The fantastic precision of this nuclear transfer between cells was performed using the most basic of equipment (see Figure 1.3).[64]

This basic science research later led to what is probably the most important significant development of the past decade relevant to stem cell gene therapy—the generation of induced pluripotent stem cells (iPSCs) by Shinya Yamanaka. He was able to take a mature cell committed to specific tissue function and turn back its biological clock, returning it to a pluripotent state. From this pluripotent state, the cell could be selectively differentiated into other cell types, an essential element in the design of an appropriate cellular vector.

1.5.4 Stem Cell Gene Therapy

Most stem cell therapies to date have used organ-specific stem cells for organ-specific diseases, such as hematopoietic stem cells for blood disorders or epidermal stem cells for diseases of skin. Nevertheless, since the foundational discoveries of nuclear and cell reprogramming by Briggs, King, Gurdon, Yamanaka, and Thomson, the culture of embryonic stem cells, and the bioengineering of iPSCs[21,62,63,65–71], it has been possible to think that these cell types may be useful as cellular targets for future stem cell gene therapy approaches and in regenerative medicine for disease and injury.[72] In this book, Mason and Kaufman discuss the history of iPSCs and the manipulation of iPSC genomes. This evaluation of efficacy and safety in the clinical translation of cells derived from gene-corrected iPSCs is especially important because the risk and benefits of stem cell therapy are balanced with those of gene therapy.

Disease-specific challenges will remain. For example, in DNA repair-deficient Fanconi anemia cells, the bone marrow cells for gene correction are few in number, extraordinarily intolerant to ex vivo manipulations, and at risk of accumulating pre-leukemic mutations, the effect of which can inadvertently be increased by gene correction.[73,74] On the other hand, the disease-specific challenges can also be viewed as opportunities. An example is lipoprotein lipase deficiency, in which enzyme replacement therapy is not effective, but intramuscular injection of the enzyme gene is, and the European Union's approval of alipogene tiparvovec for therapy of this disorder. This is correctly seen as encouraging news for all of gene therapy.[75] In this way, using these technologies in a rare and narrow disorder leads the way for other uses to follow more easily.

In all clinical applications of the gene corrective strategy, the cell type manipulated and delivery of gene-corrected cells has to be done in a disease-specific fashion. For example, in another clinically significant development, the immune system and its responses can be manipulated by transgenic means. In this book, three chapters deal with different aspects of the intersection of lymphohematopoiesis and gene therapy: Yang and Chan summarize efforts in gene therapy for diabetes; Gill and Kalos describe the great potential of T cell-based anticancer approaches; and Chen, Xie, Wu, and Chan discuss the immune response to pancreatic beta cells in type 1 diabetes and immunotherapies that can target this process. Related to this, Ning and Rabkin as well as Kratzke and Patel discuss the use of gene therapy with oncolytic viruses, induced suicide therapy, and immunotherapy for glioblastoma, one of the most lethal types of brain tumors.

1.6 ROADMAP TO THE FUTURE

These critical steps—the ability to modify the structure of DNA, the fact that all cells contain the blueprint for a complete organism, that cells can be regressed to a state of pluripotency, and that with appropriate immune manipulation cells from one individual can be transplanted into another—make it possible to transfer defined cells and designer genomes into whole organisms, thus completing the synthetic biology concept of stem cell gene therapy.

It should be noted that these discoveries were from areas of research almost entirely unconnected to each other: developmental biology, molecular crystallography, and tissue grafting. Individually, they answered important questions fundamental for their fields. Taken together, they complemented each other and represented a major turning point in biology and medicine. Furthermore, it was not obvious at that time how these milestones from three separate fields were related; as is often the case for any turning point in history, the consequences were recognizable only in retrospect. This may also be the case in the current development of gene therapy, in which critical discoveries are being made in different fields and by unconnected researchers.

Some areas of needed collaboration seem obvious. Improvements in vectorology, gene editing, and the delivery of the therapeutic genes into targeted cell populations need to connect with the development of other experimental therapies, such as the expansion of corrected cells with stem cell potential. These developments will likely result in patient-specific therapy, partial examples of which already exist. This indicates that an individualized medical approach is possible and that a combination of cell-based, gene, protein, and drug therapies may simultaneously be safer and more effective than any one treatment approach alone.[76–79]

The field of gene therapy started as a visionary and daring idea. The preclinical data were not as clear as one would have wished them to be—they never are—and it was only when the risks of therapy were compared to that of living with a genetic disease, or to the risks and incomplete efficacy of alternative therapies (e.g., hematopoietic cell therapy), that the first clinical trials were permitted.[80] The field has overcome several crises in which the public expectation that gene therapy would cure everything (and without side effects) has collided with the hard realities of science and medicine, but it has emerged as a broadly accepted new standard in several disorders.[81]

Most of the work done to date illustrates the irreversible expansion of basic molecular, cellular, and transplantation biologies and their enormous potential in the clinical application of gene therapy. It is important to note that although the outcomes of current basic and clinical gene therapy science are in the future, the actions that will lead to them are not. Accurate observation and meticulous experimentation are required to improve the odds of correctly guessing the biological mechanisms of action, but it is evident that the sequence of scientific discoveries and the path to the next turning point are unpredictable.

National and international coordination are needed to connect the flashes of discovery and progress. To make gene therapy widely available and used, there will need to be clinical trial coordination similar to historically transformative collaborations in pediatric oncology. These clinical trials need to be combined with real-time data exchange and evaluation of these data on the basis of scientific merit. Last, but not least, there must be a focused goal (i.e., rapid translation to a world-scale change in medical practice).

No person or institution can do this alone, and "eusocial" teams operating on the basis of collaboration rather than competition will prevail.[82] To recruit sufficient numbers of patients, to apply emerging technology advances (e.g., expansion of the gene-corrected stem cells or incorporation of an inducible "safety switch" apoptotic gene), and to harmonize pharmaceutical and ethical requirements will require multicenter coordination with a pragmatic "federation" rather than "pipeline" strategy. The complexity of genetic disorders and methods of gene therapy promise challenging and exciting times ahead.

As you can appreciate from the chapters of this book, those of us in this field are serious about our work, are all personally invested in our research, and have all had, or will have, someone we know and deeply care about stricken with a life-threatening illness that could be treated or even cured by gene therapy. We continue to focus on those clinically meaningful studies that combine persevering observations with smart experiments and that have the ultimate goal of converting individual and disease-specific insights into a collective understanding of the emerging gene transfer platforms and their subsequent translation to patient care.

CONFLICT OF INTEREST

No competing financial interests exist.

ACKNOWLEDGMENTS

The author apologizes to the authors whose work could not be quoted because of space constraints and thanks Nancy Griggs Morgan for editorial assistance. J.T. is supported by grants from the National Institutes of Health (R01 AR063070 and R01 AR059947), the Department of Defense (USAMRAA/DOD Department of the Army W81XWH-12-1-0609 and USAMRAA/DOD W81XWH-10-1-0874), DebRA International, the Jackson Gabriel Silver Fund, the Epidermolysis Bullosa Medical Research Fund, University of Pennsylvania grant MPS I-11-009-01, and Children's Cancer Research Fund, Minnesota.

REFERENCES

1. Matalova A. Published primary sources to Gregor Mendel's biography. Mendelianum of the Moravian Museum, Brno. *Folia Mendel* 1981;**16**:239–51.
2. Schrödinger E. *What is life? and other scientific essays*. Garden City: Doubleday; 1956.
3. Lo Ten Foe JR, Kwee ML, Rooimans MA, et al. Somatic mosaicism in Fanconi anemia: molecular basis and clinical significance. *Eur J Hum Genet* May–June 1997;**5**(3):137–48.
4. Stephan V, Wahn V, Le Deist F, et al. Atypical X-linked severe combined immunodeficiency due to possible spontaneous reversion of the genetic defect in T cells. *N Engl J Med* November 21, 1996;**335**(21):1563–7.
5. Jonkman MF, Scheffer H, Stulp R, et al. Revertant mosaicism in epidermolysis bullosa caused by mitotic gene conversion. *Cell* February 21, 1997;**88**(4):543–51.
6. Ellis NA, Lennon DJ, Proytcheva M, Alhadeff B, Henderson EE, German J. Somatic intragenic recombination within the mutated locus BLM can correct the high sister-chromatid exchange phenotype of Bloom syndrome cells. *Am J Hum Genet* November 1995;**57**(5):1019–27.
7. Hirschhorn R, Yang DR, Puck JM, Huie ML, Jiang CK, Kurlandsky LE. Spontaneous in vivo reversion to normal of an inherited mutation in a patient with adenosine deaminase deficiency. *Nat Genet* July 1996;**13**(3):290–5.

8. Ariga T, Yamada M, Sakiyama Y, Tatsuzawa O. A case of Wiskott–Aldrich syndrome with dual mutations in exon 10 of the WASP gene: an additional de novo one-base insertion, which restores frame shift due to an inherent one-base deletion, detected in the major population of the patient's peripheral blood lymphocytes. *Blood* July 15, 1998;**92**(2):699–701.

9. Gregory Jr JJ, Wagner JE, Verlander PC, et al. Somatic mosaicism in Fanconi anemia: evidence of genotypic reversion in lymphohematopoietic stem cells. *Proc Natl Acad Sci USA* February 27, 2001;**98**(5):2532–7.

10. Gross M, Hanenberg H, Lobitz S, et al. Reverse mosaicism in Fanconi anemia: natural gene therapy via molecular self-correction. *Cytogenet Genome Res* 2002;**98**(2–3):126–35.

11. Darling TN, Yee C, Bauer JW, Hintner H, Yancey KB. Revertant mosaicism: partial correction of a germ-line mutation in COL17A1 by a frame-restoring mutation. *J Clin Invest* May 15, 1999;**103**(10):1371–7.

12. Schuilenga-Hut PH, Scheffer H, Pas HH, Nijenhuis M, Buys CH, Jonkman MF. Partial revertant mosaicism of keratin 14 in a patient with recessive epidermolysis bullosa simplex. *J Invest Dermatol* April 2002;**118**(4):626–30.

13. Pasmooij AM, Garcia M, Escamez MJ, et al. Revertant mosaicism due to a second-site mutation in COL7A1 in a patient with recessive dystrophic epidermolysis bullosa. *J Invest Dermatol* October 2010;**130**(10):2407–11.

14. Mankad A, Taniguchi T, Cox B, et al. Natural gene therapy in monozygotic twins with Fanconi anemia. *Blood* April 15, 2006;**107**(8):3084–90.

15. Hirschhorn R. In vivo reversion to normal of inherited mutations in humans. *J Med Genet* October 2003;**40**(10):721–8.

16. Choate KA, Lu Y, Zhou J, et al. Mitotic recombination in patients with ichthyosis causes reversion of dominant mutations in KRT10. *Science* October 1, 2010;**330**(6000):94–7.

17. Pasmooij AM, Jonkman MF, Uitto J. Revertant mosaicism in heritable skin diseases: mechanisms of natural gene therapy. *Discov Med* September 2012;**14**(76):167–79.

18. Lai-Cheong JE, McGrath JA, Uitto J. Revertant mosaicism in skin: natural gene therapy. *Trends Mol Med* March 2011;**17**(3):140–8.

19. Watson JD, Crick FH. Molecular structure of nucleic acids; a structure for deoxyribose nucleic acid. *Nature* April 25, 1953;**171**(4356):737–8.

20. Billingham RE, Brent L, Medawar PB. Actively acquired tolerance of foreign cells. *Nature* October 3, 1953;**172**(4379):603–6.

21. Gurdon JB. The developmental capacity of nuclei taken from intestinal epithelium cells of feeding tadpoles. *J Embryol Exp Morphol* December 1962;**10**:622–40.

22. Dawkins R. *The selfish gene*. New York: Oxford University Press; 1976.

23. Kay MA. State-of-the-art gene-based therapies: the road ahead. *Nat Rev Genet* May 2011;**12**(5):316–28.

24. Maguire AM, Simonelli F, Pierce EA, et al. Safety and efficacy of gene transfer for Leber's congenital amaurosis. *N Engl J Med* May 22, 2008;**358**(21):2240–8.

25. Jacobson SG, Cideciyan AV, Ratnakaram R, et al. Gene therapy for leber congenital amaurosis caused by RPE65 mutations: safety and efficacy in 15 children and adults followed up to 3 years. *Arch Ophthalmol* January 2012;**130**(1):9–24.

26. Bennett J, Ashtari M, Wellman J, et al. AAV2 gene therapy readministration in three adults with congenital blindness. *Sci Transl Med* February 8, 2012;**4**(120). 120ra115.

27. Hufnagel RB, Ahmed ZM, Correa ZM, Sisk RA. Gene therapy for Leber congenital amaurosis: advances and future directions. *Graefes Arch Clin Exp Ophthalmol* August 2012;**250**(8):1117–28.

28. Hacein-Bey-Abina S, Garrigue A, Wang GP, et al. Insertional oncogenesis in 4 patients after retrovirus-mediated gene therapy of SCID-X1. *J Clin Invest*. September 2008;**118**(9):3132–42.

29. Moiani A, Paleari Y, Sartori D, et al. Lentiviral vector integration in the human genome induces alternative splicing and generates aberrant transcripts. *J Clin Invest* May 1, 2012;**122**(5):1653–66.

30. Cavazza A, Moiani A, Mavilio F. Mechanisms of retroviral integration and mutagenesis. *Hum Gene Ther* February 2013;**24**(2):119–31

31. Berg P, Baltimore D, Boyer HW, et al. Letter: potential biohazards of recombinant DNA molecules. *Science* July 26, 1974;**185**(4148):303.

32. Berg P, Baltimore D, Brenner S, Roblin RO, Singer MF. Summary statement of the Asilomar conference on recombinant DNA molecules. *Proc Natl Acad Sci USA* June 1975;**72**(6):1981–4.

33. Brenner S, Wolpert L, Friedberg EC, Lawrence E. *A life in science*. Rev. ed. London: BioMed Central; 2001.

34. Bush V. *Science–The Endless Frontier: A Report to the President on a Program for Postwar Scientific Research*. U.S. Office of Scientific Research and Development, Government Printing Office Washington, D.C. 1945.

35. Califano A, Butte AJ, Friend S, Ideker T, Schadt E. Leveraging models of cell regulation and GWAS data in integrative network-based association studies. *Nat Genet* August 2012;**44**(8):841–7.

36. Handel EM, Cathomen T. Zinc-finger nuclease based genome surgery: it's all about specificity. *Curr Gene Ther* February 2011;**11**(1):28–37.

37. Bogdanove AJ, Voytas DF. TAL effectors: customizable proteins for DNA targeting. *Science* September 30, 2011;**333**(6051):1843–6.

38. Joung JK, Sander JD. TALENs: a widely applicable technology for targeted genome editing. *Nat Rev Mol Cell Biol* January 2013;**14**(1):49–55.

39. Osborn MJ, Starker CG, McElroy AN, et al. TALEN-based gene correction for epidermolysis bullosa. *Mol Ther* June 2013;**21**(6):1151–9.

40. Rahman SH, Maeder ML, Joung JK, Cathomen T. Zinc-finger nucleases for somatic gene therapy: the next frontier. *Hum Gene Ther* August 2011;**22**(8):925–33.

41. Sebastiano V, Maeder ML, Angstman JF, et al. In situ genetic correction of the sickle cell anemia mutation in human induced pluripotent stem cells using engineered zinc finger nucleases. *Stem Cells* November 2011;**29**(11):1717–26.

42. Lorenz E, Uphoff D, Reid TR, Shelton E. Modification of irradiation injury in mice and guinea pigs by bone marrow injections. *J Natl Cancer Inst* August 1951;**12**(1):197–201.

43. Thomas ED, Lochte Jr HL, Lu WC, Ferrebee JW. Intravenous infusion of bone marrow in patients receiving radiation and chemotherapy. *N Engl J Med* September 12, 1957;**257**(11):491–6.

44. Medawar PB. *Memoir of a thinking radish : an autobiography*. Oxford (NY): Oxford University Press; 1986.

45. Appelbaum FR. Hematopoietic-cell transplantation at 50. *N Engl J Med* October 11, 2007;**357**(15):1472–5.
46. Fratantoni JC, Hall CW, Neufeld EF. Hurler and Hunter syndromes: mutual correction of the defect in cultured fibroblasts. *Science* November 1, 1968;**162**(3853):570–2.
47. Gatti RA, Meuwissen HJ, Allen HD, Hong R, Good RA. Immunological reconstitution of sex-linked lymphopenic immunological deficiency. *Lancet* December 28, 1968;**2**(7583):1366–9.
48. Meuwissen HJ, Gatti RA, Terasaki PI, Hong R, Good RA. Treatment of lymphopenic hypogammaglobulinemia and bone-marrow aplasia by transplantation of allogeneic marrow. Crucial role of histocompatibility matching. *N Engl J Med* September 25, 1969;**281**(13):691–7.
49. Mynarek M, Tolar J, Albert MH, et al. Allogeneic hematopoietic SCT for alpha-mannosidosis: an analysis of 17 patients. *Bone Marrow Transpl* March 2012;**47**(3):352–9.
50. Dietz AC, Orchard PJ, Baker KS, et al. Disease-specific hematopoietic cell transplantation: nonmyeloablative conditioning regimen for dyskeratosis congenita. *Bone Marrow Transpl* January 2011;**46**(1):98–104.
51. Orchard PJ, Tolar J. Transplant outcomes in leukodystrophies. *Semin Hematol* January 2010;**47**(1):70–8.
52. MacMillan ML, Auerbach AD, Davies SM, et al. Haematopoietic cell transplantation in patients with Fanconi anaemia using alternate donors: results of a total body irradiation dose escalation trial. *Br J Haematol* April 2000;**109**(1):121–9.
53. Tolar J, Teitelbaum SL, Orchard PJ. Osteopetrosis. *N Engl J Med* December 30, 2004;**351**(27):2839–49.
54. Orchard PJ, Blazar BR, Wagner J, Charnas L, Krivit W, Tolar J. Hematopoietic cell therapy for metabolic disease. *J Pediatr* October 2007;**151**(4):340–6.
55. Wagner JE, Ishida-Yamamoto A, McGrath JA, et al. Bone marrow transplantation for recessive dystrophic epidermolysis bullosa. *N Engl J Med* August 12, 2010;**363**(7):629–39.
56. Tolar J, Mehta PA, Walters MC. Hematopoietic cell transplantation for nonmalignant disorders. *Biol Blood Marrow Transpl* January 2012; **18**(Suppl 1):S166–71.
57. Cavazzana-Calvo M, Lagresle C, Hacein-Bey-Abina S, Fischer A. Gene therapy for severe combined immunodeficiency. *Annu Rev Med* 2005;**56**:585–602.
58. Rivat C, Santilli G, Gaspar HB, Thrasher AJ. Gene therapy for primary immunodeficiencies. *Hum Gene Ther* July 2012;**23**(7):668–75.
59. Boztug K, Schmidt M, Schwarzer A, et al. Stem-cell gene therapy for the Wiskott–Aldrich syndrome. *N Engl J Med* November 11, 2010;**363**(20):1918–27.
60. Aiuti A, Cattaneo F, Galimberti S, et al. Gene therapy for immunodeficiency due to adenosine deaminase deficiency. *N Engl J Med* January 29, 2009;**360**(5):447–58.
61. Naldini L. Ex vivo gene transfer and correction for cell-based therapies. *Nat Rev Genet* May 2011;**12**(5):301–15.
62. Gurdon JB, Uehlinger V. "Fertile" intestine nuclei. *Nature* June 18, 1966;**210**(5042):1240–1.
63. Briggs R, King TJ. Transplantation of living nuclei from blastula cells into enucleated frogs' eggs. *Proc Natl Acad Sci USA* May 1952;**38**(5):455–63.
64. Elsdale TR, Gurdon JB, Fischberg M. A description of the technique for nuclear transplantation in *Xenopus laevis*. *J Embryol Exp Morphol* December 1960;**8**:437–44.
65. Takahashi K, Yamanaka S. Induction of pluripotent stem cells from mouse embryonic and adult fibroblast cultures by defined factors. *Cell* August 25, 2006;**126**(4):663–76.
66. Thomson JA, Itskovitz-Eldor J, Shapiro SS, et al. Embryonic stem cell lines derived from human blastocysts. *Science* November 6, 1998;**282**(5391):1145–7.
67. Campbell KH, McWhir J, Ritchie WA, Wilmut I. Sheep cloned by nuclear transfer from a cultured cell line. *Nature* March 7, 1996;**380**(6569):64–6.
68. Wilmut I, Schnieke AE, McWhir J, Kind AJ, Campbell KH. Viable offspring derived from fetal and adult mammalian cells. *Nature* February 27, 1997;**385**(6619):810–3.
69. Evans M. Discovering pluripotency: 30 years of mouse embryonic stem cells. *Nat Rev Mol Cell Biol* October 2011;**12**(10):680–6.
70. Hirai H, Tani T, Katoku-Kikyo N, et al. Radical acceleration of nuclear reprogramming by chromatin remodeling with the transactivation domain of MyoD. *Stem Cells* 2011;**29**(9):1349–61.
71. Greder LV, Gupta S, Li S, et al. Analysis of endogenous Oct4 activation during induced pluripotent stem cell reprogramming using an inducible Oct4 lineage label. *Stem Cells* 2012;**30**(11):2596–601.
72. Garber K. Inducing translation. *Nat Biotechnol* 2013;**31**(6):483–6.
73. Tolar J, Adair JE, Antoniou M, et al. Stem cell gene therapy for Fanconi anemia: report from the 1st international Fanconi anemia gene therapy working group meeting. *Mol Ther* July 2011;**19**(7):1193–8.
74. Tolar J, Becker PS, Clapp DW, et al. Gene therapy for Fanconi anemia: one step closer to the clinic. *Hum Gene Ther* February 2012;**23**(2):141–4.
75. Miller N. Glybera and the future of gene therapy in the European Union. *Nat Rev Drug Discov* May 2012;**11**(5):419.
76. Boitano AE, Wang J, Romeo R, et al. Aryl hydrocarbon receptor antagonists promote the expansion of human hematopoietic stem cells. *Science* September 10, 2010;**329**(5997):1345–8.
77. Digiusto DL, Kiem HP. Current translational and clinical practices in hematopoietic cell and gene therapy. *Cytotherapy* August 2012;**14**(7):775–90.
78. Bernstein ID, Delaney C. Engineering stem cell expansion. *Cell Stem Cell* February 3, 2012;**10**(2):113–4.
79. Tolar J, Grewal SS, Bjoraker KJ, et al. Combination of enzyme replacement and hematopoietic stem cell transplantation as therapy for Hurler syndrome. *Bone Marrow Transpl* March 2008;**41**(6):531–5.
80. Prockop DJ. Repair of tissues by adult stem/progenitor cells (MSCs): controversies, myths, and changing paradigms. *Mol Ther* June 2009;**17**(6):939–46.
81. Goethe JWv, Constantine D. *The sorrows of young werther*. Oxford (NY): Oxford University Press; 2012.
82. Wilson EO. *The social conquest of Earth*. 1st ed. New York: Liveright Pub. Corporation; 2012.

Chapter 2

Pluripotent Stem Cells and Gene Therapy

Mathew G. Angelos[1,2], Fahad Kidwai[2,3] and Dan S. Kaufman[1,2]

[1]*Department of Medicine, University of Minnesota, Minneapolis, MN, USA;* [2]*Stem Cell Institute, University of Minnesota, Minneapolis, MN, USA;*
[3]*Department of Diagnostic and Biological Sciences, School of Dentistry, University of Minnesota, Minneapolis, MN, USA*

LIST OF ABBREVIATIONS

ALS Amyotrophic lateral sclerosis
bFGF Basic fibroblast growth factor
CRISPR Clustered regularly interspaced short palindromic repeats
DN Dopamine-releasing neurons
ESCs Embryonic stem cells
FA Fanconi anemia
FDiPSCs FD patient-specific iPSCs
FD Familial dysautonomia
GFP Green fluorescent protein
HAC Human artificial chromosome
hESCs Human embryonic stem cells
hiPSCs Human induced pluripotent stem cells
hPSC Human pluripotent stem cells
HLA Human leukocyte antigen
IKBKAP I-*k*-B Kinase complex-associated protein
iPSCs Induced pluripotent stem cells
IRES Internal ribosome entry site
LIF Leukemia inhibitory factor
MCDNA Minicircle DNA
mESCs Mouse embryonic stem cells
MEF Mouse embryonic fibroblast cells
MHC Major histocompatibility complex
miPSCs Murine induced pluripotent stem cells
miRNAs microRNA
NIH National Institutes of Health
6-OHDA 6-Hydroxydopamine
OSKM Oct3/4, Sox2, KLF4, and *c-myc*
PolyQ Polyglutamine
PD Parkinson's disease
PDiPSCs PD patient-specific iPSCs
PF4-FVIII-HAC Platelet factor-4 and human factor VIII gene in a human artificial chromosome vector
RiPSCs RNA-induced pluripotent stem cells
SCID Severe combined immunodeficiency
SCNT Somatic cell nuclear transfer
SCD Sickle-cell disease
STEMCCA Stem cell cassette
SeV Sendai virus
SB *Sleeping Beauty* transposon system
STAP Stimulus-triggered acquisition of pluripotency
SMA Spinal muscular atrophy
SOD1 Superoxide dismutase

Translating Gene Therapy to the Clinic. http://dx.doi.org/10.1016/B978-0-12-800563-7.00002-6

SMN1 Survival motor neuron gene
TALEN Transcription activator-like effector nucleases
TDP-43 Tar DNA-binding protein-43
ZFN Zinc finger endonucleases

2.1 GENETIC APPROACHES TO PLURIPOTENCY

Research on pluripotent stem cells has been instrumental for our understanding of basic developmental biology and genetic disease. The early approaches to establishing and maintaining pluripotency and self-renewal began in the 1960s (Box 2.1). In 2006, Takahashi and Yamanaka made perhaps the most stunning breakthrough by demonstrating that through the expression of a set of defined genes, murine cells could be reprogrammed to a pluripotent state.[16] In a series of elegant experiments, a set of 24 candidate genes encoding transcription factors that maintain pluripotency in early embryos, promote proliferation, and support an embryonic stem cell (ESC)-specific phenotype and proliferation were introduced into mouse embryonic and adult fibroblasts by retroviral transduction. Selective withdrawal from the candidate pool identified four transcription factors that were critical for pluripotent-like colony formation—Oct3/4, Sox2, KLF4, and *c-myc* (OSKM). The reprogrammed murine cells, termed induced pluripotent stem cells (iPSCs), were capable of forming blastocyst chimeras similar to mouse ESCs (mESCs), expressed pluripotency-specific markers (such as SSEA-1 and alkaline phosphatase), and demonstrated a similarity to mESCs in global gene expression.[16] The following year, both Yamanaka and Thomson published reports demonstrating the derivation of human iPS cells (hiPSCs) from terminally-differentiated human fibroblasts.[17,18] Takahashi et al. were able to induce pluripotency in adult human dermal fibroblasts using the OSKM factors introduced through retroviral transduction. Reprogrammed hiPSCs successfully expressed human embryonic stem cell (hESC)-specific gene and protein markers, formed teratomas in severe combined immunodeficiency (SCID) mice, and could be specifically directed to differentiate into both neural structures and cardiomyocytes under standard protocols. Interestingly, Yu et al. reported hiPSCs could be faithfully engineered using lentivirus transduction of Oct4, Sox2, Nanog, and LIN28 in both IMR90 fetal fibroblasts and adult dermal fibroblasts.[18] Unlike the Yamanaka group, they selected reprogramming factors on their basis of preferential expression in hESCs relative to myeloid precursor cells. The derived hiPSCs were also shown to express cell surface makers specific for pluripotent cells and to form teratomas.

Given the extraordinary potential of hiPSCs in either serving as a model system to study human disease or as a tool for regenerative medicine applications, there has been exponential growth in studies seeking to optimize the development of hiPSCs. However, hESCs and hiPSCs each have their own strengths as a research and possible therapeutic source. hESCs are not genetically modified and are likely a better system for studying normal human developmental biology. hiPSCs can be derived from any somatic cell isolated from patients with diverse diseases (as discussed more in this chapter) and, as a result, are more

---

Box 2.1 A Brief History of Pluripotency and the Discovery of Human Embryonic Stem Cells

In 1962, Sir John H. Gurdon provided the first evidence suggesting that terminally-differentiated somatic cells could be reprogrammed to the pluripotent state using *Xenopus laevis* as a model system.[1] In this seminal study, immature nuclei from developing oocytes were replaced with nuclei from functional intestinal epithelial cells. These modified embryos were ultimately capable of generating normal tadpoles. Gurdon's groundbreaking discovery had reversed the ideology set by Briggs and King in the 1950s that genes required for normal development are either excised or irreversibly repressed during early embryogenesis.[2] Consequently, researchers sought to engineer entire mammalian organisms through the selective transfer of terminally-differentiated, somatic cell nuclei into enucleated eggs in a process called somatic cell nuclear transfer (SCNT). Although ultimately successful in 1997 by Wilmut et al. with the development of Dolly the sheep from an adult cell, multiple reports have highlighted the technical difficulties and inefficiency of SCNT in a wide variety of mammalian species.[3–6] For example, the inability of a particular donor nucleus to support dedifferentiation and the technical challenges associated with donor cell handling are but a few obstacles preventing the wide-scale adoption of SCNT for clinical or therapeutic applications utilizing human cells.[7]

Another breakthrough for studies in pluripotency occurred in 1981 with the isolation and in vitro culture of mouse embryonic stem cells (mESCs).[8–10] These studies were the first to demonstrate that maintenance of pluripotency and self-renewal could be achieved through the addition of specific peptides, such as leukemia inhibitory factor (LIF) and subsequent activation of the gp130/Stat3 signaling cascade, at least in the murine system.[11–13] In 1998, James Thomson generated the first human pluripotent stem cell lines through the isolation and selection of the inner cell mass of donated blastocysts produced through in vitro fertilization when cocultured on feeder mouse embryonic fibroblast (MEF) cells in the presence of fetal bovine serum.[14] These hESCs fulfilled three key criteria of pluripotency in that they were derived from a preimplantation phase, proliferated in an undifferentiated state, and were capable of generating endoderm, mesoderm, and ectodermal derivatives. Subsequent studies later showed that the maintenance of hESC pluripotency is directly dependent on the presence of basic fibroblast growth factor (bFGF).[15]

suitable to study how defined (or undefined) genetic mutations affect development of specific cell populations. Additionally, hiPSCs can be derived from individual patients and can potentially be used for autologous cell replacement sources to minimize immunological barriers. Since the initial discovery of iPSCs, a variety of genetic reprogramming factors, delivery methods, starting cell sources, and culture microenvironments have been investigated, with the goal of increasing reprogramming efficiency of hiPSCs into target tissues for clinical applications. In this chapter, we will summarize the current methodology aimed at improving hiPSC engineering and their translational applications to gene therapy for correcting human disease.

2.2 TRANSCRIPTION FACTORS IMPORTANT FOR REPROGRAMMING TO PLURIPOTENCY

Both of the original hiPSC studies determined that Oct4 and Sox2 were required for genetic reprogramming. It has been well established that Oct4, also referred to as Pou5f1 as a member of the Pit1-Oct1/2-UNC-86 (POU) family, is expressed by all cells of the developing mammalian embryo up through pregastrulation and is required for the maintenance of the pluripotent state.[19,20] Oct4 is unique within the POU family in that its substitution with other family members is not sufficient to induce pluripotency because of a unique alpha-helix DNA-binding domain.[21] Oct4 can serve as a "master factor" in generating so-called 1-factor iPS cells from mouse neural stem cells (that already express high levels of Sox2) that are comparable to ESCs in morphology, global gene expression profiles, and DNA methylation patterns. These 1F iPSCs possess the ability to generate teratomas and mature into various organs upon injection by forming chimeric mouse embryos.[22,23] Sox2, a member of a superfamily of proteins that possesses a high mobility group (HMG) box DNA-binding domain that is widely expressed in pluripotent cells, is found to colocalize and cooperatively bind to DNA in conjunction with Oct4, suggesting it also is a required factor for the self-renewal of pluripotent cells through synergism.[20,24] Interestingly, using an inducible Sox2-null ES cell line, ablation of Sox2 resulted in differentiation into trophectoderm-like cells with a loss of pluripotent potential, but could be rescued by introducing an Oct3/4 transgene.[25] This established the current precedent that the predominant function of Sox2 is to maintain the appropriate expression of Oct3/4, and its role may be dispensable in maintaining pluripotency through substitution of other Sox family members.

Nanog, a homeodomain protein, is required to maintain pluripotency;[26–28] however, it is not required for the initial appearance of pluripotent colonies because of its eventual induction after Oct4 and Sox2 expression.[18] Interestingly, Nanog was originally discovered to significantly increase the reprogramming efficiency of both miPSCs and hiPSCs.[18,29] In a recent study, Nanog was found to play an integral role in maintaining pluripotency by switching from monoallelic expression in preimplantation embryos to biallelic expression in late blastocysts, providing further insight on how pluripotency is regulated during embryogenesis.[30] KLF4, a member of the Krüppel-like transcription factor family possessing zinc finger DNA-binding motifs, has been shown to be redundant for the induction of pluripotency in somatic cells.[31] However, similar to Nanog, KLF4 is found to be critical to maintain the pluripotent state, especially in the absence of a leukemia inhibitory factor (LIF).[11] One hypothesized mechanism that validates this role is that KLF4 is a central upstream regulator of large, positive feedback loops containing Oct4, Sox2, Nanog, and the *c-myc* promoter.[32] To date, the role of *c-myc*, a proto-oncogene, in reprogramming pluripotency is unclear. It has been speculated that *c-myc* may participate in chromatin remodeling very early during reprogramming through histone acetylation, facilitating Oct4/Sox2/KLF4 functionality.[20,31,33] Despite this, it is well-established that *c-myc* is not required for pluripotency induction, but is important for enhancing overall reprogramming efficiency.[31,34,35]

An important consideration in choosing transcriptional factors for pluripotent induction is their stoichiometric expression. Several studies have documented large discrepancies in reprogramming efficiency, protein expression, and even cellular senescence simply by altering the relative abundance and/or the gene order of the OSKM reprogramming factors. Carey et al. demonstrated that increasing the expression of Oct3/4 and KLF4 relative to Sox2 generated higher quality iPSCs, while Stadtfeld et al. demonstrated that altering the order of transcription factor delivery (OKSM) generated lower-quality iPSCs.[36,37] Other groups have determined Oct4 overexpression alone can maximize the efficiency of iPSC induction.[38] These studies underscore the importance of delivering appropriate stoichiometric quantities of transcription factors to maximize iPSC production.

2.3 METHODS FOR GENETIC REPROGRAMMING

Some of the earliest and to date most efficient reprogramming methods, such as lenti- and retroviruses, require integrating transgenes. Unfortunately, these methods carry an intrinsic risk of tumorigenesis and inappropriate reactivation in rapidly proliferative cells. Cassette excision and integration-free methods, such as transfection with synthetic messenger RNA (mRNA) and microRNA (miRNA), and transduction with specialized transposon systems and episomal plasmids significantly improve upon the safety profile for transplantation, but these methods have been typically less effective with lower iPSC induction efficiency. In this section, we will outline the current methods of transcription factor delivery into somatic cells while focusing on the efficiency of iPSC production (Table 2.1), long-term integration, retention in the genome

TABLE 2.1 Summary of Current Gene Delivery Methods to Engineer Human Induced Pluripotent Stem Cells

Gene Delivery Method	Risk of Insertional Mutagenesis	Induction Efficiency	Advantages	Limitations	Refs
Retrovirus	High	0.01–0.25%	Diversity in iPSC differentiation Minimal technical challenges Useful for in vitro analysis of human cells	Random, stable transgene integration Potential reactivation of reprogramming factors over time Transduction of proliferating cells only	1, 2, 25
Lentivirus	High	0.02–0.5%	Transduction of nonproliferating cells Diversity in iPSC differentiation Useful for in vitro analysis of human cells	Random, stable transgene integration Potential reactivation of reprogramming factors over time	3, 26
Cre-mediated lentiviral cassette excision	High	0.5–1.5%	Encoding transgenes removed following reprogramming, but retain pluripotency	5′ and 3′ LTR retention following excision	27–31
Synthetic mRNA	None	1.4–4.4%	Minimal number of transfections with very high efficiency Transfection maximized with serum-free culture, advantageous for clinical translation	Antiviral-like TLR activation and high immunogenicity Not validated outside of human fibroblast model	33–37
Sendai virus (SeV)	Low	0.1–2.0%	Shorter time frame to induction than integrating transgenes Temperature sensitive vectors eliminate any residual viral material from infected cells Virus never infects nucleus	Technically challenging to manipulate	32, 38–45
Synthetic miRNA	Varied	0.002% (direct)–10% (lentiviral)	Shorter time frame to induction than integrating transgenes High efficiency of iPSC induction, especially when delivered within lentiviral vector	Discrepancies in reprogramming efficiency based on miRNA clusters and delivery method Lentiviral miRNA with self-excision machinery not yet explored Limited validation using human cells	21, 46–49
piggyBac transposon	Low	0.02–1.25%	Virus-free transcriptional factor delivery; can be generated in bacteria Dramatically enhanced induction efficiency with transient presence of small molecules	Insufficient characterization of transgene excision genomic footprint in human cells	50–52
Sleeping Beauty transposon	Low	0.02%	Virus-free transcriptional factor delivery; can be generated in bacteria Does not exhibit integration bias towards transcriptional units	Not successfully demonstrated to produce a transgene-free hiPSC line Difficulty of selection if transposon is silenced or minimally expressed	53–55
Episomal plasmids	None	<0.01%	Plasmid is transiently expressed and removed with drug selection Proven success with feeder-free and nonxenogenic culture conditions	Very low induction efficiency in many different human cell types	56–61

TABLE 2.1 Summary of Current Gene Delivery Methods to Engineer Human Induced Pluripotent Stem Cells—cont'd

Gene Delivery Method	Risk of Insertional Mutagenesis	Induction Efficiency	Advantages	Limitations	Refs
Minicircle DNA	None	<0.01%	Inherently small size facilitates transfection	Very low induction efficiency Only validated in neonatal fibroblasts	62–64
Low pH (STAP-iPSC)	None	~6%	No transgenic alteration of transdifferentiated cells	To date, nonreproducible Not yet validated in human cell lines	65,66

STAP, stimulus-triggered acquisition of pluripotency; iPSC, induced pluripotent stem cells.

(Figure 2.1), and the clinical potential of each method for use in cellular replacement therapy for treating genetic diseases. We will also briefly review the potentially innovative (and controversial) notion that somatic cells can be reprogrammed to pluripotency simply by altering the microenvironmental culture conditions, exemplified by so-called stimulus triggered acquisition of pluripotency (STAP)-iPSCs.

2.3.1 Transgene Integration with Retro- and Lentiviruses

In generating the first iPSCs from mouse and human dermal fibroblasts, Takahashi and Yamanaka used retroviruses to deliver the OSKM transcription factors via constitutive promoters.[16,17] While retroviruses are relatively easy to technically manipulate, random integration into the host cell genome is required. Such integration essentially makes retroviral delivery of OKSM factors less desirable for translational therapies because of the high risk of insertional mutagenesis in the host cell genome. Furthermore, the induction efficiency with retroviruses in human cells is typically <1%, despite a high copy number delivery of the OSKM factors.[39] This low induction is directly related to the fact that retroviruses are only capable of infecting actively proliferating cells. Lentiviral vectors possessing similar pluripotent transcriptional factors were hypothesized to improve induction efficiency because they are capable of transducing nonproliferative cells, but induction efficiency remained comparable to retroviruses.[18] Moreover, the problem of multiple random transgene integration sites still persists with the lentiviral system. In an attempt to minimize these viral integration sites and to increase overall iPSC efficiency, a "stem cell cassette" (STEMCCA) containing the OSKM transcription factors and a combination of 2A peptide with internal ribosome entry sites (IRES) on a single lentiviral vector was generated.[40] Using the STEMCCA system, iPSCs could be produced with a single viral integration, which is an improvement over traditional retro- and lentivirus approaches requiring upwards of 20 integrations. Unfortunately, the kinetics and reprogramming efficiency (~0.5%) remained comparable to that of retro- and lentiviruses. Consequently, many groups have developed nonintegrating delivery methods in an attempt to improve both reprogramming efficiency and translational utility.

2.3.2 Nonintegrating Transgenes

2.3.2.1 Cre-mediated Lentiviral Cassette Excision

One of the first strategies implemented to circumvent transgene integration was iPSC induction via loxP flanking lentiviral cassettes and subsequent excision with Cre-recombinase.[41] Initially, doxycycline-inducible lentiviral vectors containing 3′ LTR loxP flanked OSKM were capable of generating iPSCs derived from fibroblasts of patients with sporadic Parkinson's disease.[41] These iPSCs were capable of maintaining all of the characteristics of a pluripotent ESC-like state after Cre-recombinase-mediated removal of each transgene and were further able to differentiate into dopaminergic neurons in vitro. Other studies expanded upon this method by modifying the STEMCCA lentiviral vector to include self-excisable loxP sites.[42] Using murine postnatal fibroblasts, STEMCCA-loxP vectors followed by Cre-mediated excision were able to generate mouse induced pluripotent stem cells (miPSCs) at the same kinetics and efficiency (~0.5%) as STEMCCA transgene insertion alone. However, the same study could not conclude that miPSCs generated using this approach were functionally equivalent to mESCs in terms of differentiation capacity. Moreover, cassette excision with Cre-recombinase was found to leave approximately 200 bp of viral LTR DNA within the genome of the miPSCs, still resulting in the potential for

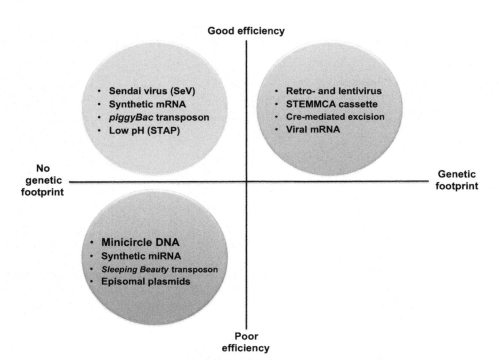

FIGURE 2.1 Categorization of gene delivery methods for generating hiPSCs. The appropriate genomic reprogramming system must be carefully assessed based on the clinical or research application. While methods yielding high hiPSC induction efficiency are desirable (e.g., lenti- and retroviruses, STEMMCA cassettes, Cre-mediated transgene excision, and viral messenger RNA (mRNA)), they often come with an intrinsic disadvantage of random integrating transgenes and establishment of a reprogramming "genomic footprint." This effect makes these systems less desirable for use in autologous transplantation, but are advantageous in recapitulating disease models for use in high-throughput pharmacological screening. Nonintegrating transgene methods (e.g., minicircle DNA, synthetic miRNA, *Sleeping Beauty* transposon system, and episomal plasmids) are ideal in improving the safety of autologous transplantation because the reprogramming factors never permanently integrate into the iPSC genome, but are relatively inefficient in generating induced pluripotent stem cells. Reprogramming via Sendai virus, synthetic mRNAs, *piggyBac* transposon, and potentially low-acid environments yield better efficiencies with low to no genomic footprint, and may ultimately be best suited for translational therapies.

oncogenic transformation and limiting their translational application. Since these initial studies, STEMCCA-loxP cassettes have been modified to generate hiPSCs from a wide variety of cell sources (e.g., bone marrow and peripheral blood) and patients across the spectrum of disease (e.g., cystic fibrosis, heart failure, and neurodegenerative disorders).[43–45] It is now possible to efficiently generate functional hiPSC clones from a CD34+-enriched population derived from a small volume of peripheral blood.[51]

2.3.2.2 Synthetic mRNA Transfection and Sendai Virus

Another method to bypass integrating transgenes is to deliver transcriptional factors directly into host cell cytoplasm as nonviral synthetic mRNA via a lipid-based transfection vehicle. Warren et al. first transfected human dermal fibroblasts with modified mRNA encoding each OSKM factor plus LIN28 to generate so-called RNA-induced pluripotent stem cells (RiPSCs).[46] Through this method, the efficiency of RiPSC reprogramming was improved to approximately 2.5%, depending on the degree of hypoxia. This increase in reprogramming efficiency was a 35-fold improvement over retroviral approaches with OSKM factors, and was applicable to four different human cell types. RiPSCs were also shown to form teratomas in vivo, clustered in global gene expression similar to hESCs, and could differentiate into myocytes when transfected with a MyoD-encoding modified RNA. Serum-free culture conditions can dramatically improve mRNA transfection efficiency, which is beneficial in minimizing the number of transfections required.[47] However, a major caveat to synthetic mRNA induction is the potential immunogenicity from toll-like receptor activation from the nucleic acids, specifically through interferon and NF-κB mediated pathways.[48–50] To attenuate this response and improve clinical applicability, mRNA molecules must be modified with synthetic nucleic acids and 5′-cap analogs prior to use.

The Sendai virus (SeV), a member of the *Paramyxoviridae* family, is an enveloped, single-stranded, negative-sense RNA virus capable of replicating within the host cell cytoplasm without nuclear transposition at any point during the host cell cycle. Consequently, SeV offers an attractive therapeutic alternative to integrating transgenes. The first reported hiPSCs engineered using SeV involved reprogramming of fibroblasts from neonatal foreskin and T cells from peripheral blood with an F-protein-deficient, nontransmissible SeV vector. This system permitted high-level expression of the OSKM transcriptional factors.[52,53] hiPSCs generated by SeV vectors expressed hESC markers, disappeared after cell growth, and formed in vivo teratomas. Favorably, SeV hiPSC induction doubled the efficiency of hiPSC production (~2%) relative to retroviruses in approximately one-half the time interval (14 days), all while avoiding a permanent genetic footprint. Temperature-sensitive mutants of SeV offer an even more attractive therapeutic option, as these viral vectors are capable of degrading when hiPSCs are incubated at 38°C for 3–5 days.[54,55] To date, SeV vectors have been successful in generating hiPSCs from peripheral blood of healthy[51] and diseased patients,[51,56,57] nasal epithelial cells,[58] and human umbilical cord blood[59] at a robust efficiency, making them an intriguing approach for broad clinical application.

2.3.2.3 Synthetic microRNA Transfection

Induction of hiPSCs with miRNA has grown out of extensive analysis of the role miRNAs play in maintenance of ESC pluripotency. High-resolution ChIP-sequencing of mESCs revealed the miR-290 cluster dominates the miRNA landscape and directly contributes to ESC cell cycle regulation.[60] Judson et al. exploited miR-291-3p, miR-294, and miR-295 from this cluster to supplement OSKM retroviral transduction in mouse embryonic fibroblast cells (MEFs).[35] Through this method, miPSCs were generated with a higher efficiency compared with OSKM alone (0.1–0.3%) and with similar characteristics as mESCs. miR302/367, which is also highly expressed in mESCs and induces Oct4 and Sox2 expression, was found to exclusively reprogram human fibroblasts independent of any additional exogenous factors when introduced within a lentiviral vector.[61] Reprogramming efficiency was twice as great as standard retroviral OSKM transduction methods. Although highly provocative, the use of miRNAs in direct reprogramming may not yield a universal technique for engineering all cell types. For example, one study was unsuccessful in generating iPSCs from MEFs when the same miR302/367 cluster was introduced via *piggyBac* transposon.[62] Moreover, miRNA302 integrated within a lentiviral vector and, when transduced in human adipose stem cells, was also unable to produce iPSCs.[63] Further mechanistic validation of miRNA reprogramming, improvement on the reproducibility of iPSC induction of multiple cell types, and an understanding of possible cytogenetic alterations of the host genome are required prior to translational uses.

2.3.2.4 PiggyBac *and* Sleeping Beauty *Transposons*

Another alternative to integrating viral transduction for generating hiPSCs are DNA transposons. Upon stable integration of transposon plasmids engineered with any gene of interest, transposases specific for repeating sequences on the flanking ends of the target gene can be utilized to excise the transposon. OSKM factors cloned into the moth-derived *piggyBac* transposon could transfect MEFs and human fibroblasts to generate iPSCs with comparable efficiencies as retro- and lentiviruses.[64–66] Interestingly, the small lipophilic molecule, butyrate, was shown to be critical to enhancing the reprogramming efficiency with the *piggyBac* system, increasing efficiency by roughly three-fold relative to retroviral transduction.[66]

The *Sleeping Beauty* (SB) transposon system is a novel, hyperactive alternative to *piggyBac* vectors that can also be exploited for use in human cells and ultimately clinical gene therapy trials.[67] A recent study has demonstrated that *SB* could be coupled with OSKM factors to reprogram MEFs into iPSCs, regardless of the underlying cell line genetic background.[68] hiPSCs have since been developed using a polycistronic OSKM SB transposon coupled with SB100X transposase at an efficiency of approximately 0.02%.[69] Although this efficiency is lower than other nonintegrating methods, such as Sendai virus or miRNA transfection, the ease of technical handling, nonrandom genomic integration, and lack of *SB*-like elements within the human genome provide advantages and may be suitable for clinical translation.[69]

2.3.2.5 Episomal Plasmids and Minicircle DNA

At the same time Cre/Lox integrating transgene excision and *piggyBac* transposons were discovered, Yu et al. also reported hiPSCs could be generated by a single transient transfection of Oct4, Sox2, Nanog, and LIN28 delivered inside episomal vectors derived from the Epstein–Barr virus.[70] Requiring only a *cis*-acting *oriP* site and a *trans*-acting *EBNA1* gene, these vectors can be simply transfected into mammalian cells without the need for viral packaging and pharmacological negative selection of nontransfected cells. Functional hiPSCs could be derived from human foreskin dermal fibroblasts using combinations of three episomal plasmids containing either 1) Oct4, Sox2, Nanog, and KLF4; 2) Oct4, Sox2, and SV40; or 3) c-myc and LIN28; however, the reprogramming efficiency was only three to six hiPSCs/10^6 human fibroblasts. Several subsequent studies have adapted similar protocols, but used instead transfected umbilical cord blood CD34$^+$ mononuclear cells, peripheral blood mononuclear cells, and neoplastic bone marrow with slight transcription factor combinatorial alterations in the *OriP/EBNA1* cassette.[71,72] In any case, reprogramming efficiency remained low, but could be enhanced 10-fold in nonfibroblasts in a shorter time. Recent studies have shown that iPSC efficiency using episomal plasmids can be enhanced to levels comparable to that of retro- and lentiviruses by *p53-p21* pathway suppression through the addition of a small hairpin RNA for *TP53*.[73–75]

Minicircle DNA (MC) vectors are small, supercoiled, episomal DNA vectors that are engineered without any bacterial plasmid backbone sequences (e.g., antibiotic resistance genes or origins of replication). MC DNA yields higher transfection efficiency because of their inherently small size and sustained levels of transgene expression because of the minimal transcriptional silencing of prokaryotic sequences.[76] The first MC DNA reprogramming study utilized a single, polycistronic cassette composed of the Oct4, Sox2, Nanog, LIN28, and a green fluorescent protein (GFP) reporter gene; this cassette was transfected into human adipose stem cells.[77,78] The MC DNA transfection yielded prolonged and a more robust ectopic transcriptional expression compared to standard plasmids. Despite a high degree of similarity between MC-derived iPSCs and hESCs and lack of MC DNA integration into iPSC clones, the reprogramming efficiency was limiting (~0.005%) and multiple series and varieties of transfection were required. Future work to increase iPSC efficiency, including evidence that MC DNA is capable of reprogramming human fibroblasts with high fidelity, is needed.

2.3.2.6 Stimulus Triggered Acquisition of Pluripotency (STAP)-iPSCs

A potential major breakthrough in iPSC engineering was made with the provocative observation that immature splenic CD45[+] hematopoietic cells can dedifferentiate and express pluripotency markers after being stressed with a 30 min., low pH (5.7) saline solution in the presence of LIF.[79] After seven days of exposure to these conditions, the cells faithfully expressed a GFP-Oct4 reporter construct in addition to other pluripotency genes, differentiated in vitro into all three embryological germ layers, and developed chimeric mice capable of germline transmission when injected into murine blastocysts. Remarkably, unlike ESCs and currently engineered iPSCs, STAP cells are additionally capable of contributing to trophoblastic tissue.[80] These so-named "stimulus triggered acquisition of pluripotency"-iPSCs (STAP iPSCs) seem to represent a unique state of pluripotency that challenges the current dogma by suggesting somatic cells are plastic and can self-induce pluripotency on the basis of their extracellular environment.

If successfully validated further by other research groups, the implications of STAP iPSCs for translation and regenerative medicine are certainly profound. For the first time, high-fidelity pluripotent cells could be engineered without the need for transgene insertion or transcription factor induction in somatic cells, essentially eliminating current iPSC hurdles. However, several questions must be addressed prior to clinical-grade STAP iPSC development. One obvious question is whether human cells can perform the same phenomenon. Moreover, it remains to be seen if pluripotent induction can be achieved using mature, adult somatic cells. A thorough understanding of the changes to the genetic and epigenetic landscape of reprogrammed cells must be attained to assess their clinical utility. Unfortunately, after the initial publications describing STAP-iPSCs, this area has rapidly become controversial as replicating these studies has not been successfully accomplished by other groups and questions regarding the quality of the published work have not been fully addressed. Indeed, both papers that initially described STAP cells have been retracted. Therefore, STAP-iPSCs likely serve as an important reminder of the need for high levels of quality control to ensure reproducibility and generalizability of potentially groundbreaking work. These issues highlight the remarkable success of iPSC technology (as developed using gene transfer by Yamanaka, Thomson, and others) that has been replicated by hundreds if not thousands of groups around the world, making iPSCs truly a remarkable and revolutionary breakthrough.

2.4 CLINICAL TRANSLATION OF iPSCs

Generating iPSCs from patients with defined genetic diseases can be used in several ways to offer new therapeutic options to patients (Figure 2.2). For example, diseased-iPSCs can be used to analyze specific cellular and developmental elements, such as signaling pathways or epigenetic changes, that can be affected by genetic mutations. Furthermore, iPSCs can be used to screen and discover small molecules that may be effective in correcting disease phenotypes. Most intriguing is the prospect of correcting mutated genes in iPSCs generated from a patient's own cells followed by autologous transplantation to effectively revert the individual to a nonaffected phenotype (Figure 2.3). In this section, we will highlight these translational applications of iPSCs, with an emphasis on reviewing several proof-of-principle studies reporting the use of human iPSCs in modeling and correcting human genetic diseases.

2.4.1 iPSCs as Models for Understanding Disease Pathophysiology

iPSCs are prime cell sources to model neurodegenerative and neurodevelopmental diseases because of the limited replicative capacity and regenerative potential of endogenous neurons in the central nervous system. Motor neuron diseases, such

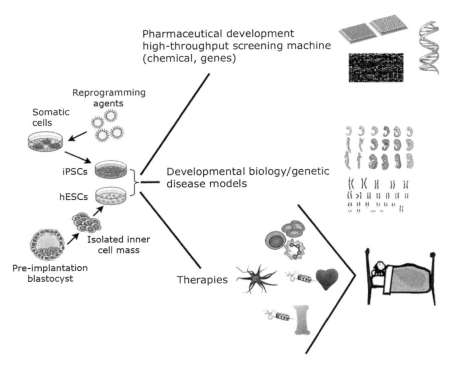

FIGURE 2.2 Potential applications of human pluripotent stem cells (hPSCs) for improving the treatment of human disease. Human embryonic stem cells (hESCs) can be directly isolated and expanded from the inner cell mass of the developing blastocyst, while human induced pluripotent stem cells can be directly engineered from any patient-derived somatic cell through introduction of defined reprogramming agents, such as OSKM. hPSCs can then be utilized directly as a cell source for high-throughput pharmacological screening, models for recapitulating human developmental biology and/or pathological disease states, or directly as cell-based therapy through autologous transplantation.

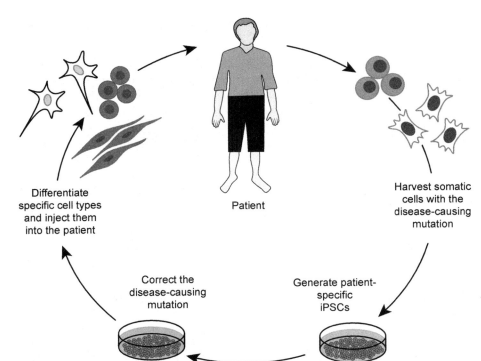

FIGURE 2.3 Induced pluripotent stem cell (iPSC)-based combined gene and cell therapy. The autologous transplantation of genetically corrected iPSC-derived cells can potentially be used to treat patients with certain genetic diseases, such as sickle cell disease, Fanconi anemia, thalassemia, Parkinson's disease, and Huntington's disease. Patient's somatic cells (e.g., skin fibroblasts or mononuclear blood cells), containing the disease-causing mutation, are in vitro transformed into iPSCs by the introduction of transcription factors. The mutation is corrected by gene-specific targeting of iPSCs. Corrected iPSCs are differentiated into the required cell type precursors or mature cells, such as hematopoietic or neural cells. Healthy differentiated cells would be injected back into the patient where they would need to integrate and provide improved function. *Used with permission from Ref. 81.*

as amyotrophic lateral sclerosis (ALS) and spinal muscular atrophy (SMA), share a common pathophysiology involving motor neuron inefficiency and destruction. Familial ALS is a motor neuron disease associated with mutations in several genes, such as *SOD1* (superoxide dismutase one) and *TDP-43* (tar DNA binding protein 43); however, the exact etiology of motor neuron dysfunction is unknown in either of these inherited cases. Dimos et al. first reported successful derivation of iPSCs from skin fibroblasts collected from elderly patients diagnosed with an *SOD1* familial form of ALS.[82] These patient-specific cells were subsequently used to produce motor neurons and glial cells that displayed characteristic neuronal transcriptional factors and intracellular markers. These iPSC-derived neurons may potentially be suitable for autologous transplantation, although the physiological function of these cells was not elucidated at the time. A recent study has shed light on a potential pathophysiological mechanism through analysis of hiPSCs generated from patients with A4V and D90A *SOD1* mutations. Chen et al. elegantly demonstrated that *SOD1* mutations caused neurofilament aggregation exclusively in spinal motor neurons with neurite swelling and degeneration in the absence of glial cells.[83] They further went on to show these intracellular changes were caused by mutant *SOD1* association with the 3'-UTR of *NF-L* mRNA. iPSC-derived motor neurons from patients with *TDP-43*-familal ALS, yielding enhanced cytoplasmic aggregation that is similar to what is found in patients with sporadic and *SOD1*-familial ALS.[84] An understanding of these mechanisms thus provides a potential therapeutic approach to target neurofilament disaggregation as a means to ultimately reverse ALS pathology.

iPSCs derived from patients with spinal muscular atrophy have also been generated and tested in preclinical models. SMA is an autosomal recessive disease caused by either a deletion or point mutation of the survival motor neuron gene (*SMN1*), resulting in motor neuron apoptosis within the anterior horn of the spinal cord. The human *SMN* locus is unique from other mammalian systems in that it is comprised of an additional inverted centromeric copy of the *SMN* gene capable of being transcribed.[85] Thus, hiPSCs offer a more replicative cellular model in understanding human SMA pathogenesis. Chang et al. reported the successful derivation of five hiPSC cell lines from an SMA Type 1 patient.[86] In this study, these patient-derived hiPSC lines showed abnormalities in neurite outgrowth and a reduced capacity to differentiate motor neurons. Although it remains to be seen how the loss of *SMN* gene expression causes attenuation in neuronal growth, it is clear that hiPSC-derived motor neurons from SMA patients can be rescued through *SMN1* transduction.[86] SMA-derived hiPSCs have also been used as a disease model for the study of mutant androgen receptor genes and how increases in polyglutamine (polyQ) result in neurodegeneration,[87] thus offering another potential avenue for disease phenotype correction.

Parkinson's disease (PD) is another highly-prevalent neurodegenerative disease caused by the loss of dopaminergic neurons in the substantia nigra. Both sporadic and familial cases of PD are symptomatically associated with skeletal muscle rigidity, tremors, postural instability and bradykinesia. Insights into the cellular mechanisms behind the selective temporospatial destruction of these neurons remain largely unknown. hiPSCs harboring familial-PD mutations have provided some insight on potential disease mechanisms. For example, hiPSC-derived neurons generated from familial-PD patients carrying a *LRRK2* or *PINK1* mutation were found to be more sensitive to reactive oxygen species than wild-type hiPSCs-derived neurons.[88] hiPSC-derived neurons from *PINK1*-carrying patients further demonstrated significant mitochondrial dysfunction, including a novel finding that such human neurons are also unable to initiate mitophagy.[89] To date, several treatment options, such as levodopa administration, deep brain stimulation, and even surgical removal of neural lesions, have all been employed for the treatment of PD, but have not yet been proven to reversibly cure the disease. Replacement of damaged neurons with fetal dopaminergic neuronal xenografts has been pursued as a potential therapeutic treatment option. However, this approach was complicated with the development of graft-induced dyskinesia in many patients.[90,91] hiPSC-derived dopamine-secreting neurons have been proposed as a more suitable therapeutic option. An early study in a rat model of PD used iPSC-derived dopamine-producing cells that were transplanted in the midbrain; these cells yielded a significant improvement in the Parkinsonian condition of rats.[92] In another study, hiPSCs were derived from PD (PDiPSCs) patients and differentiated into dopamine-releasing neurons (DN).[93] These PDiPSCs-derived DN were transplanted into 6-hydroxydopamine (6-OHDA) lesion rat models to test their survival and functionality. Results of this study demonstrated that non-DA neurons derived from PDiPSCs play a vital role in propagating neighboring axons into the remote gray and white matter. Moreover, this study demonstrates the functional effects and survival of DA neurons derived from PDiPSCs in an animal model of PD.

hiPSCs have also been derived from patents with familial dysautonomia (FD), a rare autosomal recessive neural crest lineage disorder caused by I-*k*-B kinase complex-associated protein (IKBKAP) gene mutation.[94] FD affects the survival of sympathetic, parasympathetic, and sensory neurons. The first FD-patient-specific iPSCs (FDiPSCs) validated low levels of IKBKAP transcription compared to control iPSCs, caused by a tissue-specific missplicing of the IKBKAP gene. Moreover, microarray analysis of FDiPSCs revealed significant reductions in multiple genes involved in neurogenic development. Interestingly, *ASCL1*, a gene required for autonomic neurogenesis in mice, was correlatively significantly attenuated in gene and protein expression, causing a reduction in neuronal differentiation. Thus, this FDiPSC-disease model is an example of how novel disease mechanisms can be discovered and potentially corrected through genomic editing or targeted pharmacological therapy.

2.4.2 iPSCs for High-throughput Small Molecule Screening

The success of preclinical drug development is based on substrate discovery and demonstrated efficacy in vitro, an understanding of pharmacokinetic profiles and toxicology within animal models, and ultimately evidence of efficacy with adequate safety in human clinical trials.[21] Using this current system, less than 8% of potential compounds that actually enter human clinical trials receive market approval, and the cost for the process of drug development is typically over a billion dollars per drug.[95,96] To improve upon current methods of drug discovery, hiPSC-based assays can provide an authentic, efficient, and more reliable tool for drug development via high-throughput screening. Small-molecule screening using iPSCs derived from patients with defined disorders, as previously described, may not only enhance the understanding of disease pathophysiology, but also may help to develop innovative components, such as small hairpin RNAs, to revert disease-related phenotypes to nondisease phenotypes.[97]

Various small molecules have been previously screened using mESCs and hESCs to identify critical factors important for stem cell survival and to mitigate their pluripotency to enhance the safety of transplanted therapeutic cells.[98–100] For example, marketed drugs like flurbiprofen and gatifloxacin were discovered to influence hESC self-renewal, while cardioglycosides, selegiline, and retinonic acid were found to enhance hESC differentiation.[101] With the establishment of hiPSCs, small molecules are beginning to be screened to identify factors that may increase the functional phenotype of iPSC-derived cells from disease models and improve drug efficacy directly using human cells.[102] To date, a handful of small molecules have been screened using iPSCs-derived disease models and documented pathological improvement. For example, Choi et al. used iPSCs from a patient with α_1-antitrypsin-deficiency to identify five novel compounds that inhibited mutant α_1-antitrypsin accumulation in hepatocyte-like cells and reversed disease pathology.[103] Xu et al. used iPSCs derived from healthy subjects to generate glutamatergic and GABAergic neurons, which were subsequently screened for amyloid-β(1–42) inhibitor toxicity by measuring neurite outgrowth of these differentiated cells.[104] Lee et al. analyzed roughly 7000 small molecules and found that SKF-86,466 can rescue intrinsic cell function in iPSCs derived from patients with familial dysautonomia.[94] Egawa et al. reported hiPSC generation from *TDP43*-familial ALS patients, which were subsequently differentiated into diseased motor neurons for high-throughput drug screening. This led to the discovery of anacardic acid, a histone acetyltransferase inhibitor, which was found to reverse ALS cellular phenotypes.[84] Taken together, these studies serve to highlight the useful role that iPSCs can play in the execution and development of drug-efficacy screens for identifying novel compounds.

2.4.3 Proof-of-principle iPSC Monogenetic Gene Correction for Translational Therapy

iPSC technology opens the door to potentially treating genetic and metabolic diseases that are otherwise nonresponsive to pharmacological targeting. The advent of gene-editing technologies—e.g., zinc finger endonucleases (ZFN), transcription activator-like effector nucleases (TALEN), and clustered regularly interspaced short palindromic repeats (CRISPR)-Cas9 endonucleases— have all enhanced the clinical potential of iPSCs to treat monogenetic diseases. Notably, ZFNs have been used for clinical trials to treat T cells of HIV-infected patients as a strategy to promote immune reconstitution and HIV resistance.[105] Although there are no actively-sponsored NIH clinical trials utilizing hiPSC autologous transplantation in the United States, several in vivo proof-of-principle studies using small and large-scale animal models have demonstrated the potential that gene-edited hiPSCs possess for future human trials, especially in hematological diseases. Here, we highlight several successful model systems that demonstrate the use of gene-corrected iPSCs in improving disease pathology and consider the prospect of utilizing hiPSCs for autologous transplantation.

Sickle cell disease (SCD), a red blood cell disorder characterized by hypoxia-mediated changes in red cell morphology, is caused by a well-defined glutamic acid to valine mutation in the globin β-chain. An important collaboration between the Jaenisch and Townes labs initially reported treatment of SCD using miPSCs.[106] In this study, the β^s globin gene was corrected by homologous recombination in miPSCs derived from a mouse model harboring the human sickle cell gene mutation. These iPSCs were then differentiated into hematopoietic cells, transplanted into mice with SCD, and were shown to form normal erythroid cells with improvement in total cell count and other physiological parameters. Sebastiano et al. were successful in applying ZFNs to correct the β-globin gene with iPSCs derived from dermal fibroblasts using lentivirus in two SCD patients.[107] They further highlighted the clinical applicability of such an approach by demonstrating that reprogrammed cells retained pluripotency with a normal karyotype following cassette excision. A similar ZFN gene-editing system was further applied to correct SCD with the encouraging result that patient-specific hiPSCs could be produced using Cre-mediated excisable transposons.[108] TALEN-mediated gene targeting in combination with a *piggyBac* transposon system has also been shown to correct sickle-cell mutations with no residual transgene sequence using human cells.[109]

Several studies have been conducted that suggest iPSCs can be used in hemophilia patients with the potential for use in future therapies. Hemophilia A is a bleeding disorder caused by an X-linked chromosomal-inherited defect in coagulation Factor VIII. Current pharmacological treatments involve the replacement of plasma-derived factors or recombinant Factor VIII products. However, these treatments are hampered because of high costs and frequent injections.[110] Alternatively, gene therapy using hiPSCs has been proposed and developed as an attractive strategy to treat hemophilia.[111–113] Xu et al. directly injected iPSCs generated from murine fibroblasts that were differentiated into endothelial progenitor cells into the liver of hemophilia A mice.[114] These iPSC-derived endothelial cells were able to survive more than 3 months compared to nontransplanted control mice with a significantly increased production of coagulation Factor VIII. Other recent studies also reported a new strategy to treat hemophilia A using hiPSCs engineered with an integration-free human Factor VIII gene and megakaryocyte-specific platelet factor-4 promoter within a human artificial chromosome (HAC) vector (PF4-FVIII-HAC).[115] PF4-FVIII-HAC expressing iPSCs were then differentiated into platelets that appropriately expressed Factor VIII. Although in vivo correction of hemophilia was not demonstrated in this study, this work does provide an interesting example of the diversity of genetic diseases that can be potentially treated by iPSC-derived therapies.

hiPSC technology has been implemented as a proof-of-principle model for correcting inherited bone marrow failure, particularly Fanconi anemia (FA). FA is a rare disorder of chromosomal instability caused by a defect in any one of the 13 genes of the Fanconi DNA repair pathway.[116] While the ideal treatment of FA is autologous, gene-corrected hematopoietic stem cell (HSC) transplantation, such work remains in development and in the early stages of clinical trial testing for patients with specific germline mutations.[117,118] Human leukocyte antigen (HLA)-matched HSC transplantation is also a therapeutic option, but poses the potential risk of graft-versus-host disease and recurrent infections. Consequently, patient-derived hiPSCs can be utilized to generate a theoretically limitless supply of autologous HSCs suitable for gene correction and transplantation. Raya et al. were able to genetically correct fibroblasts isolated from FA-A and FA-D2 complementation group patients using lentiviral delivery of the *FANCA* or *FANCD2* with retroviral reprogramming into patient-specific hiPSCs.[119] These FA-derived iPSCs were further validated to differentiate into CD34+CD45+ hematopoietic progenitor cells that could give rise to normal erythroid and myeloid lineages. However, studies to date have not demonstrated a system to produce HSCs from hESCs or hiPSCs that are capable of long-term multilineage engraftment when transplanted into appropriate immunodeficient mice.[120,121]

While the hematopoietic system has been a focus for hiPSC monogenetic disease correction given its accessibility and readily available supply of HSCs, there have also been successful examples involving nonhematopoietic disease processes. Using hiPSC-disease models of PD as previously described, Soldner et al. demonstrated ZFN-mediated gene editing was capable of generating point mutations in α-synuclein genes, followed by successful phenotypic correction.[122] Huntington's disease, an autosomal dominant genomic disease caused by repetitive CAG sequences within the huntingtin gene (*HTT*), could also be corrected in patient-specific hiPSCs by exchanging the expanded *HTT* allele with a normal number of polyglutamine-encoding repeats.[123] Furthermore, iPSCs produced from patients with Hutchinson–Gilford progeria syndrome (HGPS), a premature aging disorder associated with truncation of mRNA transcribed from the *LMNA* gene, could be corrected using a single-helper adenoviral vector.[124] While viable and pathologically representative patient-specific iPSCs have been generated from a number of other monogenetic diseases (e.g., cystic fibrosis,[45,46] various mucopolysaccharidoses,[125] glucose-6-phosphate deficiency,[126] Lesch–Nyhan syndrome,[125] and others[127]), an opportunity for further research remains as iPSCs have yet to be completely corrected to yield physiologically normal-derived cells in these models.

2.4.4 Important Considerations of Autologous hiPSC Transplantation

While promising, patient-specific hiPSCs most likely have several years of additional in vitro, in vivo, and preclinical study prior to wide-scale clinical adoption. As previously discussed, addressing the challenges of generating patient-specific hiPSCs with high efficiency, no residual genomic footprint of reprogramming factors, and in high cell quantity is of utmost priority. While some disease models, particularly neurodegenerative conditions, may only require less than a million differentiated iPSCs to drive phenotypic improvement, other diseases require a significantly larger scale.[120]

Transplantation with hiPSCs must also be validated for patient safety. While teratoma formation is always a concern when using heterogeneously-differentiated hiPSCs, acute immunological rejection is a more pressing limitation. Our current understanding of the immunological response against iPSCs is skewed, given that most hiPSC-derived cells are traditionally validated by transplantation into immunodeficient mice.[120] However, in vitro hESC and hiPSC models have shown significant modification to major histocompatibility complex (MHC) antigen recognition and presentation, which can ultimately initiate T cell-mediated immunological responses.[128] Natural killer (NK) cells have also been shown to target hPSCs based on decreased expression of MHC class I.[129] One potentially controllable cause of immunological rejection are peptides derived from either nonhuman "feeder" stromal layers or animal serum from the culture media that are capable of being presented on MHC and mediate effector T cell responses. Cell culture methods using high concentrations of animal-free knockout serum have been shown to alter the antigenic signature of human pluripotent stem cells (hPSCs) and could potentially enhance immunogenicity.[130,131] The host tissue epigenetic landscape of reprogrammed hiPSCs is also hypothesized to play an important role in establishing their antigenicity, which needs to be further elucidated.[132] While immunological rejection is an intrinsic obstacle to engineered iPSCs, several therapeutic methods can be employed to marginalize graft-versus-host responses. As is currently performed in allogeneic stem cell transplantation, immunosuppressive therapies can be provided to attenuate rejection. Genomic editing can also be further exploited by HLA gene-specific knockdown or overexpression of immunosuppressant molecules prior to autologous transplantation to circumvent the immunological response.[120]

Despite these hurdles, there is only one human iPSC clinical trial in preparation at the RIKEN Center for Developmental Biology and Institute of Biomedical Research and Innovation Hospital in Japan.[133,134] This trial intends to focus on the safety of hiPSC transplantation and possible correction of wet-type age-related macular degeneration, a highly prevalent form of visual impairment particularly in elderly individuals. If executed, this potential landmark study will provide important evidence regarding the safety and efficacy of iPSC therapy and will be useful in furthering autologous hiPSC transplantation to a plethora of degenerative diseases.

2.5 CONCLUSION

The remarkable discovery of iPSCs by Takahashi and Yamanaka has changed the paradigm for how we approach our understanding of pluripotency and the development of new therapies for genetic diseases. Considerable progress has been made in the genetic engineering of hiPSCs, development of representative iPSC-derived disease models, and differentiation into normal autologous tissue upon gene correction. These iPSC-derived disease models have proven to be a great platform to analyze disease pathogenesis. However, the challenges associated with generating iPSCs and their therapeutic applications still require further investigation prior to wide-scale clinical application. It remains critical to understand whether iPSC-derived disease models faithfully recapitulate their respective disease phenotype in vitro and in vivo. Additionally, correcting the diseased gene and developing therapeutically useful cells that appropriately integrate into the host tissue remains a challenge for many cell therapy models. Moreover, immunological rejection may also be problematic, as even autologously transplanted, gene-corrected cells have been found to not be immunotolerant because of expression of novel genes. Nonetheless, recent advancements using patient-specific iPSCs have helped pave the way for future preclinical studies and human clinical studies in developing solutions to otherwise historically incurable genetic diseases.

REFERENCES

1. Gurdon JB. The developmental capacity of nuclei taken from intestinal epithelium cells of feeding tadpoles. *J Embryol Exp Morphol* 1962;**10**:622–40.
2. King T, Briggs R. Serial transplantation of embryonic nuclei. *Cold Spring Harb Symp Quant Biol* 1956;**21**:271–90.
3. Wilmut I, Schnieke AE, McWhire J, Kind AJ, Campbell KHS. Viable offspring derived from fetal and adult mammalian cells. *Nature* 1997;**385**:810–3.
4. Li Z, et al. Cloned ferrets produced by somatic cell nuclear transfer. *Dev Biol* 2006;**293**:439–48.
5. Wakayama S, et al. Equivalency of nuclear transfer-derived embryonic stem cells to those derived from fertilized mouse blastocysts. *Stem Cells* 2006;**24**:2023–33.
6. Baguisi A, et al. Production of goats by somatic cell nuclear transfer. *Nat Biotechnol* 1999;**17**:456–61.
7. Meissner A, Jaenisch R. Mammalian nuclear transfer. *Dev Dyn* 2006;**235**:2460–9.
8. Evans M, Kaufman M. Establishment in culture of pluripotential cells from mouse embryos. *Nature* 1981;**292**:154–6.
9. Smith AG, et al. Inhibition of pluripotential embryonic stem cell differentiation by purified polypeptides. *Nature* 1988;**336**:688–90.
10. Martin GR. Isolation of a pluripotent cell line from early mouse embryos cultured in medium conditioned by teratocarcinoma stem cells. *Dev Biol* 1981;**78**:7634–8.
11. Niwa H, Ogawa K, Shimosato D, Adachi K. A parallel circuit of LIF signalling pathways maintains pluripotency of mouse ES cells. *Nature* 2009;**460**:118–22.
12. Williams RL, et al. Myeloid leukaemia inhibitory factor maintains the developmental potential of embryonic stem cells. *Nature* 1988;**336**:684–7.
13. Smith AG, Hooper ML. Buffalo rat liver cells produce a diffusible activity which inhibits the differentiation of murine embryonal carcinoma and embryonic stem cells. *Dev Biol* 1987;**121**:1–9.
14. Thomson JA. Embryonic stem cell lines derived from human blastocysts. *Science* 1998;**282**:1145–7.
15. Xu C, et al. Basic fibroblast growth factor supports undifferentiated human embryonic stem cell growth without conditioned medium. *Stem Cells* 2005;**23**:315–23.
16. Takahashi K, Yamanaka S. Induction of pluripotent stem cells from mouse embryonic and adult fibroblast cultures by defined factors. *Cell* 2006;**126**:663–76.
17. Takahashi K, et al. Induction of pluripotent stem cells from adult human fibroblasts by defined factors. *Cell* 2007;**131**:861–72.
18. Yu J, et al. Induced pluripotent stem cell lines derived from human somatic cells. *Science* 2007;**318**:1917–20.
19. Plachta N, Bollenbach T, Pease S, Fraser SE, Pantazis P. Oct4 kinetics predict cell lineage patterning in the early mammalian embryo. *Nat Cell Biol* 2011;**13**:117–23.

20. Chambers I, Tomlinson SR. The transcriptional foundation of pluripotency. *Development* 2009;**136**:2311–22.
21. Esch D, et al. A unique Oct4 interface is crucial for reprogramming to pluripotency. *Nat Cell Biol* 2013;**15**:295–301.
22. Kim JB, et al. Oct4-induced pluripotency in adult neural stem cells. *Cell* 2009;**136**:411–9.
23. Kim JB, Zaehres H, Araúzo-Bravo MJ, Schöler HR. Generation of induced pluripotent stem cells from neural stem cells. *Nat Protoc* 2009;**4**:1464–70.
24. Ambrosetti DC, et al. Synergistic activation of the fibroblast growth factor 4 enhancer by Sox2 and Oct-3 depends on protein-protein interactions facilitated by a specific spatial arrangement of factor binding sites. Synergistic Activation of the Fibroblast Growth Factor 4 En. *Mol Cell Biol* 1997;**17**:6321.
25. Masui S, et al. Pluripotency governed by Sox2 via regulation of Oct3/4 expression in mouse embryonic stem cells. *Nat Cell Biol* 2007;**9**:625–35.
26. Chambers I, et al. Nanog safeguards pluripotency and mediates germline development. *Nature* 2007;**450**:1230–4.
27. Hatano S-Y, et al. Pluripotential competence of cells associated with Nanog activity. *Mech Dev* 2005;**122**:67–79.
28. Ivanova N, et al. Dissecting self-renewal in stem cells with RNA interference. *Nature* 2006;**442**:533–8.
29. Silva J, Chambers I, Pollard S, Smith A. Nanog promotes transfer of pluripotency after cell fusion. *Nature* 2006;**441**:997–1001.
30. Miyanari Y, Torres-Padilla M-E. Control of ground-state pluripotency by allelic regulation of Nanog. *Nature* 2012;**483**:470–3.
31. Huangfu D, et al. Induction of pluripotent stem cells from primary human fibroblasts with only Oct4 and Sox2. *Nat Biotechnol* 2008;**26**:1269–75.
32. Kim J, Chu J, Shen X, Wang J, Orkin SH. An extended transcriptional network for pluripotency of embryonic stem cells. *Cell* 2008;**132**:1049–61.
33. Knoepfler PS, et al. Myc influences global chromatin structure. *EMBO J* 2006;**25**:2723–34.
34. Nakagawa M, et al. Generation of induced pluripotent stem cells without Myc from mouse and human fibroblasts. *Nat Biotechnol* 2008;**26**:101–6.
35. Judson RL, Babiarz JE, Venere M, Blelloch R. Embryonic stem cell-specific microRNAs promote induced pluripotency. *Nat Biotechnol* 2009;**27**:459–61.
36. Carey BW, et al. Reprogramming factor stoichiometry influences the epigenetic state and biological properties of induced pluripotent stem cells. *Cell Stem Cell* 2011;**9**:588–98.
37. Stadtfeld M, et al. Aberrant silencing of imprinted genes on chromosome 12qF1 in mouse induced pluripotent stem cells. *Nature* 2010;**465**:175–81.
38. Papapetrou EP, et al. Stoichiometric and temporal requirements of Oct4, Sox2, Klf4, and c-Myc expression for efficient human iPSC induction and differentiation. *Proc Natl Acad Sci USA* 2009;**106**:12759–64.
39. Rao MS, Malik N. Assessing iPSC reprogramming methods for their suitability in translational medicine. *J Cell Biochem* 2012;**113**:3061–8.
40. Sommer CA, et al. Induced pluripotent stem cell generation using a single lentiviral stem cell cassette. *Stem Cells* 2009;**27**:543–9.
41. Soldner F, et al. Parkinson's disease patient-derived induced pluripotent stem cells free of viral reprogramming factors. *Cell* 2009;**136**:964–77.
42. Sommer CA, et al. Excision of reprogramming transgenes improves the differentiation potential of iPS cells generated with a single excisable vector. *Stem Cells* 2010;**28**:64–74.
43. Zwi-Dantsis L, et al. Derivation and cardiomyocyte differentiation of induced pluripotent stem cells from heart failure patients. *Eur Heart J* 2013;**34**:1575–86.
44. Camnasio S, et al. The first reported generation of several induced pluripotent stem cell lines from homozygous and heterozygous Huntington's disease patients demonstrates mutation related enhanced lysosomal activity. *Neurobiol Dis* 2012;**46**:41–51.
45. Somers A, et al. Generation of transgene-free lung disease-specific human induced pluripotent stem cells using a single excisable lentiviral stem cell cassette. *Stem Cells* 2010;**28**:1728–40.
46. Warren L, et al. Highly efficient reprogramming to pluripotency and directed differentiation of human cells with synthetic modified mRNA. *Cell Stem Cell* 2010;**7**:618–30.
47. Tavernier G, et al. Activation of pluripotency-associated genes in mouse embryonic fibroblasts by non-viral transfection with in vitro-derived mRNAs encoding Oct4, Sox2, Klf4 and cMyc. *Biomaterials* 2012;**33**:412–7.
48. Lester SN, Li K. Toll-like receptors in antiviral Innate Immunity. *J Mol Biol* 2013:1–19. http://dx.doi.org/10.1016/j.jmb.2013.11.024.
49. Diebold SS, Kaisho T, Hemmi H, Akira S, Reis e Sousa C. Innate antiviral responses by means of TLR7-mediated recognition of single-stranded RNA. *Science* 2004;**303**:1529–31.
50. Drews K, et al. The cytotoxic and immunogenic hurdles associated with non-viral mRNA-mediated reprogramming of human fibroblasts. *Biomaterials* 2012;**33**:4059–68.
51. Merling RK, et al. Transgene-free iPSCs generated from small volume peripheral blood nonmobilized CD34+ cells. *Blood* 2013;**121**:e98–107.
52. Seki T, et al. Generation of induced pluripotent stem cells from human terminally differentiated circulating T cells. *Cell Stem Cell* 2010;**7**:11–4.
53. Fusaki N, Ban H, Nishiyama A, Saeki K, Hasegawa M. Efficient induction of transgene-free human pluripotent stem cells using a vector based on Sendai virus, an RNA virus that does not integrate into the host genome. *Proc Jpn Acad Ser B* 2009;**85**:348–62.
54. Ban H, et al. Efficient generation of transgene-free human induced pluripotent stem cells (iPSCs) by temperature-sensitive Sendai virus vectors. *Proc Natl Acad Sci USA* 2011;**108**:14234–9.
55. Seki T, Yuasa S, Fukuda K. Generation of induced pluripotent stem cells from a small amount of human peripheral blood using a combination of activated T cells and Sendai virus. *Nat Protoc* 2012;**7**:718–28.
56. Chen I-P, et al. Induced pluripotent stem cell reprogramming by integration-free Sendai virus vectors from peripheral blood of patients with craniometaphyseal dysplasia. *Cell Reprogram* 2013;**15**:503–13.
57. Jin Z-B, Okamoto S, Xiang P, Takahashi M. Integration-free induced pluripotent stem cells derived from retinitis pigmentosa patient for disease modeling. *Stem Cells Transl Med* 2012;**1**:503–9.
58. Ono M, et al. Generation of induced pluripotent stem cells from human nasal epithelial cells using a Sendai virus vector. *PLoS One* 2012;**7**:e42855.

59. Nishishita N, Takenaka C, Fusaki N, Kawamata S. Generation of human induced pluripotent stem cells from cord blood cells. *J Stem Cells* 2011;**6**:101–8.

60. Marson A, et al. Connecting microRNA genes to the core transcriptional regulatory circuitry of embryonic stem cells. *Cell* 2008;**134**:521–33.

61. Anokye-Danso F, et al. Highly efficient miRNA-mediated reprogramming of mouse and human somatic cells to pluripotency. *Cell Stem Cell* 2011;**8**:376–88.

62. Hu S, et al. MicroRNA-302 increases reprogramming efficiency via repression of NR2F2. *Stem Cells* 2013;**31**:259–68.

63. Lu D, et al. MiR-25 regulates Wwp2 and Fbxw7 and promotes reprogramming of mouse fibroblast cells to iPSCs. *PLoS One* 2012;**7**:e40938.

64. Kaji K, et al. Virus-free induction of pluripotency and subsequent excision of reprogramming factors. *Nature* 2009;**458**:771–5.

65. Woltjen K, et al. piggyBac transposition reprograms fibroblasts to induced pluripotent stem cells. *Nature* 2009;**458**:766–70.

66. Mali P, et al. Butyrate greatly enhances derivation of human induced pluripotent stem cells by promoting epigenetic remodeling and the expression of pluripotency-associated genes. *Stem Cells* 2010;**28**:713–20.

67. Hackett PB, Largaespada DA, Cooper LJN. A transposon and transposase system for human application. *Mol Ther* 2010;**18**:674–83.

68. Muenthaisong S, et al. Generation of mouse induced pluripotent stem cells from different genetic backgrounds using Sleeping beauty transposon mediated gene transfer. *Exp Cell Res* 2012;**318**:2482–9.

69. Davis RP, et al. Generation of induced pluripotent stem cells from human foetal fibroblasts using the Sleeping Beauty transposon gene delivery system. *Differentiation* 2013;**86**:30–7.

70. Yu J, et al. Human induced pluripotent stem cells free of vector and transgene sequences. *Science* 2009;**324**:797–801.

71. Chou B-K, et al. Efficient human iPS cell derivation by a non-integrating plasmid from blood cells with unique epigenetic and gene expression signatures. *Cell Res* 2011;**21**:518–29.

72. Hu K, et al. Efficient generation of transgene-free induced pluripotent stem cells from normal and neoplastic bone marrow and cord blood mononuclear cells. *Blood* 2011;**117**:e109–19.

73. Hong H, et al. Suppression of induced pluripotent stem cell generation by the p53-p21 pathway. *Nature* 2009;**460**:1132–5.

74. Okita K, et al. A more efficient method to generate integration-free human iPS cells. *Nat Methods* 2011;**8**:409–12.

75. Okita K, et al. An efficient nonviral method to generate integration-free human-induced pluripotent stem cells from cord blood and peripheral blood cells. *Stem Cells* 2013;**31**:458–66.

76. Dietz WM, et al. Minicircle DNA is superior to plasmid DNA in eliciting antigen-specific CD8+ T-cell responses. *Mol Ther* 2013;**21**:1526–35.

77. Jia F, et al. A nonviral minicircle vector for deriving human iPS cells. *Nat Methods* 2010;**7**:197–9.

78. Narsinh KH, et al. Generation of adult human induced pluripotent stem cells using nonviral minicircle DNA vectors. *Nat Protoc* 2011;**6**:78–88.

79. Obokata H, et al. Stimulus-triggered fate conversion of somatic cells into pluripotency. *Nature* 2014;**505**:641–7. Retracted.

80. Obokata H, et al. Bidirectional developmental potential in reprogrammed cells with acquired pluripotency. *Nature* 2014;**505**:676–80. Retracted.

81. Simara P, Motl J, Kaufman DS. Pluripotent stem cells and gene therapy. *Transl Res* 2013;**161**(4):284–92. http://dx.doi.org/10.1016/j.trsl.2013.01.001.

82. Dimos JT, et al. Induced pluripotent stem cells generated from patients with ALS can be differentiated into motor neurons. *Science* 2008;**321**:1218–21.

83. Chen H, et al. Modeling ALS with iPSCs reveals that mutant SOD1 misregulates neurofilament balance in motor neurons. *Cell Stem Cell* 2014:1–14. http://dx.doi.org/10.1016/j.stem.2014.02.004.

84. Egawa N, et al. Drug screening for ALS using patient-specific induced pluripotent stem cells. *Sci Transl Med* 2012;**4**:145ra104.

85. Cai S, Chan Y, Shum DK. Induced pluripotent stem cells and neurological disease models. *Acta Physiol Sin* 2014;**66**:55–66.

86. Chang T, et al. Brief report: phenotypic rescue of induced pluripotent stem cell-derived motoneurons of a spinal muscular atrophy patient. *Stem Cells* 2011;**29**:2090–3.

87. Nihei Y, et al. Enhanced aggregation of androgen receptor in induced pluripotent stem cell-derived neurons from spinal and bulbar muscular atrophy. *J Biol Chem* 2013;**288**:8043–52.

88. Cooper O, et al. Pharmacological rescue of mitochondrial deficits in iPSC-derived neural cells from patients with familial Parkinson's disease. *Sci Transl Med* 2012;**4**:141ra90.

89. Rakovic A, et al. Phosphatase and tensin homolog (PTEN)-induced putative kinase 1 (PINK1)-dependent ubiquitination of endogenous Parkin attenuates mitophagy: study in human primary fibroblasts and induced pluripotent stem cell-derived neurons. *J Biol Chem* 2013;**288**:2223–37.

90. Freed CR, et al. Transplantation of embryonic dopamine neurons for Severe Parkinson's disease. *N. Engl J Med* 2001;**344**:710–9.

91. Olanow CW, et al. Expedited Publication A Double-blind Controlled Trial of Bilateral Fetal Nigral Transplantation in Parkinson 's Disease. *Ann Neurol* 2003;**54**:403–14.

92. Wernig M, et al. Neurons derived from reprogrammed fibroblasts functionally integrate into the fetal brain and improve symptoms of rats with Parkinson's disease. *Proc Natl Acad Sci USA* 2008;**105**:5856–61.

93. Hargus G, et al. Differentiated Parkinson patient-derived induced pluripotent stem cells grow in the adult rodent brain and reduce motor asymmetry in Parkinsonian rats. *Proc Natl Acad Sci USA* 2010;**107**:15921–6.

94. Lee G, et al. Large-scale screening using familial dysautonomia induced pluripotent stem cells identifies compounds that rescue IKBKAP expression. *Nat Biotechnol* 2012;**30**:1244–8.

95. Kaitin KI. Obstacles and opportunities in new drug development. *Clin Pharmacol Ther* 2008;**83**:210–2.

96. Sollano J a, Kirsch JM, Bala MV, Chambers MG, Harpole LH. The economics of drug discovery and the ultimate valuation of pharmacotherapies in the marketplace. *Clin Pharmacol Ther* 2008;**84**:263–6.

97. Marchetto MCN, et al. A model for neural development and treatment of Rett syndrome using human induced pluripotent stem cells. *Cell* 2010;**143**:527–39.

98. Xu Y, et al. Revealing a core signaling regulatory mechanism for pluripotent stem cell survival and self-renewal by small molecules. *Proc Natl Acad Sci USA* 2010;**107**:8129–34.

99. Desbordes SC, Studer L. Adapting human pluripotent stem cells to high-throughput and high-content screening. *Nat Protoc* 2013;**8**:111–30.

100. Ben-David U, et al. Selective elimination of human pluripotent stem cells by an oleate synthesis inhibitor discovered in a high-throughput screen. *Cell Stem Cell* 2013;**12**:167–79.

101. Desbordes SC, et al. High-throughput screening assay for the identification of compounds regulating self-renewal and differentiation in human embryonic stem cells. *Cell Stem Cell* 2009;**2**:602–12.

102. Kunisada Y, Tsubooka-Yamazoe N, Shoji M, Hosoya M. Small molecules induce efficient differentiation into insulin-producing cells from human induced pluripotent stem cells. *Stem Cell Res* 2012;**8**:274–84.

103. Choi SM, et al. Efficient drug screening and gene correction for treating liver disease using patient-specific stem cells. *Hepatology* 2013;**57**:2458–68.

104. Xu X, Lei Y, Luo J, Wang J, Zhong Z. Prevention of β-amyloid induced toxicity in human iPS cell-derived neurons by inhibition of cyclin-dependent kinases and associated cell cycle events. *Stem Cell Res* 2013;**10**:213–27.

105. Tebas P, et al. Gene editing of CCR5 in autologous CD4 T cells of persons infected with HIV. *N. Engl J Med* 2014;**370**:901–10.

106. Hanna J, et al. Treatment of sickle cell anemia mouse model with iPS cells generated from autologous skin. *Science* 2007;**318**:1920–3.

107. Sebastiano V, et al. In situ genetic correction of the sickle cell anemia mutation in human induced pluripotent stem cells using engineered zinc finger nucleases. *Stem Cells* 2011;**29**:1717–26.

108. Zou J, Mali P, Huang X, Dowey SN, Cheng L. Site-specific gene correction of a point mutation in human iPS cells derived from an adult patient with sickle cell disease. *Blood* 2011;**118**:4599–608.

109. Sun N, Zhao H. Seamless correction of the sickle cell disease mutation of the HBB gene in human induced pluripotent stem cells using TALENs. *Biotechnol Bioeng* 2013:1–6. http://dx.doi.org/10.1002/bit.25018.

110. High K a. Update on progress and hurdles in novel genetic therapies for hemophilia. *Hematology Am Soc Hematol Educ Program* 2007:466–72. http://dx.doi.org/10.1182/asheducation-2007.1.466.

111. Hasbrouck NC, High K a. AAV-mediated gene transfer for the treatment of hemophilia B: problems and prospects. *Gene Ther* 2008;**15**:870–5.

112. Kren BT, et al. Nanocapsule-delivered Sleeping Beauty mediates therapeutic factor VIII expression in liver sinusoidal endothelial cells of hemophilia A mice. *J Clin Invest* 2009;**119**:2086–99.

113. Scott DW, Lozier JN. Gene therapy for haemophilia: prospects and challenges to prevent or reverse inhibitor formation. *Br J Haematol* 2012;**156**:295–302.

114. Xu D, et al. Phenotypic correction of murine hemophilia A using an iPS cell-based therapy. *Proc Natl Acad Sci USA* 2009;**106**:808–13.

115. Yakura Y, et al. An induced pluripotent stem cell-mediated and integration-free factor VIII expression system. *Biochem. Biophys Res Commun* 2013;**431**:336–41.

116. Moldovan G-L, D'Andrea AD. How the fanconi anemia pathway guards the genome. *Annu. Rev Genet* 2009;**43**:223–49.

117. Tolar J, et al. Stem cell gene therapy for fanconi anemia: report from the 1st international Fanconi anemia gene therapy working group meeting. *Mol Ther* 2011;**19**:1193–8.

118. Tolar J, et al. Gene therapy for Fanconi anemia: one Step closer to the clinic. *Hum Gene Ther* 2012;**23**(2):141–4.

119. Raya A, et al. Disease-corrected haematopoietic progenitors from Fanconi anaemia induced pluripotent stem cells. *Nature* 2009;**460**:53–9.

120. Kaufman DS. Toward clinical therapies using hematopoietic cells derived from human pluripotent stem cells. *Blood* 2009;**114**:3513–23.

121. Slukvin II. Hematopoietic specification from human pluripotent stem cells: current advances and challenges toward de novo generation of hematopoietic stem cells. *Blood* 2013;**122**(23):4035–56.

122. Soldner F, et al. Generation of isogenic pluripotent stem cells differing exclusively at two early onset Parkinson point mutations. *Cell* 2011;**146**:318–31.

123. An MC, et al. Genetic correction of Huntington's disease phenotypes in induced pluripotent stem cells. *Cell Stem Cell* 2012;**11**:253–63.

124. Liu G-H, et al. Targeted gene correction of laminopathy-associated LMNA mutations in patient-specific iPSCs. *Cell Stem Cell* 2011;**8**:688–94.

125. Park I-H, et al. Disease-specific induced pluripotent stem cells. *Cell* 2008;**134**:877–86.

126. Rashid ST, et al. Technical Advance Modeling inherited metabolic disorders of the liver using human induced pluripotent stem cells. *J Clin Invest* 2010;**120**:3127–36.

127. Garate Z, Davis BR, Quintana-Bustamante O, Segovia JC. New frontier in regenerative medicine: site-specific gene correction in patient-specific induced pluripotent stem cells. *Hum Gene Ther* 2013;**24**:571–83.

128. Tang C, Weissman IL, Drukker M. Embryonic stem cell immunobiol. Methods Protoc. In: Zavazava N, editor. *Methods Mol. Biol,* ;vol. 1029. Humana Press; 2013. pp. 17–31.

129. Dressel R, et al. Pluripotent stem cells are highly susceptible targets for syngeneic, allogeneic, and xenogeneic natural killer cells. *FASEB J* 2010;**24**:2164–77.

130. Tangvoranuntakul P, et al. Human uptake and incorporation of an immunogenic nonhuman dietary sialic acid. *Proc Natl Acad Sci USA* 2003;**100**:12045–50.

131. Chung T-L, et al. Ascorbate promotes epigenetic activation of CD30 in human embryonic stem cells. *Stem Cells* 2010;**28**:1782–93.

132. Kim K, et al. Epigenetic memory in induced pluripotent stem cells. *Nature* 2010;**467**:285–90.

133. Kamao H, et al. Characterization of human induced pluripotent stem cell-derived retinal pigment epithelium cell sheets aiming for clinical application. *Stem Cell Reports* 2014;**2**:205–18.

134. Kanemura H, et al. Tumorigenicity studies of induced pluripotent stem cell (iPSC)-derived retinal pigment epithelium (RPE) for the treatment of age-related macular degeneration. *PLoS One* 2014;**9**:e85336.

Chapter 3

Genome Engineering for Therapeutic Applications

Pratiksha I. Thakore[1] and Charles A. Gersbach[1,2,3]

[1]Department of Biomedical Engineering, Duke University, Durham, North Carolina, USA; [2]Institute for Genome Sciences and Policy, Duke University, Durham, North Carolina, USA; [3]Department of Orthopedic Surgery, Duke University Medical Center, Durham, North Carolina, USA

LIST OF ABBREVIATIONS

AAV Adeno-associated vector
CFTR Cystic fibrosis transmembrane conductance regulator
CoDA Context-dependent assembly
CRISPR Clustered, regularly interspaced short palindromic repeats
crRNA CRISPR RNA
dCas9 Catalytically inactive, "dead" Cas9
DMD Duchenne muscular dystrophy
gRNA Guide RNA
HDR Homology-directed repair
Indels Small insertions or deletions
iPSCs Induced pluripotent stem cells
KRAB Krüppel-associated box
MASPIN Mammary serine protease inhibitor
NHEJ Nonhomologous end joining
NmCas9 Cas9 protein derived from *N. meningitidis*
OPEN Oligomerized pool engineering
PAM Protospacer-adjacent motif
RNAi RNA interference
RVD Repeat-variable diresidue
SCID Severe combined immune deficiency
SpCas9 Cas9 protein derived from *S. pyogenes*
StCas9 Cas9 protein derived from *S. thermophilus*
TALE Transcription activator-like effector
tracrRNA Trans-activating CRISPR RNA
ZFP Zinc finger proteins

3.1 INTRODUCTION

Genome engineering technologies based on synthetic nucleases and transcription factors enable the targeted modification of the sequence and expression of genes. These engineered nucleases and transcription factors typically consist of a DNA-binding domain linked to an effector module. Zinc finger proteins (ZFPs) and transcription activator-like effector (TALE) DNA-binding domains were discovered in nature and systems have been developed to engineer synthetic versions of these proteins with the potential to recognize any nucleotide sequence in the genome. More recently, RNA-guided targeting of DNA sequences through the clustered regularly interspaced short palindromic repeats (CRISPR)/Cas system has further simplified custom genome engineering, obviating the need for the complex protein engineering necessary for the ZFP- and TALE-based systems and therefore enabling widespread use of genome engineering. Effector modules that can be attached to targeted DNA-binding domains include endonuclease catalytic domains, transcriptional activators, and transcriptional repressors. With the potential to control the expression of any gene or modify any sequence in the human genome, these proteins have promising therapeutic potential, with zinc finger proteins in phase II clinical trials and multiple biotechnology companies emerging with a focus on genome engineering-based medicine. Synthetic nucleases have been used to guide targeted therapeutic gene addition, correct pathogenic mutations, and knock out disease-associated genes. Synthetic transcription factors have the potential to activate cellular reprogramming for regenerative medicine and modulate aberrant gene expression to treat cancer and other genetic diseases. This chapter reviews each of these technologies and their therapeutic applications (Table 3.1).

3.2 CUSTOMIZABLE DNA-TARGETING PROTEINS

Zinc fingers and TALEs are modular DNA-binding proteins isolated from nature whose amino acid composition can be modified to bind custom, contiguous DNA sequences with high specificity. In contrast, the RNA-guided CRISPR/Cas9 system exploits RNA:DNA base pair complementarity for DNA targeting. Presented here is a brief overview of the DNA-binding mechanism, target site design, assembly methods, and delivery strategies for each DNA-targeting protein (Table 3.2).

3.2.1 Zinc Finger Proteins (ZFPs)

ZFPs are the oldest and most established of the three classes of engineered DNA-binding proteins.[39–42] Cys_2His_2 ZFPs are the most common DNA-binding domains in the human proteome. Each zinc finger domain is 30 amino acids long and forms a $\beta\beta\alpha$ configuration, where individual amino acids in the α-helix interact with three successive nucleotide bases in the major groove of DNA (Figure 3.1(a)).[39,41] Arrays of several engineered zinc finger domains can be generated using a highly conserved linker sequence to connect the individual DNA-binding modules.[39–42] This method allows the creation of polydactyl ZFPs that recognize long sequences of DNA with high specificity.

Multiple methods are available for selecting, designing, and building ZFPs targeted to new sites. Several modular assembly libraries have been generated to target each of the 64 possible nucleotide triplets.[36,43–45] Depending on the flexibility of target site selection, it may be necessary to build and design several proteins to find one with high activity at its genomic target site.[46–48] Adjacent modules in a polydactyl zinc finger have cooperative effects that can affect activity of the entire DNA-binding array.[49] ZFP assembly strategies that account for this limitation include oligomerized pool engineering (OPEN), a library selection-based method, and context-dependent assembly (CoDA), an open-source library of zinc finger duos that have been validated to bind in concert.[50,51] Libraries of prevalidated two-finger modules can also be used to address this issue.[52,53] Alternatively, custom, validated ZFPs are also available commercially from Sigma–Aldrich.

3.2.2 Transcriptional Activator-like Effectors (TALEs)

TALEs are naturally occurring proteins from the pathogenic bacteria genus *Xanthomonas* that bind and regulate genes in host plants. The DNA-binding domains of these proteins are composed of 33–35 amino acid repeats in which the 12th and 13th positions, termed the repeat variable diresidue (RVD), determine single nucleotide specificity (Figure 3.1(b)).[54,55] The base pair recognition by the RVDs follows a simple code, allowing for this technology to be reconfigured as an alternative platform for customizable DNA-binding proteins. The original RVD amino acid code for recognition was described as NI for adenine, HD for cytosine, NG for thymine, and NN for guanine or adenine, but alternatives such as NH or NK have since been developed for guanine recognition.[56,57]

TALE subunits recognize single nucleotides independently, and do not exhibit context-dependent effects like zinc fingers, lending them great flexibility in target site selection. One target sequence limitation for TALEs is that the first nucleotide must be a T,[54,55] although this requirement may become obsolete as TALEs have been developed through directed evolution to recognize any nucleotide in the first position.[58] The tremendous versatility of TALE DNA-binding domains has been demonstrated by the creation of a library of TALE nucleases targeted to 18,740 human protein-coding genes.[59]

A truncated version of TALE proteins is used for genome engineering applications, most typically consisting of a 152 amino acid truncation on the N terminus to remove plant cell-targeting domains and 63 amino acids remaining on the C terminus of the DNA-binding domain to optimize cleavage activity.[60] TALE activators also operated efficiently with a minimal C terminal region, but were less tolerant of truncations to the N terminus.[61]

Rapid assembly of custom TALEs is possible with Golden Gate assembly,[37] solid-phase synthesis,[62,63] or ligation-independent assembly.[64] Many of these methods employ distinct, position-dependent single-stranded DNA overhangs to concatemerize TALE domains with various RVDs in a predefined order. Solid-phase synthesis and ligation-independent strategies have been particularly configured for high-throughput, automated TALE production using specialized protocols

TABLE 3.1 Examples of Preclinical Development of Genome Engineering for Therapeutic Applications

Type of Modification	Target Gene	Therapeutic Application	DNA-Targeting Platform	Reference
Gene addition	ROSA26	Safe-harbor site	Zinc finger nuclease	1
	AAVS1	Safe-harbor insertion of gp91phox minigene for X-linked chronic granulomatous disease	Zinc finger nuclease	2
Gene correction	IL2Rγ	HDR-based correction in primary T cells or CD34+ stem cells for X-linked SCID	Zinc finger nuclease	3,4
	β-globin	HDR-based correction in iPSCs for sickle cell anemia	Zinc finger nuclease	5,6
	CFTR	HDR-based correction in intestinal stem cells for cystic fibrosis	CRISPR/Cas9	7
	SERPINA1	HDR-based correction in hepatic cells for α_1-antitrypsin deficiency	Zinc finger nuclease	8
	Factor IX	In vivo correction in the liver for hemophilia B	Zinc finger nuclease	9,10
	Dystrophin	Reading frame restoration by NHEJ in primary myoblasts for DMD	TALE nuclease	11
	Crygc	In vivo correction of dominant negative cataract disorder	CRISPR/Cas9	12
	Fah	In vivo correction of hereditary tyrosinemia	CRISPR/Cas9	13
Gene disruption	CCR5	Gene knockout in T cells for resistance to HIV infection	Zinc finger nuclease	14
	CCR5	Gene knockout in iPSCs for resistance to HIV infection	CRISPR/Cas9	15
	mtDNA	Disrupt mtDNA expressing mutations for Leber's hereditary optic neuropathy and dystonia	TALE nuclease	16
	T-cell receptor, α- and β-chains	Disrupt native T-cell receptor gene for cancer immunotherapy	Zinc finger nucleases	17,18
	Myostatin	Increase muscle mass and fiber size for treatment of sarcopenia and DMD	TALE nuclease	19
	HLA-A2	Disrupted HLA genes in T cells and embryonic stem cells to prevent immune rejection for allogeneic cell therapy	Zinc finger nuclease	20
	PCSK9	In vivo gene knockout to increase LDL receptor levels and decrease cholesterol levels	CRISPR/Cas9	21
Gene activation	MASPIN	Activate tumor suppressor to suppress breast cancer tumor growth	ZFP-VP64 fusion	22–24
	γ-globin	Upregulate compensatory embryonic globin to treat sickle cell anemia and β-thalassemia	ZFP-VP64, CRISPR/dCas9-VP64	25–27
	Utrophin	Activate compensatory extracellular matrix protein to treat DMD	ZFP-VP16	28,29
	GDNF	Treat neural cell degeneration in Parkinson's disease	ZFP-VP64	30
	VEGF	Activate all isoforms of VEGF to treat ischemia in diabetic neuropathy	ZFP-VP64	31–33
Gene repression	BCR-ABL	Inhibit oncogene expression	ZFP, without effector	34
	htt	Repress mutant htt expression to treat Huntington's disease	ZFP-KRAB	35

TABLE 3.2 Comparison of Genome Engineering Platforms

	Zinc Finger Proteins	Transcriptional Activator-like Effectors	CRISPR/Cas9 RNA-Guided DNA Targeting
Mechanism for targeting specificity	Protein–DNA interaction, ~3:1 nucleotides to zinc finger subunit	Protein–DNA interaction, 1:1 nucleotide to RVD subunit recognition	RNA:DNA base pair complementarity
Relative cost to generate	+++	++	+
Relative time to generate using standard cloning techniques	>7 days[36]	~5 days[37]	~3 days[38]
Target site guidelines	Rich in 5′-GNN-3′ nucleotide triplets	Target sequence must follow a T	Target sequence 18–20 base pairs and must precede a proto-spacer-adjacent motif (PAM)
Size of DNA-binding subunits	30 amino acids per nucleotide triplet	34 amino acids per nucleotide	1372 amino acid SpCas9[+] 100 base pairs gRNA molecule to target 20 nucleotides
Entered clinical trials	Yes	No	No

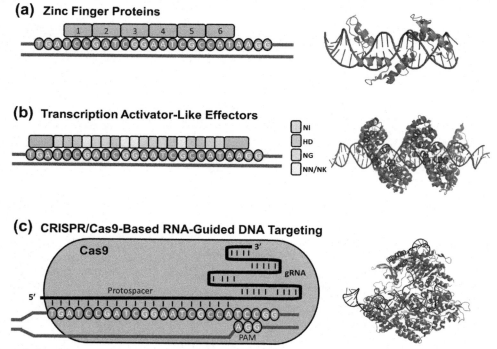

FIGURE 3.1 Three platforms for DNA-targeting. (a) Zinc fingers and (b) TALEs are modular DNA-binding proteins that interact with the major groove of DNA to recognize specific base pairs. Each zinc finger module interacts with a nucleotide triplet, whereas TALE subunits bind single base pairs. Zinc fingers have been designed to target almost all possible nucleotide triplets, although individual zinc fingers in an array can exhibit context-dependent effects that can affect targeting affinity of the entire DNA-binding domain. (c) In the CRISPR system, the Cas9 protein (purple) is directed to the target site by a guide RNA (gRNA, black), consisting of a 20 base pair protospacer that binds the complementary strand of the target sequence (blue) and a constant region that interacts with the Cas9 protein. Cas9 also requires the presence of a protospacer-adjacent motif (PAM) immediately adjacent to the target sequence. SpCas9 recognizes a 5′-NGG-3′ PAM immediately following the target sequence. Example crystal structures are shown on the right (PDB files 2I13, 3UGM, and 4OO8 for the zinc finger protein, TALE, and CRISPR/Cas9 structures, respectively). (For interpretation of the references to color in this figure legend, the reader is referred to the online version of this book.)

or equipment. Online protocols and starter kits from nonprofit plasmid repositories such as Addgene are available for these methods. Alternatively, custom TALEs can also be ordered from a variety of commercial sources.

3.2.3 RNA-Guided DNA Targeting

Multiple groups have adapted components of the type II prokaryotic CRISPR/Cas adaptive immune system to facilitate RNA-guided site-specific DNA cleavage in eukaryotic cells.[65–68] In bacteria, the CRISPR/Cas system is an adaptive defense to viral infection, in which short sequences of invading foreign DNA are incorporated into the CRISPR genomic locus, flanked by direct repeats.[69] These short sequences, termed protospacers, and flanking direct repeats are transcribed and processed into pre-crRNA. The constant region of the pre-crRNA anneals to the trans-activating crRNA (tracrRNA), and this RNA complex guides the endonuclease Cas9 to cleave foreign DNA based on sequence complementarity with the protospacer. To minimize the number of required components for genome engineering applications, the tracrRNA and crRNA have been combined into a single chimeric guide RNA (gRNA) molecule, consisting of a protospacer fused to a partial tracrRNA.[70] The use of a chimeric gRNA allows genome engineering with delivery of a single RNA molecule and a Cas9-based effector (Figure 3.1(c)). Additionally, multiplexing gRNAs allow the modification of multiple genomic loci with a single Cas9 effector.[65,66] Because specificity is determined by DNA complementarity and does not require multistep protein engineering, the CRISPR/Cas system has emerged as a straightforward, faster, and more affordable method for genome engineering compared to traditional ZFP- and TALE-based approaches.

DNA targeting by the CRISPR/Cas system is achieved by selecting a protospacer complementary to the genomic sequence of interest. Protospacers are typically 18–20 base pairs in length, and the target sequence must be immediately followed by a protospacer-adjacent motif (PAM) for recognition by Cas9 (Figure 3.1(c)). Although gRNAs delivered using a U6 RNA Pol III promoter must also contain a G at the first position for efficient transcript expression,[65] it has been widely observed that a mismatch at the 5′ end of the protospacer does not substantially reduce targeting efficiency and therefore this is not a significant restriction on site selection. The primary determinant of target site flexibility is the PAM. The *Streptococcus pyogenes* Cas9 (SpCas9), the most widely used Cas9 to date in genome editing, gene activation, and gene repression systems,[71] recognizes a 5′NGG PAM.[72] Therefore, SpCas9 protospacers can be selected an average of every eight base pairs in the genome when considering both DNA strands. Two Cas9 genes from other species, *Neisseria meningitidis* Cas9 (NmCas9) and *S. thermophilus* Cas9 (StCas9), have also been reconfigured for genome engineering in eukaryotic cells.[73,74] NmCas9 and StCas9 recognize PAMs of 5′NNNNGATT and 5′NNAGAAW, respectively, where W represents A or T.

Additionally, Cas9 systems have demonstrated the ability to target sequences in heterochromatin, cleaving at highly methylated sites such as the human SERPINB5 locus[75] and targeting DNase-inaccessible sites for activation.[25] This, combined with the flexibility conferred by multiple versions of Cas9 for target selection, lends the CRISPR/Cas system exciting versatility for genome engineering applications.

SpCas9 generates a blunt-end, double-strand break three base pairs upstream of the PAM.[70] The nuclease domains of the protein are the HNH and RuvC-like domains, which cleave the complementary and noncomplementary strands of target DNA, respectively.[76] These catalytic domains have been mutated to abolish endonuclease activity, generating a dead Cas9 (dCas9) that retains its RNA-based DNA-binding activity. Multiple effector domains, including the VP64 activator domain and the Krüppel-associated box (KRAB) repressor domain, have been fused to dCas9 to generate artificial transcription factors.[25,76–78]

3.2.4 Delivery Strategies for Targeted DNA-Binding Proteins

One of the primary barriers to clinical application of genome engineering technologies is the development of optimal delivery strategies for each type of DNA-binding domain. Transfection of plasmids encoding ZFPs, TALEs, and the CRISPR/Cas9 system is sufficient to screen the activity of artificial nucleases and transcription factors in model cell lines. For cell-based, ex vivo genome editing therapies, transfection can also be appropriate, although selection of transfected cells may be required as high efficiencies of transfection are not possible for all relevant primary cell types. Additionally, electroporation and cationic lipid-based transfection methods can be highly toxic. For applications that may require stable expression, such as endogenous gene regulation with engineered transcription factors, lentiviral delivery methods present a reliable means to achieve potent, long-term expression. Zinc finger transcription factor libraries have been generated using retroviral vectors that allow for high-throughput screening of gene activation or repression.[79,80] Cas9-based effectors and gRNAs have been delivered in lentiviral vectors,[76] and similar high-throughput knockout screening strategies have been developed with gRNA libraries.[81,82] The repeated sequences in TALE subunits make lentiviral delivery of the protein challenging, as the subunits tend to recombine during viral transduction.[83] Taking advantage of codon degeneracy, TALE proteins have been recoded to minimize repetitive sequences while maintaining amino acid sequence in order to facilitate TALE lentiviral delivery.[84]

Adeno-associated vectors (AAV) provide a promising alternative for stable expression of genome engineering proteins. AAVs are nonpathogenic, are capable of transducing dividing and nondividing cells, and are already being tested in several gene therapy clinical trials. Furthermore, an AAV-based gene therapy product, Glybera, has been approved in Europe for treatment of lipoprotein lipase deficiency. Zinc finger nucleases and transcription factors delivered by AAVs have been successful in preclinical in vivo studies.[9,10,35] The limited packaging capacity of AAVs, about 4.7 kilobases, is the primary drawback of this delivery system. Although a single AAV vector can accommodate both units of a zinc finger nuclease dimer,[85] the larger TALE nuclease monomers must be delivered on separate vectors due to the limited packaging capacity. Similarly, the gene encoding SpCas9 is 4.2 kilobases, precluding delivery of a gRNA and SpCas9 vector along with the necessary regulatory elements on a single vector. However, StCas9 and NmCas9 are orthogonal Cas9 proteins that have demonstrated gene regulation and editing capabilities in human cells and are 3.4 and 3.3 kilobases in length, respectively, which is short enough to package with a gRNA expression cassette in an AAV.[73] Further studies on the specificity of these versions of Cas9 need to be performed. Integrase-defective lentiviruses (IDLVs) are another potential clinically relevant viral delivery method that have a larger carrying capacity of greater than 10 kilobases.[86] TALE protein repeats present similar difficulties as with traditional lentiviral vectors, but IDLVs have been used to efficiently express active zinc finger nucleases in human cells.[87]

Direct delivery of purified protein is another strategy that avoids the potential regulatory complications of viral vectors. Because of their highly cationic charge, purified zinc finger nucleases can cross cell membranes in culture and have demonstrated specific activity with minimal cytotoxicity in human cells.[88] A similar approach has been used for TALE nucleases by chemical conjugation of cationic cell-penetrating peptides onto purified proteins.[89] Delivery of Cas9 protein in a complex with the gRNA has been achieved by injection of single cells,[90] electroporation,[91] or as conjugates to cell-penetrating peptides.[92]

3.3 GENOME EDITING WITH ENGINEERED NUCLEASES

Genome editing harnesses endogenous DNA repair pathways and the precise gene targeting capabilities of zinc finger proteins, TALEs, and the CRISPR/Cas9 system in order to create sequence modifications in the genome. Zinc finger nucleases, TALE nucleases, and Cas9 generate double-strand breaks in the genome that can be repaired via two pathways: nonhomologous end-joining (NHEJ) and homology-directed repair (HDR) (Figure 3.2). In NHEJ, double-strand DNA breaks are repaired by a stochastic, error-prone ligation process that often results in small insertions or deletions. NHEJ can be used to generate gene knockouts and correct out-of-frame mutations. In HDR, a donor sequence is inserted at the site of DNA fracture through homologous recombination, mediated by homology arms that flank the target site and donor sequence. Homologous recombination is rare in human cells, occurring approximately one in every 10^6–10^9 cells, but this frequency can be increased to as much as more than 10% of cells with the introduction of a double-strand break.[93,94] The actual frequency of HDR is dependent on many factors, including cell type, donor template concentration, cell cycle state, the activity of the nuclease, and the level and the duration of expression of the nuclease.[95,96] As a result of these high levels of gene editing, HDR is a powerful tool for targeted gene addition and gene sequence correction.

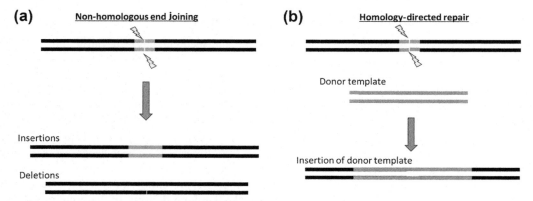

FIGURE 3.2 Double-strand breaks are repaired via two pathways. (a) Nonhomologous end joining is an error-prone pathway that often results in small insertions or deletions. (b) Homology-directed repair can result in insertion of a donor template that has significant homology to the DNA break site.

The most common nuclease effector fused to zinc finger proteins and TALEs is the catalytic domain of FokI.[60,97,98] FokI is an endonuclease that dimerizes to generate double-stranded breaks with a 5′ overhang. Cleavage only occurs when two independent zinc finger nucleases or TALE nucleases bind adjacently to opposite DNA strands, colocalizing the FokI heterodimers in head-to-head orientation. FokI has been mutated to require heterodimer formation for enzymatic activity, in order to reduce the probability of off-target events.[99–101] For zinc finger nucleases, a designed spacer length of four to seven base pairs between target sites is recommended for peak enzyme activity, depending on the length and identity of the amino acid linker used to fuse FokI to the polydactyl protein.[94,102] For TALE nucleases, spacer lengths between heterodimer binding sites of 12–20 base pairs are optimal for activity, with significantly shorter spacers tending toward reduced cleavage efficiency.[60,64]

Cas9 is an endonuclease protein that generates blunt-end double-strand breaks in order to stimulate NHEJ and HDR. Cleavage occurs on the 3′ end of the protospacer; for SpCas9, the double-strand break is located three base pairs upstream of the PAM. Cas9 has two nuclease domains, the HNH and RuvC domains, which cleave the strand complementary and noncomplementary to the gRNA strand, respectively. A D10A mutation that renders the HNH nuclease domain inactive transforms Cas9 into a nickase that generates 5′ overhangs when directed by gRNAs on opposing strands.[103,104] A Cas9-based cooperative nickase strategy for genome editing improves specificity as any off-target single-stranded DNA nicks are repaired with high fidelity.[103]

For the successful therapeutic application of genome editing, sequence-targeting fidelity is critical to prevent mutagenesis and minimize the cytotoxicity of artificial nucleases. Zinc finger and TALE specificity has been hypothesized to depend on sequence identity and binding energy of the protein as a whole.[104–106] Several clinically relevant, active zinc finger nucleases have been found to exhibit substantial off-target activity, and a minimum of three mismatches to all other possible target sequences is recommended to prevent nonspecific enzyme activity.[105,107] Similarly, TALEs are generally more sensitive to 5′ end mismatches, and 18-mer TALEs can exhibit activity with one or two mutations in target sequences.[106,108,109]

Studies have also explored the specificity of CRISPR/Cas-based targeting. Off-target CRISPR/Cas activity is dependent on protospacer sequence and varies based on the combination of mismatch number and the position of mismatches.[75,104,110,111] Single mismatches that disrupt the PAM sequence greatly diminish Cas9 activity.[112] Generally, mismatches in the PAM-distal region on the 5′ end of the protospacer are tolerated more than PAM-proximal, 3′ mismatches. Tolerance of single mismatches anywhere in the protospacer region, however, has the potential to generate unintended cleavage events. For multiple mismatches in the target sequence, adjacent mismatches are more likely to result in off-target nuclease activity than nonadjacent mismatches.

Methods for robustly predicting off-target activity of artificial nucleases are under development. Several online resources are available for sequence selection, some of which include scoring of off-target sites.[37,75,113] Strategies to increase sequence fidelity include reducing the amount of enzyme delivered, but this also reduces on-target activity.[104,105,110] Decreasing the length of the protospacer target may also be an effective means for selectively decreasing off-target activity by reducing excess binding energy.[114] Directed evolution of the catalytic domains for Cas9 nucleases is a promising future approach for improving specificity. For clinical applications, deep sequencing of predicted off-target sites in relevant cells treated with the engineered nuclease can reveal potential unintended mutations to the genome.

3.3.1 Targeted Gene Addition

Conventional gene addition methods such as retroviral and lentiviral delivery that rely on random integration have the potential to disrupt coding regions or local gene regulation, which can lead to adverse effects when applied clinically.[115] Additionally, uncontrolled random integration is variable in copy number and is affected by the local chromatin state at the site of insertion. As a result, expression profiles of the transgene are highly heterogeneous within a modified cell population. HDR can be harnessed to target gene addition at double-strand break locations induced by artificial nucleases. In human and mice genomes, "safe-harbor" loci have been identified where inserted genes are expressed ubiquitously in the organism without observable undesirable side effects. ROSA26 has been identified as one such locus on chromosome 6 in the murine genome. Zinc finger nucleases directed to the ROSA26 genome were used to integrate a GFP transgene via HDR.[1] Targeted integration at the ROSA26 locus resulted in uniform transgene expression compared to a variable profile exhibited by clones derived from random donor plasmid integration. In the human genome, the AAVS1 locus, a common integration site of AAV, has been identified as a nonpathogenic safe-harbor location for robust transgene expression.[116,117] The AAVS1 locus is generally epigenetically accessible and is shielded from trans-activation or repression by flanking insulator elements. Zinc finger nucleases were used to insert a therapeutic gp91[phox] minigene at the AAVS1 locus in induced pluripotent stem cells (iPSCs) derived from X-linked chronic granulomatous patients, achieving functional correction.[2]

3.3.2 Gene Correction

The ability of engineered nucleases to invoke HDR and NHEJ pathways has been exploited to restore the expression of mutated genes for monogenic disease therapies. In the presence of a donor template, targeted nuclease-facilitated HDR is a powerful tool for scarless gene correction. Optimized zinc finger nucleases targeted to exon 5 of IL2Rγ, the most commonly mutated region in X-linked severe combined immune deficiency (SCID), achieved 20% HDR-based correction rates in primary human CD4+ T cells when delivered with a donor template.[3] This seminal work presented a model for the generation of therapeutic nucleases for autologous ex vivo treatment strategies. More recently, this approach has been extended to human CD34+ hematopoietic stem cells that are capable of reconstituting the complete immune system.[4] An autologous, broadly applicable cell source for such strategies are iPSCs. Sickle cell anemia patient-derived iPSCs have been treated with zinc finger nucleases and donor vectors for correction of β-globin gene mutations.[5,6] Sequencing of selected clones elucidated the specificity of these ZFP pairs: no sites with a single nucleotide mismatch, including sites in adjacent δ- and γ-globin genes, demonstrated off-target cleavage events.[6] Autologous ex vivo strategies driven by HDR are not limited to iPSCs. The repair of cystic fibrosis-causing mutations in the cystic fibrosis transmembrane conductance regulator (CFTR) gene was exhibited in patient-derived intestinal stem cells using the CRISPR/Cas9 system.[7] The two active gRNAs, targeted to treat the most common CFTR mutation in cystic fibrosis, demonstrated minimal off-target cleavage. Gene correction with the CRISPR/Cas9 system has also been demonstrated in vivo following delivery to the liver in a mouse model of hereditary tyrosinemia.[13] Plasmid-based delivery of Cas9 and the gRNA, combined with an oligonucleotide donor to direct gene correction, led to correction of the disease phenotype in these mice.

Because of low recombination efficiencies, donor templates in HDR-based correction strategies often include an antibiotic resistance cassette to facilitate clonal selection.[5,6,8] The selection marker can be excised from verified, corrected clones through Cre-mediated recombination or through transposition by the piggybac system.[8] Piggybac offers the potential for scarless integration, since no sequences are left behind following excision of the selectable marker. This is in contrast to Cre-mediated excision that leaves a 34 base pair loxP site in the genome. The *piggyBac*-based approach was demonstrated in combination with zinc finger nucleases designed for biallelic correction of the Glu342Lys point mutation that is the basis of α$_1$-antitrypsin deficiency in hepatic cells.[8]

HDR with custom nucleases can be used to replace entire mutation hotspot regions to treat a variety of mutations for a particular monogenic disease. In hemophilia B, 95% of disease-causing mutations occur across the coding sequence of Factor IX in exons 2–8. Zinc finger nucleases were used to insert a target vector encoding WT exons 2–8 preceded by a slice acceptor.[9,10] When zinc finger nucleases were delivered via a liver-tropic AAV8 to a humanized mouse model of hemophilia, approximately 40% of alleles in liver cells were modified by the nuclease. When a donor AAV containing Factor IX exons 2–8 was also delivered, 2–3% gene correction resulted, as assayed by direct sequencing. This is a useful method to treat monogenic diseases that can be caused by a variety of mutations to the same gene.

Some mutated genes can be corrected by NHEJ without the need for delivery of a donor repair template. For example, Duchenne muscular dystrophy (DMD) is an X-linked disorder caused by deleterious mutations in the dystrophin gene that often disrupt the reading frame of this critical skeletal muscle protein. Importantly, many of the internal regions of the dystrophin protein are dispensable, and a truncated protein may be partially functional so long as the reading frame and terminal regions are intact.[118] Repair of double-strand breaks by NHEJ is an error-prone process that often results in small insertions or deletions (indels) at the fracture site. TALE nucleases (TALENs) targeted to exon 51, a commonly mutated region of the dystrophin gene, created indels that corrected the disrupted reading frame and restored protein expression in myoblasts isolated from DMD patients.[11] Alternatively, exons missing from the deleted dystrophin gene have been inserted into the appropriate intron by HDR.[119] Another strategy is to use the wild-type allele to serve as a repair template for the mutant allele. For this approach, the nuclease must only cut the mutated allele, which can be accomplished by designing nucleases to specifically target sequences containing single nucleotide polymorphisms or by incorporating the mutation into the CRISPR PAM sequence. For example, a one base pair deletion in exon 3 of the Crygc gene is known to cause a dominant cataract disorder by resulting in expression of a truncated γC-crystallin protein.[12] Cas9 targeted to cleave upstream of this deletion corrected the Crygc reading frame at a rate of 20% in mouse models.[12] One of the tested gRNAs contained a mutation-contingent PAM, which served as an elegant means to target only the disease-causing allele for correction. Additionally, because this was a heterozygous model of the cataract disorder, HDR directed by the wild-type allele as the repair template resulted in an additional 40% rate of gene correction in vivo.

3.3.3 Gene Knockout

Targeted gene disruption combines engineered, targeted nucleases and endogenous NHEJ repair pathways. NHEJ is an error-prone process that often results in indels at the site of nuclease activity. These mutations can cause a frameshift or premature stop codon that abolishes wild-type gene expression. The application of zinc finger nucleases targeted to disrupt the CCR5 gene in order to confer resistance to HIV infection in T cells is the pioneering paradigm of this therapeutic strategy.[14] Sangamo Bioscience is currently conducting Phase II clinical trials for these zinc finger nucleases for CCR5 modification and re-engraftment of autologous CD4+ T cells in HIV-infected patients (NCT01252641, NCT00842634, NCT01044654).[120] Preclinical work is advancing the same strategy to the modification of autologous hematopoietic stem cells,[121] and a study produced similar results in targeting the CCR5 gene in iPSCs with the CRISPR/Cas9 system.[15]

Gene disruption has also been proposed as a possible treatment of mitochondrial DNA-based diseases, which are unique for their often heteroplasmic nature. TALE nucleases were designed to specifically cleave mitochondrial DNA that exhibited the mutations characteristic of Leber's hereditary optic neuropathy and dystonia.[16] Both of these TALE nuclease pairs increased the ratio of wild-type to mutated mitochondrial DNA expression in vitro. These are promising proof-of-principle results for the use of engineered nucleases for mitochondrial disease therapy.

Custom nucleases can also be used to create therapeutic gene knockouts. For example, TALE nucleases have been used to disrupt the myostatin gene in multiple cell lines across different species.[19] Myostatin inhibits muscle cell proliferation, and animal models of myostatin knockout have increased muscle mass. Blocking myostatin has been proposed as a possible treatment for muscle-weakening diseases such as muscular dystrophy and sarcopenia. In another study, the CRISPR/Cas9 system was used to knock out the PCSK9 gene in the mouse liver in order to increase LDL receptor levels and decrease cholesterol levels.[21] This approach has therapeutic potential for preventing cardiovascular disease.

Targeted gene knockouts created by engineered nucleases can also be employed as a complement to T cell-based cancer immunotherapy strategies.[17,18] T cells were treated with zinc finger nucleases designed to disrupt endogenous T cell receptor α- and β-chain genes, preventing the expression of the cell's native T cell receptor. These primed T cells were then transduced with a lentivirus encoding a leukemia-specific T cell receptor, generating a re-engineered lymphocyte that demonstrated specific in vivo tumor-specific targeting. By disrupting the endogenous T cell receptors, this approach ensured that the cells expressed solely the tumor-specific antigens and that there was no cross-reactivity between the endogenous and engineered receptors.

Artificial nucleases have also been developed to improve donor–host HLA compatibility for allogeneic cell therapy. Primary human T cells and human embryonic cells were modified with TALE nucleases to disrupt HLA-A genes and, thus, prevent host T cell-recognition and immune rejection.[20] This work proposes a potential strategy for creating a widely compatible, "off-the-shelf" donor cell source for allogeneic T cell therapies without the need for HLA-type matching or immunosuppression.

3.4 SYNTHETIC TRANSCRIPTION FACTORS FOR THERAPEUTIC APPLICATIONS

Customizable DNA-binding domains can be fused to transcriptional regulators in order to control expression of a target gene. Artificial transcription factors can turn on gene expression through effectors that recruit transcriptional complexes, such as VP64. Conversely, gene expression can be suppressed through the use of DNA-binding proteins without effector modules, which act through steric blocking of transcription, or through the fusion of a repressor domain, such as the KRAB domain. Synthetic transcription factors have been developed for many gene therapy applications, including treatments for cancer, Parkinson's disease, sickle cell anemia, HIV, and Huntington's disease.

3.4.1 Activators of Gene Expression

Custom gene activators are generated by fusing a transcriptional activation domain to a DNA-binding domain, such as ZFPs,[122,123] TALEs,[60,61] or the catalytically inactive dCas9 coexpressed with a gRNA.[25,77,104,124] The most commonly used effector is VP64, a concatemer of four copies of the minimal VP16 activation domain.[122,123] Robust, synergistic activation of gene expression can be attained with multiple TALEs[125,126] or gRNAs targeted to a single gene (Figure 3.3(a)).[25,77] This multiplexing strategy is particularly well-suited for the CRISPR/Cas system, in which the single effector module, the dCas9-VP64 activator, can be delivered distinctly from multiple gRNAs, reducing the overall delivery load of this system. Using multiple, weak transcriptional activators can also increase specificity, as a single gRNA alone would be unlikely to have significant off-target effects.[25]

Statistically, target sequences for synthetic transcription factors should be at least 16 base pairs in length to be specific in the context of the three billion base pair haploid human genome. Target sites selected both upstream and downstream of the transcriptional start site have resulted in successful endogenous gene activation, but are typically within 500 base pairs of transcriptional initiation.[125,126] For TALE and dCas9-based activators, targeting is not necessarily limited to DNase-hypersensitive regions; activation of silenced genes that are located in heterochromatic regions of the genome has been possible, particularly when multiple binding sites are targeted.[25,125]

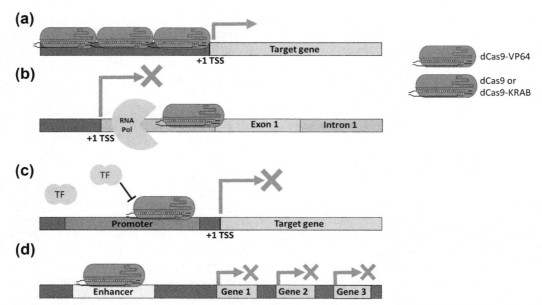

FIGURE 3.3 Mechanisms for transcriptional regulation. (a) Multiple synthetic transcription factors created with a VP64 effector domain and targeted upstream of the transcription start site (TSS) result in robust activation of a target gene. Synthetic transcription factors can repress transcription of a target gene by (b) interfering with RNA polymerase activity or (c) binding within the promoter and blocking the binding sites of endogenous transcription factors (TF). (d) Targeting regulatory elements such as enhancers could also potentially block the expression of multiple distal genes.

Targeted, synthetic transcriptional activators have been applied in cancer gene therapy to stimulate expression of endogenous tumor suppressors in cancer cells to halt tumor growth and progression. Mammary serine protease inhibitor (MASPIN) is a tumor suppressor that is epigenetically silenced in aggressive epithelial tumors. Artificial zinc finger activators targeting 18 base pair sequences in the MASPIN promoter induced apoptosis in breast cancer cells in vitro and suppressed tumor growth in a xenograft breast cancer model in nude mice.[22] Combining zinc finger activators with chromatin-modifying drugs resulted in greater activation of MASPIN and 95% inhibition of tumor cell proliferation.[23] In vivo delivery of these repressors to the tumor by nanoparticles also repressed tumor growth.[24]

Artificial transcription factor activators can also be used to upregulate compensatory genes in disease therapies. For example, a suggested treatment for sickle cell anemia and β-thalassemia is to activate expression of the embryonic γ-globin gene in order to counteract loss of β-globin expression. Zinc finger transcription factors designed for γ-globin promoter activation for the treatment of sickle cell anemia and beta-thalassemia upregulated γ-globin expression in the K562 leukemia cell line and in primary cells.[26,27] Potent activation of γ-globin expression has also been demonstrated using CRISPR/dCas-VP64.[25] Similarly, utrophin upregulation has demonstrated such a compensatory effect in mouse models of DMD. Artificial zinc finger transcription factors designed to bind a nine base pair target sequence upregulated utrophin expression in vitro[28] or when delivered in vivo to a mouse model of muscular dystrophy.[29]

Synthetic transcription factors developed for the treatment of symptoms associated with disease have also demonstrated some success. A promising zinc finger activator in development by Sangamo Biosciences is targeted to upregulate glial cell line-derived neurotrophic factor (GDNF) to treat neural cell degeneration in Parkinson's disease.[30] In rat models, GDNF was upregulated with AAV-mediated ZFP delivery, and treated animals were functionally protected from dopaminergic neural damage that was induced by chemical treatment.[30] Interestingly, the authors hypothesized that artificial transcription factor-based activation may circumvent some of the toxic side effects observed in ectopic GDNF factor treatments because targeting the endogenous promoter limited GDNF expression to physiological levels.

For the treatment of ischemia associated with diabetic neuropathy, Sangamo Biosciences developed a zinc finger activator for vascular endothelial growth factor (VEGF) upregulation.[31–33] In vivo results showing the formation of intact, functional vessels in mouse ears treated with the therapeutic zinc finger protein highlighted a key advantage of VEGF-targeted zinc finger activators versus traditional delivery of VEGF cDNA. By inducing the expression of all isoforms of VEGF from the endogenous gene, the zinc finger protein-based approach avoided generating the leaky blood vessels typically associated with overexpression of a single VEGF isoform. The VEGF-targeted zinc finger transcription factors progressed to Phase IIb clinical trials for treating ischemia in diabetic neuropathy, but did not meet critical endpoints (NCT01079325), possibly due to the choice of plasmid DNA as the delivery vehicle.

For some gene therapy applications, such as tumor suppressor activation, permanent upregulation of the target gene may be undesirable once the therapy has been completed. Conditional systems have been developed for temporal and spatial control of activation. Ligand-inducible zinc finger protein- and TALE-based activators have been created based on fusing hormone receptors to the artificial transcription factors.[127–130] Light-inducible artificial transcription factors also have the potential to provide spatiotemporal control.[131,132] In these optogenetic systems, the DNA-binding and effector domains of artificial transcription factors are separated and fused to light-responsive heterodimers. Only in the presence of light of a specified wavelength do the heterodimers interact, bringing the effector domain to the DNA target and inciting gene activation. Light-inducible TALE activators have been used to upregulate endogenous neural genes in vitro and in vivo using a Cry2/Cib light-dependent dimerization system.[132]

3.4.2 Repressors of Gene Expression

Artificial transcription factor-based repressors can provide many advantages over RNA interference (RNAi), the traditional technique for targeted silencing of gene expression. RNAi relies on the endogenous microRNA pathway, and important cellular functions may be disrupted by overloading the system's components with ectopic small interfering RNAs.[133] Artificial transcription factors also provide additional flexibility for target selection, as a gene's regulatory elements, coding regions, and introns can be targeted (Figure 3.3(b), (c)). Lastly, transcriptional control allows for genome-wide regulation beyond coding genes, including manipulation of enhancers, insulators, noncoding RNAs, nuclear localized RNAs and other Pol III-dependent transcripts. Regulatory elements such as distal enhancers may be targeted to modulate multiple genes (Figure 3.3(d)).

Steric hindrance of the RNA polymerase II machinery by DNA protein interactions has been demonstrated to result in transcriptional repression using engineered DNA-binding domains. Early ZFP work demonstrated in vitro silencing of the BCR-ABL oncogene expression by a polydactyl ZFP sans effector domain.[34] Catalytically inactive dCas9 has been used for endogenous protein repression when targeted to the 5′UTR of CD71 and CXCR4 in HeLa cells.[124]

In order to increase the potency of gene inhibition, repressor effector domains have been linked to zinc finger proteins, TALEs, and dCas9.[56,76,122,134–136] The KRAB repressor domain is conserved in many naturally occurring repressors in the human genome, although the form most commonly used in engineered transcription factors is derived from the human Kox1 transcriptional repressor. The KRAB domain has been widely used to generate artificial repressors and functions by recruiting heterochromatin-forming proteins to the genomic target site.[137] Fusion of a KRAB repressor domain to zinc finger proteins targeted to Huntington's disease-causing CAG repeats decreased expression of mutant *htt* more potently than the targeted zinc finger alone, achieving 90% silencing of mutant *htt* levels in vitro.[35] AAV delivery of the *htt* zinc finger repressor resulted in 40% reduction of mutant *htt* aggregates in an in vivo mouse model.

Zinc finger repressors have also been developed for the inhibition of viral infection and replication.[136] A six-finger zinc finger repressor targeted to the HIV-1 5′ LTR promoter inhibited HIV-1 replication by 75% compared to controls in an in vitro infection assay.[134] Additionally, zinc finger repressors have been shown to inhibit expression of the herpes simplex virus and reduce viral titer by 90%.[138]

Endogenous repression has been demonstrated but not yet employed for therapeutic applications with newer TALE and CRISPR/dCas9-based technologies. Expression of TALEs specific to the Sox2 locus led to 80% gene repression in HEK293T cells when the TALEs were attached to an SID or KRAB repressor effector domain.[56] dCas9-KRAB fusions have demonstrated potent gene regulation with single gRNAs targeted to CD31 and CXCR4 in HeLa cells.[124] In some cases, the dCas9-KRAB fusion led to a greater degree of repression than delivery of dCas9 alone, which may be a result of active recruitment of heterochromatin-forming proteins, in addition to the steric hindrance effects created by the large dCas9 protein. As this technology is relatively new, it is likely that many improvements and optimizations will soon follow. Indeed, a published alternative gRNA design increased repression potency up to five-fold in a reporter system, putatively due to increased RNA stability.[139] Modifying dCas9-KRAB linker size and composition, multiplexing gRNAs for repression of a single gene, and designing targets to DNase I hypersensitivity regions are all areas for potential improvement.

The distance over which repressor domains such as KRAB exert their effects is not known and is likely highly dependent on local chromatin structure and function, although some studies suggest that the repressive effects may spread quite far from the target site.[140] This could pose a concern for the potential specificity of synthetic transcriptional repressors. In the study of the zinc finger repressors targeted to the *htt* gene, expression of neighboring genes was not affected.[35] Furthermore, CRISPR/dCas9-KRAB studies found no significant repression of off-target genes by RNA-seq analysis when validating a gRNA against a GFP reporter in a human cell line.[124] Collectively, these studies indicate that gene repression with CRISPR/Cas9 will be a useful tool to complement traditional approaches such as RNAi.

3.5 CONCLUSION

Multiple methods now exist for any laboratory with standard molecular cloning capabilities to manipulate the genome at will. From the advent of zinc finger proteins to the recent emergence of the CRISPR/Cas9 system, each subsequent technology has rendered greater target site flexibility and facility of use, enabling the precipitous expansion of genome engineering strategies for gene therapy. These new technologies allow the rapid generation of custom genome engineering strategies in the laboratory aimed at treating a bevy of diseases that previously had no cure, from Duchenne muscular dystrophy to HIV infection to cancer. Research advances are outpacing clinical results: the earliest zinc finger proteins have demonstrated promising effects in clinical trials, but are not yet approved and critical end points still need to be attained. As for TALE and CRISPR/Cas genome engineering strategies, it is as yet unknown whether promising preclinical results for these new technologies will translate to usable therapies. Further work on optimizing methods for delivery and characterizing and improving specificity will likely determine the success of genome engineering treatments for each individual disease and target tissue or cell type. For genome editing strategies, further characterization of toxicity and off-target events is essential to the application of therapies. For TALE and CRISPR/Cas-based systems, which have been imported from bacteria, potential immunogenicity events warrant further investigation. Although the particular diseases for which genome engineering will have the greatest impact are not yet known, it is certain that these new technologies will have a tremendous impact on advancing the long-awaited goal of widespread clinical translation of gene therapy.

GLOSSARY

Adeno-associated vector (AAV) Nonintegrating, DNA-based viral vector used in gene therapy applications

Cas9 Nuclease protein derived from a bacterial immune system. Cas9 cleaves DNA in a site-specific manner when directed by a chimeric guide RNA containing the protospacer target sequence or by CRISPR and CRISPR trans-activating RNA

CRISPR Clustered, regularly interspaced short palindromic repeats; an adaptive immunity system in bacteria

crRNA CRISPR RNAs containing the protospacer region

dCas9 Cas9 that has been mutated to abrogate nuclease activity

Guide RNA (gRNA) Chimeric RNA used in genome engineering for CRISPR/Cas system, combining the tracrRNA and crRNA to complex with and guide Cas9 to a target sequence using a single RNA molecule

Homology-directed repair (HDR) A double-strand break repair pathway, in which a donor strand with significant homology is inserted at the site of DNA cleavage

Indels Small insertions or deletions in DNA

KRAB Repressor domain commonly linked to DNA-binding domains in order to create synthetic transcriptional repressors

Lentivirus Integrating gene delivery viral vector that can transduce dividing and nondividing cells

Nonhomologous end-joining (NHEJ) A double-strand break DNA repair pathway in which DNA is ligated together at the site of fracture. This is an error-prone process that can result in small insertions or deletions.

Protospacer Short sequences of ~20 base pairs that encode the recognition sequence for CRISPR targets and are transcribed as part of the CRISPR RNA or chimeric guide RNA

Protospacer-adjacent motif (PAM) Short sequence contained on the 3′ end of the protospacer in the genomic target which is necessary for Cas9 recognition. The PAM sequence is present in the genome but not in the CRISPR RNA or guide RNA molecule. SpCas9 recognizes a 5′-NGG-3′ PAM

Repeat variable diresidue (RVD) Transcription activator-like effector subunit comprised of 33–35 amino acids, in which the 12th and 13th amino acid confer nucleotide specificity. Each RVD in a transcription activator-like effector targets a single nucleotide.

X-linked severe combined immune deficiency (SCID) Heritable disease caused by mutations in the Il2-Rγ gene, affecting T cells, NK cells, and B cells and severely inhibiting the immune system

Transcription activator-like effector (TALE) DNA-binding domain adapted from a plant virus, consisting of an array of repeat variable diresidue subunits that complex with DNA. TALEs can be linked to effector domains to create nucleases or synthetic transcription factors

tracrRNA Trans-activating CRISPR RNA

VP64 Activator domain consisting of four tandem copies of VP16 and commonly linked zinc fingers, TALEs, or dCas9 to create targeted activator proteins

Zinc fingers DNA-binding domain commonly found in the human proteome, consisting of ~30 amino acids that form a ββα configuration, where individual amino acids in the α-helix interact with three successive nucleotide bases. Zinc fingers can be linked to target longer, contiguous nucleotide sequences and fused to nuclease or other effector domains

ACKNOWLEDGMENTS

This work was supported by a US National Institutes of Health (NIH) Director's New Innovator Award (DP2OD008586), National Science Foundation (NSF) Faculty Early Career Development (CAREER) Award (CBET-1151035), NIH R01DA036865, NIH R21AR065956, NIH R03AR061042, the Muscular Dystrophy Association (MDA277360), a Hartwell Foundation Individual Biomedical Research Award, a

Basil O'Connor Starter Scholar Award from the March of Dimes Foundation, and an American Heart Association Scientist Development Grant (10SDG3060033) to C.A.G. P.I.T. was supported by a National Science Foundation Graduate Research Fellowship and an American Heart Association Mid-Atlantic Affiliate Predoctoral Fellowship.

REFERENCES

1. Perez-Pinera P, Ousterout DG, Brown MT, Gersbach CA. Gene targeting to the ROSA26 locus directed by engineered zinc finger nucleases. *Nucleic Acids Res* 2012;**40**(8):3741–52.
2. Zou J, Sweeney CL, Chou BK, et al. Oxidase-deficient neutrophils from X-linked chronic granulomatous disease iPS cells: functional correction by zinc finger nuclease-mediated safe harbor targeting. *Blood* 2011;**117**(21):5561–72.
3. Urnov FD, Miller JC, Lee YL, et al. Highly efficient endogenous human gene correction using designed zinc-finger nucleases. *Nature* 2005;**435**(7042):646–51.
4. Genovese P, Schiroli G, Escobar G, et al. Targeted genome editing in human repopulating haematopoietic stem cells. *Nature* 2014;**510** (7504):235–40.
5. Sebastiano V, Maeder ML, Angstman JF, et al. In situ genetic correction of the sickle cell anemia mutation in human induced pluripotent stem cells using engineered zinc finger nucleases. *Stem Cells* 2011;**29**(11):1717–26.
6. Zou J, Mali P, Huang X, Dowey SN, Cheng L. Site-specific gene correction of a point mutation in human iPS cells derived from an adult patient with sickle cell disease. *Blood* 2011;**118**(17):4599–608.
7. Schwank G, Koo BK, Sasselli V, et al. Functional repair of CFTR by CRISPR/Cas9 in intestinal stem cell organoids of cystic fibrosis patients. *Cell Stem Cell* 2013;**13**(6):653–8.
8. Yusa K, Rashid ST, Strick-Marchand H, et al. Targeted gene correction of alpha1-antitrypsin deficiency in induced pluripotent stem cells. *Nature* 2011;**478**(7369):391–4.
9. Li H, Haurigot V, Doyon Y, et al. In vivo genome editing restores haemostasis in a mouse model of haemophilia. *Nature* 2011;**475**(7355):217–21.
10. Anguela XM, Sharma R, Doyon Y, et al. Robust ZFN-mediated genome editing in adult hemophilic mice. *Blood* 2013;**122**(19):3283–7.
11. Ousterout DG, Perez-Pinera P, Thakore PI, et al. Reading frame correction by targeted genome editing restores dystrophin expression in cells from Duchenne muscular dystrophy patients. *Mol Ther* 2013;**21**(9):1718–26.
12. Wu Y, Liang D, Wang Y, et al. Correction of a genetic disease in mouse via use of CRISPR-Cas9. *Cell Stem Cell* 2013;**13**(6):659–62.
13. Yin H, Xue W, Chen S, et al. Genome editing with Cas9 in adult mice corrects a disease mutation and phenotype. *Nat Biotechnol* 2014;**32**(6): 551–3.
14. Perez EE, Wang J, Miller JC, et al. Establishment of HIV-1 resistance in CD4+ T cells by genome editing using zinc-finger nucleases. *Nat Biotechnol* 2008;**26**(7):808–16.
15. Ye L, Wang J, Beyer AI, et al. Seamless modification of wild-type induced pluripotent stem cells to the natural CCR5Delta32 mutation confers resistance to HIV infection. *Proc Natl Acad Sci USA* 2014;**111**(26):9591–6.
16. Bacman SR, Williams SL, Pinto M, Peralta S, Moraes CT. Specific elimination of mutant mitochondrial genomes in patient-derived cells by mito-TALENs. *Nat Med* 2013;**19**(9):1111–3.
17. Provasi E, Genovese P, Lombardo A, et al. Editing T cell specificity towards leukemia by zinc finger nucleases and lentiviral gene transfer. *Nat Med* 2012;**18**(5):807–15.
18. Torikai H, Reik A, Liu PQ, et al. A foundation for universal T-cell based immunotherapy: T cells engineered to express a CD19-specific chimeric-antigen-receptor and eliminate expression of endogenous TCR. *Blood* 2012;**119**(24):5697–705.
19. Xu L, Zhao P, Mariano A, Han R. Targeted myostatin gene editing in multiple mammalian species directed by a single pair of TALE nucleases. *Mol Ther Nucleic Acids* 2013;**2**:e112.
20. Torikai H, Reik A, Soldner F, et al. Toward eliminating HLA class I expression to generate universal cells from allogeneic donors. *Blood* 2013;**122**(8):1341–9.
21. Ding Q, Strong A, Patel KM, et al. Permanent alteration of PCSK9 with in vivo CRISPR-Cas9 genome editing. *Circ Res* 2014; http://dx.doi.org/ 10.1161/CIRCRESAHA.115.304351 [Epub ahead of print].
22. Beltran A, Parikh S, Liu Y, et al. Re-activation of a dormant tumor suppressor gene maspin by designed transcription factors. *Oncogene* 2007;**26**(19):2791–8.
23. Beltran AS, Sun X, Lizardi PM, Blancafort P. Reprogramming epigenetic silencing: artificial transcription factors synergize with chromatin remodeling drugs to reactivate the tumor suppressor mammary serine protease inhibitor. *Mol Cancer Ther* 2008;**7**(5):1080–90.
24. Lara H, Wang Y, Beltran AS, et al. Targeting serous epithelial ovarian cancer with designer zinc finger transcription factors. *J Biol Chem* 2012;**287**(35):29873–86.
25. Perez-Pinera P, Kocak DD, Vockley CM, et al. RNA-guided gene activation by CRISPR-Cas9-based transcription factors. *Nat Methods* 2013;**10**(10):973–6.
26. Graslund T, Li X, Magnenat L, Popkov M, Barbas 3rd CF. Exploring strategies for the design of artificial transcription factors: targeting sites proximal to known regulatory regions for the induction of gamma-globin expression and the treatment of sickle cell disease. *J Biol Chem* 2005;**280**(5):3707–14.
27. Wilber A, Tschulena U, Hargrove PW, et al. A zinc-finger transcriptional activator designed to interact with the gamma-globin gene promoters enhances fetal hemoglobin production in primary human adult erythroblasts. *Blood* 2010;**115**(15):3033–41.
28. Corbi N, Libri V, Fanciulli M, Tinsley JM, Davies KE, Passananti C. The artificial zinc finger coding gene "Jazz" binds the utrophin promoter and activates transcription. *Gene Ther* 2000;**7**(12):1076–83.

29. Strimpakos G, Corbi N, Pisani C, et al. Novel adeno-associated viral vector delivering the utrophin gene regulator jazz counteracts Dystrophic pathology in mdx mice. *J Cell Physiol* 2014;**229**(9):1283–91.

30. Laganiere J, Kells AP, Lai JT, et al. An engineered zinc finger protein activator of the endogenous glial cell line-derived neurotrophic factor gene provides functional neuroprotection in a rat model of Parkinson's disease. *J Neurosci Off J Soc Neurosci* 2010;**30**(49):16469–74.

31. Rebar EJ, Huang Y, Hickey R, et al. Induction of angiogenesis in a mouse model using engineered transcription factors. *Nat Med* 2002; **8**(12):1427–32.

32. Yu J, Lei L, Liang Y, et al. An engineered VEGF-activating zinc finger protein transcription factor improves blood flow and limb salvage in advanced-age mice. *FASEB J* 2006;**20**(3):479–81.

33. Liu PQ, Rebar EJ, Zhang L, et al. Regulation of an endogenous locus using a panel of designed zinc finger proteins targeted to accessible chromatin regions. Activation of vascular endothelial growth factor A. *J Biol Chem* 2001;**276**(14):11323–34.

34. Choo Y, Sanchez-Garcia I, Klug A. In vivo repression by a site-specific DNA-binding protein designed against an oncogenic sequence. *Nature* 1994;**372**(6507):642–5.

35. Garriga-Canut M, Agustin-Pavon C, Herrmann F, et al. Synthetic zinc finger repressors reduce mutant huntingtin expression in the brain of R6/2 mice. *Proc Natl Acad Sci USA* 2012;**109**(45):E3136–45.

36. Gonzalez B, Schwimmer LJ, Fuller RP, Ye Y, Asawapornmongkol L, Barbas 3rd CF. Modular system for the construction of zinc-finger libraries and proteins. *Nat Protoc* 2010;**5**(4):791–810.

37. Cermak T, Doyle EL, Christian M, et al. Efficient design and assembly of custom TALEN and other TAL effector-based constructs for DNA targeting. *Nucleic Acids Res* 2011;**39**(12):e82.

38. Ran FA, Hsu PD, Wright J, Agarwala V, Scott DA, Zhang F. Genome engineering using the CRISPR-Cas9 system. *Nat Protoc* 2013;**8**(11): 2281–308.

39. Gersbach CA, Gaj T, Barbas 3rd CF. Synthetic zinc finger proteins: the advent of targeted gene regulation and genome modification technologies. *Acc Chem Res* 2014;**47**(8):2309–18.

40. Urnov FD, Rebar EJ, Holmes MC, Zhang HS, Gregory PD. Genome editing with engineered zinc finger nucleases. *Nat Rev Genet* 2010; **11**(9):636–46.

41. Klug A. The discovery of zinc fingers and their applications in gene regulation and genome manipulation. *Annu Rev Biochem* 2010;**79**:213–31.

42. Perez-Pinera P, Ousterout DG, Gersbach CA. Advances in targeted genome editing. *Curr Opin Chem Biol* 2012;**16**(3–4):268–77.

43. Wright DA, Thibodeau-Beganny S, Sander JD, et al. Standardized reagents and protocols for engineering zinc finger nucleases by modular assembly. *Nat Protoc* 2006;**1**(3):1637–52.

44. Carroll D, Morton JJ, Beumer KJ, Segal DJ. Design, construction and in vitro testing of zinc finger nucleases. *Nat Protoc* 2006;**1**(3):1329–41.

45. Kim HJ, Lee HJ, Kim H, Cho SW, Kim JS. Targeted genome editing in human cells with zinc finger nucleases constructed via modular assembly. *Genome Res* 2009;**19**(7):1279–88.

46. Ramirez CL, Foley JE, Wright DA, et al. Unexpected failure rates for modular assembly of engineered zinc fingers. *Nat Methods* 2008;**5**(5):374–5.

47. Lam KN, van Bakel H, Cote AG, van der Ven A, Hughes TR. Sequence specificity is obtained from the majority of modular C_2H_2 zinc-finger arrays. *Nucleic Acids Res* 2011;**39**(11):4680–90.

48. Bhakta MS, Henry IM, Ousterout DG, et al. Highly active zinc-finger nucleases by extended modular assembly. *Genome Res* 2013;**23**(3):530–8.

49. Isalan M, Choo Y, Klug A. Synergy between adjacent zinc fingers in sequence-specific DNA recognition. *Proc Natl Acad Sci USA* 1997; **94**(11):5617–21.

50. Maeder ML, Thibodeau-Beganny S, Osiak A, et al. Rapid "open-source" engineering of customized zinc-finger nucleases for highly efficient gene modification. *Mol Cell* 2008;**31**(2):294–301.

51. Sander JD, Dahlborg EJ, Goodwin MJ, et al. Selection-free zinc-finger-nuclease engineering by context-dependent assembly (CoDA). *Nat Methods* 2011;**8**(1):67–9.

52. Isalan M, Klug A, Choo Y. A rapid, generally applicable method to engineer zinc fingers illustrated by targeting the HIV-1 promoter. *Nat Biotechnol* 2001;**19**(7):656–60.

53. Gupta A, Christensen RG, Rayla AL, Lakshmanan A, Stormo GD, Wolfe SA. An optimized two-finger archive for ZFN-mediated gene targeting. *Nat Methods* 2012;**9**(6):588–90.

54. Boch J, Scholze H, Schornack S, et al. Breaking the code of DNA binding specificity of TAL-type III effectors. *Science* 2009;**326**(5959):1509–12.

55. Moscou MJ, Bogdanove AJ. A simple cipher governs DNA recognition by TAL effectors. *Science* 2009;**326**(5959):1501.

56. Cong L, Zhou R, Kuo YC, Cunniff M, Zhang F. Comprehensive interrogation of natural TALE DNA-binding modules and transcriptional repressor domains. *Nat Commun* 2012;**3**:968.

57. Streubel J, Blucher C, Landgraf A, Boch J. TAL effector RVD specificities and efficiencies. *Nat Biotechnol* 2012;**30**(7):593–5.

58. Lamb BM, Mercer AC, Barbas 3rd CF. Directed evolution of the TALE N-terminal domain for recognition of all 5' bases. *Nucleic Acids Res* 2013;**41**(21):9779–85.

59. Kim Y, Kweon J, Kim A, et al. A library of TAL effector nucleases spanning the human genome. *Nat Biotechnol* 2013;**31**(3):251–8.

60. Miller JC, Tan S, Qiao G, et al. A TALE nuclease architecture for efficient genome editing. *Nat Biotechnol* 2011;**29**(2):143–8.

61. Zhang F, Cong L, Lodato S, Kosuri S, Church GM, Arlotta P. Efficient construction of sequence-specific TAL effectors for modulating mammalian transcription. *Nat Biotechnol* 2011;**29**(2):149–53.

62. Reyon D, Tsai SQ, Khayter C, Foden JA, Sander JD, Joung JK. FLASH assembly of TALENs for high-throughput genome editing. *Nat Biotechnol* 2012;**30**(5):460–5.

63. Reyon D, Maeder ML, Khayter C, et al. Engineering customized TALE nucleases (TALENs) and TALE transcription factors by fast ligation-based automatable solid-phase high-throughput (FLASH) assembly. In: Ausubel Frederick M, et al., editor. *Current protocols in molecular biology*; 2013. [Chapter 12]: Unit 12 6.

64. Schmid-Burgk JL, Schmidt T, Kaiser V, Honing K, Hornung V. A ligation-independent cloning technique for high-throughput assembly of transcription activator-like effector genes. *Nat Biotechnol* 2013;**31**(1):76–81.

65. Mali P, Yang L, Esvelt KM, et al. RNA-guided human genome engineering via Cas9. *Science* 2013;**339**(6121):823–6.

66. Cong L, Ran FA, Cox D, et al. Multiplex genome engineering using CRISPR/Cas systems. *Science* 2013;**339**(6121):819–23.

67. Cho SW, Kim S, Kim JM, Kim JS. Targeted genome engineering in human cells with the Cas9 RNA-guided endonuclease. *Nat Biotechnol* 2013;**31**(3):230–2.

68. Jinek M, East A, Cheng A, Lin S, Ma E, Doudna J. RNA-programmed genome editing in human cells. *eLife* 2013;**2**:e00471.

69. Wiedenheft B, Sternberg SH, Doudna JA. RNA-guided genetic silencing systems in bacteria and archaea. *Nature* 2012;**482**(7385):331–8.

70. Jinek M, Chylinski K, Fonfara I, Hauer M, Doudna JA, Charpentier E. A programmable dual-RNA-guided DNA endonuclease in adaptive bacterial immunity. *Science* 2012;**337**(6096):816–21.

71. Mali P, Esvelt KM, Church GM. Cas9 as a versatile tool for engineering biology. *Nat Methods* 2013;**10**(10):957–63.

72. Sternberg SH, Redding S, Jinek M, Greene EC, Doudna JA. DNA interrogation by the CRISPR RNA-guided endonuclease Cas9. *Nature* 2014;**507**(7490):62–7.

73. Esvelt KM, Mali P, Braff JL, Moosburner M, Yaung SJ, Church GM. Orthogonal Cas9 proteins for RNA-guided gene regulation and editing. *Nat Methods* 2013;**10**(11):1116–21.

74. Hou Z, Zhang Y, Propson NE, et al. Efficient genome engineering in human pluripotent stem cells using Cas9 from *Neisseria meningitidis*. *Proc Natl Acad Sci USA* 2013;**110**(39):15644–9.

75. Hsu PD, Scott DA, Weinstein JA, et al. DNA targeting specificity of RNA-guided Cas9 nucleases. *Nat Biotechnol* 2013;**31**(9):827–32.

76. Qi LS, Larson MH, Gilbert LA, et al. Repurposing CRISPR as an RNA-guided platform for sequence-specific control of gene expression. *Cell* 2013;**152**(5):1173–83.

77. Maeder ML, Linder SJ, Cascio VM, Fu Y, Ho QH, Joung JK. CRISPR RNA-guided activation of endogenous human genes. *Nat Methods* 2013;**10**(10):977–9.

78. Cheng AW, Wang H, Yang H, et al. Multiplexed activation of endogenous genes by CRISPR-on, an RNA-guided transcriptional activator system. *Cell Res* 2013;**23**(10):1163–71.

79. Blancafort P, Magnenat L, Barbas 3rd CF. Scanning the human genome with combinatorial transcription factor libraries. *Nat Biotechnol* 2003;**21**(3):269–74.

80. Park KS, Lee DK, Lee H, et al. Phenotypic alteration of eukaryotic cells using randomized libraries of artificial transcription factors. *Nat Biotechnol* 2003;**21**(10):1208–14.

81. Shalem O, Sanjana NE, Hartenian E, et al. Genome-scale CRISPR-Cas9 knockout screening in human cells. *Science* 2014;**343**(6166):84–7.

82. Wang T, Wei JJ, Sabatini DM, Lander ES. Genetic screens in human cells using the CRISPR-Cas9 system. *Science* 2014;**343**(6166):80–4.

83. Holkers M, Maggio I, Liu J, et al. Differential integrity of TALE nuclease genes following adenoviral and lentiviral vector gene transfer into human cells. *Nucleic Acids Res* 2013;**41**(5):e63.

84. Yang L, Guell M, Byrne S, et al. Optimization of scarless human stem cell genome editing. *Nucleic Acids Res* 2013;**41**(19):9049–61.

85. Ellis BL, Hirsch ML, Porter SN, Samulski RJ, Porteus MH. Zinc-finger nuclease-mediated gene correction using single AAV vector transduction and enhancement by Food and Drug Administration-approved drugs. *Gene Ther* 2012;**20**:35–42.

86. Banasik MB, McCray Jr PB. Integrase-defective lentiviral vectors: progress and applications. *Gene Ther* 2010;**17**(2):150–7.

87. Lombardo A, Genovese P, Beausejour CM, et al. Gene editing in human stem cells using zinc finger nucleases and integrase-defective lentiviral vector delivery. *Nat Biotechnol* 2007;**25**(11):1298–306.

88. Gaj T, Guo J, Kato Y, Sirk SJ, Barbas 3rd CF. Targeted gene knockout by direct delivery of zinc-finger nuclease proteins. *Nat Methods* 2012;**9**(8):805–7.

89. Liu J, Gaj T, Patterson JT, Sirk SJ, Barbas III CF. Cell-penetrating peptide-mediated delivery of TALEN proteins via bioconjugation for genome engineering. *PLoS One* 2014;**9**(1):e85755.

90. Cho SW, Lee J, Carroll D, Kim JS, Lee J. Heritable gene knockout in Caenorhabditis elegans by direct injection of Cas9-sgRNA ribonucleoproteins. *Genetics* 2013;**195**(3):1177–80.

91. Kim S, Kim D, Cho SW, Kim J, Kim JS. Highly efficient RNA-guided genome editing in human cells via delivery of purified Cas9 ribonucleoproteins. *Genome Res* 2014;**24**(6)1012–19.

92. Liu J, Gaj T, Patterson JT, Sirk SJ, Barbas 3rd CF. Cell-penetrating peptide-mediated delivery of TALEN proteins via bioconjugation for genome engineering. *PLoS One* 2014;**9**(1):e85755.

93. Jasin M. Genetic manipulation of genomes with rare-cutting endonucleases. *Trends Genetics TIG* 1996;**12**(6):224–8.

94. Bibikova M, Carroll D, Segal DJ, et al. Stimulation of homologous recombination through targeted cleavage by chimeric nucleases. *Mol Cell Biol* 2001;**21**(1):289–97.

95. Certo MT, Ryu BY, Annis JE, et al. Tracking genome engineering outcome at individual DNA breakpoints. *Nat Methods* 2011;**8**(8):671–6.

96. Branzei D, Foiani M. Regulation of DNA repair throughout the cell cycle. *Nat Rev Mol Cell Biol* 2008;**9**(4):297–308.

97. Kim YG, Cha J, Chandrasegaran S. Hybrid restriction enzymes: zinc finger fusions to Fok I cleavage domain. *Proc Natl Acad Sci USA* 1996;**93**(3):1156–60.

98. Mussolino C, Morbitzer R, Lutge F, Dannemann N, Lahaye T, Cathomen T. A novel TALE nuclease scaffold enables high genome editing activity in combination with low toxicity. *Nucleic Acids Res* 2011;**39**(21):9283–93.

99. Doyon Y, Vo TD, Mendel MC, et al. Enhancing zinc-finger-nuclease activity with improved obligate heterodimeric architectures. *Nat Methods* 2011;**8**(1):74–9.

100. Miller JC, Holmes MC, Wang J, et al. An improved zinc-finger nuclease architecture for highly specific genome editing. *Nat Biotechnol* 2007;**25**(7):778–85.

101. Szczepek M, Brondani V, Buchel J, Serrano L, Segal DJ, Cathomen T. Structure-based redesign of the dimerization interface reduces the toxicity of zinc-finger nucleases. *Nat Biotechnol* 2007;**25**(7):786–93.

102. Handel EM, Alwin S, Cathomen T. Expanding or restricting the target site repertoire of zinc-finger nucleases: the inter-domain linker as a major determinant of target site selectivity. *Mol Ther* 2009;**17**(1):104–11.

103. Ran FA, Hsu PD, Lin CY, et al. Double nicking by RNA-guided CRISPR Cas9 for enhanced genome editing specificity. *Cell* 2013;**154**(6):1380–9.

104. Mali P, Aach J, Stranges PB, et al. CAS9 transcriptional activators for target specificity screening and paired nickases for cooperative genome engineering. *Nat Biotechnol* 2013;**31**(9):833–8.

105. Pattanayak V, Ramirez CL, Joung JK, Liu DR. Revealing off-target cleavage specificities of zinc-finger nucleases by in vitro selection. *Nat Methods* 2011;**8**(9):765–70.

106. Guilinger JP, Pattanayak V, Reyon D, et al. Broad specificity profiling of TALENs results in engineered nucleases with improved DNA-cleavage specificity. *Nat Methods* 2014;**11**(4):429–35.

107. Gabriel R, Lombardo A, Arens A, et al. An unbiased genome-wide analysis of zinc-finger nuclease specificity. *Nat Biotechnol* 2011;**29**(9):816–23.

108. Sanjana NE, Cong L, Zhou Y, Cunniff MM, Feng G, Zhang F. A transcription activator-like effector toolbox for genome engineering. *Nat Protoc* 2012;**7**(1):171–92.

109. Meckler JF, Bhakta MS, Kim MS, et al. Quantitative analysis of TALE-DNA interactions suggests polarity effects. *Nucleic Acids Res* 2013;**41**(7):4118–28.

110. Pattanayak V, Lin S, Guilinger JP, Ma E, Doudna JA, Liu DR. High-throughput profiling of off-target DNA cleavage reveals RNA-programmed Cas9 nuclease specificity. *Nat Biotechnol* 2013;**31**(9):839–43.

111. Fu Y, Foden JA, Khayter C, et al. High-frequency off-target mutagenesis induced by CRISPR-Cas nucleases in human cells. *Nat Biotechnol* 2013;**31**(9):822–6.

112. Cradick TJ, Fine EJ, Antico CJ, Bao G. CRISPR/Cas9 systems targeting beta-globin and CCR5 genes have substantial off-target activity. *Nucleic Acids Res* 2013;**41**(20):9584–92.

113. Doyle EL, Booher NJ, Standage DS, et al. TAL effector-nucleotide targeter (TALE-NT) 2.0: tools for TAL effector design and target prediction. *Nucleic Acids Res* 2012;**40**(Web Server issue):W117–22.

114. Fu Y, Sander JD, Reyon D, Cascio VM, Joung JK. Improving CRISPR-Cas nuclease specificity using truncated guide RNAs. *Nat Biotechnol* 2014;**32**(3)279–84.

115. Baum C, Modlich U, Gohring G, Schlegelberger B. Concise review: managing genotoxicity in the therapeutic modification of stem cells. *Stem Cells* 2011;**29**(10):1479–84.

116. DeKelver RC, Choi VM, Moehle EA, et al. Functional genomics, proteomics, and regulatory DNA analysis in isogenic settings using zinc finger nuclease-driven transgenesis into a safe harbor locus in the human genome. *Genome Res* 2010;**20**(8):1133–42.

117. Lombardo A, Cesana D, Genovese P, et al. Site-specific integration and tailoring of cassette design for sustainable gene transfer. *Nat Methods* 2011;**8**(10):861–9.

118. England SB, Nicholson LV, Johnson MA, et al. Very mild muscular dystrophy associated with the deletion of 46% of dystrophin. *Nature* 1990;**343**(6254):180–2.

119. Popplewell L, Koo T, Leclerc X, et al. Gene correction of a duchenne muscular dystrophy mutation by meganuclease-enhanced exon knock-in. *Hum Gene Ther* 2013;**24**(7):692–701.

120. Tebas P, Stein D, Tang WW, et al. Gene editing of CCR5 in autologous CD4 T cells of persons infected with HIV. *N Engl J Med* 2014;**370**(10):901–10.

121. Holt N, Wang J, Kim K, et al. Human hematopoietic stem/progenitor cells modified by zinc-finger nucleases targeted to CCR5 control HIV-1 in vivo. *Nat Biotechnol* 2010;**28**(8):839–47.

122. Beerli RR, Segal DJ, Dreier B, Barbas 3rd CF. Toward controlling gene expression at will: specific regulation of the erbB-2/HER-2 promoter by using polydactyl zinc finger proteins constructed from modular building blocks. *Proc Natl Acad Sci USA* 1998;**95**(25):14628–33.

123. Beerli RR, Dreier B, Barbas 3rd CF. Positive and negative regulation of endogenous genes by designed transcription factors. *Proc Natl Acad Sci USA* 2000;**97**(4):1495–500.

124. Gilbert LA, Larson MH, Morsut L, et al. CRISPR-mediated modular RNA-guided regulation of transcription in eukaryotes. *Cell* 2013;**154**(2):442–51.

125. Perez-Pinera P, Ousterout DG, Brunger JM, et al. Synergistic and tunable human gene activation by combinations of synthetic transcription factors. *Nat Methods* 2013;**10**(3):239–42.

126. Maeder ML, Linder SJ, Reyon D, et al. Robust, synergistic regulation of human gene expression using TALE activators. *Nat Methods* 2013;**10**(3):243–5.

127. Mercer AC, Gaj T, Sirk SJ, Lamb BM, Barbas 3rd CF. Regulation of endogenous human gene expression by ligand-inducible TALE transcription factors. *ACS Synth Biol* 2013; http://dx.doi.org/10.1021/sb400114p [Epub ahead of print].

128. Magnenat L, Schwimmer LJ, Barbas 3rd CF. Drug-inducible and simultaneous regulation of endogenous genes by single-chain nuclear receptor-based zinc-finger transcription factor gene switches. *Gene Ther* 2008;**15**(17):1223–32.

129. Schwimmer LJ, Gonzalez B, Barbas 3rd CF. Benzoate X receptor zinc-finger gene switches for drug-inducible regulation of transcription. *Gene Ther* 2012;**19**(4):458–62.

130. Beerli RR, Schopfer U, Dreier B, Barbas 3rd CF. Chemically regulated zinc finger transcription factors. *J Biol Chem* 2000;**275**(42):32617–27.

131. Polstein LR, Gersbach CA. Light-inducible spatiotemporal control of gene activation by customizable zinc finger transcription factors. *J Am Chem Soc* 2012;**134**(40):16480–3.

132. Konermann S, Brigham MD, Trevino AE, et al. Optical control of mammalian endogenous transcription and epigenetic states. *Nature* 2013;**500** (7463):472–6.

133. Grimm D, Streetz KL, Jopling CL, et al. Fatality in mice due to oversaturation of cellular microRNA/short hairpin RNA pathways. *Nature* 2006;**441**(7092):537–41.

134. Reynolds L, Ullman C, Moore M, et al. Repression of the HIV-1 5′ LTR promoter and inhibition of HIV-1 replication by using engineered zinc-finger transcription factors. *Proc Natl Acad Sci USA* 2003;**100**(4):1615–20.

135. Larson MH, Gilbert LA, Wang X, Lim WA, Weissman JS, Qi LS. CRISPR interference (CRISPRi) for sequence-specific control of gene expression. *Nat Protoc* 2013;**8**(11):2180–96.

136. Segal DJ, Goncalves J, Eberhardy S, et al. Attenuation of HIV-1 replication in primary human cells with a designed zinc finger transcription factor. *J Biol Chem* 2004;**279**(15):14509–19.

137. Barde I, Laurenti E, Verp S, et al. Regulation of episomal gene expression by KRAB/KAP1-mediated histone modifications. *J Virol* 2009; **83**(11):5574–80.

138. Papworth M, Moore M, Isalan M, Minczuk M, Choo Y, Klug A. Inhibition of herpes simplex virus 1 gene expression by designer zinc-finger transcription factors. *Proc Natl Acad Sci USA* 2003;**100**(4):1621–6.

139. Chen B, Gilbert LA, Cimini BA, et al. Dynamic imaging of genomic loci in living human cells by an optimized CRISPR/Cas system. *Cell* 2013;**155**(7):1479–91.

140. Groner AC, Meylan S, Ciuffi A, et al. KRAB-zinc finger proteins and KAP1 can mediate long-range transcriptional repression through heterochromatin spreading. *PLoS Genet* 2010;**6**(3):e1000869.

Chapter 4

Immune System Obstacles to In vivo Gene Transfer with Adeno-Associated Virus Vectors

Hildegard Büning[1,2,3] and Federico Mingozzi[4,5]

[1]Center for Molecular Medicine Cologne (CMMC), University of Cologne, Cologne, Germany; [2]German Center for Infection Research (DZIF), Partner Site Bonn-Cologne, Germany; [3]Department I of Internal Medicine, University Hospital of Cologne, Cologne, Germany; [4]University Pierre and Marie Curie Paris, Institute of Myology, Paris, France; [5]Genethon, Evry, France

LIST OF ABBREVIATIONS

AAV Adeno-associated virus
AAP Assembly activating protein
BLyS B-lymphocyte stimulator
EGFR Epidermal growth factor receptor
FGFR Fibroblast growth factor receptor
HGFR Hepatocyte growth factor receptor
HSPG Heparan sulfate proteoglycan
hNPCs Human non-parenchymal liver cells
ITRs Inverted terminal repeats
IFNs Interferons
KCs Kupffer cells
LN Lupus nephritis
LSECs Liver sinusoidal endothelial cells
MVM Minute virus of mice
NAbs Neutralizing antibodies
NHP Nonhuman primates
NPC Nuclear pore complex
PAMPs Pathogen-associated molecular patterns
PBMC Peripheral blood mononuclear cells
pDCs Plasmacytoid dendritic cells
PDGFR Platelet-derived growth factor receptor
PLA Phospholipase
PRR Pattern recognition receptors
TLRs Toll-like-receptors

4.1 INTRODUCTION

The goal of gene therapy for inherited diseases is the long-term correction of the disease phenotype in affected patients. Several clinical studies document the therapeutic potential of the approach in attaining this goal[1-9] using integrating vectors (typically retroviral or lentiviral vectors) ex vivo to introduce the therapeutic gene into a stem cell[10] or non-integrating vectors to deliver the gene into a long-lived post-mitotic tissue in vivo.[11]

In particular, over the past decade, viral vectors derived from adeno-associated virus (AAV) have become the platform of choice for in vivo gene transfer,[11] mainly because of their superior transduction efficiency demonstrated in preclinical

Translating Gene Therapy to the Clinic. http://dx.doi.org/10.1016/B978-0-12-800563-7.00004-X

studies,[12–18] their tropism for a broad variety of tissues,[12–16] and their poor immunogenicity profile compared to other vectors.[19,20] Experience with in vivo administration of AAV vectors in human trials confirmed promising preclinical data, leading to some of the most exciting results in the field of gene therapy.[11] Moreover, the recent market approval of the first gene therapy in Europe,[21] an AAV-based drug, is evidence that the field is progressing from proof-of-concept studies towards a more mature stage. However, some of the limitations of the approach, not entirely identified in preclinical studies, became obvious in clinical studies. In particular, it is now clear that the host immune system represents one of the most important obstacles to be overcome in terms of both safety and efficacy of in vivo gene transfer with viral vectors, including AAV. This was well exemplified in a number of clinical studies, in which lack of efficacy, or short-lived transgene expression, were documented.[22]

Over the past 10 years, gene therapists learned a lot about the complexity of immune responses triggered by in vivo gene transfer and, at least to some extent, how to modulate these responses.[6] Studies in humans have been the main source of this knowledge, mainly because experimental animal models failed to predict outcomes of AAV vector application in humans.[23–25] In this chapter, we will summarize the progress on the understanding of immune responses to the AAV vector capsid, with special attention to the clinical development of AAV-based therapeutics and possible strategies to overcome the limitations posed by the host immune system.

4.2 AAV VECTORS

Recombinant AAV vectors are engineered from a naturally occurring parvovirus, which was first isolated as a contaminant of an adenovirus preparation.[26] AAV are small (20–25 nm in diameter), non-enveloped viruses with a single-stranded DNA genome of ~4.7 kilobases (Figure 4.1). They contain two sets of genes: the *rep* genes encoding a set of proteins required for replication, transcription control, and packaging; and the *cap* genes encoding the three capsid proteins (VP1, VP2, and VP3) and the assembly activating protein (AAP) (Figure 4.1). While the three capsid proteins that share most of the amino acid sequences form the 60-mer icosahedral viral capsid with a stoichometry of 1:1:10, respectively,[28] AAP is required to initiate capsid assembly in the nucleus.[29] The *rep* and *cap* genes are flanked by two inverted terminal repeats (ITRs) that serve as packaging signals[30]; the ITRs are the only viral sequences retained in AAV vectors, which are otherwise devoid of all viral coding sequences.

AAV is not an autonomously replicating virus, but is rather dependent on a helper virus such as adenovirus or herpes simplex virus for progeny production. Although AAV vectors are not dependent on helper virus factors for cell transduction, this dependency impacts the efficiency and safety of AAV as a delivery tool in human clinical trials. Specifically, humans typically become infected by AAV one year after birth, with a peak at three years of age,[31] a pattern that closely follows that of adenovirus and is consistent with acquisition of AAV as a consequence of an adenovirus infection.[31] This initial exposure during a co-infection may account for the generation of both antibody and memory T-cell responses to AAV (which will be discussed later). Aside from a recent report suggesting an association between spontaneous abortions and AAV infections in the genital tract,[32] AAV has never been associated with any known illnesses in the human population. In addition, studies in experimental animal models revealed a unique immune profile—i.e., low immunogenicity and induction of tolerance—valuable features for obtaining long-term gene correction. This accounts for both the wide adoption of the AAV vector platform for in vivo gene transfer and for the initial lack of extensive immunology studies on AAV.

The genome of wild-type AAV is known to integrate in a specific site of human chromosome 19[33–35]; this specificity of integration is lost in AAV vectors, whose genomes persist mainly episomally as concatemers and, to some extent, integrate non-actively and randomly in the host genome with preference for expressed genes.[36] Whether vector genome integration can lead to an increased frequency of tumor formation is a matter of debate[37–39]; thus far, no evidence of the genotoxicity of AAV vectors has emerged from long-term large animal studies[40,41] or in human trials in which subjects have been followed for years after gene transfer.

Since the initial reports of long-term expression of a transgene in immunocompetent animals following AAV vector gene delivery[42,43] and the first trials of AAV vectors in human subjects in the setting of cystic fibrosis,[44] about 100 clinical trials of in vivo gene transfer with AAV vectors have been conducted (http://www.abedia.com/wiley/vectors.php) and many more are planned.

4.2.1 Early Steps in Host–Vector Interaction: AAV Infection Biology

Comprehensive knowledge of early steps in cell transduction is required to decipher mechanisms of immune recognition and to develop strategies that overcome host-related barriers to initial cell infection and/or long-term modification. The

(a)

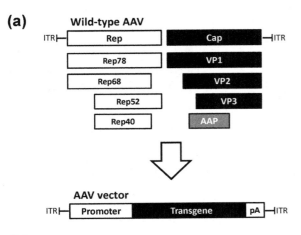

(b)

Molecular weight	
Protein (74 % or MW)	M$_r$ ~ 3750 kDa
DNA (26% of MW)	M$_r$ ~ 1350 kDa
Total MW virus	M$_r$ ~ 5100 kDa

Particle diameter:
25nm

(c)

Serotype	Host	Genome size (nt)	Homology to AAV2 (%)
1	Monkey	4718	80
2	Human	4681	100
3	Human	4722	82
4	Monkey	4767	75
5	Human	4642	55
6	Human	4683	82
7	Monkey	4721	84
8	Monkey	4393	84

FIGURE 4.1 (a) Genomic organization of wild-type AAV2 (top) and of an AAV vector (bottom). The inverted terminal repeats (ITRs) flank the two open reading frames (ORFs) encoding for four nonstructural proteins (Rep gene) and three structural proteins (Cap gene). The Cap gene encodes also for the Assembly Activating Protein (AAP). (b) Left, molecular model of the AAV capsid; right, dimensions and molecular weight of AAV particles. (c) Comparison of AAV capsid serotypes 1–8. (*Adapted from Grimm and Kay.*[27])

most comprehensive knowledge on AAV's infection biology is available for AAV serotype 2, the prototype AAV vector, and was mainly obtained on highly permissive cell types such as the cervix carcinoma cell line HeLa (Figure 4.2).

Initiating AAV's cell entry seems to require multiple contacts with receptors on the cell membrane, each lasting for approximately 62 ms, before the viral vector particles are rapidly endocytosed.[45] Knowledge of receptors is still sparse, in particular regarding the more recently identified serotypes but, given the broad tropism, receptors shared by multiple cell types are exploited. In addition, AAV—as most viruses—follows a two-step kinetic process, i.e., cell attachment through binding of a primary receptor that fosters secondary receptor binding and as a consequence induces virus/viral vector internalization.[46] In the case of AAV2, the glycan moiety of heparan sulfate proteoglycan (HSPG) serves as primary receptor.[47] The specific sequence of sugar residues was shown to fit between two neighboring protrusions at the threefold axes of the capsid formed by two capsid monomers[48] contacting R484, R487, K527, K532, R585, and R588[49,50] and being influenced by G512 and R729.[51] Binding to HSPG induces conformational changes in the capsid that are believed to prepare AAV for cell infection. Specifically, they may (1) enable binding or increased affinity for AAV2's internalization receptors (αvβ5 or α5β1 integrin, respectively), (2) prepare the viral capsid for a second conformational change required to extrude the

FIGURE 4.2 Key steps of cell infection by AAV. (a) Primary receptor binding (heparan sulfate proteoglycan (HSPG) in case of AAV2) induces a conformational change in capsid enabling binding to internalization receptors ($\alpha v\beta 5$ or $\alpha 5\beta 1$ in case of AAV2). Cell binding can be stabilized by binding to further co-receptors (for details see text). (b) AAV vectors are internalized through endocytosis. (c) Vector-containing endosomes are transported along the cytoskeleton towards the nuclear area. (d) Maturation of the endosomes induces a second conformational change that enables release of vector particles from the endosomal compartment. (e) This is followed by nuclear delivery and release of vector genomes from the capsid through mechanisms that are not yet understood.

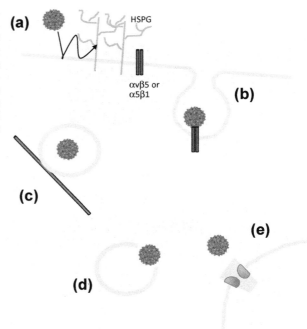

N-terminus of VP1 and maybe VP2 (discussed later), and (3) induce a more favorable interaction between the packed DNA and the inner capsid surface, perhaps for genome uncoating.[48] Besides HSPG and integrins, hepatocyte growth factor receptor (HGFR), fibroblast growth factor receptor (FGFR), and 37/67 kDa laminin receptor have been reported as co-receptors presumably supporting or stabilizing AAV2's cell membrane contacts.[52] Glycans, either again as HSPG (AAV3, AAV13) or with sialic acids (α 2–6: AAV4, α 2–3: AAV1, AAV5, AAV6) or with terminal galactose moieties (AAV9), serve also as primary receptors for other AAV serotypes.[53] Co-receptors are less well described for serotypes other than AAV2. As of yet, FGFR, 37/67 kDa laminin receptor, and $\alpha v\beta 5$ integrin have been reported for AAV3, platelet-derived growth factor receptor (PDGFR) for AAV5, epidermal growth factor receptor (EGFR) for AAV6, and 37/67 kDa laminin receptor for AAV8 and AAV9.

After the initial cell contact, receptor-mediated endocytosis of AAV occurs rapidly, i.e., in the range of 64 ms[45] via clathrin-coated vesicles initiating an endosomal trafficking process that ultimately leads to AAV's release into the cytoplasm.[52,54,55] Besides this main route, evidence for unspecific uptake[56] as well as specific uptake through alternative routes[52,57] has been provided. Endosomal vesicles can contain more than one particle and are transported along the cytoskeleton towards the nuclear area by the cellular transport system. On this "journey," the viral capsid encounters a second conformational change triggered by the constantly decreasing pH[58] and probably through the activity of cathepsin D and L.[59] It is an essential step for productive cell infection as indicated by the negligible transduction obtained after microinjection of AAV vectors directly into the cytoplasm.[58] As a consequence of this second capsid modification, N-termini of VP1 and maybe VP2 are externalized through the channels at the five-fold symmetry axis. Both N-termini contain basic regions that may function as nuclear localization sequence of incoming particles (in addition to their key role in progeny production).[60,61] Moreover, in the N'-terminus of VP1, a further sorting signal[62] and a conserved phospholipase (PLA) A2 domain is located.[63,64] These parvoviral PLAs form a separate class of phospholipases that share the catalytic domain (HDXXY motif) and the calcium-binding GXG motif with the cellular counterparts[65] and—as suggested by their name—hydrolyze phospholipid substrates at the 2-acyl-ester position.[64] Through this activity, pores are introduced into the endosomal membrane through which presumably still intact AAV particles are released into the cytoplasm.[63,64,66] Once released, AAV particles may become subjects of ubiquitination and proteasomal degradation, a process that hampers cell transduction and may result in MHC class I loading (discussed later), or succeed in delivering their payload to the nucleus. The mechanism of nuclear entry is still not understood. Similarly, it is so far unknown where and how uncoating (i.e., the release of viral/vector genome) takes place. Viral genomes were detected in cells infected with a low AAV particle per cell ratio in AAV-permissive cells in the presence of adenovirus, but no intact viral capsids were detected in the nucleus; in contrast, co-localization of capsids and genomes was observed in the perinuclear area and the cytoplasm.[67] While these results are in favor of occurrence of viral uncoating before or during nuclear entry, blockage of cell transduction by intranuclear injection of neutralizing antibodies argues for uncoating as a postnuclear entry event.[58] For the latter, capsids have to enter the nucleus. Indeed, capsids are detectable in nuclear invaginations, which are tubular structures protruding into the nuclear area,[67] as well as within the nucleus in cells infected with high particle-per-cell ratios.[68] The cluster of basic regions in VP1/VP2, the small size of the particles, and the interaction with the nuclear shuttle protein nucleolin[69] argue in favor of an intranuclear delivery of AAV particles through the nuclear pore complex (NPC); however, blockage of NPC activity was unable to impair nuclear accumulation of AAV particles.[70,71] Furthermore, the N'-terminus of VP1 does contain three PDZ-binding motifs, required for viral infectivity which, in the context of the autonomous parvovirus MVM, induced formation of holes within the nuclear envelope.[62]

Nevertheless, once released, single-stranded AAV vector genomes have to be converted to double-stranded DNA that serves as a template for transcription and for episome formation. Second-strand synthesis is a rate-limiting step and sometimes an inefficient process. Therefore, artificial vector genomes, termed self-complementary, are exploited to enhance onset and level of in vivo transgene expression.[72] These self-complementary vector genomes encode the sense and anti-sense DNA sequences on a single-stranded DNA molecule separated by an additional ITR structure that initiates DNA hybridization upon uncoating.[72]

4.3 INNATE IMMUNITY IN AAV GENE TRANSFER

Intrinsic host defense mechanisms, evolved to protect against invading pathogens including viruses, are a continuous challenge to vector application. This first line of defense does not discriminate between a virus and a viral vector, and thus induces inflammation and activates and shapes adaptive immune responses upon sensing of shared patterns. Such patterns, termed pathogen-associated molecular patterns (PAMPs), are evolutionarily conserved structures which mark the respective "source" as foreign. Recognition occurs through a limited set of germ-line encoded receptors (pattern recognition receptors, PRRs), which are either present in body fluids, activating the complement cascade upon pathogen recognition, or in cells.[73] Not only classical immune cells, but also nonimmune cells such as hepatocytes, endothelial cells, or epithelial cells express a cell-type specific set of these PRRs and are thus capable of mounting a host–defense reaction upon pathogen recognition. The expression pattern is, however, not static. Ongoing infection or other environmental stress may result in up- or downregulation of specific PRRs.[74]

Four families of cellular PPRs have been described: toll-like-receptors (TLRs), NOD-like receptors, RIG-like receptors, and C-type lectin receptors. Of these, only TLRs, specifically TLR-2 and TLR-9, have been identified as molecular sensors of AAV.[75] TLR-2 and -9, like all family members, are type I transmembrane proteins that recognize PAMPs through a leucine-rich repeat-containing ectodomain, while a toll-interleukin-1 receptor homology domain is responsible for downstream signaling.[76] Through the latter, cell autonomous immune responses are activated. In addition, secretion of cytokines/chemokines and upregulation of cell-surface molecules induce or modulate—as mentioned above—the adaptive arm of the immune system.

The two TLRs thus far identified as PRRs for AAV differ in the vector component that is sensed as well as in the cell types involved in AAV recognition. Specifically, TLR-2 is a cell-surface located PRR initially described as a sensor of pathogens displaying triacyl lipopeptides (TLR-1/2 heterodimer) and diacyl lipopeptides (TLR-2/6 heterodimer).[76] More recently, however, its role in viral recognition has become evident.[77] With regard to AAV, it was demonstrated that TLR-2 functions in sensing the AAV capsid in primary human non-parenchymal liver cells (hNPCs), specifically Kupffer cells (KCs) and liver sinusoidal endothelial cells (LSECs), eliciting a transient and vector-dose dependent NFκB-mediated inflammatory response.[75] Interestingly, and in contrast to hNPCs, challenges with AAV vectors do not alter expression of key innate immune genes in primary human hepatocytes, even though they were successfully transduced. These results confirmed earlier reports in mice on the function of KCs, the liver-resident macrophages, in anti-AAV innate immune responses,[78] but were the first to characterize LSECs as sentinel cells for AAV. LSECs both structurally and functionally differ from the macrovascular endothelial cells lining the interior surface of blood vessels. Subsequent experiments showed that the latter are also capable of sensing AAV upon receiving inflammatory signals which, as a consequence, results in upregulation of TLR-2 expression up to 300-fold.[74] While this finding confirmed involvement of TLR-2 in sensing of AAV, it may also call for a careful risk assessment prior to use of AAV vectors in patients suffering from inflammatory diseases.

Among the antigen-presenting cells, plasmacytoid dendritic cells (pDCs) are the key players in antiviral immunity.[79] In contrast to myeloid DC and monocytes, pDCs selectively express TLR-7 and TLR-9, which sense viral RNA and DNA, respectively, within the endosomal–lysosomal compartment, rapidly inducing production and release of large amounts of interferons (IFNs). Stimulating these cells with AAV vectors resulted—analogous to adenovirus vectors—in IFN-α and IFN-β production.[80] Interestingly, optimal responses required long-term stimulation (18 h).[80] Subsequent experiments in specific knockout cells confirmed involvement of TLR-9 and MyD88 in AAV recognition, while in vivo mouse models demonstrated the impact of TLR-9/MyD88 sensing of AAV vector genomes for mounting both AAV-specific humoral and T-cell responses.[80] Interestingly, in contrast to AAV vectors, neither wild-type AAV nor the autonomous parvovirus minute virus of mice (MVM) induced IFN production in pDCs, suggesting that parvoviruses have evolved escape strategies to innate immune recognition, which are not present in AAV vectors.[81] Because pDC-mediated innate immune responses occur independent of the transgene delivered by the vectors, either the single-stranded DNA and/or the viral packaging signals are sensed as PAMP. If the ITR is the target, innate immune responses ought to be increased when vectors with self-complementary genome conformations are used. Indeed, comparison of single-stranded and self-complementary AAV vectors revealed that the latter elicit increased immunogenicity.[82] Nevertheless, the transgene expression cassette may also impact on immune response against the vector and/or the transgene product.[83] Specifically, CpG motifs contained within the transgene expression cassette augmented TLR-9-dependent induction of T-cell responses against the capsid and the transgene product in the context of AAVrh32.33, but not of AAV8 vectors, after intramuscular injection in C57BL/6 mice.[83] This finding is interesting given the fact that AAVrh32.33 vectors deliver their genome within a nonnatural capsid (i.e., a hybrid capsid formed of VP protein of two AAV serotypes isolated from rhesus macaque).[84] They are unique among AAV vectors for their high immunogenicity, due to which this artificial AAV variant is currently considered a promising backbone in DNA vaccine approaches.[84,85]

In addition to cellular PRR, which either induce a TLR-2/MyD88-driven and NFκB-dependent inflammatory response or a TLR-9/MyD88-dependent interferon response, factors in body fluids also enhance anti-AAV innate immune responses. In vitro transduction experiments revealed, for example, a pivotal function of iC3b, a protein of the complement cascade, in enhancing uptake of AAV vectors into murine and human macrophages, which could be correlated with increased macrophage activation.[86] These in vitro findings are in line with mouse experiments demonstrating impaired humoral anti-AAV responses in mice deficient for the complement receptor 1/2 or the complement factor C3. The latter is cleaved upon activation of the complement cascade into C3a and C3b. C3b is catabolized into the above-mentioned iC3b, which inhibits amplification of C3 cleavage and thus results in downregulation of the complement system.[86] In addition, iC3b is deposited on the surface of pathogens, a mechanism that leads—presumably as in the case of AAV—to pathogen opsonization and enhanced phagocytosis through binding to complement receptors.[86] AAV capsids (serotypes 1, 5, and 6) also interact with human galectin 3-binding protein (hu-G3BP), a soluble scavenger receptor in sera, which seems to impair cell transduction by forming aggregates.[87] In contrast to the inhibitory effect of hu-G3BP, mouse C-reactive protein, an acute phase protein, significantly enhances AAV vector-mediated transduction (AAV-1 and AAV-6) of skeletal muscles, heart, lung, and liver.[87] Thus, species- and serotype-specific differences do exist, which may impact on the outcome of AAV's interaction with the host immune system.

4.4 T-CELL RESPONSES TO VECTORS

An important concept to keep in mind when discussing the immune responses to AAV vectors is that these recombinant viruses are complex multicomponent biological entities, and each of these components, as already exemplified in the discussion of the innate immune responses, may contribute to shaping the host immune response to gene transfer. The viral capsid in AAV vectors is, in most cases, identical or nearly identical to the capsid of the wild-type virus, to which humans are exposed (vide infra). Thus, it is expected that the immune responses triggered by vector administration will be similar to those associated with a natural infection with AAV, although the quantity of viral particles administered and the route(s) of administration itself are different and may contribute to fundamental differences in the magnitude and kinetics of immune responses. Another fundamental difference between wild-type AAV and AAV vectors is that the vector-encoded transgene product may be recognized as foreign, especially if the recipient of gene transfer is not tolerant against the protein encoded by the vector, thus contributing to the overall immunogenicity of AAV vectors.

4.4.1 Intracellular Processing and MHC I Presentation of the AAV Capsid Antigen

As outlined above, cell transduction with AAV vectors begins with binding of the virion to the cell-surface proteins and carbohydrates followed by endocytosis. Processing of AAV particles within the endosome appears to be important for nuclear transport because vector injected directly into the cytoplasm fails to accumulate in the nucleus. The steps of endosomal escape, nuclear transport, and vector uncoating are not well understood in terms of timing or mechanism, and differences among serotypes at these steps are also not well-defined because of the poor transduction efficiency of most serotypes, except AAV2, in vitro. Several studies indicate that the capsid is a substrate for ubiquitination,[88,89] and that proteasome inhibitors, or mutation of surface exposed tyrosine residues that normally undergo phosphorylation and ubiquitination, enhance transduction by enhancing nuclear uptake of the virus.[90,91] Compelling data supporting the hypothesis that AAV capsid antigen is processed by the proteasome and presented on MHC class I comes from CTL assays performed using AAV-transduced human hepatocytes as targets, and HLA-matched human peripheral blood mononuclear cells (PBMCs) expanded using AAV whole capsid or specific peptides as effectors.[92] The fact that lysis can be inhibited specifically using a capsid antigen-specific soluble T-cell receptor to block T-cell access to the target cell further strengthens this conclusion. Presentation of antigen in the context of MHC I was also demonstrated by our lab using a T-cell line engineered to express luciferase when recognizing the capsid antigen.[93] In this study, levels of antigen presentation were directly correlated with the multiplicity of infection used in the assay, and the proteasome inhibitor bortezomib blocked antigen presentation. These results are in agreement with results from clinical studies, which suggest a correlation between vector dose and magnitude of T-cell responses to the capsid (which will be discussed later).

4.4.2 T-Cell Responses to AAV in Human Studies

In parallel to the study of capsid antigen presentation on MHC class I, investigators have analyzed the frequency of AAV capsid-specific T cells in healthy subjects and in humans undergoing gene transfer. Surveys of T-cell reactivity to the AAV2 capsid conducted in humans undergoing splenectomy for nonmalignant indications show that about two-thirds of adults >25 years old carry a pool of T cells reactive to AAV2.[94] Similarly, recent work from Veron and colleagues shows that about one-third of adult humans present T-cell reactivity against AAV1 in PBMC, measured by IFN-γ ELISpot and intracellular cytokine staining.[95] The discrepancy in frequency of T-cell responses directed against the AAV capsid in spleen[94] versus PBMC[95] may simply reflect serotype-specific differences, or else could be because of preferential homing of capsid T cells to lymphoid organs such as the spleen. Owing to the high degree of conservation of the AAV's capsid amino acid sequence,[96] a high degree of cross-reactivity of T-cell responses was observed in these subjects towards other serotypes, like AAV1 and AAV8, a result in agreement with published data from human trials[97] and other studies in PBMC from healthy donors looking at AAV1 T-cell reactivity.[95]

The importance of T-cell immunity to the AAV capsid in terms of both safety and efficacy of AAV gene transfer in humans was evidenced in the first clinical trial in which an AAV2 vector was introduced into the liver of severe hemophilia B subjects.[98] In this study, upon AAV gene transfer to liver, two subjects developed transient elevation of liver enzymes and loss of coagulation Factor IX (F.IX) transgene expression around week 4 after vector delivery because of immune rejection of transduced hepatocytes mediated by capsid-specific CD8[+] T cells.[97] A similar set of observations was made in the context of a clinical trial of AAV8 gene transfer to the liver of subjects affected by severe hemophilia B.[6] This study showed that AAV8 vector administration in humans resulted in activation of capsid-specific CD8[+] T cells and increased liver enzymes in four out of six subjects (A. Davidoff, American Society of Gene and Cell Therapy 2014 meeting oral communication, and Ref. 6) from the high-dose cohort, who received 2×10^{12} vg/kg of vector, ~8–9 weeks after vector delivery. Results from this AAV8 clinical study suggest that a short course of steroids, administered at the time of liver enzyme elevation, can rescue transgene expression. Timely intervention with steroids can fully ablate the detrimental effect of the immune response on transgene expression. This study also highlights differences in kinetics of T-cell responses between the AAV2 and the AAV8 liver trials, as liver enzyme elevation in the AAV2 trial was observed around week 4 following vector delivery,[98] as opposed to 8–9 weeks in the AAV8 trial.[6] It is not clear at this point what the likely explanation is for this difference. The lack of in vitro studies comparing the behavior of different AAV serotypes is a limitation to the understanding of the underlying mechanism(s) driving these differences; however, a recent in vivo model of capsid T-cell immune responses to AAV effectively recapitulated the human findings, showing that AAV8-transduced hepatocytes remain susceptible to CD8[+] T-cell-mediated lysis longer than those transduced with AAV2 vectors.[99] Additional studies in mice suggest that the genome of AAV8 vectors is released faster than that of AAV2,[100] highlighting possible differences between the two serotypes. Recent work published on the differences in transduction efficiency of AAV vectors in mouse versus human hepatocytes[101] suggests that humanized animal models may help better model capsid T-cell responses in humans.

Results from the AAV8-F.IX clinical trial indicate that, upon AAV vector delivery to humans, capsid-specific T-cell responses are detected in a dose-dependent fashion, a result consistent with published in vitro antigen presentation data.[92] Above a certain threshold of capsid antigen load, the activation of capsid-specific T cells results in loss of transduced hepatocytes. It is not clear at this point whether all subjects will mount a T-cell response that will result in loss of transgene expression at higher vector doses than those tested thus far, and whether steroids will effectively control capsid T cells at all vector doses. Emerging data from clinical trials indicate that not all subjects dosed with AAV will require immunosuppression (A. Davidoff, American Society of Gene and Cell Therapy 2014 meeting oral communication, and Ref. 6); however, the individual differences (HLA type, exposure to the wild-type virus, etc.) that may account for the different outcome of gene transfer among subjects are unknown.

The issue of AAV vector-induced T-cell immunity is not unique to liver-directed gene transfer. Monitoring of capsid T-cell responses has been performed in the context of several muscle-directed gene transfer clinical studies.[102–110] In agreement with the findings in AAV liver gene-transfer studies, results accumulated for muscle gene transfer suggest that the magnitude of T-cell responses directed against the AAV capsid correlates with the dose of vector administered.[22,110,111] After intramuscular AAV vector delivery, an increase in frequency of circulating reactive T cells in PBMCs is observed at higher vector doses.[109] In some cases, detection of capsid T-cell activation in PBMCs correlated with lack of transgene expression in vector-injected muscle,[107,109] while other groups reported no consequences on transduced cells despite the detection of capsid-specific T cells in PBMCs.[111] Immunosuppression has been used in some of the muscle gene transfer studies conducted thus far to modulate capsid immunogenicity.[112] Whether this helped the persistence of transgene expression is not completely clear because of the lack of readily detectable efficacy end-points and the fact that immunosuppression itself complicates immunomonitoring as it is likely to modify capsid-directed T-cell responses.

In a recent study of intramuscular gene transfer for α_1-antitrypsin deficiency,[110] vector administration was associated with detection of capsid-specific T-cell reactivity and increase in serum creatine kinase in some subjects around day 30 after vector administration. Furthermore, activation of both CD4+ and CD8+ T cells in peripheral blood and T-cell infiltrates in muscle biopsies were detected. T-cell reactivity against the AAV capsid antigen did not seem to result in clearance of transduced muscle fibers or loss of transgene expression, a finding that may be explained by the fact that significant amounts of CD4+CD25+FoxP3+ regulatory T cells (Tregs) were found in muscle biopsies.[111] Whether the local induction of Tregs is a phenomenon unique to muscle or to the α_1-antitrypsin transgene[113] remains to be established. However, these findings indicate that pro-inflammatory and tolerogenic signals may be concomitantly elicited by vector administration and may co-determine the outcome of gene transfer.

The results in the AAV8 hemophilia trial[6] represent an important stepping-stone in the management of unwanted immune responses in AAV gene therapy, as they show that it is possible to monitor liver enzymes and administer transient immunosuppression with steroids only if required. However, it should be kept in mind that the ease of end-point monitoring characteristic of this trial (i.e., follow-up of liver enzymes to guide intervention with steroids) is unlikely to apply for all gene therapy scenarios. For certain disease indications (e.g., Duchenne muscular dystrophy), muscle and/or liver enzymes are often constitutively elevated; thus they are not useful as surrogate end-points of tissue damage. Similarly, for gene transfer strategies targeted to the central nervous system, markers of inflammation are not always easy to follow, as they would require frequent access to the cerebrospinal fluid or the use of complex imaging technologies. This leaves monitoring of T-cell responses as the main end-point that can offer a measure of an ongoing immune response. Given the limit of the current assays used in the clinic for immunomonitoring (Table 4.1) and the fact that immunomonitoring is conventionally performed on peripheral blood and not on lymphocytes homing to the tissue target of gene transfer, it will be important to complement immunomonitoring data generated by implementing more up-to-date, yet robust, technologies (e.g., polyfunctional assays analyzed with flow cytometry) and by collecting and analyzing vector-injected tissues, when feasible.

Another solution to the issue of immune responses to AAV could be to administer an immunosuppression regimen up front for all subjects enrolled in a given AAV trial. However, this is not an ideal solution to the problem, as not all individuals will have an immune response to the vector; the timing of immune responses may vary with the vector dose, serotype, etc.[6,98,109] and immunosuppression may change the outcome of gene transfer by decreasing transduction efficiency[114] or triggering unwanted reactions to the donated transgene.[115]

Future immunosurveillance studies conducted as part of clinical gene transfer studies will provide the basis for a better understanding of the determinants of T-cell responses in AAV-mediated gene transfer, for example, the role of target tissue inflammation due to the underlying disease,[116] and conversely, the role of the immune privilege characteristic of some tissues such as the eye.[117] Standardization of technologies used for monitoring of T-cell responses[118,119] will help to correlate them with clinical outcomes and, eventually, devise strategies around the issue. Finally, hyperactive variants of therapeutic proteins,[120–122] AAV capsid mutants that escape proteasomal processing and have higher transduction efficiency,[91,99] and other maneuvers will help maintain the total capsid dose below a threshold critical for T-cell activation.

4.5 HUMORAL IMMUNITY

Antibodies to AAV can neutralize large doses of vector, even when present at low titers.[123,124] The impact of neutralizing antibodies (NAbs) directed against AAV on vector transduction was well evidenced in an AAV2-F.IX clinical trial.[98] In this trial, one subject who enrolled in the high vector dose cohort (2×10^{12} vg/kg) had an NAb titer to AAV2 of 1:2 and expressed peak levels of F.IX transgene of ~11% of normal. Another subject with a pre-treatment NAb titer of 1:17 did not have any detectable circulating F.IX despite the fact that he received the same vector dose. After this initial observation,

TABLE 4.1 Technologies Used in Gene Transfer to Monitor Anti-Capsid Immunogenicity

Assay	Advantages	Disadvantages
Anti-AAV Antibodies		
Binding assay (ELISA): Capsid particles are coated onto microplates and test samples added to the plates	• Easy to set up • Works well with all serotypes • Allows looking at antibody subclasses binding to AAV capsids	• Does not discriminate between neutralizing and non-neutralizing antibodies
Dot-blot assay: Capsid particles are spotted on blot membranes that are incubated with test samples	• Easy to set up • Works well with all serotypes • Allows looking at antibody subclasses binding to AAV capsids • Highly sensitive	• Does not discriminate between neutralizing and non-neutralizing antibodies
Neutralization assay in vitro: Neutralization activity of antibodies is measured with a reporter vector used to transduced cells in vitro	• Relatively easy to set up • Directly reflects the neutralization activity of anti-AAV antibodies	• Most serotypes do not efficiently transduce cells in vitro
Neutralization assay in vivo: Neutralization activity of antibodies is measured by passively immunizing mice with a test sample and then delivering a reporter vector in vivo	• The assay performs equally with all serotypes • More faithfully represents the effect of antibodies on the transduction of a given target tissue	• Not very sensitive • High variability • Time consuming and expensive
T-Cell Responses to AAV		
ELISpot: PBMC are cultured in the presence of antigen in filter plates coated with antibodies specific for cytokines. Each reactive T cell will leave a spot on the well, which can be enumerated to determine the magnitude of the response	• Gold standard for immunomonitoring • Extensive experience with the technology • Sensitive and relatively easy to set up • Reproducible, solid technique	• Does not discriminate which cell is responding (e.g., CD4$^+$ vs CD8$^+$ T cells) • Can analyze only very few markers at a time
Flow cytometry: PBMC are stimulated with an antigen and then stained with fluorescently conjugated antibodies (intracellularly and on the surface)	• Can analyze several markers at the same time, better defining the nature of the immune response • Virtually endless numbers of markers available • Highly flexible	• Requires skilled operator to avoid mistakes/bias in sample preparation, acquisition, and analysis of results • Hard to standardize

Scallan and colleagues showed that even low NAb titers (1:4) can prevent liver transduction in mice passively immunized intraperitoneally with intravenous immunoglobulin.[124] Murphy and colleagues further explored the impact of NAbs on vector transduction by administering NAb-containing serum to animals at defined time-points after vector administration, demonstrating that vector remains susceptible to antibody-mediated neutralization for several hours after delivery.[125] These results were also confirmed by experiments in nonhuman primates, a natural host for AAV,[96] which showed that NAb titers as low as ~1:5 can completely block transduction of the liver following AAV8-F.IX vector administration at doses of 5×10^{12} vg/kg.[123]

Neutralizing antibodies to AAV do not only influence liver transduction. Data in pre-immunized dogs suggest that anti-AAV antibodies are likely to affect transduction efficiency when the vector is administered through the vasculature under hydrostatic pressure,[126] a technique developed to transduce large areas of muscle and tested as a delivery strategy for hemophilia B[126,127] and muscular dystrophy.[128,129] Similarly, recent data show that significant amounts of antibodies against the AAV capsid are present in the joint space.[130,131] Measurement of the NAb titer in matched serum and synovial fluid samples showed a generally good correlation in titers, with the synovial fluid displaying a slightly lower titer than serum.[130]

As expected, the effect of NAbs to the AAV capsid on the efficacy of vector transduction of the target organ is highly dependent on the route of administration. For example, the presence of baseline NAb titers to AAV2 does not affect the efficacy of AAV2 gene transfer for RPE65 after injection of the vector into the subretinal space in experimental animal models[132] and humans.[2,5,9,133,134] Similarly, anti-AAV antibodies do not block muscle transduction when the vector is directly injected intramuscularly.[102,135,136] For gene transfer applications targeting the brain, focal delivery of vectors does not appear to be influenced by anti-AAV NAbs;[137] conversely, systemic delivery of AAV vectors to target the central nervous system through the blood–brain barrier[14,138,139] is likely to be affected by circulating NAbs to AAV. Finally, experiments in large animal models suggest that NAbs partially affect the efficiency of brain transduction after delivery of AAV vectors into the cerebrospinal fluid.[140,141]

4.5.1 Prevalence and Significance of Anti-AAV Antibodies

After exposure to wild-type AAV, a significant proportion of individuals develop humoral immunity against the capsid early in life, starting around 2 years of age;[31,142,143] additionally, maternal anti-AAV antibodies can be found in newborns, which decrease a few months after birth before increasing again.[31,143] Thus, the window of time in which the majority of humans appears to be naïve to anti-AAV antibodies is narrow.

Several groups published extensive surveys on the frequency of antibodies to AAV in humans.[31,130,143–145] Among the AAV serotypes isolated in nature and currently considered for clinical use, AAV2 is the most seroprevalent, which restricts the use of this serotype to gene transfer applications such as gene delivery to the eye[2,5,9,133,134] and the brain parenchyma,[137] for which NAbs are not a concern. AAV5, carrying one of the least-conserved capsid sequences,[146] and the nonhuman primate serotype AAV8,[96] are among the least prevalent in humans.[130,144,145]

Antibody subclass analysis[147] shows that anti-AAV IgG1 antibodies are highly prevalent among humans exposed to the wild-type virus, although lower levels of IgG2, 3, and 4 are also found. After AAV vector delivery to muscle and liver, IgG1 antibodies remain the preponderant subclass,[147] with some subjects developing a robust anti-capsid IgG3 antibody response.[109,147] The significance of these responses in terms of their possible impact on vector transduction or their possible prognostic value for capsid-triggered T-cell immunity is not fully understood. Nonhuman primate studies suggest that low levels of antibodies redirect AAV vector tropism to spleen, perhaps influencing T-cell responses to the capsid antigen;[148] however, the relevance of this finding to the outcome of gene transfer and, therefore, to the clinical setting remains unclear. Anecdotal evidence from human studies suggests that NAbs, when present, simply block AAV transduction without leading to any sequelae.[98]

Studies in healthy humans also show that seropositivity for anti-AAV antibodies is not predictive of T-cell reactivity to the antigen, as it appears that humoral and cellular responses to AAV are mostly decoupled.[95] One possible important implication of this finding is that it is not possible to predict who is at higher risk of developing a T-cell response to the capsid after gene transfer on the basis of baseline serology.

Finally, after gene transfer in animal models and humans, anti-AAV antibodies develop at very high titers (>1:1000) and persist for >10 years (F.M. unpublished observation), preventing vector re-administration.

4.5.2 Anti-AAV Antibody Titering Methods

As a result of the profound effect of anti-AAV antibodies on the efficiency of transduction, even when present at low levels, the development of sensitive and reliable methods to measure NAb titers is critical to screen subjects prior to enrollment in gene-transfer studies and to ensure correlation of baseline titers with clinical outcome (see Table 4.1).

ELISA-based capture assays, in which whole AAV capsids or peptides are coated onto a plate and serum anti-AAV Ig are detected with a secondary antibody, are the easiest assay to set up and give a relatively sensitive measurement of antibody levels. However, the total amount of anti-AAV antibodies is not always proportional to their neutralizing activity, especially at low titers, which makes this assay poorly reliable in determining subjects' eligibility for enrollment in AAV trials.

Cell-based in vitro assays are among the most broadly used to screen samples for anti-AAV neutralizing antibodies.[130,144,145,149,150] These assays typically measure the residual expression of a reporter gene in cells after transduction with an AAV vector pre-incubated with a test serum. The output of the measure is the dilution of the test sample at which <50% of the reporter signal is detected. While in vitro NAb assays are relatively easy to set up and give consistent results, they suffer from some limitations. The most important derives from the fact that most AAV serotypes are highly inefficient in transducing cells in vitro, forcing the use of higher MOIs in the assay and thus underestimating NAb titers. This limitation can be at least partially overcome with the use of reporter genes with high sensitivity of detection (e.g., luciferase) and cell lines more permissive for transduction by AAV vectors. Second, several parameters related to the cell culture conditions are likely to contribute to the variability of the assay, such as the cell line used, the cell density, and the number of passages the cells have been subjected to.

The limitations of in vitro methods to determine anti-AAV NAb can be overcome using in vivo assays in which mice are pre-immunized with test serum a few hours before vector delivery.[6,148,151] However, in vivo measurements of anti-AAV NAb is variable, time-consuming, more expensive than in vitro methods, and may not be highly sensitive.

The most important consideration in selecting an AAV antibody test assay in the context of an AAV clinical trial is whether the baseline NAb titer is an exclusion criteria, which is highly dependent on the target tissue for gene transfer and the route of administration of the vector (described earlier). Additional considerations on the assays used to measure anti-AAV antibodies may include the need for standardized assays to compare results across studies and the need to validate them as the field moves towards late-stage clinical development.

4.5.3 Strategies to Overcome Humoral Immunity to AAV

Gene transfer to subjects with pre-existing immunity to AAV is a major challenge to systemic vector delivery. Patients who present detectable levels of NAbs are currently excluded from trials; however, this is not an acceptable solution to the problem, as it prevents up to two-thirds of the population (depending on the serotype) from accessing in vivo gene-based therapies with AAV vectors. Thus, better solutions must be found. One perhaps obvious solution to the issue could be to increase the dose of vector administered to overcome antibody-mediated neutralization; however, high vector doses face the challenge of activation of T-cell immunity to the AAV capsid, which may limit the duration of transgene expression. Several additional strategies have been tested with some evidence of efficacy; what appears evident from the published studies is that high-titer NAbs are harder to tackle. The optimal strategy for the delivery of AAV vectors in the presence of high NAbs may require a combination of physical and/or pharmacological methods.

1. Selection of naïve subjects

Selecting young subjects for enrollment in gene transfer studies may increase the likelihood of finding naïve individuals, or at least subjects with relatively low NAb titers. Limitations to the approach include the early appearance of anti-AAV NAbs in humans (discussed earlier), which essentially restricts the selection of subjects to those aged 1–2 years. Furthermore, research in pediatric subjects has additional obvious limitations.

2. Identification of less seroprevalent AAV serotypes and modification of the viral capsid to escape antibodies

Studies in humans show that some AAV serotypes are less prevalent than others (discussed earlier). Granted that serotypes with different antibody reactivity profiles and similar affinity for target tissue are available, switching AAV serotype is a potentially feasible approach to the issue of NAbs, which showed efficacy in nonhuman primate[152,153] models of gene transfer. However, the main limitation of this strategy is that anti-AAV antibodies are highly cross-reactive,[144] a fact that reflects the high level of sequence homology among most AAV serotypes.

Development of novel, artificial AAV serotypes with different technologies is also a promising approach.[154] The use of a library of AAV capsids created with error-prone PCR techniques and/or DNA shuffling followed by selection for resistance to NAbs has been used to create novel AAV serotypes with significantly decreased sensitivity to NAbs.[155,156] One strategy proposed for capsid engineering to evade humoral immunity consists in the identification of regions at the surface of the AAV capsid where B-cell epitopes are found. For example, Gurda and colleagues studied in silico the binding between AAV8 and the neutralizing monoclonal antibody ADK8, identifying a variable region on AAV8 important for vector neutralization and using site-specific mutagenesis to escape antibody binding without altering vector tropism.[157] This approach could be useful, although a careful characterization of the main B-cell epitopes found within the AAV capsid will be needed to drive capsid engineering.

Finally, when approaching the problem of capsid engineering, it is important to choose the optimal model to select the mutants developed. While in vitro models may not accurately predict efficiency of transduction in vivo, one limitation of animal models is that they may overestimate efficiency of transduction in humans.[101] Possible solutions to the problem are the use of primary human cells, humanized mouse models, or studies in large animal models such as nonhuman primates. The main advantage of capsid switching and engineering strategies is that no immunosuppression or any other kind of intervention is needed to decrease the antibody titer. Conversely, implementing the use of a novel serotype in the clinic requires extensive biodistribution studies and development of tailored GMP production processes, which are expensive and time-consuming.

3. Plasmapheresis

A recent report[158] shows that repeated cycles of plasmapheresis are effective in lowering anti-AAV NAbs. Plasmapheresis is an established and safe technique used to isolate or eliminate specific components from blood. In this study, investigators measured neutralization titers after each of five cycles of plasma exchange. In a subset of subjects with baseline NAb titers of <1:20, a drop in NAb titers to undetectable levels was observed, indicating that the approach is useful. However, the approach suffers from the limitation that it is not very effective in reducing higher antibody titers, even after repeated cycles.[158] These results are in agreement with another study showing that a single cycle of plasmapheresis was effective in decreasing NAb titers to <1:20 only in subjects with baseline titers of <1:100.[159] These results suggest that additional maneuvers may be needed to overcome high antibody titers.

The initial observations on the effect of plasmapheresis on anti-AAV titers in humans have recently been translated to nonhuman primates, showing that with this technique it is possible to decrease anti-AAV antibodies.[160]

One key question about plasmapheresis is how long the decrease in NAb titers will persist after a plasma exchange cycle, and whether this will be enough to allow the vector to reach and transduce target tissue.

4. Transient immunosuppression

One potential advantage of immunosuppression used in the context of AAV gene transfer is that, unlike organ transplant or autoimmune disease, the duration of the intervention is relatively short. Several drugs more or less specifically targeting B cells are available for clinical use; one of the most widely used is rituximab, a chimeric mouse–human monoclonal antibody specific to human CD20, which was first approved in 1997 for the treatment of B-cell malignancies and now is used in the management of several other nonmalignant diseases owing to its efficacy and safety profile.[161]

We tested the effect of B-cell depletion with rituximab on anti-AAV NAbs in nonhuman primates (NHP) and humans. In NHPs, we administered a combination of rituximab and cyclosporine A to eradicate anti-human F.IX transgene antibodies that developed after AAV8-F.IX gene transfer. After administration of the drug combination, a transient profound depletion of circulating B cells (CD20+) was observed, which was accompanied by disappearance of anti-transgene antibodies and a decrease in anti-AAV NAb titers, which was more pronounced when the baseline NAb titers were only moderately elevated. In this study, we successfully performed vector re-administration with an AAV6-F.IX vector, showing that pharmacologic immunosuppression combined with serotype switching may allow for vector re-administration. The effect of a single course of rituximab (two intravenous administrations at a two-week interval) and methotrexate on anti-AAV NAb titers in a cohort of rheumatoid arthritis patients who received the drug combination as part of the management of their autoimmune disease was recently analyzed. In these subjects, a drop in NAb titers was observed in subjects with titers <1:400, compared with subjects with higher titers who did not experience any change in NAbs.[130] Taken together, results in nonhuman primates and humans are consistent with what was observed in acquired hemophilia patients, where B-cell depletion was less effective in patients with high-titer inhibitors,[162] and with the observation that rituximab does not completely deplete B cells and has virtually no effect on plasma cells, which do not express CD20.

Other drugs targeting B cells could be used to reduce anti-AAV antibodies; many of them are already licensed for clinical use or in late stage clinical development. Among the drugs of interest, the clinically approved drug bortezomib seems promising, as it efficiently targets plasma cells[163] while reducing capsid antigen presentation and enhancing AAV vector transduction.[93,164,165] Karman and colleagues tested the use of bortezomib to decrease the anti-AAV titers after vector administration in mice.[166] In this study, an 8- to 10-fold drop in anti-AAV antibody titer was observed, but the decrease was not sufficient to allow for vector re-administration, and toxicity was observed in some animals. Nonetheless, these results are promising, as typically animals develop very high antibody titers after vector administration, which are hard to eradicate. Thus, additional testing is needed to assess whether bortezomib can eradicate anti-AAV titers that are found in humans naturally exposed to AAV, who typically present lower NAb titers. Additionally, carfilzomib, another proteasome inhibitor with fewer side effects,[167,168] could be an interesting alternative to bortezomib.

Drugs targeting the B-cell-stimulating factors APRIL (a proliferation-inducing ligand) and BLyS (B-lymphocyte stimulator)/BAFF pathways[169] could also be interesting. For example, the combination of atacicept (a fully-human chimeric protein inhibiting the APRIL/BLyS pathway) with mycolphenolate mofetil and corticosteroids led to a profound decline in serum IgG in patients affected by lupus nephritis (LN);[170] while this was considered a severe adverse event in the context of an LN clinical trial, it also demonstrates that pharmacologic reduction of IgG titers is feasible.

One major limitation to the use of immunosuppression to achieve complete eradication of anti-AAV antibodies is the lack of antigen specificity of the approach, which raises concerns with the risk of inducing hypogammaglobulinemia and increasing the risk of serious infections.[170] Antigen-specific B-cell tolerance strategies[171,172] or combination with mild immunosuppressive regimens with other strategies may help overcome this limitation.

5. Delivery of the vector with isolated perfusion and saline flushing

The use of delivery techniques such as balloon catheters or tourniquets to isolate target tissue from the systemic circulation and thereby avoid vector dilution in blood and exposure to NAbs has been previously explored in the context of adenoviral vector delivery.[173–175] Mimuro and colleagues[176] recently tested the efficacy of saline flushing, with or without balloon catheter isolation of the hepatic circulation, to minimize the inhibitory effect of NAbs. Administration of an AAV8 vector encoding for human F.IX under the control of a liver-specific promoter resulted in low transgene expression (<0.2% of normal), when injected via the mesenteric vein in rhesus macaques displaying low levels of NAbs against AAV8 at baseline. To minimize the inhibitory effect of NAbs, the authors blocked the portal vein circulation by clamping it with vascular forceps or by using a balloon catheter. The procedure resulted in F.IX transgene expression levels between 4% and 10% of normal, even in the presence of a low titer of NAbs. Both clamping and catheterization resulted in equivalent avoidance of the inhibitory effect of NAbs; however, the use of vascular forceps to clamp the portal vein is a more invasive strategy, which is potentially associated with vascular damage, making the balloon catheterization more suitable for clinical translation.[177]

6. Capsid decoys

One obvious solution to the problem of anti-AAV NAbs is to increase the vector dose to overcome vector neutralization. While the approach clearly has the potential to achieve transduction of the target organ in the presence of antibodies, it is not ideal, as it would require the infusion of high amounts of the genome-containing viral vectors, with variable results depending on the baseline titer. One alternative to this approach is the use of an excess of AAV capsid antigen devoid of a genome (empty capsids) to achieve consistent transduction of target tissue in the presence of antibodies.

This strategy was recently described to be effective in mice and nonhuman primates undergoing liver gene transfer with AAV8 vectors.[178] Formulation of AAV8 vectors in an excess of AAV8 empty capsids was sufficient to overcome anti-AAV NAb titers of up to 1:100 in mice passively immunized with human IgG. One limitation of this approach is that empty capsids are a replica of AAV vectors and therefore can be recognized by the immune system and trigger CTL-mediated lysis of target cells.[92] Alternative solutions to the use of native empty capsid decoys include the use of noninfectious decoys,[178] the use of peptides corresponding to the B-cell epitopes, or the use of domains of the AAV capsid that are recognized by neutralizing antibodies.[157,179]

4.6 CONCLUSIONS

Since the initial proof-of-concept studies in experimental models, the field of in vivo gene transfer has made enormous progress. Experience from clinical translation of AAV vector-based gene transfer strategies has helped optimize the choice of serotype and delivery methods and has highlighted some of the limitations of the approaches tested. In particular, the interactions between the different arms of the immune system and all the components of gene therapy vectors seem to represent one of the major limitations to long-lasting therapeutic efficacy.

Challenges to understanding the determinants of immune responses to AAV vector gene transfer include the small scale of gene transfer trials and the fact that these studies are very diverse in vector serotype and dose, therapeutic transgene, and route of vector delivery. The underlying disease, the ability of measuring the end-points of therapeutic efficacy, and individual variability associated with some HLA haplotypes are among the additional challenges to data interpretation. Nevertheless, studies conducted thus far in humans have yielded a tremendous amount of information and have given us the opportunity to test strategies to address the immune responses to AAV vectors. Moving forward, it will be important to standardize some of the key assays used in immunomonitoring, such as the ones used for neutralizing antibody titer determination and the ELISpot assay to measure T-cell responses.

The administration of a course of steroids effectively blunted T-cell responses directed against the AAV8 capsid in the setting of liver gene transfer for hemophilia. Whether immunosuppression will be effective and safe in the context of higher vector doses and gene transfer with other AAV serotypes remains to be tested. An alternative solution to immunosuppression is to decrease the therapeutic vector dose; to this end, the use of codon-optimized transgenes and hyperactive variants of therapeutic enzymes may provide a solution. Additionally, recent data emerging from liver-directed gene therapy trials suggest that the immunogenicity of different AAV serotypes is not identical, with some serotypes possibly being less immunogenic[180] and others more immunogenic.[181]

Neutralizing antibodies to AAV pose more of a challenge to the success of in vivo gene therapy than T-cell reactivity to the capsid, particularly when vectors are delivered systemically. The current approach of excluding seropositive patients is far from ideal, and a variety of approaches to the problem are being tested in preclinical models. These strategies, alone or in combination, will have to be translated to the clinic to allow seropositive patients to be enrolled in clinical trials and, when necessary, to re-administer the therapeutic vector to subjects previously treated.

Despite the remaining challenges, results are promising, and long-term follow-up data from AAV gene transfer trials suggest that once transgene expression is established, and immune responses avoided, multiyear therapeutic efficacy is a goal attainable in humans.

GLOSSARY

Adaptive immunity Arm of the immune system found in vertebrates acting in an antigen-specific manner. The adaptive immune system includes both humoral and cell-mediated immunity.

B cells Cells that are part of the adaptive immune system involved in the production of antibodies and presentation of antigen.

Cell-mediated immunity Type of adaptive immunity that is mediated by cells such as lymphocytes and macrophages.

CTL Cytotoxic T lymphocyte that can eliminate cells presenting foreign antigens in the form of epitopes complexed with MHC class I molecules. Typically CTLs express the glycoprotein CD8 at their surface, which mediates binding to the cognate MHC class I molecule. CD8+ T cells are not the only lymphocytes that present cytotoxic activity.

ELISpot Enzyme-Linked ImmunoSpot assay, a technique commonly used to monitor T-cell responses specific to an antigen.

Episome Extrachromosomal circular DNA.

Epitope Also called antigenic determinant, an epitope is the part of a protein that is recognized by the immune system. T-cell epitopes are linear sequences within the primary amino acid sequence of a given protein, while B-cell epitopes are conformational or structural and not necessarily linear.

GMP Good Manufacturing Practices. Manufacturing of a drug product following the guidelines of the regulatory agencies.

HLA Human leukocyte antigen, which is the locus corresponding to the major histocompatibility complex in humans.

Humoral immunity Compartment of the adaptive immunity that is mediated by antibodies located in the vascular and extravascular space. Antibodies are produced by B lymphocytes.

Hypogammaglobulinemia Condition associated with abnormally low levels of circulating gamma globulins, which requires supplementation due to a high risk of infections.

Innate immunity Arm of the immune system and first line of defense against invading pathogens. Innate immunity is also known for not being antigen specific. The innate immune system recognizes through pathogen recognition receptors like Toll-like-receptor (TLR) specific so-called pathogen-associated molecular patterns (PAMPs). Sensing of these PAMPs indicates pathogen invasion.

MHC class I Major histocompatibility complex class I. A set of cell-surface molecules mediating interactions between lymphocytes and other cells of the body, MHC class I molecules are expressed by all cells of the body except for erythrocytes and have the role of presenting endogenous epitopes (self and those produced by invaders) to lymphocytes. Through this mechanism, cells are constantly checked for being healthy and noninfected by pathogens.

NAb Neutralizing antibody titer.

Parvovirus Family of viruses with a non-enveloped protein capsid carrying a single-stranded DNA genome. Parvoviruses are among the smallest viruses in nature (18–28 nm in diameter).

PCR Polymerase chain reaction, a technology developed in 1983 to amplify DNA sequences.

Plasmapheresis Technique used in the clinic to remove components of plasma via extracorporeal circulation through specific binding columns.

Proteasome Protein complex involved in the proteolytic degradation of ubiquitinated proteins.

T cells T cells belong to the adaptive immune system and recognize antigens via the T-cell receptor. Typically, MHC molecules mediate antigen recognition by T cells. T-helper cells and regulatory T cells are important regulators of the immune responses, while cytotoxic T cells have mainly lytic functions.

Ubiquitination Posttranslational modification of proteins consisting of their conjugation to ubiquitin, which results in targeting for degradation via proteasome or lysosome.

Uncoating Commonly referred as the process of release of the vector/viral genome from the protein capsid following entry of the particle into the cell. This step is essential to achieve successful transduction of the target cell.

Vg Vector genome, a commonly adopted measure of titer of AAV vectors, usually obtained using real-time quantitative PCR.

ACKNOWLEDGMENT

This work was support by the German Research Foundation (CRC670 and SPP1230) and the Center for Molecular Medicine Cologne (CMMC) to H.B. It was also supported by Genethon, by the European Union FP7-PEOPLE-2012-CIG grant 333628 and ERC-2013-CoG grant 617432, and by the Bayer grant ECIA to F.M.

REFERENCES

1. Aiuti A, Cattaneo F, Galimberti S, Benninghoff U, Cassani B, Callegaro L, et al. Gene therapy for immunodeficiency due to adenosine deaminase deficiency. *N Engl J Med* 2009;**360**:447–58.
2. Bainbridge JW, Smith AJ, Barker SS, Robbie S, Henderson R, Balaggan K, et al. Effect of gene therapy on visual function in Leber's congenital amaurosis. *N Engl J Med* 2008;**358**:2231–9.
3. Cartier N, Hacein-Bey-Abina S, Bartholomae CC, Veres G, Schmidt M, Kutschera I, et al. Hematopoietic stem cell gene therapy with a lentiviral vector in X-linked adrenoleukodystrophy. *Science* 2009;**326**:818–23.
4. Cavazzana-Calvo M, Payen E, Negre O, Wang G, Hehir K, Fusil F, et al. Transfusion independence and HMGA2 activation after gene therapy of human beta-thalassaemia. *Nature* 2010;**467**:318–22.

5. Maguire AM, Simonelli F, Pierce EA, Pugh Jr EN, Mingozzi F, Bennicelli J, et al. Safety and efficacy of gene transfer for Leber's congenital amaurosis. *N Engl J Med* 2008;**358**:2240–8.

6. Nathwani AC, Tuddenham EG, Rangarajan S, Rosales C, McIntosh J, Linch DC, et al. Adenovirus-associated virus vector-mediated gene transfer in hemophilia B. *N Engl J Med* 2011;**365**:2357–65.

7. Biffi A, Montini E, Lorioli L, Cesani M, Fumagalli F, Plati T, et al. Lentiviral hematopoietic stem cell gene therapy benefits metachromatic leukodystrophy. *Science* 2013;**341**:1233158.

8. Aiuti A, Biasco L, Scaramuzza S, Ferrua F, Cicalese MP, Baricordi C, et al. Lentiviral hematopoietic stem cell gene therapy in patients with Wiskott–Aldrich syndrome. *Science* 2013;**341**:1233151.

9. Cideciyan AV, Aleman TS, Boye SL, Schwartz SB, Kaushal S, Roman AJ, et al. Human gene therapy for RPE65 isomerase deficiency activates the retinoid cycle of vision but with slow rod kinetics. *Proc Natl Acad Sci USA* 2008;**105**:15112–7.

10. Naldini L. Ex vivo gene transfer and correction for cell-based therapies. *Nat Rev Genet* 2011;**12**:301–15.

11. Mingozzi F, High KA. Therapeutic in vivo gene transfer for genetic disease using AAV: progress and challenges. *Nat Rev Genet* 2011;**12**:341–55.

12. Acland GM, Aguirre GD, Ray J, Zhang Q, Aleman TS, Cideciyan AV, et al. Gene therapy restores vision in a canine model of childhood blindness. *Nat Genet* 2001;**28**:92–5.

13. Arruda VR, Stedman HH, Haurigot V, Buchlis G, Baila S, Favaro P, et al. Peripheral transvenular delivery of adeno-associated viral vectors to skeletal muscle as a novel therapy for hemophilia B. *Blood* 2010;**115**:4678–88.

14. Foust KD, Nurre E, Montgomery CL, Hernandez A, Chan CM, Kaspar BK. Intravascular AAV9 preferentially targets neonatal neurons and adult astrocytes. *Nat Biotechnol* 2009;**27**:59–65.

15. Limberis MP, Wilson JM. Adeno-associated virus serotype 9 vectors transduce murine alveolar and nasal epithelia and can be readministered. *Proc Natl Acad Sci USA* 2006;**103**:12993–8.

16. Mount JD, Herzog RW, Tillson DM, Goodman SA, Robinson N, McCleland ML, et al. Sustained phenotypic correction of hemophilia B dogs with a factor IX null mutation by liver-directed gene therapy. *Blood* 2002;**99**:2670–6.

17. Gregorevic P, Allen JM, Minami E, Blankinship MJ, Haraguchi M, Meuse L, et al. rAAV6-microdystrophin preserves muscle function and extends lifespan in severely dystrophic mice. *Nat Med* 2006;**12**:787–9.

18. Herzog RW, Yang EY, Couto LB, Hagstrom JN, Elwell D, Fields PA, et al. Long-term correction of canine hemophilia B by gene transfer of blood coagulation factor IX mediated by adeno-associated viral vector. *Nat Med* 1999;**5**:56–63.

19. Jooss K, Yang Y, Fisher KJ, Wilson JM. Transduction of dendritic cells by DNA viral vectors directs the immune response to transgene products in muscle fibers. *J Virol* 1998;**72**:4212–23.

20. Fields PA, Kowalczyk DW, Arruda VR, Armstrong E, McCleland ML, Hagstrom JN, et al. Role of vector in activation of T cell subsets in immune responses against the secreted transgene product factor IX. *Mol Ther* 2000;**1**:225–35.

21. Buning H. Gene therapy enters the pharma market: the short story of a long journey. *EMBO Mol Med* 2013;**5**:1–3.

22. Mingozzi F, High KA. Immune responses to AAV in clinical trials. *Curr Gene Ther* 2011;**11**:321–30.

23. Li C, Hirsch M, Asokan A, Zeithaml B, Ma H, Kafri T, et al. Adeno-associated virus type 2 (AAV2) capsid-specific cytotoxic T lymphocytes eliminate only vector-transduced cells coexpressing the AAV2 capsid in vivo. *J Virol* 2007;**81**:7540–7.

24. Li H, Murphy SL, Giles-Davis W, Edmonson S, Xiang Z, Li Y, et al. Pre-existing AAV capsid-specific CD8+ T cells are unable to eliminate AAV-transduced hepatocytes. *Mol Ther* 2007;**15**:792–800.

25. Wang L, Figueredo J, Calcedo R, Lin J, Wilson JM. Cross-presentation of adeno-associated virus serotype 2 capsids activates cytotoxic T cells but does not render hepatocytes effective cytolytic targets. *Hum Gene Ther* 2007;**18**:185–94.

26. Atchison RW, Casto BC, Hammon WM. Adenovirus-associated defective virus particles. *Science* 1965;**149**:754–6.

27. Grimm D, Kay MA. From virus evolution to vector revolution: use of naturally occurring serotypes of adeno-associated virus (AAV) as novel vectors for human gene therapy. *Curr Gene Ther* 2003;**3**:281–304.

28. Muzyczka N, Berns KI. *Parvoviridae: the viruses and their replication.* Philadelphia: Lippincott, Williams and Wilkins; 2001.

29. Sonntag F, Kother K, Schmidt K, Weghofer M, Raupp C, Nieto K, et al. The assembly-activating protein promotes capsid assembly of different adeno-associated virus serotypes. *J Virol* 2011;**85**:12686–97.

30. Zhou X, Muzyczka N. In vitro packaging of adeno-associated virus DNA. *J Virol* 1998;**72**:3241–7.

31. Calcedo R, Morizono H, Wang L, McCarter R, He J, Jones D, et al. Adeno-associated virus antibody profiles in newborns, children, and adolescents. *Clin Vaccine Immunol* 2011;**18**:1586–8.

32. Arechavaleta-Velasco F, Gomez L, Ma Y, Zhao J, McGrath CM, Sammel MD, et al. Adverse reproductive outcomes in urban women with adeno-associated virus-2 infections in early pregnancy. *Hum Reprod* 2008;**23**:29–36.

33. Kotin RM, Menninger JC, Ward DC, Berns KI. Mapping and direct visualization of a region-specific viral DNA integration site on chromosome 19q13-qter. *Genomics* 1991;**10**:831–4.

34. Samulski RJ, Zhu X, Xiao X, Brook JD, Housman DE, Epstein N, et al. Targeted integration of adeno-associated virus (AAV) into human chromosome 19. *EMBO J* 1991;**10**:3941–50.

35. Kotin RM, Siniscalco M, Samulski RJ, Zhu XD, Hunter L, Laughlin CA, et al. Site-specific integration by adeno-associated virus. *Proc Natl Acad Sci USA* 1990;**87**:2211–5.

36. Nakai H, Montini E, Fuess S, Storm TA, Grompe M, Kay MA. AAV serotype 2 vectors preferentially integrate into active genes in mice. *Nat Genet* 2003;**34**:297–302.

37. Donsante A, Miller DG, Li Y, Vogler C, Brunt EM, Russell DW, et al. AAV vector integration sites in mouse hepatocellular carcinoma. *Science* 2007;**317**:477.

38. Li H, Malani N, Hamilton SR, Schlachterman A, Bussadori G, Edmonson SE, et al. Assessing the potential for AAV vector genotoxicity in a murine model. *Blood* 2011;**117**:3311–9.

39. Wang PR, Xu M, Toffanin S, Li Y, Llovet JM, Russell DW. Induction of hepatocellular carcinoma by in vivo gene targeting. *Proc Natl Acad Sci USA* 2012;**109**:11264–9.

40. Nathwani AC, Rosales C, McIntosh J, Rastegarlari G, Nathwani D, Raj D, et al. Long-term safety and efficacy following systemic administration of a self-complementary AAV vector encoding human FIX pseudotyped with serotype 5 and 8 capsid proteins. *Mol Ther* 2011;**19**:876–85.

41. Niemeyer GP, Herzog RW, Mount J, Arruda VR, Tillson DM, Hathcock J, et al. Long-term correction of inhibitor-prone hemophilia B dogs treated with liver-directed AAV2-mediated factor IX gene therapy. *Blood* 2009;**113**:797–806.

42. Kessler PD, Podsakoff GM, Chen X, McQuiston SA, Colosi PC, Matelis LA, et al. Gene delivery to skeletal muscle results in sustained expression and systemic delivery of a therapeutic protein. *Proc Natl Acad Sci USA* 1996;**93**:14082–7.

43. Xiao X, Li J, Samulski RJ. Efficient long-term gene transfer into muscle tissue of immunocompetent mice by adeno-associated virus vector. *J Virol* 1996;**70**:8098–108.

44. Wagner JA, Reynolds T, Moran ML, Moss RB, Wine JJ, Flotte TR, et al. Efficient and persistent gene transfer of AAV-CFTR in maxillary sinus. *Lancet* 1998;**351**:1702–3.

45. Seisenberger G, Ried MU, Endress T, Buning H, Hallek M, Brauchle C. Real-time single-molecule imaging of the infection pathway of an adeno-associated virus. *Science* 2001;**294**:1929–32.

46. Marsh M, Helenius A. Virus entry: open sesame. *Cell* 2006;**124**:729–40.

47. Summerford C, Samulski RJ. Membrane-associated heparan sulfate proteoglycan is a receptor for adeno-associated virus type 2 virions. *J Virol* 1998;**72**:1438–45.

48. Levy HC, Bowman VD, Govindasamy L, McKenna R, Nash K, Warrington K, et al. Heparin binding induces conformational changes in adeno-associated virus serotype 2. *J Struct Biol* 2009;**165**:146–56.

49. Opie SR, Warrington Jr KH, Agbandje-McKenna M, Zolotukhin S, Muzyczka N. Identification of amino acid residues in the capsid proteins of adeno-associated virus type 2 that contribute to heparan sulfate proteoglycan binding. *J Virol* 2003;**77**:6995–7006.

50. Kern A, Schmidt K, Leder C, Muller OJ, Wobus CE, Bettinger K, et al. Identification of a heparin-binding motif on adeno-associated virus type 2 capsids. *J Virol* 2003;**77**:11072–81.

51. Lochrie MA, Tatsuno GP, Christie B, McDonnell JW, Zhou S, Surosky R, et al. Mutations on the external surfaces of adeno-associated virus type 2 capsids that affect transduction and neutralization. *J Virol* 2006;**80**:821–34.

52. Nonnenmacher M, Weber T. Intracellular transport of recombinant adeno-associated virus vectors. *Gene Ther* 2012;**19**:649–58.

53. Mietzsch M, Broecker F, Reinhardt A, Seeberger PH, Heilbronn R. Differential adeno-associated virus serotype-specific interaction patterns with synthetic heparins and other glycans. *J Virol* 2014;**88**:2991–3003.

54. Bartlett JS, Wilcher R, Samulski RJ. Infectious entry pathway of adeno-associated virus and adeno-associated virus vectors. *J Virol* 2000;**74**:2777–85.

55. Sanlioglu S, Benson PK, Yang J, Atkinson EM, Reynolds T, Engelhardt JF. Endocytosis and nuclear trafficking of adeno-associated virus type 2 are controlled by rac1 and phosphatidylinositol-3 kinase activation. *J Virol* 2000;**74**:9184–96.

56. Uhrig S, Coutelle O, Wiehe T, Perabo L, Hallek M, Buning H. Successful target cell transduction of capsid-engineered rAAV vectors requires clathrin-dependent endocytosis. *Gene Ther* 2012;**19**:210–8.

57. Nonnenmacher M, Weber T. Adeno-associated virus 2 infection requires endocytosis through the CLIC/GEEC pathway. *Cell Host Microbe* 2011;**10**:563–76.

58. Sonntag F, Bleker S, Leuchs B, Fischer R, Kleinschmidt JA. Adeno-associated virus type 2 capsids with externalized VP1/VP2 trafficking domains are generated prior to passage through the cytoplasm and are maintained until uncoating occurs in the nucleus. *J Virol* 2006;**80**:11040–54.

59. Akache B, Grimm D, Shen X, Fuess S, Yant SR, Glazer DS, et al. A two-hybrid screen identifies cathepsins B and L as uncoating factors for adeno-associated virus 2 and 8. *Mol Ther* 2007;**15**:330–9.

60. Grieger JC, Snowdy S, Samulski RJ. Separate basic region motifs within the adeno-associated virus capsid proteins are essential for infectivity and assembly. *J Virol* 2006;**80**:5199–210.

61. Popa-Wagner R, Sonntag F, Schmidt K, King J, Kleinschmidt JA. Nuclear translocation of adeno-associated virus type 2 capsid proteins for virion assembly. *J Gen Virol* 2012;**93**:1887–98.

62. Popa-Wagner R, Porwal M, Kann M, Reuss M, Weimer M, Florin L, et al. Impact of VP1-specific protein sequence motifs on adeno-associated virus type 2 intracellular trafficking and nuclear entry. *J Virol* 2012;**86**:9163–74.

63. Girod A, Wobus CE, Zadori Z, Ried M, Leike K, Tijssen P, et al. The VP1 capsid protein of adeno-associated virus type 2 is carrying a phospholipase A2 domain required for virus infectivity. *J Gen Virol* 2002;**83**:973–8.

64. Zadori Z, Szelei J, Lacoste MC, Li Y, Gariepy S, Raymond P, et al. A viral phospholipase A2 is required for parvovirus infectivity. *Dev Cell* 2001;**1**:291–302.

65. Canaan S, Zadori Z, Ghomashchi F, Bollinger J, Sadilek M, Moreau ME, et al. Interfacial enzymology of parvovirus phospholipases A2. *J Biol Chem* 2004;**279**:14502–8.

66. Stahnke S, Lux K, Uhrig S, Kreppel F, Hosel M, Coutelle O, et al. Intrinsic phospholipase A2 activity of adeno-associated virus is involved in endosomal escape of incoming particles. *Virology* 2011;**409**:77–83.

67. Lux K, Goerlitz N, Schlemminger S, Perabo L, Goldnau D, Endell J, et al. Green fluorescent protein-tagged adeno-associated virus particles allow the study of cytosolic and nuclear trafficking. *J Virol* 2005;**79**:11776–87.

68. Johnson JS, Samulski RJ. Enhancement of adeno-associated virus infection by mobilizing capsids into and out of the nucleolus. *J Virol* 2009;**83**: 2632–44.

69. Qiu J, Brown KE. A 110-kDa nuclear shuttle protein, nucleolin, specifically binds to adeno-associated virus type 2 (AAV-2) capsid. *Virology* 1999;**257**:373–82.

70. Hansen J, Qing K, Srivastava A. Infection of purified nuclei by adeno-associated virus 2. *Mol Ther* 2001;**4**:289–96.

71. Xiao W, Warrington Jr KH, Hearing P, Hughes J, Muzyczka N. Adenovirus-facilitated nuclear translocation of adeno-associated virus type 2. *J Virol* 2002;**76**:11505–17.

72. McCarty DM, Fu H, Monahan PE, Toulson CE, Naik P, Samulski RJ. Adeno-associated virus terminal repeat (TR) mutant generates self-complementary vectors to overcome the rate-limiting step to transduction in vivo. *Gene Ther* 2003;**10**:2112–8.

73. Lee MS, Kim YJ. Signaling pathways downstream of pattern-recognition receptors and their cross talk. *Annu Rev Biochem* 2007;**76**:447–80.

74. Satta N, Kruithof EK, Reber G, de Moerloose P. Induction of TLR2 expression by inflammatory stimuli is required for endothelial cell responses to lipopeptides. *Mol Immunol* 2008;**46**:145–57.

75. Hosel M, Broxtermann M, Janicki H, Esser K, Arzberger S, Hartmann P, et al. Toll-like receptor 2-mediated innate immune response in human nonparenchymal liver cells toward adeno-associated viral vectors. *Hepatology* 2012;**55**:287–97.

76. Newton K, Dixit VM. Signaling in innate immunity and inflammation. *Cold Spring Harb Perspect Biol* 2012;**4**.

77. Boo KH, Yang JS. Intrinsic cellular defenses against virus infection by antiviral type I interferon. *Yonsei Med J* 2010;**51**:9–17.

78. Zaiss AK, Liu Q, Bowen GP, Wong NC, Bartlett JS, Muruve DA. Differential activation of innate immune responses by adenovirus and adeno-associated virus vectors. *J Virol* 2002;**76**:4580–90.

79. Gilliet M, Cao W, Liu YJ. Plasmacytoid dendritic cells: sensing nucleic acids in viral infection and autoimmune diseases. *Nat Rev Immunol* 2008;**8**:594–606.

80. Zhu J, Huang X, Yang Y. The TLR9-MyD88 pathway is critical for adaptive immune responses to adeno-associated virus gene therapy vectors in mice. *J Clin Invest* 2009;**119**:2388–98.

81. Mattei LM, Cotmore SF, Li L, Tattersall P, Iwasaki A. Toll-like receptor 9 in plasmacytoid dendritic cells fails to detect parvoviruses. *J Virol* 2013;**87**:3605–8.

82. Martino AT, Suzuki M, Markusic DM, Zolotukhin I, Ryals RC, Moghimi B, et al. The genome of self-complementary adeno-associated viral vectors increases Toll-like receptor 9-dependent innate immune responses in the liver. *Blood* 2011;**117**:6459–68.

83. Faust SM, Bell P, Cutler BJ, Ashley SN, Zhu Y, Rabinowitz JE, et al. CpG-depleted adeno-associated virus vectors evade immune detection. *J Clin Invest* 2013;**123**:2994–3001.

84. Lin J, Calcedo R, Vandenberghe LH, Bell P, Somanathan S, Wilson JM. A new genetic vaccine platform based on an adeno-associated virus isolated from a rhesus macaque. *J Virol* 2009;**83**:12738–50.

85. Mays LE, Vandenberghe LH, Xiao R, Bell P, Nam HJ, Agbandje-McKenna M, et al. Adeno-associated virus capsid structure drives CD4-dependent CD8+ T cell response to vector encoded proteins. *J Immunol* 2009;**182**:6051–60.

86. Zaiss AK, Cotter MJ, White LR, Clark SA, Wong NC, Holers VM, et al. Complement is an essential component of the immune response to adeno-associated virus vectors. *J Virol* 2008;**82**:2727–40.

87. Denard J, Beley C, Kotin R, Lai-Kuen R, Blot S, Leh H, et al. Human galectin 3 binding protein interacts with recombinant adeno-associated virus type 6. *J Virol* 2012;**86**:6620–31.

88. Ding W, Zhang L, Yan Z, Engelhardt JF. Intracellular trafficking of adeno-associated viral vectors. *Gene Ther* 2005;**12**:873–80.

89. Yan Z, Zak R, Luxton GW, Ritchie TC, Bantel-Schaal U, Engelhardt JF. Ubiquitination of both adeno-associated virus type 2 and 5 capsid proteins affects the transduction efficiency of recombinant vectors. *J Virol* 2002;**76**:2043–53.

90. Duan D, Yue Y, Yan Z, Yang J, Engelhardt JF. Endosomal processing limits gene transfer to polarized airway epithelia by adeno-associated virus. *J Clin Invest* 2000;**105**:1573–87.

91. Zhong L, Li B, Mah CS, Govindasamy L, Agbandje-McKenna M, Cooper M, et al. Next generation of adeno-associated virus 2 vectors: point mutations in tyrosines lead to high-efficiency transduction at lower doses. *Proc Natl Acad Sci USA* 2008;**105**:7827–32.

92. Pien GC, Basner-Tschakarjan E, Hui DJ, Mentlik AN, Finn JD, Hasbrouck NC, et al. Capsid antigen presentation flags human hepatocytes for destruction after transduction by adeno-associated viral vectors. *J Clin Invest* 2009;**119**:1688–95.

93. Finn JD, Hui D, Downey HD, Dunn D, Pien GC, Mingozzi F, et al. Proteasome inhibitors decrease AAV2 capsid derived peptide epitope presentation on MHC class I following transduction. *Mol Ther* 2010;**18**:135–42.

94. Hui DJ, Basner-Tschakarjan E, Chen Y, Edmonson SC, Maus MV, Podsakoff GM, et al. Characterization of AAV T cell epitopes presented by Splenocytes from Normal human donors. *Mol Ther ASGCT Annu Meet Abstracts* 2012:S557.

95. Veron P, Leborgne C, Monteilhet V, Boutin S, Martin S, Moullier P, et al. Humoral and cellular capsid-specific immune responses to adeno-associated virus type 1 in randomized healthy donors. *J Immunol* 2012;**188**:6418–24.

96. Gao GP, Alvira MR, Wang L, Calcedo R, Johnston J, Wilson JM. Novel adeno-associated viruses from rhesus monkeys as vectors for human gene therapy. *Proc Natl Acad Sci USA* 2002;**99**:11854–9.

97. Mingozzi F, Maus MV, Hui DJ, Sabatino DE, Murphy SL, Rasko JE, et al. CD8(+) T-cell responses to adeno-associated virus capsid in humans. *Nat Med* 2007;**13**:419–22.

98. Manno CS, Pierce GF, Arruda VR, Glader B, Ragni M, Rasko JJ, et al. Successful transduction of liver in hemophilia by AAV-Factor IX and limitations imposed by the host immune response. *Nat Med* 2006;**12**:342–7.

99. Martino AT, Basner-Tschakarjan E, Markusic DM, Finn JD, Hinderer C, Zhou S, et al. Engineered AAV vector minimizes in vivo targeting of transduced hepatocytes by capsid-specific CD8+ T cells. *Blood* 2013;**121**:2224–33.

100. Thomas CE, Storm TA, Huang Z, Kay MA. Rapid uncoating of vector genomes is the key to efficient liver transduction with pseudotyped adeno-associated virus vectors. *J Virol* 2004;**78**:3110–22.

101. Lisowski L, Dane AP, Chu K, Zhang Y, Cunningham SC, Wilson EM, et al. Selection and evaluation of clinically relevant AAV variants in a xenograft liver model. *Nature* 2014;**506**:382–6.

102. Brantly ML, Chulay JD, Wang L, Mueller C, Humphries M, Spencer LT, et al. Sustained transgene expression despite T lymphocyte responses in a clinical trial of rAAV1-AAT gene therapy. *Proc Natl Acad Sci USA* 2009;**106**:16363–8.

103. Flotte TR, Schwiebert EM, Zeitlin PL, Carter BJ, Guggino WB. Correlation between DNA transfer and cystic fibrosis airway epithelial cell correction after recombinant adeno-associated virus serotype 2 gene therapy. *Hum Gene Ther* 2005;**16**:921–8.

104. Herson S, Hentati F, Rigolet A, Behin A, Romero NB, Leturcq F, et al. A phase I trial of adeno-associated virus serotype 1-gamma-sarcoglycan gene therapy for limb girdle muscular dystrophy type 2C. *Brain* 2012;**135**:483–92.

105. Jaski BE, Jessup ML, Mancini DM, Cappola TP, Pauly DF, Greenberg B, et al. Calcium upregulation by percutaneous administration of gene therapy in cardiac disease (CUPID trial), a first-in-human phase 1/2 clinical trial. *J Card Fail* 2009;**15**:171–81.

106. Mendell JR, Campbell K, Rodino-Klapac L, Sahenk Z, Shilling C, Lewis S, et al. Dystrophin immunity in Duchenne's muscular dystrophy. *N Engl J Med* 2010;**363**:1429–37.

107. Mendell JR, Rodino-Klapac LR, Rosales XQ, Coley BD, Galloway G, Lewis S, et al. Sustained alpha-sarcoglycan gene expression after gene transfer in limb-girdle muscular dystrophy, type 2D. *Ann Neurol* 2010;**68**:629–38.

108. Mendell JR, Rodino-Klapac LR, Rosales-Quintero X, Kota J, Coley BD, Galloway G, et al. Limb-girdle muscular dystrophy type 2D gene therapy restores alpha-sarcoglycan and associated proteins. *Ann Neurol* 2009;**66**:290–7.

109. Mingozzi F, Meulenberg JJ, Hui DJ, Basner-Tschakarjan E, Hasbrouck NC, Edmonson SA, et al. AAV-1-mediated gene transfer to skeletal muscle in humans results in dose-dependent activation of capsid-specific T cells. *Blood* 2009;**114**:2077–86.

110. Flotte TR, Trapnell BC, Humphries M, Carey B, Calcedo R, Rouhani F, et al. Phase 2 clinical trial of a recombinant adeno-associated viral vector expressing alpha1-antitrypsin: interim results. *Hum Gene Ther* 2011;**22**:1239–47.

111. Mueller C, Chulay JD, Trapnell BC, Humphries M, Carey B, Sandhaus RA, et al. Human Treg responses allow sustained recombinant adeno-associated virus-mediated transgene expression. *J Clin Invest* 2013;**123**:5310–8.

112. Gaudet D, Methot J, Dery S, Brisson D, Essiembre C, Tremblay G, et al. Efficacy and long-term safety of alipogene tiparvovec (AAV1-LPLS447X) gene therapy for lipoprotein lipase deficiency: an open-label trial. *Gene Ther* 2013;**20**:361–9.

113. Ozeri E, Mizrahi M, Shahaf G, Lewis EC. Alpha-1 antitrypsin promotes semimature, IL-10-producing and readily migrating tolerogenic dendritic cells. *J Immunol* 2012;**189**:146–53.

114. Montenegro-Miranda PS, ten Bloemendaal L, Kunne C, de Waart DR, Bosma PJ. Mycophenolate mofetil impairs transduction of single-stranded adeno-associated viral vectors. *Hum Gene Ther* 2011;**22**:605–12.

115. Mingozzi F, Hasbrouck NC, Basner-Tschakarjan E, Edmonson SA, Hui DJ, Sabatino DE, et al. Modulation of tolerance to the transgene product in a nonhuman primate model of AAV-mediated gene transfer to liver. *Blood* 2007;**110**:2334–41.

116. Boisgerault F, Gross DA, Ferrand M, Poupiot J, Darocha S, Richard I, et al. Prolonged gene expression in muscle is achieved without active immune tolerance using microRNA 142.3p-regulated rAAV gene transfer. *Hum Gene Ther* 2013;**24**:393–405.

117. Willett K, Bennett J. Immunology of AAV-mediated gene transfer in the eye. *Front Immunol* 2013;**4**:261.

118. Britten CM, Janetzki S, van der Burg SH, Huber C, Kalos M, Levitsky HI, et al. Minimal information about T cell assays: the process of reaching the community of T cell immunologists in cancer and beyond. *Cancer Immunol Immunother* 2011;**60**:15–22.

119. Britten CM, Walter S, Janetzki S. Immunological monitoring to rationally guide AAV gene therapy. *Front Immunol* 2013;**4**:273.

120. Finn JD, Nichols TC, Svoronos N, Merricks EP, Bellenger DA, Zhou S, et al. The efficacy and the risk of immunogenicity of FIX Padua (R338L) in hemophilia B dogs treated by AAV muscle gene therapy. *Blood* 2012;**120**:4521–3.

121. Siner JI, Iacobelli NP, Sabatino DE, Ivanciu L, Zhou S, Poncz M, et al. Minimal modification in the factor VIII B domain sequence ameliorates the murine hemophilia A phenotype. *Blood* 2013;**121**:4396–403.

122. Cantore A, Nair N, Della Valle P, Di Matteo M, Matrai J, Sanvito F, et al. Hyperfunctional coagulation factor IX improves the efficacy of gene therapy in hemophilic mice. *Blood* 2012;**120**:4517–20.

123. Jiang H, Couto LB, Patarroyo-White S, Liu T, Nagy D, Vargas JA, et al. Effects of transient immunosuppression on adenoassociated, virus-mediated, liver-directed gene transfer in rhesus macaques and implications for human gene therapy. *Blood* 2006;**108**:3321–8.

124. Scallan CD, Jiang H, Liu T, Patarroyo-White S, Sommer JM, Zhou S, et al. Human immunoglobulin inhibits liver transduction by AAV vectors at low AAV2 neutralizing titers in SCID mice. *Blood* 2006;**107**:1810–7.

125. Murphy SL, Li H, Zhou S, Schlachterman A, High KA. Prolonged susceptibility to antibody-mediated neutralization for adeno-associated vectors targeted to the liver. *Mol Ther* 2008;**16**:138–45.

126. Arruda VR, Stedman HH, Haurigot V, Buchlis G, Baila S, Favaro P, et al. Peripheral transvenular delivery of adeno-associated viral vectors to skeletal muscle as a novel therapy for hemophilia B. *Blood* 2010;**115**:4678–88.

127. Haurigot V, Mingozzi F, Buchlis G, Hui D, Chen Y, Basner-Tschakarjan E, et al. Safety of AAV factor IX peripheral transvenular gene delivery to muscle in hemophilia B dogs. *Mol Ther* 2010;**18**:1318–29.

128. Su LT, Gopal K, Wang Z, Yin X, Nelson A, Kozyak BW, et al. Uniform scale-independent gene transfer to striated muscle after transvenular extravasation of vector. *Circulation* 2005;**112**:1780–8.

129. Toromanoff A, Cherel Y, Guilbaud M, Penaud-Budloo M, Snyder RO, Haskins ME, et al. Safety and efficacy of regional intravenous (r.i.) versus intramuscular (i.m.) delivery of rAAV1 and rAAV8 to nonhuman primate skeletal muscle. *Mol Ther* 2008;**16**:1291–9.

130. Mingozzi F, Chen Y, Edmonson SC, Zhou S, Thurlings RM, Tak PP, et al. Prevalence and pharmacological modulation of humoral immunity to AAV vectors in gene transfer to synovial tissue. *Gene Ther* 2012;**20**:417–24.

131. Boissier MC, Lemeiter D, Clavel C, Valvason C, Laroche L, Begue T, et al. Synoviocyte infection with adeno-associated virus (AAV) is neutralized by human synovial fluid from arthritis patients and depends on AAV serotype. *Hum Gene Ther* 2007;**18**:525–35.

132. Amado D, Mingozzi F, Hui D, Bennicelli JL, Wei Z, Chen Y, et al. Safety and efficacy of subretinal readministration of a viral vector in large animals to treat congenital blindness. *Sci Transl Med* 2010;**2**:21ra16.

133. Bennett J, Ashtari M, Wellman J, Marshall KA, Cyckowski LL, Chung DC, et al. AAV2 gene therapy readministration in three adults with congenital blindness. *Sci Transl Med* 2012;**4**:120ra115.

134. Maguire AM, High KA, Auricchio A, Wright JF, Pierce EA, Testa F, et al. Age-dependent effects of RPE65 gene therapy for Leber's congenital amaurosis: a phase 1 dose-escalation trial. *Lancet* 2009;**374**:1597–605.

135. Manno CS, Chew AJ, Hutchison S, Larson PJ, Herzog RW, Arruda VR, et al. AAV-mediated factor IX gene transfer to skeletal muscle in patients with severe hemophilia B. *Blood* 2003;**101**:2963–72.

136. Stroes ES, Nierman MC, Meulenberg JJ, Franssen R, Twisk J, Henny CP, et al. Intramuscular administration of AAV1-lipoprotein lipase S447X lowers triglycerides in lipoprotein lipase-deficient patients. *Arterioscler Thromb Vasc Biol* 2008;**28**:2303–4.

137. Kaplitt MG, Feigin A, Tang C, Fitzsimons HL, Mattis P, Lawlor PA, et al. Safety and tolerability of gene therapy with an adeno-associated virus (AAV) borne GAD gene for Parkinson's disease: an open label, phase I trial. *Lancet* 2007;**369**:2097–105.

138. Duque S, Joussemet B, Riviere C, Marais T, Dubreil L, Douar AM, et al. Intravenous administration of self-complementary AAV9 enables transgene delivery to adult motor neurons. *Mol Ther* 2009;**17**:1187–96.

139. Ruzo A, Marco S, Garcia M, Villacampa P, Ribera A, Ayuso E, et al. Correction of pathological accumulation of glycosaminoglycans in central nervous system and peripheral tissues of MPSIIIA mice through systemic AAV9 gene transfer. *Hum Gene Ther* 2012;**23**:1237–46.

140. Haurigot V, Marco S, Ribera A, Garcia M, Ruzo A, Villacampa P, et al. Whole body correction of mucopolysaccharidosis IIIA by intracerebrospinal fluid gene therapy. *J Clin Invest* 2013 pii: 66778. doi: 10.1172/JCI66778. [Epub ahead of print].

141. Gray SJ, Matagne V, Bachaboina L, Yadav S, Ojeda SR, Samulski RJ. Preclinical differences of intravascular AAV9 delivery to neurons and glia: a comparative study of adult mice and nonhuman primates. *Mol Ther* 2011;**19**:1058–69.

142. Erles K, Sebokova P, Schlehofer JR. Update on the prevalence of serum antibodies (IgG and IgM) to adeno-associated virus (AAV). *J Med Virol* 1999;**59**:406–11.

143. Li C, Narkbunnam N, Samulski RJ, Asokan A, Hu G, Jacobson LJ, et al. Neutralizing antibodies against adeno-associated virus examined prospectively in pediatric patients with hemophilia. *Gene Ther* 2011;**19**:288–94.

144. Boutin S, Monteilhet V, Veron P, Leborgne C, Benveniste O, Montus MF, et al. Prevalence of serum IgG and neutralizing factors against adeno-associated virus (AAV) types 1, 2, 5, 6, 8, and 9 in the healthy population: implications for gene therapy using AAV vectors. *Hum Gene Ther* 2010;**21**:704–12.

145. Calcedo R, Vandenberghe LH, Gao G, Lin J, Wilson JM. Worldwide epidemiology of neutralizing antibodies to adeno-associated viruses. *J Infect Dis* 2009;**199**:381–90.

146. Chiorini JA, Kim F, Yang L, Kotin RM. Cloning and characterization of adeno-associated virus type 5. *J Virol* 1999;**73**:1309–19.

147. Murphy SL, Li H, Mingozzi F, Sabatino DE, Hui DJ, Edmonson SA, et al. Diverse IgG subclass responses to adeno-associated virus infection and vector administration. *J Med Virol* 2009;**81**:65–74.

148. Wang L, Calcedo R, Wang H, Bell P, Grant R, Vandenberghe LH, et al. The pleiotropic effects of natural AAV infections on liver-directed gene transfer in macaques. *Mol Ther* 2010;**18**:126–34.

149. Halbert CL, Miller AD, McNamara S, Emerson J, Gibson RL, Ramsey B, et al. Prevalence of neutralizing antibodies against adeno-associated virus (AAV) types 2, 5, and 6 in cystic fibrosis and normal populations: implications for gene therapy using AAV vectors. *Hum Gene Ther* 2006;**17**:440–7.

150. Moskalenko M, Chen L, van Roey M, Donahue BA, Snyder RO, McArthur JG, et al. Epitope mapping of human anti-adeno-associated virus type 2 neutralizing antibodies: implications for gene therapy and virus structure. *J Virol* 2000;**74**:1761–6.

151. Sun L, Tu L, Gao G, Sun X, Duan J, Lu Y. Assessment of a passive immunity mouse model to quantitatively analyze the impact of neutralizing antibodies on adeno-associated virus-mediated gene transfer. *J Immunol Methods* 2013;**387**:114–20.

152. Mingozzi F, Chen Y, Murphy SL, Edmonson SC, Tai A, Price SD, et al. Pharmacological modulation of humoral immunity in a nonhuman primate model of AAV gene transfer for hemophilia B. *Mol Ther* 2012;**20**:1410–6.

153. Nathwani AC, Gray JT, Ng CY, Zhou J, Spence Y, Waddington SN, et al. Self-complementary adeno-associated virus vectors containing a novel liver-specific human factor IX expression cassette enable highly efficient transduction of murine and nonhuman primate liver. *Blood* 2006;**107**:2653–61.

154. Bartel M, Schaffer D, Buning H. Enhancing the clinical potential of AAV vectors by capsid engineering to evade pre-existing immunity. *Front Microbiol* 2011;**2**:204.

155. Stone D, Koerber JT, Mingozzi F, Podsakoff GM, High KA, Schaffer DV. Associated virus variants that evade human neutralizing antibodies. *Mol Ther ASGCT Annu Meet Abstracts* 2010;S8.

156. Perabo L, Endell J, King S, Lux K, Goldnau D, Hallek M, et al. Combinatorial engineering of a gene therapy vector: directed evolution of adeno-associated virus. *J Gene Med* 2006;**8**:155–62.

157. Gurda BL, Raupp C, Popa-Wagner R, Naumer M, Olson NH, Ng R, et al. Mapping a neutralizing epitope onto the capsid of adeno-associated virus serotype 8. *J Virol* 2012;**86**:7739–51.

158. Monteilhet V, Saheb S, Boutin S, Leborgne C, Veron P, Montus MF, et al. A 10 patient case report on the impact of plasmapheresis upon neutralizing factors against adeno-associated virus (AAV) types 1, 2, 6, and 8. *Mol Ther* 2011;**19**:2084–91.

159. Hurlbut GD, Ziegler RJ, Nietupski JB, Foley JW, Woodworth LA, Meyers E, et al. Preexisting immunity and low expression in primates highlight translational challenges for liver-directed AAV8-mediated gene therapy. *Mol Ther* 2010;**18**:1983–94.

160. Chicoine LG, Montgomery CL, Bremer WG, Shontz KM, Griffin DA, Heller KN, et al. Plasmapheresis eliminates the negative impact of AAV antibodies on microdystrophin gene expression following vascular delivery. *Mol Ther* 2014;**22**:338–47.

161. Tony HP, Burmester G, Schulze-Koops H, Grunke M, Henes J, Kotter I, et al. Safety and clinical outcomes of rituximab therapy in patients with different autoimmune diseases: experience from a national registry (GRAID). *Arthritis Res Ther* 2011;**13**:R75.

162. Stasi R, Provan D. Management of immune thrombocytopenic purpura in adults. *Mayo Clin Proc* 2004;**79**:504–22.

163. Neubert K, Meister S, Moser K, Weisel F, Maseda D, Amann K, et al. The proteasome inhibitor bortezomib depletes plasma cells and protects mice with lupus-like disease from nephritis. *Nat Med* 2008;**14**:748–55.

164. Monahan PE, Lothrop CD, Sun J, Hirsch ML, Kafri T, Kantor B, et al. Proteasome inhibitors enhance gene delivery by AAV virus vectors expressing large genomes in hemophilia mouse and dog models: a strategy for broad clinical application. *Mol Ther* 2010;**18**:1907–16.

165. Nathwani AC, Cochrane M, McIntosh J, Ng CY, Zhou J, Gray JT, et al. Enhancing transduction of the liver by adeno-associated viral vectors. *Gene Ther* 2009;**16**:60–9.

166. Karman J, Gumlaw NK, Zhang J, Jiang JL, Cheng SH, Zhu Y. Proteasome inhibition is partially effective in attenuating pre-existing immunity against recombinant adeno-associated viral vectors. *PLoS One* 2012;**7**:e34684.

167. Arastu-Kapur S, Anderl JL, Kraus M, Parlati F, Shenk KD, Lee SJ, et al. Nonproteasomal targets of the proteasome inhibitors bortezomib and carfilzomib: a link to clinical adverse events. *Clin Cancer Res* 2011;**17**:2734–43.

168. Thompson JL. Carfilzomib: a second-generation proteasome inhibitor for the treatment of relapsed and refractory multiple myeloma. *Ann Pharmacother* 2013;**47**:56–62.

169. Chan VS, Tsang HH, Tam RC, Lu L, Lau CS. B-cell-targeted therapies in systemic lupus erythematosus. *Cell Mol Immunol* 2013;**10**:133–42.

170. Ginzler EM, Wax S, Rajeswaran A, Copt S, Hillson J, Ramos E, et al. Atacicept in combination with MMF and corticosteroids in lupus nephritis: results of a prematurely terminated trial. *Arthritis Res Ther* 2012;**14**:R33.

171. Dimitrov JD, Dasgupta S, Navarrete AM, Delignat S, Repesse Y, Meslier Y, et al. Induction of heme oxygenase-1 in factor VIII-deficient mice reduces the immune response to therapeutic factor VIII. *Blood* 2010;**115**:2682–5.

172. Lei TC, Scott DW. Induction of tolerance to factor VIII inhibitors by gene therapy with immunodominant A2 and C2 domains presented by B cells as Ig fusion proteins. *Blood* 2005;**105**:4865–70.

173. Brunetti-Pierri N, Ng T, Iannitti DA, Palmer DJ, Beaudet AL, Finegold MJ, et al. Improved hepatic transduction, reduced systemic vector dissemination, and long-term transgene expression by delivering helper-dependent adenoviral vectors into the surgically isolated liver of nonhuman primates. *Hum Gene Ther* 2006;**17**:391–404.

174. Brunetti-Pierri N, Palmer DJ, Mane V, Finegold M, Beaudet AL, Ng P. Increased hepatic transduction with reduced systemic dissemination and proinflammatory cytokines following hydrodynamic injection of helper-dependent adenoviral vectors. *Mol Ther* 2005;**12**:99–106.

175. Hodges BL, Taylor KM, Chu Q, Scull SE, Serriello RG, Anderson SC, et al. Local delivery of a viral vector mitigates neutralization by antiviral antibodies and results in efficient transduction of rabbit liver. *Mol Ther* 2005;**12**:1043–51.

176. Mimuro J, Mizukami H, Hishikawa S, Ikemoto T, Ishiwata A, Sakata A, et al. Minimizing the inhibitory effect of neutralizing antibody for efficient gene expression in the liver with adeno-associated virus 8 vectors. *Mol Ther* 2013;**21**:318–23.

177. Raper SE, Wilson JM, Nunes FA. Flushing out antibodies to make AAV gene therapy available to more patients. *Mol Ther* 2013;**21**:269–71.

178. Mingozzi F, Anguela XM, Pavani G, Chen Y, Davidson RJ, Hui DJ, et al. Overcoming preexisting humoral immunity to AAV using capsid decoys. *Sci Transl Med* 2013;**5**:194ra192.

179. Tseng YS, Agbandje-McKenna M. Mapping the AAV capsid host antibody response toward the development of second generation gene delivery vectors. *Front Immunol* 2014;**5**(9).

180. D'Avola D, Lopez-Franco E, Harper P, Fontanellas A, Grosios N, Sangro B, et al. Phase 1 clinical trial of liver directed gene therapy with rAAV5-PBGD in acute intermittent porphyria: preliminary safety data. *Mol Ther ASGCT Annu Meet Abstr* 2014;S14.

181. Mays LE, Wang L, Tenney R, Bell P, Nam HJ, Lin J, et al. Mapping the structural determinants responsible for enhanced T cell activation to the immunogenic adeno-associated virus capsid from isolate rhesus 32.33. *J Virol* 2013;**87**:9473–85.

Chapter 5

Risks of Insertional Mutagenesis by DNA Transposons in Cancer Gene Therapy

Perry B. Hackett[1], Timothy K. Starr[1,2] and Laurence J.N. Cooper[3]

[1]*Department of Genetics, Cell Biology and Development, Center for Genome Engineering and Masonic Cancer Center, University of Minnesota, Minneapolis, MN, USA;* [2]*Department of Obstetrics, Gynecology and Women's Health, Center for Genome Engineering and Masonic Cancer Center, University of Minnesota Medical School, Duluth, MN, USA;* [3]*Division of Pediatrics and Department of Immunology, University of Texas MD Anderson Cancer Center, Houston, TX, USA*

LIST OF ABBREVIATIONS

AAV Adeno-associated viruses
aAPCs Antigen-presenting cells
ADA Adenosine deaminase
CISs Common insertion sites
CGD Chronic granulomatous disease
DNF Dominant-negative function
GOF Gain-of-function
HERVs Human endogenous retroviral sequences
HSCs Hematopoietic stem cells
IL Interleukin
LOF Loss-of-function
LTR Long terminal repeat
mRNA Messenger RNA
MSCV Murine stem cell virus
pA Polyadenylation
PCR Polymerase chain reaction
SA Splice acceptors
TAL Transcription activator-like
TG Therapeutic gene

5.1 INSERTIONAL MUTAGENESIS—THE DOWNSIDE OF GENE THERAPY?

After a long probationary period, human gene therapy has entered the Western commercial world with the approval of Glybera to treat lipoprotein lipase deficiency using an adeno-associated virus gene vector that has a largely random pattern of integration into host chromosomes. The potential of gene therapy to treat disease has always been obvious, but so are the risks of adverse outcomes associated with altering the genome. Accordingly, the principal objective now is to reduce risk as the applications are broadened to the point that gene therapy becomes commonplace in clinical practice. Although recent clinical trials have demonstrated the increasing promise of gene therapy, they have also illustrated the difficulties of assessing risks given the inherent uncertainty of trial outcomes.[1]

This chapter is focused on evaluating the risks of insertional mutagenesis when using *Sleeping Beauty* (SB) transposons in genetically modified T cells to treat cancer. The genetic modification of T cells was recognized as part of The Scientific Breakthrough of the Year in 2013, thus the risks associated with introducing new genes into human chromosomes will take on added importance as adoptive immunotherapy enters into clinical practice. Chromosomal integration is necessary when outgrowth of the genetically engineered cells is a required condition for effective clinical treatment, as is the case with adoptive immunotherapy. Here we discuss the concerns of insertional mutagenesis in the light of our recent increased understanding of the plasticity of the human genome, the newly discovered levels of background transposition by endogenous retrotransposable elements, and the consequences of genetically modifying oncogenes to induce neoplastic transformation of cells.

5.1.1 Retrovirus-Associated Adverse Events in Gene Therapy Trials

Whenever any DNA sequence "randomly" integrates into a genome, there is a chance that the relationship of endogenous genes and their regulatory components will be affected. This is called *insertional mutagenesis*. There are many ways that endogenous genetic information can be corrupted by insertional mutagenesis, of which, the most commonly discussed are shown in Figure 5.1. Genetic alterations that are associated with adverse health consequences are (1) *Loss-of-function*

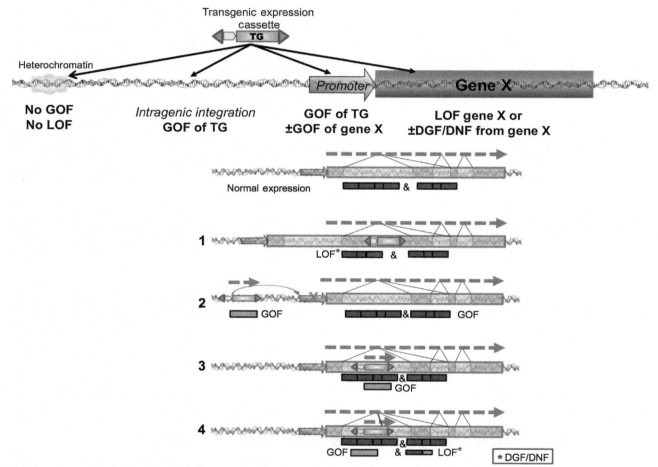

FIGURE 5.1   Insertional mutagenesis by integration of therapeutic vectors. This figure uses a DNA transposon as a model vector, denoted by the inverted terminal repeats (blue triangles), but the same concepts apply to viral vectors. Transposon or viral integration can occur in four general types of chromatin. Integration into heterochromatin, essentially a black hole for expression cassettes, normally results in silencing of the vector. Integration into intergenic regions is desirable because a therapeutic gene (TG) will be expressed without affecting endogenous genes. Integration into or proximal to a transcriptional regulatory region that contains a protein-encoding gene (gene X); this can have several outcomes including gain-of-function (GOF) of the transgene, as well as GOF or loss-of-function (LOF) of gene X. Normal expression of gene X is shown by the dashed green arrow representing RNA synthesis, light black lines representing splicing of exons (with alternative splicing of the first exon to either the second or third exon being indicated), and alternative translational products represented by the purple boxes. Four examples of insertional mutagenesis are shown: (1) integration of a vector into an exon, which can lead to formation of a hybrid polypeptide (purple/blue product) similar to normal proteins that come from alternative splicing; (2) integration of a vector with a strong promoter that leads to expression of the transgene, but also may activate a silenced promoter (the red X) for gene X, thereby leading to unwanted expression of an endogenous gene; (3) integration into an intron of gene X, which generally results in expression of the transgene with little effect on gene X because the transgenic sequences are spliced out of gene X's pre-mRNAs; and (4) integration of a vector with cryptic splicing sites (red line) that can lead to hybrid proteins (purple/gold product), resulting in LOF of gene X. In both examples 1 and 4, as well as other possibilities not shown, the hybrid polypeptides could have dominant-negative function (DNF) and/or dominant gain-of-function (DGF) that need not be "negative" per se. Many other types of genomic alterations, including induction or repression of epigenetic markings are possible. (For interpretation of the references to color in this figure legend, the reader is referred to the online version of this book.)

(LOF) of a critical gene because of interruption by the insertion of an exogenous DNA sequence into an exon. (2) *Gain-of-function* (GOF) of an oncogene, wherein a gene associated with cellular expansion that should not be expressed in a differentiated cell is activated by integration of strong transcriptional regulatory elements that were designed to drive the incoming transgene. (3) LOF of a critical gene because of insertions into introns. Insertion of a vector into an intron generally is more likely than into an exon because introns are on average six- to 10-fold longer than exons, although vectors that prefer to integrate proximal to transcriptional start sites may have a preference for the first exon.[2] As long as there are no cryptic splice-acceptor or splice-donor sites in the vector, normal splicing should occur, but the level of transcription may be reduced resulting in low expression of the endogenous gene. (4) Should there be *cryptic* (low level) splice-donor or splice-acceptor sites in the transgenic vector, then there is the possibility of synthesis of chimeric polypeptides that may have dominant-negative function or dominant gain-of-function, either of which could have adverse effects on cellular growth. Other potential mutagenic processes by transposons have been extensively described.[3] The important lesson of Figure 5.1 is that integration of transgenic expression cassettes inside a gene *does not* irrevocably lead to loss of normal gene expression. This point is discussed further in terms of recently discovered plasticity of regulation of gene expression in human cells.

The potential of unpredictable outcomes from insertional vectors was observed when recombinant retroviruses were employed to deliver genes semirandomly into mouse genomes.[4] Two decades later, similar results were seen in humans undergoing gene therapy in which retroviral vectors were used to introduce a therapeutic interleukin 2γc gene to treat X-linked severe combined immunodeficiency disease. Five of twenty boys developed leukemia three or more years after their treatments.[5] Two boys died, the other three were successfully treated for their leukemias as well as immunodeficiency. In each case, integration of a gamma-retroviral vector was found upstream of the endogenous LMO2 proto-oncogene (Figure 5.1, example 2), which led to the hypothesis that in hematopoietic stem cells (HSCs) integrated retroviral vectors and the enhancer/promoter cassettes they carry could activate an oncogene.[6] In subsequent studies, therapeutic retroviruses containing the gp91Phox gene for treatment of chronic granulomatous disease (CGD) resulted in leukemia and one death.[7] In the CGD patients, retroviral integrations proximal to growth-related genes have been associated with enhanced proliferation.[8] Consequently, viral vectors that integrated into chromosomes were associated with leukemias that, in some cases, could be successfully treated. These early events in retroviral transduction of HSCs cemented the notion of the dangers of insertional mutagenesis in the minds of many researchers.[9]

In contrast, retroviral vectors carrying either the adenosine deaminase (ADA) gene for treatment of ADA-linked immunodeficiency[10,11] or WASp for Wiskott–Aldrich syndrome[12] have not been associated with similar adverse events. Currently, by using altered retroviruses and appropriate doses, adverse events have been avoided.[13–15] Recently, successful use of lentiviral therapies has been demonstrated for metachromatic leukodystrophy, an inherited lysosomal storage disease caused by arylsulfatase A deficiency,[16] and CGD in which vector integration analyses revealed highly polyclonal and multilineage hematopoiesis that resulted from the gene-corrected HSCs. These lentiviral gene therapies have not induced selection of integrations near oncogenes nor aberrant clonal expansion.[17] Thus, although vector integration can be mutagenic, consequential adverse events are not a *fait accompli*. In some cases, there is limited, temporary clonal expansion of treated cells that actually increases the likelihood of successful gene therapy. Moreover, in contrast to early gene therapy of HSCs, viral-mediated transduction of T cells for adoptive immunotherapy has not been associated with adverse outcomes; comparable treatments of T cells for adoptive immunotherapy has not resulted in undesirable clinical outcomes.[13,14,18,19]

It is unclear why certain gene therapy treatments led to negative outcomes whereas others did not. There may be differences between cell types for integration sites as well as the availability of endogenous genes for activation. Analyses of populations of gamma retrovirus-modified HSCs show that many transgene integration sites were proximal to genes involved in cell survival and growth control (e.g., 40% of insertions were found in only a very small proportion of the genome, about 0.5%, that had nearly a 100-fold affinity for integrants compared with the genome as a whole[20,21]). These and other studies make it clear that only a small subset of all the integrations are responsible for most of the outgrowths of therapeutically treated cells. Retroviral parameters associated with integration effects and genotoxicity are still poorly understood.[22–25]

The deliberate use of insertional vectors to uncover oncogenes by inducing leukemia and solid tumors has exacerbated the concerns of insertional mutagenesis when the very same vectors are used for gene therapy. Retroviruses[26] as well as transposons[27] are used for this purpose, although predisposing mutations are often included in studies to raise the numbers of oncogenic events per mouse.[28] These studies have identified a number of mechanisms by which integrations of specialized retroviruses induce mutations that lead to cancer.

AIDS patients are at increased risk for some cancers because of their immune state,[29] but lentiviral vectors have not been associated with adverse outcomes stemming from either activation, inactivation, and/or alteration of splicing of endogenous genes.[30] The absence of cancers/leukemias after treatments with lentiviruses might be due either to the ability of these vectors to infect nondividing cells for therapeutic benefit[31] or to their selection of a different set of genetic loci for integration.[32]

5.2 *SLEEPING BEAUTY* TRANSPOSON/TRANSPOSASE SYSTEM ADAPTED FOR GENE THERAPY

Nonviral vectors have distinct advantages over viral vectors for gene therapy.[33,34] Transposons, specifically, have two major improvements over viruses as gene therapy vectors. First, clinical grade manufacture and quality control of purified transposon DNA is easier, more reliable, and less expensive than employing recombinant clinical-grade virus. Second, unlike viral vectors that have an integration bias either into or proximal to transcriptional units, SB transposons have few known preferences for integration. Nevertheless, insertional mutagenesis is always a concern.

During the past 2 years, there have been significant findings in areas that pertain to genotoxicity and its impact on gene therapy. First, the results from the 1000 Genomes Project and ENCODE Project[35] have demonstrated that the interactions of genetic elements in our chromosomes are far more variable and complex than previously thought. Second, it is evident that endogenous transposons are far more active in human cells than was surmised until very recently.[36] Two questions arise from these findings: Do these elements induce similar genetic consequences as therapeutic transposons and do cells have mechanisms to cope with insertional mutagenic agents? In light of these recent discoveries, we have three objectives in this review that pertain to the use of SB transposons to engineer human T cells for adoptive immunotherapy in the clinic: (1) present the context of SB transposition in an environment of relatively high retro-transposition activity in human cells; (2) evaluate SB transposon-mediated induction of cancer in mouse studies and their pertinence to patients treated with SB transposons; and (3) present approaches to ameliorate potential risks of transposon-mediated events in patients.

5.2.1 DNA Transposons

At least 45% of the human genome is composed of clearly identifiable transposable elements,[37] and up to two-thirds of chromosomes may be derived from transposons that have been adapted to support cell function.[38] Transposons are divided into two classes (Figure 5.2). Class I transposons are retroelements that spread by a *copy-and-paste* mechanism, whereby the transposon is transcribed and the RNA transcript reverse transcribed for insertion elsewhere in the genome. Class II transposons, which include the SB system, are DNA sequences that can "hop/jump" from one site to the next via a *cut-and-paste* mechanism. The distinctive features of class I and class II transposons are important considerations in evaluating risks associated with any type of gene therapy that involves integration of genetic sequences into chromosomes.

Class II DNA transposons, unlike class I transposons, are DNA sequences that can be excised from the genome for insertion elsewhere in the same, or different chromosome, by a cut-and-paste mechanism (Figure 5.2). The transposon sequence that is excised is precise, in that the termini of the transposed sequence are exact. However, with many transposons, when the donor DNA sequence is repaired, there is commonly a "footprint" that is left that, in the case of SB, appears to vary according to cell type, but often is a 5-bp AC(A/T)GT insertion[39] flanked by the TA sites that flank all SB transposons. Approximately 300,000 class II transposons comprise about 3% of the human genome, but all are inactive because of a lack of a functional transposase gene.[37] *Active* class II transposons are defined by inverted terminal repeats that flank a transposase gene, which does not have a promoter. Thus, remobilization is dependent on integration proximal to an active, endogenous promoter, which is infrequent because of evolutionary pressure to suppress insertional mutagenesis. However, DNA transposons under some circumstances can spread when the transposase gene is activated.[40] Transposons, like viruses, are natural mechanisms used in nature to introduce new genetic sequences into cells. However, the host-cell responses to viruses and transposons, both parasitic invaders, differ. The general strategy of most viruses is to *hit-and-run* (i.e., infect and replicate regardless of consequences to the cell or whole organism). As a result, cells have elaborate defenses against viruses. In contrast, transposons are naked DNA sequences that lack the infective ability of viruses; they rarely penetrate the membrane barriers that shield genomes from the outer world, and so the defenses against transposons are generally, but not always, minimal. In nature, transposons are *hit-and-hide* vectors that are maintained over evolutionary periods in all the offspring of the cell.

5.2.2 *Sleeping Beauty* Transposon System

The potential of DNA transposons for use as vectors for transgene delivery was realized with the resurrection of an extinct DNA transposon in current salmonid genomes.[41] DNA transposons accommodate genetic cargos of up to, and on occasion, more than 10 kbp.[42] There are not any active class II transposons (transposases) in the human genome. The SB system (SB transposon plus SB transposase) derives its name from its resurrection from an evolutionary sleep of more than 10 million years.[41] Owing to approximately 500 million years of evolutionary separation, the salmonid-based SB transposase

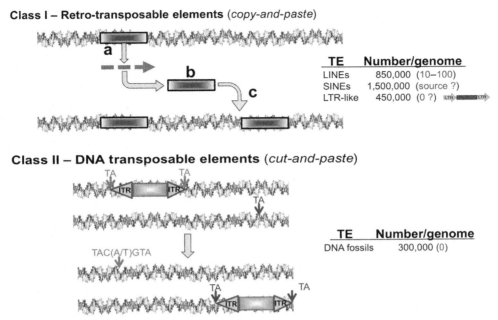

Class I – Retro-transposable elements (*copy-and-paste*)

TE	Number/genome	
LINEs	850,000	(10–100)
SINEs	1,500,000	(source ?)
LTR-like	450,000	(0 ?)

Class II – DNA transposable elements (*cut-and-paste*)

TAC(A/T)GTA

TE	Number/genome
DNA fossils	300,000 (0)

FIGURE 5.2   Transposable elements in the human genome. Transposable elements, which comprise nearly 50% of the human genome, are divided into two classes: class I retrotransposable elements spread by a copy-and-paste mechanism whereby the transposon is transcribed by either RNA polymerase II (long interspersed elements (LINEs) and LTR-like) or RNA polymerase III (short interspersed elements (SINEs)), and the RNA transcript (dashed green line) is then reverse-transcribed for insertion as a double-stranded DNA elsewhere in the genome. The LTR-like transposons have long terminal repeat sequences (illustrated in the table to the right) that flank a set of genes that are related to retroviral genomes; the genes generally are defective. Only a very small proportion of class I transposons are active and even these rarely transpose. Class II transposons, which include the synthetic *Sleeping Beauty* (SB) system, are DNA sequences that can "hop/jump" from one site to the next via a cut-and-paste mechanism when an active transposase is available that recognizes their inverted terminal repeat (ITR) sequences. No active DNA transposase genes in the human genome have been identified, and the SB transposase does not bind and transpose the inactive human Class II sequences. Blue entries in the tables indicate the numbers of *active* elements of each type in humans; there are an unknown number of active SINEs, generally *Alu* elements. (For interpretation of the references to color in this figure legend, the reader is referred to the online version of this book.)

does not recognize remnants of class II transposons in the human genome. Since its first use as a vector for gene therapy in mice,[43] SB transposase has been reengineered for increased activity (e.g., SB11, SB100X).[42,44] Likewise, the inverted-terminal-repeat structure of original transposon, *T*, also has been reengineered for greater activity to produce *T2* and other versions[33]. DNA-based transposons have been employed as vectors for transgene delivery since 1997 (Box 5.1). Maximal DNA transposon activity is not always desirable for gene therapy because multiple integrations increase the probability of unwanted mutations and unwanted effects on the target cells. Currently, the intermediate strength of the enzymatic activity of SB11[42] has been most extensively used in human trials.

For current clinical applications of the SB system, the SB11 expression cassette is supplied in a separate plasmid (Figure 5.3) rather than being incorporated into the same plasmid that carries the therapeutic transposon. There are two reasons for using two plasmids. First, small plasmids are more efficiently introduced into target cells for transposition.[42] Second, for clinical trials, once an SB transposase plasmid has been certified, it can be used with other plasmids that carry various therapeutic transposons. That said, a single plasmid with both the transposon and SB transposase gene (*cis*) could reduce cost as only one clinical-grade DNA plasmid would be needed, a consideration if transposons are used widely to treat cancer or other genetic diseases.

There are multiple advantages to using transposons as gene-delivery vectors. They can be straightforwardly delivered to single cells, the entire transposon is precisely integrated, and each integration reaction is independent so that concatemers are essentially never formed.[67] A further advantage of the SB system, but not all other transposons, is that it only requires a TA dinucleotide sequence for integration, with very few preferences for integration, unlike most integrating viruses[2,68–70] (Figure 5.4, Table 5.1). There are about 200 million TA sites in the human genome, which is appealing for obtaining a wide distribution of integrated vectors. Although about 10% of TA sites can be considered *preferential* because of an inherent flexibility of their flanking base pairs,[71] the SB system has the least preference for integration in and around transcriptional units (Table 5.1). As a consequence of its overall lack of preference for promoters and actively transcribed genes, unlike other transposons adapted for use in mammals, the SB transposon system has become the leading nonviral vector for gene

Box 5.1 A brief history of the *Sleeping Beauty* transposon system used for gene transfer

The *Sleeping Beauty* (SB) system was developed in 1997 as a nonviral vector for introducing genes into fish. The goal was to generate faster growing animals without using viral vectors, which at the time were considered too dangerous to use in animals for human consumption. However, it became apparent that the SB system was most active in human (HeLa) cells, which opened up consideration for its use as a nonviral vector for human gene therapy. The perceived advantages of using the SB system carried in plasmids were the relative ease and low cost of producing the required numbers of vectors to meet a large patient population, storage stability, and lack of immunogenicity.[45–47] Yant et al.[43] first demonstrated the efficacy of the SB system for gene therapy achieving sustained expression of α1-antitrypsin in normal mice and of factor IX clotting factor in hemophilic mice. Successful treatments of other genetic diseases in mice followed, including tyrosinemia,[48] hemophilia A and B,[49–52] mucopolysaccharidosis types I and VII,[53–56] epidermolysis bullosa,[57] glioblastoma multiforme,[58] sickle cell anemia,[59] and B-cell lymphoma.[60–62] In rats, SB transposons have been used to treat pulmonary hypertension[63] and jaundice.[64]

All of these in vivo studies were conducted by introducing transposons into the entire animal, most commonly using the hydrodynamic (high pressure, short interval of less than 9s) that directs transgene uptake and expression in the liver.[65] Two issues face hydrodynamic delivery to the liver. First, only about 10%, at most, of hepatocytes take up and stably express transgenes in SB transposons. This rather low efficiency can be overcome in two ways—either use the more active SB100X transposase rather than SB10 or SB11 to achieve multiple integrants per cell (e.g., references 43 and 44) or use an adenovirus as a carrier for the SB system.[52] The second issue impeding systemic delivery of SB for gene therapy in humans is that no one has reported successful scale-up of effective delivery to livers of larger animals such as the dog, suggesting that hydrodynamic impulses required for successful breach of the plasma membranes of hepatocytes are attenuated in the larger and more complex vasculatures of organs in animals weighing 1 kg or more.[66] The ex vivo approaches described in this chapter, wherein isolated hematopoietic cells are transduced by electroporation, avoids the issues with systemic delivery.

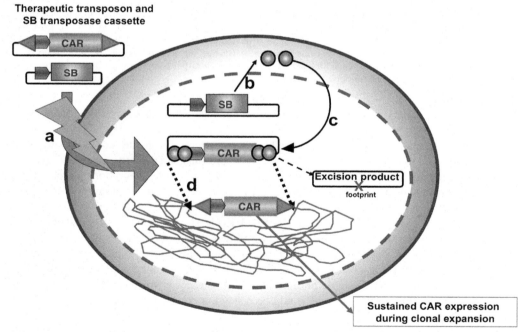

FIGURE 5.3 DNA transposition. DNA transposition, as exemplified by the *Sleeping Beauty* (SB) transposon system, is a cut-and-paste reaction in which a transposon containing an expression cassette with a therapeutic gene (in this case a chimeric antigen receptor (CAR)) and its promoter (green pentagon) is delivered through the plasma membrane of target cells; electroporation (yellow bolt) is the method of choice for delivery to T cells (a). After entry into the nucleus (dashed oval), the SB transposase is expressed (b) and SB transposase cuts the transposon out of the plasmid (c) and inserts it into a chromosome (d). The inverted terminal repeats (inverted set of double arrowheads) define the transposon. The second part of the SB system is SB transposase, which in this example is carried in a separate plasmid. The remaining plasmid, called an excision product, will have a footprint (red X) that is a sequence resulting from the repair of the cleavages on either side of the transposon. Excision products are a convenient way of evaluating the extent of transposition reactions. The SB transposon will almost always integrate into a TA-dinucleotide base pair. There are about 200 million in a diploid human genome, which allows nearly random insertion in the genome. Integration into a chromosome can confer sustained expression of the gene of interest through many rounds of cell amplification. (For interpretation of the references to color in this figure legend, the reader is referred to the online version of this book.)

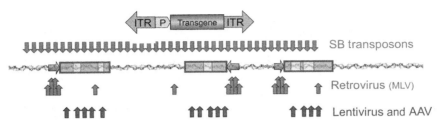

FIGURE 5.4 Comparison of integration preferences of vectors used in gene therapy. The schematic illustrates the nearly random integration profile of *Sleeping Beauty* transposons (blue arrows) that can integrate at any of about 200 million TA sites in the human genome more or less regardless of genetic activity, compared with retroviruses that prefer transcriptional regulatory motifs and lentiviruses and adeno-associated viruses (AAV) that prefer integration into active transcriptional units (see also Table 5.1). (For interpretation of the references to color in this figure legend, the reader is referred to the online version of this book.)

TABLE 5.1 Comparison of Insertion-Site Preferences of HIV, MLV, and SB[a]

Agent	% of Insertions in RefSeq Genes	% of Insertions ± 5 kb of a TSS
HIV	83%	11%
MLV	51%	20%
SB	39%	8%
Expect (random)	33%	5%

MLV, murine leukemia virus; SB, Sleeping Beauty; TSS, transcriptional start site.
[a]Adapted from data presented in references 2,68–70.

therapy. The SB vector has been validated for ex vivo gene delivery to stem cells, including T cells for the treatment of B-cell malignancies,[72] and SB transposons have been delivered to liver for treatment of various diseases in mice[33].

When used in a clinical setting, the possibility of remobilization of a transposon leading to an adverse event, as a result of residual transposase activity, is a theoretical concern that is ameliorated by the understanding that SB-mediated transposition is based on the conserved nature of DDE transposases[73] that are designed to be extremely inefficient. Remobilization of an SB transposon would have two consequences. First, remobilization would leave a *footprint* (Figure 5.3) that generally is an addition of 5 bp, although sometimes deletions and larger insertions occur as a result of repair at the excision site.[39] Hence, remobilization out of a protein-encoding exon would likely induce a frameshift mutation resulting in an abnormal polypeptide. Because exons compose no more than 2% of the human genome and the rate of excision of a transposon in a cell is about $1/10^4$, we estimate the chance of remobilization of a transposon into an exon is about $1/10^6$, and the chances of integration into an exon in a haplo-insufficient oncogene to induce an adverse event would be more than 100-fold smaller. Of course, the chance of an adverse event from reintegration is the same as for the initial delivery wherein far more sites are hit.

Nevertheless, there are two ways to deal with residual SB activity. First, use a transiently active promoter, such as the cytomegalovirus early promoter, to control SB transposase expression. This method results in SB activity being reduced about 10,000-fold over a couple of days.[74] Nevertheless, because the possibility of a few cells continuing to express transposase cannot be ruled out, a second alternative is to deliver the SB transposase as messenger RNA (mRNA) instead of a DNA plasmid.[75] Theoretically, this is sound, but the instability of mRNA raises quality control issues that could hinder widespread use for gene therapy.

Experience suggests that more elaborate constructions may not be needed. After more than a decade of using SB transposons for gene therapy in mice, there have not been any reports of tumor formation or leukemia. However, as discussed in the following section, SB transposons are notable for their role in identifying putative oncogenes.

5.2.3 SB Transposon-Mediated Induction of Cancer in Mice

Although it was developed as a vector for therapeutic gene transfer into genomes, the SB transposon system is more widely known as a tool to cause cancer and thereby identify oncogenes and pathways that lead to cancer.[27] Clearly this raises issues for its use as a gene therapy vector. On the basis of early studies that showed murine leukemia virus can cause cancer by acting as an insertional mutagen,[76] SB transposons were designed to mimic the abilities of retroviruses to cause both GOF and LOF after integration. T/Onc, the original SB transposon, and a number of derivatives such as T2/Onc, have been used in these experiments; herein, we collectively refer to them as T/Onc transposons. These transposons contain splice acceptors (SA) followed by polyadenylation (pA) signals in both orientations (Figure 5.5). Upon insertion into introns, these elements are designed to intercept upstream splice donors and elicit premature transcript truncation. Between the two SAs are sequences from the 5′LTR of the murine stem cell virus (MSCV) that contain strong, methylation-resistant, transcriptional motifs that will initiate expression from the site of integration.[77] Immediately downstream of the LTR is a splice donor for splicing of a transcript initiated from the LTR into a neighboring gene. Many variations on this theme have been developed.[27] As shown in Figure 5.5, integration of a T2/Onc transposon into a gene, exon or intron can lead to its nearly complete disruption with termination of transcription regardless of orientation of insertion (right-to-left or left-to-right). Additionally, the strong MSCV promoter can cause overexpression of the entire gene or a truncated portion of the gene. This alters the expression of endogenous genes and can direct the synthesis of dominant-negative gain of function polypeptides that can affect any of a number of pathways leading to cancer.

Figure 5.6 shows the general strategy for using SB transposons to disrupt genes. Because the transposon sequence is known, it is possible to identify the genomic insertion location using various polymerase chain reaction (PCR) techniques. This means that the mutated genes can be fairly easily identified. The two genetic constructs required for SB activity are the transposon and the transposase. Additional constructs can be used to limit SB activity to specific cell types or to predispose the mouse to cancer using a known cancer gene (Figure 5.6). The first required element is a concatemer of T/Onc transposons that is introduced into mice using standard pro-nuclear injection techniques to generate transgenic mice. This technique results in multiple transposons inserted into a single location in the mouse genome. The multiple transposons, referred to as a concatemer, can range from 25 to 200 copies. The second required element is the SB transposase expression cassette. The expression cassette can be designed to be ubiquitously expressed or conditionally expressed using a third element. Ubiquitous expression is commonly generated by "knocking in" an SB complementary DNA to the ubiquitous Rosa26 promoter in mice. This method has been used to induce or accelerate tumors.[78] Conditional activity is achieved by including a *lox-stop-lox* signal between the Rosa26 promoter and the SB complementary DNA. The *lox-stop-lox* signal blocks expression of SB unless the cell also expresses Cre recombinase, a bacteriophage enzyme capable of removing the *lox-stop-lox* signal. Many strains of mice expressing Cre recombinase under the control of different tissue-specific

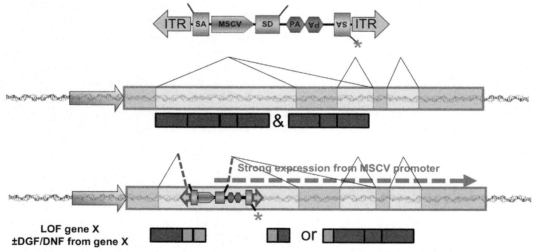

FIGURE 5.5 **T/Onc transposon.** The T/Onc series of *Sleeping Beauty* mutagenic vectors is illustrated by T2/Onc. T2/Onc contains elements designed to elicit either transcriptional activation from the mouse stem cell virus 5′-LTR (green murine stem cell virus (MSCV) pentagon) and splice donor (purple SD box) or inactivation (splice acceptors (gold SA boxes) and polyadenylation signals (red pA hexagons)). The inverted terminal repeats are indicated by the blue arrows labeled ITR. T2/Onc is effective in disrupting gene expression regardless of the orientation in which it integrates. Potential consequences of integration are shown in the bottom panel. Transcriptional regulatory regions (promoters and enhancers) are shown as green arrows; transcriptional units are in pink, with exon sequences darker than intron sequences and with lines showing possible splicing of exons. Normally encoded polypeptides are shown in purple, with the contributions by each exon indicated by the divisions within each box; and abnormal polypeptides with contributions from viral (green boxes) or intron sequences (orange boxes) indicated below the DNAs. The MSCV LTR promoter directs unusually high levels of RNA synthesis and the splice acceptor and donor sites in the T2/Onc vector can create unusual splicing, as indicated in the lower panel by the dashed green arrow and the dashed purple lines, respectively. (For interpretation of the references to color in this figure legend, the reader is referred to the online version of this book.)

promoters are available and can be used to direct the expression of SB to specific cell types. This method has been used to generate intestinal epithelial adenocarcinoma, hepatocellular carcinoma and many other cancer types.[79,80] Finally, a known cancer predisposing mutation, such as mutant *p53* or *Apc*, can be introgressed into the SB mice to identify mutations that cooperate with these predisposing mutations.[81,82] Figure 5.6 shows a representative experiment using a predisposing mutation (Onc[+/−]) and a conditional SB expression system using Cre recombinase. In the outlined experiment, four genotypes are produced: the experimental group (active SB and Onc[+/−]), the active SB-only group (active SB, no Onc[+/−]), the predisposed group (Onc[+/−], no Cre), and the control group (no Onc[+/−], no Cre). The expected outcome of this experimental design is a high rate of tumorigenesis in the experimental group, attenuated rates in the SB-only and Onc[+/−] groups, and few or no tumors in the control group.

To identify the genomic location of the transposon insertions, DNA is extracted from a tumor, and linker-mediated PCR is used to amplify fragments of DNA where the transposon is adjacent to genomic DNA. These PCR amplicons are sequenced using next generation sequencing technology, and the sequences are mapped to the mouse genome. In a typical

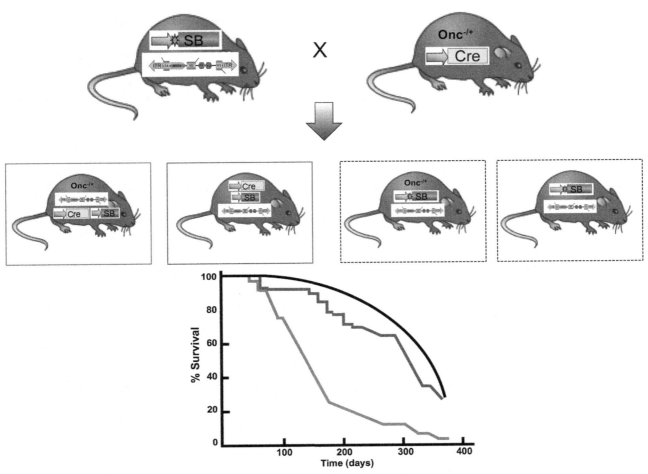

FIGURE 5.6 T2/Onc transposon-mediated cancer-gene screens in mice. The steps in mouse breeding for a tissue-specific cancer-gene screen are depicted. The steps include a mouse with a predisposing mutation (Onc[+/−]). A double transgenic mouse (top left) is generated to contain a concatemer of T2/Onc transposons (20–250 in a single locus) and a *Sleeping Beauty* (SB) transposase expression cassette that contains a blocking *lox-stop-lox* signal (gold star) between the promoter (green arrowhead) and the SB transpose gene (red box). The blocking *lox-stop-lox* signal permits tissue-specific expression of SB when the mice are bred to another mouse line that harbors a *Cre recombinase* transgene driven by a tissue-specific promoter. A second double-transgenic mouse (top right) is generated to contain the tissue-specific *Cre* transgene and another mutant tumor suppressor or oncogene. The two double-transgenic mice are mated resulting in offspring containing various combinations of these genes. Animals are genotyped and offspring with the four possible different genotypes (four middle boxes) are aged and monitored for tumor growth. Transposition in somatic cells causes random insertion mutations in mice that have an active transposase expression cassette and a T2/Onc concatemer (red and blue boxed mice on the left); owing to the presence of the Onc[+/−] predisposition (red box) there should be more oncogenic events than without it (blue box). Typical survival curves are shown at the bottom; generally, there is no significant difference between control mice without any part of the SB system and mice that have either a T2/Onc concatemer or SB transposase gene. (For interpretation of the references to color in this figure legend, the reader is referred to the online version of this book.)

experiment, a single tumor will have several hundred transposon insertions that can be identified. By comparing the insertions in all the tumors generated in the experiment, it is possible to identify common insertion sites (Figure 5.7). Clustering of insertions in certain regions, beyond what is expected by chance, indicates selection for these rare insertion events, among many neutral or negative insertion events, that can give the cell a selective growth or survival advantage. The genes present in these common insertion sites are, therefore, candidate cancer genes.

Five overall observations are important for evaluation of SB transposons used for gene therapy. First, although there is efficient induction of T-cell malignancy[78] and myeloid leukemia[83] in mice that are not genetically predisposed to tumor formation, the experience of performing many of these studies *in mice* has shown that unless many conditions are met, tumor induction by transposon mobilization is not efficient. Second, the presence of either a concatemer of T/Onc transposons or expression of the SB transposase by itself does not result in tumorigenesis in mice, even when the transposase is expressed ubiquitously over the lifetime of the mice. These findings suggest that SB transposase does not induce chromosomal instability and/or mobilize endogenous mammalian transposons at a detectable rate. Third, an intermediate to strong promoter sequence to drive transposition is required for efficient tumor induction, consistent with transpositional studies in somatic tissues.[74] Fourth, a promoter/splice donor sequence appears to be required to direct efficient oncogenesis. Fifth, in aggregate, all of the various studies using SB transposons to identify oncogenes by inducing tumors and leukemias highlight the major influence that genetic background has on genotoxicity by integrating vectors.

In rough numbers, assuming a mouse organ has about 1 million cells, with each cell containing a total of 100 transposons, a single tumor per animal would mean that the incidence of an adverse event is about $1/10^8$ transposons. This is a high number that does not reflect the actual incidence of a single transposon hit because in these animals there are hundreds of insertional mutations per tumor. This adverse event rate will be further reduced in gene therapy because therapeutic transposons will not contain the elements specifically designed to cause tumors (LTRs, splicing signals and transcription termination sequences) that are present in the mutagenesis experiments.

Thus, although the use of SB transposons to cause cancer may suggest danger, more careful consideration of the differences between the transposons used in mice to tag cancer genes and the SB transposons used in a clinical setting reveals that cancer is a highly unlikely outcome after therapeutic transposon delivery. To summarize, in the mouse experiments: (1) transposons are designed to maximally disrupt endogenous genetic pathways in multiple ways; (2) multiple copies of the disruptive transposon are mobilized in every targeted cell; (3) transposase enzyme is expressed continuously in all cells of a given tissue throughout the lifetime of the animal, which can lead to multiple remobilizations; and (4) efficient cancer induction is achieved in mice genetically predisposed to cancer. Thus, given the limitations of using mice as a model system (fewer cells per organ and considerably shorter lives) in the setting of human gene therapy, we do not expect cancer to be a likely side effect of using of transposon vectors.

A central point of this discussion is that for all the diagrams and models, we have to explain mutagenesis that leads to cancer: it takes an extraordinarily high number of events that employ insertion of a collection of transcriptional and RNA

FIGURE 5.7 Analysis of T2/Onc-induced solid tumors and leukemias. Cells from tumors (colored objects in the top right mouse) caused by T/Onc transposition are collected, the genomes are isolated and the transposon integration sites (colored arrows on the green lines representing the genome) are identified and mapped to chromosomal sites (lower left corner; star color relates to the transposons shown in the lower right corner). Genes that are repeatedly mutated in multiple, independent tumors are designated as *common insertion sites* (CISs). CISs can be analyzed to determine what genes and genetic pathways contribute to cancer. (For interpretation of the references to color in this figure legend, the reader is referred to the online version of this book.)

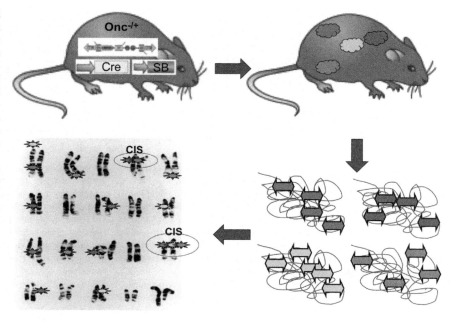

processing motifs to elicit adverse events. Recent investigations into the regulation of gene expression, the unexpected variations in human genomes and the natural instability of genomes may explain the actual resistance to transformation events. These observations are important considerations in evaluating risks of insertional mutagenesis by SB transposons when used clinically.

5.3 PLASTICITY OF GENOMES AND GENE EXPRESSION IN HUMANS

Sequencing of human genomes from various normal tissue and many tumors and leukemias has revealed unexpected variations in chromosomal sequences including large alterations and complex rearrangements.[84,85] Deep sequencing has found that there are approximately 22,000 insertions and deletions (indels), which are equivalent to insertions of transposons, in a given human genome with up to 50% being novel for any individual.[86] Moreover, the average human has about 250–300 LOF mutations in his or her genome, 50–100 of which are in disease-related genes.[87,88] Included in the variation are copy number variations, especially of L1 and Alu retroelements found in protein-encoding genes, including growth-promoting genes, in cells of somatic tissues.[89–91] Thus, the human genome is not only highly variable, but it can sustain genetic hits from transposons without apparent phenotypic consequences.

5.3.1 Retrotransposable Elements

DNA transposons are relatively minor contributors to the transposon load in mammalian genomes, and they all appear to be inactive.[92] Class I transposons (Figure 5.2) comprise about 42% of the genome and are divided into four types. The most important with respect to gene therapy are the 500,000–800,000 long interspersed elements (LINEs) that are up to 6–8 kb long. About 100 LINE-1 elements are active and about 10 are considered highly active (hot) because they encode all of the necessary products for integration.[93] The second subtype of class I transposons are the approximate 1.5 million short interspersed elements (SINEs) that are about 100–300 bp and are transcribed by RNA polymerase III. SINEs are found everywhere in the genome. A small subset of one prominent SINE family, called *source elements*, also contributes to genomic instability.[94] The third category comprises the nearly half-million retrovirus-like, human endogenous retroviral sequences (HERVs). HERVs have hallmarks of retroviruses that persistently invaded primate genomes over eons and since have lost their activities to mobilize. HERVs account for about 8% of the human genome, but their integrations over millions of years have contributed substantially to primate-specific gene regulation.[95] Resurrection of a consensus HERV-K sequence indicated that these elements preferentially integrated into or proximal to transcriptional units.[96] The fourth category comprises a heterogeneous set of repetitive sequences that are several hundred base pairs in length and probably are transcribed by RNA polymerase II.

Of the nearly 3 million retrotransposable elements, only a small proportion of LINEs, those designated L1Hs, and SINEs, primarily *Alu* elements, are active and responsible for insertional mutagenic events that are associated with genetic disease,[93,97] including cancer.[90,98,99] For transpositional activity, the LINE or SINE first must be transcribed, generally from an endogenous promoter, and the RNA must be reverse transcribed to form a double-stranded DNA that can integrate somewhere into the genome. L1 sequences have two open reading frames that support the integration step. The origin of the reverse transcriptase activity in cells is poorly understood. Regardless, the polypeptides encoded in the *hot* L1 elements can transmobilize both inactive LINE as well as SINE sequences[100] because both depend on reverse transcriptase and integration enzymes for their spread. With more than 2 million LINEs and SINEs in the genome, on average they will occur about every 2000 base pairs. However, they are not randomly distributed because evolutionary selection eliminates overactive transposons and transposons in regions that lower the fitness of the cell, but this calculation indicates that many will be well within the influence of endogenous promoters with uncharacterized leakiness.

Decades of analyses of mutations that cause genetic diseases in humans have documented the ability of LINEs to inflict deleterious mutations in human genomes.[101] Nevertheless, owing to the huge number of retroelements in genomes and the paucity of mutagenic events, retroelements were thought to be largely inactive, mainly playing roles in genome evolution.[102] This notion was reconsidered when whole-genome sequencing became available.[101,103] Mobilization of L1 transposons were found in many somatic neuronal cells[89,104] and in cancers.[105,106] The association of the loss of methyl-CpG-binding protein-2 activity in some cells with mobilization suggested that a failure in epigenetic silencing accounted for rare cases of L1 transposition.[107,108] Investigations into the genome-wide DNA methylation status in human embryonic and adult tissues have identified unexpected tissue-specific and subfamily-specific hypomethylation signatures that can support transposition of retrotransposons.[109] Epigenetic silencing is relaxed in germline cells[110]; hence, it is not surprising that epigenetic silencing fails to keep all retrotransposable elements quiet all of the time in all cells.

5.3.2 Complexity of Transcription and Gene Expression in Human Genomes

The ENCODE project revealed that about 80% of the human genome could be attributed with a functional role in one cell type or another and that 95% of the genome was subject to transcription. Expression is highly variable: transcription of different sequences spans six orders of magnitude for polyadenylated RNA, which is generally associated with pre-mRNAs, with a large proportion of the transcripts being initiated from repetitive elements. The average protein-encoding gene has four splicing isoforms that are produced in unequal amounts.[111] The insertion of transgenic DNA into chromosomes can alter epigenetic marks that affect gene expression[108] and, conversely, retroviral vectors are subject to epigenetic silencing.[112] Moreover, enhancer elements can work over long distances via chromosomal looping, with only about 7% of the looping interactions being to the nearest gene.[113] These findings support earlier findings that murine leukemia virus integration up to 100 kbp upstream of the c-*myb* locus could activate a linearly distal locus through chromosomal looping loci.[114] Hence, physical proximity of an integrated vector is not a simple predictor for activation (or repression) of an endogenous gene. Transcriptional regulatory proteins selectively interact with each other and specific promoters,[115] which may explain the pronounced lack of activation of endogenous genes that lead to cancer by enhancers in lentiviral vectors and the repetitive findings of LMO2 activations with some murine retroviral vectors.

These findings are important for risk assessment of integrating vectors used in gene therapy. The interactions between regions of the genome that affect gene expression are complex and not based entirely on physical (linear) proximity. Cells have a bewildering level of variety in gene expression at both the transcriptional and RNA processing levels. Moreover, the genome appears to have the ability to assimilate transposable elements and even make use of them. All of these features strongly suggest that the genome is unexpectedly flexible in terms of events that would be expected to destabilize its panoply of functions. Recent findings from the 1000 Genomes Project and other whole-genome sequencing projects support this assertion. The bottom line is that transcriptional activation is far more complex than previously thought, and the possible consequences of insertional mutagenesis are clearly more complex than shown in Figure 5.1, which serves more as a model than a guide. In fact, experiments conducted decades ago analyzing transposition in nematodes showed that despite a strong preference for integration into transcriptional units, most Tc1-transposon integrations did not have an obvious phenotypic effect because when they integrated into introns (Figure 5.1, example 3) their sequences were removed during pre-mRNA processing, thereby avoiding detrimental consequences.[116]

5.3.3 Genetic Consequences of Natural Transposition and Therapeutic Transposition

One way of addressing insertional mutagenesis by SB transposon vectors is to compare the nearly random insertion profile of transposons with natural variation and the germline mutation rate. For example, variations in transcriptional regulation in cells normally lead to expression of millions of transposable elements, including the $\sim 10^{14}$ *hot* retrotransposable elements and an unknown number of active *Alu* elements. Combined with the background mutation rate in normal cells, the potential for deleterious mutations is potentially millions of fold greater than the number of mutations resulting from single integrations of SB transposons in several thousand cells. Superficially, with the histories of gamma retrovirus- and lentivirus-mediated gene therapies in mind, it would appear that the consequences of insertions of SB transposons carrying a therapeutic cassette with enhancers that are *not* designed to interact promiscuously with all promoters should be relatively small. Specifically, because only approximately 1.2% of the genome consists of protein-encoding exons,[35] almost 99% of integrations are unlikely to affect protein sequences in any way other than their rates of expression, which can vary widely (Figure 5.1, example 3). The remaining 1% of integrations that might occur into exons must be viewed in terms of other conclusions from deep sequencing of human genomes. The average gene is multiply spliced giving isoforms that, for the most part, have undefined specific roles. Thus, in many cases only one of the isoforms of multiply spliced genes would be affected by an exonic integration, which may be the reason that background integration of retrotransposable elements does not cause as many adverse events as might be expected. Although retrotransposition and DNA transposition occur by different mechanisms (copy-and-paste compared with cut-and-paste), the phenotypic results will be similar when integration interrupts a genomic sequence that either encodes proteins or regulates their expression (Figure 5.1). These controls over insertional damage by endogenous transposons will likely be relevant to transposons used for gene therapy.

In the clinical studies discussed next, a chimeric antigen receptor (CAR) is used to redirect the specificity of T cells to bind to a tumor-associated antigen. It is possible that a small subset of cancer patients treated with CAR-modified T cells carry inherited cancer-predisposing genetic alterations.[117] Thus, potential mutations arising from insertions of SB transposons may be more likely to produce an undesirable effect in these individuals than in cells from healthy individuals.

5.4 TRANSPOSON-MEDIATED GENE THERAPY IN THE CLINIC

Adoptive transfer of CAR+ T cells with a CD-19 targeted CAR in early clinical trials resulted in the desired outcome, the loss of normal CD19+ B cells and the destruction of malignant B cells in some patients.[118,119] At present, most trials targeting CD19 have used viruses to deliver the CAR to T cells. There have not been any reports of insertional mutagenesis leading to either persistent clonal expansion or enrichment for integration sites near genes implicated in growth control and/or transformation.[13] These clinical data, generated at MD Anderson Cancer Center, provide the foundation for the first human application of the SB system to redirect T-cell specificity against B-cell leukemias and lymphomas. In these trials, a CAR (Figure 5.8(a)) was constructed to recognize the CD19-lineage antigen, a common polypeptide found on the surface of normal and malignant B cells. In preparation for these trials, we accumulated data regarding the ability of the SB system to enforce expression of a second-generation CAR that activates T cells via chimeric CD3-ζ and CD28.[72] These two chimeric signaling motifs achieve a fully competent T-cell activation event, which is defined, at a minimum, as CAR-mediated proliferation, cytokine production, as well as serial and specific lysis. The introduction of the CAR was achieved by co-electroporation (Figure 5.3) of two supercoiled DNA plasmids, one containing the CAR-transposon and the second encoding SB11.[60] T cells stably expressing the CAR could be expanded by coculture with designer artificial antigen-presenting cells (aAPCs) in the presence of the soluble recombinant cytokines interleukin (IL)-2 and IL-21 (Figure 5.8(b)). The aAPCs used for the initial clinical trials were derived from the human K-562 immortalized cells that were genetically modified using lentivirus to coexpress CD19 as well as CD64 and the T-cell costimulatory molecules CD86, CD137L, and a membrane-bound mutant form of human IL-15. By recursively adding γ-irradiated aAPCs to the stably modified CAR+ T cells, we achieved outgrowth of CAR+ T cells. The electroporation method employs defined ratios of mononuclear cells and components of the SB transposon system that result in between 1 and 2 CAR insertions per T-cell genome without integration of SB11 transgene, based on the absence of a PCR signature for SB11. The combination of SB and aAPC platforms routinely results in the outgrowth of T cells, with the majority expressing CAR within a few weeks after electroporation.[120]

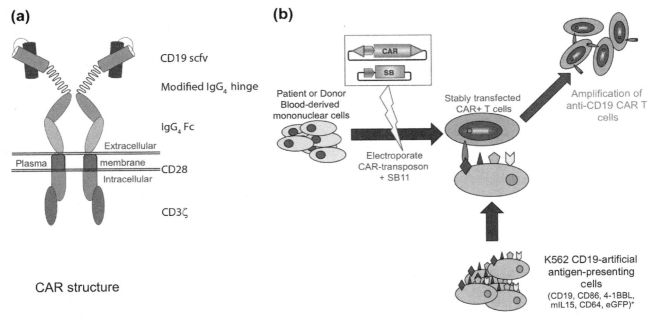

(a)

CD19 scfv

Modified IgG$_4$ hinge

IgG$_4$ Fc

Extracellular

Plasma membrane

Intracellular

CD28

CD3ζ

CAR structure

(b)

CAR

SB

Patient or Donor
Blood-derived
mononuclear cells

Electroporate
CAR-transposon
+ SB11

Stably transfected
CAR+ T cells

Amplification of
anti-CD19 CAR T
cells

K562 CD19-artificial
antigen-presenting
cells
(CD19, CD86, 4-1BBL,
mIL15, CD64, eGFP)+

FIGURE 5.8   Design and manufacture of chimeric antigen receptor (CAR+) T cells for nonviral gene therapy. (a) Structure of a prototypical CAR used at MD Anderson Cancer Center. A second-generation CAR is shown as a homodimer. Antigen recognition independent of human leukocyte antigen is achieved by scFv region derived from a CD19-specific mouse monoclonal antibody. The scFv is fused in frame to a modified hinge and Fc region derived from human IgG4. These extracellular structures are bound to the cell surface via chimeric CD28 and CD3-zeta, which activate T cells upon docking with CD19. (b) Blood from either a recipient or a healthy donor is collected by venipuncture. Supercoiled DNA plasmids from *Sleeping Beauty* system coding for transposon (CAR) and transposase (SB11) are synchronously electroporated into mononuclear cells using a commercial electroporator. The electroporation conditions and transposon to transposase ratio are calibrated to deliver between one to two CAR integration events per T-cell genome. Intranuclear and intracellular events that follow are illustrated in Figure 5.3. T cells that have stably integrated the CD19-specific CAR are retrieved by coculture with irradiated artificial antigen-presenting cells (aAPC) derived from K562 cells. The K562 cells function as aAPC because they are genetically modified to coexpress CD19 and costimulatory molecules (CD86, CD137L, and a membrane-bound variant of IL-15). The CD19-specific CAR+ T cells are numerically expanded before cryopreservation, release (and in-process) testing, and infusion into a recipient. (For interpretation of the references to color in this figure legend, the reader is referred to the online version of this book.)

5.4.1 Risks Associated with T Cells Modified to Stably Express CARs from Transposons

Adverse events may be associated with (1) lympho-depletion of the recipient to improve engraftment of infused T cells, (2) massive cell death and the accompanying cytokine storm triggered by the CAR T-cell killing of B cells, and (3) rare adverse effects from either the integration of the CAR transgene into a particular locus or the remobilization of a transposon into a deleterious locus as a result of lingering SB transposase activity.

Recipients that receive myelosuppressive or myeloablative therapy to induce lymphopenia are at risk from opportunistic infections and toxicity from the administered drugs. Indeed, there has been one death attributed to administering cyclophosphamide in an elderly patient with bulky CLL that received CD19-specific CAR⁺ T cells[121]; but exceptional events of this sort should be manageable by modifications in caring for medically fragile patients. There have been expected and unexpected adverse events due to on-target toxicities as the CAR recognizes tumor-associated antigen on normal cells.[122,123] For example, CAR⁺ T cells will not distinguish between CD19 on normal and cancerous B cells. This results in damage and even loss of humoral immunity in patients that benefit from T-cell therapy for B-lineage leukemias and lymphomas.[124–127] Destruction of normal B cells can be managed by repeated administrations of intravenous immunoglobulin, which appears tolerable in most patients at risk from dying from progressive B-cell malignancies. In addition, the synchronous activation of CAR⁺ T cells engaging large numbers of CD19⁺ malignant B cells can result in supraphysiologic elevations of cytokines and attendant systemic side effects. In severe cases, patients experience fever and degrading vital signs that can be complicated by tumor lysis syndrome associated with the desired destruction of large numbers of tumor cells. This type of adverse event can be managed by timely application of IL-6 receptor blockade and intravenous administration of corticosteroids.

Unforeseen adverse events can occur because of on-target, but off-tissue, binding of the CAR to normal cells. For example, one patient that received a large number of ERBB2-specific CAR⁺ T cells died as a result of off-tissue effects.[123] Expected toxicities may be ameliorated using an interpatient T-cell dose-escalation schema and pausing between cohorts of patients before dose escalation to evaluate for emergence of delayed adverse events. Whether unforeseen toxicities can be managed with a similar approach remains to be seen.

The third type of risk associated with infusing CAR⁺ T cells stems from insertional mutagenesis either directly or indirectly following remobilization of the therapeutic transposon. As noted earlier, unlike gene transfer into hematopoietic stem cells[9] there has not been a reported genotoxic event in genetically modified T cells. Indeed, hundreds of infusions have been delivered using mostly recombinant retrovirus to transduce T cells.[13] Extensive in-process testing and release testing is undertaken before infusion of the SB-modified T cells. The in-process evaluations include performing PCR to exclude integration of SB11 transgene, limiting the number of CAR integrants per T-cell genome to 1 or 2, and assessing the diversity of the T-cell receptor to exclude the emergence of either a mono- or an oligo-clonal population of genetically modified T cells that might herald unwanted uncontrolled growth in vivo. To guard against this possibility we undertake a culturing assay to measure the potential for autonomous proliferation. Other release tests include measurements of (1) sterility (absence of bacteria, fungal, *Mycoplasma*, endotoxin), (2) viability, (3) enumeration of CAR⁺ T cells, (4) chain of custody (measurement of human leukocyte antigen class I to validate identity between CAR⁺ T cells and the intended recipient), (5) identity (CD3 expression to validate the presence of T cells), (6) contamination to exclude B cells and aAPC, and (7) expression of the CAR. The SB-modified clinical-grade T cells are prepared in compliance with current good manufacturing practice for phase 1/2 trials at MD Anderson Cancer Center.

Remobilization of SB transposons resulting from residual SB transposase activity should not be a major safety issue. Transposition after co-electroporation of plasmids carrying transposons and transposase occurs because the appropriate levels of transposase to transposon are delivered; transposons are designed by nature to be autoregulated such that too much or too little transposase will not support transposition.[128] Thus, in this context, the data from clinical quality-control criteria studies that suggest there is no more than about 1 SB-mRNA/10^8 T cells[67] strongly suggest that remobilization of SB transposons will not occur at a measurable level in preparations of cells for gene therapy. Moreover, the studies using SB transposons to identify oncogenes described earlier as well as transformation of hepatocytes[74] suggest that approximately 100 SB-mRNAs per cell are required to support transposition. Hence, the rate of remobilization is below detectable limits. In any case, a remobilization event does not imply adverse consequences to a greater extent than the initial delivery of the integrating vector. That is, if remobilization occurs at a rate of $1/10^4$, then the chances of an adverse event would be 0.01% of a similar event happening after the initial delivery, which so far has been nil.

Four phase 1 studies are under way to establish safety, feasibility, and persistence of the infused T cells. Three of these trials infuse patient- and donor-derived CAR⁺ T cells in a setting of minimal residual disease after autologous and allogeneic hematopoietic stem-cell transplantation, including after umbilical cord blood transplantation. The fourth trial infuses autologous CAR⁺ T cells after lympho-depleting chemotherapy. Safety is a primary concern to be balanced with the medical condition of the patient for these early-phase clinical trials employing a new approach to gene therapy.

5.4.2 Future Directions: Directing SB Vectors to "Safe-Harbor" Sites in Genomes

The fears of insertional mutagenesis during the course of adoptive cell therapy will likely persist—the memories of the X-linked severe combined immunodeficiency disease adverse events will not be forgotten. A recurrent idea has been to guide integrations into "safe harbors" in the genome.[129,130] Site-specific nucleases—including zinc-finger nucleases, TAL-effector nucleases, and the CRISPR/Cas9 system—are all feasible for efficiently supporting homologous recombination into specific genomic sites.[131] The SB system can also be combined with gene-editing tools such as designer zinc-finger nucleases and TAL-effector nucleases.[132] Gene-editing technology allows placements of genes into precise locations in genomes with efficiencies that appear to be at least 80%, depending on the specific locus and the number of related sequences in the genome.[133] Although off-target activity will occur in a minority of cells, the unselected sites will be somewhat predictable so that selective screening can be done to determine the level of insertions into nontargeted sites after cleavage by whatever gene-editing tool is used. Because gene-editing technologies have been in widespread use for less than a decade, we expect that the targeting efficiencies and our understanding of the rules governing specificity will continue to improve.

5.5 CONCLUSIONS

The barriers to widespread adoption of gene therapy are being lowered as we achieve greater understanding of how gene expression is regulated and the genomic environments into which vectors insert. The promise of genetically modifying the human genome to overcome cancer has never been greater. The SB system provides a clear alternative to viral vectors to achieve safe and efficient gene therapy. The low cost of manufacturing clinical-grade SB plasmids and availability of expertise to manufacture these expression vectors in a commercial setting contrasts with the expense, uncertainty, and lengthy timeline to manufacture recombinant clinical-grade virus. Lowering the cost and improving the reliability of the gene delivery method will be essential for widespread application of gene therapy. Several cohorts of patients have successfully received T cells genetically modified with the SB system. Facilitating the use of SB transposons is the availability of methods such as electroporation to introduce immunologically neutral naked DNA molecules into cells. Now it is possible to adopt the SB system for further human applications beyond T cells and CARs, as described in Box 7.1.

ACKNOWLEDGMENT

We thank Drs Judy S. Moyes, Harjeet Singh (MD Anderson Cancer Center) and Dr Elena Aronovich (University of Minnesota) for careful reading and editing of the manuscript and our colleagues in the Center for Genome Engineering for many insightful discussions. We acknowledge the financial support of National Institutes of Health grants 1R01DK082516 and P01-HD32652 (PH), CA16672, CA124782, CA120956, CA141303, CA163587, and CA148600 (LJNC), and 5R00CA151672-04 and P30 CA77598 (TKS).

REFERENCES

1. Deakin CT, Alexander IE, Hooker CA, Kerridge IH. Gene therapy researchers' assessments of risks and perceptions of risk acceptability in clinical trials. *Mol Ther* 2013;**21**:806–15.
2. Berry C, Hannenhalli S, Leipzig J, Bushman FD. Selection of target sites for mobile DNA integration in the human genome. *PLoS Comp Biol* 2006;**2**:e157.
3. Huang CR, Burns KH, Boeke JD. Active transposition in genomes. *Ann Rev Genet* 2012;**46**:651–75.
4. Wagner EF, Covarrubias L, Stewart TA, Mintz B. Prenatal lethalities in mice homozygous for human growth hormone gene sequences integrated in the germ line. *Cell* 1983;**35**:647–55.
5. Hacein-Bey-Abina S, Hauer J, Lim A, Picard C, Wang GP, Berry CC, et al. Efficacy of gene therapy for X-linked severe combined immunodeficiency. *New Eng J Med* 2010;**363**:355–64.
6. Gaspar HB, Cooray S, Gilmour KC, et al. Long-term persistence of a polyclonal T cell repertoire after gene therapy fo rX-linked severe combined immunodeficiency. *Sci Transl Med* 2011;**3**:97ra79.
7. Kang EM, Choi U, Theobald N, Linton G, Long Priel DA, Kuhns D, et al. Retrovirus gene therapy for X-linked chronic granulomatous disease can achieve stable long-term correction of oxidase activity in peripheral blood neutrophils. *Blood* 2010;**115**:783–91.
8. Biasco L, Baricordi C, Aiuti A. Retroviral integrations in gene therapy trials. *Mol Ther* 2012;**20**:709–16.
9. Baum C, von Kalle C, Staal FJ, et al. Chance or necessity? insertional mutagenesis in gene therapy and its consequences. *Mol Ther* 2004;**9**:5–13.
10. Aiuti A, Cassani B, Andolfi G, et al. Multilineage hematopoietic reconstitution without clonal selection in ADA-SCID patients treated with stem cell gene therapy. *J Clin Invest* 2007;**117**:2233–40.
11. Cassani B, Montini E, Maruggi G, et al. Integration of retroviral vectors induces minor changes in the transcriptional activity of T cells from ADA-SCID patients treated with gene therapy. *Blood* 2009;**114**:3546–56.
12. Boztug K, Schmidt M, Schwarzer A, Banerjee PP, Díez IA, Dewey RA, et al. Stem-cell gene therapy for the Wiskott-Aldrich Syndrome. *New Eng J Med* 2010;**363**:1918–27.

13. Scholler J, Brady TL, Binder-Scholl G, et al. Decade-long safety and function of retroviral-modified chimeric antigen receptor T cells. *Sci Transl Med* 2012;**4**:132ra53.

14. Schwarzwaelder K, Howe SJ, Schmidt M, Brugman MH, Deichmann A, Glimm H, et al. Gammaretrovirus-mediated corrrection of SCID-X1 is associated with skewed vector integration site distribution in vivo. *J Clin Invest* 2007;**117**:2241–9.

15. van der Loo JC, Swaney WP, Grassman E, et al. Critical variables affecting clinical-grade production of the self-inactivating gamma-retroviral vector for the treatment of X-linked severe combined immunodeficiency. *Gene Ther* 2012;**19**:872–6.

16. Biffi A, Montini E, Lorioli L, et al. Lentiviral hematopoietic stem cell gene therapy benefits metachromatic leukodystrophy. *Science* 2013;**341**:1233158.

17. Aiuti A, Biasco L, Scaramuzza S, et al. Lentiviral hematopoietic stem cell gene therapy in patients with Wiskott-Aldrich syndrome. *Science* 2013;**341**:1233151.

18. Newrzela S, Cornils K, Li Z, et al. Resistance of mature T cells to oncogene transformation. *Blood* 2008;**112**:2278–86.

19. Cattoglio C, Maruggi G, Bartholomae C, et al. High-definition mapping of retroviral integration sites defines the fate of allogeneic T cells after donor lymphocyte infusion. *PLoS One* 2010;**5**:e15688.

20. Wang GP, Berry CC, Malani N, et al. Dynamics of gene-modified progenitor cells analyzed by tracking retroviral integration sites in a human SCID-X1 gene therapy trial. *Blood* 2010;**115**:4356–66.

21. Deichmann A, Brugman MH, Bartholomae CC, et al. Insertion sites in engrafted cells cluster within a limited repertoire of genomic areas after gammaretroviral vector gene therapy. *Mol Ther* 2011;**19**:2031–9.

22. Baum C. Parachuting in the epigenome: the biology of gene vector insertion profiles in the context of clinical trials. *EMBO Mol Med* 2011;**3**:75–7.

23. Corrigan-Curay J, Cohen-Haguenauer O, O'Reilly M, Ross SR, Fan H, Rosenberg N, et al. Challenges in vector and trial design using retroviral vectors for long-term gene correction in hematopoietic stem cell gene therapy. *Mol Ther* 2012;**20**:1084–94.

24. Montini E, Cesana D, Schmidt M, et al. The genotoxic potential of retroviral vectors is strongly modulated by vector design and integration site selection in a mouse model of HSC gene therapy. *J Clin Invest* 2009;**119**:964–75.

25. Kustikova O, Brugman M, Baum C. The genomic risk of somatic gene therapy. *Semin Cancer Biol* 2010;**20**:269–78.

26. Uren AG, Kool J, Berns A, van Lohuizen M. Retroviral insertional mutagenesis: past, present and future. *Oncogene* 2005;**24**:7656–72.

27. Copeland NG, Jenkins NA. Harnessing transposons for cancer gene discovery. *Nat Rev Genet* 2010;**10**:696–706.

28. Bergerson RJ, Collier LS, Sarver AL, Been RA, Lugthart S, Diers MD, et al. An insertional mutagenesis screen identifies genes that cooperate with Mll-AF9 in a murine leukemogenesis model. *Blood* 2012;**119**:4512–23.

29. Oliveira-Cobucci RN, Saconato H, Lima PH, Rodrigues HM, et al. Comparative incidence of cancer in HIV-AIDS patients and transplant recipients. *Cancer Epidemiol* 2012;**36**:e69–73.

30. Cesana D, Squaldino J, Rudilosso L, Merella S, Naldini L, Montini E. Whole transcriptome characterization of aberrant splicing events induced by lentivirus vector integrations. *J Clin Invest* 2012;**122**:1667–76.

31. Brady T, Bushman FD. Nondividing cells: a safer bet for integrating vectors? *Mol Ther* 2011;**19**:640–1.

32. Biffi A, Bartolomae CC, Cesana D, et al. Lentiviral vector common integration sites in preclinical models and a clinical trial reflect a benign integration bias and not oncogenic selection. *Blood* 2012;**117**:5332–9.

33. Aronovich EL, McIvor RS, Hackett PB. The *Sleeping Beauty* transposon system: a non-viral vector for gene therapy. *Hum Mol Genet* 2011;**20**(R1): 14–20.

34. Zhang Y, Satterlee A, Huang L. In vivo gene delivery by nonviral vectors: overcoming hurdles? *Mol Ther* 2012;**20**:1298–304.

35. Consortium ENCODE P, Bernstein BE, Birney E, et al. An integrated encyclopedia of DNA elements in the human genome. *Nature* 2012;**489**: 57–74.

36. Kaer K, Speek M. Retroelements in human disease. *Gene* 2013;**518**:231–41.

37. Lander ES, et al. Initial sequencing and analysis of the human genome. *Nature* 2001;**409**:860–921.

38. de Koning AP, Gu W, Castoe TA, Batzer MA, Pollock DD. Repetitive elements may comprise over two-thirds of the human genome. *PLoS Genet* 2011;**7**:e1002384.

39. Liu G, Aronovich EL, Cui Z, Whitley CB, Hackett PB. Excision of *Sleeping Beauty* transposons: parameters and applications to gene therapy. *J Gene Med* 2004;**6**:574–83.

40. Plasterk RHA, Izsvák Z, Ivics Z. Resident aliens: the Tc1/mariner superfamily of transposable elements. *Trends Genet* 1999;**15**:326–32.

41. Ivics Z, Hackett PB, Plasterk RH, Izsvak Z. Molecular reconstruction of *Sleeping Beauty*, a Tc1-like transposon from fish, and its transposition in human cells. *Cell* 1997;**91**:501–10.

42. Geurts AM, Yang Y, Clark KJ, et al. Gene transfer into genomes of human cells by the *Sleeping Beauty* transposon system. *Mol Ther* 2003;**8**:108–17.

43. Yant SR, Meuse L, Chiu W, Ivics Z, Izsvak Z, Kay MA. Somatic integration and long-term transgene expression in normal and haemophilic mice using a DNA transposon system. *Nat Genet* 2000;**25**:35–41.

44. Mátés L, Chuah MK, Belay E, et al. Molecular evolution of a novel hyperactive *Sleeping Beauty* transposase enables robust stable gene transfer in vertebrates. *Nat Genet* 2009;**41**:753–61.

45. Hodges BL, Cheng SH. Cell and gene-based therapies for the lysosomal storage diseases. *Curr Gene Ther* 2006;**6**:227–41.

46. Essner JJ, McIvor RS, Hackett PB. Awakening of gene therapy with *Sleeping Beauty* transposons. *Curr Opin Pharmacol* 2005;**5**:513–9.

47. Hackett PB, Largaespada DA, Cooper LJN. A transposon and transposase system for human application. *Mol Ther* 2010;**18**:674–83.

48. Montini E, Held PK, Noll M, et al. In vivo correction of murine tyrosinemia type I by DNA-mediated transposition. *Mol Ther* 2002;**6**(6):759–69.

49. Ohlfest JR, Frandsen JL, Fritz S, et al. Phenotypic correction and long-term expression of factor VIII in hemophilic mice by immunotolerization and nonviral gene transfer using the *Sleeping Beauty* transposon system. *Blood* 2005;**105**:2691–8.

50. Liu L, Mah C, Fletcher BS. Sustained FVIII expression and phenotypic correction of hemophilia A in neonatal mice. *Mol Ther* 2006;**13**:1006–15.

51. Kren BT, Unger GM, Sjeklocha L, et al. Nanocapsule-delivered *Sleeping Beauty* mediates therapeutic factor VIII expression in liver sinusoidal endothelial cells of hemophilia A mice. *J Clin Invest* 2009;**119**:2086–99.

52. Zhang W, Muck-Hausl M, Wang J, et al. Integration profile and safety of an adenovirus hybrid-vector utilizing hyperactive sleeping beauty transposase for somatic integration. *PLoS One* 2013;**8**:e75344.

53. Aronovich EL, Bell JB, Belur LR, et al. Prolonged expression of a lysosomal enzyme in mouse liver after *Sleeping Beauty* transposon-mediated gene delivery: implications for non-viral gene therapy of mucopolysaccharidoses. *J Gene Med* 2007;**9**:403–15.

54. Aronovich EL, Bell JB, Kahn SA, et al. Systemic correction of storage disease in MPS I NOD/SCID mice using the *Sleeping Beauty* transposon system. *Mol Ther* 2009;**17**:1136–44.

55. Aronovich EL, Bell JB, Koniar BL, Hackett PB. A liver-specific promoter enhances expression of alpha-L-iduronidase in Sleeping Beauty-mediated gene therapy for murine MPS I. *Mol Ther* 2010;**18**:S284.

56. Aronovich EL, Bell JB, Podetz-Pedersen K, Koniar BL, McIvor RS, Hackett PB. Quantitative analysis of α-L-Iduronidase expression in immunocompetent mice Treated with the *Sleeping Beauty* transposon system. *PLoS One* 2013;**8**:e78161.

57. Ortiz-Urda S, Lin Q, Yant SR, Keene D, Kay MA, Khavari PA. Sustainable correction of junctional epidermollysis bullosa via transposon-mediated nonviral gene transfer. *Gene Ther* 2003;**10**:1099–104.

58. Ohlfest JR, Lobitz PD, Perkinson SG, Largaespada DA. Integration and long-term expression in xenografted human glioblastoma cells using a plasmid-based transposon system. *Mol Ther* 2004;**10**:260–8.

59. Belcher JD, Vineyard JV, Bruzzone CM, et al. Heme oxygenase-1 gene delivery by *Sleeping Beauty* inhibits vascular stasis in a murine model of sickle cell disease. *J Mol Med* 2010;**88**:665–75.

60. Singh H, Manuri PR, Olivares S, et al. Redirecting specificity of T-cell populations for CD19 using the *Sleeping Beauty* system. *Cancer Res* 2008;**68**:2961–71.

61. Huang X, Guo H, Kang J, et al. *Sleeping Beauty* transposon-mediated engineering of human primary T cells for therapy of CD19(+) lymphoid malignancies. *Mol Ther* 2008;**16**:580–9.

62. Jena B, Dotti G, Cooper LJ. Redirecting T-cell specificity by introducing a tumor-specific chimeric antigen receptor. *Blood* 2010;**116**:1035–44.

63. Liu L, Liu H, Visner G, Fletcher BS. *Sleeping Beauty*-mediated eNOS gene therapy attenuates monocrotaline-induced pulmonary hypertension in rats. *FASEB J* 2006;**20**:2594–6.

64. Wang X, Sarkar DP, Mani P, Steer CJ, Chen Y, Guha C, et al. Long-term reduction of jaundice in Gunn rats by nonviral liver-targeted delivery of *Sleeping Beauty* transposon. *Hepatology* 2009;**50**:815–24.

65. Podetz-Pedersen K, Bell JB, Steele TW, et al. Gene expression in lung and liver after intravenous infusion of polyethylenimine complexes of *Sleeping Beauty* transposons. *Gene Ther* 2010;**21**:210–20.

66. Hackett PB, Aronovich EL, Hunter D, et al. Efficacy and safety of *Sleeping Beauty* transpson-mediated gene transfer in preclinical animal studies. *Curr Gene Ther* 2011;**11**:341–9.

67. Hackett PB, Largaespada DA, Switzer KC, Cooper LJ. Evaluating risks of insertional mutagenesis by DNA transposons in gene therapy. *Transl Res* 2013;**161**:265–83.

68. Wu X, Li Y, Crise B, Burgess SM. Transcription start regions in the human genome are favored targets for MLV integration. *Science* 2003;**300**:1749–51.

69. Yant SR, Wu X, Huang Y, Garrison B, Burgess SM, Kay MA. High-resolution genome-wide mapping of transposon integration in mammals. *Mol Cell Biol* 2005;**25**:2085–94.

70. Yant SR, Wu X, Huang Y, et al. Nonrandom insertion site preferences for the SB transposon in vitro and in vivo. *Mol Ther* 2004;**9**:S309.

71. Geurts AM, Hackett CS, Bell JB, et al. DNA structural patterns influence integration site preferences for mobile elements. *Nucl Acids Res* 2006;**34**:2803–11.

72. Singh H, Huls H, Kebriaei P, Cooper LJN. A new approach to gene therapy using Sleeping Beauty to genetically modify clinical-grade T cells to target CD19. *Immunol Rev* 2014;**257**:181–90.

73. Nesmelova IV, Hackett PB. DDE Transposases: structural similarity and diversity. *Adv Drug Del Rev* 2010;**62**:1187–95.

74. Bell JB, Aronovich EL, Schreifels JM, Beadnell TC, Hackett PB. Duration of expression of *Sleeping Beauty* transposase in mouse liver following hydrodynamic DNA delivery. *Mol Ther* 2010;**18**:1792–802.

75. Wilber AC, Frandsen JL, Geurts AM, Largaespada DA, H PB, McIvor RS. RNA as a source of transposase for *Sleeping Beauty*-mediated gene insertion and expression in somatic cells and tissues. *Mol Ther* 2006;**13**:625–30.

76. Mikkers H, Berns A. Retroviral insertional mutagenesis: tagging cancer pathways. *Adv Cancer Res* 2003;**88**:53–99.

77. Cherry SR, Biniszkiewicz D, van Parijs L, Baltimore D, Jaenisch R. Retroviral expression in embryonic stem cells and hematopoietic stem cells. *Mol Cell Biol* 2000;**20**:7419–26.

78. Dupuy AJ, Akagi K, Largaespada DA, Copeland NG, Jenkins NA. Mammalian mutagenesis using a highly mobile somatic *Sleeping Beauty* transposon system. *Nature* 2005;**436**:221–6.

79. Starr TK, Allaei R, Silverstein KA, et al. A transposon-based genetic screen in mice identifies genes altered in colorectal cancer. *Science* 2009;**323**:1747–50.

80. Keng VW, Villanueva A, Chiang DY, et al. A conditional transposon-based insertional mutagenesis screen for genes associated with mouse hepatocellular carcinoma. *Nat Biotech* 2009;**27**:264–74.

81. Starr TK, Scott PM, Marsh BM, Zhao L, Than BL, O'Sullivan MG, et al. A Sleeping Beauty transposon-mediated screen identifies murine susceptibility genes for adenomatous polyposis coli (Apc)-dependent intestinal tumorigenesis. *Proc Natl Acad Sci USA* 2011;**108**:5765–70.

82. Rahrmann EP, Watson AL, Keng VW, et al. Forward genetic screen for malignant peripheral nerve sheath tumor formation identifies new genes and pathways driving tumorigenesis. *Nat Genet* 2013;**45**:756–66.

83. Vassiliou GS, Cooper JL, Rad R, et al. Mutant nucleophosmin and cooperating pathways drive leukemia initiation and progression in mice. *Nat Genet* 2011;**43**:470–5.

84. Abyzov A, Mariani J, Palejev D, et al. Somatic copy number mosaicism in human skin revealed by induced pluripotent stem cells. *Nature* 2013;**492**:438–42.

85. Handsaker RE, Korn JM, Nemesh J, McCarroll SA. Discovery and genotyping of genome structural polymorphism by sequencing on a population scale. *Nat Genet* 2011;**43**:269–76.

86. Alkan C, Coe BP, Eichler EE. Genome structural variation discovery and genotyping. *Nat Rev Genet* 2011;**12**:363–76.

87. Durbin RM, Abecasis GR, Altshuler DL, et al. A map of human genome variation from population-scale sequencing. *Nature* 2010;**467**:1061–73.

88. MacArthur DG, Balasubramanian S, Frankish A, et al. A systematic survey of loss-of-function variants in human protein-coding genes. *Science* 2012;**335**:823–8.

89. Baillie JK, Barnett MW, Upton KR, Gerhardt DJ, Richmond TA, De Sapio F, et al. Somatic retrotransposition alters the genetic landscape of the human brain. *Nature* 2011;**479**:534–7.

90. Lee E, Iskow R, Yang L, et al. Landscape of somatic retrotransposition in human cancers. *Science* 2012;**337**:967–71.

91. O'Huallachain M, Karczewski KJ, Weissman SM, Urban AE, Snyder MP. Extensive genetic variation in somatic human tissues. *Proc Natl Acad Sci USA* 2012;**109**:18018–23.

92. Pace II JK, Feschotte C. The evolutionary history of human DNA transposons: evidence for intense activity in the primate lineage. *Genome Res* 2007;**17**:422–32.

93. Beck CR, Collier P, Macfarlane C, et al. LINE-1 retrotransposition activity in human genomes. *Cell* 2010;**141**:1159–70.

94. Ade C, Roy-Engel AM, Deininger PL. Alu elements: an intrinsic source of human genome instability. *Curr Opin Virol* 2013;**3**:639–45.

95. Faulkner GJ, Kimura Y, Daub CO, et al. The regulated retrotransposon transcriptome of mammalian cells. *Nat Genet* 2009;**40**:563–71.

96. Brady T, Lee YN, Ronen K, et al. Integration target site selection by a resurrected human endogenous retrovirus. *Genes Dev* 2009;**23**:633–42.

97. Brouha B, Schustak J, Badge RM, et al. Hot L1s account for the bulk of retrotransposition in the human population. *Proc Natl Acad Sci USA* 2003;**100**:5280–8.

98. Solyom S, Ewing AD, Rahrmann EP, et al. Extensive somatic L1 retrotransposition in colorectal tumors. *Genome Res* 2012;**22**:2328–38.

99. Shukla R, Upton KR, Muñoz-Lopez M, et al. Endogenous retrotransposition activates oncogenic pathways in hepatocellular carcinoma. *Cell* 2013;**153**:101–11.

100. Wagstaff BJ, Hedges DJ, Derbes RS, et al. Rescuing Alu: recovery of new inserts shows LINE-1 preserves Alu activity through A-tail expansion. *PLoS Genet* 2012;**8**:e1002842.

101. Ewing AD, Kazazian HHJ. Whole-genome resequencing allows detection of many rare LINE-1 insertion alleles in humans. *Genome Res* 2011;**21**:985–90.

102. Cordaux R, Batzer MA. The impact of retrotransposons on human genome structure. *Nat Rev Genet* 2009;**10**:691–703.

103. Burns KH, Boeke JD. Human transposon tectonics. *Cell* 2012;**149**:740–52.

104. Coufal NG, Garcia-Perez JL, Peng GE, et al. L1 retrotransposition in human neural progenitor cells. *Nature* 2009;**460**:1127–31.

105. Iskow RC, McCabe MT, Mills RE, et al. Natural mutagenesis of human genomes by endogenous retrotransposons. *Cell* 2010;**141**:1253–61.

106. Rodić N, Burns KH. Long interspersed element-1 (LINE-1): passenger or driver in human neoplasms? *PLoS Genet* 2013;**9**:e1003402.

107. Garcia-Perez JL, Morell M, Scheys JO, et al. Epigenetic silencing of engineered L1 retrotransposition events in human embryonic carcinoma cells. *Nature* 2010;**466**:769–73.

108. Doerfler W. Impact of foreign DNA integration on tumor biology and on evolution via epigenetic alterations. *Epigenomics* 2012;**4**:41–9.

109. Xie M, Hong C, Zhang B, et al. DNA hypomethylation within specific transposable element families associates with tissue-specific enhancer landscape. *Nat Genet* 2013;**45**:836–41.

110. Bao J, Yan W. Male germline control of transposable elements. *Biol Reprod* 2012;**86**(162):1–14.

111. Djebali S, Davis CA, Merkel A, et al. Landscape of transcription in human cells. *Nature* 2012;**489**:101–8.

112. Poleshko A, Palagin I, Zhang R, et al. Identification of cellular proteins that maintain retroviral epigenetic silencing: evidence for an antiviral response. *J Virol* 2008;**82**:2313–23.

113. Sanyal A, Lajoie BR, Jain G, Dekker J. The long-range interaction landscape of gene promoters. *Nature* 2012;**489**:109–13.

114. Zhang J, Markus J, B J, Paul T, Wolff L. Three murine leukemia virus integration regions within 100 kb upstream of c-myb are proximal to the 5' regulatory region of the gene through DNA looping. *J Virol* 2012;**86**:10524–32.

115. Zhang Y, Wong CH, Birnbaum RY, et al. Chromatin connectivity maps reveal dynamic promoter-enhancer long-range associations. *Nature* 2013;**504**:306–10.

116. Rushforth AM, Anderson P. Splicing removes the Caenorhabditis elegans transposon Tc1 from most mutant pre-mRNAs. *Mol Cell Biol* 1996;**16**:422–9.

117. Vogelstein B, Papadopoulos N, Velculescu VE, Zhou S, Diaz LAJ, Kinzler KW. Cancer genome landscapes. *Science* 2013;**339**:1446–58.

118. Kershaw MH, Westwood JA, Darcy PK. Gene-engineered T cells for cancer therapy. *Nat Rev Cancer* 2013;**18**:525–41.

119. Themeli M, Kloss CC, Ciriello G, et al. Generation of tumor-targeted human T lymphocytes from induced pluripotent stem cells for cancer therapy. *Nat Biotech* 2013;**31**:928–33.

120. Manuri PV, Wilson MH, Maiti SN, et al. piggyBac transposon/transposase system to generate CD19-specific T cells for the treatment of B-lineage malignancies. *Hum Gene Ther* 2010;**21**:427–37.

121. Brentjens R, Yeh R, Bernal Y, Riviere I, Sadelain M. Treatment of chronic lymphocytic leukemia with genetically targeted autologous T cells: case report of an unforeseen adverse event in a phase I clinical trial. *Mol Ther* 2010;**18**:666–8.

122. Lamers CH, Sleijfer S, Vulto AG, et al. Treatment of metastatic renal cell carcinoma with autologous T-lymphocytes genetically retargeted against carbonic anhydrase IX: first clinical experience. *J Clin Oncol* 2006;**24**:e20–2.

123. Morgan RA, Yang JC, Kitano M, Dudley ME, Laurencot CM, R S/A. Case report of a serious adverse event following the administration of T cells transduced with a chimeric antigen receptor recognizing ERBB2. *Mol Ther* 2010;**18**:843–51.

124. Kochenderfer JN, Dudley ME, Feldman SA, et al. B-cell depletion and remissions of malignancy along with cytokine-associated toxicity in a clinical trial of anti-CD19 chimeric-antigen-receptor-transduced T cells. *Blood* 2012;**119**:2709–20.

125. Porter DL, Levine BL, Kalos M, Bagg A, June CH. Chimeric antigen receptor-modified T cells in chronic lympholid leukemia. *New Eng J Med* 2011;**365**:725–33.

126. Brentjens RJ, Rivière I, Park JH, et al. Safety and persistence of adoptively transferred autologous CD19-targeted T cells in patients with relapsed or chemotherapy refractory B-cell leukemias. *Blood* 2011;**118**:4817–28.

127. Kalos M, Levine BL, Porter DL, et al. T cells with chimeric antigen receptors have potent antitumor effects and can establish memory in patients with advanced leukemia. *Sci Transl Med* 2011;**3**:95ra73.

128. Claeys-Bouuaert C, Lipkow K, Andrews SS, Liu D, Chalmers R. The autoregulation of a eukaryotic DNA transposon. *eLIFE* 2013;**2**:00668.

129. Sadelain M, Papapetrou EP, Bushman FD. Safe harbours for the integration of new DNA in the human genome. *Nat Rev Cancer* 2011;**12**:51–8.

130. Papapetrou EP, Lee G, Malani N, et al. Genomic safe harbors permit high beta-globin transgene expression in thalassemia induced pluirpotent stem cells. *Nat Biotech* 2011;**29**:73–8.

131. Gaj T, Gersbach CA, Barbas CFr. ZFN, TALEN, and CRISPR/Cas-based methods for genome engineering. *Trends Biotechnol* 2013;**31**:397–404.

132. Torikai H, Reik A, Liu PQ, et al. A foundation for universal T-cell based immunotherapy: T cells engineered to express a CD19-specific chimeric-antigen-receptor and eliminate expression of endogenous TCR. *Blood* 2012;**119**:5697–705.

133. Tan S, Carlson DF, Walton MW, Fahrenkrug SC, Hackett PB. Precision editing of large animal genomes. *Adv Genet* 2012;**80**:37–97.

Chapter 6

Arthritis Gene Therapy: A Brief History and Perspective

Christopher H. Evans[1], Steven C. Ghivizzani[2] and Paul D. Robbins[3]

[1]Mayo Clinic, Rehabilitation Medicine Research Center, Rochester, MN, USA; [2]Department of Orthopedics and Rehabilitation, University of Florida College of Medicine, Gainesville, FL, USA; [3]Department of Metabolism and Aging, The Scripps Research Institute, Jupiter, FL, USA

LIST OF ABBREVIATIONS

AAV Adeno-associated virus
CMV Cytomegalovirus
HSV Herpes simplex virus
IFN Interferon
IL Interleukin
IL-1Ra Interleukin-1 receptor antagonist
IRAP Interleukin-1 receptor antagonist protein
MCP Metacarpophalangeal
MoMLV Moloney murine leukemia virus
NF-κB Nuclear factor kappa B
OA Osteoarthritis
PVNS Pigmented villonodular synovitis
RA Rheumatoid arthritis
TGF-β$_1$ Transforming growth factor-β$_1$
TNF Tumor necrosis factor

6.1 INTRODUCTION

Arthritis is not an obvious target for gene therapy because no common form of arthritis results from a single gene defect. Rather than compensate for a genetic abnormality, the goal of arthritis gene therapy is to deliver, in a sustained and focal fashion, therapeutic gene products to the joints of subjects with arthritis.[1] These individuals may well be genetically unremarkable. Because of the unique physiology and pharmacokinetics of the joint,[2,3] sustained intra-articular delivery of such products is almost impossible to achieve by any other clinically reasonable means. Not only do proteins have restricted entry to joints from the circulation, but their exit via the lymphatics is rapid.

As described in the next section, the concept of using gene transfer as a delivery system to joints originated at the University of Pittsburgh in the late 1980s, at the same time as the first human gene transfer trials in subjects with cancer and severe combined immunodeficiency disease were getting underway. Arthritis was the first nongenetic, nonlethal disease targeted by gene therapy, a radical departure at the time that brought new types of common, everyday diseases within the scope of gene therapy.

Subsequent progress with arthritis gene therapy has been fitful.[4–9] Nevertheless, from this beginning have emerged a few clinical trials, with a pivotal Phase III study in the advanced planning stage and four biotechnology companies devoted to the commercial development of arthritis gene therapy. This article describes the sequence of events leading to the present state and identifies matters in need of resolution. To avoid duplication, subjects that were described in detail in a previous review[5] are not addressed in depth.

6.2 CONCEPTION AND STRATEGIES

The original strategy was, and is, conceptually very simple—to transfer cDNAs encoding therapeutic products to the synovial linings of individual joints with arthritis. Figures 6.1(a) and (b) are reproduced from the original concept paper by Bandara et al.[1] Several other strategies soon emerged. For polyarticular arthritides (e.g., rheumatoid arthritis (RA)), systemic delivery and facilitated local delivery, using genetically modified cells with the ability to home to joints, were suggested[10,11] (Figure 6.2). Although impressive results in animal models of disease were reported,[12–14] none of these alternative approaches to therapy were developed further. This was partly because of the emergence of successful, alternative therapies for RA that were based on systemically blocking tumor necrosis factor-α (TNF-α).[15] Thus, the unique ex vivo and in vivo intra-articular approaches proposed by Bandara et al.[1] remain the strategies of choice for clinical and commercial development. They have changed surprisingly little in the quarter-century since being proposed. The two main modifications are the inclusion of allogeneic cells for ex vivo delivery[16] and the realization that chondrocytes can also be considered a target for gene transfer when using small vectors such as adeno-associated virus (AAV).[17]

It was originally proposed to transfer cDNAs encoding secreted protein products. There are several reasons for this. First, when this approach was initially conceptualized, gene transfer vectors were not as efficient as they now are. The chances of therapeutic success were considered greater with a secreted product because it would not be necessary to transduce synovial cells with high efficiency. It would simply be necessary to transduce a sufficient number of cells with sufficient efficiency to achieve, in aggregate, a therapeutic concentration of the transgene within the joint. This resonated with the authors' interest in using the interleukin-1 receptor antagonist (IL-1Ra) as the transgene product. It should also be

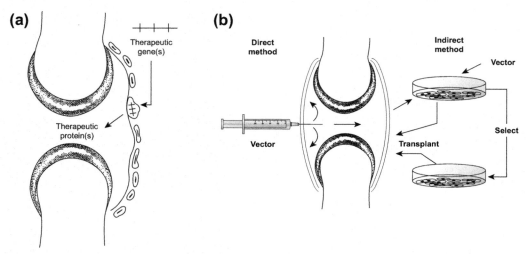

FIGURE 6.1 Original concept behind arthritis gene therapy. (a) cDNAs encoding therapeutic products are transferred to the synovial linings of individual joints with arthritis. (b) Therapeutic genes can be transferred to the synovial linings of joints by ex vivo or in vivo methods. *From Ref. 1.*

FIGURE 6.2 Comprehensive strategies for arthritis gene therapies. In local gene therapy, genes are transferred to individual diseased joints, as in Figure 6.1. In systemic gene therapy, genes are transferred to extra-articular locations where secreted gene products can enter the circulation to treat multiple joints and address extra-articular manifestations of disease. In facilitated local delivery, genes are transferred to cells, such as dendritic cells, with the ability to home selectively to diseased joints. *Modified from Refs 10,11.*

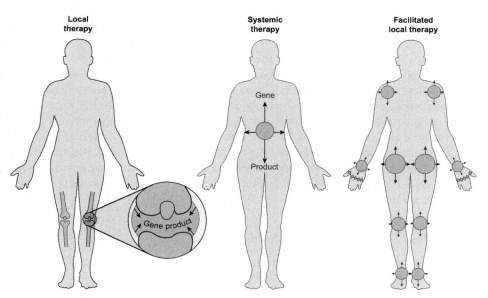

remembered that at this time Moloney-based retroviruses were the most developed of the vectors for human use. Because of their requirement for host cell division, they needed to be deployed in an ex vivo fashion.

With the development of more powerful vectors for in vivo delivery, it is now possible to envisage using transgenes encoding proteins with an intracellular site of therapeutic action as well as noncoding RNA molecules such as small interfering RNAs and micro RNAs. However, such developments remain at the early preclinical stage.

Using genes to kill cells within synovium became another addition to the conceptual armamentarium.[18] In RA and certain other intractable diseases where the synovium becomes severely and uncontrollably inflamed, there is a long history of surgically removing synovium.[19] This can provide symptomatic relief for an extended period. Nonsurgical synovectomy using toxins[20] or irradiation[21] has also been used, especially in diseases in which surgery is counter-indicated; the latter include hemophilia where hemoarthrosis commonly occurs as a result of repeated bleeding into the joint.[22,23] In many cases, approaches toward achieving a genetic synovectomy converge with those of cancer gene therapy where the aim is to kill offending cells.

6.3 TECHNOLOGY DEVELOPMENT

6.3.1 Retroviruses and Ex vivo Gene Delivery

When research into arthritis gene therapy was initiated, vectors based on the Moloney murine leukemia virus (MoMLV) were the most advanced for clinical use. Indeed, the first 12 human trials approved by the Recombinant DNA Advisory Committee of the National Institutes of Health used this type of vector. Thus, it was the obvious choice for arthritis studies. The theoretical occurrence of insertional mutagenesis was recognized, but it was not considered an insurmountable barrier to use at this time.

Robbins, Guild, and Mulligan engineered a novel version of MoMLV-based vectors known as MFG.[24] This simplified vector contains a region of the viral *gag* gene to increase viral titer, the *env* splice acceptor, and 5′ Ncol and 3′ BamH1 cloning sites for insertion of transgene(s). Expression of the transgene is driven by the viral 5′ long terminal repeat. Translational efficiency is high because the start codon for translation of the transgene is coincident with the natural start codon of the viral *env* gene. The construct containing human IL-1Ra cDNA is known as MFG-IRAP (Figure 6.3), IRAP being a former acronym for IL-1Ra. MFG-IRAP was used in the first two human arthritis gene therapy trials.

Ex vivo gene delivery to joints required cells that could be recovered readily from the patient, expanded easily in culture, genetically modified with MFG-IRAP and, after being returned by intra-articular injection, would stably colonize the host synovium and express the transgene for an extended period. Our laboratory had extensive experience with the harvest and culture expansion of synovial fibroblasts from the knee joints of rabbits[25,26] as well as human synovial biopsies,[27] which were readily available as a result of surgical synovectomy or joint replacement surgery. These cells were readily grown in culture, transduced with retroviral vectors, and continued to express transgenes intra-articularly after intra-articular injection.[28,29] Independent, parallel experiments by Makarov et al.[30,31] made equivalent observations. In later research, we confirmed that skin fibroblasts were also suitable for this role but, by then, we had moved on to in vivo delivery.

Subsequent clinical trials using MFG-IRAP for the ex vivo delivery of IL-1Ra to human joints using autologous synovial fibroblasts were successful,[32–34] but they also highlighted the cumbersome, costly, and complicated nature of the process. Using an allograft cell line offers one way to avoid the need to harvest and expand under good manufacturing practice conditions, genetically modify, and safety-test cells from each individual patient. The notion of using allograft cells for ex vivo gene transfer to joints was not immediately attractive for fear of an allograft response that would cause inflammation, eliminate the transplanted cells, and thus curtail transgene expression. However, TissueGene, Inc. established a line of primary human chondrocytes recovered from the articular cartilage of a newborn with polydactyly and used this cell line as a universal donor for transforming growth factor-β_1 (TGF-β_1) gene transfer to the joints of subjects with osteoarthritis (OA).[16] As noted later, this protocol is about to enter Phase III trials.

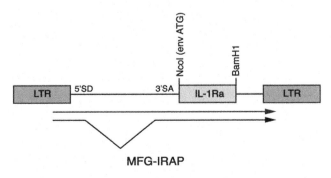

MFG-IRAP

FIGURE 6.3 Structure of MFG-IRAP. This retrovirus was the first taken into clinical trials for arthritis gene therapy. Expression of the full coding sequence of human interleukin-1 receptor antagonist (IL-1Ra) is driven by the 5′ long terminal repeat (LTR) of the virus, which retains endogenous splice sites and initiation codon. SD, splice donor; SA, splice acceptor.

6.3.2 In vivo Gene Delivery

Although attention was initially focused on ex vivo delivery to joints, it was obvious that in vivo delivery held advantages of simplicity and lower total cost.[35] At about the same time, recombinant adenovirus vectors were becoming available[36] and they proved very effective in transferring genes to cells within the synovial lining, but not cartilage, in various experimental species,[37] including nonhuman primates.[38] These vectors allowed for rapid screening of different candidate cDNAs as antiarthritic agents,[39–43] but they were not good candidates for human application because early generation adenovirus vectors continue to express viral proteins in transduced cells, thus targeting them for destruction by the immune system with ensuing loss of transgene expression.

Transient transgene expression is not an impediment when attempting genetic synovectomy, and adenovirus has been used extensively and successfully to deliver various cytocidal genes to the joints of experimental animals.[18,44–46] Clinical development of this approach has not occurred, but the strategy using gene transfer of herpes simplex virus (HSV) thymidine kinase in combination with ganciclovir treatment has been evaluated in monkeys, leading to one related clinical trial as a means of dealing with the aseptic loosening of prosthetic joints.[47]

Removal of nearly all viral sequences from adenovirus produces a "gutted," helper-dependent version that does not express viral proteins in transduced cells and has a high DNA carrying capacity of approximately 37 kDa. Because they no longer express viral proteins, transduced cells are not cleared by the immune system, which allows extended periods of transgene expression in nondividing cells. Robust transgene expression from synovium has been reported.[48]

Recombinant lentiviruses derived from human immunodeficiency virus are very effective vectors for gene transfer to synovium, but not cartilage, after intra-articular injection.[49] Transgene expression is extremely high and is amplified upon an inflammatory flare because the transduced synovial fibroblasts proliferate under inflamed conditions.[50] This advantage reflects the ability of such vectors to integrate their DNA into the chromosomal DNA of the host cells, something that raises many prohibitive safety concerns for human use. Nonintegrating lentivirus vectors would eliminate these safety concerns but at the cost of losing the intrinsic expansion of transgene during a flare. However, if the field as a whole makes sufficient progress, then it may only be a matter of time before nonintegrating lentivirus is advanced toward clinical trials for the local, genetic treatment of arthritis.

Much interest surrounds the use of AAV as a vector for local arthritis gene therapy. Its advantages as a safe, injectable virus with the ability to transduce nondividing cells were appreciated from the outset, but its path to the clinic was retarded by production problems and the unreliability of second strand DNA synthesis.[51] Moreover, one of the first reported intra-articular uses of AAV to transfer transgenes to synovium[52] was later modified to suggest that the virus had actually transduced adjacent muscle instead of synovium.[53]

Advances in AAV production technology, the development of self-complementing vectors, and the expanded choice of serotypes have greatly accelerated progress in developing AAV for intra-articular gene transfer.[17,54–56] Moreover, studies in the horse confirm that, after intra-articular injection, AAV indeed transduces synovium and, unexpectedly, articular chondrocytes.[17] As described below, AAV has advanced to Phase II clinical trials in arthritis and two additional trials are planned. In conjunction with Dr Jude Samulski of the University of North Carolina, our group has generated the vector sc-rAAV2.5IL-1Ra (Figure 6.4). This is a self-complementing, recombinant AAV, serotype 2.5 containing human IL-1Ra cDNA that is at the pre-investigational new drug stage of development toward a clinical trial in patients with OA.

Other vectors screened for their ability to transduce synovium in vivo in a sustained fashion include those based upon HSV,[57] foamy virus,[58] and a multitude of nonviral systems.[59] High-titer, MoMLV-based vectors also have been tested in vivo under conditions of synovial inflammation and proliferation.[60] Although they showed promise in animal models, their use is limited to conditions of acute inflammation; in any case, they are unlikely to find clinical application because of insurmountable safety issues. Vectors based upon HSV proved cytotoxic. A very wide variety of different nonviral vectors have been screened, but transgene expression was always low, transient, or both. Moreover, many nonviral formulations were inflammatory.[59]

Most information concerning the utility of various vectors for delivering genes intra-articularly has been obtained empirically. The investigation by Gouze et al.[61] is one of the few studies to examine the biology of intra-articular gene transfer in detail. Their data, obtained using the rat knee as the model system, indicate that long-term, intra-articular transgene expression can be obtained with immunologically silent vectors and transgenes using uncomplicated constitutive promoters such as the cytomegalovirus (CMV) immediate early promoter or the elongation factor 1-α promoter. Persistence of transgene expression resided in a sub-set of synovial cells within specific anatomical locations. With AAV, especially in diseased joints, chondrocytes also could serve as a reservoir of long-term transgene expression.[17]

FIGURE 6.4   Structure of s.c.-rAAV2.5IL-1Ra. This self-complementing, serotype 2.5 adeno-associated virus (AAV) carries the full coding sequence of human interleukin-1 receptor antagonist (IL-1Ra) driven by the human cytomegalovirus (CMV) immediate early promoter. TR, terminal repeat.

6.3.3 Regulated Transgene Expression

Regulated transgene expression is attractive because arthritis is often accompanied by exacerbations and remissions. Integrating viruses may achieve this spontaneously because exacerbations are usually accompanied by synovial cell proliferation.[50] However, safety issues limit the clinical development of such viruses in this context. Although there has been some research into the use of drug-inducible promoters,[55] most interest has centered on the use of promoters that are sensitive to ambient inflammatory conditions.

According to early literature reports, in vivo synovial expression of a transgene driven by the CMV immediate early promoter within an AAV genome was enhanced under inflammatory conditions.[62,63] Expression fell when the inflammation resolved, but it was restored when inflammation returned. This effect has been difficult to replicate,[56] but Traister et al.[64] reported a strong and reversible increase in transgene expression from AAV-transduced human synovial fibroblasts exposed to IL-1β, TNF-α, and IL-6 in vitro. This effect was replicated when cells were exposed to synovial fluids recovered from inflamed joints, but not with fluids from noninflamed joints. Because amplification of expression occurred with single- and double-stranded vectors, it was presumably unrelated to second strand synthesis. It was also independent of the promoter used to drive transgene expression. The authors suggested that inflammatory cytokines affected intracellular trafficking of AAV mediated by phosphatidylinositol-3-kinase. The same group also demonstrated a marked and reversible stimulation of transgene expression in the presence of proteasome inhibitors.[65]

Miagkov et al.[66] introduced the concept of endogenous regulation of transgene expression using promoters that respond to ambient inflammatory stimuli. Subsequent adaptation of this technology has produced promoters responsive to nuclear factor-kappa B (NF-κB) that are being proposed for human clinical trials in RA.[67,68] Additional disease-activated promoters are also under investigation.[69]

6.3.4 Summary

Armed with these technologies, investigators have generated an abundance of data confirming unequivocally the ability to transfer genes to joints of multiple species, to express them intra-articularly at therapeutic concentrations, and to successfully treat animal models of arthritis. These findings have been extensively reviewed.[5,6,11,70,71] Although subsequent studies continue to repeat and confirm these initial proof-of-concept experiments, the main challenge is now to bring this technology into the clinic while addressing unresolved issues.

6.4 UNRESOLVED ISSUES

6.4.1 Immune Responses to Vectors and Their Significance

Very little detailed research has been performed on the immune response to viral vectors delivered by intra-articular injection. Most humans already contain neutralizing antibodies against adenovirus5[72] and certain AAV serotypes[73] in their sera and synovial fluids.[72–74] The degree to which this limits intra-articular transduction efficiency has not been examined empirically, but it can be assumed to be a barrier.

One way to obviate pre-existing immunity is to use serotypes, such as adenovirus35[72] and AAV5 or AAV8,[73] to which human populations do not normally have high natural immunity. However, in immunologically naïve animals, intra-articular administration of vector generates immune responses that would be expected to inhibit repeat dosing, but this also has been little studied. Because the joint is an accessible, isolated body cavity, it lends itself to various strategies for mitigating the effects of immune reactivity; lavage is the simplest of these. Mingozzi et al.[75] using anti-CD20 antibodies (rituximab) as an example, have pointed out that many treatments for RA are immunosuppressive and could be used as combination therapy to reduce immune responses to viral vectors.

6.4.2 Which Arthritis and Which Transgene?

Arthritis is derived from Greek words meaning "inflamed joint," a wide designation. According to rheumatologists, there are over 100 forms of arthritis. Most, if not all, of these could benefit from suitable intra-articular gene therapy.

The greatest clinical need occurs in OA, the most common cause of disability in the elderly. OA affects 27 million Americans, and this number will increase because of demographic changes; there is no cure and treatment options are very limited. Many individuals progress to the point of needing a joint replacement, something that is expensive and requires major surgery with extended postsurgical rehabilitation. A successful gene therapy would certainly address a major unmet clinical need.

TABLE 6.1 Transgenes Explored as Possible Therapeutic Agents in Experimental Models of Osteoarthritis

Transgene	Model	References
Interleukin-1 receptor antagonist	Rabbit: meniscectomy Horse: osteochondral defect Dog: ligament transection	76 77 78
Lubricin	Mouse: aging, ligament transection	48
Interleukin-10	Rabbit: meniscectomy	76
RNA Knockdown		
Macrophage inflammatory protein-1α NF-κB	Mouse: ligament transection Rat: ligament transection	79 80
Hemoxygenase-1	Mouse: spontaneous	81
Thrombospondin-1	Rat: ligament transection	82
Kallistatin	Rat: ligament transection	83

One reason that OA remains incurable and difficult to treat is our poor knowledge of the etiopathophysiology of the disease. As a consequence, the choice of a therapeutic transgene is not obvious. We have focused on interleukin (IL)-1 as an important mediator of symptoms and the disease process in OA, but there are many other candidates. Table 6.1 lists transgenes that have been evaluated in experimental models of OA. However, animal models are inadequate surrogates for human OA, with the possible exception of post-traumatic OA; therefore, their predictive value is weak.

RA is the second most common form of arthritis. Although this was the first form of arthritis to be targeted by gene therapy,[32] its urgency has been diminished by progress in nongenetic therapies. For example, many cases of RA are treated in a satisfactory fashion with proteins, delivered by injection or infusion, that block TNF-α or IL-6 or target T or B lymphocytes.[84] Such therapies are expensive, inconvenient, and invasive, but inhibitors of janus kinases have recently been approved as orally active drugs for RA.[85] More progress along these lines can be expected.[86] Thus, unless gene therapy proves to be dramatically more effective, safe, or economical than these existing treatments, it may be relegated to the fringes of therapy for RA.

Many other forms of arthritis, such as psoriatic arthritis and the crystal-induced arthropathies, are good candidates for gene therapy, but they have not been well studied. The hemarthrosis of hemophiliacs is discussed later in this review.

6.5 CLINICAL TRIALS

There have been four clinical trials of gene therapy in subjects with RA (Table 6.2) and four in subjects with OA (Table 6.3).

6.5.1 Ex vivo Gene Delivery

The first arthritis gene therapy trial,[32] using MFG-IRAP (Figure 6.3) in an ex vivo fashion, was only the 74th human gene therapy study to gain approval by the Recombinant DNA Advisory Committee, which at last count (November 25, 2013), has approved 1258 human gene transfer protocols.[90] Because of its novelty, the study raised considerable issues with regard to safety and the risk:benefit ratio. The main safety concerns were insertional mutagenesis triggered by retroviral insertion, and germ-line transmission of the transgene. The latter concern was addressed by restricting recruitment to postmenopausal females. Insertional mutagenesis was addressed by targeting the metacarpophalangeal (MCP; "knuckle") joints of patients with RA who were to receive prosthetic MCP joint replacements. This provided the opportunity to deliver the genetically modified cells to arthritic joints 1 week before joint replacement surgery. We knew from our preclinical studies that the cells did not migrate in large numbers from the site of application. This ensured that they would be removed surgically when the joints were replaced. Such a strategy also provided biopsy material for analysis. One week was selected as the time between injection and removal of the cells because this allowed sufficient opportunity to confirm that the cells had survived injection, colonized the synovium, and continued to secrete transgenic IL-1Ra (Figure 6.5).

Entry criteria also included the need for at least one additional joint surgery required as part of the surgical management of the patients' RA. This provided tissue for the recovery of autologous synovial cells. These were expanded in vitro and divided into

TABLE 6.2 Clinical Trials of Gene Therapy for Rheumatoid Arthritis

Transgene	Method of Delivery	Phase	Principal Investigator, Institution or Sponsor	NIH OBA Protocol Number	Number of Subjects	References
IL-1Ra	Retrovirus ex vivo	I	Evans and Robbins University of Pittsburgh	9406-074	9	32
IL-1Ra	Retrovirus ex vivo	I	Wehling, University of Düsseldorf, Germany	NA	2	33
Etanercept	AAV In vivo	I	Mease, Targeted Genetics	0307-588[a]	15	87
Etanercept	AAV In vivo	I/II	Mease, Targeted Genetics	0503-705[b]	127	88
IFN-β	AAV In vivo	Preclinical	Tak, Arthrogen, Netherlands			

OBA, Office of Biotechnology Activities; NIH, National Institutes of Health; NA, not applicable; AAV, adeno-associated virus; IL-1Ra, interleukin-1 receptor antagonist; IFN, interferon.
[a]Included one subject with ankylosing spondylitis.
[b]Included subjects with ankylosing spondylitis and psoriatic arthritis.
Adapted from Ref. 5.

TABLE 6.3 Clinical Trials of Gene Therapy for Osteoarthritis

Transgene	Method of Delivery	Phase	Principal Investigator, Institution or Sponsor	NIH OBA Protocol Number	Number of Subjects	References
TGF-β$_1$	Retrovirus, ex vivo	I	Ha, Kolon Life Sciences, Korea	NA	12	16
TGF-β$_1$	Retrovirus, ex vivo	I	Mont, TissueGene	0307-594	9	
TGF-β$_1$	Retrovirus, ex vivo	IIa	Ha, Kolon Life Sciences, Korea	NA	28	89
TGF-β$_1$	Retrovirus, ex vivo	II	Mont, TissueGene	0912-1016	100	
IL-1Ra	AAV, in vivo	Preclinical	Evans, Mayo Clinic			

OBA, Office of Biotechnology Activities; NA, not applicable; NIH, National Institutes of Health; AAV, adeno-associated virus; IL-1Ra, interleukin-1 receptor antagonist; TGF-β, transforming growth factor-β$_1$.
Adapted from Ref. 5.

two cultures, one of which was transduced with MFG-IRAP. Release criteria required the secretion of at least 30 ng IL-1Ra/10^6 cells/48 h. Before injection, cells were tested for replication-competent retrovirus, sterility, mycoplasma, and endotoxin.

Because all four MCP joints on one hand were replaced, it was possible to inject two MCP joints with transduced cells and two MCP joints with unmodified, control cells. These were administered in a blinded, dose-escalation fashion, with three groups of three patients receiving 10^6 cells (low dose), $1.5–5 \times 10^6$ (medium dose), or $6.5 \times 10^6–10^7$ (high dose) cells per joint. One week later, the MCPs were surgically replaced and tissues analyzed.

This first-in-human trial confirmed that it is possible to transfer genes to joints safely and express them intra-articularly. There were no patient-related issues and several subjects reported symptomatic improvement, although this was attributed to a placebo effect. Transduced cells were detected by immunohistochemistry and in situ hybridization in the retrieved synovia, where they continued to express IL-1Ra. Activity of IL-1Ra was confirmed by reduced production of IL-6 and prostaglandin E2 by synovia receiving the transgene.

A similar clinical trial took place subsequently in Germany.[33] This also introduced IL-1Ra cDNA into MCP joints of patients with RA using MFG-IRAP as the vector in conjunction with autologous synovial fibroblasts. However, instead of

FIGURE 6.5 Protocol for the first gene transfer to human joints. Surgery of the joints of the hand or foot (step one) provided autologous synovium, which was used to establish cultures of synovial fibroblasts (step two). Half of the cells were transduced with the retroviral vector (step three), and all cells were tested for replication-competent retrovirus and adventitious agents (step four) before injection into metacarpophalangeal (MCP) joints numbers two to five on one hand (step five). In a double-blinded fashion, two joints received transduced cells, and two received control cells. One week later, the injected joints were surgically removed during total joint replacement surgery (step six), and the retrieved tissues were analyzed for evidence of successful gene transfer and gene expression (step seven). *Redrawn from Ref. 32 with permission.*

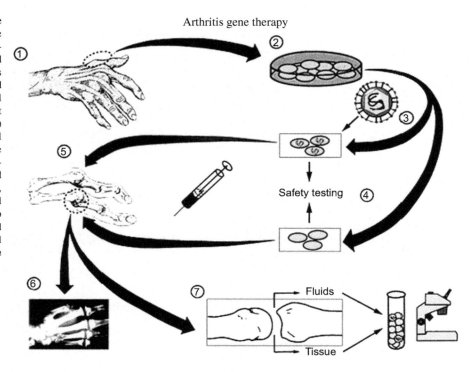

FIGURE 6.6 Protocol for use of genetically modified, allogeneic chondrocytes in osteoarthritis. Transduced chondrocytes expressing transforming growth factor-β_1 (TGF-β_1) are irradiated and mixed with unmodified cells in a 1:3 ratio. They are delivered to the clinic for intra-articular injection. *Redrawn from Ref. 16, with permission.*

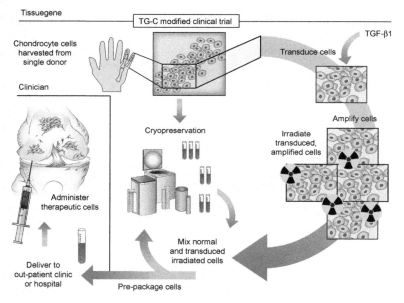

MCP joint replacement surgery, the injected joints were subjected to surgical synovectomy. Moreover, a period of 1 month was allowed between injection of the genetically modified cells and surgery.

Although the ethics committee gave permission for six subjects to be treated in this manner, safety issues arising in a separate trial elsewhere involving MFG caused the RA study to be curtailed after only two procedures had been completed. Nevertheless, the results were highly encouraging and one patient experienced dramatic reductions in pain.[33]

These ex vivo studies were highly successful[32,33] but revealed the cumbersome, laborious, and expensive nature of ex vivo gene therapy when using expanded autologous cells. TissueGene has obviated this problem by using allograft cells that serve as universal donors. Their protocol[16] uses retrovirally transduced allograft chondrocytes expressing TGF-β_1 (Figure 6.6). To address the issue of insertional mutagenesis, the transduced cells are irradiated before injection. Because this limits their survival, the irradiated cells are combined with untransduced, unirradiated cells before injection into knee joints with OA. This combination of chondrocytes and TGF-β_1 is expected to curtail cartilage damage and reduce symptoms. The protocol has completed Phase II testing in Korea and the United States; a Phase III study should begin in 2015.

6.5.2 In vivo Gene Delivery

Injectable vectors hold many obvious advantages for arthritis gene therapy. AAV was the first safe vector to offer in vivo delivery and the prospect of long-term transgene expression. Its clinical use for arthritis gene therapy was initiated by Targeted Genetics, Inc., who undertook Phase I and Phase II studies in subjects with RA (Table 6.2). Certain subjects with ankylosing spondylitis and psoriatic arthritis were also included. Etanercept, a TNF antagonist, was selected as the transgene, using AAV serotype 2. Although etanercept is widely administered for the treatment of RA, its effects are often incomplete, and certain patients have persistent symptoms in one or more joints. Gene delivery was used as a way of augmenting the concentration of etanercept in persistently symptomatic joints. Because of the size of etanercept cDNA, it was not possible to use self-complementing AAV.

The Phase I trial was accomplished without incident, but one subject died in the Phase II study. After an investigation, the U.S. Food and Drug Administration allowed the study to continue with modifications to the consent form, suggesting that gene transfer was not held responsible for the patient's death. Although certain therapeutic trends were noted, any improvements in the signs and symptoms of RA did not reach statistical significance. Possible reasons for this include the low levels of etanercept expression from the single-copy AAV vector and the possibility that additional intra-articular etanercept is of little value in subjects mounting incomplete responses to TNF antagonists.

As far as we know, the only other protocols in late preclinical development are the ones listed in Table 6.2 using AAV2.5 to deliver IL-1Ra in OA and AAV5 to deliver IFN-β in RA. The latter construct is notable for using an inducible promoter based on NF-κB responsiveness.

6.6 VETERINARY APPLICATIONS

OA is found commonly in horses and dogs, suggesting veterinary applications for arthritis gene therapy. Proof of principle was obtained in horses, in which a first-generation recombinant adenovirus vector was used to deliver equine IL-1Ra intra-articularly to horses with experimental OA.[77] This procedure lowered inflammation, protected the articular cartilage and, importantly, reduced lameness. The reduction in lameness is highly significant because it indicates that the gene therapy reduced pain and increased function—two very important considerations when treating OA.

Recombinant AAV is being used for clinical development of this strategy, carrying an equine IL-1Ra cDNA equivalent to the one shown in Figure 6.4.[17] In this way, the equine product will not only provide a novel veterinary therapy, but it will also help regulatory approval of the human construct.

6.7 OTHER APPLICATIONS OF INTRA-ARTICULAR GENE THERAPY

6.7.1 Mucopolysaccharidosis Type VI

This disease provides a striking example of the unique power of gene transfer to deliver proteins to joints in a sustained fashion. Mucopolysaccharidosis type VI (MPS VI; Maroteaux-Lamy syndrome) is a lysosomal storage disease resulting from loss of N-acetyl galactosamine-4-sulfatase (aryl sulfatase B). Once-per-week intravenous infusion of patients with recombinant N-acetyl galactosamine-4-sulfatase successfully treats all extra-articular manifestations of disease. Because this protein does not accumulate in joints to a sufficient degree, intra-articular pathology is more resistant to therapy. Glycosylated, recombinant N-acetyl galactosamine-4-sulfatase has a molecular weight of only 56kDa, less than that of serum albumin, indicating the severe limitations on protein accumulation in joints. Treatment of a feline model of MPS VI by in vivo intra-articular gene therapy using a lentivirus vector successfully treated intra-articular disease that was resistant to systemic protein therapy.[91]

6.7.2 Hemophilia

In similar fashion, administration of clotting factors to hemophiliacs, while controlling bleeding at extra-articular locations, is limited in its ability to control bleeding into the joint. This leads to a form of arthritis known as hemarthrosis. Studies in a murine model of hemophilia B showed that local gene delivery was superior to systemic introduction of recombinant factor IX as a means of preventing hemarthrosis.[92] The transition of this technology into human clinical trials will be facilitated by the previous use of AAV to deliver factor IX to the livers of patients with hemophilia and by progress in using AAV intra-articularly to treat OA and RA.

TABLE 6.4 Companies Commercializing Arthritis Gene Therapy

Company	Location	Year Founded/ Incorporated	Lead Product/Technology Platform	Web sites
TissueGene, Inc.	Rockville, MD	1999	Allogeneic chondrocytes expressing TGF-β_1	www.tissuegene.com
Arthrogen BV	Amsterdam, The Netherlands	2005	AAV/IFN-β	www.arthrogen.nl
Molecular Orthopaedics, Inc.	Chapel Hill, NC	2007	AAV/IL-1Ra	N.A.
GeneQuine Biotherapeutics GmbH	Hamburg, Germany	2012	Helper-dependent adenovirus	www.genequine.com

IL-1Ra, interleukin-1 receptor antagonist; AAV, adeno-associated virus; IFN, interferon; N.A., not applicable; TGF-β_1, transforming growth factor-β_1.

6.7.3 Pigmented Villonodular Synovitis

Pigmented villonodular synovitis (PVNS) is a rare condition resulting from hyperproliferation and fibrosis of the synovial lining. Surgical synovectomy is used to remove the hyperplastic tissue, but effectiveness is limited by regrowth. Moreover, surgical synovectomy is not effective in the diffuse form of PVNS. The disease is debilitating and, in rare cases, has led to amputation.[93] In many ways, the lesion in PVNS resembles a tumor and should be susceptible to genetic synovectomy. This could be particularly useful in eliminating residual hyperproliferative cells after surgical debulking of the gross tissue. Using vectors and strategies that are already in clinical trials for cancer should expedite translation.

6.8 COMMERCIALIZATION

The path to commercial development was initiated when a gene therapy company, TissueGene, Inc., licensed a suite of patents on arthritis gene therapy owned by the University of Pittsburgh (Table 6.4). As noted, this company is developing therapies for OA using genetically modified allogeneic chondrocytes.[16] These cells also show promise in the related area of cartilage repair, and a clinical trial of this application has successfully completed a Phase IIb trial.[94] Their parent company, Kolon Industries, Inc., is pursuing parallel clinical trials in arthritis and cartilage repair in Korea.

Arthrogen B.V. is based in the Netherlands with funding from Dubai. Their lead product uses AAV serotype 5 to deliver IFN-β to individual joints with RA.[95] According to the company's website, visited February 25, 2014, the pipeline also includes the delivery of neutralizing antibodies and small hairpin RNA.

Two of us (S.C.G., P.D.R.) co-founded Molecular Orthopaedics, Inc., a virtual company created around the use of AAV technology to deliver IL-1Ra intra-articularly in human and veterinary medicine.

GeneQuine Biotherapeutics was recently formed to exploit the use of helper-dependent adenovirus for delivery to joints. Their website, visited February 25, 2014, does not specify the lead product, but they have published preclinical studies using a lubricin transgene.[48]

6.9 PERSPECTIVES

Although progress in arthritis gene therapy has been slow and tortuous,[5] there are increasing grounds for optimism. Recent data from equine studies confirm unequivocally that it is possible to put a therapeutic gene into a human-sized joint, have it express at very high levels for at least 6 months,[17] and produce beneficial effect in a model of OA (Ghivizzani et al., unpublished). Several human trials have been performed. An additional protocol should enter a Phase III trial next year, and one or two additional Phase I protocols may be initiated soon. There are now four companies devoted to this area of development, and three of them appear to have closed successful rounds of financing. Thanks to recent clinical successes in treating certain genetic diseases, the field of gene therapy is now infused with renewed optimism; although U.S. federal funding is stagnant, the venture capital market for gene therapy is regaining momentum and several corporate deals have been concluded in recent months.

Many of the immediate technology hurdles to arthritis gene therapy appear to have been successfully addressed, although a full understanding of the immune responses to vectors and how this might affect dosing and re-dosing is lacking.

Information on the required duration and level of transgene expression in the joints of large animals is incomplete. However, possibly the biggest deficiency is the lack of a clear therapeutic target, especially in OA, a complex and heterogeneous group of diseases. It is quite likely that different groups of patients will respond differently to interventions, but there are no data on this point. The selection of target determines, among other things, the necessary level and duration of therapeutic gene expression. Matters are not helped by the inadequacies of animal models, a circumstance that encourages early human trials after safety has been established.

Cost is an additional matter that will factor into the clinical success of an arthritis gene therapy. For comparison, the annual cost of the anti-TNF agent Humira is approximately $44,000, and the new janus kinase inhibitors are unlikely to be much cheaper. However, the patents protecting the first TNF blockers, such as enbrel, are expiring and, with new laws allowing biosimilars, the price may drop considerably. Intra-articular injection of hyaluronic acid, commonly and controversially used to treat OA of the knee, costs approximately $1500. Knee joint replacement surgery costs approximately $50,000. Where a genetic therapy would fall in the price-efficacy continuum is not yet predictable.

ACKNOWLEDGMENTS

The authors' work in this area has been funded by NIH grants R01 AR43623, R21 AR049606, R01 AR048566, R01 AR057422, R01 AR051085, X01 NS066865, and by Orthogen AG.

REFERENCES

1. Bandara G, Robbins PD, Georgescu HI, Mueller GM, Glorioso JC, Evans CH. Gene transfer to synoviocytes: prospects for gene treatment of arthritis. *DNA Cell Biol* April 1992;**11**(3):227–31.
2. Evans CH, Kraus VB, Setton LA. Progress in intra-articular therapy. *Nat Rev Rheumatol* January 2014;**10**(1):11–22.
3. Simkin PA. Synovial perfusion and synovial fluid solutes. *Ann Rheum Dis* May 1995;**54**(5):424–8.
4. Evans CH, Ghivizzani SC, Herndon JH, et al. Clinical trials in the gene therapy of arthritis. *Clin Orthop Relat Res* October 2000;(Suppl. 379):S300–7.
5. Evans CH, Ghivizzani SC, Robbins PD. Arthritis gene therapy and its tortuous path into the clinic. *Transl Res* April 2013;**161**(4):205–16.
6. Evans CH, Ghivizzani SC, Robbins PD. Gene therapy of the rheumatic diseases: 1998 to 2008. *Arthritis Res Ther* 2009;**11**(1):209.
7. Evans CH, Ghivizzani SC, Robbins PD. Getting arthritis gene therapy into the clinic. *Nat Rev Rheumatol* April 2013;**7**(4):244–9.
8. Evans CH, Ghivizzani SC, Robbins PD. Orthopedic gene therapy–lost in translation? *J Cell Physiol* Feburary 2012;**227**(2):416–20.
9. Evans CH. Arthritis gene therapy at an inflection point. *Future Rheumatol* 2008;**3**:207–10.
10. Evans CH, Ghivizzani SC, Kang R, et al. Gene therapy for rheumatic diseases. *Arthritis Rheum* January 1999;**42**(1):1–16.
11. Evans CH, Ghivizzani SC, Robbins PD. Gene therapy for arthritis: what next? *Arthritis Rheum* June 2006;**54**(6):1714–29.
12. Kim SH, Evans CH, Kim S, Oligino T, Ghivizzani SC, Robbins PD. Gene therapy for established murine collagen-induced arthritis by local and systemic adenovirus-mediated delivery of interleukin-4. *Arthritis Res* 2000;**2**(4):293–302.
13. Kim SH, Kim S, Evans CH, Ghivizzani SC, Oligino T, Robbins PD. Effective treatment of established murine collagen-induced arthritis by systemic administration of dendritic cells genetically modified to express IL-4. *J Immunol* March 1, 2001;**166**(5):3499–505.
14. Tarner IH, Slavin AJ, McBride J, et al. Treatment of autoimmune disease by adoptive cellular gene therapy. *Ann NY Acad Sci* September 2003;**998**: 512–9.
15. Thalayasingam N, Isaacs JD. Anti-TNF therapy. *Best Pract Res Clin Rheumatol* August 2011;**25**(4):549–67.
16. Ha CW, Noh MJ, Choi KB, Lee KH. Initial phase I safety of retrovirally transduced human chondrocytes expressing transforming growth factor-beta-1 in degenerative arthritis patients. *Cytotherapy* Feburary 2012;**14**(2):247–56.
17. Watson RS, Broome TA, Levings PP, et al. scAAV-mediated gene transfer of interleukin-1-receptor antagonist to synovium and articular cartilage in large mammalian joints. *Gene Ther* June 2013;**20**(6):670–7.
18. Zhang H, Gao G, Clayburne G, Schumacher HR. Elimination of rheumatoid synovium in situ using a Fas ligand 'gene scalpel'. *Arthritis Res Ther* 2005;**7**(6):R1235–43.
19. London PS. Synovectomy of the knee in rheumatoid arthritis; an assay in surgical salvage. *J Bone Joint Surg Br* August 1955;**37-B**(3):392–9.
20. Schaumburger J, Trum S, Anders S, et al. Chemical synovectomy with sodium morrhuate in the treatment of symptomatic recurrent knee joint effusion. *Rheumatol Int* October 2012;**32**(10):3113–7.
21. Schneider P, Farahati J, Reiners C. Radiosynovectomy in rheumatology, orthopedics, and hemophilia. *J Nucl Med* January 2005;**46**(Suppl. 1): 48S–54S.
22. Teyssler P, Kolostova K, Bobek V. Radionuclide synovectomy in haemophilic joints. *Nucl Med Commun* April 2013;**34**(4):291–7.
23. Luck Jr JV, Silva M, Rodriguez-Merchan EC, Ghalambor N, Zahiri CA, Finn RS. Hemophilic arthropathy. *J Am Acad Orthop Surg* July–August 2004;**12**(4):234–45.
24. Robbins PD, Tahara H, Mueller G, et al. Retroviral vectors for use in human gene therapy for cancer, Gaucher disease, and arthritis. *Ann NY Acad Sci* May 31, 1994;**716**:72–88. discussion 88–79.
25. Georgescu HI, Mendelow D, Evans CH. HIG-82: an established cell line from rabbit periarticular soft tissue, which retains the "activatable" phenotype. *In Vitro Cell Dev Biol* October 1988;**24**(10):1015–22.

26. Stefanovic-Racic M, Stadler J, Georgescu HI, Evans CH. Nitric oxide synthesis and its regulation by rabbit synoviocytes. *J Rheumatol* October 1994;**21**(10):1892–8.
27. Del Vecchio MA, Georgescu HI, McCormack JE, Robbins PD, Evans CH. Approaches to enhancing the retroviral transduction of human synovio-cytes. *Arthritis Res* 2001;**3**(4):259–63.
28. Bandara G, Mueller GM, Galea-Lauri J, et al. Intraarticular expression of biologically active interleukin 1-receptor-antagonist protein by ex vivo gene transfer. *Proc Natl Acad Sci USA* November 15, 1993;**90**(22):10764–8.
29. Hung GL, Galea-Lauri J, Mueller GM, et al. Suppression of intra-articular responses to interleukin-1 by transfer of the interleukin-1 receptor antago-nist gene to synovium. *Gene Ther* January 1994;**1**(1):64–9.
30. Makarov SS, Olsen JC, Johnston WN, et al. Suppression of experimental arthritis by gene transfer of interleukin 1 receptor antagonist cDNA. *Proc Natl Acad Sci USA* January 9, 1996;**93**(1):402–6.
31. Makarov SS, Olsen JC, Johnston WN, et al. Retrovirus mediated in vivo gene transfer to synovium in bacterial cell wall-induced arthritis in rats. *Gene Ther* August 1995;**2**(6):424–8.
32. Evans CH, Robbins PD, Ghivizzani SC, et al. Gene transfer to human joints: progress toward a gene therapy of arthritis. *Proc Natl Acad Sci USA* June 14, 2005;**102**(24):8698–703.
33. Wehling P, Reinecke J, Baltzer AW, et al. Clinical responses to gene therapy in joints of two subjects with rheumatoid arthritis. *Hum Gene Ther* February 2009;**20**(2):97–101.
34. Evans CH, Robbins PD, Ghivizzani SC, et al. Clinical trial to assess the safety, feasibility, and efficacy of transferring a potentially anti-arthritic cytokine gene to human joints with rheumatoid arthritis. *Hum Gene Ther* June 20, 1996;**7**(10):1261–80.
35. Nita I, Ghivizzani SC, Galea-Lauri J, et al. Direct gene delivery to synovium. An evaluation of potential vectors in vitro and in vivo. *Arthritis Rheum* May 1996;**39**(5):820–8.
36. Crystal RG. Adenovirus: the first effective in vivo gene delivery vector. *Hum Gene Ther* January 2014;**25**(1):3–11.
37. Roessler BJ, Allen ED, Wilson JM, Hartman JW, Davidson BL. Adenoviral-mediated gene transfer to rabbit synovium in vivo. *J Clin Invest.* August 1993;**92**(2):1085–92.
38. Goossens PH, Schouten GJ, t Hart BA, et al. Feasibility of adenovirus-mediated nonsurgical synovectomy in collagen-induced arthritis-affected rhesus monkeys. *Hum Gene Ther* May 1, 1999;**10**(7):1139–49.
39. Ghivizzani SC, Lechman ER, Kang R, et al. Direct adenovirus-mediated gene transfer of interleukin 1 and tumor necrosis factor alpha soluble receptors to rabbit knees with experimental arthritis has local and distal anti-arthritic effects. *Proc Natl Acad Sci USA* April 14, 1998;**95**(8):4613–8.
40. Lechman ER, Jaffurs D, Ghivizzani SC, et al. Direct adenoviral gene transfer of viral IL-10 to rabbit knees with experimental arthritis ameliorates disease in both injected and contralateral control knees. *J Immunol* August 15, 1999;**163**(4):2202–8.
41. Mi Z, Ghivizzani SC, Lechman E, Glorioso JC, Evans CH, Robbins PD. Adverse effects of adenovirus-mediated gene transfer of human transforming growth factor beta 1 into rabbit knees. *Arthritis Res Ther* 2003;**5**(3):R132–9.
42. Mi Z, Ghivizzani SC, Lechman ER, et al. Adenovirus-mediated gene transfer of insulin-like growth factor 1 stimulates proteoglycan synthesis in rab-bit joints. *Arthritis Rheum* November 2000;**43**(11):2563–70.
43. Whalen JD, Lechman EL, Carlos CA, et al. Adenoviral transfer of the viral IL-10 gene periarticularly to mouse paws suppresses development of collagen-induced arthritis in both injected and uninjected paws. *J Immunol* March 15, 1999;**162**(6):3625–32.
44. Yao Q, Glorioso JC, Evans CH, et al. Adenoviral mediated delivery of FAS ligand to arthritic joints causes extensive apoptosis in the synovial lining. *J Gene Med* May–June 2000;**2**(3):210–9.
45. Yao Q, Wang S, Gambotto A, et al. Intra-articular adenoviral-mediated gene transfer of trail induces apoptosis of arthritic rabbit synovium. *Gene Ther* June 2003;**10**(12):1055–60.
46. Yao Q, Wang S, Glorioso JC, et al. Gene transfer of p53 to arthritic joints stimulates synovial apoptosis and inhibits inflammation. *Mol Ther* June 2001;**3**(6):901–10.
47. de Poorter JJ, Hoeben RC, Hagendoorn S, et al. Gene therapy and cement injection for restabilization of loosened hip prostheses. *Hum Gene Ther* January 2008;**19**(1):83–95.
48. Ruan MZ, Erez A, Guse K, et al. Proteoglycan 4 expression protects against the development of osteoarthritis. *Sci Transl Med* March 13, 2013;**5**(176):176 ra134.
49. Gouze E, Pawliuk R, Pilapil C, et al. In vivo gene delivery to synovium by lentiviral vectors. *Mol Ther* April 2002;**5**(4):397–404.
50. Gouze E, Pawliuk R, Gouze JN, et al. Lentiviral-mediated gene delivery to synovium: potent intra-articular expression with amplification by inflam-mation. *Mol Ther* April 2003;**7**(4):460–6.
51. Goater J, Muller R, Kollias G, et al. Empirical advantages of adeno associated viral vectors in vivo gene therapy for arthritis. *J Rheumatol* April 2000;**27**(4):983–9.
52. Watanabe S, Imagawa T, Boivin GP, Gao G, Wilson JM, Hirsch R. Adeno-associated virus mediates long-term gene transfer and delivery of chondro-protective IL-4 to murine synovium. *Mol Ther* August 2000;**2**(2):147–52.
53. Katakura S, Jennings K, Watanabe S, et al. Recombinant adeno-associated virus preferentially transduces human, compared to mouse, synovium: implications for arthritis therapy. *Mod Rheumatol* 2004;**14**(1):18–24.
54. Takahashi H, Kato K, Miyake K, Hirai Y, Yoshino S, Shimada T. Adeno-associated virus vector-mediated anti-angiogenic gene therapy for collagen-induced arthritis in mice. *Clin Exp Rheumatol* July–August 2005;**23**(4):455–61.
55. Apparailly F, Millet V, Noel D, Jacquet C, Sany J, Jorgensen C. Tetracycline-inducible interleukin-10 gene transfer mediated by an adeno-associated virus: application to experimental arthritis. *Hum Gene Ther* July 1, 2002;**13**(10):1179–88.

56. Kay JD, Gouze E, Oligino TJ, et al. Intra-articular gene delivery and expression of interleukin-1Ra mediated by self-complementary adeno-associated virus. *J Gene Med* July 2009;**11**(7):605–14.

57. Oligino T, Ghivizzani S, Wolfe D, et al. Intra-articular delivery of a herpes simplex virus IL-1Ra gene vector reduces inflammation in a rabbit model of arthritis. *Gene Ther* October 1999;**6**(10):1713–20.

58. Weber C, Armbruster N, Scheller C, et al. Foamy virus-adenovirus hybrid vectors for gene therapy of the arthritides. *J Gene Med* March–April 2013;**15**(3–4):155–67.

59. Ghivizzani SC, Oligino TJ, Glorioso JC, Robbins PD, Evans CH. Direct gene delivery strategies for the treatment of rheumatoid arthritis. *Drug Discov Today* March 1, 2001;**6**(5):259–67.

60. Ghivizzani SC, Lechman ER, Tio C, et al. Direct retrovirus-mediated gene transfer to the synovium of the rabbit knee: implications for arthritis gene therapy. *Gene Ther* September 1997;**4**(9):977–82.

61. Gouze E, Gouze JN, Palmer GD, Pilapil C, Evans CH, Ghivizzani SC. Transgene persistence and cell turnover in the diarthrodial joint: implications for gene therapy of chronic joint diseases. *Mol Ther* June 2007;**15**(6):1114–20.

62. Pan RY, Xiao X, Chen SL, et al. Disease-inducible transgene expression from a recombinant adeno-associated virus vector in a rat arthritis model. *J Virol* April 1999;**73**(4):3410–7.

63. Pan RY, Chen SL, Xiao X, Liu DW, Peng HJ, Tsao YP. Therapy and prevention of arthritis by recombinant adeno-associated virus vector with delivery of interleukin-1 receptor antagonist. *Arthritis Rheum* February 2000;**43**(2):289–97.

64. Traister RS, Fabre S, Wang Z, Xiao X, Hirsch R. Inflammatory cytokine regulation of transgene expression in human fibroblast-like synoviocytes infected with adeno-associated virus. *Arthritis Rheum* July 2006;**54**(7):2119–26.

65. Jennings K, Miyamae T, Traister R, et al. Proteasome inhibition enhances AAV-mediated transgene expression in human synoviocytes in vitro and in vivo. *Mol Ther* April 2005;**11**(4):600–7.

66. Miagkov AV, Varley AW, Munford RS, Makarov SS. Endogenous regulation of a therapeutic transgene restores homeostasis in arthritic joints. *J Clin Invest*. May 2002;**109**(9):1223–9.

67. Khoury M, Adriaansen J, Vervoordeldonk MJ, et al. Inflammation-inducible anti-TNF gene expression mediated by intra-articular injection of serotype 5 adeno-associated virus reduces arthritis. *J Gene Med* July 2007;**9**(7):596–604.

68. van de Loo FA. Inflammation-responsive promoters for fine-tuned gene therapy in rheumatoid arthritis. *Curr Opin Mol Ther* October 2004;**6**(5):537–45.

69. Broeren MG, Vermeij EA, Arntz OJ, et al. The validation of disease-inducible promoter constructs for gene therapy in rheumatoid arthritis and osteoarthritis in human THP-1 cells. *Ann Rheum Dis* March 1, 2014;**73**(Suppl. 1):A69.

70. Fabre S, Apparailly F. Gene therapy for rheumatoid arthritis: current status and future prospects. *BioDrugs* December 1, 2011;**25**(6):381–91.

71. Ghivizzani SC, Gouze E, Gouze JN, et al. Perspectives on the use of gene therapy for chronic joint diseases. *Curr Gene Ther* August 2008;**8**(4):273–86.

72. Goossens PH, Vogels R, Pieterman E, et al. The influence of synovial fluid on adenovirus-mediated gene transfer to the synovial tissue. *Arthritis Rheum* January 2001;**44**(1):48–52.

73. Boissier MC, Lemeiter D, Clavel C, et al. Synoviocyte infection with adeno-associated virus (AAV) is neutralized by human synovial fluid from arthritis patients and depends on AAV serotype. *Hum Gene Ther* June 2007;**18**(6):525–35.

74. Cottard V, Valvason C, Falgarone G, Lutomski D, Boissier MC, Bessis N. Immune response against gene therapy vectors: influence of synovial fluid on adeno-associated virus mediated gene transfer to chondrocytes. *J Clin Immunol* March 2004;**24**(2):162–9.

75. Mingozzi F, Chen Y, Edmonson SC, et al. Prevalence and pharmacological modulation of humoral immunity to AAV vectors in gene transfer to synovial tissue. *Gene Ther* April 2013;**20**(4):417–24.

76. Zhang X, Mao Z, Yu C. Suppression of early experimental osteoarthritis by gene transfer of interleukin-1 receptor antagonist and interleukin-10. *J Orthop Res* July 2004;**22**(4):742–50.

77. Frisbie DD, Ghivizzani SC, Robbins PD, Evans CH, McIlwraith CW. Treatment of experimental equine osteoarthritis by in vivo delivery of the equine interleukin-1 receptor antagonist gene. *Gene Ther* January 2002;**9**(1):12–20.

78. Pelletier JP, Caron JP, Evans C, et al. In vivo suppression of early experimental osteoarthritis by interleukin-1 receptor antagonist using gene therapy. *Arthritis Rheum* June 1997;**40**(6):1012–9.

79. Shen PC, Lu CS, Shiau AL, Lee CH, Jou IM, Hsieh JL. Lentiviral small hairpin RNA knockdown of macrophage inflammatory protein-1gamma ameliorates experimentally induced osteoarthritis in mice. *Hum Gene Ther* October 2013;**24**(10):871–82.

80. Chen LX, Lin L, Wang HJ, et al. Suppression of early experimental osteoarthritis by in vivo delivery of the adenoviral vector-mediated NF-κBp65-specific siRNA. *Osteoarthr Cartil* February 2008;**16**(2):174–84.

81. Kyostio-Moore S, Bangari DS, Ewing P, et al. Local gene delivery of heme oxygenase-1 by adeno-associated virus into osteoarthritic mouse joints exhibiting synovial oxidative stress. *Osteoarthr Cartil* February 2013;**21**(2):358–67.

82. Hsieh JL, Shen PC, Shiau AL, et al. Intraarticular gene transfer of thrombospondin-1 suppresses the disease progression of experimental osteoarthritis. *J Orthop Res* October 2010;**28**(10):1300–6.

83. Hsieh JL, Shen PC, Shiau AL, et al. Adenovirus-mediated kallistatin gene transfer ameliorates disease progression in a rat model of osteoarthritis induced by anterior cruciate ligament transection. *Hum Gene Ther* February 2009;**20**(2):147–58.

84. Emery P, Sebba A, Huizinga TW. Biologic and oral disease-modifying antirheumatic drug monotherapy in rheumatoid arthritis. *Ann Rheum Dis* December 2013;**72**(12):1897–904.

85. Riese RJ, Krishnaswami S, Kremer J. Inhibition of JAK kinases in patients with rheumatoid arthritis: scientific rationale and clinical outcomes. *Best Pract Res Clin Rheumatol* August 2010;**24**(4):513–26.

86. Simmons DL. Targeting kinases: a new approach to treating inflammatory rheumatic diseases. *Curr Opin Pharmacol* June 2013;**13**(3):426–34.

87. Mease PJ, Hobbs K, Chalmers A, et al. Local delivery of a recombinant adenoassociated vector containing a tumour necrosis factor alpha antagonist gene in inflammatory arthritis: a phase 1 dose-escalation safety and tolerability study. *Ann Rheum Dis* August 2009;**68**(8):1247–54.

88. Mease PJ, Wei N, Fudman EJ, et al. Safety, tolerability, and clinical outcomes after intraarticular injection of a recombinant adeno-associated vector containing a tumor necrosis factor antagonist gene: results of a phase 1/2 Study. *J Rheumatol* April 2010;**37**(4):692–703.

89. Ha CW, Lee KH, Lee BS, et al. Efficacy of TissueGene-C (TG-C), a cell-mediated gene therapy, in patients with osteoarthritis: a phase IIa clinical study. *J Tissue Eng Regen Med* 2012;**6**(Suppl. 1). 287 (Abstract 248.205).

90. Ginn SL, Alexander IE, Edelstein ML, Abedi MR, Wixon J. Gene therapy clinical trials worldwide to 2012 – an update. *J Gene Med* February 2013;**15**(2):65–77.

91. Byers S, Rothe M, Lalic J, Koldej R, Anson DS. Lentiviral-mediated correction of MPS VI cells and gene transfer to joint tissues. *Mol Genet Metab* June 2009;**97**(2):102–8.

92. Sun J, Hakobyan N, Valentino LA, Feldman BL, Samulski RJ, Monahan PE. Intraarticular factor IX protein or gene replacement protects against development of hemophilic synovitis in the absence of circulating factor IX. *Blood* December 1, 2008;**112**(12):4532–41.

93. Tyler WK, Vidal AF, Williams RJ, Healey JH. Pigmented villonodular synovitis. *J Am Acad Orthop Surg* June 2006;**14**(6):376–85.

94. Evans CH. Advances in regenerative orthopedics. *Mayo Clin Proc* November 2013;**88**(11):1323–39.

95. Adriaansen J, Fallaux FJ, de Cortie CJ, Vervoordeldonk MJ, Tak PP. Local delivery of beta interferon using an adeno-associated virus type 5 effectively inhibits adjuvant arthritis in rats. *J Gen Virol* June 2007;**88**(Pt 6):1717–21.

Chapter 7

Type 1 Diabetes Mellitus: Immune Modulation as a Prerequisite for Successful Gene Therapy Strategies

Wenhao Chen, Aini Xie, Jie Wu and Lawrence Chan

Diabetes Research Center, Division of Diabetes, Endocrinology, and Metabolism, Departments of Medicine and Molecular & Cellular Biology, Baylor College of Medicine, Houston, TX, USA

LIST OF ABBREVIATIONS

Ad Adenovirus
Aire Autoimmune regulator
APC Antigen-presenting cell
ATG Antithymocyte globulin
CsA Cyclosporine A
CTLA-4 Cytotoxic T-lymphocyte antigen
DCs Dendritic cells
Flt3L Fms-like tyrosine kinase-3 ligand
Gad65 Glutamate decarboxylase 65
G-CSF Granulocyte colony-stimulating factor
HLA Human leukocyte antigen
ICOS Inducible costimulator
Idd3 Insulin dependent diabetes susceptibility 3
IFIH1 Interferon induced with helicase C domain 1
IGRP Islet-specific glucose-6-phosphatase catalytic subunit-related protein
MHC Major histocompatibility complex
mTECs Medullary thymic epithelial cells
mTOR Mammalian target of rapamycin
NOD Nonobese diabetic
PD-1 Programmed death 1
PPI Preproinsulin
STZ Streptozotocin
TCR T cell receptor
Th T helper
Treg Regulatory T
T1D Type 1 diabetes
VNTR Variable number tandem repeat

7.1 INTRODUCTION

Type 1 diabetes (T1D) is an autoimmune disease. T1D patients present with absolute insulin deficiency; hyperglycemia occurs when the vast majority of pancreatic beta cells in the body are destroyed by autoimmunity. To date, insulin replacement remains the mainstay of treatment. In the past three decades, the introduction of multiple forms of insulin formulations that exhibit widely varying durations of action has significantly added to our armamentarium for the clinical management of diabetes. The continued development of better insulin pumps and glucose monitoring systems will further improve glycemic control among diabetics. These advances notwithstanding, insulin injection therapy is not a cure for diabetes, as exogenously administered insulin cannot mimic the pattern of natural glucose-stimulated insulin secretion from the normal pancreas.

The induction of new beta cells, or beta-like cells, in T1D subjects could lead to a cure. Strategies for the generation of new beta cells include the in vivo approaches of directed differentiation/transdifferentiation of pancreatic nonbeta cells or the directed transdetermination of liver progenitor cells into beta cells; in contrast to these two in vivo approaches, ex vivo approaches include directed differentiation of embryonic stem cells and induced pluripotent stem cells into beta cells in vitro (recently reviewed in *Translational Research* by Wagner et al.[1]). For instance, we showed that gene transfer of the lineage-defining transcription factor neurogenin 3 leads to reversal of diabetes and restoration of normal glucose and insulin dynamics in streptozotocin-induced diabetic mice. Neurogenin 3 gene therapy induces the reprogramming of hepatic progenitor cells (also known as oval cells) into neo-beta cells that closely mimic normal pancreatic beta cells functionally, biochemically, and ultrastructurally.[2,3] Even if we assume that a gene therapy regimen succeeds in inducing new beta cells, the latter would be destroyed immediately by ongoing autoimmunity. As shown in Figure 7.1, any effective therapy that enables the acquisition of new beta cells must be supported by a potent immunotherapy to reverse T1D. Herein, we will review the mechanistic background information on the autoimmune process that underlies T1D and the current status of immunotherapy that addresses specific T1D autoimmune processes.

7.2 EFFECTIVE IMMUNE THERAPY/MODULATION: A PREREQUISITE FOR SUCCESSFUL GENE THERAPY OF TYPE 1 DIABETES

In the past 20 years, there has been considerable progress toward the therapeutic induction of neo-beta cells in vitro and in experimental animals in vivo. Overt T1D is preceded by autoimmune destruction of pancreatic beta cells such that, at presentation, T1D patients have lost 70–90% of their beta cell mass. Gene therapy-based beta cell neogenesis has the capacity to restore insulin production and reverse hyperglycemia. However, for a sustained therapeutic response, the neo-beta cells must be protected from the ongoing autoimmune process. Over the past decades, many laboratories have tested different immunomodulatory therapies with the objective of preventing, postponing, or reversing the onset of T1D. The same principle applies if, and when, diabetes gene therapy-mediated neo-islet regeneration becomes a reality. The islet neogenesis regimen must be supported by an immunotherapy that allows the new beta cell to survive the onslaught of autoimmunity, so it can continue to do its job.

T1D occurs when the body's T cells attack and destroy insulin-producing beta cells in the pancreatic islets (Figure 7.2). If the T cell-mediated immune attack could be halted, residual beta cells (or neo-beta cells produced by gene therapy) in T1D patients would be preserved and perhaps be given a chance to proliferate and produce sufficient insulin to normalize blood glucose. The fact that T cells that directly attack beta cells are islet antigen autoreactive[4] forms the basis for the many antigen-based therapies for T1D that have been devised over the years. Unfortunately, clinical trials using such antigen-specific strategies have been largely unsuccessful in delaying human T1D progression.[5] Numerous other clinical trials have depended on the use of immune-modifying drugs to treat T1D. Treatments using such drugs have often been borrowed from effective therapeutic

FIGURE 7.1 Combining β cell replacement with immunotherapy to treat T1D. Insulin-producing β cells in late-stage T1D subjects are destroyed by T and B lymphocytes. To restore insulin secretion in these patients, effective therapies that produce new beta (neo-β) cells must be supported by a potent immunotherapy that halts T and B cell attack. Subjects with new-onset T1D have some residual beta cells. Potential immunotherapies may reverse disease in these subjects by protecting the residual beta cells.

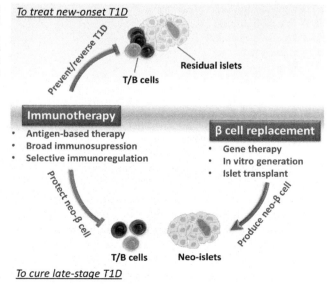

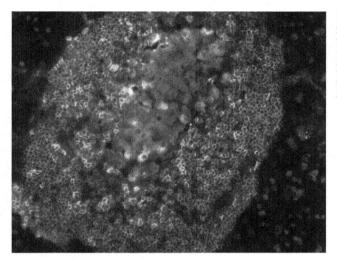

FIGURE 7.2 **Immune cells attack islet cells in the pancreas of NOD mice.** Pancreatic tissue was obtained from a diabetic NOD mouse, followed by immunofluorescence staining of beta cells with anti-insulin Ab (white), B cells with anti-B220 mAb (red), and T cells with anti-CD3 mAb (green). Nuclei were stained with 4,6-diamidino-2-phenylindole (DAPI, blue). (For interpretation of the references to color in this figure legend, the reader is referred to the online version of this book.)

regimens for transplant rejection or other T cell-mediated autoimmune diseases. However, the use of these drugs in T1D patients only transiently delays (but does not abrogate) beta cell loss.[5] Therefore, rather than assuming that T cell-mediated beta cell destruction can be halted by common immunosuppressive drugs, the rational development of innovative immunotherapy for T1D depends on a clear understanding of the molecular basis of the T cell response against a patient's own pancreatic islets.

To identify optimal therapeutic targets for T1D immunotherapies, it is necessary to develop a mechanistic framework for understanding T1D immune pathogenesis, including a detailed understanding of the (1) genetic factors and mechanisms responsible for the generation of islet antigen-specific T cells and the success (or failure) of antigen-based therapies for T1D; (2) T cell-mediated pathogenesis in T1D and its modulation and control by broad immunosuppressants; and (3) immunotherapies that selectively target the chain of events that enable and sustain the effector function of diabetogenic T cells. These mechanistic underpinnings of T1D immune pathogenesis are described below.

7.3 TARGETED ISLET ANTIGEN RECOGNITION AND ANTIGEN-BASED THERAPIES

7.3.1 Altered Self-Antigen Recognition Generates Diabetogenic T Cells

T cells are the master regulators that control the quality and quantity of adaptive immune responses. A majority of mature T cells express their own unique T cell receptor (TCR) consisting of an α and a β chain, which recognize specific "foreign" or "self" peptide antigens presented by major histocompatibility complex (MHC) molecules. To ensure immune tolerance to self-antigens, developing T cells in the thymus that have a TCR that recognizes a self-antigen/MHC molecule with high affinity generally undergo apoptosis, a process called negative selection or central tolerance. However, not all self-reactive developing T cells are eliminated during development. Some of them may leave the thymus and circulate in the periphery. To maintain peripheral tolerance to self-antigens, self-reactive T cells entering the periphery can be eliminated, become anergic (a lack of reactivity against specific antigen), or be actively suppressed by regulatory T (Treg) cells.[6] The development of autoimmune T1D is the consequence of impaired thymic deletion and peripheral tolerance of diabetogenic T cells (Figure 7.3).[6]

To efficiently delete self-reactive developing T cells, self-antigens need to be displayed in the thymus and presented to developing T cells. The transcription factor autoimmune regulator (Aire) influences ectopic expression of thousands of self-tissue-specific antigens (including pancreatic beta cell-restricted insulin) in medullary thymic epithelial cells (mTECs). Disruption of Aire function eliminates the expression of insulin and other tissue-specific antigens in the thymus, which leads to lethal autoimmunity that often includes T1D as part of the disease phenotype.[7] Hence, Aire mutations may cause a T1D phenotype by impairing negative selection of insulin-reactive T cells. Aire mutations are rare in the general population. Altered insulin expression levels in the human thymus are mainly attributed to the variable number of tandem repeat (VNTR) polymorphisms that lie upstream of the INS promoter. While the class III INS-VNTR genotype provides protection from T1D and is associated with high expression of insulin in mTECs, the class I INS-VNTR genotype contributes to the risk of T1D and is associated with lower thymic insulin expression.[8,9] Recently, Fan et al.[10] genetically inactivated insulin expression in the mTECs of C57BL/6 background mice, while the animals were allowed to maintain pancreatic insulin production. These mice developed diabetes spontaneously around 3 weeks after birth. Therefore, a defect in negative selection of insulin-reactive T cells in the thymus indeed causes T1D.

FIGURE 7.3 Development of T1D is the consequence of impaired thymic deletion and peripheral tolerance of diabetogenic T cells. Certain polymorphisms of HLA and genotypes of insulin gene variable number of tandem repeats (INS-VNTR) impair the central tolerance (thymic deletion) of potential diabetogenic T cells, which enter the peripheral tissues and lymphoid organs. Treg suppression and other peripheral tolerance mechanisms are breached by genetic factors, such as gene polymorphisms in *IL2RA*, *PTPN22*, and *CTLA4*. Diabetogenic T cells are then activated in the pancreatic lymph nodes upon encounter with T1D antigens (e.g., insulin, IGRP, and Gad65) presented by DCs, and infiltrate into the pancreas to destroy islets and in turn induce T1D. B cells also infiltrate the pancreas but their pathogenic role is still undefined.

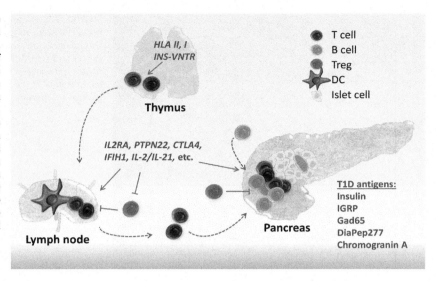

Self-antigen is presented to both developing and mature T cells by MHC molecules. Therefore, polymorphisms of human MHC, the human leukocyte antigen (HLA), may affect the generation and effector function of diabetogenic T cells. Indeed, polymorphisms of HLA represent the primary etiological pathway of T1D.[11] For instance, the predisposing HLA class II haplotypes—DRB1*03, DRB1*04, and DQB1*0302—are present in 39% of patients who develop T1D before age 20 and confer the highest T1D risk. The risk of developing T1D reaches 50% in children who have a predisposing HLA genotype and two or more relatives with T1D. By contrast, children with a T1D family history but with protective HLA-DQB alleles (e.g., DQB1*0602) have only about a 1% risk of developing T1D.[12] Genome-wide association studies have verified that the major T1D susceptibility loci indeed map to HLA-DQB1 and HLA-DRB1 and that several HLA class I regions are also associated with T1D.[13]

Knowledge of the MHC molecules that contribute to T1D susceptibility in the generation and function of diabetogenic T cells is still in its infancy. It has been suggested that the particular polymorphic peptide-binding pockets of these HLA molecules may contribute to T1D susceptibility by presenting a unique set of peptides.[14] Interestingly, nonobese diabetic (NOD) mice express only one MHC class II allele, I-A^{g7}, which has a substantially wider peptide-binding pocket that accounts for its distinct peptide preferences compared to other MHC class II alleles. The unusual peptide-binding pocket of I-A^{g7} leads to its weak interaction with proinsulin peptides, which may in turn provide low binding affinity and avidity to autoreactive T cells.[15] Autoreactive T cells with low avidity are capable of escaping thymic negative selection and peripheral tolerance mechanisms. To drive the chronic progression of T1D, polyclonal autoreactive T cells that escape into the periphery with low avidity for MHC antigens may transit into a set of oligoclonal T cells with higher avidity for cognate antigens.[16] Taken together, the generation of diabetogenic T cells is driven by both susceptibility to MHC molecules and altered thymic expression of self-antigens.

7.3.2 Insulin Immunotherapy

Insulin was the first identified human T1D antigen. Insulin antibodies can be detected in T1D patients prior to exogenous insulin treatment.[17] Several preproinsulin (PPI)/insulin epitopes can also be recognized by human T cells.[18,19] For instance, certain CD4+ T cell clones obtained from pancreatic lymph nodes recognize the insulin A 1–15 peptide restricted by HLA-DRB1*0401 allele,[18] and some circulating CD8+ T cells react to PPI$_{15-24}$ bound to the HLA-A*0201 molecule.[19] In NOD mice, insulin B chain 9–23 peptide (insulin B:9–23) is a primary autoantigen that initiates T1D. Hence, auto-immune diabetes can be prevented in transgenic NOD mice that express, in place of native insulin, a functional insulin with a Tyr→Ala substitution in position 16 of its B chain.[20] Insulin B:9–23 reactive CD4 T cells and B:15–23 reactive CD8 T cells are the predominant infiltrating cells in the islets with initial insulitis from 4-week-old NOD mice. However, the proportion of those cells is dramatically reduced at or near disease onset (usually >12-week old).[21] Pathogenic T cells recognizing other antigens further accumulate in the pancreas during the early progression of T1D. The progressive involvement of other PPI epitopes or other pancreatic beta cell molecules is called "antigen spreading."[21]

Mucosal (oral or nasal) administration of insulin has been used clinically in an attempt to restore immune tolerance to insulin antigens. In particular, antigens crossing from the intestine into the gut-associated lymphoid tissue will be embedded in an environment enriched with suppressive cytokines (e.g., TGF-β and IL-10) as well as unique CD103+ gut dendritic cells (DCs) that express indoleamine 2,3-dioxygenase and retinoic acid. Such an environment favors the induction of tolerance to antigens by increasing Treg cells or deleting antigen-specific T cells.[22] Indeed, mucosal insulin inhibits insulin-specific T cell and antibody responses in humans. Nevertheless, it has proven difficult to delay or prevent human T1D by mucosal insulin immunotherapy.[23] The reason for the lack of success may be due to the occurrence of antigen spreading prior to oral insulin administration. Moreover, in contrast to inbred NOD mice, there is no conclusive evidence that PPI/insulin is "the initiating antigen" in human T1D. Further studies are required to define the role of PPI/insulin antigenicity in distinct human populations.

In NOD mice, intranasal administration of insulin peptides prevents the development of T1D.[24] When combined with a low dose of systemic anti-CD3 mAb treatment, oral or intranasal administration of proinsulin can even enhance remission from recent-onset autoimmune diabetes in mice.[25] Other PPI/insulin delivery strategies have also been found to reduce the T1D incidence in NOD mice, such as intramuscular injection of a plasmid DNA vaccine encoding proinsulin II,[26] intravenous administration of syngeneic splenocytes ECDI cross-linked to intact insulin or specific peptides (B:9–23 or B:15–23),[27] and administration of an agonistic insulin mimetope (replacing the arginine residue [R22] in the natural peptide B:12–23 with glutamic acid).[28] The clinical utility of these strategies remains to be determined.

7.3.3 Other Antigen-Based Therapies

Islet-specific glucose-6-phosphatase catalytic subunit-related protein (IGRP) is a diabetic antigen that is expressed exclusively in pancreatic islets. Because IGRP is not expressed in thymic mTECs, it regulates the response or tolerance of IGRP-reactive T cells mainly in the periphery.[29] TCR transgenic mice with CD8+ T cells specific for the $IGRP_{206-214}$ epitope (NOD8.3 mice) develop accelerated diabetes, but the development of such a phenotype requires the presence of autoimmunity against proinsulin.[30] Thus, in NOD mice, IGRP may be one of the many "spreading" antigens following pathogenic proinsulin-specific autoimmunity. In humans, several HLA-A2-restricted IGRP epitopes have been identified as diabetic antigens. Circulating CD8+ T cells from patients with recent-onset T1D but not from healthy controls respond to the identified IGRP epitopes,[31] indicating that the presence/absence of IGRP-reactive T cells may be a marker for monitoring the progression of T1D.

Glutamate decarboxylase 65 (Gad65) is predominantly expressed in the pancreas and brain, and is an essential autoantigen in both T1D and stiff-person syndrome.[32] Autoimmune diabetes in NOD mice can be prevented not only by intrathymic injection of GAD65 at 3 weeks of age,[33] but also by intravenous injection of large amounts of GAD65 during later stages when autoimmunity has been initiated.[34] Hence, GAD65 antigen that has arisen very early in life could continuously mediate disease progression. In humans, immunization with GAD65 formulated with alum was initially shown to delay the progressive loss of c-peptide (a measure of beta cell function) in new-onset diabetic individuals, but a recent phase 3 trial of GAD65-alum did not show any clinical benefits.[35]

The list of diabetic antigens is constantly growing, and approximately 20 antigens have now been identified to be targeted by diabetogenic T cells (Table 7.1). For instance, an extensively studied diabetogenic CD4+ T cell clone, BDC2.5, has recently been shown to recognize the peptide epitope from chromogranin A.[36] Many antigen-based T1D therapies are under active clinical trial. For example, trials using heat shock protein 60 DiaPep277 peptide have shown some promise in preserving residual beta cell function.[37] Moreover, innovative methods of antigen administration (e.g., injection of nanoparticles coated with peptide-MHC) are under investigation for the treatment of T1D.[38] Overall, antigen-based therapy for T1D remains attractive because it offers the advantage of affecting only diabetogenic T cells. The success of this approach relies not only on the identification of primary antigen targets that initiate or aggravate T1D in distinct "outbred" human subjects, but also on the development of innovative and efficient antigen delivery systems.

7.4 BROAD IMMUNOSUPPRESSIVE THERAPIES

7.4.1 Pathogenic Role of T Cells in T1D

A recent study of pancreas autopsy samples removed from 29 patients with recent-onset T1D showed that CD8+ cytotoxic T cells were the most abundant cell population identified in insulitis specimens. Macrophages and CD4+ cells were also prominent in the islet infiltrate of T1D patients.[39] However, it remains difficult to define the pathogenic role of T cells in human T1D because of the difficulty in accessing pancreatic tissues from patients. By contrast, in NOD mice, there is overwhelming evidence that T1D is a T cell-mediated autoimmune disease. Priming of islet-reactive T cells occurs in the pancreatic lymph nodes,[40] where islet antigens are present and initially associated with a ripple of physiological beta cell death that occurs at 2 weeks of age.[41] Primed T cells then infiltrate into islets causing further islet damage and the

TABLE 7.1 Antigens Recognized by Human or Murine Diabetogenic T Cells

Antigen	Human T Cell	Murine T Cell
Insulin	+	+
Islet-specific glucose-6-phosphatase catalytic subunit-related protein (IGRP)	+	+
Glutamate decarboxylase 65 (Gad65)	+	+
Glutamate decarboxylase 67 (Gad67)	+	+
Heat shock protein 60 (Hsp60)	+	+
Insulinoma-associated protein 2 (IA-2)	+	+
Insulinoma-associated protein 2β (IA-2β)	+	+
Carboxypeptidase E (CPE)	+	+
Glucose transporter 2 (GLUT2)	+	+
S100β	+	+
Glial fibrillary acidic protein (GFAP)	+	+
Islet cell autoantigen 69 kDa (ICA69)	+	+
Islet amyloid polypeptide (IAPP)	+	+
Heat shock protein 70 (Hsp70)	+	N/A[a]
Mitochondrial 38 kDa islet antigen (Imogen 38)	+	N/A
Chromogranin A (Chg A)	N/A	+
Pancreatic duodenal homeobox 1 (PDX1)	N/A	+
Peripherin	N/A	+
Hepatocarcinoma–intestine–pancreas/pancreatic associated protein (HIP/PAP)	N/A	+
Dystrophia myotonica kinase (DMK)	N/A	+
Regenerating gene II (REG II)	N/A	+

[a]N/A indicates that positive results are not available.

subsequent release of different peptides from islet proteins that may contribute to early antigen spreading.[21] Amplified T cell responses, together with the reaction of other immune cells, eventually induce T1D onset at around 12 weeks of age or later. It is noteworthy that T cell entry to the pancreas is an antigen-specific process, and the majority of T cells recruited to the pancreatic islets in the course of T1D are islet autoreactive.[4] Destruction of islet beta cells by diabetogenic T cells is mediated via perforin and granzyme, the Fas/Fas ligand pathway, and the TNF-α-dependent pathway.[42]

A central step to start a T cell response is the activation of antigen-specific T cells. Naïve T cell activation requires at least TCR/CD3 and costimulatory signals, which are delivered by antigen-MHC and other molecules expressed on the surface of the antigen-presenting cell (APC). TCR/CD3 signaling activates various intracellular pathways (e.g., PKC/NF-κB, AP-1, and calcium/calcineurin/NFAT) to fulfill the needs of T cell proliferation and effector function.[43] Thus, in theory, decreased TCR/CD3 signal strength may prevent T1D autoimmunity. Nevertheless, the *PTPN22* gene encodes a negative regulator of TCR/CD3 signaling, lymphoid tyrosine phosphatase (LYP). The T1D susceptibility gene *PTPN22* increases LYP activity and suppresses TCR/CD3 signaling, which results in a profound deficit in T cell responsiveness to antigen stimulation.[44] Therefore, it is not clear how changes in TCR/CD3 signal strength affect T1D autoimmunity, because TCR/CD3 signaling controls multiple processes involved in thymic T cell development, Treg cell generation, and naïve T cell activation and function.

7.4.2 Small-Molecule Immunosuppressants

Immunosuppressive drugs have been used to treat T1D, a T cell-mediated autoimmune disease. Cyclosporin A (CsA) is a calcineurin inhibitor that prevents T cell activation and function. Continuous CsA treatment in patients with new-onset T1D was shown to eliminate the need for exogenous insulin, but the effect was not lasting.[45] Moreover, it caused renal toxicity, which considerably dampened enthusiasm for its use in T1D patients.

Rapamycin (also known as sirolimus) inhibits the mammalian target of the rapamycin (mTOR) pathway, which is critical for cell growth, proliferation, motility, and survival. A recent trial examined islet beta cell function in patients with long-term T1D and showed that rapamycin monotherapy leads to an increase in serum c-peptide and a reduction in exogenous insulin requirements.[46] By contrast, another trial showed that rapamycin/IL-2 combination therapy in T1D patients actually impaired beta cell function.[47] The clinical benefit of rapamycin therapy, if any, is debatable.

7.4.3 Anti-CD3 mAbs

Binding of cell surface receptors by mAbs in vivo could result in the depletion of cells or in the agonistic/antagonistic effects mimicking/blocking the action of the receptor's natural ligands. Therefore, selecting mAbs against distinct receptors on immune cells is a promising approach for developing new immunotherapies. The first mAb approved for clinical use was a mouse anti-human CD3ε (OKT3) mAb, which is a potent mitogenic agent for T cells by providing agonist effects via the TCR/CD3 complex.[48] OKT3 mAb treatment elicits almost immediate massive cytokine release leading to mild-to-life threatening symptoms. To reduce such side effects, FcR-nonbinding humanized CD3 mAbs (teplizumab and otelixizumab) were developed to reduce the quantity and quality of TCR signals and attenuate toxic reactions. FcR-nonbinding CD3 mAb treatment marginally and transiently depletes T cells and exerts long-lasting immune regulatory effects.[49,50]

In initial clinical trials, brief administration of FcR-nonbinding CD3 mAbs in new-onset T1D patients exhibited substantial therapeutic benefits. The cytokine release syndrome was largely eliminated, and the circulating T cell number recovered to pretreatment levels by one month of treatment. Most importantly, transient therapy with FcR-nonbinding CD3 mAbs preserved residual beta cell function in patients for more than two years.[51] Depending on patient age and baseline residual beta cell mass, transient treatment with otelixizumab even preserved insulin production for up to 3–4 years.[52] In a small group of patients, brief teplizumab treatment preserved residual beta cell function as long as 5 years.[53] Nevertheless, in a recent phase 3 trial using teplizumab in 516 T1D patients within 12 weeks of diagnosis, the composite endpoint (glycated hemoglobin A1c (HbA1c) <6.5% and insulin dose <0.5 U/kg per day at 1 year) did not differ significantly between the teplizumab-treated groups and the placebo group.[54] A phase 3 trial of otelixizumab also did not meet the primary efficacy endpoint of change in C-peptide at month 12 in T1D patients.[5] It is unclear whether further optimization of therapeutic antibody concentration and timing of treatment would lead to better outcomes.

7.4.4 Antithymocyte Globulin

Antithymocyte globulin (ATG) is a rabbit polyclonal gamma-immunoglobulin against human thymocytes. ATG has multiple specificities for various molecules displayed on T cells and other immune cells and is a potent immune-depleting agent. Transient ATG therapy delays the loss of beta cell function in T1D patients,[55] but does not induce long-lasting remission.[5] Recently, a group of new-onset T1D patients were transplanted with autologous nonmyeloablative hematopoietic stem cells. The majority of these patients achieved insulin independence after a mean follow-up of 29.8 months.[56] Importantly, these patients were also treated with cyclophosphamide and granulocyte colony-stimulating factor (G-CSF) for hematopoietic stem cell isolation, as well as with ATG and cyclophosphamide before cell transplantation.[57] Thus, a combination of potent immune depletion and suppression may forestall beta cell loss; however, potential adverse side effects may complicate such an aggressive approach.

7.5 IMMUNOTHERAPIES THAT TARGET EVENTS IN T CELL RESPONSE

7.5.1 Environmental Triggers of T1D

An effective T cell response requires not only the presence of antigens and antigen-specific T cells, but also a proinflammatory environment. In the context of autoimmune T1D, the diabetogenic T cell response may be mainly triggered by undefined "sterile" inflammation. On the other hand, T1D onset could also be influenced by microbial infections and the intestinal microbial environment. For instance, inoculation of encephalomyocarditis virus induces diabetes in wild-type SJL mice by triggering an immune attack on beta cells.[58] In NOD mice, interaction of intestinal microbes with the innate immune system was found to be a critical factor modifying T1D predisposition.[59] In humans, rare variants of IFIH1 (interferon induced with helicase C domain 1) have been shown to provide protection against T1D. These variants are predicted to alter the expression and structure of IFIH1 and disrupt the normal function of IFIH1's response to viral RNA.

Hence, viral stimuli through IFIH1 may be an environmental trigger of human T1D.[60] Over the last 50 years, a several-fold increase has been observed in the incidence of human T1D, indicating an urgent need to identify possible diabetogenic environmental triggers.

7.5.2 Antigen-Presenting Cells

Local inflammation induces maturation of APCs, which in turn migrate to lymphoid tissues and activate antigen-specific T cells. DCs are a type of professional APC and can be divided into several subsets. For instance, conventional DCs are specialized for antigen processing and presentation, whereas plasmacytoid DCs express CD45RA and have poor antigen-presenting capacity. Conventional DCs can be further classified into CD4$^+$ DCs, CD8α^+ DCs, and CD4$^-$CD8α^- DCs. Interestingly, NOD mice exhibit a defect in CD8α^+ conventional DCs.[61] Some studies indicate that enhancing DC development or function predisposes to autoimmune disease.[62] By contrast, others showed that Fms-like tyrosine kinase-3 ligand treatment prevents diabetes in NOD mice by upregulating the frequency of CD8α^+ conventional DCs.[61] G-CSF treatment also prevents diabetes by recruiting plasmacytoid DCs.[63] To date, the exact role of DCs in T1D remains elusive, and the therapeutic use of DCs needs to be further investigated.

B cells are crucial APCs in the pathogenesis of T1D. Although B cell-derived autoantibodies against islet antigens are useful markers for predicting T1D disease activity, these autoantibodies may not be pathogenic.[64] By contrast, B cells themselves infiltrate the pancreatic islets[39] and present autoantigens to T cells.[65] NOD mice deficient in B cells are insulitis resistant and protected from developing diabetes.[66] Moreover, in a transgenic NOD mouse expressing human CD20 (hCD20) on their B cells, transient depletion of B cells by an anti-hCD20 mAb delays the onset of diabetes and even reverses overt diabetes.[67] A clinical trial with a course of an anti-hCD20 mAb (rituximab) showed that, at 1 year of therapy, the rituximab group had a significantly higher level of C-peptide and required less exogenous insulin for treatment than the placebo group. However, B cell depletion persists for months and is associated with some adverse events.[68] A shortcoming of anti-CD20 mAb therapy is the risk of lowered systemic immunity.

7.5.3 Positive and Negative Costimulatory Molecules

APCs express various costimulatory molecules to regulate T cell response. For instance, CD28 expressed on T cells is the receptor for B7.1 (CD80) and B7.2 (CD86) expressed on APCs. TCR/CD3 stimulation plus CD28 signaling are sufficient to activate naïve T cells. In the absence of a CD28 costimulatory signal, TCR/CD3 stimulation induces functional anergy rather than activation of naïve T cells. T cell activation induces surface expression of cytotoxic T-lymphocyte antigen 4 (CTLA-4), which also binds to CD80 and CD86 on APCs. However, CTLA-4 transmits an inhibitory signal to "turn off" T cell activation and serves as a physiologic antagonist of CD28. Mice deficient in CTLA-4 die within the first month of life of severe pancreatitis and mycocarditis, demonstrating the critical regulatory function of this negative costimulatory receptor.[69] In the context of murine T1D, specific CTLA-4 deletion in autoreactive T cells facilitates the induction of autoimmunity,[70] whereas artificial expression of a CTLA-4 agonist on B cells protects NOD mice from spontaneous auto-immune diabetes.[71] In humans, an association of the CTLA-4 locus with susceptibility to T1D has been reported.[72] Moreover, a clinical trial of abatacept (CTLA4-Ig; a fusion protein of the CTLA4 and an IgG1 Fc domain) was conducted in T1D patients to block the CD28 pathway. Continued administration of abatacept resulted in an estimated 9.6 months delay in C-peptide reduction; however, beta cell mass and function continued to deteriorate.[73]

Inducible costimulator (ICOS) is a CD28 superfamily costimulatory molecule expressed on activated T cells. ICOS serves as a positive mediator for activated T cells by upregulating cell proliferation and cytokine secretion. NOD mice deficient in ICOS or ICOSL (ICOS ligand) are protected from developing diabetes by limiting the activation of diabetogenic T cells.[74] Combined treatment of mAbs blocking ICOS and CD40 ligand signaling also prevents T1D in adult NOD mice.[75] By contrast, abrogation of the ICOS pathway in BDC2.5 NOD mice exacerbates T1D, which may be due to the critical role of ICOS in homeostasis and functional stability of Treg cells in prediabetic islets.[76] Therefore, caution should be exercised in immunotherapeutic use of ICOS blockade.

Activated T cells express programmed death 1 (PD-1). Binding of PD-1 with its two ligands, PD-L1 and PD-L2, maintains peripheral T cell tolerance by not only inhibiting primary T cell activation but also triggering exhaustion of activated T cells.[77] Thus, blockade of the PD-1–PD-L1 pathway with mAbs rapidly precipitates diabetes in prediabetic NOD mice.[78] PD-1 deficiency in NOD mice also accelerates the onset and frequency of T1D development.[79] It is worth noting that PD-L1 is expressed broadly on hematopoietic and parenchymal cells. PD-L1 expression on parenchymal cells rather than hematopoietic cells protects against T cell-mediated autoimmune diabetes.[80] PD-L2 is predominantly expressed on hematopoietic cells and is less potent in controlling T1D in NOD mice.[80] Lymphocyte activation gene 3

(LAG-3) is another inhibitory coreceptor expressed on activated T cells. LAG3 acts synergistically with PD-1 to prevent autoimmunity in mice,[81] possibly by inducing exhaustion of activated T cells.[77] Moreover, CD137 (4-1BB), a member of the tumor necrosis factor receptor superfamily, is primarily expressed on activated T cells and may prevent their exhaustion.[82] However, potential risks may associate with anti-4-1BB-based immunotherapy for autoimmune diseases,[83] which requires further study.

The development of immunotherapies targeting costimulatory pathways is an attractive but challenging undertaking. Activated CD4+ T helper (Th) cells express high levels of CD154 (CD40L) to control the activation and function of various CD40-expressing immune cell subsets. Blocking CD40–CD154 interaction by using an anti-CD154 mAb proved effective in inducing peripheral T cell tolerance in various animal models. However, clinical trials of anti-CD154 mAb had to be prematurely terminated, because the mAb also binds to CD154 expressed on platelets and stimulates coagulation.[84] It is important to note that attempts to target costimulatory pathways (e.g., the use of CD28 superagonist antibody TGN1412) in humans have led to life-threatening complications.[85] Thorough preclinical investigations are required for understanding the therapeutic potential and possible toxicity of targeting costimulatory molecules.

7.5.4 Cytokines in Treg and Effector T Cell Function

Activated T cells produce many cytokines that have a variety of actions on themselves and other immune cells. In particular, recent studies highlight that impairment of the IL-2 pathway plays a central role in the pathogenesis of T1D. IL-2 is predominantly produced by activated T cells and supports the expansion of activated T cells. However, IL-2 is also a critical cytokine for the development and homeostasis of Treg cells. In murine models, insulin-dependent diabetes susceptibility 3 (Idd3) on NOD mouse chromosome 3 is the strongest single non-MHC T1D susceptibility locus, which contains the *Il2* gene and transcribes much less IL-2 mRNA as compared to protective Idd3 alleles. This genetic deficiency in IL-2 production correlates with reduced Treg cell function in NOD mice.[86] Moreover, transient low-dose IL-2 administration in diabetic NOD mice can reverse T1D progression by a local effect on pancreatic Treg cells.[87] Therefore, in NOD mice, the defect in the IL-2–Treg pathway is critical for T1D development, and such a defect can be corrected by IL-2 therapy.

The IL-2–Treg pathway in "outbred" human T1D subjects should be broadly interrogated. First, the Foxp3 transcription factor is the master regulator for maintaining Treg cell phenotype and function. Humans with a rare Foxp3 mutation develop severe T1D and other lethal autoimmune disorders, demonstrating that a defect in Treg cell function triggers human T1D.[7] Second, Treg cells isolated from patients with recent-onset T1D exhibit markedly reduced suppressor function in vitro.[88] Finally, polymorphisms in genes encoding IL-2 receptor alpha chain (IL-2Rα or CD25) and IL-2 are strongly associated with human T1D.[11] In particular, individuals with *IL2RA* susceptibility genotypes are associated with lower IL-2 secretion by T cells as well as lower expression of CD25 on Treg cells, both of which may impair Foxp3 expression and Treg cell function.[89] Taken together, the IL-2–Treg pathway could be a regulator controlling the development of human T1D. Nevertheless, effector T cells isolated from T1D patients are also resistant to suppression by Treg cells,[90] which may be a barrier for Treg cell-based therapy (i.e., low dose IL-2 administration). Indeed, in a recent clinical trial, IL-2 therapy combined with rapamycin in T1D patients causes transient beta cell dysfunction rather than protection, though treated patients exhibit increased Treg cell frequency.[47]

Recent studies demonstrated that IL-21 plays an essential role in T1D onset in NOD mice.[91–94] IL-21 and IL-2 are closely related cytokines, both of which are predominantly produced by activated Th cells. The genes encoding IL-21 and IL-2 are adjacent to each other, and receptors for IL-21 and IL-2 share a common γ chain. However, IL-21 and IL-2 play distinct roles in regulating the immune system. While both cytokines can support expansion and survival of antigen-specific T cells during acute viral infection, IL-21 is the major cytokine required for sustaining effector T cell function in chronic viral infection.[95] Moreover, IL-21 is not required for Treg cell homeostasis, but provides pleiotropic effects on many other immune cell types. Therefore, in contrast to IL-2-deficient mice that suffer from lethal autoimmunity due to Treg cell defect, IL-21 or IL-21 receptor (IL-21R) deficiency in mice prevents the induction of several autoimmune diseases.[95] This has proved true in NOD mice as well. Deficiency in IL-21 or IL-21R renders NOD mice resistant to the onset of T1D.[91,92] Recent studies further showed that CCR9+ Th cells in the pancreas produce IL-21[93] and that the IL-21R signal controls autoreactive T cell infiltration in pancreatic islets.[94] Such thorough investigations of IL-21 actions will facilitate the design of immunotherapy targeting the IL-21 signal.

IL-21R is expressed on CD4+ Th cells, CD8+ T cells, B cells, NK cells, DCs, macrophages, and keratinocytes. Thus, IL-21 produced by activated Th cells has a broad range of actions on various immune cells.[95] For instance, Th17 cells produce IL-21, but IL-21 also serves as an autocrine cytokine for Th17 cell development and upregulates IL-23R expression to maintain the Th17 cell phenotype.[95] Interestingly, enhanced Th17 immunity is observed in T1D patients.[96] Inhibition of

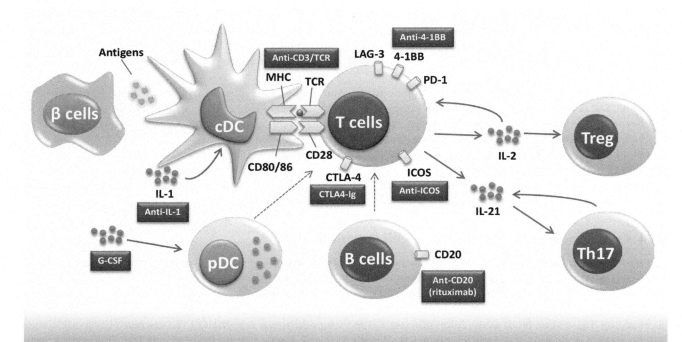

FIGURE 7.4   Developing immunotherapies by targeting molecules that regulate T cell response. A central step in initiating a T cell response is the activation of T cells upon MHC-restricted antigen presentation by APCs. In the context of T1D, cDCs are professional APCs specified for processing beta cell antigens and presenting them to T cells. Proinflammatory cytokines (e.g., IL-1) trigger "sterile" inflammation and activate cDCs. IL-1 also directly causes beta cell destruction. Hence, IL-1 blocker may synergize with other immunotherapies in reversing T1D. By contrast, G-CSF prevents the T1D onset in mice by recruiting pDCs, which have poor antigen-presenting capacity. B cells are also crucial APCs in the pathogenesis of T1D. A clinical trial of an anti-hCD20 mAb showed some protective effects on beta cells by depleting B cells. APCs activate T cells through at least TCR/CD3 and CD28 costimulatory signals. In recent trials using FcR-nonbinding CD3 mAbs to alter TCR signals or CTLA4-Ig to block the interaction between CD80/CD86 and CD28, the beta cell loss was delayed but not abrogated. Other costimulatory or coinhibitory receptors (e.g., ICOS, 4-1BB, LAG-3, PD-1) are also potent regulators for T cell activation and function. Moreover, activated T cells produce many cytokines (e.g., IL-2 and IL-21) that have a variety of actions on themselves and other immune cells. IL-2 controls the generation and survival of Treg cells; however, IL-2 therapy in T1D patients did not protect beta cells from destruction. IL-21 provides pleiotropic effects on many immune cell types. Promoting Th17 cell response may be one of many actions by which IL-21 contributes to the pathogenesis of T1D. Targeting costimulatory pathways and cytokine signals remains an attractive but challenging therapeutic approach to treat T1D.

Th17 cell function even prevents diabetes in NOD mice.[97] Therefore, enhancing the Th17 cell response may be one of many actions by which IL-21 underlies the pathogenesis of T1D.

Inflamed tissues contain many types of immune cytokines, all of which play distinct roles in regulating immune responses. In the context of T1D, polymorphisms in genes encoding IL-7 receptor alpha, IL-10, and IL-27 also predispose to islet autoimmunity.[11] The biology of these cytokines in T1D should be investigated to evaluate their therapeutic potential. Moreover, proinflammatory cytokines (e.g., IL-1) can directly cause beta cell destruction.[98] Results from a recent clinical trial indicated that use of the IL-1 blocker anakinra lowers the requirement for exogenous insulin for months in children with new-onset T1D.[99] Although IL-1 blocker alone may be insufficient to induce long-term remission of T1D, it should synergize with other immunotherapies in reversing diabetes (Figure 7.4).[100]

7.6 PROSPECTS FOR IMMUNOTHERAPY IN PROTECTING NEO-BETA CELLS

In the previous sections, we have reviewed many of the approaches that have been tested either in laboratory animals or in T1D patients in clinical trials (Table 7.2). Here, we offer additional thoughts on possible strategies for immunotherapy. We note that similar immune mechanisms that have led to beta cell destruction in a T1D subject will also destroy neo-beta cells that are produced in response to novel gene and cell therapy. Hence, a safe and effective immunotherapy is not available for the long-term protection of neo-beta cells. A possible approach is to halt the autoimmune attack first with potent short-term general immunosuppression (e.g., FcR-nonbinding anti-CD3 mAb, ATG, or anti-CD20 mAb) and prevent or ameliorate late recurrence of autoimmunity by instituting prolonged or intermittent

TABLE 7.2 Clinical Attempts Described in This Chapter That Target Antigens and Cells Involved in Immune-Mediated Beta Cell Destruction

Antigen/Cell Types	Mechanism of Action in T1D	Therapeutic Strategy	Patient Selection	Clinical Outcomes and Limitations
Insulin	A major antigen that may trigger T1D onset	Oral/nasal insulin administration for tolerance induction	T1D-risk children with HLA genotypes and autoantibodies	Did not prevent or delay T1D[23]
GAD65	A major antigen in T1D	Subcutaneous (s.c.) injection of alum-formulated GAD65	T1D children; within 3 months after diagnosis	Did not reduce the loss of beta cell function or improve clinical outcomes over a 15 month period[35]
Hsp60 peptide p277	An antigen in T1D	s.c. injection of a synthetic Hsp60 p277, DiaPep277	Adult and pediatric patients with new-onset T1D	Significantly preserved insulin synthesis in adult, but not pediatric, T1D patients[37]
T cells	Effector cells that attack beta cells	CsA inhibits T cell function	Patients with new-onset T1D	Required continued use of CsA to protect beta cells; associated with renal toxicity[45]
T cells	Effector cells that attack beta cells	FcR-nonbinding CD3 mAbs alter TCR signals	Patients with new-onset T1D	Did not meet the primary efficacy endpoint of a phase 3 trial[54]; associated with transient cytokine release and EBV reactivation
T cells	Effector cells that attack beta cells	ATG depletes T cells	T1D patients within 4 weeks of diagnosis	Delayed the loss of beta cell function; transient fever and symptoms of serum sickness[55]
T cells	Effector cells that attack beta cells	CTLA4-Ig blocks costimulatory signal of T cell activation	Patients with new-onset T1D	Slowed reduction in beta cell function; no increase in infections or neutropenia[73]
B cells	APC and secrete autoantibodies	Anti-CD20 mAb depletes B cells	Patients with new-onset T1D	Partially preserved beta cell function over a period of 1 year; grade 1 or 2 reactions after the first infusion[68]
Innate immune cells	APC and release inflammatory mediators	Anti-IL1 mAb blocks inflammatory mediator IL-1	Patients with new-onset T1D	Safe but were not effective as a single immunomodulatory drug[99]

immunotherapies that target costimulatory signals or cytokines (e.g., CTLA4-Ig or IL-1 blocker). Each of those therapies delays beta cell loss for months. It would be interesting to determine the safety and efficacy of such combined therapies as a treatment strategy.

As immunologists continue their research on the mechanistic basis of the cause and progression of T1D, they should pay attention also to the precise pathogenic roles of diabetogenic T cells. Undoubtedly, they will uncover new treatment targets that will help refine future immunotherapies. For example, it has been noted recently that IL-21 produced by activated Th cells is required to sustain the chronic immune response in different autoimmune diseases.[95] IL-21 and its related signaling molecules are thus attractive targets for T1D immunotherapy. Moreover, other innovative technologies for immunotherapy should be explored. A plasmid DNA vaccine encoding mouse proinsulin II has been shown to reduce the incidence of diabetes in NOD mice,[26] which demonstrates the potential of DNA-based immunotherapy for T1D.

Immune response against neo-beta cells is only one of the challenges for the clinical application of beta cell regeneration. Efficient gene therapy to induce neo-beta cell formation in vivo still relies mostly on viral vectors, which themselves exert potent immunogenicity as well as toxic and other adverse effects. The generation of mature beta cells in vitro from pluripotent stem cells remains inefficient for obtaining clinically relevant cell numbers for therapy. The tumorigenicity of pluripotent cells also poses a potential threat that must be neutralized or circumvented before clinical application. Extensive preclinical investigations are required to resolve these challenges.

7.7 CONCLUSION

The past decade has witnessed steady progress in experimental in vivo reprogramming of nonbeta cells or adult stem cells in situ, and ex vivo-targeted differentiation of embryonic and adult stem cells, to mature beta cells as a therapy for diabetes. To prepare for the clinical translation of beta cell creation strategies in T1D patients in the future, we must add to our armamentarium an effective treatment method that counters the autoimmunity that underlies T1D. T1D patients harbor activated diabetogenic T cells that specifically attack beta cells. None of the current immunosuppressive agents can selectively eliminate diabetogenic T cells to protect neo-beta cells in the long term. A better understanding of the T cell response in T1D would greatly improve our chances of developing a safe and effective immunotherapy that protects de novo-generated beta cells and enables a cure of this devastating disease.

ACKNOWLEDGMENT

The research in the authors' laboratories discussed in this manuscript was supported by National Institutes of Health Grants P30-DK079638 for a Diabetes Research Center at Baylor College of Medicine, R01-HL51586 (to L. Chan), the Charles and Barbara Close Foundation (to L.Chan) and the Juvenile Diabetes Research Foundation (46-2010-752 to L. Chan), the American Heart Association 11SDG7690000 (to W. Chen) and the Fondation de la Recherche en Transplantation (IIG201101 to W. Chen), the Betty Rutherford Chair in Diabetes Research and St. Luke's Medical Center, Houston, Texas, USA.

REFERENCES

1. Wagner RT, Lewis J, Cooney A, Chan L. Stem cell approaches for the treatment of type 1 diabetes mellitus. *Trans Res* September 2010;**156**(3):169–79.
2. Yechoor V, Liu V, Espiritu C, Paul A, Oka K, Kojima H, et al. Neurogenin3 is sufficient for transdetermination of hepatic progenitor cells into neo-islets in vivo but not transdifferentiation of hepatocytes. *Dev Cell* March 2009;**16**(3):358–73.
3. Yechoor V, Liu V, Paul A, Lee J, Buras E, Ozer K, et al. Gene therapy with neurogenin 3 and betacellulin reverses major metabolic problems in insulin-deficient diabetic mice. *Endocrinology* November 2009;**150**(11):4863–73.
4. Lennon GP, Bettini M, Burton AR, Vincent E, Arnold PY, Santamaria P, et al. T cell islet accumulation in type 1 diabetes is a tightly regulated, cell-autonomous event. *Immunity* October 16, 2009;**31**(4):643–53.
5. Waldron-Lynch F, Herold KC. Immunomodulatory therapy to preserve pancreatic beta-cell function in type 1 diabetes. *Nat Rev Drug Discov* June 2011;**10**(6):439–52.
6. Bluestone JA. Mechanisms of tolerance. *Immunol Rev* May 2011;**241**(1):5–19.
7. Husebye ES, Anderson MS. Autoimmune polyendocrine syndromes: clues to type 1 diabetes pathogenesis. *Immunity* April 23, 2010;**32**(4):479–87.
8. Pugliese A, Zeller M, Fernandez Jr A, Zalcberg LJ, Bartlett RJ, Ricordi C, et al. The insulin gene is transcribed in the human thymus and transcription levels correlated with allelic variation at the INS VNTR-IDDM2 susceptibility locus for type 1 diabetes. *Nat Genet* March 1997;**15**(3):293–7.
9. Vafiadis P, Bennett ST, Todd JA, Nadeau J, Grabs R, Goodyer CG, et al. Insulin expression in human thymus is modulated by INS VNTR alleles at the IDDM2 locus. *Nat Genet* March 1997;**15**(3):289–92.
10. Fan Y, Rudert WA, Grupillo M, He J, Sisino G, Trucco M. Thymus-specific deletion of insulin induces autoimmune diabetes. *EMBO J* September 16, 2009;**28**(18):2812–24.
11. Todd JA. Etiology of type 1 diabetes. *Immunity* April 23, 2010;**32**(4):457–67.
12. Ziegler AG, Nepom GT. Prediction and pathogenesis in type 1 diabetes. *Immunity* April 23, 2010;**32**(4):468–78.
13. Nejentsev S, Howson JM, Walker NM, Szeszko J, Field SF, Stevens HE, et al. Localization of type 1 diabetes susceptibility to the MHC class I genes HLA-B and HLA-A. *Nature* December 6, 2007;**450**(7171):887–92.
14. Suri A, Walters JJ, Gross ML, Unanue ER. Natural peptides selected by diabetogenic DQ8 and murine I-A(g7) molecules show common sequence specificity. *J Clin Invest* August 2005;**115**(8):2268–76.
15. Levisetti MG, Lewis DM, Suri A, Unanue ER. Weak proinsulin peptide-major histocompatibility complexes are targeted in autoimmune diabetes in mice. *Diabetes* July 2008;**57**(7):1852–60.
16. Amrani A, Verdaguer J, Serra P, Tafuro S, Tan R, Santamaria P. Progression of autoimmune diabetes driven by avidity maturation of a T-cell population. *Nature* August 17, 2000;**406**(6797):739–42.
17. Palmer JP, Asplin CM, Clemons P, Lyen K, Tatpati O, Raghu PK, et al. Insulin antibodies in insulin-dependent diabetics before insulin treatment. *Science* December 23, 1983;**222**(4630):1337–9.
18. Kent SC, Chen Y, Bregoli L, Clemmings SM, Kenyon NS, Ricordi C, et al. Expanded T cells from pancreatic lymph nodes of type 1 diabetic subjects recognize an insulin epitope. *Nature* May 12, 2005;**435**(7039):224–8.
19. Skowera A, Ellis RJ, Varela-Calvino R, Arif S, Huang GC, Van-Krinks C, et al. CTLs are targeted to kill beta cells in patients with type 1 diabetes through recognition of a glucose-regulated preproinsulin epitope. *J Clin Invest* October 2008;**118**(10):3390–402.
20. Nakayama M, Abiru N, Moriyama H, Babaya N, Liu E, Miao D, et al. Prime role for an insulin epitope in the development of type 1 diabetes in NOD mice. *Nature* May 12, 2005;**435**(7039):220–3.
21. Wong FS, Karttunen J, Dumont C, Wen L, Visintin I, Pilip IM, et al. Identification of an MHC class I-restricted autoantigen in type 1 diabetes by screening an organ-specific cDNA library. *Nat Med* September 1999;**5**(9):1026–31.

22. Weiner HL, da Cunha AP, Quintana F, Wu H. Oral tolerance. *Immunol Rev* May 2011;**241**(1):241–59.

23. Nanto-Salonen K, Kupila A, Simell S, Siljander H, Salonsaari T, Hekkala A, et al. Nasal insulin to prevent type 1 diabetes in children with HLA genotypes and autoantibodies conferring increased risk of disease: a double-blind, randomised controlled trial. *Lancet* November 15, 2008;**372**(9651):1746–55.

24. Martinez NR, Augstein P, Moustakas AK, Papadopoulos GK, Gregori S, Adorini L, et al. Disabling an integral CTL epitope allows suppression of autoimmune diabetes by intranasal proinsulin peptide. *J Clin Invest* May 2003;**111**(9):1365–71.

25. Takiishi T, Korf H, van Belle TL, Robert S, Grieco FA, Caluwaerts S, et al. Reversal of autoimmune diabetes by restoration of antigen-specific tolerance using genetically modified *Lactococcus lactis* in mice. *J Clin Invest* May 1, 2012;**122**(5):1717–25.

26. Solvason N, Lou YP, Peters W, Evans E, Martinez J, Ramirez U, et al. Improved efficacy of a tolerizing DNA vaccine for reversal of hyperglycemia through enhancement of gene expression and localization to intracellular sites. *J Immunol* December 15, 2008;**181**(12):8298–307.

27. Prasad S, Kohm AP, McMahon JS, Luo X, Miller SD. Pathogenesis of NOD diabetes is initiated by reactivity to the insulin B chain 9–23 epitope and involves functional epitope spreading. *J Autoimmun* May 28, 2012;**39**(4):347–53.

28. Daniel C, Weigmann B, Bronson R, von Boehmer H. Prevention of type 1 diabetes in mice by tolerogenic vaccination with a strong agonist insulin mimetope. *J Exp Med* July 4, 2011;**208**(7):1501–10.

29. Krishnamurthy B, Chee J, Jhala G, Fynch S, Graham KL, Santamaria P, et al. Complete diabetes protection despite delayed thymic tolerance in NOD8.3 TCR transgenic mice due to antigen-induced extrathymic deletion of T cells. *Diabetes* February 2012;**61**(2):425–35.

30. Krishnamurthy B, Dudek NL, McKenzie MD, Purcell AW, Brooks AG, Gellert S, et al. Responses against islet antigens in NOD mice are prevented by tolerance to proinsulin but not IGRP. *J Clin Invest* December 2006;**116**(12):3258–65.

31. Jarchum I, Nichol L, Trucco M, Santamaria P, Dilorenzo TP. Identification of novel IGRP epitopes targeted in type 1 diabetes patients. *Clin Immunol* June 2008;**127**(3):359–65.

32. Hanninen A, Soilu-Hanninen M, Hampe CS, Deptula A, Geubtner K, Ilonen J, et al. Characterization of CD4+ T cells specific for glutamic acid decarboxylase (GAD65) and proinsulin in a patient with stiff-person syndrome but without type 1 diabetes. *Diabetes Metab Res Rev* May 2010;**26**(4):271–9.

33. Tisch R, Yang XD, Singer SM, Liblau RS, Fugger L, McDevitt HO. Immune response to glutamic acid decarboxylase correlates with insulitis in non-obese diabetic mice. *Nature* November 4, 1993;**366**(6450):72–5.

34. Tian J, Clare-Salzler M, Herschenfeld A, Middleton B, Newman D, Mueller R, et al. Modulating autoimmune responses to GAD inhibits disease progression and prolongs islet graft survival in diabetes-prone mice. *Nat Med* December 1996;**2**(12):1348–53.

35. Ludvigsson J, Krisky D, Casas R, Battelino T, Castano L, Greening J, et al. GAD65 antigen therapy in recently diagnosed type 1 diabetes mellitus. *N Engl J Med* February 2, 2012;**366**(5):433–42.

36. Stadinski BD, Delong T, Reisdorph N, Reisdorph R, Powell RL, Armstrong M, et al. Chromogranin A is an autoantigen in type 1 diabetes. *Nat Immunol* March 2010;**11**(3):225–31.

37. Luo X, Herold KC, Miller SD. Immunotherapy of type 1 diabetes: where are we and where should we be going? *Immunity* April 23, 2010;**32**(4):488–99.

38. Tsai S, Shameli A, Yamanouchi J, Clemente-Casares X, Wang J, Serra P, et al. Reversal of autoimmunity by boosting memory-like autoregulatory T cells. *Immunity* April 23, 2010;**32**(4):568–80.

39. Willcox A, Richardson SJ, Bone AJ, Foulis AK, Morgan NG. Analysis of islet inflammation in human type 1 diabetes. *Clin Exp Immunol* February 2009;**155**(2):173–81.

40. Gagnerault MC, Luan JJ, Lotton C, Lepault F. Pancreatic lymph nodes are required for priming of beta cell reactive T cells in NOD mice. *J Exp Med* August 5, 2002;**196**(3):369–77.

41. Turley S, Poirot L, Hattori M, Benoist C, Mathis D. Physiological beta cell death triggers priming of self-reactive T cells by dendritic cells in a type-1 diabetes model. *J Exp Med* November 17, 2003;**198**(10):1527–37.

42. Roep BO, Peakman M. Diabetogenic T lymphocytes in human type 1 diabetes. *Curr Opin Immunol* December 2011;**23**(6):746–53.

43. Smith-Garvin JE, Koretzky GA, Jordan MS. T cell activation. *Annu Rev Immunol* 2009;**27**:591–619.

44. Rieck M, Arechiga A, Onengut-Gumuscu S, Greenbaum C, Concannon P, Buckner JH. Genetic variation in PTPN22 corresponds to altered function of T and B lymphocytes. *J Immunol* October 1, 2007;**179**(7):4704–10.

45. Bougneres PF, Landais P, Boisson C, Carel JC, Frament N, Boitard C, et al. Limited duration of remission of insulin dependency in children with recent overt type I diabetes treated with low-dose cyclosporin. *Diabetes* October 1990;**39**(10):1264–72.

46. Piemonti L, Maffi P, Monti L, Lampasona V, Perseghin G, Magistretti P, et al. Beta cell function during rapamycin monotherapy in long-term type 1 diabetes. *Diabetologia* February 2011;**54**(2):433–9.

47. Long SA, Rieck M, Sanda S, Bollyky JB, Samuels PL, Goland R, et al. Rapamycin/IL-2 combination therapy in patients with type 1 diabetes augments Tregs yet transiently impairs beta-cell function. *Diabetes* September 2012;**61**(9):2340–8.

48. Kjer-Nielsen L, Dunstone MA, Kostenko L, Ely LK, Beddoe T, Mifsud NA, et al. Crystal structure of the human T cell receptor CD3 epsilon gamma heterodimer complexed to the therapeutic mAb OKT3. *Proc Natl Acad Sci USA* May 18, 2004;**101**(20):7675–80.

49. Belghith M, Bluestone JA, Barriot S, Megret J, Bach JF, Chatenoud L. TGF-beta-dependent mechanisms mediate restoration of self-tolerance induced by antibodies to CD3 in overt autoimmune diabetes. *Nat Med* September 2003;**9**(9):1202–8.

50. Bisikirska B, Colgan J, Luban J, Bluestone JA, Herold KC. TCR stimulation with modified anti-CD3 mAb expands CD8+ T cell population and induces CD8+CD25+ Tregs. *J Clin Invest* November 2005;**115**(10):2904–13.

51. Herold KC, Gitelman SE, Masharani U, Hagopian W, Bisikirska B, Donaldson D, et al. A single course of anti-CD3 monoclonal antibody hOKT3gamma1(Ala-Ala) results in improvement in C-peptide responses and clinical parameters for at least 2 years after onset of type 1 diabetes. *Diabetes* June 2005;**54**(6):1763–9.

52. Keymeulen B, Walter M, Mathieu C, Kaufman L, Gorus F, Hilbrands R, et al. Four-year metabolic outcome of a randomised controlled CD3-antibody trial in recent-onset type 1 diabetic patients depends on their age and baseline residual beta cell mass. *Diabetologia* April 2010;**53**(4):614–23.

53. Herold KC, Gitelman S, Greenbaum C, Puck J, Hagopian W, Gottlieb P, et al. Treatment of patients with new onset type 1 diabetes with a single course of anti-CD3 mAb Teplizumab preserves insulin production for up to 5 years. *Clin Immunol* August 2009;**132**(2):166–73.

54. Sherry N, Hagopian W, Ludvigsson J, Jain SM, Wahlen J, Ferry Jr RJ, et al. Teplizumab for treatment of type 1 diabetes (Protege study): 1-year results from a randomised, placebo-controlled trial. *Lancet* August 6, 2011;**378**(9790):487–97.

55. Saudek F, Havrdova T, Boucek P, Karasova L, Novota P, Skibova J. Polyclonal anti-T-cell therapy for type 1 diabetes mellitus of recent onset. *Rev Diabet Stud* 2004;**1**(2):80–8.

56. Couri CE, Oliveira MC, Stracieri AB, Moraes DA, Pieroni F, Barros GM, et al. C-peptide levels and insulin independence following autologous nonmyeloablative hematopoietic stem cell transplantation in newly diagnosed type 1 diabetes mellitus. *JAMA* April 15, 2009;**301**(15):1573–9.

57. Voltarelli JC, Couri CE, Stracieri AB, Oliveira MC, Moraes DA, Pieroni F, et al. Autologous nonmyeloablative hematopoietic stem cell transplantation in newly diagnosed type 1 diabetes mellitus. *JAMA* April 11, 2007;**297**(14):1568–76.

58. Yoon JW, McClintock PR, Onodera T, Notkins AL. Virus-induced diabetes mellitus. XVIII. Inhibition by a nondiabetogenic variant of encephalomyocarditis virus. *J Exp Med* November 1, 1980;**152**(4):878–92.

59. Wen L, Ley RE, Volchkov PY, Stranges PB, Avanesyan L, Stonebraker AC, et al. Innate immunity and intestinal microbiota in the development of type 1 diabetes. *Nature* October 23, 2008;**455**(7216):1109–13.

60. Nejentsev S, Walker N, Riches D, Egholm M, Todd JA. Rare variants of IFIH1, a gene implicated in antiviral responses, protect against type 1 diabetes. *Science* April 17, 2009;**324**(5925):387–9.

61. O'Keeffe M, Brodnicki TC, Fancke B, Vremec D, Morahan G, Maraskovsky E, et al. Fms-like tyrosine kinase 3 ligand administration overcomes a genetically determined dendritic cell deficiency in NOD mice and protects against diabetes development. *Int Immunol* March 2005;**17**(3):307–14.

62. Poligone B, Weaver Jr DJ, Sen P, Baldwin Jr AS, Tisch R. Elevated NF-kappaB activation in nonobese diabetic mouse dendritic cells results in enhanced APC function. *J Immunol* January 1, 2002;**168**(1):188–96.

63. Kared H, Masson A, Adle-Biassette H, Bach JF, Chatenoud L, Zavala F. Treatment with granulocyte colony-stimulating factor prevents diabetes in NOD mice by recruiting plasmacytoid dendritic cells and functional CD4(+)CD25(+) regulatory T-cells. *Diabetes* January 2005;**54**(1):78–84.

64. Mallone R, Brezar V. To B or not to B: (anti)bodies of evidence on the crime scene of type 1 diabetes? *Diabetes* August 2011;**60**(8):2020–2.

65. Silveira PA, Johnson E, Chapman HD, Bui T, Tisch RM, Serreze DV. The preferential ability of B lymphocytes to act as diabetogenic APC in NOD mice depends on expression of self-antigen-specific immunoglobulin receptors. *Eur J Immunol* December 2002;**32**(12):3657–66.

66. Serreze DV, Chapman HD, Varnum DS, Hanson MS, Reifsnyder PC, Richard SD, et al. B lymphocytes are essential for the initiation of T cell-mediated autoimmune diabetes: analysis of a new "speed congenic" stock of NOD.Ig mu null mice. *J Exp Med* November 1, 1996;**184**(5):2049–53.

67. Hu CY, Rodriguez-Pinto D, Du W, Ahuja A, Henegariu O, Wong FS, et al. Treatment with CD20-specific antibody prevents and reverses autoimmune diabetes in mice. *J Clin Invest* December 2007;**117**(12):3857–67.

68. Pescovitz MD, Greenbaum CJ, Krause-Steinrauf H, Becker DJ, Gitelman SE, Goland R, et al. Rituximab, B-lymphocyte depletion, and preservation of beta-cell function. *N Engl J Med* November 26, 2009;**361**(22):2143–52.

69. Zotos D, Coquet JM, Zhang Y, Light A, D'Costa K, Kallies A, et al. IL-21 regulates germinal center B cell differentiation and proliferation through a B cell-intrinsic mechanism. *J Exp Med* February 15, 2010;**207**(2):365–78.

70. Eggena MP, Walker LS, Nagabhushanam V, Barron L, Chodos A, Abbas AK. Cooperative roles of CTLA-4 and regulatory T cells in tolerance to an islet cell antigen. *J Exp Med* June 21, 2004;**199**(12):1725–30.

71. Fife BT, Griffin MD, Abbas AK, Locksley RM, Bluestone JA. Inhibition of T cell activation and autoimmune diabetes using a B cell surface-linked CTLA-4 agonist. *J Clin Invest* August 2006;**116**(8):2252–61.

72. Ueda H, Howson JM, Esposito L, Heward J, Snook H, Chamberlain G, et al. Association of the T-cell regulatory gene CTLA4 with susceptibility to autoimmune disease. *Nature* May 29, 2003;**423**(6939):506–11.

73. Orban T, Bundy B, Becker DJ, DiMeglio LA, Gitelman SE, Goland R, et al. Co-stimulation modulation with abatacept in patients with recent-onset type 1 diabetes: a randomised, double-blind, placebo-controlled trial. *Lancet* July 30, 2011;**378**(9789):412–9.

74. Prevot N, Briet C, Lassmann H, Tardivel I, Roy E, Morin J, et al. Abrogation of ICOS/ICOS ligand costimulation in NOD mice results in autoimmune deviation toward the neuromuscular system. *Eur J Immunol* August 2010;**40**(8):2267–76.

75. Nanji SA, Hancock WW, Luo B, Schur CD, Pawlick RL, Zhu LF, et al. Costimulation blockade of both inducible costimulator and CD40 ligand induces dominant tolerance to islet allografts and prevents spontaneous autoimmune diabetes in the NOD mouse. *Diabetes* January 2006;**55**(1):27–33.

76. Kornete M, Sgouroudis E, Piccirillo CA. ICOS-dependent homeostasis and function of Foxp3⁺ regulatory T cells in islets of nonobese diabetic mice. *J Immunol* February 1, 2012;**188**(3):1064–74.

77. Wherry EJ. T cell exhaustion. *Nat Immunol* June 2011;**12**(6):492–9.

78. Ansari MJ, Salama AD, Chitnis T, Smith RN, Yagita H, Akiba H, et al. The programmed death-1 (PD-1) pathway regulates autoimmune diabetes in nonobese diabetic (NOD) mice. *J Exp Med* July 7, 2003;**198**(1):63–9.

79. Wang J, Yoshida T, Nakaki F, Hiai H, Okazaki T, Honjo T. Establishment of NOD-Pdcd1−/− mice as an efficient animal model of type I diabetes. *Proc Natl Acad Sci USA* August 16, 2005;**102**(33):11823–8.

80. Keir ME, Liang SC, Guleria I, Latchman YE, Qipo A, Albacker LA, et al. Tissue expression of PD-L1 mediates peripheral T cell tolerance. *J Exp Med* April 17, 2006;**203**(4):883–95.

81. Okazaki T, Okazaki IM, Wang J, Sugiura D, Nakaki F, Yoshida T, et al. PD-1 and LAG-3 inhibitory co-receptors act synergistically to prevent autoimmunity in mice. *J Exp Med* February 14, 2011;**208**(2):395–407.

82. Wang C, McPherson AJ, Jones RB, Kawamura KS, Lin GH, Lang PA, et al. Loss of the signaling adaptor TRAF1 causes CD8+ T cell dysregulation during human and murine chronic infection. *J Exp Med* January 16, 2012;**209**(1):77–91.

83. Sytwu HK, Lin WD, Roffler SR, Hung JT, Sung HS, Wang CH, et al. Anti-4-1BB-based immunotherapy for autoimmune diabetes: lessons from a transgenic non-obese diabetic (NOD) model. *J Autoimmun* November 2003;**21**(3):247–54.

84. Roth GA, Zuckermann A, Klepetko W, Wolner E, Ankersmit HJ, Moser B, et al. Thrombophilia associated with anti-CD154 monoclonal antibody treatment and its prophylaxis in nonhuman primates. *Transplantation* October 27, 2004;**78**(8):1238–9.

85. Hunig T. The storm has cleared: lessons from the CD28 superagonist TGN1412 trial. *Nat Rev Immunol* May 2012;**12**(5):317–8.

86. Tang Q, Adams JY, Penaranda C, Melli K, Piaggio E, Sgouroudis E, et al. Central role of defective interleukin-2 production in the triggering of islet autoimmune destruction. *Immunity* May 2008;**28**(5):687–97.

87. Grinberg-Bleyer Y, Baeyens A, You S, Elhage R, Fourcade G, Gregoire S, et al. IL-2 reverses established type 1 diabetes in NOD mice by a local effect on pancreatic regulatory T cells. *J Exp Med* August 30, 2010;**207**(9):1871–8.

88. Lindley S, Dayan CM, Bishop A, Roep BO, Peakman M, Tree TI. Defective suppressor function in CD4(+)CD25(+) T-cells from patients with type 1 diabetes. *Diabetes* January 2005;**54**(1):92–9.

89. Long SA, Cerosaletti K, Bollyky PL, Tatum M, Shilling H, Zhang S, et al. Defects in IL-2R signaling contribute to diminished maintenance of FOXP3 expression in CD4(+)CD25(+) regulatory T-cells of type 1 diabetic subjects. *Diabetes* February 2010;**59**(2):407–15.

90. Schneider A, Rieck M, Sanda S, Pihoker C, Greenbaum C, Buckner JH. The effector T cells of diabetic subjects are resistant to regulation via CD4+ FOXP3+ regulatory T cells. *J Immunol* November 15, 2008;**181**(10):7350–5.

91. Sutherland AP, Van BT, Wurster AL, Suto A, Michaud M, Zhang D, et al. Interleukin-21 is required for the development of type 1 diabetes in NOD mice. *Diabetes* May 2009;**58**(5):1144–55.

92. McGuire HM, Walters S, Vogelzang A, Lee CM, Webster KE, Sprent J, et al. Interleukin-21 is critically required in autoimmune and allogeneic responses to islet tissue in murine models. *Diabetes* March 2011;**60**(3):867–75.

93. McGuire HM, Vogelzang A, Ma CS, Hughes WE, Silveira PA, Tangye SG, et al. A subset of interleukin-21+ chemokine receptor CCR9+ T helper cells target accessory organs of the digestive system in autoimmunity. *Immunity* April 22, 2011;**34**(4):602–15.

94. van Belle TL, Nierkens S, Arens R, von Herrath MG. Interleukin-21 receptor-mediated signals control autoreactive T cell infiltration in pancreatic islets. *Immunity* June 29, 2012;**36**(6):1060–72.

95. Xie A, Buras E, Xia J, Chen W. The emerging role of interleukin-21 in transplantation. *J Clin Cell Immunol* August 24, 2013;Suppl 9(2):1–7.

96. Arif S, Moore F, Marks K, Bouckenooghe T, Dayan CM, Planas R, et al. Peripheral and islet interleukin-17 pathway activation characterizes human autoimmune diabetes and promotes cytokine-mediated beta-cell death. *Diabetes* August 2011;**60**(8):2112–9.

97. Emamaullee JA, Davis J, Merani S, Toso C, Elliott JF, Thiesen A, et al. Inhibition of Th17 cells regulates autoimmune diabetes in NOD mice. *Diabetes* June 2009;**58**(6):1302–11.

98. Bendtzen K, Mandrup-Poulsen T, Nerup J, Nielsen JH, Dinarello CA, Svenson M. Cytotoxicity of human pI 7 interleukin-1 for pancreatic islets of Langerhans. *Science* June 20, 1986;**232**(4757):1545–7.

99. Sumpter KM, Adhikari S, Grishman EK, White PC. Preliminary studies related to anti-interleukin-1beta therapy in children with newly diagnosed type 1 diabetes. *Pediatr Diabetes* November 2011;**12**(7):656–67.

100. Ablamunits V, Henegariu O, Hansen JB, Opare-Addo L, Preston-Hurlburt P, Santamaria P, et al. Synergistic reversal of type 1 diabetes in NOD mice with anti-CD3 and interleukin-1 blockade: evidence of improved immune regulation. *Diabetes* January 2012;**61**(1):145–54.

Chapter 8

Gene Therapy for Diabetes

Yisheng Yang[1,2] **and Lawrence Chan**[1,2,3]

[1]*Diabetes Research Center, Houston, TX, USA;* [2]*Division of Diabetes, Endocrinology and Metabolism, Department of Medicine, Baylor College of Medicine, Houston, TX, USA;* [3]*Department of Molecular and Cellular Biology, Baylor College of Medicine, Houston, TX, USA*

LIST OF ABBREVIATIONS

ABT Adenosine kinase inhibitors
Arx Aristaless-related homeobox
BMP Active bone morphogenetic protein
CAII Carbonic anhydrase II
cdk6 Cyclin-dependent kinase 6
Cpa1 Carboxypeptidase A1
DDC Dihydrocollidine
ESCs Embryonic stem cells
FBS Fetal bovine serum
FGAd First generation adenovirus
Fgf10 Fibroblast growth factor 10
GSIS Glucose-stimulated insulin secretion
GWAS Genome-wide association studies
HDAd Helper-dependent adenovirus
Hes1 Hairy and enhancer of split1
HFD High-fat diet
HGF Hepatocyte growth factor
iPSCs Induced pluripotent stem cells
KAAD Keto-N-aminoethyl-aminocaproyl-dihydrocinnamoyl
Klf4 Krüppel-like factor 4
MODY Maturity onset diabetes of the young
MSCs Mesenchymal stem cells
mtDNA Mitochondrial DNA
mTOR Mammalian target of rapamycin
Oct4 Octamer-4
Pax4 Paired box gene 4
PDL Pancreatic duct ligation
RA Retinoic acid
Shh Sonic hedgehog
Stat3 Signal transducer and activator of transcription 3
STZ Streptozotocin
TCF7 Transcription factor 7
TGFβ Transforming growth factor-β

8.1 INTRODUCTION

The prevalence of diabetes mellitus has been on the rise globally. It is a major health issue in well-developed countries and a major threat to the health and welfare of developing countries. Diabetes occurs when there is absence or insufficient circulating insulin to drive glucose into the cell and maintain euglycemia. Pancreatic β cells are the source of circulating insulin. Type 1 diabetes (T1D) is caused by absolute insulin deficiency, and type 2 diabetes (T2D) by relative insulin

Translating Gene Therapy to the Clinic. http://dx.doi.org/10.1016/B978-0-12-800563-7.00008-7

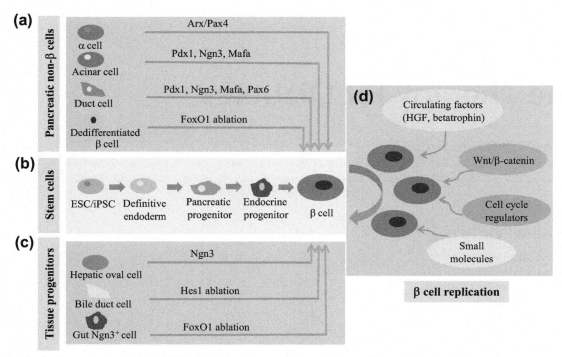

FIGURE 8.1 **Strategies to generate new β cells.** New β cells can be generated from (a) pancreatic non-β cells, (b) pluripotent stem cells including ESCs and iPSCs, (c) tissue progenitor cells, and (d) β cells via replication.

deficiency in the presence of insulin resistance. Replacement of the missing insulin by the administration of exogenous insulin is the standard treatment for T1D; it is also an important adjunct therapy for T2D. However, insulin therapy cannot mimic the secretion pattern of endogenous insulin in response to changes in blood glucose. Diabetic individuals who are treated with exogenous insulin invariably develop hyper- and hypoglycemia at least some of the time, and they are not completely protected from chronic diabetic complications. Despite some limited initial success,[1] islet transplantation is not a cure; most transplanted islets eventually fail during long-term follow-up. Furthermore, the availability of islet donors is severely limiting, and transplant recipients must receive lifetime toxic immunosuppressive therapy to prevent rejection.[2] These limitations have prompted research on the generation of new β cells for replacement therapy. This approach is applicable to T1D individuals, who produce little or no insulin. The knowledge is also potentially useful in the treatment of subgroups of T2D individuals who exhibit severe β cell loss and dysfunction. Functional β cells can be generated by a number of strategies outlined in Figure 8.1.

8.2 GENERATION OF β CELLS FROM PANCREATIC MATURE NON-β CELLS

Reprogramming the existing pancreatic non-β cells to transdifferentiate into new β cells has been widely explored. The rationale behind this approach is that these cells share close developmental lineage with β cells, and the reprogramming barrier may be lower than in the developmentally less well-related cells.

8.2.1 Pancreatic α Cells

A close lineage relationship provides an opportunity for directed transdifferentiation between α cells and β cells. The antagonistic functions of the transcription factors aristaless-related homeobox (Arx) and paired box gene 4 (Pax4) serve as a switch in the maintenance of α cell and β cell identity, respectively.[3] The repression of Arx by DNA methyltransferase 1-dependent methylation is required for the maintenance of β cell identity in mice.[4] Forced expression of Arx is sufficient to transdifferentiate β cells to α cells.[5] Ablation of *Pax4* leads to the loss of β cells and a concomitant increase in the number of α cells,[6] which raises the possibility that forced expression of Pax4 in α cells may have the reciprocal effect. Indeed, Collombat et al.[7] discovered that ectopic expression of *Pax4* is sufficient to convert α cells to β cells in vivo. Forced Pax4 expression in α cells induces hypoglucagonemia, which induces the conversion of Ngn3+ progenitor cells into α cells that subsequently acquire a β cell phenotype upon Pax4 ectopic expression. The animals display a 10-fold increase in β cell mass after 20 months but, inexplicably, eventually develop diabetes.[7]

The potential utility of α-to-β cell conversion has been reinforced by a report that conditions of near complete β cell depletion are sufficient to lead to a similar change in cell identity. In a transgenic mouse model that permits α cell lineage tracing, near total β cell ablation using the diphtheria toxin receptor system leads to the regeneration of β cells but, unexpectedly, β cell replenishment is mediated by α cell transdifferentiation rather than by β cell replication.[8] In another β cell ablation model using simultaneous pancreatic duct ligation (PDL) and alloxan, both α-to-β cell conversion and β cell replication are involved in the restoration of β cell mass.[9]

In 2013, a recent genome-wide study showed that α cells harbor bivalent chromatin signatures marked by the activating H3K4me3 and repressing H3K27me3 histone modifications, particularly at the loci of genes that are active in β cells, such as *Pdx1* and *Mafa*.[10] The presence of bivalent signatures is characteristic of progenitor cells and may suggest that α cells carry the plasticity to be reprogrammed into β cells. Interestingly, treatment of cultured human islets with a histone methyltransferase inhibitor results in insulin and Pdx1 expression in glucagon+ cells, suggesting that epigenetic manipulation may provide an avenue for α-to-β cell reprogramming.[10]

8.2.2 Pancreatic Acinar Cells

Pancreatic acinar cells are the major cell type in the pancreas; they share a similar lineage history with β cells during embryogenesis. Therefore, it is reasonable to expect that these cells may be amenable to reprogramming to β cells. Zhou et al.[11] discovered that transduction of acinar cells with the three transcription factors, Pdx1, Ngn3, and Mafa reprograms these cells to β cells in vivo. Lineage tracing confirms that the insulin-positive cells have been derived from carboxypeptidase A1 (Cpa1)-positive mature acinar cells. They also coexpress mature β cell markers such as Nkx6.1, Glut2, and glucokinase, but no longer express acinar cell markers such as Ptf1a and amylase. Sufficient reprogrammed insulin-positive cells can be produced by transduction in situ to improve the glycemia of streptozotocin (STZ)-induced diabetes in mice.[11]

While adenovirus-mediated transduction with the three transcription factors is sufficient to reprogram acinar cells to functional β cells in vivo, the same manipulation fails to fully reprogram the acinar cells to β cells in vitro.[12] One concern in testing the in vivo strategy to generate new β cells in humans is the risk of pancreatitis caused by the manipulation.

A recent study claimed that Ptf1a+ acinar cells can transdifferentiate into Ngn3+ progenitor cells after PDL injury; they would further differentiate into endocrine β cells that express mature β cell markers, such as Pdx1, Nkx6.1, and Mafa.[13] According to this report, acinar cells may spontaneously retrieve embryonic multipotency after injury and transdifferentiate into mature β cells. However, other groups have failed to identify any newly generated β cells in the PDL injury models.[14,15] The reason for these discrepancies is unknown.

Baeyens et al.[16] recently reported that transient simultaneous administration of epidermal growth factor and ciliary neurotrophic factor to adult mice with alloxan-induced diabetes efficiently stimulates the reprogramming of acinar cells to β-like cells. These newly generated β-like cells durably reinstate normal glycemic control and display a degree of glucose-stimulated insulin secretion (GSIS). Activation of signal transducer and activator of transcription 3 (Stat3) signaling and Ngn3 expression in acinar cells are required for the complete transdifferentiation of acinar cells to β-like cells in this model.

As acinar cells can be reprogrammed to β cells by genetic manipulation in vivo, it is possible that one can treat cultured pancreatic acinar cells with different cocktails and accomplish reprogramming in vitro. Unfortunately, it has been challenging to use cultured pancreatic acinar cells for this type of experiment, partly because the exocrine component undergoes rapid dedifferentiation (see section 8.2.4 for details) under standard culture conditions. Lima et al.[17] reported that addition of a combination of rho-kinase inhibitor (Y27632) and transforming growth factor β1 inhibitor (SB431542) to exocrine enriched pancreatic cells cultured in vitro slows down the dedifferentiation process.

8.2.3 Pancreatic Ductal Cells

During embryonic development, the pancreatic epithelium undergoes dynamic structural segregation of tip and trunk domains at the onset of the secondary transition. Lineage-tracing experiments showed that the tip domain (marked by the expression of Ptf1a, c-Myc, and Cpa) is quickly restricted to an acinar cell fate, whereas the trunk domain (marked by the expression of Nkx6.1/6.2, Sox9, Hnf1β, onecut1, Prox1, and Hes1) predominantly gives rise to the endocrine and ductal cell lineages.[18]

Although pancreatic β cells are known to expand by self-replication in postnatal life,[19] the contribution of the ductal progenitors in the process remains controversial, discrepant data being reported by investigators using different lineage-tracing animal models.[20] Inada et al.[21] used an inducible human carbonic anhydrase II (CAII)-Cre to cross with ROSA26 loxP-Stop-loxP LacZ reporter mice to trace the lineage of ductal progenitors and found that CAII-expressing cells within the pancreas act as progenitors that give rise to both new islets and acini normally after birth and after PDL injury. In contrast, Solar et al.,[22] who used Hnf1β as a ductal-specific lineage tracing marker, concluded that Hnf1β+ cells contribute to β cells during embryogenesis, but make no significant contribution to acinar or endocrine cells postnatally under basal conditions or after PDL injury. In another study using mucin gene *Muc1* expression to track exocrine and ductal cells, Kopinke et al.[23] observed that Muc1+ cells give rise to β cells and other islet cells in utero, but do not detectably contribute to islet cells after birth, whereas Muc1 lineage-labeled cells contribute to the exocrine compartment even after birth. It is noteworthy that the lineage tracing systems used in all these studies have limitations, which may explain some of the discrepancies between laboratories.[20]

Whether ductal cells are a postnatal source of β cells remains an open question. It is clear, however, that under appropriate conditions, ductal cells can be converted into insulin-producing cells in vitro. Recently, Lee et al.[24] reported that CD133+ human adult pancreatic duct cells can be successfully transdifferentiated into insulin-producing cells after transduction with the four islet developmental factors Pdx1, Ngn3, Mafa, and Pax6. These human insulin-producing cells exhibit features of functional neonatal β cells including high-level preproinsulin expression, proinsulin processing and dense-core granule formation as well as secretion of insulin and C-peptide in response to glucose and ATP-sensitive potassium channel stimulants. Despite the fact that many of these cells created in vitro die shortly after transplantation in NOD-*SCID* gamma mice, a substantial proportion undergo further maturation in the host animal and secrete insulin when the animal receives an acute glucose challenge in vivo.[24]

8.2.4 Pancreatic Dedifferentiated β Cells

In 2004, Gershengorn et al.[25] observed that human insulin-positive cells maintained in serum-containing media undergo dedifferentiation, characterized by expression of the mesenchymal cell marker vimentin, exponential expansion, and retention of the ability to return to hormone-positive cells in serum-free media. Lentiviral-based lineage tracing showed a substantial proliferative capacity in dedifferentiated human islet β cells, whereas a much lower proliferative potential is observed in dedifferentiated mouse islet β cells.[26] In another lineage tracing study, Weinberg et al.[27] came to the same conclusion for mouse islet β cells, indicating that significant differences exist in β cell proliferative behavior in a species-specific manner.[26]

It has been suggested that pancreatic β cells may dedifferentiate to progenitor-like cells, a process that may contribute to the pathogenesis of β cell failure in T2D.[28] Lineage tracing experiments demonstrated that loss of β cell mass in *FoxO1* deficient mice is caused by β cell dedifferentiation.[29] Dedifferentiated β cells can revert to progenitor-like cells that express Ngn3, Oct4, Nanog, and L-Myc and fail to produce insulin.[30] Interestingly, it is proposed that these dedifferentiated β cells could be used to reconstitute functional β cells by expanding these progenitor-like cells and redifferentiating them. Efrat and colleagues identified that inhibition of Notch pathway or epithelial–mesenchymal transition regulatory factor SLUG (SNAI2) facilitates redifferentiation of expanded human pancreatic β cell-derived cells.[31]

8.3 GENERATION OF β CELLS FROM TISSUE PROGENITOR CELLS

Progenitor cells have been widely identified in different tissues in adults and have a tendency to differentiate into a specific type of cell under specific conditions. Unlike stem cells, progenitor cells have more limited proliferative capacity.[32] An alternate strategy to generate new β cells is by nudging these progenitor cells toward β cell lineage by gene transfer.

8.3.1 Transdetermination of Liver Progenitor Cells into β Cells

The liver and pancreas both arise from the endodermal cells in the embryonic foregut and share common progenitors. The adult liver harbors facultative progenitors, called "oval cells," that can be activated in response to certain types of injury, especially under conditions of impaired hepatocyte proliferation, for example, when rodents are fed a 3,5-diethoxycarbonyl-1,4-dihydrocollidine (DDC) diet.[33] The oval cells, presenting as a heterogeneous population, are bipotential in that they can differentiate into both biliary epithelial cells and hepatocytes.[33] Hepatic oval cells have been shown to become insulin-producing cells under appropriate circumstances. When cultured in vitro in a high glucose environment for prolonged periods, oval cells acquire the capability to produce insulin.[34] STZ-induced diabetes in rodents fed a DDC diet can lead to low-level insulin production by the newly formed oval cells.[35]

Adenovirus-mediated transfer of Pdx1 was shown to induce the expression of insulin in the liver of STZ-induced diabetic mice.[36,37] This was true whether the gene was transferred by first generation adenovirus (FGAd) or helper-dependent adenovirus (HDAd), an adenovirus vector that is devoid of all adenoviral protein genes. FGAd caused a severe toxic reaction, and the experiments were terminated eight days later.[36] In mice treated with HDAd-Pdx1, high-level Pdx1 transgene expression also led to severe fulminant hepatitis four weeks later.[37] Further analysis revealed that hepatic Pdx1 expression led to the appearance of cells that express both insulin and pancreatic acinar enzymes such as trypsin. It appears that, as a pancreas lineage-defining transcription factor, Pdx1 induces the production of both pancreatic exocrine and endocrine cells in the liver of adult diabetic mice, though the cellular origin of these ectopic pancreatic cells was not determined in this study.

To circumvent the undesirable exocrine-inducing action of Pdx1, Kojima et al.[37] transduced Beta2/NeuroD1, a transcription factor downstream of Pdx1, into the liver and achieved robust islet neogenesis that encompassed cells that produced all four major islet hormones (insulin, glucagon, somatostatin, and pancreatic polypeptide). The insulin secretion from the liver exhibited GSIS and stably reversed diabetes in STZ-treated mice. However, the origin of the newly formed insulin-producing cells after HDAd-Beta2/NeuroD1 treatment was not addressed in this study.

Subsequently, Yechoor et al.[38] used HDAd to deliver the genes for *Ngn3* and betacellulin (a β cell growth factor) to the liver of STZ-diabetic mice, and found neoislet formation in the periportal region of the treated animals. The newly formed islets encompassed cells that produced the four major islet hormones, especially insulin in a robust fashion. The periportal neoislets exhibit GSIS and stably reverse diabetes in STZ-diabetic mice. They also exhibit an ultrastructure and a transcription profile that is essentially identical to those of pancreatic β cells. Lineage tracing in combination with histological analyses indicate that the insulin-producing cells arose from a progenitor cell population, the hepatic oval cells.[38] The process of differentiation of an adult stem cell or progenitor cell population into developmentally related cells is called transdetermination.[39] These proof-of-concept experiments in mice could lay the foundation for future research exploiting the process of transdetermination to treat diabetes.

8.3.2 Generation of β Cells from Hepatic Bile Ducts

In pancreatic progenitor cells, the protein hairy and enhancer of split1 (Hes1), a mediator of Notch signaling, inhibits the expression of Ptf1a, which controls exocrine cell differentiation, and of Ngn3, which drives islet differentiation.[18] Foci of ectopic pancreatic tissue in the extrahepatic bile ducts have been detected in *Hes1*-deficient mice.[40] Dutton et al.[41] found that a small population of endocrine cells spontaneously appears in the extrahepatic bile ducts, suggesting that the latter may be a new source from which to generate β cells for the treatment of diabetes. Using an adenoviral polycistronic vector, Banga et al.[42] transferred the *Pdx1*, *Ngn3*, and *Mafa* genes in the livers of STZ-diabetic NOD-*SCID* mice and successfully reversed hyperglycemia of these animals. Interestingly, marked ectopic duct-like structures that express a variety of β cell markers were observed in the livers of these mice. The insulin-positive cells display GSIS and dense-core granules characteristic of rodent β cells. Lineage tracing experiments showed that they are derived from Sox9+ cells,[42] which are normally associated with small bile ducts in the periportal region, strongly suggesting in vivo reprogramming of cells in the biliary tree to a β cell-like phenotype. Recently, multipotent progenitor cells have been isolated from human biliary ducts that differentiate toward β cell-like cells in conditional media. These insulin-producing cells also express C-peptide, Pdx1, glucagon, and somatostatin, and can have glucose-lowering effects in STZ-diabetic mice.[43,44]

8.3.3 Generation of β Cells from Gut Neuroendocrine Cells

Despite their common endodermal origin, pancreatic and gut Ngn3+ progenitor cells have distinct developmental fates and lifespans.[45] In the pancreas, Ngn3+ progenitors are formed during embryonic development and their numbers are drastically decreased to a very low level after birth.[46] In the gut, Ngn3+ progenitors instead continually arise from gut stem cells and contribute to the repopulation of the high-turnover enteroendocrine population.[45] A recent study showed that ablation of *FoxO1* in Ngn3+ enteroendocrine progenitor cells redirects them toward gut insulin-positive cells.[29] These cells display the characteristics of mature β cells, e.g., expression of mature β cell markers, the ability to secrete bioactive insulin and C-peptide in response to glucose and sulfonylureas. The ectopic gut-produced insulin was shown to ameliorate diabetes in STZ-treated *FoxO1* deficient mice.[29] The plasticity of enteroendocrine cells demonstrated in *FoxO1* ablated cells raises the possibility for gut cell-based therapies for diabetes.

8.4 GENERATION OF β CELLS FROM STEM CELLS

Different types of stem cells including embryonic stem cells (ESCs), induced pluripotent stem cells (iPSCs) and mesenchymal stem cells (MSCs) have been exploited as starting material for directed differentiation to a β cell phenotype. The major benefit of using stem cells to generate functional islet cells is their unique ability to divide indefinitely in culture while maintaining an undifferentiated fate in a process known as self-renewal.[47]

A set of criteria have been proposed that define MSCs: (1) adherence to plastic in standard culture conditions; (2) the simultaneous presence of CD105, CD73, CD90, and absence of CD45, CD34, CD14, or CD11b, CD79a, or CD19 and HLA class II; (3) the ability to differentiate to osteoblasts, adipocytes, and chondroblasts under standard in vitro differentiation conditions.[48] Numerous attempts have been made to differentiate MSCs from distinct tissue origins to insulin-producing cells; unfortunately, the majority of reports have used insulin production as the sole criterion in their assessment and as evidence of success. Furthermore, considerable uncertainty remains with respect to the identity and heterogeneity of MSCs (reviewed by Dominguez-Bendala et al.[49]). Recently, Carlotti et al.[50] characterized a subset of human islet-derived MSCs, possibly related to pericytic CD105+/CD90+ progenitors. When these endocrine hormone-negative cells are transferred to a serum-free medium, they aggregate to form clusters and develop the capacity to produce insulin, glucagon, and somatostatin. These cells appear to fall short of what is required for efficient GSIS and therapy in vivo. Pluripotent stem cells, particularly iPSCs, have received much more attention than MSCs as a potential source of pancreatic β cells.[51]

8.4.1 Embryonic Stem Cells

Establishing executable differentiation protocols to direct pluripotent ESCs to functional somatic cells has been a major focus of research in the field of regenerative medicine (Figure 8.2). Baetge and colleagues established an efficient protocol for directed differentiation of ESCs to pancreatic endoderm, which was recapitulated as four stages of embryonic development to obtain glucose-responsive, insulin-secreting cells in vitro.[52,53]

8.4.1.1 Nodal and Wnt Signals Drive Definitive Endoderm Formation from ESCs

During gastrulation, Nodal and Wnt signals coordinately regulate patterning of the streak and the formation of midline organizing tissues.[54] Nodal, a transforming growth factor-β (TGFβ) ligand, displays antagonism in mesoderm and endoderm induction within the embryo. High levels of Nodal signal direct mesendoderm cells toward an endoderm fate, and to a mesoderm fate at lower doses.[55] Activin A, another TGFβ ligand that mimics the action of Nodal, was successfully used in the first efficient protocol for generating definitive endoderm in vitro from ESCs.[52] The differentiation of ESCs to definitive endoderm is achieved by culturing pluripotent cells at a low concentration of fetal bovine serum (FBS) for three days in the presence of a high level of Activin A (100 ng/mL). Additional Wnt3a on the first day of Activin A exposure facilitates mesendoderm specification and definitive endoderm formation, characterized by the expression of Sox17, Foxa2, CER, and Cxcr4. The differentiation of pluripotent cells to definitive endoderm is a critical step for the efficient differentiation of pancreatic endoderm; therefore, future protocols may benefit from incorporating the small molecules IDE1 and IDE2, which are known inducers of definitive endoderm via activating the Activin/Nodal signaling pathway in mouse and human ESCs.[56]

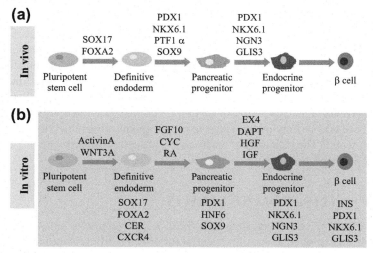

FIGURE 8.2 Schematic representation of differentiation procedure and the expression of key markers of (a) normal β cell development and (b) directed differentiation of ESCs/iPSCs toward β-like cells.

8.4.1.2 Removal of Activin A Promotes Transition to Primitive Gut Endoderm

During embryogenesis, sonic hedgehog (Shh) plays an essential role in primitive gut tube endoderm development; however, Shh must be repressed later in the posterior foregut for pancreatic buds to be specified.[57,58] In vitro, this process could be mimicked by returning FBS and removing Activin A from the culture for three days. The removal of Activin A is critical for transition from definitive endoderm to gut tube endoderm. At this stage, the expression of gut tube markers such as Hnf1β and Hnf4α is increased; in contrast, the expression of definitive endoderm markers CER and Cxcr4 is reduced.[52] Notably, Sox17 is persistently expressed in Hnf1β-positive cells, indicating their endodermal origin. Although Activin A removal alone is sufficient to induce the expression of gut tube markers Hnf1β and Hnf4α, D'Amour et al.[52] observed that the addition of the fibroblast growth factor 10 (Fgf10) and hedgehog signaling inhibitor 3-keto-N-aminoethyl-aminocaproyl-dihydrocinnamoyl (KAAD)-cyclopamine at this stage significantly promotes β cell differentiation as manifested in end stage differentiated cells marked by a massive increase in insulin production.

8.4.1.3 Noggin and Retinoic Acid Promote Pancreas Specification

The foregut endoderm gives rise to the pancreas, stomach, lungs, liver, and biliary apparatus in embryos. Active bone morphogenetic protein (BMP) signaling is associated with hepatic induction from foregut endoderm, while BMP inhibition in the dorsal endoderm favors a pancreatic fate.[59] Retinoic acid (RA) is sufficient to induce the expression of the pancreatic marker gene Pdx1 in pancreatic endoderm from *Xenopus* to humans.[60,61] The signaling environment could be recreated in vitro by supplementing the previously differentiated cells with RA, Fgf10, and KAAD-cyclopamine. Recent studies indicate that Noggin, a BMP inhibitor, combined with RA, succeeded in producing massive Pdx1+ pancreatic progenitor cells.[62,63] Properly differentiated cells at this stage express significant levels of Pdx1, Hnf6, and Sox9. Importantly, RA treatment at this stage is required for later expression of insulin and Ngn3.[52]

8.4.1.4 Amplification and Maturation of Pancreatic Endoderm and Endocrine Precursors

During pancreas development, the expansion of Pdx1+ pancreatic progenitor cells is driven by signals from the pancreatic mesenchyme. Mesenchymal Fgf10 plays a crucial role in inducing progenitor cell proliferation and arresting their terminal differentiation.[64,65] Both Fgf10 and epidermal growth factor treatments facilitate the expansion of human ESC-derived Pdx1+ pancreatic progenitor cells.[62,63]

It has been a major challenge to achieve the subsequent transition from pancreatic Pdx1+ progenitors to endocrine Ngn3+ progenitors. The paucity of data on proper signals to specify β cell fate underlies some of this difficulty. Recent genome-wide association studies (GWAS) in adult populations identified GLIS family zinc finger 3 (GLIS3) as a candidate gene for T1D,[66] as well as a gene associated with T2D.[67,68] The *Glis3* gene appears to control the specification of endocrine progenitors in mice by transactivating *Ngn3*, collaborating with Hnf6 and Foxa2 during early pancreas development.[69] This finding opens an avenue for a translational effort to promote the formation of endocrine progenitors by overexpressing *Glis3* and *Hnf6* or *Foxa2* properly.

Sander and colleagues reported that aberrant histone modification patterns are associated with inefficient induction of endocrine genes during terminal differentiation into endocrine cells in vitro. Notably, several transcription factors such as the Hnf1α, transcription factor 7 (TCF7) and hematopoietically-expressed homeobox protein HHEX, are poorly induced, which could underlie the inappropriate chromatin remodeling of β cell genes including insulin and urocotin 3 during in vitro differentiation.[70] It is unclear whether direct manipulations of these transcription factors or chromatin-modifying enzymes would facilitate production of fully differentiated β cells in vitro.

8.4.2 Induced Pluripotent Stem Cells

The term "induced pluripotency" refers to the reprogramming of a somatic cell into a cell with pluripotent potential. Shinya Yamanaka first showed that somatic cells can be reprogrammed into iPSCs directly in culture through forced expression of four defined factors: Octamer-4 (Oct4), Sox2, Krüppel-like factor 4 (Klf4), and c-Myc.[71] Shortly afterwards, studies demonstrated that somatic cells can be reprogrammed by substituting Nanog and Lin28 for Klf4 and c-Myc.[72,73] Recently, Yamanaka and colleagues discovered that GLIS1 can replace c-Myc and markedly enhances the generation of iPSCs from both mouse and human fibroblasts.[74]

The iPSCs exhibit many ESC-like properties such as high telomerase activity, hypomethylation of endogenous pluripotency gene promoters, as well as reactivation of the X-chromosome in XX cell lines.[75] In the future, the generation of iPSCs in vitro likely will be favored over ESCs for tissue transplantation for several reasons. First, iPSCs are patient-specific, thereby minimizing the likelihood of immune rejection after transplantation. Second, the induction process is becoming increasingly efficient with fewer technical limitations. Finally, the production of iPSCs does not require the use of oocytes or the destruction of an embryo, circumventing major ethical concerns associated with ESC production.

Importantly, iPSCs can be generated from patients with T1D by reprogramming their adult fibroblasts with three transcription factors (Oct4, Sox2, and Klf4). T1D-specific iPSCs have the hallmarks of pluripotency and can be differentiated into insulin-producing cells,[76] although efficiency is low and clonal variance has been noted.[77] In addition, patient-specific iPSCs have been generated from patients with maturity onset diabetes of the young (MODY) mutations[78] and mitochondrial DNA (mtDNA) mutations.[79] In time, these and other patient-derived iPSC lines may enable personalized medicine for diabetic patients at a level not previously achievable.

One must overcome some hurdles before iPSCs can be used to treat T1D patients. First, reprogramming typically requires genomic integration to drive exogenous gene expression. Three out of the four reprogramming factors—c-Myc, Oct4, and Klf4—initially identified are known oncogenes or have oncogenic potential.[80,81] It is unlikely that they will be used to generate iPSCs for reimplantation in a patient. It has been reported that the transgenes undergo silencing after reprogramming has occurred[71]; nonetheless, there is continued concern over the risk of reactivation and subsequent transformation of transplanted tissue. Furthermore, viral-mediated genomic integration may also cause insertional mutagenesis. Currently, nonintegrating or excisable systems, RNA delivery, or protein transduction methods have been developed to circumvent this potential drawback.[82–84]

The second obstacle is that iPSCs, like ESCs, exhibit enormous tumorigenic potential.[85] Different strategies (e.g., optimizing the directed-differentiation protocols and use of cell-sorting techniques) are being developed to optimize the production of a more homogeneous β cell population produced in vitro for transplantation.[51,86]

Finally, although there are major advances in the induction of insulin production capacity from iPSCs, the generation of authentic β cells in vitro has not been achieved.[87] Several factors seem to underlie the difficulty. First, the presence of epigenetic memory may influence the propensity of cells to differentiate toward β cell lineage. Bar-Nur et al.[88] found that β cell-derived iPSCs maintain open chromatin structure at key β cell-specific genes, conferring them with an enhanced capacity to differentiate to insulin-producing cells both in vitro and in vivo, as compared with ESCs and non-β cell-derived iPSCs. Noncoding RNAs, particularly microRNAs, are also involved in the regulation of iPSC generation and morphological, as well as functional, β cell differentiation.[89] Recent studies showed that miR302 can replace all the formerly defined factors to reprogram human and mouse somatic cells to iPSCs. Mechanistically, miR302, as a cytoplasmic gene silencer, represses the translation of multiple key epigenetic regulators to induce global DNA demethylation, which subsequently triggers the previously defined factors such as Oct4, Sox2, and Nanog to complete the reprogramming process.[90–92] Inhibition of mouse embryonic fibroblast-enriched miRNAs such as miR-21 or miR-29a enhances reprogramming efficiency to iPSCs.[92] Deletion of the miRNA processing enzyme Dicer1 in the early pancreas affects all subsequent pancreatic lineage development, suggesting that a unique profile of miRNAs is required for pancreas development and β cell formation.[89] Therefore, in addition to identifying the proteins needed for conferring true β cell identity, we must keep in mind the importance of epigenetic regulation and microRNA modulation in the process.[89,92,93]

8.5 GENERATION OF NEW β CELLS BY INDUCING THEIR REPLICATION

Dor et al. reported in 2004 that newly generated β cells in adult pancreas arise directly from preexisting mature β cells rather than from specialized progenitor cells.[19] Teta et al.[94] used an innovative DNA double-labeling system to trace the origins of new β cells and came to the same conclusion. Whether adult human β cells arise from preexisting β cells remains to be determined. Formation of a human β-cell population within pancreatic islets is established before the age of five years or even earlier.[95] Early in life, β cell mass increases as β cell replication rates are increased, and islet sizes increase but islet number remains stable; thus, the predominant source of new β cells in humans in early life is from previously existing β cells.[96]

A recent GWAS implicated growth-associated loci (e.g., *CDKN2 A/B* and *CCND2*) as contributors to the hereditary risk of T2D.[97] It should be noted that most patients with long-duration (greater than five years) T1D retain some β cells,[98] suggesting that they may benefit from therapies that induce proliferation of these residual cells. The fact that self-duplication as the primary mechanism for β cell growth and regeneration after birth has heightened the interest in finding approaches to stimulate β cell replication to treat diabetes.

8.5.1 Insulin Resistance-Induced Circulating Factors Stimulate β Cell Replication

The insulin resistant state is a strong stimulant for β cell replication in rodents. The connection between insulin resistance and β cell replication is well-established in the liver-specific insulin receptor knockout mouse model, which fails to suppress hepatic glucose production in response to insulin. The inability to suppress hepatic gluconeogenesis causes marked hyperinsulinemia and robust compensatory β cell expansion.[99] In humans, whether insulin resistance induces β cell replication is controversial. Hanley et al.[100] discovered that β cell proliferation is increased in obese nondiabetic patients and decreased in obese diabetic patients, compared to that in lean nondiabetic controls. In contrast, Saisho et al.[101] and Kou et al.[102] reported that β cell replication is not significantly different between obese nondiabetic patients and lean nondiabetic controls. Recently, Mezza et al.[103] found that increased islet neogenesis, but not β cell replication, contributes to increased β cell mass in insulin-resistant living subjects, compared to that in body weight-matched insulin sensitive controls. Low proliferation rate, sample origins, ethnic differences, and poorly matched comparisons may explain some of the discrepancies.

Glucose has been identified as a stimulant for β cell replication in vitro and in vivo for decades.[104] This notion was challenged by the observation that β cell replication is increased in islets transplanted into euglycemic insulin resistant animals,[105] suggesting that circulating factors other than glucose are responsible for driving β cell replication. In the past 2 years, two hormones—hepatocyte growth factor (HGF)[106] and angiopoietin-like protein 8/betatrophin[107]—have been identified to be responsible for the compensatory expansion of β cells in the face of insulin resistance in rodents.

Circulating levels of HGF are elevated in obesity-associated insulin resistance. The physiological role of HGF in the pancreatic β cell was examined in conditional knockout mice in which HGF/c-Met signaling was disrupted in β cells. These mice display impaired GSIS without significant alterations in β cell mass, suggesting that HGF is important for controlling normal insulin secretion but dispensable for normal β cell growth.[106] On the other hand, HGF administration or overexpression in β cells increases β cell proliferation, survival, and differentiation. Interestingly, HGF gene transfer to islets ex vivo enhances islet graft survival and function and reduces the number of islets needed for successful transplantation, suggesting a potential therapeutic future for HGF in islet transplantation in humans.[106]

Betatrophin, predominantly expressed in liver and adipose tissue, is upregulated in mouse models of insulin resistance and during pregnancy in mice.[107] Overexpression of betatrophin in mouse liver markedly promotes pancreatic β cell proliferation, expands β cell mass, and improves glucose tolerance. Betatrophin has been highlighted as a β cell selective growth factor with therapeutic potential.[107] This notion was challenged by the recent observation that circulating levels of betatrophin are doubled in individuals with long-standing T1D, compared to normal individuals. An increased level of betatrophin does not protect against progressive loss of functional β cells.[108] More recently, Jiao et al.[109] reported that elevated betatrophin fails to stimulate the proliferation of transplanted human islet β cells in animals. These recent data indicate that manipulation of betatrophin may not have therapeutic value in human diabetes as it may in insulin-resistant diabetic rodents.

8.5.2 Wnt/GSK-3/β-catenin Pathway

The conserved Wnt/GSK-3/β-catenin pathway plays an essential role in cell proliferation and differentiation.[110] Pharmacologic inhibition of GSK-3 in combination with nutrient activation of mammalian target of rapamycin (mTOR) leads to activation of the Wnt/GSK-3/β-catenin pathway, which in turn modestly stimulates DNA synthesis, cell cycle progression, and proliferation of adult human β cells in vitro.[111] However, a study utilizing this approach showed that the expansion of human islets is associated with a loss of insulin content and secretory function.[111] Aly et al.[112] discovered that cadaver-derived intact human islets treated with a conditioned medium from intestinal L cells that constitutively produce Wnt-3a, R-spondin-3, and Noggin, in combination with inhibitors of ROCK and RhoA, display significantly increased DNA synthesis by six-fold, and are associated with a 20-fold increase in proliferation of the human β cells beyond that with glucose alone. Importantly, this approach appears not to impair GSIS or decrease insulin content of the human β cells.[112] This novel strategy might have a potential for use in the in vitro expansion of human islet β cells to treat diabetes.

8.5.3 Cell Cycle Regulators

A number of studies have focused on developing strategies to expand or restore β cell mass by exploring pathways that drive β cell proliferation.[104,113] Delivery of cell cycle regulators such as cyclin-dependent kinase 6 (cdk6), c-Myc, and cyclin D1 can induce the replication of adult human β cells in vitro or in transplanted mouse models.[104,113] Notably, therapeutics targeting the cell cycle must be highly selective for β cells and have a reversible effect on β cell growth to avoid neoplastic transformation.

Haploinsufficiency of *Glis3*, a potent insulin gene transactivatior,[114] has been shown to make adult *Glis3*$^{+/-}$ mice much more prone to develop high-fat diet (HFD)-induced diabetes, because of the impairment of β cell proliferation and β cell mass expansion during HFD feeding. Mechanistically, *Glis3* controls β cell proliferation in part by directly regulating *Ccnd2* transcription.[115] GWAS showed both *GLIS3* and *CCND2* loci are closely associated with T2D,[67,68,97] which highlights the interest to drive β cell replication by manipulating *Glis3* expression.

Although many strategies can successfully drive β cell replication in rodents, induction of adult human β cell replication has proven to be much more challenging.[113] Recently, Stewart and colleagues found that almost all of the critical G1/S cell cycle control molecules, including cyclins or cdks, necessary to drive human β cell proliferation are located in the cytoplasm of the quiescent human β cell, whereas the cell cycle inhibitors, pRb, p57, and variably, p21 are present in the nuclear compartment.[116] This difference between humans and rodents may be one of the reasons why human β cells are resistant to some stimuli that seem to work in rodents.

8.5.4 Small-Molecule Inducers

Identification of small molecules to stimulate β cell proliferation, instead of making use of endogenous factors that promote β cell replication, is an alternative strategy to generate more β cells. By high-throughput screening, adenosine kinase inhibitors (ABT-702, 5-IT), phorbol esters, thiophene-pyrimidines, and dihydropyridine derivatives have been shown to stimulate β cell proliferation in immortalized β cell lines or primary rodent islets.[117,118] Notably, adenosine kinase inhibitors selectively promote β cell replication, but not replication of other cell types such as exocrine cells, hepatocytes, and myocytes.[117] It is unclear whether adenosine kinase inhibitors stimulate human β cell replication. Stainier and colleagues confirmed that adenosine kinase inhibitors promote β cell regeneration in a diabetic zebrafish model.[119] More recently, the small molecule WS6 was shown to potently stimulate β cell proliferation in rodents and humans.[120]

8.6 CLOSING REMARKS

A major challenge facing cell transplantation for treating diabetes is the host immune system via allo- and autoimmunity.[121] An updated summary on the immune process underlying T1D and current immunotherapies to protect the newly transplanted islets or β cells from autoimmune attack is the subject matter in Chapter 7 in this volume. One alternative approach of bypassing the immune system is to encapsulate the islets or β cells within a semipermeable membrane barrier, allowing diffusion of insulin, glucose, and other nutrients, but not of larger molecules, cells, or antibodies.[122] Although islet encapsulation has been proven a success, the variability amongst small-scale clinical trials in T1D highlights a variety of issues, such as homogeneity, stability, purity, porosity, and biocompatibility of the materials.[123] Improvement in the efficacy of islet encapsulation has been achieved by combining with a short-term or low-dose immunosuppression regimen.[124]

In summary, tremendous progress has been made in the areas of β cell differentiation, replication, and reprogramming of non-β cells to β cells. The pros and cons of current approaches to produce β cells have been briefly summarized in Table 8.1. Current protocols for deriving pancreatic endocrine cells have to be further optimized to achieve efficient differentiation and production of homogeneous β cell populations that lack residual multipotent cells. Additional research

TABLE 8.1 Comparisons of Current Approaches to Generate Pancreatic β Cells

Approach	Pros	Cons
Pancreatic mature non-β cells	Closely related developmental lineage, lower reprogramming barrier	Acinar cells: risk of pancreatitis; ductal cells: absence of a unique lineage tracing system makes validation difficult; dedifferentiated β cells: limited available data
Tissue progenitor cells	Common endodermal origin, single or few transcription factors are introduced	Difficult to isolate and expand
Stem cells	Essentially infinite proliferative capacity, no ethical concerns (iPSCs), iPSCs present great translational potential	Tumorigenic potential; full differentiation to authentic β cells still a challenge
Induction of β cell proliferation	Adult β cells form by self-duplication; clinical potential	Human β cells seem resistant to the known stimuli

into the mechanisms of endocrine specification will be needed to eventually yield the necessary tools to allow for direct differentiation of non-β cells toward endocrine β cells. In any event, new discoveries on β cell biology and manipulation have been appearing in rapid succession, and we are optimistic that β cell replacement therapy will become a practical strategy for treating diabetes in the not too distant future.

ACKNOWLEDGMENTS

The research in this chapter that deals with gene therapy experiments performed in the authors' laboratory was supported by grants from the National Institutes of Health (P30-DK796380), the Juvenile Diabetes Foundation, the T.T. & W.F. Chao Global Foundation, the Frank and Cindy Liu Family Foundation, the Betty Rutherford Chair in Diabetes Research at Baylor St. Luke's Medical Center, and the Amercian Heart Association (13SDG17090096, to Y.Y.).

REFERENCES

1. Shapiro AM, Lakey JR, Ryan EA, et al. Islet transplantation in seven patients with type 1 diabetes mellitus using a glucocorticoid-free immunosuppressive regimen. *N Engl J Med* 2000;**343**(4):230–8.
2. Vantyghem MC, Defrance F, Quintin D, et al. Treating diabetes with islet transplantation: lessons from the past decade in Lille. *Diabetes Metab* 2014;**40**(2):108–19.
3. Collombat P, Mansouri A, Hecksher-Sorensen J, et al. Opposing actions of Arx and Pax4 in endocrine pancreas development. *Genes Dev* 2003;**17**(20):2591–603.
4. Dhawan S, Georgia S, Tschen SI, et al. Pancreatic beta cell identity is maintained by DNA methylation-mediated repression of Arx. *Dev Cell* 2011;**20**(4):419–29.
5. Collombat P, Hecksher-Sorensen J, Krull J, et al. Embryonic endocrine pancreas and mature beta cells acquire alpha and PP cell phenotypes upon Arx misexpression. *J Clin Invest* 2007;**117**(4):961–70.
6. Sosa-Pineda B, Chowdhury K, Torres M, et al. The Pax4 gene is essential for differentiation of insulin-producing beta cells in the mammalian pancreas. *Nature* 1997;**386**(6623):399–402.
7. Collombat P, Xu X, Ravassard P, et al. The ectopic expression of Pax4 in the mouse pancreas converts progenitor cells into alpha and subsequently beta cells. *Cell* 2009;**138**(3):449–62.
8. Thorel F, Nepote V, Avril I, et al. Conversion of adult pancreatic alpha-cells to beta-cells after extreme beta-cell loss. *Nature* 2010;**464**(7292): 1149–54.
9. Chung CH, Hao E, Piran R, et al. Pancreatic beta-cell neogenesis by direct conversion from mature alpha-cells. *Stem Cells* 2010;**28**(9):1630–8.
10. Bramswig NC, Everett LJ, Schug J, et al. Epigenomic plasticity enables human pancreatic alpha to beta cell reprogramming. *J Clin Invest* 2013;**123**(3):1275–84.
11. Zhou Q, Brown J, Kanarek A, et al. In vivo reprogramming of adult pancreatic exocrine cells to beta-cells. *Nature* 2008;**455**(7213):627–32.
12. Akinci E, Banga A, Greder LV, et al. Reprogramming of pancreatic exocrine cells towards a beta (beta) cell character using Pdx1, Ngn3 and MafA. *Biochem J* 2012;**442**(3):539–50.
13. Pan FC, Bankaitis ED, Boyer D, et al. Spatiotemporal patterns of multipotentiality in Ptf1a-expressing cells during pancreas organogenesis and injury-induced facultative restoration. *Development* 2013;**140**(4):751–64.
14. Cavelti-Weder C, Shtessel M, Reuss JE, et al. Pancreatic duct ligation after almost complete beta-cell loss: exocrine regeneration but no evidence of beta-cell regeneration. *Endocrinology* 2013;**154**(12):4493–502.
15. Rankin MM, Wilbur CJ, Rak K, et al. beta-Cells are not generated in pancreatic duct ligation-induced injury in adult mice. *Diabetes* 2013;**62**(5):1634–45.
16. Baeyens L, Lemper M, Leuckx G, et al. Transient cytokine treatment induces acinar cell reprogramming and regenerates functional beta cell mass in diabetic mice. *Nat Biotechnol* 2014;**32**(1):76–83.
17. Lima MJ, Muir KR, Docherty HM, et al. Suppression of epithelial-to-mesenchymal transitioning enhances ex vivo reprogramming of human exocrine pancreatic tissue toward functional insulin-producing beta-like cells. *Diabetes* 2013;**62**(8):2821–33.
18. Shih HP, Wang A, Sander M. Pancreas organogenesis: from lineage determination to morphogenesis. *Annu Rev Cell Dev Biol* 2013;**29**:81–105.
19. Dor Y, Brown J, Martinez OI, et al. Adult pancreatic beta-cells are formed by self-duplication rather than stem-cell differentiation. *Nature* 2004;**429**(6987):41–6.
20. Kushner JA, Weir GC, Bonner-Weir S. Ductal origin hypothesis of pancreatic regeneration under attack. *Cell Metab* 2010;**11**(1):2–3.
21. Inada A, Nienaber C, Katsuta H, et al. Carbonic anhydrase II-positive pancreatic cells are progenitors for both endocrine and exocrine pancreas after birth. *Proc Natl Acad Sci USA* 2008;**105**(50):19915–9.
22. Solar M, Cardalda C, Houbracken I, et al. Pancreatic exocrine duct cells give rise to insulin-producing beta cells during embryogenesis but not after birth. *Dev Cell* 2009;**17**(6):849–60.
23. Kopinke D, Murtaugh LC. Exocrine-to-endocrine differentiation is detectable only prior to birth in the uninjured mouse pancreas. *BMC Dev Biol* 2010;**10**:38.
24. Lee J, Sugiyama T, Liu Y, et al. Expansion and conversion of human pancreatic ductal cells into insulin-secreting endocrine cells. *Elife* 2013; **2**:e00940.

25. Gershengorn MC, Hardikar AA, Wei C, et al. Epithelial-to-mesenchymal transition generates proliferative human islet precursor cells. *Science* 2004;**306**(5705):2261–4.

26. Russ HA, Bar Y, Ravassard P, et al. In vitro proliferation of cells derived from adult human beta-cells revealed by cell-lineage tracing. *Diabetes* 2008;**57**(6):1575–83.

27. Weinberg N, Ouziel-Yahalom L, Knoller S, et al. Lineage tracing evidence for in vitro dedifferentiation but rare proliferation of mouse pancreatic beta-cells. *Diabetes* 2007;**56**(5):1299–304.

28. Dor Y, Glaser B. beta-cell dedifferentiation and type 2 diabetes. *N Engl J Med* 2013;**368**(6):572–3.

29. Talchai C, Xuan S, Kitamura T, et al. Generation of functional insulin-producing cells in the gut by Foxo1 ablation. *Nat Genet* 2012;**44**(4): 406–12. S1.

30. Talchai C, Xuan S, Lin HV, et al. Pancreatic beta cell dedifferentiation as a mechanism of diabetic beta cell failure. *Cell* 2012;**150**(6):1223–34.

31. Russ HA, Sintov E, Anker-Kitai L, et al. Insulin-producing cells generated from dedifferentiated human pancreatic beta cells expanded in vitro. *PLoS ONE* 2011;**6**(9):e25566.

32. Weissman IL, Anderson DJ, Gage F. Stem and progenitor cells: origins, phenotypes, lineage commitments, and transdifferentiations. *Annu Rev Cell Dev Biol* 2001;**17**:387–403.

33. Fausto N, Campbell JS. The role of hepatocytes and oval cells in liver regeneration and repopulation. *Mech Dev* 2003;**120**(1):117–30.

34. Yang L, Li S, Hatch H, et al. In vitro trans-differentiation of adult hepatic stem cells into pancreatic endocrine hormone-producing cells. *Proc Natl Acad Sci USA* 2002;**99**(12):8078–83.

35. Kim S, Shin JS, Kim HJ, et al. Streptozotocin-induced diabetes can be reversed by hepatic oval cell activation through hepatic transdifferentiation and pancreatic islet regeneration. *Lab Invest* 2007;**87**(7):702–12.

36. Ferber S, Halkin A, Cohen H, et al. Pancreatic and duodenal homeobox gene 1 induces expression of insulin genes in liver and ameliorates strepto-zotocin-induced hyperglycemia. *Nat Med* 2000;**6**(5):568–72.

37. Kojima H, Fujimiya M, Matsumura K, et al. NeuroD-betacellulin gene therapy induces islet neogenesis in the liver and reverses diabetes in mice. *Nat Med* 2003;**9**(5):596–603.

38. Yechoor V, Liu V, Espiritu C, et al. Neurogenin3 is sufficient for transdetermination of hepatic progenitor cells into neo-islets in vivo but not trans-differentiation of hepatocytes. *Dev Cell* 2009;**16**(3):358–73.

39. Manohar R, Lagasse E. Transdetermination: a new trend in cellular reprogramming. *Mol Ther* 2009;**17**(6):936–8.

40. Fukuda A, Kawaguchi Y, Furuyama K, et al. Ectopic pancreas formation in Hes1-knockout mice reveals plasticity of endodermal progenitors of the gut, bile duct, and pancreas. *J Clin Invest* 2006;**116**(6):1484–93.

41. Dutton JR, Chillingworth NL, Eberhard D, et al. Beta cells occur naturally in extrahepatic bile ducts of mice. *J Cell Sci* 2007;**120**(Pt 2):239–45.

42. Banga A, Akinci E, Greder LV, et al. In vivo reprogramming of Sox9+ cells in the liver to insulin-secreting ducts. *Proc Natl Acad Sci USA* 2012;**109**(38):15336–41.

43. Cardinale V, Wang Y, Carpino G, et al. Multipotent stem/progenitor cells in human biliary tree give rise to hepatocytes, cholangiocytes, and pancre-atic islets. *Hepatology* 2011;**54**(6):2159–72.

44. Wang Y, Lanzoni G, Carpino G, et al. Biliary tree stem cells, precursors to pancreatic committed progenitors: evidence for possible life-long pan-creatic organogenesis. *Stem Cells* 2013;**31**(9):1966–79.

45. Schonhoff SE, Giel-Moloney M, Leiter AB. Minireview: development and differentiation of gut endocrine cells. *Endocrinology* 2004;**145**(6): 2639–44.

46. Gu G, Dubauskaite J, Melton DA. Direct evidence for the pancreatic lineage: NGN3+ cells are islet progenitors and are distinct from duct progeni-tors. *Development* 2002;**129**(10):2447–57.

47. He S, Nakada D, Morrison SJ. Mechanisms of stem cell self-renewal. *Annu Rev Cell Dev Biol* 2009;**25**:377–406.

48. Dominici M, Le BK, Mueller I, et al. Minimal criteria for defining multipotent mesenchymal stromal cells. The international society for cellular therapy position statement. *Cytotherapy* 2006;**8**(4):315–7.

49. Dominguez-Bendala J, Lanzoni G, Inverardi L, et al. Concise review: mesenchymal stem cells for diabetes. *Stem Cells Transl Med* 2012;**1**(1): 59–63.

50. Carlotti F, Zaldumbide A, Loomans CJ, et al. Isolated human islets contain a distinct population of mesenchymal stem cells. *Islets* 2010;**2**(3): 164–73.

51. Wagner RT, Lewis J, Cooney A, et al. Stem cell approaches for the treatment of type 1 diabetes mellitus. *Transl Res* 2010;**156**(3):169–79.

52. D'Amour KA, Bang AG, Eliazer S, et al. Production of pancreatic hormone-expressing endocrine cells from human embryonic stem cells. *Nat Biotechnol* 2006;**24**(11):1392–401.

53. Kroon E, Martinson LA, Kadoya K, et al. Pancreatic endoderm derived from human embryonic stem cells generates glucose-responsive insulin-secreting cells in vivo. *Nat Biotechnol* 2008;**26**(4):443–52.

54. Tam PP, Loebel DA. Gene function in mouse embryogenesis: get set for gastrulation. *Nat Rev Genet* 2007;**8**(5):368–81.

55. Vincent SD, Dunn NR, Hayashi S, et al. Cell fate decisions within the mouse organizer are governed by graded Nodal signals. *Genes Dev* 2003;**17**(13):1646–62.

56. Borowiak M, Maehr R, Chen S, et al. Small molecules efficiently direct endodermal differentiation of mouse and human embryonic stem cells. *Cell Stem Cell* 2009;**4**(4):348–58.

57. Hebrok M, Kim SK, Melton DA. Notochord repression of endodermal Sonic hedgehog permits pancreas development. *Genes Dev* 1998;**12**(11): 1705–13.

58. Hebrok M, Kim SK, St JB, et al. Regulation of pancreas development by hedgehog signaling. *Development* 2000;**127**(22):4905–13.

59. Wandzioch E, Zaret KS. Dynamic signaling network for the specification of embryonic pancreas and liver progenitors. *Science* 2009;**324**(5935): 1707–10.

60. Stafford D, Hornbruch A, Mueller PR, et al. A conserved role for retinoid signaling in vertebrate pancreas development. *Dev Genes Evol* 2004;**214**(9):432–41.

61. Martin M, Gallego-Llamas J, Ribes V, et al. Dorsal pancreas agenesis in retinoic acid-deficient Raldh2 mutant mice. *Dev Biol* 2005;**284**(2):399–411.

62. Zhang D, Jiang W, Liu M, et al. Highly efficient differentiation of human ES cells and iPS cells into mature pancreatic insulin-producing cells. *Cell Res* 2009;**19**(4):429–38.

63. Mfopou JK, Chen B, Mateizel I, et al. Noggin, retinoids, and fibroblast growth factor regulate hepatic or pancreatic fate of human embryonic stem cells. *Gastroenterology* 2010;**138**(7):2233–45.

64. Bhushan A, Itoh N, Kato S, et al. Fgf10 is essential for maintaining the proliferative capacity of epithelial progenitor cells during early pancreatic organogenesis. *Development* 2001;**128**(24):5109–17.

65. Norgaard GA, Jensen JN, Jensen J. FGF10 signaling maintains the pancreatic progenitor cell state revealing a novel role of Notch in organ development. *Dev Biol* 2003;**264**(2):323–38.

66. Barrett JC, Clayton DG, Concannon P, et al. Genome-wide association study and meta-analysis find that over 40 loci affect risk of type 1 diabetes. *Nat Genet* 2009;**41**(6):703–7.

67. Dupuis J, Langenberg C, Prokopenko I, et al. New genetic loci implicated in fasting glucose homeostasis and their impact on type 2 diabetes risk. *Nat Genet* 2010;**42**(2):105–16.

68. Cho YS, Chen CH, Hu C, et al. Meta-analysis of genome-wide association studies identifies eight new loci for type 2 diabetes in east Asians. *Nat Genet* 2012;**44**(1):67–72.

69. Yang Y, Chang BH, Yechoor V, et al. The Kruppel-like zinc finger protein GLIS3 transactivates neurogenin 3 for proper fetal pancreatic islet differentiation in mice. *Diabetologia* 2011;**54**(10):2595–605.

70. Xie R, Everett LJ, Lim HW, et al. Dynamic chromatin remodeling mediated by polycomb proteins orchestrates pancreatic differentiation of human embryonic stem cells. *Cell Stem Cell* 2013;**12**(2):224–37.

71. Takahashi K, Yamanaka S. Induction of pluripotent stem cells from mouse embryonic and adult fibroblast cultures by defined factors. *Cell* 2006;**126**(4):663–76.

72. Park IH, Zhao R, West JA, et al. Reprogramming of human somatic cells to pluripotency with defined factors. *Nature* 2008;**451**(7175):141–6.

73. Yu J, Vodyanik MA, Smuga-Otto K, et al. Induced pluripotent stem cell lines derived from human somatic cells. *Science* 2007;**318**(5858):1917–20.

74. Maekawa M, Yamaguchi K, Nakamura T, et al. Direct reprogramming of somatic cells is promoted by maternal transcription factor Glis1. *Nature* 2011;**474**(7350):225–9.

75. Puri MC, Nagy A. Concise review: embryonic stem cells versus induced pluripotent stem cells: the game is on. *Stem Cells* 2012;**30**(1):10–4.

76. Maehr R, Chen S, Snitow M, et al. Generation of pluripotent stem cells from patients with type 1 diabetes. *Proc Natl Acad Sci USA* 2009;**106**(37): 15768–73.

77. Thatava T, Kudva YC, Edukulla R, et al. Intrapatient variations in type 1 diabetes-specific iPS cell differentiation into insulin-producing cells. *Mol Ther* 2013;**21**(1):228–39.

78. Teo AK, Windmueller R, Johansson BB, et al. Derivation of human induced pluripotent stem cells from patients with maturity onset diabetes of the young. *J Biol Chem* 2013;**288**(8):5353–6.

79. Fujikura J, Nakao K, Sone M, et al. Induced pluripotent stem cells generated from diabetic patients with mitochondrial DNA A3243G mutation. *Diabetologia* 2012;**55**(6):1689–98.

80. Knoepfler PS. Myc goes global: new tricks for an old oncogene. *Cancer Res* 2007;**67**(11):5061–3.

81. Rowland BD, Bernards R, Peeper DS. The KLF4 tumour suppressor is a transcriptional repressor of p53 that acts as a context-dependent oncogene. *Nat Cell Biol* 2005;**7**(11):1074–82.

82. Yu J, Hu K, Smuga-Otto K, et al. Human induced pluripotent stem cells free of vector and transgene sequences. *Science* 2009;**324**(5928):797–801.

83. Woltjen K, Michael IP, Mohseni P, et al. piggyBac transposition reprograms fibroblasts to induced pluripotent stem cells. *Nature* 2009;**458**(7239): 766–70.

84. Kim D, Kim CH, Moon JI, et al. Generation of human induced pluripotent stem cells by direct delivery of reprogramming proteins. *Cell Stem Cell* 2009;**4**(6):472–6.

85. Miura K, Okada Y, Aoi T, et al. Variation in the safety of induced pluripotent stem cell lines. *Nat Biotechnol* 2009;**27**(8):743–5.

86. Maehr R. iPS cells in type 1 diabetes research and treatment. *Clin Pharmacol Ther* 2011;**89**(5):750–3.

87. Pagliuca FW, Melton DA. How to make a functional beta-cell. *Development* 2013;**140**(12):2472–83.

88. Bar-Nur O, Russ HA, Efrat S, et al. Epigenetic memory and preferential lineage-specific differentiation in induced pluripotent stem cells derived from human pancreatic islet beta cells. *Cell Stem Cell* 2011;**9**(1):17–23.

89. Lynn FC, Skewes-Cox P, Kosaka Y, et al. MicroRNA expression is required for pancreatic islet cell genesis in the mouse. *Diabetes* 2007;**56**(12): 2938–45.

90. Subramanyam D, Lamouille S, Judson RL, et al. Multiple targets of miR-302 and miR-372 promote reprogramming of human fibroblasts to induced pluripotent stem cells. *Nat Biotechnol* 2011;**29**(5):443–8.

91. Lin SL, Chang DC, Chang-Lin S, et al. Mir-302 reprograms human skin cancer cells into a pluripotent ES-cell-like state. *RNA* 2008;**14**(10): 2115–24.

92. Li Z, Yang CS, Nakashima K, et al. Small RNA-mediated regulation of iPS cell generation. *EMBO J* 2011;**30**(5):823–34.
93. Robinton DA, Daley GQ. The promise of induced pluripotent stem cells in research and therapy. *Nature* 2012;**481**(7381):295–305.
94. Teta M, Rankin MM, Long SY, et al. Growth and regeneration of adult beta cells does not involve specialized progenitors. *Dev Cell* 2007;**12**(5):817–26.
95. Gregg BE, Moore PC, Demozay D, et al. Formation of a human beta-cell population within pancreatic islets is set early in life. *J Clin Endocrinol Metab* 2012;**97**(9):3197–206.
96. Meier JJ, Butler AE, Saisho Y, et al. Beta-cell replication is the primary mechanism subserving the postnatal expansion of beta-cell mass in humans. *Diabetes* 2008;**57**(6):1584–94.
97. Morris AP, Voight BF, Teslovich TM, et al. Large-scale association analysis provides insights into the genetic architecture and pathophysiology of type 2 diabetes. *Nat Genet* 2012;**44**(9):981–90.
98. Oram RA, Jones AG, Besser RE, et al. The majority of patients with long-duration type 1 diabetes are insulin microsecretors and have functioning beta cells. *Diabetologia* 2014;**57**(1):187–91.
99. Michael MD, Kulkarni RN, Postic C, et al. Loss of insulin signaling in hepatocytes leads to severe insulin resistance and progressive hepatic dysfunction. *Mol Cell* 2000;**6**(1):87–97.
100. Hanley SC, Austin E, ssouline-Thomas B, et al. {beta}-Cell mass dynamics and islet cell plasticity in human type 2 diabetes. *Endocrinology* 2010;**151**(4):1462–72.
101. Saisho Y, Butler AE, Manesso E, et al. β-cell mass and turnover in humans: effects of obesity and aging. *Diabetes Care* 2013;**36**(1):111–7.
102. Kou K, Saisho Y, Satoh S, et al. Change in beta-cell mass in Japanese nondiabetic obese individuals. *J Clin Endocrinol Metab* 2013;**98**(9):3724–30.
103. Mezza T, Muscogiuri G, Sorice GP, et al. Insulin resistance alters islet morphology in nondiabetic humans. *Diabetes* 2014;**63**(3):994–1007.
104. Heit JJ, Karnik SK, Kim SK. Intrinsic regulators of pancreatic beta-cell proliferation. *Annu Rev Cell Dev Biol* 2006;**22**:311–38.
105. Flier SN, Kulkarni RN, Kahn CR. Evidence for a circulating islet cell growth factor in insulin-resistant states. *Proc Natl Acad Sci USA* 2001;**98**(13):7475–80.
106. Araujo TG, Oliveira AG, Carvalho BM, et al. Hepatocyte growth factor plays a key role in insulin resistance-associated compensatory mechanisms. *Endocrinology* 2012;**153**(12):5760–9.
107. Yi P, Park JS, Melton DA. Betatrophin: a hormone that controls pancreatic beta cell proliferation. *Cell* 2013;**153**(4):747–58.
108. Espes D, Lau J, Carlsson PO. Increased circulating levels of betatrophin in individuals with long-standing type 1 diabetes. *Diabetologia* 2014;**57**(1):50–3.
109. Jiao Y, Le LJ, Yu M, et al. Elevated mouse hepatic betatrophin expression does not increase human beta-cell replication in the transplant setting. *Diabetes* 2014;**63**(4):1283–8.
110. Willert K, Jones KA. Wnt signaling: is the party in the nucleus? *Genes Dev* 2006;**20**(11):1394–404.
111. Liu Y, Mziaut H, Ivanova A, et al. beta-Cells at the crossroads: choosing between insulin granule production and proliferation. *Diabetes Obes Metab* 2009;**11**(Suppl. 4):54–64.
112. Aly H, Rohatgi N, Marshall CA, et al. A novel strategy to increase the proliferative potential of adult human beta-cells while maintaining their differentiated phenotype. *PLoS ONE* 2013;**8**(6):e66131.
113. Kulkarni RN, Mizrachi EB, Ocana AG, et al. Human beta-cell proliferation and intracellular signaling: driving in the dark without a road map. *Diabetes* 2012;**61**(9):2205–13.
114. Yang Y, Chang BH, Samson SL, et al. The Kruppel-like zinc finger protein Glis3 directly and indirectly activates insulin gene transcription. *Nucleic Acids Res* 2009;**37**(8):2529–38.
115. Yang Y, Chang BH, Chan L. Sustained expression of the transcription factor GLIS3 is required for normal beta cell function in adults. *EMBO Mol Med* 2013;**5**(1):92–104.
116. Fiaschi-Taesch NM, Kleinberger JW, Salim FG, et al. Cytoplasmic-nuclear trafficking of G1/S cell cycle molecules and adult human beta-cell replication: a revised model of human beta-cell G1/S control. *Diabetes* 2013;**62**(7):2460–70.
117. Annes JP, Ryu JH, Lam K, et al. Adenosine kinase inhibition selectively promotes rodent and porcine islet beta-cell replication. *Proc Natl Acad Sci USA* 2012;**109**(10):3915–20.
118. Wang W, Walker JR, Wang X, et al. Identification of small-molecule inducers of pancreatic beta-cell expansion. *Proc Natl Acad Sci USA* 2009;**106**(5):1427–32.
119. Andersson O, Adams BA, Yoo D, et al. Adenosine signaling promotes regeneration of pancreatic beta cells in vivo. *Cell Metab* 2012;**15**(6):885–94.
120. Shen W, Tremblay MS, Deshmukh VA, et al. Small-molecule inducer of beta cell proliferation identified by high-throughput screening. *J Am Chem Soc* 2013;**135**(5):1669–72.
121. Chen W, Xie A, Chan L. Mechanistic basis of immunotherapies for type 1 diabetes mellitus. *Transl Res* 2013;**161**(4):217–29.
122. Tuch BE, Keogh GW, Williams LJ, et al. Safety and viability of microencapsulated human islets transplanted into diabetic humans. *Diabetes Care* 2009;**32**(10):1887–9.
123. O'Sullivan ES, Vegas A, Anderson DG, et al. Islets transplanted in immunoisolation devices: a review of the progress and the challenges that remain. *Endocr Rev* 2011;**32**(6):827–44.
124. Toso C, Mathe Z, Morel P, et al. Effect of microcapsule composition and short-term immunosuppression on intraportal biocompatibility. *Cell Transpl* 2005;**14**(2–3):159–67.

Chapter 9

Gene Therapy for Neurological Diseases

Michele Simonato[1], Lars U. Wahlberg[2], William F. Goins[3] and Joseph C. Glorioso[3]

[1]Section of Pharmacology, Department of Medical Sciences, University of Ferrara, Ferrara, Italy; [2]NsGene A/S, Ballerup, Denmark; [3]Department of Microbiology and Molecular Genetics, University of Pittsburgh School of Medicine, Pittsburgh, PA, USA

LIST OF ABBREVIATIONS

AADC Aromatic amino acid decarboxylase
BBB Blood–brain barrier
BDNF Brain-derived neurotrophic factor
CED Convection-enhanced delivery
FGF2 Fibroblast growth factor 2
GABA γ-Aminobutyric acid
GAL Galanin
GDNF Glial cell line-derived neurotrophic factor
GAD Glutamic acid decarboxylase
GlyR Glycine receptor
GTP Guanosine triphosphate
HSV Herpes simplex virus
IL Interleukin
NMDA N-methyl-D-aspartic acid
NPY Neuropeptide Y
NTF Neurotrophic factor
PD Parkinson's disease
SE Status epilepticus

9.1 INTRODUCTION

No other organ system can reach the complexity of the nervous system. Not only are the principal cells (the neurons) morphologically and physiologically complex and heterogeneous, but they are also highly interconnected to generate circuits and regions organizing and regulating all kinds of functions: from the vegetative, to the sensory-motor, to the so-called "higher" functions (perception, memory, and emotion). Many other cell types, such as astrocytes and microglia, participate in and sustain neuronal activity. Achieving normal neuronal function is a balance between correct developmental pathways and maintenance of the cells in the adult brain. Contrary to a dogma that prevailed for about a century, it was demonstrated that neural stem cells persist even in the adult brain, which ensures continuous replacement of some neuronal elements and increases their activity in conjunction with physiological needs (such as learning and memory) or damage.[1]

Because of the complexity, it is not surprising that the nervous system is the site where many of the most elusive and pervasive disease processes arise. These diseases encompass a broad spectrum of pathologic states (neurodegeneration, abnormal cell activity, cancer, neuroinflammation, metabolic, and developmental disorders) and can have global or local effects on development, metabolism, and function. Because of the complexity and limited understanding of the relevant neuropathophysiology, conventional therapeutic approaches (drug treatments and neurosurgical procedures) provide limited and often unsatisfactory effects. Because the continuous aging of the population leads to a progressive increase in the incidence of nervous system diseases, there is an urgent need to improve the currently available treatments and to develop effective alternatives.

In this review, we will present and discuss data supporting the notion that gene therapy may become a realistic option for the treatment of nervous system diseases. The field has matured, with the development of sophisticated gene delivery vehicles and their successful application to treatment of animal models. Based on these preclinical studies, early human clinical trials were carried out to test the safety and the efficacy of gene therapy and these studies showed encouraging results.

Translating Gene Therapy to the Clinic. http://dx.doi.org/10.1016/B978-0-12-800563-7.00009-9

A single review cannot cover all these data in detail. Therefore, we will focus on some of the most representative cases and lead the reader to other recent papers for further information (see Ref. 2). Here, we will first briefly describe the advantages and disadvantages of the different vectors for application to nervous system diseases. We will then discuss data on three diseases in detail: (1) neuropathic pain, representative of peripheral nervous system diseases; (2) epilepsy, characterized by abnormal hyperactivity; and (3) Parkinson's disease, representative of neurodegenerative diseases. It should be noted that this choice was dictated by the will to demonstrate the widespread potential applicability of gene therapy to heterogeneous groups of nervous system diseases. Spectacular advances were made in other areas of neurological disorders, like sensory organs diseases and brain cancer,[2] but these will not be discussed in detail. The authors believe that gene therapy has tremendous potential to prevent development and progression of many nervous system diseases and possibly to restore normal function. Optimistically, some of these approaches will become approved medications for patients in the not too distant future.

9.2 VIRAL VECTORS FOR NEUROLOGICAL DISEASES

The delivery of genes in the nervous system has employed a large array of vectors. Nonviral vectors composed of naked DNA have been attempted by a number of investigators. These vectors suffer from difficulties (Table 9.1) related to delivery, short-term expression and immune reactivity to the vector itself. Viral vectors were also extensively employed for pain gene therapy with considerable success (Table 9.1). Viruses can be delivered locally and many, but not all, can be transported down the axons by a retrograde transport mechanism to the cell body following infection of the nerve termini. This retrograde transport may be particularly attractive for the therapy of neuropathic pain, as described in subsequent sections. It occurs with different efficiencies depending on the virus type and strain but the most highly evolved virus type for this purpose in human infections is herpes simplex virus (HSV). Retrograde transport provides a means of targeting specific ganglia following intradermal infection or other delivery routes.

Other important considerations for the choice of vector relate to dosing, mechanism of persistence (e.g., integrated or episomal), induction of local inflammation, payload capacity and complexity, level and duration of transgene expression, and the degree to which the vector platform can be delivered to specific neuronal subtypes. In general, vectors such as those derived from retroviruses (Figure 9.1(a)) and lentiviruses (Figure 9.1(b)) can deliver durable transgene expression but require integration (Table 9.1). Axonal transport of these vectors is inefficient and standard retroviruses require cell division for gene delivery. Lentiviruses can integrate into nondividing cells and can be pseudotyped with rabies G protein to enhance retrograde delivery[3,4]; however, they run the risk of disruption of normal genes including tumor suppressor functions or activation of oncogenes.[5,6] Vectors that stably persist as episomes are preferred for peripheral nerve applications. Adenoviral vectors (Figure 9.1(c)) do not require integration and undergo retrograde transport, but can recruit inflammatory cells because of viral gene expression[7,8] and have not been shown to persist long-term in sensory neurons (Table 9.1). The immunogenicity of adenoviral vectors has relegated their exploitation to vaccine development and as oncolytic vectors to treat cancer. Adeno-associated (Figure 9.1(d)) viruses (AAV) have proven attractive for gene delivery to the nervous system. They are dependent on helper virus such as HSV and adenovirus for replication and are not in themselves pathogenic. Several clades have been isolated based on isolation from primates[9,10] and pigs,[11] and these different virus serotypes have shown proclivity for infection of different cell types (Table 9.2) compared with standard AAV serotypes.[9–13] The new serotypes, as well as AAV serotypes 1–11, transduce neurons with varying efficiencies, while some serotypes, such as AAV1, 5, and 8, can readily transduce glial cells. However, various neuronal-specific promoters can be employed to achieve transgene expression exclusively in neurons.[10,13] AAV can efficiently transduce nerve tissue, can be retrogradely transported efficiently at least in mice[14–16] and express transgenes long term. A limitation of these vectors is payload capacity (3–4 kb) and thus only single moderate-sized genes are generally accommodated and virus doses to achieve effective transduction are very high compared with other vectors. Moreover, targeting of AAV vectors for infection of specific cell types, especially neurons, has thus far not been adequately demonstrated. AAV is readily neutralized by antibody and repeat dosing is usually not possible. There have been concerns that high vector dosing in the periphery and loading of genomes into cells can induce DNA damage responses that lead to inflammation.[17,18] Besides transduction of neurons of the brain, vectors are needed for the transfer of therapeutic genes to the peripheral nervous system to treat diseases such as pain and peripheral neuropathies. HSV (Figure 9.1(e)) is probably the most logical choice for peripheral nerve delivery because it only establishes latency in human peripheral ganglia accounting for its neurotropism and persists for life in humans without integration or apparent harm to sensory neurons.

The biology of HSV is complex and the genome is large (152 kb) compared with other vectors (Figure 9.1(e)). Among the approximately 80 encoded gene products, about half are essential for virus replication. The viral genes (Figure 9.1(e)) are classified as immediate early (IE), early, and late and they are expressed in a cascade manner during the replication cycle. Deletion of any of the essential IE genes blocks virus replication at a very early stage, sidestepping the need to make extensive deletions in the genome to make highly defective vectors suitable for gene transfer. The propagation of deletion mutants

TABLE 9.1 Vector Characteristics

Vector	Plasmid	RV	LV	AdV	AAV	HSV
(1) Genome size	Varies	~10 kb	~10 kb	~40 kb	~5 kb	~150 kb
(2) Payload size	Varies	Medium	Medium	Large	Very small	Very large
(a) Size	Varies	~7 kb	~6.5 kb	~7–36 kb	~3–4.5 kb	~40 kb+
(b) Genes	Varies	1–2	1–2	1–many	1	1–many
(3) Host range	Varies	Limited	Limited[a]	Broad	Broad	Broad
(a) Dividing	Yes	Yes	Yes	Yes	Yes	Yes
(b) Nondividing	Yes	No	Yes	Yes	Yes	Yes
(4) Transduction efficiency	Low	High	High	Medium	Low-medium	High
(5) Genome stability	Low	High	High	Low	High	Medium-high
(a) Episomal	Yes	No	No	No	Yes	Yes[c]
(b) Integrated	Rare[d]	Yes	Yes	No	Yes	No
(6) Expression levels	Medium	Medium	Medium	High	Medium	Medium
(a) Short term	Low	Low	Low	High	Low	Medium
(b) Long term	Poor-rare	High	High	Poor-rare	High	Low[e]
(7) Production	Easy	Easy	Easy	Easy	Easy-hard	Hard
(a) Cell line needed	No	Yes	Yes	Yes	Yes	Yes
(b) Kits available	Yes	Yes	Yes	Yes	Yes	No
(c) Cost	High	Low	Low	Low	Medium	Low
(8) Titers (TU/mL)	[f]	10^5–10^7	10^6–10^8	10^{10}–10^{13}	10^8–10^{12}	10^9–10^{11}
(9) Safety	High	Poor-low	Poor-low	Poor	Medium-poor	Low
(a) Tumor potential	No	Medium	Medium	No	Medium-low	No
(b) Recombination	No	Medium	Medium	Low	Low	Low-rare
(c) Immune response	Med	Rare	Rare	Very high	High	Low-rare
(d) Cytotoxicity	Low-rare	No	No	High	No	Medium-rare
(10) Repeat dosing	No	No	No	No	No (eye)[b]	Yes

AdV, adenovirus; AAV, adeno-associated virus; HSV, herpes simplex virus; kb, kilobases; LV, lentivirus; mL, milliliters; RV, retrovirus; TU, transducing units.
[a]Host range of pseudotyped LV varies with the glycoproteins employed, which affects transduction efficiency.
[b]AAV repeat dosing has been achieved during vector delivery to the eye/retina.
[c]Only maintained as an episome.
[d]Marker drug selection can enable integration.
[e]Extended expression using HSV latency promoter system.
[f]Plasmid DNA preparations in mg/mL rather than TU/mL.

defective for the essential IE genes ICP4 and ICP27 require engineered cell lines that complement these mutant genes by providing these essential products in trans. An important component of HSV is the latency locus that expresses an 8.3-kb latency transcript that is noncoding and unstable. However, this region encodes a number of microRNAs that are likely to play a role in establishing or maintaining the latent state.[19,20] In addition, there is a large 2-kb intron (LAT) that is spliced from the 8.3-kb latency transcript. LAT is highly stable and accumulates in most latently infected neurons. The function of the latency gene noncoding RNA is unclear but some evidence shows that it plays a role in preventing virus-mediated apoptosis and aids

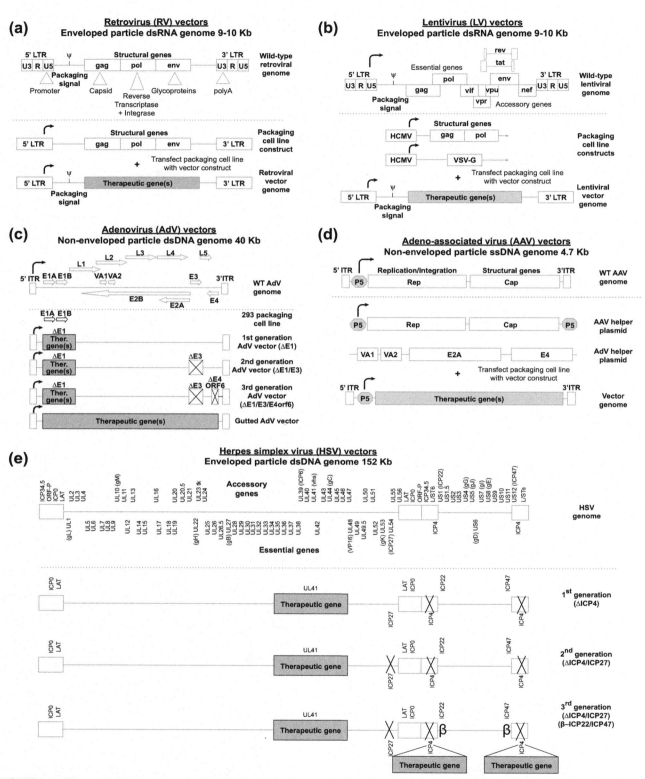

FIGURE 9.1 Diagrams of the genomes of various viral vectors used in gene therapy approaches. The diagram for each vector depicts the genome of the wild-type (wt) virus above the dotted lines and the genome of the corresponding viral vectors below the dotted lines. The diagrams show viral structural genes, some essential transcriptional regulatory genes and viral polymerases, and nonstructural accessory genes. Accessory genes are not required for the virus to replicate in cell culture, but they may play a role in the virus life cycle in vivo. All of these replication-defective virus vectors are produced using standard cell lines that express the essential viral genes that are deleted from the viral vector genome *in trans* to enable vector propagation in vitro that

TABLE 9.2 Adeno-Associated Virus (AAV) Vector Serotype-Specific Transduction of the Brain, Eye, and Peripheral Nervous System

AAV Serotype	Central Nervous System[d] (Str, Ctx, Hip, SN)	Eye[e]	Peripheral Nervous System (Aβ-A™ >> C)
AAV1	Medium[b,c]	High (PR/HC/MGC)	Medium
AAV2	Low	High (HC/GCL)	Medium
AAV3	Low-poor	ND	Low-poor
AAV4	Low-poor	ND	Low
AAV5	High[a]	Medium (PR/ONL/INL)	Medium
AAV6	Low	ND	Medium
AAV7	Medium	ND	ND
AAV8	High-medium[a,b]	High-medium[a] (PR/HC/GCL)	High
AAV9	High-medium	High (PR/ONL/INL/HC/GCL)	ND
AAV10	ND	Medium (PR/ONL/INL/HC/GCL)	ND
Rh43	Low-medium	ND	ND

AAV, adeno-associated virus; Ctx, cortex; GCL, ganglion cell layer; Hip, hippocampus; HC, horizontal cells; INL, inner nuclear layer; MGC, Mueller glial cells; ND, not determined; ONL, outer nuclear layer; PR, photoreceptors; SN, substantia nigra; Str, striatum.
[a]High-level transduction of astrocytes.
[b]High-level transduction of oligodendrocytes.
[c]High level transduction of microglia.
[d]Rhesus macaque rh2, rh8, rh10, rh20, rh39, Cynomolgus macaque cy5, and Baboon bb2 serotypes also tested in the central nervous system with varying transduction efficiencies.
[e]Porcine AAV serotypes po1, po2, po4, po5, and po6 also tested in eye with varying transduction efficiencies.

reactivation.[21,22] Deletion of the latency gene has little effect on the establishment of latency, allowing the manipulation of this genomic region to express genes long term in neurons. Replication-defective HSV genomic vectors have been developed for gene transfer (Figure 9.1(e)) as these vectors are easier to engineer and produce to high titer (Table 9.1). Although wild-type virus is capable of reactivation from the latent state, replication within the host and spread to others, replication-defective HSV vectors enter a permanent state of "latency" incapable of reactivation.

9.3 GENE THERAPY FOR CHRONIC PAIN

Chronic pain remains a disease state in need of effective therapies that lack the unwanted side effects and tolerance or abuse problems associated with the various current pain drug therapies. Gene therapy represents a viable approach for the treatment of chronic pain that will avoid the unwanted drug-associated whole body side effects because the vectors can be injected into patients only at sites where they are experiencing pain. This section covers the various strides made in preclinical studies using gene therapy approaches to treat various forms of chronic pain such as arthritis pain, interstitial cystitis-painful bladder syndrome, bone cancer pain, spinal cord injury pain, infectious agent-induced pain, and even chemotherapeutic agent-induced pain. Although lentivirus, adenovirus, and AAV vectors have been used, the majority of these studies have used HSV vectors that cumulated in the first phase 1 clinical trial to treat chronic cancer pain associated with bone metastases.[23]

are also depicted below the dotted line above the vector genome diagram. The more complex vectors such as adenovirus and HSV, show various stages of vectors that over time have decreased toxicity and immunogenicity because of the deletion of more viral gene products. Viral genes that are transcribed in the 5–3′ direction (rightward arrow) are depicted above the viral genome; those transcribed in the opposite direction (leftward arrow) are depicted below the genome. The location of the therapeutic gene cassettes within the vector genomes is shown in gray and genes covered by an "X" are deleted from the vector genome. AAV, adeno-associated virus; AdV, adenovirus; β, beta or early gene; ds, double-stranded; E, early gene; HCMV, human cytomegalovirus promoter; HSV, herpes simplex virus; ICP, infected cell polypeptide; IE, immediate early gene; IRL, inverted repeat long; IRS, inverted repeat short; ITR, inverted terminal repeat; kb, kilobase; L, late gene; LAT, latency-associated transcript; LTR, long terminal repeat; LV, lentivirus; ORF, open reading frame; P5, AAV promoter; PolyA, polyadenylation sequences; RV, retrovirus; ss, single-stranded; TR, terminal repeat; U3, retroviral; U5, retroviral LTR promoter; UL, unique long; US, unique short.

9.3.1 Chronic Pain and Normal Pain Signaling

Detection and withdrawal from pain is a natural protective mechanism for avoidance of tissue damage and injury. Pain is a cognitive experience that is derived from activation of pain sensing neurons, nociceptors, in the peripheral nervous system. The pain pathway is well-described and involves neurotransmitter release from primary nociceptor afferent termini into the dorsal horn of the spinal cord, thereby activating second-order neurons that transmit pain signals to the thalamus to third-order neurons that activate cognitive centers in the brain.[24] Activation of nociceptors can result from a variety of events including nerve damage, inflammation, contact with caustic chemicals, and extreme pressure and temperature. Two types of nociceptive neurons exist which are involved in pain signaling, the medium-sized Aδ-fibers that are fast-firing because of being myelinated and the smaller C-fibers, which fire very slowly because they are non myelinated. Pain can be classified as being either acute-sharp in nature being primarily mediated by myelinated Aδ-fibers, whereas the slow chronic prolonged pain is thought to be mediated by thinly myelinated C-fibers, most likely involved in chronic pain states. Although acute pain is adaptive, chronic pain represents a pathologic state that results from long-term inflammation or nerve damage. Chronic pain is defined as severe and ever-present pain that persists for at least 3 months after initial injury or tissue damage. The mechanism(s) that underlie chronic pain are incompletely understood, but clearly the biology of nociceptors is altered with the establishment of low threshold patterns of activation. Common causes of chronic pain include cancer, diabetes, arthritis, pancreatitis, cystitis, infections such as shingles, irritable bowel syndrome, surgery, and amputation. It is estimated that 80 million Americans will experience chronic pain at some time in their lives and with an estimated 90 billion in health care costs and loss of work time.[25] As the population ages, the toll of chronic pain is expected to increase.

The treatment of acute pain is manageable and involves the use of drugs that range from anti-inflammatory substances such as nonsteroidal anti-inflammatory drugs that inhibit cyclooxygenase, opioid analgesics, and corticosteroids. Chronic pain is far more difficult to treat and may include combinations of anti-inflammatory substances, opioids, and adjuvant analgesics that include tricyclic antidepressants, γ-aminobutyric acid (GABA), and N-methyl-D-aspartic acid (NMDA) agonists.[26] Although these drugs can be effective for some patients they are not without systemic toxicities and have unpleasant sedative-like side affects on patients. Opioids have been widely used for treatment of chronic pain and include hydrocodone, morphine, oxycodone, and fentanyl. Although initial analgesia is quite good, use beyond a few weeks is accompanied by growing tolerance and lack of efficacy. Because these drugs act systemically, they can have profound effects on brain activity, bowel function, and respiration. Addiction is commonly associated with opioid abuse. Vicodin has become one of the most highly dispensed drugs in the United States, with estimates of exceeding 15–20 million people.[27] Clearly safer, more effective strategies lacking the undesired side effects are needed to combat chronic pain.

9.3.2 Genetic Intervention in Pain

Gene therapy for chronic pain has many attractive features that suggest a potential broad-based utility. The intent of gene therapy for chronic pain is to introduce genes into the ganglia where pain arises. Pain-ameliorating gene products block the activation of nociceptors and/or the release of neurotransmitters or binding to second-order neurons in the spinal cord. The achievement of this goal will block pain at its source thereby representing a local rather than systemic therapy. Both viral and nonviral vectors have been used for pain gene therapy and the therapeutic genes are wide ranging.[28] Because HSV establishes latency in peripheral nerve ganglia, we focused our research efforts on developing gene therapies for peripheral neuropathies[29,30] and chronic pain.[31,32] Introduction into the skin allows vector uptake by sensory nerve terminals and transport of the defective virus to the nerve cell body where the viral genome persists and can function as a platform for transgene expression. Third-generation HSV vectors are suitable for gene delivery to sensory neurons. Just like wild-type HSV, replication-defective HSV vectors are readily taken up by nerve terminals following intradermal injection or by other delivery routes. The efficiency of retrograde axonal transport exceeds other vector systems and the virus genome is delivered to the nerve cell nucleus where it persists for life as a stable circular episome.

Our initial work to exploit HSV vectors to treat pain involved the local expression of opiate peptides. Preproenkephalin gene delivery results in protein processing to produce met and leu opiate peptides that impede pain-related neurotransmitter release through binding to the mu and delta receptors found on both nociceptors and second-order neurons in the spinal cord. Thus, we reasoned that HSV expression of preproenkephalin in the primary afferent could substantially dampen pain signaling at the site where nociceptor activation was occurring; in other words, treating pain at its source. This strategy proved effective in blocking both neuropathic and inflammatory pain in multiple animal models of chronic pain.[31–43] Transient expression of enkephalin prevented pain and longer-term treatment could be achieved by repeat dosing.[32] Importantly, enkephalin gene therapy did not lead to tolerance and could be observed even when tolerance to morphine was induced by high-dose administration of the opioid drug.[34] These features were essential because early-phase human clinical trials for treatment of cancer-related pain would encounter these two issues. Data from the phase 1 trial looked very promising[23] (Table 9.3) and the results of the phase 2 trial are in review (Darren Wolfe, personal communication, Periphagen Inc., Pittsburgh, PA).

A similar approach has involved the engineering of an enkephalin gene cassette that encodes both endomorphin and enkephalin peptides.[49,50] Other genes such as glutamic acid decarboxylase (GAD) produce GABA that will dampen pain signaling by binding the GABA receptor.[51] Other strategies have involved the use of genes that express neurotrophins[52] or growth factors (nerve growth factor, brain-derived neurotrophic factor (BDNF), glial cell line-derived neurotrophic factor (GDNF), vascular endothelial growth factor, erythropoietin, fibroblast growth factor 2 (FGF2), and hepatocyte growth factor) that offer the ability to reduce nerve cell damage and neuropathic pain. In cases of pain largely because of inflammation, immune modulatory factors such as interleukin (IL)-10, IL-2, IL-4, tumor necrosis factor soluble receptor,[53] and IkB have been used to reduce the root cause of pain that relies on the presence of an inflammatory soup to induce nociceptor activation. The finding that chronic pain signaling can result from increased expression of certain sodium channels such as Nav1.7, 1.8, and 1.9 has led to the use of short hairpin RNAs to reduce channel expression.[54] Genes that prevent activation (PP1α) or transport (dominant negative PKCε) of the calcium channel TRPV1 have been shown to prevent inflammatory pain responses (Reinhart et al., unpublished).[55]

Despite the utility of enkephalin and other therapies,[31–43,49–56] the consequences of long-term pain-relieving transgene expression using a form of the latency promoter remained uncertain and untested in patients. Therefore, regulated therapy may prove essential for pain gene therapy to become a practical solution for chronic pain treatment. A possible solution may involve the use of a ligand-mediated chloride channel that can hyperpolarize the nerve cell and prevent the formation of a pain signal-induced action potential. The glycine receptor (GlyR), a common chloride channel that is expressed only in the central nervous system and the alpha subunit alone (GlyRα) could be distributed along axons and was sufficient for channel activation by glycine.[57] Activation resulted in hyperpolarization of the neuron with subsequent silencing, raising the question of whether delivery of the GlyRα gene to sensory nerves with subsequent glycine administration to the skin would block pain. We recently showed that both neuropathic and inflammatory pain (i.e., regardless of the cause of pain), could be blocked by GlyRα activation following HSV-mediated gene transfer to sensory nerves.[57] Thus, we had created a "nerve-silencing" tool that could be regulated by a drug, providing a novel drug-regulated approach to gene therapy. We also found that glycine could be administered systemically to treat bladder pain.[57] These studies have now been extended to the use of a mutant form of the GlyR that is exclusively activated by ivermectin,[58] an approved drug for treatment of helminth infections in humans[59] and it appears feasible that ivermectin-inducible GlyR activation will create a powerful and highly specific means of controlled, long-term pain treatment.

9.4 GENE THERAPY FOR EPILEPSY

9.4.1 Clinical Manifestation of Epilepsy

Epilepsy is the most common serious neurological disorder. It affects an estimated 1% of the population (i.e., about 60 million individuals worldwide and about 10 million between Europe and North America); approximately one in 25 people will develop epilepsy at some point in his or her lifetime [60,61]. The term "epilepsy" actually includes a large group of genetic and acquired chronic neurological disorders whose single common feature is a persistent increase of neuronal excitability occasionally and unpredictably expressed as a seizure. According to previous work,[62] epilepsy comprises "the transient occurrence of signs and/or symptoms due to abnormal, excessive or synchronous neuronal activity." Seizures can be of two types: generalized, when occurring in and rapidly engaging bilaterally distributed networks, or focal, when occurring within networks limited to one hemisphere.[63] Etiologically, epileptic syndromes are classified as genetic (when resulting from a known genetic defect), structural/metabolic (when resulting from a structural or metabolic lesion), and of unknown cause.[63] Genetic epilepsies are most often associated with generalized seizures, whereas most structural epilepsies are associated with focal seizures that originate within or around the lesion area.

There is a significant unmet medical need in epilepsy. First, no truly antiepileptogenic therapy is available. None of the antiepileptic drugs in clinical use can prevent the development of epilepsy in cases in which the cause of the epileptogenic lesion is identifiable (e.g., lesional epilepsies following for example a head trauma, an episode of status epilepticus (SE), a stroke, a brain infection). Second, pharmacological therapy is unsatisfactory: one-third of the patients treated with antiepileptic drugs continue to experience seizures. Furthermore, in patients in whom seizures are well controlled, drugs may exert debilitating side effects and, in time, refractoriness to their therapeutic effects may develop. For some patients with focal seizures that are refractory or become refractory to pharmacological therapy, one final option is the surgical resection of the epileptogenic

TABLE 9.3 Gene Therapy Clinical Trials for Chronic Pain and Parkinson Disease

Disease	Anatomical Target	Phase	Sponsor	Viral Vector	Transgene—Therapeutic Mechanism	Reference	Status
Chronic bone cancer pain	Intradermal injection of vector into dermatome innervated by neurons showing pain	Phase 1 trial	Diamyd Inc.	HSV-1	Preproenkephalin synthesized in dorsal root ganglion secreted by vesicles into synaptic cleft within spinal dorsal horn	23	Completed. Phase 2 trial results currently under review.
Parkinson's disease	Putamen	Controlled phase 2	Ceregene	AAV2	Neurturin (NRTN) for neuroprotection, nigrostriatal regeneration, and upregulation of dopamine production	48	Completed. Met safety but failed efficacy endpoints.
Parkinson's disease	Putamen and substantia nigra	Controlled phase 2	Ceregene	AAV2	Same as above but added substantia nigra target	47	Completed. Met safety but failed efficacy endpoints. No further clinical development expected at this point.
Parkinson's disease	Putamen	Open phase 1b	NIH/NINDS	AAV2	GDNF: similar to NRTN mechanism, but convection-enhanced delivery (CED) expected to increase putaminal coverage	Bankiewicz, personal communication	Ongoing and recruiting NCT01973543
Parkinson's disease	Subthalamic nucleus	Controlled phase 2	Neurologix	AAV2	GAD: converts glutamate to GABA to increase synaptic inhibition of the subthalamic nucleus	46	Completed and met endpoints. Significance not enough to attract funding for further development.
Parkinson's disease	Putamen	Open phase 1b	Genzyme	AAV2	AADC: converts systemic L-dopa to dopamine	44	Completed and met endpoints. Additional study is recruiting NCT01973543.
Parkinson's disease	Putamen	Open phase 1/2	Oxford Biomedica	Equine infectious anemia virus	AADC, tyrosine hydroxylase, guanosine triphosphate (GTP) cyclohydrolase. Intraputamenal autonomous dopamine production	45	Completed and met endpoints. Additional clinical development expected.

region. Overall, we do need easily tolerated therapies and therapies that are effective in pharmacoresistant patients. Third, there is a need for disease-modifying therapies. Antiepileptic drugs do not prevent the progressive worsening of seizures observed in many patients, and we lack therapies that can ameliorate or prevent the associated cognitive, neurological, and psychiatric comorbidities, or epilepsy-related mortality. The annual direct and indirect cost attributable to epilepsy in the United States is $12.5 billion, and 80% of this is accounted for by patients whose seizures are not adequately controlled by antiepileptic drugs.

9.4.2 Gene Therapy Intervention Strategies for Epilepsy

More than one-third of the epilepsies are of genetic origin. One may think that these are the best candidates for gene therapy, but this is not the case. Very rarely a single mutant gene causes epilepsy, whereas more commonly it results from the inheritance of two or more susceptibility genes.[64] Moreover, the pathology in these cases often affects a large part of the brain and, thus, would require widespread gene transfer. These features pose two big obstacles against gene therapy. First, there is the need to transfer multiple genes into diseased cells, when most viral vectors (with the notable exception of HSV) have a small genome that can host only one gene at a time. Second, widespread expression of the therapeutic gene(s) is needed, and currently available gene therapy methods mostly provide localized effects.

Researchers are attempting to overcome these obstacles of multiple gene transfer and widespread expression. As noted, HSV can host multiple transgenes in a single viral particle and this feature may be exploited for multi-predisposing genes caused epilepsies. Moreover, strategies are under development for the global delivery of genes to the brain by crossing the blood–brain barrier (BBB) after administering vectors in the peripheral blood. One such strategy is to employ a pathway used by circulating endogenous molecules, such as transferrin or insulin, to reach neurons and glia.[65,66] These molecules bind to their specific receptors on the luminal side of the endothelial cell membrane, and a vesicle is then formed that contains the bound receptor and ligand. This vesicle is then carried across the endothelial cell cytoplasm (from the luminal to the abluminal side) in a transport mechanism called transcytosis. For gene therapy, a vector can be conjugated to a ligand (a peptide or a single-chain antibody against the transcytosis receptor) that mimics the natural ligand (e.g., transferrin, insulin). The vector–ligand conjugate remains intact and unmodified while in transit and is released intact into the interstitial space. Although some AAV vectors have been shown to undergo transcytosis of the rodent BBB,[67,68] much work remains to be done to prove that this approach can be applied to the treatment of genetic epilepsies in humans.

For now, epileptic syndromes characterized by a focal lesion appear to be much better candidates for gene therapy because, in these cases, seizures originate in an often-restricted brain area, allowing local delivery of the vector using stereotaxic surgery. Epilepsy caused by focal lesions often has an identifiable cause in a damaging insult (head trauma, SE, stroke, brain infection) that sets a cascade of neurobiological events in motion. This eventually leads to spontaneous seizures and to the diagnosis of epilepsy. The development and extension of tissue capable of generating spontaneous seizures, resulting in (1) development of an epileptic condition and/or (2) progression of the epilepsy after it is established is termed "epileptogenesis". Accordingly, antiepilepsy treatments can be symptomatic (i.e., antiseizure), and disease-modifying, which include antiepileptogenic and comorbidity-modifying therapies. Lesional epilepsies offer opportunities for intervention at all levels.

9.4.3 Administering Gene Therapy in Epilepsy

The vector types employed thus far in epilepsy studies have been AAV (different serotypes), lentivirus, and HSV. Typically, the route of delivery has been the stereotactic injection of the vector directly into the epileptogenic region (the hippocampus, in most instances). This approach ensures high levels of transgene expression and a limited immune response (even if the surgical procedure may induce some breakdown of the BBB and some penetration of lymphocytes). The extent of spread of the viral particles can be calibrated to ensure covering the target epileptogenic area and limit the volume injected and the number of injections. Two examples: the retrograde transport of HSV can allow bilateral delivery of therapeutic genes after injection into one hippocampus, HSV being retrogradely transported to the contralateral hippocampus by commissural fibers[69]; different AAV serotypes have been reported to have different attitudes to spread around the site of injection.[70]

Researchers have tested other routes of administration to obtain specific accumulation of the transgene in the region of interest avoiding a direct intracerebral administration. For example, intranasal delivery has been tested using a replication-defective HSV-2 vector to deliver the antiapoptotic gene *ACP10PK*,[71] but the resulting transgene expression was not specific to the area of interest and level of expressions were low. A more promising approach may be to take advantage of the BBB breakdown that occurs following an epileptogenic insult or seizures. An AAV clone has been identified that can cross the seizure-compromised, and not the intact, BBB,[72] suggesting that it may be feasible to create vectors that, after peripheral administration, selectively target the brain areas involved in seizure activity.

9.4.4 Genetic Intervention in the Prevention of Epilepsy

In both humans and animals, epileptogenesis (that is, the transformation of a normal brain to epileptic) is associated with focal pathological abnormalities, like cell death (the most prominent is a loss of neurons in the hippocampus termed "hippocampal sclerosis"), axonal and dendritic plasticity, neurogenesis, neuroinflammation, alterations in ion channel, and synaptic properties. The molecular mechanisms underlying these abnormalities are still incompletely understood, but impairment in the levels of neurotrophic factors (NTFs) like FGF-2 and BDNF may be one factor.[73] Both FGF-2 and BDNF protect neurons from damage and, moreover, FGF-2 is a potent proliferation factor for neural stem cells, whereas BDNF favors their differentiation into neurons.[74]

We used a replication-defective HSV-1 vector expressing FGF-2 and BDNF to test the hypothesis that local supplementation of these NTFs can attenuate seizure-induced damage and contrast epileptogenesis.[69] This vector was injected into one hippocampus 4 days after a chemically induced SE in rats (i.e., during latency and after the establishment of damage). These conditions are similar to those of a person who, after the occurrence of an epileptogenic insult, is in the latency period preceding the beginning of spontaneous seizures. Expression of the transgenes was bilateral (because of retrograde transport of the vector to the contralateral hippocampus) and transient (approximately 2 weeks). However, short-term expression is an advantage in these specific settings because NTFs can trigger plastic changes that remain detectable when they are no longer expressed, whereas their long-term expression may be detrimental for brain function. Thus, this approach allowed modification of the microenvironment by generating cells capable of constitutively but transiently secreting FGF-2 and BDNF. The treatment slightly attenuated the ongoing cell loss, favored the proliferation of early progenitors, and led to the production of cells that entered the neuronal lineage of differentiation. One month after SE, all untreated animals displayed hippocampal sclerosis and spontaneous seizures. Treated animals, in contrast, had a highly significant reduction of cell loss in the hippocampus and of the frequency and severity of spontaneous seizures[69,75,76] (i.e., the FGF-2 and BDNF expressing vector exerted a disease-modifying anti-epileptogenic effect).

In summary, these studies provide evidence that gene therapy may become an option in epilepsy prevention in at-risk patients. The disease-modifying effect was striking and could be interpreted as truly antiepileptogenic, an effect that is not achieved by the currently available therapy. Moreover, virus injection was performed under conditions compatible with the clinical settings of patient observation (after the epileptogenic insult), conditions that reproduce those that may allow therapeutic intervention. However, clinical translation is premature: data must be replicated in other models and, most importantly, invasive techniques are not justified because only 10% of at-risk patients develop epilepsy.[77–79] Before moving to the clinics, therefore, it will be essential to develop biomarkers to identify patients that will actually develop epilepsy after a potentially epileptogenic insult.

9.4.5 Using Gene Therapy as an Antiseizure Treatment

A logical approach to stop seizures in drug-resistant patients is modulation of excitability (i.e., to increase the strength of inhibitory signals or reduce the strength of excitatory signals). One target for such approach is the $GABA_A$ receptor, the main receptor subtype for the principal inhibitory neurotransmitter. This is a pentameric chloride-permeable channel. In the granule cells of the hippocampus of epileptic rats, the expression of $GABA_A$ alpha-1 subunits is decreased and that of alpha-4 subunits is increased,[80] generating receptors that rapidly desensitize, thus favoring the generation of seizures.[81] An AAV2 vector containing the alpha-4 subunit promoter driving alpha-1 expression was used to increase alpha-1 expression in the hippocampal granule cells after SE, obtaining an increase in latency duration and a decreased number of rats developing spontaneous seizures.[82]

A reduction in the strength of excitatory signals was obtained by downregulating a receptor for glutamate, the principal excitatory neurotransmitter. Thus, an essential subunit for the functioning of NMDA receptors (NR1) was cloned in antisense in two AAV vectors, one in which the transgene was under control of a promoter expressed in inhibitory interneurons, the other under control of a promoter expressed in excitatory neurons.[83] The two different vectors had opposite effects: reducing excitatory input onto inhibitory neurons favored seizures, whereas doing the same onto excitatory neurons reduced seizures.[83] It is therefore critical to transduce a specific cell population anytime the transgene codes for a receptor (or a channel) that is expressed on both inhibitory and excitatory neurons. This concept has more recently found support from a study showing that inhibition of excitatory neurons via lentivirus vector-mediated overexpression of a potassium channel had a therapeutic effect in a model of neocortical epilepsy (Figure 9.2).[84]

To avoid the problem of targeting specific neuronal populations, one could express an inhibitory factor that is constitutively secreted from the transduced cells. If the receptors for that factor are present in the injected area, seizure control may be achieved without the need to target specific cells. Indeed, antiseizure effects have been obtained by overexpressing the NTF GDNF in the hippocampus[85] and increasing hippocampal levels of the endogenous anticonvulsant adenosine.[86] The most promising results have been obtained with the inhibitory neuropeptides galanin (GAL) and neuropeptide Y (NPY).

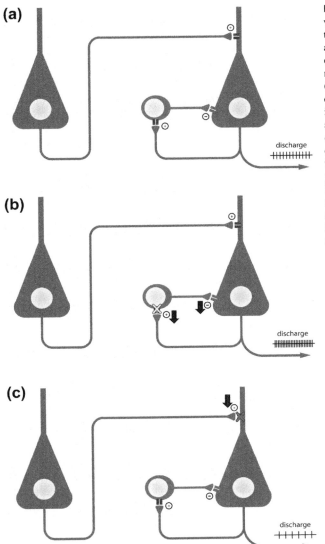

(a)

discharge
++++++++++

(b)

discharge
+++++++++++++++

(c)

discharge
++++++

FIGURE 9.2 Consequences of transduction of a specific cell population when the transgene codes for a receptor that is expressed on both inhibitory and excitatory neurons. (a) A typical interaction between excitatory and inhibitory neurons, as occurs for example in the hippocampus. Excitatory output neurons (right blue neuron) receive input from other excitatory neurons (left blue neuron) and feedback inhibition from inhibitory interneurons (red) that are driven by their own collaterals. The same receptors mediate the excitatory input onto the output neuron and the activation of the inhibitory interneurons by the collaterals. These settings ensure an output discharge of specific frequency. (b) The effect of a gene therapy intervention that knocks down expression of the excitatory receptor in the interneuron. The consequence is a decreased activation of the interneuron, with decreased inhibitory feedback and increased output discharge (that may translate in increased probability of seizures). (c) The effect of a gene therapy intervention that knocks down expression of the excitatory receptor in the output neuron. The consequence is a decreased activation of the output neuron, with decreased output discharge (that may translate in reduced probability of seizures). (For interpretation of the references to color in this figure legend, the reader is referred to the online version of this book).

Increased GAL signal (obtained in GAL transgenic mice or with GAL receptor agonists) attenuates seizures, whereas its blockade (in GAL knockout mice or with receptor antagonists) exerts proconvulsant effects.[87] To obtain constitutive secretion of GAL from transduced cells in the seizure-generating area, an AAV vector was constructed in which the GAL coding sequence was preceded by the secretory signal sequence of fibronectin, a protein that is constitutively secreted.[88] After injection into the hippocampus, this vector attenuated seizures and prevented seizure-induced cell death in two experimental models.[88] Similar results were obtained using different GAL-expressing vectors in different acute and chronic seizure models.[89,90] Notably, these effects were independent of the promoter driving GAL expression. Together, these studies support the notion that vector-derived GAL expression and constitutive secretion suppress epileptic seizure activity.

Activation of the NPY Y2 and Y5 receptors inhibits glutamate release in the hippocampus and attenuates seizures. Transgenic rats overexpressing NPY show reduced seizure susceptibility, whereas knockout mice lacking NPY or the Y2 or Y5 receptor gene are more vulnerable to seizures. Moreover, NPY potently inhibits excitatory responses in hippocampal slices from epileptic patients.[91] AAV vector-mediated overexpression of NPY in the hippocampus reduced seizures in several acute and chronic models of seizures and epilepsy,[70,92,93] and this effect was potentiated if the Y2 receptor was coexpressed with NPY.[94] In chronic models, a remarkable decrease in the progressive increase of seizure frequency (i.e., a disease-modifying effect) was also observed.[93] Moreover, AAV-mediated NPY supplementation did not produce significant side effects such as impairment in learning and memory, anxiety, or alterations in locomotor activity.[95,96] The overall evidence supports the application of NPY gene therapy for human epilepsy.

9.4.6 Opportunities for Gene Therapy Clinical Studies in Epilepsy

No clinical trial has been conducted so far in epilepsy. However, patients with partial epilepsies selected for surgical resection of the epileptogenic area are ideal candidates for gene therapy. There are an estimated 100,000–200,000 patients in the United States alone that could benefit from surgery. Although surgical treatment for epilepsy is effective in about 60–70% of the patients, only about 3000 surgeries per year are performed in the United States. The reasons for the lack of referral to surgery by the neurology community may be partly explained by fear (by both patients and neurologists) of the idea of surgical removal of partially normal brain and the fact that this surgery is associated with death or major disability in 4% of the patients.[97] Gene therapy may offer a more cost-effective and much less invasive alternative, with higher benefit-risk ratio than surgery. Tissue resection may be viewed as the most extreme form of cellular "silencing," and gene therapy may provide a realistic alternative by ensuring the sustained release of inhibitory molecules (such as GAL or NPY) that "silence" hyperactivity.

Gene transfer of inhibitory factors into the epileptogenic area in patients who are planning to undergo surgery does not require ad hoc stereotaxic intervention when these patients undergo implantation of depth electrodes for diagnostic purposes before surgery. Also, a rescue procedure would be in place because, should gene therapy fail to produce any advantage, patients would simply undergo resective surgery as originally planned. Therefore, the only deviation from standard care will be the postponement of the surgery from the typical 2–4 months following intracranial recording to 6 months, to allow assessment of the gene therapy effects. In summary, we believe that the latest vector technologies (including their human safety) and the specific features of these patients will soon allow the start of phase 1 gene therapy studies for epilepsy.

9.5 PARKINSON'S DISEASE

9.5.1 Parkinson's Disease and the Current Standard of Care

Parkinson's disease (PD) is a degenerative disorder of the central nervous system characterized by progressive, debilitating motor impairments resulting from the dysfunction and loss of dopamine-secreting neurons in the substantia nigra. The disease affects more than four million people worldwide[98] and pharmacologic treatments such as levodopa (L-dopa) and dopaminergic agonists provide symptomatic relief of the motor symptoms early in the disease by compensating for the loss of endogenous dopamine activity. Although L-dopa is remarkably effective for many years in PD patients, its therapeutic effects are only temporary and continued loss of dopaminergic neurons eventually leads to a point when drugs become ineffective. Therapies that could halt or reverse the degenerative process or replace or modulate the dopaminergic synthesis would therefore be a great advance in the treatment of the disease. Even though the dopaminergic neuronal loss is variable from patient to patient, the loss occurs in well-defined anatomical regions making it suitable for targeted gene therapy approaches. In human trials, gene therapy has been applied to promote dopaminergic neuron survival and regeneration, restore striatal dopamine levels, or modulate the basal ganglia neurotransmission. As the underlying genetic and pathogenic causes of the dopaminergic degeneration are becoming better understood, the modification of the dopaminergic neurons themselves is also being explored in the laboratory but no clinical trial has yet been started. Both in vivo and ex vivo gene therapy approaches have been applied to the treatment of PD. So far, only the in vivo approaches have been applied in the clinic using AAV or lentivirus vectors.

9.5.2 Preclinical Data for Neuroprotection and Restoration

An ideal treatment for PD would be one that restores the function of dopaminergic neurons while also protecting them from further damage and loss. Accordingly, trophic protein factors have been under intensive investigation as treatments from their inherent roles in promoting neuronal survival, growth, and function. In particular, the GDNF family of proteins has been investigated as these factors are involved in survival and maturation of dopaminergic neurons under embryonic development and have shown both safety and efficacy when applied in several PD models and species.[99] Because growth factors such as GDNF do not pass the BBB, direct delivery approaches such as gene therapy are needed. Experiments have shown that GDNF and its family member neurturin need to be applied to the dopaminergic nerve terminals in the striatum to have both neuroprotective and regenerative effects whereas application to the dopaminergic neuronal cell bodies in the substantia nigra alone only confers neuroprotection.[99,100] Toxin-induced dopaminergic neuronal degeneration has been particularly responsive to the treatment effects of the GDNF family, but it is currently unclear if these models are predictive of clinical translation. With the increased understanding of PD and the important role of misfolded α-synuclein in the pathogenesis, new models have been developed where the AAV-mediated overexpression of α-synuclein in dopaminergic neurons causes dopaminergic neurodegeneration unresponsive to striatal GDNF treatment.[101] Furthermore, the overexpression of α-synuclein downregulates the dopaminergic neuronal survival factor Nurr-1, which in turn appears to decrease the responsiveness to GDNF by lowering the levels of its signaling receptor Ret.[102] It is therefore possible that α-synuclein-induced, reduced GDNF responsiveness in the clinic can explain the discrepant treatment results in the toxin-induced animal models and actual PD patients. However, it should be noted that AAV overexpression of α-synuclein in dopaminergic neurons may not be representative of the clinical situation and data suggest that the GDNF family of neurotrophic factors may still be effective if applied earlier in the disease process in the right amounts. Current preclinical work aims at elucidating the interaction between α-synuclein and survival signaling and how different species of misfolded α-synuclein induce dopaminergic toxicity and responsiveness to growth factors.

9.5.3 Neuromodulation in PD

In patients with severe degeneration of the dopaminergic system, growth factor therapy becomes ineffective as the loss of dopaminergic neurons and their nerve endings becomes too severe to reverse. In these patients, the restoration of striatal dopamine levels may confer symptomatic therapeutic effects. Because systemic dopamine cannot enter the BBB and would be associated with systemic side effects, long-term localized striatal delivery is necessary, making gene therapy an optimal therapeutic choice. Because dopamine is enzymatically synthesized, genetic enzyme replacement in striatal cells has been tried and shown to be effective in rodent and nonhuman primate models. An elegant way of making dopamine in the striatum has been to combine systemic L-dopa therapy with a striatal gene therapy enzyme replacement of aromatic amino acid decarboxylase (AADC). AADC converts L-dopa into dopamine and AAV vectors making AADC targeting striatal nondopaminergic neurons did indeed restore the responsiveness to L-dopa therapy in MPTP-treated nonhuman primates and safety and tolerability and evidence of clinical improvement was seen in a phase 1b study.[44]

The full replacement of the enzyme machinery converting the amino acid tyrosine into dopamine has also been explored. A tri-cistronic lentiviral vector based on a replication-incompetent equine infectious anemia virus lentiviral vector coding for the dopamine synthesizing enzymes AADC, GCH1, and TH has shown reduction in motor impairment as well as L-dopa-induced dyskinesias and clinical translation has been successful meeting both safety and efficacy endpoints in a phase 1/2 clinical trial.[45] A variant gene therapy enzyme replacement approach where L-dopa is synthesized locally in the striatum by overexpression of TH and GCH1 has also been explored and shown to improve motor deficits and, importantly, reduce systemically administered L-dopa-induced dyskinesias. This approach is currently being translated toward clinical application.[103]

With the loss of dopaminergic signaling in the basal ganglia neuronal circuitry, the inhibitory control exerted by the internal segment of the globus pallidus on the subthalamic nucleus is reduced, resulting in subthalamic hyperactivity and Parkinsonian motor symptoms. The inhibition of the hyperactivity of the subthalamic nucleus is a target for deep brain stimulation, where high-frequency stimulation inhibits neuronal activity and modulates the basal ganglia circuitry toward more normal functioning. A similar modulatory effect can be achieved by increasing the inhibitory neurotransmitter GABA by overexpressing the enzyme GAD with gene therapy in the subthalamic nucleus. This approach has been shown to reduce motor deficits in both rodent and nonhuman primate models[104] and was successfully translated into clinical trials based on preclinical safety and efficacy.

9.5.4 Dopaminergic Neuron Modification

With an increasing understanding of the molecular basis of dopaminergic neuronal degeneration in PD and potential early detection of the disease by biomarkers, early intervention, and genetic modification of the dopaminergic neurons themselves could stop the degeneration and disease progression. The role and understanding of α-synuclein in the pathogenesis of PD has become increasingly important and a reduction of misfolded α-synuclein should result in neuroprotection and retardation of the disease process. Several gene therapy vectors have therefore been designed to interfere with α-synuclein levels in dopaminergic neurons. In some experiments, α-synuclein silencing with RNA interference led to reduced levels of endogenous or overexpressed α-synuclein and positive effects were shown in vivo and in vitro.[105] However, other experiments have shown that the reduction of endogenous wild-type α-synuclein can lead to dopaminergic neuronal degeneration[106] and it appears that retained levels of α-synuclein are necessary for survival and normal functioning. Therefore, a potential gene therapy vector targeting α-synuclein may be difficult to design and translate into clinical development that reduces pathological levels of misfolded α-synuclein while sustaining normal α-synuclein function.

Parkin is another protein in which mutations have been linked to PD. The overexpression of normal Parkin appears to confer neuroprotective effects in vivo[107] but few studies have explored translation toward the clinic.

With the recent data suggesting a link between α-synuclein and Nurr1 downregulation, the overexpression of Nurr1 in dopaminergic neurons may provide a simple manner to confer neuroprotection to dopaminergic neurons, independent of underlying cause.[108] The exploration of this approach in animal models is currently ongoing.

9.5.5 Clinical Trials for PD

The discovery of neurotrophic factors like GDNF and neurturin with survival and regenerative effect on dopaminergic neurons in the aforementioned toxin-induced PD models brought hope that if applied to patients with PD, neuroprotection and restoration could be readily attained. However, the translation of GDNF and neurturin into therapeutics for PD has been a mixed bag of hope and failure where unfortunately several clinical trials have been unable to demonstrate efficacy.[47,48,109] However, refinements in pump and gene therapy technologies provide hope that factors like GDNF may become a therapeutic reality in the not-too-distant future. It appears that the trials to date have been plagued by "too little too late" in that the appropriate target tissue therapeutic levels of growth factor have not been achieved and that the patients selected were too far in their disease progression. New approaches are therefore aimed at optimizing growth factor levels in the target tissue in patients and applied earlier in their disease process. Currently, only GDNF is undergoing clinical trials using convection enhanced striatal delivery of an AAV vector coding for human GDNF (Bankiewicz, personal communication). An ex vivo gene therapy approach using GDNF expressing encapsulated cells is in late preclinical development and should enter clinical trials in the near future. This ex vivo approach uses a slender implantable device to house the modified cells allowing for removal if necessary.[110,111] This technology combines the safety of a retrievable device with the potency of gene therapy allowing for entry into patients earlier in the disease. It is envisioned that this approach will therefore stand a better chance to show efficacy of GDNF than was possible in earlier trials.

With respect to neuromodulation, the AAV-GAD approach was successfully translated into clinical trials and statistically significant efficacy was demonstrated in a phase 2 proof-of-concept trial.[46] However, as DBS is targeting the same structure (subthalamic nucleus) and with similar or better results, the AAV-GAD approach was not competitive enough and additional clinical development has been stopped. The aforementioned dopamine-enhancing enzyme replacement strategies are currently undergoing clinical trials and both initial safety and efficacy have been demonstrated warranting additional clinical development.

As described in this section and summarized in Table 16.3, gene therapy for PD has successfully undergone clinical translation and even though some trials have failed from an efficacy standpoint, important safety and feasibility results have been established enabling refinements in methods and more rapid translation of new approaches in future trials. It is likely that both protective/restorative and dopamine replacement approaches will be successfully developed in the next few years.

9.6 CONCLUSIONS

The goal of this chapter was to provide some examples of promising gene therapy strategies for diseases of the nervous system. The successful development of gene therapy strategies for neurological diseases will require both advancements in vector technology and in the understanding of the mechanisms of disease. Regarding mechanisms of disease, as illustrated in this chapter, intervention will primarily target pathogenic mechanisms and not defective genes; therefore, a correct understanding of these mechanisms will be instrumental for gene therapy. Regarding vector technology, vectors should be further improved in terms of efficacy, safety, duration, and regulation of transgene expression. Another major issue that will need careful consideration is the targeting of specific neuronal populations: as described in the epilepsy section, the consequences of targeting one or another neuronal population may be huge. More in general, out-of-target expression of transgenes may cause unwanted side effects.

Peripheral nervous system diseases are the most advanced toward approval for clinical use. These include neuropathic pain but also deficits of vision, hearing, and smell that were not discussed here. Gene therapy for central nervous system diseases is far more difficult, but preclinical and early clinical data like those described here for epilepsy and PD prompt the optimistic view that the scientific and technical hurdles that remain on the path to standard clinical application will be overcome in the near future.

REFERENCES

1. Gage FH, Temple S. Neural stem cells: generating and regenerating the brain. *Neuron* 2013;**80**:588–601.
2. Simonato M, Bennett J, Boulis NM, Castro MG, Fink DJ, Goins WF, et al. Progress in gene therapy for neurological disorders. *Nat Rev Neurol* 2013;**9**:277–91.
3. Mazarakis ND, Azzouz M, Rohll JB, Ellard FM, Wilkes FJ, Olsen AL, et al. Rabies virus glycoprotein pseudotyping of lentiviral vectors enables retrograde axonal transport and access to the nervous system after peripheral delivery. *Hum Mol Genet* 2001;**10**:2109–21.
4. Wong LF, Azzouz M, Walmsley LE, Askham Z, Wilkes FJ, Mitrophanous KA, et al. Transduction patterns of pseudotyped lentiviral vectors in the nervous system. *Mol Ther* 2004;**9**:101–11.
5. Beard BC, Keyser KA, Trobridge GD, Peterson LJ, Miller DG, Jacobs M, et al. Unique integration profiles in a canine model of long-term repopulating cells transduced with gammaretrovirus, lentivirus, or foamy virus. *Hum Gene Ther* 2007;**18**:423–34.

6. Derse D, Crise B, Li Y, Princler G, Lum N, Stewart C, et al. Human T-cell leukemia virus type 1 integration target sites in the human genome: comparison with those of other retroviruses. *J Virol* 2007;**81**:6731–41.

7. Yang Y, Li Q, Ertl H, Wilson J. Cellular and humoral immune responses to viral antigens create barriers to lung-directed gene therapy with recombinant adenoviruses. *J Virol* 1995;**69**:2004–15.

8. Varnavski AN, Calcedo R, Bove M, Gao G, Wilson JM. Evaluation of toxicity from high-dose systemic administration of recombinant adenovirus vector in vector-naive and pre-immunized mice. *Gene Ther* 2005;**12**:427–36.

9. Swain GP, Prociuk M, Bagel JH, O'Donnell P, Berger K, Drobatz K, et al. Adeno-associated virus serotypes 9 and rh10 mediate strong neuronal transduction of the dog brain. *Gene Ther* 2014;**21**:28–36.

10. Lawlor PA, Bland RJ, Mouravlev A, Young D, During MJ. Efficient gene delivery and selective transduction of glial cells in the mammalian brain by AAV serotypes isolated from nonhuman primates. *Mol Ther* 2009;**17**:1692–702.

11. Puppo A, Bello A, Manfredi A, Cesi G, Marrocco E, Della Corte M, et al. Recombinant vectors based on porcine AAV serotypes transduce the murine and pig retina. *PLOS ONE* 2013;**8**:e59025.

12. Mason MRJ, Ehlert EME, Eggers R, Pool CW, Hermening S, Huseinovic A, et al. Comparison of AAV serotypes for gene delivery to dorsal root ganglion neurons. *Mol Ther* 2010;**18**:715–24.

13. Watanabe S, Sanuki R, Ueno S, Koyasu T, Hasegawa T, Furukawa T. Tropisms of AAV for subretinal delivery to the neonatal mouse retina and its application for *in vivo* rescue of developmental photoreceptor disorders. *PlOS One* 2013;**8**:e54146.

14. Castle MJ, Perlson E, Holzbaur EL, Wolfe JH. Long-distance axonal transport of AAV9 is driven by dynein and Kinesin-2 and is trafficked in a highly motile Rab7-positive compartment. *Mol Ther* 2013;**22**:554–66.

15. Towne C, Schneider BL, Kieran D, Redmond Jr DE, Aebischer P. Efficient transduction of non-human primate motor neurons after intramuscular delivery of recombinant AAV serotype 6. *Gene Ther* 2010;**17**:141–6.

16. Zheng H, Qiao C, Wang CH, Li J, Yuan Z, Zhang C, et al. Efficient retrograde transport of adeno-associated virus type 8 to spinal cord and dorsal root ganglion after vector delivery in muscle. *Hum Gene Ther* 2010;**21**:87–97.

17. Lovric J, Mano M, Zentilin L, Eulalio A, Zacchigna S, Giacca M. Terminal differentiation of cardiac and skeletal myocytes induces permissivity to AAV transduction by relieving inhibition imposed by DNA damage response proteins. *Mol Ther* 2012;**20**:2087–97.

18. Rosas LE, Grieves JL, Zaraspe K, La Perle KM, Fu H, McCarty DM. Patterns of scAAV vector insertion associated with oncogenic events in a mouse model for genotoxicity. *Mol Ther* 2012;**20**:2098–110.

19. Tang S, Bertke AS, Patel A, Margolis TP, Krause PR. Herpes simplex virus 2 microRNA miR-H6 is a novel latency-associated transcript-associated microRNA, but reduction of its expression does not influence the establishment of viral latency or the recurrence phenotype. *J Virol* 2011;**85**:4501–9.

20. Tang S, Patel A, Krause PR. Novel less-abundant viral microRNAs encoded by herpes simplex virus 2 latency-associated transcript and their roles in regulating ICP34.5 and ICP0 mRNAs. *J Virol* 2009;**83**:1433–42.

21. Du T, Zhou G, Roizman B. HSV-1 gene expression from reactivated ganglia is disordered and concurrent with suppression of latency-associated transcript and miRNAs. *Proc Natl Acad Sci USA* 2011;**108**:18820–4.

22. Nicoll MP, Proenca JT, Connor V, Efstathiou S. Influence of herpes simplex virus 1 latency-associated transcripts on the establishment and maintenance of latency in the ROSA26R reporter mouse model. *J Virol* 2012;**86**:8848–58.

23. Fink DJ, Wechuck J, Mata M, Glorioso JC, Goss J, Krisky D, et al. Gene therapy for pain: results of a phase I clinical trial. *Ann Neurol* 2011;**70**:207–12.

24. Woolf CJ. Windup and central sensitization are not equivalent. *Pain* 1996;**66**:105–8.

25. Martin BI, Turner JA, Mirza SK, Lee MJ, Comstock BA, Deyo RA. Trends in health care expenditures, utilization, and health status among US adults with spine problems, 1997-2006. *Spine (Phila Pa 1976)* 2009;**34**:2077–84.

26. Chou R, Fanciullo GJ, Fine PG, Adler JA, Ballantyne JC, Davies P, et al. Clinical guidelines for the use of chronic opioid therapy in chronic non-cancer pain. *J Pain* 2009;**10**:113–30.

27. Fishbain DA, Cole B, Lewis J, Rosomoff HL, Rosomoff RS. What percentage of chronic nonmalignant pain patients exposed to chronic opioid analgesic therapy develop abuse/addiction and/or aberrant drug-related behaviors? A structured evidence-based review. *Pain Med* 2008;**9**:444–59.

28. Goins WF, Cohen JB, Glorioso JC. Gene therapy for the treatment of chronic peripheral nervous system pain. *Neurobiol Dis* 2012;**48**:255–70.

29. Goins WF, Lee KA, Cavalcoli JD, O'Malley ME, DeKosky ST, Fink DJ, et al. Herpes simplex virus type 1 vector-mediated expression of nerve growth factor protects dorsal root ganglion neurons from peroxide toxicity. *J Virol* 1999;**73**:519–32.

30. Chattopadhyay M, Goss J, Wolfe D, Goins WC, Huang S, Glorioso JC, et al. Protective effect of herpes simplex virus-mediated neurotrophin gene transfer in cisplatin neuropathy. *Brain* 2004;**127**:929–39.

31. Goss JR, Harley CF, Mata M, O'Malley ME, Goins WF, Hu X, et al. Herpes vector-mediated expression of proenkephalin reduces bone cancer pain. *Ann Neurol* 2002;**52**:662–5.

32. Goss JR, Mata M, Goins WF, Wu HH, Glorioso JC, Fink DJ. Antinociceptive effect of a genomic herpes simplex virus-based vector expressing human proenkephalin in rat dorsal root ganglion. *Gene Ther* 2001;**8**:551–6.

33. Braz J, Beaufour C, Coutaux A, Epstein AL, Cesselin F, Hamon M, et al. Therapeutic efficacy in experimental polyarthritis of viral-driven enkephalin overproduction in sensory neurons. *J Neurosci* 2001;**21**:7881–8.

34. Hao S, Mata M, Goins W, Glorioso JC, Fink DJ. Transgene-mediated enkephalin release enhances the effect of morphine and evades tolerance to produce a sustained antiallodynic effect in neuropathic pain. *Pain* 2003a;**102**:135–42.

35. Lu Y, McNearney TA, Lin W, Wilson SP, Yeomans DC, Westlund KN. Treatment of inflamed pancreas with enkephalin encoding HSV-1 recombinant vector reduces inflammatory damage and behavioral sequelae. *Mol Ther* 2007;**15**:1812–9.

36. Lu Y, McNearney TA, Wilson SP, Yeomans DC, Westlund KN. Joint capsule treatment with enkephalin-encoding HSV-1 recombinant vector reduces inflammatory damage and behavioural sequelae in rat CFA monoarthritis. *Eur J Neurosci* 2008;**27**:1153–65.

37. Meunier A, Latremoliere A, Mauborgne A, Bourgoin S, Kayser V, Cesselin F, et al. Attenuation of pain-related behavior in a rat model of trigeminal neuropathic pain by viral-driven enkephalin overproduction in trigeminal ganglion neurons. *Mol Ther* 2005;**11**:608–16.

38. Pinto M, Castro AR, Tshudy F, Wilson SP, Lima D, Tavares I. Opioids modulate pain facilitation from the dorsal reticular nucleus. *Mol Cell Neurosci* 2008;**39**:508–18.

39. Wilson SP, Yeomans DC, Bender MA, Lu Y, Goins WF, Glorioso JC. Antihyperalgesic effects of infection with a preproenkephalin-encoding herpes virus. *Proc Natl Acad Sci USA* 1999;**96**:3211–6.

40. Yang H, McNearney TA, Chu R, Lu Y, Ren Y, Yeomans DC, et al. Enkephalin-encoding herpes simplex virus-1 decreases inflammation and hotplate sensitivity in a chronic pancreatitis model. *Mol Pain* 2008;**4**:8.

41. Yeomans DC, Jones T, Laurito CE, Lu Y, Wilson SP. Reversal of ongoing thermal hyperalgesia in mice by a recombinant herpesvirus that encodes human preproenkephalin. *Mol Ther* 2004;**9**:24–9.

42. Yeomans DC, Lu Y, Laurito CE, Peters MC, Vota-Vellis G, Wilson SP, et al. Recombinant herpes vector-mediated analgesia in a primate model of hyperalgesia. *Mol Ther* 2006;**13**:589–97.

43. Yokoyama H, Sasaki K, Franks ME, Goins WF, Goss JR, de Groat WC, et al. Gene therapy for bladder overactivity and nociception with herpes simplex virus vectors expressing preproenkephalin. *Hum Gene Ther* 2009;**20**:63–71.

44. Mittermeyer G, Chadwick CW, Rosenbluth KH, Baker SL, Starr P, Larson P, et al. Long-term evaluation of a phase 1 study of AADC gene therapy for Parkinson's disease. *Hum Gene Ther* April 2012;**23**(4):377–81.

45. Palfi S, Gurruchaga JM, Ralph GS, Lepetit H, Lavisse S, Buttery PC, et al. Long-term safety and tolerability of ProSavin, a lentiviral vector-based gene therapy for Parkinson's disease: a dose escalation, open-label, phase 1/2 trial, *Lancet* **383**:1138–1146.

46. LeWitt P, Rezai AR, Leehey MA, Ojemann SG, Flaherty AW, Eskandar EN, et al. AAV2-GAD gene therapy for advanced Parkinson's disease: a double-blind, sham-surgery controlled, randomised trial. *Lancet Neurol* 2011;**10**:309–19.

47. Bartus RT, Baumann TL, Siffert J, Herzog CD, Alterman R, Boulis N, et al. Safety/feasibility of targeting the substantia nigra with AAV2-neurturin in Parkinson patients. *Neurology* April 30, 2013;**80**(18):1698–701.

48. Marks WJ, Bartus RT, Siffert J, et al. Gene delivery of AAV2-neurturin for Parkinson's disease: a double-blind, randomised, controlled trial. *Lancet Neurol* 2010;**9**:1164–72.

49. Hao S, Wolfe D, Glorioso JC, Mata M, Fink DJ. Effects of transgene-mediated endomorphin-2 in inflammatory pain. *Eur J Pain* 2009;**13**:380–6.

50. Wolfe D, Hao S, Hu J, Srinivasan R, Goss J, Mata M, et al. Engineering an endomorphin-2 gene for use in neuropathic pain therapy. *Pain* 2007;**133**:29–38.

51. Liu J, Wolfe D, Hao S, Huang S, Glorioso JC, Mata M, et al. Peripherally delivered glutamic acid decarboxylase gene therapy for spinal cord injury pain. *Mol Ther* 2004;**10**:57–66.

52. Hao S, Mata M, Wolfe D, Huang S, Glorioso JC, Fink DJ. HSV-mediated gene transfer of the glial cell-derived neurotrophic factor provides an antiallodynic effect on neuropathic pain. *Mol Ther* 2003b;**8**:367–75.

53. Hao S, Mata M, Glorioso JC, Fink DJ. Gene transfer to interfere with TNFalpha signaling in neuropathic pain. *Gene Ther* 2007;**14**:1010–6.

54. Chattopadhyay M, Zhou Z, Hao S, Mata M, Fink DJ. Reduction of voltage gated sodium channel protein in DRG by vector mediated miRNA reduces pain in rats with painful diabetic neuropathy. *Mol Pain* 2012;**8**:17.

55. Srinivasan R, Wolfe D, Goss J, Watkins S, de Groat WC, Sculptoreanu A, et al. Protein kinase C epsilon contributes to basal and sensitizing responses of TRPV1 to capsaicin in rat dorsal root ganglion neurons. *Eur J Neurosci* 2008;**28**:1241–54.

56. Hao S, Mata M, Wolfe D, Huang S, Glorioso JC, Fink DJ. Gene transfer of glutamic acid decarboxylase reduces neuropathic pain. *Ann Neurol* 2005;**57**:914–8.

57. Goss JR, Cascio M, Goins WF, Huang S, Krisky DM, Clarke RJ, et al. HSV delivery of a ligand-regulated endogenous ion channel gene to sensory neurons results in pain control following channel activation. *Mol Ther* 2011;**19**:500–6.

58. Lynagh T, Lynch JW. An improved ivermectin-activated chloride channel receptor for inhibiting electrical activity in defined neuronal populations. *J Biol Chem* 2010;**285**:14890–7.

59. Omura S. Ivermectin: 25 years and still going strong. *Int J Antimicrob Agents* 2008;**31**:91–8.

60. England MJ, Liverman CT, Schultz AM, Strawbridge LM. *Epilepsy across the spectrum*. Washington, DC: National Academy Press; 2011.

61. ILAE/IBE/WHO. *Epilepsy: out of the shadows*. Hoofddorp, The Netherlands: Global Campaign against Epilepsy; 2011.

62. Fisher RS, van Emde Boas W, Blume W, Elger C, Genton P, Lee P, et al. Epileptic seizures and epilepsy: definitions proposed by the International League Against Epilepsy (ILAE) and the International Bureau for Epilepsy (IBE). *Epilepsia* 2005;**46**:470–2.

63. Berg AT, Berkovic SF, Brodie MJ, Buchhalter J, Cross JH, van Emde Boas W, et al. Revised terminology and concepts for organization of seizures and epilepsies: report of the ILAE Commission on Classification and Terminology, 2005–2009. *Epilepsia* 2010;**51**:676–85.

64. Berkovic SF, Mulley JC, Scheffer IE, Petrou S. Human epilepsies: interaction of genetic and acquired factors. *Trends Neurosci* 2006;**29**:391–7.

65. de Boer AG, Gaillard PJ. Drug targeting to the brain. *Annu Rev Pharmacol Toxicol* 2007;**47**:323–5.

66. Simionescu M, Popov D, Sima A. Endothelial transcytosis in health and disease. *Cell Tissue Res* 2009;**335**:27–40.

67. Di Pasquale G, Chiorini JA. AAV transcytosis through barrier epithelia and endothelium. *Mol Ther* 2006;**13**:506–16.

68. Foust KD, Nurre E, Montgomery CL, Hernandez A, Chan CM, Kaspar BK. Intravascular AAV9 preferentially targets neonatal neurons and adult astrocytes. *Nat Biotechnol* 2009;**27**:59–65.

69. Paradiso B, Marconi P, Zucchini S, Berto E, Binaschi A, Bozac A, et al. Localized delivery of fibroblast growth factor-2 and brain-derived neurotrophic factor reduces spontaneous seizures in an epilepsy model. *Proc Natl Acad Sci USA* 2009;**106**:7191–6.

70. Richichi C, Lin EJ, Stefanin D, Colella D, Ravizza T, Grignaschi G, et al. Anticonvulsant and antiepileptogenic effects mediated by adeno-associated virus vector neuropeptide Y expression in the rat hippocampus. *J Neurosci* 2004;**24**:3051–9.

71. Laing JM, Gober MD, Golembewski EK, Thompson SM, Gyure KA, Yarowsky PJ, et al. Intranasal administration of the growth-compromised HSV-2 vector DeltaRR prevents kainate-induced seizures and neuronal loss in rats and mice. *Mol Ther* 2006;**13**:870–81.

72. Gray SJ, Blake BL, Criswell HE, Nicolson SC, Samulski RJ, McCown TJ. Directed evolution of a novel adeno-associated virus (AAV) vector that crosses the seizure-compromised blood–brain barrier (BBB). *Mol Ther* 2010;**18**:570–8.

73. Simonato M, Tongiorgi E, Kokaia M. Angels and demons: neurotrophic factors and epilepsy. *Trends Pharmacol Sci* 2006;**27**:631–8.

74. Simonato M, Zucchini S. Are the neurotrophic factors a suitable therapeutic target for the prevention of epileptogenesis? *Epilepsia* 2010;**51**(Suppl. 3):48–51.

75. Bovolenta R, Zucchini S, Paradiso B, Rodi D, Merigo F, Navarro Mora G, et al. Hippocampal FGF-2 and BDNF overexpression attenuates epileptogenesis-associated neuroinflammation and reduces spontaneous recurrent seizures. *J Neuroinflammation* 2010;**7**:81.

76. Paradiso B, Zucchini S, Su T, Bovolenta R, Berto E, Marconi P, et al. Localized overexpression of FGF-2 and BDNF in hippocampus reduces mossy fiber sprouting and spontaneous seizures up to 4 weeks after pilocarpine-induced status epilepticus. *Epilepsia* 2011;**52**:572–8.

77. Annegers JF, Hauser WA, Beghi E, Nicolosi A, Kurland LT. The risk of unprovoked seizures after encephalitis and meningitis. *Neurology* 1988;**38**:1407–10.

78. Annegers JF, Hauser WA, Coan SP, Rocca WA. A population-based study of seizures after traumatic brain injuries. *N Engl J Med* 1998;**338**:20–4.

79. Burn J, Dennis M, Bamford J, Sandercock P, Wade D, Warlow C. Epileptic seizures after a first stroke: the Oxfordshire Community Stroke Project. *BMJ* 1997;**315**:1582–7.

80. Brooks-Kayal AR, Shumate MD, Jin H, Rikhter TY, Coulter DA. Selective changes in single cell $GABA_A$ receptor subunit expression and function in temporal lobe epilepsy. *Nat Med* 1998;**4**:1166–72.

81. Lagrange AH, Botzolakis EJ, Macdonald RL. Enhanced macroscopic desensitization shapes the response of alpha4 subtype-containing GABAA receptors to synaptic and extrasynaptic GABA. *J Physiol* 2007;**578**:655–76.

82. Raol YH, Lund IV, Bandyopadhyay S, Zhang G, Roberts DS, Wolfe JH, et al. Enhancing $GABA_A$ receptor α1 subunit levels in hippocampal dentate gyrus inhibits epilepsy development in an animal model of temporal lobe epilepsy. *J Neurosci* 2006;**26**:11342–6.

83. Haberman R, Criswell H, Snowdy S, Ming Z, Breese G, Samulski R, et al. Therapeutic liabilities of *in vivo* viral vector tropism: adeno-associated virus vectors, NMDAR1 antisense, and focal seizure sensitivity. *Mol Ther* 2002;**6**:495–500.

84. Wykes RC, Heeroma JH, Mantoan L, Zheng K, MacDonald DC, Deisseroth K, et al. Optogenetic and potassium channel gene therapy in a rodent model of focal neocortical epilepsy. *Sci Transl Med* 2012;**4**:161ra152.

85. Kanter-Schlifke I, Georgievska B, Kirik D, Kokaia M. Seizure suppression by GDNF gene therapy in animal models of epilepsy. *Mol Ther* 2007;**15**:1106–13.

86. Theofilas P, Brar S, Stewart KA, Shen HY, Sandau US, Poulsen D, et al. Adenosine kinase as a target for therapeutic antisense strategies in epilepsy. *Epilepsia* 2011;**52**:589–601.

87. Lerner JT, Sankar R, Mazarati AM. Galanin and epilepsy. *Cell Mol Life Sci* 2008;**65**:1864–71.

88. Haberman RP, Samulski RJ, McCown TJ. Attenuation of seizures and neuronal death by adeno-associated virus vector galanin expression and secretion. *Nat Med* 2003;**9**:1076–80.

89. Lin EJ, Richichi C, Young D, Baer K, Vezzani A, During MJ. Recombinant AAV-mediated expression of galanin in rat hippocampus suppresses seizure development. *Eur J Neurosci* 2003;**18**:2087–92.

90. McCown TJ. Adeno-associated virus-mediated expression and constitutive secretion of galanin suppresses limbic seizure activity *in vivo*. *Mol Ther* 2006;**14**:63–8.

91. Noè F, Frasca A, Balducci C, Carli M, Sperk G, Ferraguti F, et al. Neuropeptide Y overexpression using recombinant adeno-associated viral vectors. *Neurotherapeutics* 2009;**6**:300–6.

92. Foti S, Haberman RP, Samulski RJ, McCown TJ. Adeno-associated virus-mediated expression and constitutive secretion of NPY or NPY13-36 suppresses seizure activity *in vivo*. *Gene Ther* 2007;**14**:1534–6.

93. Noè F, Pool AH, Nissinen J, Gobbi M, Bland R, Rizzi M, et al. Neuropeptide Y gene therapy decreases chronic spontaneous seizures in a rat model of temporal lobe epilepsy. *Brain* 2008;**131**:1506–15.

94. Woldbye DP, Angehagen M, Gøtzsche CR, Elbrønd-Bek H, Sørensen AT, Christiansen SH, et al. Adeno-associated viral vector–induced overexpression of neuropeptide Y Y2 receptors in the hippocampus suppresses seizures. *Brain* 2010;**133**:2778–88.

95. Noè F, Vaghi V, Balducci C, Fitzsimons H, Bland R, Zardoni D, et al. Anticonvulsant effects and behavioural outcomes of rAAV serotype 1 vector–mediated neuropeptide Y overexpression in rat hippocampus. *Gene Ther* 2010;**17**:643–52.

96. Sørensen AT, Nikitidou L, Ledri M, Lin EJ, During MJ, Kanter-Schlifke I, et al. Hippocampal NPY gene transfer attenuates seizures without affecting epilepsy-induced impairment of LTP. *Exp Neurol* 2009;**215**:328–33.

97. Ellis TL, Stevens A. Deep brain stimulation for medically refractory epilepsy. *Neurosurg Focus* 2008;**25**:E11.

98. Alves G, Forsaa EB, Pedersen KF, Dreetz Gjerstad M, Larsen JP. Epidemiology of Parkinson's disease. *J Neurol* 2008;**255**(Suppl. 5):18–32.

99. Kirik D, Georgievska B, Björklund A. Localized striatal delivery of GDNF as a treatment for parkinson disease. *Nat Neurosci* 2004;**7**:105–10.

100. Fjord-Larsen L, Johansen JL, Kusk P, Tornøe J, Grønborg M, Rosenblad C, et al. Efficient in vivo protection of nigral dopaminergic neurons by lentiviral gene transfer of a modified neurturin construct. *Exp Neurol* 2005;**195**:49–60.

101. Decressac M, Ulusoy A, Mattsson B, Georgievska B, Romero-Ramos M, Kirik D, et al. GDNF fails to exert neuroprotection in a rat α-synuclein model of Parkinson's disease. *Brain* 2011;**134**:2302–11.

102. Decressac M, Kadkhodaei B, Mattsson B, Laguna A, Perlmann T, Björklund A. α-Synuclein-induced down-regulation of Nurr1 disrupts GDNF signaling in nigral dopamine neurons. *Sci Transl Med* 2012;**4**:163ra156.
103. Carlsson T, Björklund T, Kirik D. Restoration of the striatal dopamine synthesis for parkinson's disease: viral vector-mediated enzyme replacement strategy. *Curr Gene Ther* 2007;**7**:109–20.
104. Emborg ME, Carbon M, Holden JE, During MJ, Ma Y, Tang C, et al. Subthalamic glutamic acid decarboxylase gene therapy: changes in motor function and cortical metabolism. *J Cereb Blood Flow Metab* 2007;**27**:501–9.
105. Sapru MK, Yates JW, Hogan S, Jiang L, Halter J, Bohn MC. Silencing of human alpha-synuclein in vitro and in rat brain using lentiviral-mediated RNAi. *Exp Neurol* 2006;**198**:382–90.
106. Gorbatyuk OS, Li S, Nash K, Gorbatyuk M, Lewin AS, Sullivan LF, et al. In vivo RNAi-mediated alpha-synuclein silencing induces nigrostriatal degeneration. *Mol Ther* 2010;**18**:1450–7.
107. Vercammen L, Van der Perren A, Vaudano E, Gijsbers R, Debyser Z, Van den Haute C, et al. Parkin protects against neurotoxicity in the 6-hydroxydopamine rat model for Parkinson's disease. *Mol Ther* 2006;**14**:716–23.
108. Decressac M, Volakakis N, Björklund A, Perlmann T. NURR1 in Parkinson disease–from pathogenesis to therapeutic potential. *Nat Rev Neurol* 2013;**9**:629–36.
109. Douglas MR. Gene therapy for Parkinson's disease: state-of-the-art treatments for neurodegenerative disease. *Exp Rev Neurother* 2013;**13**:695–705.
110. Emerich DF, Orive G, Thanos C, Tornoe J, Wahlberg LU. Encapsulated cell therapy for neurodegenerative diseases: from promise to product. *Adv Drug Deliv Rev* 2013;**67-68**:131–41.
111. Lindvall O, Wahlberg LU. Encapsulated cell biodelivery of GDNF: a novel clinical strategy for neuroprotection and neuroregeneration in Parkinson's disease? *Exp Neurol* 2008;**209**:82–8.

Chapter 10

Genetic and Cell-Mediated Therapies for Duchenne Muscular Dystrophy

Jacopo Baglieri and Carmen Bertoni

Department of Neurology, David Geffen School of Medicine, University of California Los Angeles, CA, USA

LIST OF ABBREVIATIONS

2′OMeRNA 2′-O-methyl phosphorothioate oligoribonucleotides
6MWD 6 Minute walking distance
AAV Adeno-associated virus
ActRIIB Activin type II receptor
Ad Adenoviral
AONs Antisense oligonucleotides
BER Base excision repair
BIV Bovine immunodeficiency virus
BM Bone marrow
BMD Becker muscular dystrophy
CCND2 Cyclin D2
CDMRP Congressionally Directed Medical Research Programs
cDNA Complementary DNA
chODNs Chimeric oligonucleotides
CPPs Cell-penetrating peptides
DAPC Dystrophin-associated protein complex
DMD Duchenne muscular dystrophy
dmd The dystrophin gene
DNA Deoxyribonucleic acid
DoD Department of Defense
DSBs Double-strand breaks
ECM Extracellular matrix
EIA Equine infectious anemia
ES Embryonic stem
FDA Food and Drug Administration
FIV Feline immunodeficiency virus
GFP Green fluorescent protein
GRMD Golden retriever muscular dystrophy
GSK GlaxoSmithKline
H&E Hematoxylin and eosin
HDR Homology-directed repair
HIV Human immunodeficiency virus
HR Homologous recombination
HTS High-throughput screening
iPS Induced pluripotent stem
ITR Inverted terminal repeat
Lv Lentiviral
MDA Muscular Dystrophy Association
MGNs Meganucleases or homing endonucleases
MLV Moloney murine leukemia virus

Translating Gene Therapy to the Clinic. http://dx.doi.org/10.1016/B978-0-12-800563-7.00010-5

MMR Mismatch repair
mRNA Messenger RNA
MSCs Mesenchymal stem cells
NER Nucleotide excision repair
NHEJ Nonhomologous end-joining
NIH National Institutes of Health
ODNs Oligodeoxynucleotides
PA Polyadenylation
PMOs Phoshorodiamidate morpholino oligonucleotides
PNAs Peptide nucleic acids
PNA-ssODNs ssODNs made of PNA
pre-mRNA Precursor mRNA
rAAV Recombinant AAV
RT Read through
SCs Satellite cells
siRNAs Small interfering RNAs
ssODNs Single-stranded oligodeoxynucleotides
TALENs Transcription activator-like effector nucleases
ZFNs Zinc finger nucleases

10.1 INTRODUCTION

To date, more than 30 different types of muscular dystrophies have been described and each of these disorders varies considerably in its age of onset, clinical manifestations, and severity. Among those, Duchenne muscular dystrophy (DMD) is the most severe and among the most prevalent of the genetic disorders affecting childhood. The disease manifests within the first 2–5 years of life and is characterized by progressive muscle wasting. The prognosis is premature death, by cardiac or respiratory failure, within the second to third decade of the patient's life. The first report of DMD dates from 1834 and was presented at a meeting in Naples by the physician Giovanni Semmola. Another physician, Gaetano Conte, more clearly described the symptoms observed in two brothers and characterized the disease as presenting with progressive muscle wasting, muscle hypertrophy, spontaneous contractures, and cardiomyopathy. However, the disease is named after the French neurologist Guillaume Benjamin Amand Duchenne who described in detail a case of a boy with this condition in 1861.

The dystrophin (*dmd*) gene is one of the largest human genes known, occupying 2.2 megabases of the genome and encompassing 79 exons that encode for a complementary DNA (cDNA) of 14 kilobases (kb) in length. Its complexity is further evidenced by the presence of at least eight independent, tissue-specific promoters that control its expression in different tissues and by the presence of differentially-spliced dystrophin RNA transcripts that encode for a large number of isoforms. In skeletal muscles, the prevalent form of the protein is Dp427, which is encoded by all 79 exons and is herein referred to as full-length dystrophin. The full-length protein consists of over 3500 amino acid residues that, once assembled, reach 427 kDa in molecular weight. Due to its complexity, mutations in *dmd* (both sporadic and hereditary) can occur throughout its length and can vary from large deletions, single point mutations, frameshift mutations, insertions, duplications, and translocations. Mutations in the *dmd* gene can lead to either a complete absence of dystrophin expression (DMD) or a milder form of the disease known as Becker muscular dystrophy (BMD). In general, DMD patients carry mutations that cause premature termination of messenger RNA (mRNA) or protein translation (nonsense or frameshift mutations), while in BMD patients, dystrophin is reduced either in molecular weight (derived from in-frame deletions) or in expression level.

Dystrophin is part of the dystrophin-associated protein complex (DAPC), which bridges the inner cytoskeleton (F-actin) and the extracellular matrix (ECM) (Figure 10.1). While the dystrophin function has not yet been completely understood, its elastic and flexible characteristics along with its involvement in linking the ECM and the cytoskeleton suggest that dystrophin may serve a role in protecting muscle fibers from the stresses generated during muscle contraction, although other functions are likely to be present. In DMD, the absence of dystrophin causes destabilization of the DAPC which leads to an increase in sarcolemma fragility and muscle damage in response to injury or even during normal physical activity. As a result, skeletal muscles of DMD patients are characterized by inflammation, weakness, and myofiber necrosis, which lead to continuous rounds of degeneration and regeneration. As it advances, the disease is accompanied by progressive loss of muscle tissue and muscle strength that ultimately results in wheelchair dependence by the age of 11. The loss of muscle mass has been attributed primarily to the loss of regenerative capacity of satellite cells (SCs), which are muscle stem cells that reside underneath the basal lamina that surrounds myofibers and that are thought to be the main source of cells responsible for the maintenance and repair of muscle. Though the diaphragm is the most affected skeletal

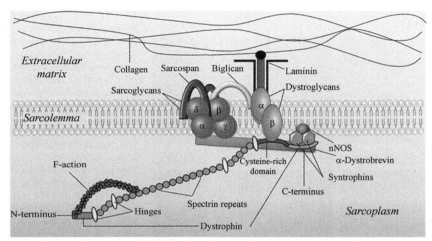

FIGURE 10.1 Structural organization of the four domains of the dystrophin protein and their interacting partners. The dystrophin-associated protein complex (DAPC) is composed of a number of different proteins that interact, directly or indirectly, with dystrophin and connect the intracellular cytoplasm (sarcoplasm) with the extracellular matrix (ECM) (a dense and complex structure that surrounds individual myofibers and that is composed of laminin, collagen, and other proteins). The N-terminal, the middle rod, the cysteine-rich, and the C-terminal domains that compose the dystrophin protein all play critical functions. Dystrophin binds cytoskeletal F-actin through its N-terminus and the DAPC through its C-terminus, acting as an important link between the internal cytoskeleton and the ECM. The central rod domain is formed by 24 triple-helical segments similar to the repeat domains of spectrin which are interrupted by four hinge regions. The cysteine-rich domain binds β-dystroglycan and links the sarcoplasm to the ECM while the C-terminal domain binds to α-dystrobrevin, syntrophins, and nNOS. Maintenance of the linkage between the ECM and the cytoskeleton is critical to skeletal muscle integrity, and mutations in the genes of the DAPC alter the expression of the majority of the DAPC proteins leading to degeneration of muscle fibers.

muscle in DMD, more than 90% of DMD patients show cardiomyopathy indicating that the disease also affects cardiac muscle. It is the involvement of the diaphragm and cardiac muscles that leads to premature death of DMD patients by respiratory or cardiac failure.

Current treatment options for DMD include the use of corticosteroids that have been shown to provide temporary benefits. Corticosteroids slow down the loss of muscle strength and improve respiratory and muscle function compared to age-matched patients who are off medication. However, these treatments are associated with severe side effects, limiting their use over prolonged periods of time. Physical therapy and assisted ventilation have proven beneficial in improving muscle function and on prolonging survival respectively, but are unable to slow down the progression of the disease.

Efforts made by the scientific community, family advocates, as well as private and government agencies toward the identification of effective therapeutic approaches for the disease have enabled the field of DMD to experience an exponential growth during the last few years in the number of applications being pursued for the treatment of this disease. Many of those approaches have already reached the stage of clinical trials with promising results.

Animal models for the disease have been instrumental in the quest for a cure. Among those, the *mdx* mouse is the most used primarily due to its availability through commercial sources, although other mouse models have also been employed and assessed in preclinical testing. They all display features typical of DMD such as increased susceptibility to myofiber instability and absence of dystrophin expression (Figure 10.2), but exhibit a milder phenotype and close to normal lifespan. The canine model (golden retriever muscular dystrophy (GRMD)) for DMD exhibits disease features similar to DMD, including progressive loss of muscle function, muscle membrane fragility, cardiomyopathy, and premature death. It has been used extensively in studies at a more advanced stage of development as an additional animal species to demonstrate the feasibility of pharmacological, gene, and molecular approaches to DMD. Although it appears to more accurately resemble the pathophysiological changes detected in DMD patients, phenotypical differences between individual animals have been observed and breeding of the GRMD colony is challenging. Additional dog models for DMD have been described, as well as a cat model, but their use and their availability for research purposes has been limited.

Initial studies aimed at developing therapeutic approaches to DMD have focused on establishing the minimum amount of dystrophin expression required to achieve clinical benefits in patients. Transgenic animals expressing different levels of dystrophin and crossed to the *mdx* background have demonstrated that as little as 20% of the levels of dystrophin expression detected in wild-type mice is sufficient to ameliorate the disease.[1] These values seem to correlate well with what has been observed in BMD patients, suggesting that even low levels of dystrophin expression achieved within each individual fiber may be therapeutically relevant. However, as we began to assess the potential of different therapeutic approaches, it became more and more evident that factors like the longitudinal distribution of dystrophin along the length of the fiber, the number of dystrophin positive fibers restored following intervention, and even the number of myonuclei successfully

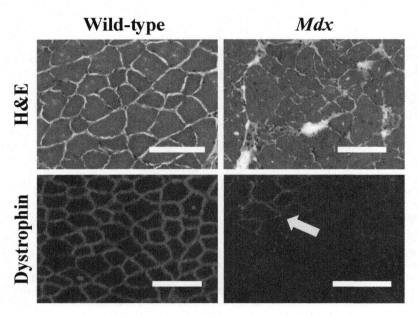

Wild-type **Mdx**

H&E

Dystrophin

FIGURE 10.2 Histological features of normal and dystrophic muscles. Micrographs of transverse sections of the quadriceps muscle of wild-type and *mdx* mice at 12 months of age. Hematoxylin and eosin (H&E) staining shows clear pathological changes in dystrophin-deficient mice typical of Duchenne muscular dystrophy (DMD) patients, which include myofibers of irregular size, accumulation of fibrotic and adipose tissue, muscle necrosis, regeneration, and inflammation. Immunostaining analysis using an antibody specific to dystrophin (labeled in red) demonstrates the absence of dystrophin expression in *mdx* mice with the exception of rare isolated dystrophin-positive fibers known as revertant fibers (indicated by an arrow) common to all DMD models and also present in patients. The origin of these so-called revertant fibers is not completely understood, but is thought to be the result of mutations in the dystrophin gene that restore the coding reading frame of the messenger RNA (mRNA) and that therefore result in the expression of shorter proteins. (Scale bar: 100 μm). (For interpretation of the references to color in this figure legend, the reader is referred to the online version of this book.)

targeted within multisyncytial fibers influence the stability of the therapeutic approach and the persistence of the beneficial effects achieved.[2,3]

Since the initial identification of the *dmd* gene, therapeutic approaches to the disorder have focused primarily on restoring the expression of a functional dystrophin protein through the delivery of vectors capable of efficiently distributing a new gene throughout the muscle. These approaches are commonly referred to as gene therapies or gene replacement strategies. More recently, new approaches have been pursued capable of directly correcting the dystrophin gene defect and of restoring its expression (molecular therapies) or capable of regenerating tissue lost as the result of the disease (cell-mediated regenerative approaches). Of equal value, are strategies capable of compensating for the lack of the dystrophin protein. Each approach, although valuable, presents limitations which may one day preclude it from entering into the clinic. Nonetheless, the results shown to date have clearly demonstrated the possibility of treating the disorder or, at least, substantially improving the quality of life of DMD patients. The therapeutic approaches to DMD are outlined in Table 10.1.

10.2 GENE REPLACEMENT THERAPIES

One of the major challenges of any application aimed at introducing a newly functional gene in order to replace the defective one is represented by the necessity to deliver a sufficient amount of vector to achieve therapeutic levels of protein expression and ensure that the levels of expression obtained are sustained over prolonged periods of time. Treatments aimed at restoring dystrophin expression face an even bigger obstacle as the vector needs to target the vast majority of the muscles that compose the human body in order to be clinically relevant.

While viral-mediated gene delivery has dominated the field through the development of efficient systemic delivery systems capable of achieving sustained levels of expression of the therapeutic vector, the use of plasmid vectors has faced hurdles in terms of both delivery and sustainability of transgene expression. Nonetheless, plasmid-mediated gene therapy approaches still remain a valid therapeutic application due to the ease of production of vectors and their cloning capacity that allows the delivery of the full-length dystrophin gene.

10.2.1 Viral Gene Therapy for DMD

Gene therapy for DMD using viral vectors has employed mainly adenoviral (Ad), lentiviral (Lv) and adeno-associated viral (AAV) vectors as delivery systems. Viral vectors have proven to be good candidates for gene therapy approaches into muscle due, primarily, to their excellent transduction efficiency in both cell cultures, as well as living organisms. However, only a few viral vectors possess sufficient cloning capacity to carry the full-length dystrophin cDNA.[4] Therefore, increasing efforts have been made toward the optimization of internally-deleted mini- or micro-dystrophin constructs expressing regions of the dystrophin protein required to maintain its functional activity into muscle (Figure 10.3). The possibility that

TABLE 10.1 Therapeutic Approaches to Duchenne Muscular Dystrophy (DMD)

Approach	Method of Delivery	DMD Models Tested	Latest Clinical Trials in DMD/BMD
Gene replacement therapies			
Viral gene therapy for DMD	IM, IP	Mouse and dog	Phase 1: rAAV2.5-CMV-minidystrophin (d3990) in DMD (ClinicalTrials.gov Id: NCT00428935)
Plasmid-mediated gene therapy	IM, IV[a]	Mouse and dog	Phase 1: Full-length dystrophin (trial Id number not available)
Correction of the dystrophin gene			
Antisense-mediated therapies using oligonucleotides	IM, IP, SC	Mouse and dog	Phase 2: Efficacy, safety, and tolerability rollover study of eteplirsen (AVI-4568) in DMD (ClinicalTrials.gov Id: NCT01540409)
			Phase 2: Open label escalating dose pilot study of PRO051 (GSK2402968) in DMD (ClinicalTrials.gov Id: NCT01910649)
Oligonucleotide-mediated gene editing of the dystrophin gene	IM	Mouse and dog	Not tested in humans
Endonuclease-mediated gene editing	IM	Mouse	Not yet tested in humans
Suppression of nonsense mutations mediated by stop codon read through compounds	IM, IP, oral	Mouse	Phase 1: Six-month study of gentamicin in DMD (ClinicalTrials. gov Id: NCT00451074)
			Phase 3: Extension study of Ataluren (PTC124) in DMD (ClinicalTrials.gov Id: NCT02090959)
Cell-based therapies for DMD			
Cell-based therapies using muscle progenitor or muscle stem cells	IM	Mouse and dog	Phase 1A: High-density injection of myogenic cells throughout large volumes in a 26-year-old DMD patient (trial Id number not available)
			Phase 1: Autologous transplantation of muscle-derived CD113 + cells in DMD (trial Id number not available)
Non muscle-derived stem cells therapies	IM, IV	Mouse and dog	Phase 1: Intra-arterial delivery of HLA-identical allogeneic mesoangioblasts in DMD. (EudraCT number: 2011-000176-33)
Upregulation of utrophin	IM, IP, oral	Mouse and dog	Phase 1b: Study of SMT C1100 in DMD subjects (ClinicalTrials.gov Id: NCT02056808)
Myostatin blockers	IP, SC, IV	Mouse and dog	Phase 1: MYO-029 in BMD and other muscular dystrophies (ClinicalTrials.gov Id: NCT00104078)
			Phase 2: Extension study of ACE-031 in subjects with DMD (ClinicalTrials.gov Id: NCT01239758)

IM, intramuscular; IP, intraperitoneal; SC, subcutaneous; IV, intravenous; DMD, Duchenne muscular dystrophy.
[a]*vascular delivery was restricted to isolated limbs.*

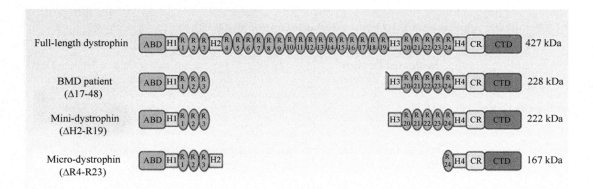

FIGURE 10.3   Proteins encoded by the vectors most commonly used for therapeutic purposes. This schematic representation shows the structure of the domains and their arrangement in full-length, mini- (ΔH2-R19) and micro-dystrophin (ΔR4-R23) containing the regions of the dystrophin protein believed to be required to restore partial function in patients. The arrangement of the domains of the protein predicted to be expressed in a BMD patient with a milder form of muscular dystrophy and characterized by a deletion of the coding sequence of the dystrophin gene encoded by exons 17 through 48 (Δ17-48) is also shown for comparison. Δ indicates deletion of spectrin-like repeats: ABD, actin-binding domain; H, hinge region; R, spectrin-like repeat; CR, cysteine-rich domain; CTD, C-terminal domain. Molecular weights of full-length dystrophin and truncated proteins are indicated on the right.

large portions of the dystrophin central region could be deleted without significantly affecting the role of the protein in myofibers is based on clinical data obtained from a BMD patient characterized by a deletion of the central part of the dystrophin gene lacking exons 17 through 48 (Δ17-48) who was still ambulant at the age of 61. When expressed into *mdx* mice, the Δ17-48 mini-dystrophin gene was shown to effectively prevent the dystrophic pathology.[5] These results have opened a new venue in the field of gene therapy for DMD by prompting subsequent studies aimed at better characterizing the regions of the dystrophin gene that are required for the proper function of the protein in skeletal muscle. Several mini- and micro-dystrophin genes that were small enough to be incorporated into most viral vectors were generated and a few of those were shown to be able to prevent and partially reverse the dystrophic phenotype.[6,7] Among those, constructs expressing the ΔH2-R19 and ΔR4-R23 proteins appear the most effective in reversing the dystrophic pathology (Figure 10.3).[7]

10.2.1.1 Gene Therapy Based on Ad Vectors

Adenoviruses are nonenveloped icosohedral viruses containing a 36 kb double-stranded DNA genome that remains episomal once into the cell nuclei. Since their initial isolation in the early 1950s, tremendous progress has been made toward understanding the adenovirus lifecycle, its interaction with the host immune response, and its role in diseases.

Adenoviruses are extremely stable against chemical or physical agents, they can be produced in sufficient quantities to satisfy clinical needs, and have been shown to infect a large number of cell types, rendering these viruses particularly attractive in gene therapy applications. The studies conducted to date have focused on optimizing the structure of the adenovirus genome to allow maximal transgene capacity and optimal transduction efficiency while minimizing immunogenicity of the proteins encoded by the virus following introduction into host organisms. First generation Ad vectors lack the E1 region that is responsible for allowing the expression of other viral genes and for modulating the host cell cycle, rendering the vector unable to replicate.[8] In addition, a deletion of the E3 region that is responsible for encoding several proteins that modulate host antiviral and inflammatory responses[9] allows introduction of the therapeutic gene.

Early reports demonstrated that a mini-dystrophin human cDNA could be successfully delivered into *mdx* after intramuscular injection of a first generation Ad vector. Up to 50% of fibers in the injected muscle showed correct localization of a mini-dystrophin protein at the sarcolemma of muscle fibers.[10] Similar studies using the same mini-dystrophin construct demonstrated that expression was stable for up to six months following delivery and was able to ameliorate some of the pathological features associated with the disease.[11] Subsequent studies also established the beneficial effects that could be achieved by delivering a mini-dystrophin gene into newborn *mdx* muscles, as well as immunosuppressed adult *mdx* mice by demonstrating a diminished susceptibility of myofibers to damage in response to force contractions.[12,13] However, despite the deletion of the regions of the Ad vector responsible for eliciting an immune response, first generation Ad vectors have been shown to cause a significant immune response in vivo, mainly as the result of the synthesis of viral proteins. To overcome these limitations, a second generation of Ad vectors was developed to be devoid of additional genes necessary for viral DNA replication. Although second generation Ad vectors have been shown to better sustain the level of expression of transgenes in immunocompetent mice, their use to permanently treat disorders is hampered by the strong immune response detected following a second administration and, therefore, has precluded them from entering testing in animal models for DMD.

In order to overcome the Ad vectors' immunogenicity and in order to increase their packaging capacity, "gutted" (also known as gutless, high-capacity, or helper-dependent Ad vectors) have been developed by several groups.[14,15] The gutted vectors lack all the viral coding sequences and therefore, Ad replication and packaging require the presence of a helper virus which carries all the viral proteins necessary for its production in cells. As a result, their assembly, and therefore titer obtained in cultures are reduced. Promising results have been obtained in neonatal, as well as juvenile *mdx* mice, with levels of dystrophin expression sustained for at least one year following in vivo administration.[16–22] Despite the progress made, the use of third generation Ad vectors still face major hurdles, namely the humoral response which results in the generation of antibodies to Ad proteins and that limits multiple administrations.[23] Furthermore, because infections from adenoviruses are fairly common in humans, most individuals are likely to have already developed an immunoresponse to proteins present in the Ad virion. The efficiency of transduction that can be achieved in patients may therefore vary substantially among individuals and, in certain subjects, could lead to a failure to express the desired gene. The use of approaches that target muscles during the early stages of embryo development, when the immune system is not fully developed and where muscle stem cells are actively replicating to induce muscle growth, has been used to overcome these problems and to facilitate the persistence of transgene expression into muscle.[24,25]

10.2.1.2 Gene Therapies for DMD Using Retroviral and Lv Vectors

Retroviruses were among the first to be developed and used in gene therapy trials. Most of the clinical trials approved by the Food and Drug Administration (FDA) have employed the Moloney murine leukemia virus (MLV) that has one of the simplest genomes of all the retroviruses and is therefore easier to manipulate and engineer for therapeutic applications. Retroviruses are composed of two identical copies of single-stranded RNA encapsulated into a capsid that is surrounded by a lipid membrane from which the enveloped proteins extrude. The enveloped proteins confer virus specificity as they are responsible for attachment to the target cells. Once in the cell, the RNA is reverse transcribed into double-stranded cDNA that translocates into the nucleus and integrates into the cell genome thanks to the integrase encoded by the retrovirus. The ability to integrate into the genome allows sustained gene expression and this is, perhaps, the best feature of these vectors and the reason why they have been successfully employed in clinical trials. The basic structure of retroviruses has undergone genetic modification to devoid the replicative ability of the vector and to allow up to 8 kb of cloning capacity. As with third generation Ad vectors, production of retroviral vectors requires the use of special packaging cells to allow assembly of the virion and release into the media. The process is technically challenging and usually results in a lower titer than that usually achieved using other types of viral vectors.

The use of retroviral vectors for the treatment of DMD has been limited to testing the feasibility of restoring dystrophin expression in culture in vitro as a proof-of-principle of the applicability of gene replacement strategies using this type of viruses.[26–28] Among the limitations that have precluded retroviral vectors from advancing into clinical applications for DMD is the rapid loss of transgene expression detected following delivery, the inability to control the integration site of the retrovirus into the host genome and, in the case of approaches aimed at targeting muscle progenitor cells or muscle stem cells, the necessity of using cells that are actively dividing due to the inability of the reverse transcribed genome to efficiently enter the nucleus unless the nuclear membrane is disrupted as usually occurs during mitosis.

These problems have largely been overcome by the development of Lv vectors.[29] They are a class of retroviral vectors capable of stably integrating transgenes into the host genome of nonreplicating cells, as well as postmitotic cells at random positions and capable of achieving long-term expression of the transgene.[30,31] The first class, and to this date the most widely used, was initially derived from the human immunodeficiency virus (HIV) although other classes of vectors such as the feline immunodeficiency virus (FIV), bovine immunodeficiency virus (BIV) and equine infectious anemia (EIA) have also been successfully used for therapeutic applications. During the years, Lv vectors have evolved tremendously and extensive studies have led to the development of safer vectors that can be grown in culture at high titers. Currently, the vectors used for most of the applications into muscles have been based on third generation Lv vectors.

Kafri et al have demonstrated that a human-lentiviral (HIV)-based vector can sustain expression of green fluorescent protein (GFP) for more than 22 weeks in the liver and more than 8 weeks in transduced muscle without sufficient humoral or cellular response to preclude repeated injection, suggesting that this system may be safe to use in patients following repeated dosing.[30] Intramuscular delivery of Lv vectors have also been shown to result in expression of a β-galactosidase reporter system, although the results showed poor distribution of the Lv vectors in the injected muscle with expression of the reporter being localized at the injection site.[32] The limited effects achieved following intramuscular injection of Lv has precluded these vectors from gaining momentum in the field of gene therapy for muscle disorders.

Better results have been achieved following delivery of Lv vectors into fetuses and newborn mice.[33,34] MacKenzie et al demonstrated that delivery of Lv vectors encoding a reporter gene into the liver or hind limbs of mouse embryos at an age of gestation between 14 and 15 days resulted in the detection of transgene expression in cardiac and skeletal muscles for at least nine months following delivery. Furthermore, the study demonstrated that systemic administration of Lv vectors could achieve widespread distribution into muscle progenitor cells or muscle stem cells with significant beneficial effects.[33] More recently, it has been demonstrated that a lentiviral vector expressing a fusion protein encoded by a micro-dystrophin and the GFP coding sequences and delivered intramuscularly in neonatal *mdx4cv* mice was able to successfully transduce myogenic stem cells, as well as myofibers, resulting in widespread and stable expression of dystrophin for at least two years.[34] These results, although promising, have clearly demonstrated that future applications of Lv vectors in DMD patients are likely to require the use of immune suppressors and that further optimization may still be necessary before these vectors can be used in a clinical setting.[31]

10.2.1.3 AAV Vectors for the Treatment of DMD

Gene therapy using AAV vectors has gained attention in the field due to their muscle tropism, long-lasting expression and relatively low, although still present, immunogenicity. The adeno-associated virus is a nonpathogenic virus that belongs to the Parvovirus family. Its genome consists of approximately 4.7 kb of single-stranded DNA flanked by inverted terminal repeats (ITRs) and is organized into two open reading frames: the *rep* region, which encodes for replication-related proteins and the *cap* region which encodes for three viral capsid proteins (VP1, VP2 and VP3). AAV have been found in many species such as nonhuman primates, humans, canines, and fowl. To date, 12 serotypes of adeno-associated viruses (AAV1–AAV12) have been identified in primates alone. Each serotype has capsid sequences with a unique serological profile resulting in different transduction efficiencies. AAV vectors can infect both dividing, as well as nondividing cells and, once in the cell, they are known to integrate into the genome. Since AAV vectors can transduce all muscles in the body with only a mild immunogenicity compared to other vectors, they appear to be good candidates for gene therapy applications in DMD.[35,36] Recombinant AAV (rAAV) vectors are generated by replacing the wild-type viral genome with any gene or DNA sequence of interest that is up to 5 kb in length. Production of these viral vectors can be tedious, although a variety of methods have been developed to facilitate the process.[37]

Vectors packaged with AAV1, AAV6, AAV8, and AAV9 capsids have been shown to induce levels of transgene expression that are significantly higher than those obtained with rAAV2 vectors.[38] Successful delivery of a functional micro-dystrophin has been achieved in both cardiac and skeletal muscles of *mdx* mice by single low pressure intravenous administration of rAAV pseudotyped 6 (rAAV6) vectors.[38,39] Among the serotypes tested for their ability to transduce muscles, rAAV1, rAAV8, and rAAV9 appear to be the most promising as they can transduce heart and skeletal muscles in both neonatal and adult mice, although at different efficiencies.[40–42]

Among the drawbacks of rAAV technologies is the limited packaging capacity. This issue has been addressed through the development of transplicing or recombining AAV vectors.[43,44] The basic principle of these strategies is to split the dystrophin gene into two or more rAAV vectors that, once into the cell, can recombine to reconstitute the dystrophin coding sequence (Figure 10.4).

Discordant results have been observed with rAAV vectors in terms of immune response. Cellular immune responses against the capsid proteins have been rarely observed in rodents after systemic delivery of rAAV1, rAAV6, rAAV8, and rAAV9,[35,39] whereas intramuscular injection of rAAV2 and rAAV6 in dogs has been shown to elicit a cell-mediated immune response[45,46] which could be prevented by immunosuppression.[47] Clinical trials using rAAV1 and rAAV2 vectors confirmed the presence of an immune response against the capsid proteins, as well as the transgene expression in humans, raising issues of safety and efficacy of rAAV-mediated technologies[48,49] and suggesting that further optimization of rAAV vectors may be necessary to guarantee their success in the clinic.

10.2.2 Plasmid-Mediated Gene Therapy

Plasmids have the advantage of being able to accommodate large genes and therefore can be used to express full-length dystrophin. However, their use in the clinical setting has been limited by low transduction efficiency achieved in muscle following administration and low persistence of transgene expression following delivery. Injection of plasmids expressing reporter genes into mouse skeletal muscle in vivo was reported for the first time in 1990. Delivery of a plasmid expressing luciferase resulted in enzyme activity in skeletal muscles of mice for at least 2 months,[50] but similar experiments using a β-galactosidase cDNA in the GRMD dog muscles resulted in low transfection efficiency.[51] Results were confirmed in a clinical trial in Duchenne and Becker patients injected with a full-length human dystrophin plasmid into the radialis muscle.[52] Although delivery of the plasmid vector was reported to be well tolerated in patients, only low levels of dystrophin expression were found in some, but not all, of the subjects that received the vector.[52]

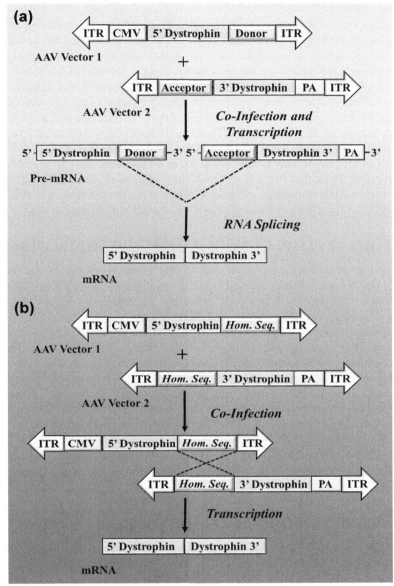

FIGURE 10.4 Strategy used to increase packaging capacity of adeno-associated virus (AAV) vectors. (a) Schematic outline of the strategy used to induce *trans*-splicing of dystrophin messenger RNA (mRNA) transcripts expressed by two independent AAV vectors. The AAV donor vector carries the promoter, the 5′ sequence of the transgene, and the splicing donor signal. The AAV acceptor vector carries the splicing acceptor signal, the 3′ region of the transgene, and the polyadenylation (PA) sequence. Infection of AAV virions leads to the expression of two mRNA containing the splicing signals necessary to remove the noncoding components of the mRNA to produce a functional mini-dystrophin gene. (b) Schematic illustration of the homologous-based approach that could be used to reconstitute the expression of functional mini-dystrophin protein. Like the *trans*-splicing system, this approach requires the coinfection of two AAV vectors, one vector containing the promoter, the 5′ sequence of the transgene, and a sequence of DNA where the homologous recombination should occur. The second AAV vector carries a copy of the homologous region, the 3′ sequence of the mini-dystrophin gene, and the PA site. Pairing of the homologous sequences and recombination events result in the reconstitution of a functional expression cassette. Both approaches are capable of expressing a gene of approximately 8 kb in length.

The use of cationic lipids, polyplexes, or similar systems that can be coupled to DNA to enhance their ability to cross the ECM together with the use of intravascular injection under elevated pressure has been shown to significantly increase the delivery of plasmid DNA into myofibers.[53-55] One of the major limitations of using plasmid vectors as carriers to introduce a new dystrophin gene into muscles is the progressive loss of transgene expression detected following delivery, which has been attributed primarily to silencing of highly methylated sequences that are particularly enriched in plasmids and rapid degradation of episomal DNA characteristic of nonintegrating vectors. Silencing effects can be in part limited by the use of vectors devoid of methylated sequences,[56] while the loss of episomal plasmid has been addressed by designing plasmid vectors capable of integrating at specific sites of the genome.[57,58] Generally, the use of systems capable of integrating a cDNA into the genome require the coadministration of two distinct sequences, a cDNA expression construct of the therapeutic gene flanked by specific sequences that can recombine and integrate, and a plasmid containing the enzyme required to activate integration of the gene into the mammalian genome. In muscle, the applicability of these approaches has been limited to proof-of-concept studies in *mdx* mice using the φC31 integrase system.[2] While the results have demonstrated that plasmid integration is capable of achieving higher levels of dystrophin expression compared to those obtained using nonintegrating plasmid vectors, they have also evidenced a loss of transgene expression over prolonged periods of time, although at a lower rate than that observed in muscles that expressed constructs unable to integrate.[2] These studies have clearly evidenced limitations in the applicability of plasmid-mediated gene replacement strategies for DMD and have demonstrated the need to further optimize the structure of the vectors used to deliver dystrophin or other therapeutic genes in muscle. Although the possibility of having neutralizing antibodies toward plasmid vectors is low, the effects of repeated administration of plasmids, which would be required to guarantee sustained level of gene expression in patients, are unknown and, at the present time, it is difficult to determine whether plasmid-mediated technologies for muscle disorders will enter clinical studies in the near future.

10.3 STRATEGIES AIMED AT CORRECTING THE DEFECTIVE DYSTROPHIN GENE

Among the therapeutic approaches currently being pursued for the treatment of DMD, the use of oligonucleotides or small molecules to correct the dystrophin gene defect is gaining attention due to the possibility of restoring expression of functional dystrophin without introducing a new gene in muscle. There are several advantages that render molecular therapies appealing for the treatment of genetic diseases and DMD in particular. First, they allow the gene to remain under its own regulatory mechanisms. Second, oligonucleotides or small molecules are relatively easy to produce at costs that are generally lower than those associated with the production of viral vectors. Third, production can be scaled up to satisfy the needs associated with treatments in humans. Issues of immune responses toward oligonucleotides or small molecules are rarely seen, easing a few of the issues associated with certain types of vectors used in gene therapy. Current applications to DMD using molecular therapies use antisense oligonucleotides (AONs) to restore the open-reading frame of the dystrophin mRNA by inducing skipping of mutated exons. Similarly, chimeric oligonucleotides (chODNs) or single stranded oligodeoxynucleotides (ssODNs) have been shown to be able to correct single base mutations and to restore dystrophin expression in vitro, as well as in vivo in DMD animal models. Furthermore, the use of small compounds like gentamicin, or more recently, PTC124, capable of promoting read through (RT) of a premature stop codon, has shown promise. Altogether, the encouraging results obtained thus far have allowed molecular therapeutic approaches for this disease to grow considerably in the last few years.

10.3.1 Antisense-Mediated Therapies Using Oligonucleotides

The use of antisense technology has proven to be one of the most promising approaches to the treatment of DMD caused by large deletions. Although the original reports of the feasibility of this approach for the treatment of DMD came from studies conducted by the groups of Matsuo and Dickson, respectively,[59,60] it is the pioneer work of Wilton and colleagues[61,62] along with the results achieved by van Deutekom et al[63,64] that has allowed this technology to reach the stage of conducting clinical trials in humans. The technology employs small, chemically-modified RNA or DNA sequences complementary to a specific region of the dystrophin pre-mRNA and capable of interfering with the splicing machinery to induce skipping of one or more exons (Figure 10.5). The result is restoration of the open reading frame and expression of a dystrophin protein that, although smaller than full-length dystrophin, conserves some of the key domains of the protein necessary for its interaction with members of the DAPC and therefore believed to be functional.

The initial testing was conducted in muscle cells isolated from the *mdx* mouse and transfected with 2′-O-methyl phosphorothioate oligoribonucleotides (2′OMeRNA) which are known to be more resistant to endonucleases and have a higher binding affinity than oligoribonucleotides made of regular bases (Figure 10.6).[60] AONs complementary to the 3′ splice site of murine dystrophin intron 22 induced splicing of exon 22 to exon 30 and restored dystrophin expression in *mdx* myotubes.[60] These results were confirmed by another study in *mdx* cultures which also demonstrated the feasibility of precisely skipping only the mutated exon when 2′OMeRNA-AONs were designed to target the donor site of exon 23.[61] Shortly thereafter it was shown that 2′OMeRNA could restore dystrophin expression in DMD patient-derived muscle cells, carrying a deletion of exon 45 by inducing skipping of exon 46.[63] When tested in vivo, 2′OMeRNA were able to efficiently restore dystrophin expression in *mdx* mice following intramuscular injection,[65] as well as after systemic administration.[62] These results have established the therapeutic relevance of the approach and have served as foundations for all subsequent studies aimed at testing and optimizing AONs' structure in animal models and in human cells isolated from DMD patients.

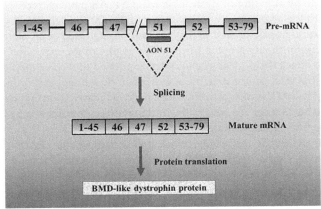

FIGURE 10.5   Antisense-mediated exon skipping strategies used to correct Duchenne muscular dystrophy (DMD) defects due to large deletions. Schematic illustration of the approach used by antisense oligonucleotide (AON)-mediated strategies to restore the reading frame of a dystrophin gene containing a deletion of exon 48 through 50. AONs can be designed to target regulatory regions of one or more exons responsible for the assembly of mature RNA transcripts to induce skipping of the target exon. The result is the production of a messenger RNA (mRNA) that, although shorter, is still in frame, thus allowing the expression of a functional dystrophin protein.

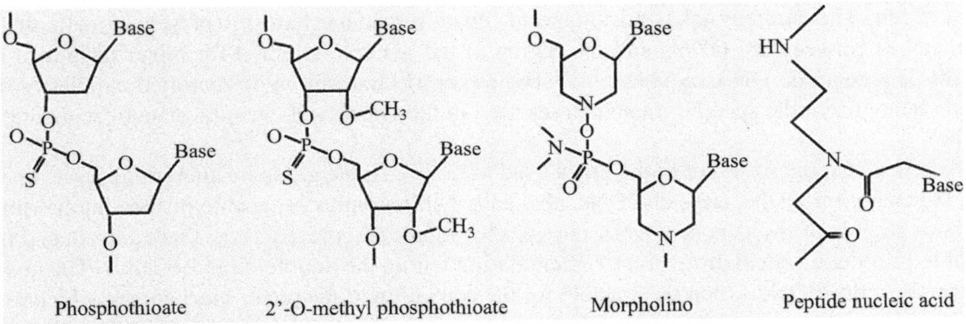

Phosphothioate	2'-O-methyl phosphothioate	Morpholino	Peptide nucleic acid

FIGURE 10.6   Chemical modifications used in RNA and DNA based approaches to Duchenne muscular dystrophy (DMD). The modifications introduced in the oligonucleotides usually consist of changes in the oligonucleotide backbone chemistry to enhance nuclease resistance by inhibiting RNase activity and by conferring tighter binding affinity to their DNA or RNA target.

Since then, the field of AON-mediated molecular therapies for DMD has focused primarily on identifying and testing new chemistries that could be used to enhance the efficacy of the AONs in redirecting splicing of the dystrophin gene. Phosphorodiamidate morpholino oligonucleotides (PMOs) in which the phosphodiester bond is replaced by phosphoroamidate linkage and the ribose replaced by a morpholino moiety (Figure 10.6), have proven to be a valid alternative to 2′OMeRNA, due primarily to their higher affinity for their target nucleic acid sequences and greater resistance to degradation than conventional nucleic acids. Weekly intravenous injections of a PMO targeting the donor site of exon 23 in *mdx* mice administered for up to seven weeks was shown to restore levels of dystrophin expression to up to 50% of those detected in wild-type mice in gastrocnemii and quadriceps throughout the majority of the muscle cross sectional area analyzed, suggesting that PMO may be more efficient than 2′OMeRNA in correcting the dystrophin gene defect.[66] More recently, it has been demonstrated that AONs made of peptide nucleic acids (PNAs), which are artificially-generated homologs of nucleic acids in which the phosphate–sugar backbone is replaced by a pseudo-peptide polymer (Figure 10.6), were more efficient than 2′OMeRNA or PMOs in inducing skipping of the dystrophin exon 23 in *mdx* mice.[67]

Clinical trials are currently under way for the treatment of DMD characterized by deletions of the dystrophin gene that could be treatable by skipping of exon 51 using 2′OMeRNA-based AONs (drisapersen, also known as PR0051) developed by Prosensa in collaboration with GlaxoSmithKline (GSK), as well as PMO-based AONs (eteplirsen, also known as AVI-4568) developed by Sarepta Therapeutics. While the initial results of the phase I and II studies of drisapersen showed great promise,[64] recent results of GSK's Phase III clinical study were somewhat disappointing as they failed to show a statistically significant improvement in the 6 minute walking distance (6MWD) test in subjects that received AONs compared to the placebo-treated group.

Better outcomes have been achieved using the PMO-AON developed and optimized by Sarepta Therapeutics.[68] The Phase IIb clinical study of eteplirsen was conducted on 12 DMD boys between the ages of seven and 10 years with deletions of the dystrophin gene correctable by skipping of dystrophin exon 51. All patients received eteplirsen for a maximum of 48 weeks, although treatment with the active drug was initiated at different times following enrollment. A statistically significant increase in dystrophin expression compared to the baseline was detected in all patients, demonstrating the efficacy of the AON in redirecting splicing for the dystrophin mRNA The results of the extension study released in June 2012 on subjects that were enrolled in the Phase IIb clinical trial and that continued to receive eteplirsen for 84 weeks, although did not show an improvement in the 6MWD, demonstrated a stabilization of walking ability in patients treated with the drug.

It is unclear at this point why two AONs designed to work on the same mutation can result in such different outcomes, but it is possible that the different chemistries, as well as the different dosages utilized may have played a role. If approved, eteplirsen may become one of the first antisense drugs on the market.

One of the limitations faced by antisense-mediated exon-skipping therapies for DMD is the inability of AONs to induce skipping of the dystrophin mRNA in cardiac tissue following systemic delivery. This issue has been recently addressed by studies that have demonstrated that PMOs, as well as PNAs conjugated with short peptides that facilitate cellular uptake such as cell-penetrating peptides (CPPs), are able to facilitate uptake of AONs into muscle including the heart with functional effects.[69–71] However, the use of these new molecules may require further optimization before testing in the clinic.

10.3.2 Oligonucleotide-Mediated Gene Editing of the Dystrophin Gene

In the last decade, gene editing strategies mediated by oligodeoxynucleotides (ODNs) have achieved notable accomplishments in the field of molecular therapeutics for DMD. The approach is particularly appealing because it acts at the genomic level and therefore has the potential to permanently correct the genetic defect allowing for stable expression of the gene being targeted for repair. This strategy takes advantage of innate repair mechanisms present in cells that are capable of recognizing mismatches between the ODNs and the region of the genome targeted for repair to induce single base pair changes in virtually any genomic DNA sequence.[72–74] The repair mechanism being recruited can vary widely depending on the type of cells being used, the specific mismatch inserted in the ODN, and even the state of replication of cells being targeted for repair.[75]

During the past few years, the structure of the ODN used to induce single genomic alterations has evolved considerably (Figure 10.7). First generation ODNs were chODNs, also called chimeraplasts, capable of forming hairpin structures and interacting with both strands of the genomic DNA targeted for repair (Figure 10.7(a)). Once introduced into the cell, the chODN is thought to introduce helical distortion by intercalating within the double-stranded DNA. The mismatch, which is present in both strands of the ODN, is then responsible for the activation of the repair mechanism, which results in the conversion of the targeted base at the genomic level using the information provided by the chODN. Since their first application,

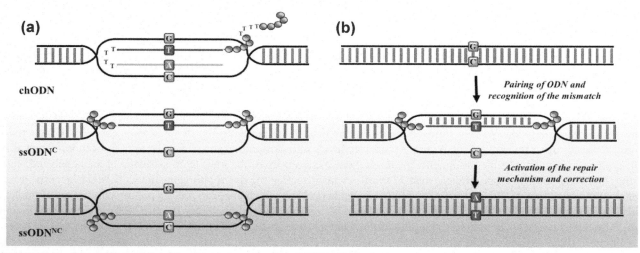

FIGURE 10.7 Gene editing mediated by oligodeoxynucleotides (ODNs). (a) The basic structure of targeting ODNs consists of a stretch of DNA perfectly homologous to the region of the gene that is targeted for correction with the exception of a single base mismatch. A chODN contains a complementary region composed of 2′-O-methyl RNA interrupted by a pentameric block of DNA bases (chODN). Newer generation of ODNs known as ssODNs consist of unmodified DNA bases that can be complementary to the coding strand (ssODN^C) or complementary to the noncoding strand (ssODN^{NC}) of the gene targeted for repair. Phosphorothioate bases are added to each end of the oligonucleotides to increase their stability to endonucleases. (b) Correction is mediated by the pairing of ODN with the genomic sequence targeted for repair. Recognition of the mismatch present on the ODNs activates innate repair mechanisms naturally present in the cell nuclei and capable of directing the correction based on the information provided by the ODN template.

chODNs have been investigated for their ability to target and induce genomic modifications in a number of different cell types and have been successfully applied in both eukaryotic and prokaryotic cells.[75]

The process of gene repair, although still not completely understood, appears to involve multiple steps (Figure 10.7(b)). Different repair mechanisms have been implicated in the process that leads to the correction of the genetic defect. Among those, the mismatch repair (MMR), homologous recombination (HR), homology-directed repair (HDR), nonhomologous end-joining (NHEJ), nucleotide excision repair (NER) pathway, and the base excision repair (BER) mechanism have all been shown to play a role throughout the different stages that lead to the replacement of the mismatched base in the genomic DNA.[75]

In muscle, chODNs have been successfully used to correct the dystrophin gene defect in *mdx* mice and the GRMD dog model.[76–78] Furthermore, chODNs have been shown to induce single point mutations at the intron/exon boundaries of the dystrophin gene defect to disrupt consensus sequences of the dystrophin gene and induce skipping of the mutation responsible for the lack of dystrophin in *mdx* mice, demonstrating the broad applicability of this approach for the treatment of DMD.[75,79] Second generation ODNs were developed based on studies aimed at better characterizing the function of the RNA and moieties of each of the strands of the chODN during the repair process. The results indicated that correction was mediated primarily by the DNA strand of the chODN. When ssODNs made of unmodified bases were tested in cells in culture or in prokaryotic systems, they were shown to be as efficient in correcting the genetic defect as chODNs. Furthermore, a strand bias was observed in the correction abilities of linear DNA ODNs depending on whether the ssODNs were targeting the coding or the noncoding strand of the gene (Figure 10.7(a)), suggesting the implication of transcription in the processes that take place in gene repair mediated by ssODNs.[75] A similar strand bias was also observed when ssODNs complementary or homologous to the dystrophin gene were tested in an mdx^{5cv} mouse, a model for DMD that has proven to be a valuable system to test and quantitate the level of gene repair achieved in muscles.[80] The use of ssODN has substantially advanced the field of gene editing for the treatment of diseases like DMD due to the feasibility of synthesizing large amounts of ODNs that would be required to administer in patients in prospective clinical trials.

Better results have been achieved using ssODNs made of PNA (PNA-ssODNs) and designed to target the mdx^{5cv} mutation. Kayali et al. showed that PNA-ssODNs can improve correction frequencies up to ten-fold compared to those achieved by older generation ssODNs.[3] The increase in efficacy has been attributed primarily to the improved stability of the ODNs toward endonucleases allowing the PNA-ssODN to reach its target more efficiently than ssODNs made of unmodified bases. More importantly, the study evidenced the presence of important factors that contributed to the stability of the repair process in muscle following intramuscular delivery of ODNs. The progressive decrease in dystrophin expression detected following intramuscular administration of ssODNs demonstrated that correction of mature myofibers alone is not sufficient to guarantee long-lasting effects.[3] The issue of the stability of the repair process has recently been tackled through transplantation studies aimed at testing the feasibility of using PNA-ssODNs to target and correct the mdx^{5cv} mutation in SCs (See Section 10.4.1). Nik-Ahd et al demonstrated that SCs explanted ex vivo from mdx^{5cv} muscles and transfected with PNA-ssODNs targeting the mdx^{5cv} mutation were amenable to gene repair mediated by oligonucleotides. When transplanted into skeletal muscles of dystrophin-deficient mice, cells that had undergone correction were able to restore dystrophin expression. Importantly, dystrophin was shown to increase over time suggesting that SCs that homed into engrafted muscles were able to activate in response to myofiber degeneration or normal muscle turnover and regenerated new myofibers expressing normal, full-length dystrophin.[81] These studies have opened new possibilities for the treatment of DMD and have demonstrated that the correction of SCs is a key to the success of oligonucleotide-mediated gene approaches to DMD.

Despite the encouraging results obtained, the field of gene editing using ODNs is still far from entering a clinical scenario. Among the major hurdles that ODN-mediated therapies for the treatment of DMD face is the possible immune response towards either the ODNs themselves or the modifications which have been generated to improve the stability and efficiency of ODNs. Issues of delivery, as for any other approach currently being developed for the treatment of this disease, is also a concern, although it is believed that the use of coupling cargo, like CPP or similar structures, may overcome this limitation.

10.3.3 Endonuclease-Mediated Gene Editing

Genome editing using engineered nucleases has emerged as a powerful tool for correcting genetic mutations and eventually treating a variety of genetic disorders. This approach uses artificial proteins capable of recognizing specific sequences in the genomic DNA. Each nuclease consists of a DNA-binding domain that is customized to recognize a specific DNA sequence fused to an enzyme capable of cutting the DNA in a nonsequence-specific manner. To date, four major classes of engineered endonucleases have been described in the literature (reviewed in more detail in accompanying chapters). However, testing

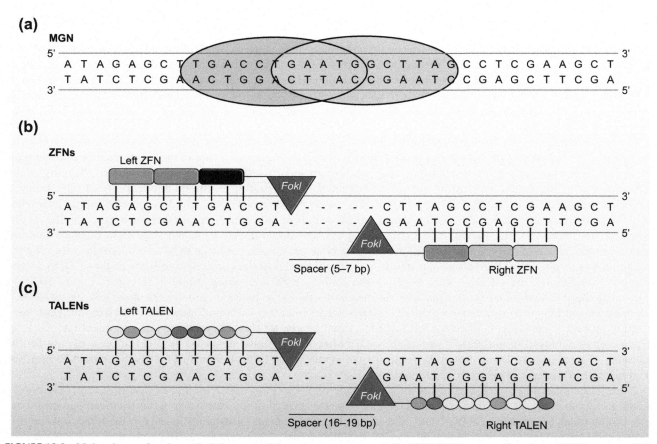

FIGURE 10.8   Major classes of engineered nucleases. (a) Schematic representation of a LAGLIDADG homing endonuclease (LHE) bound to a DNA target through its catalytic motif. Sequence specificity toward its DNA target is mediated through the association or fusion of protein domains from different enzymes to generate chimeric MGN. MGNs have the unique property of recognizing long DNA target sequences (>12 bp), making them very specific. (b) A representation of a zinc finger nuclease (ZFN) dimer bound to its target. Each ZFN contains the cleavage domain of *FokI* linked to zinc fingers designed to recognize specific trinucleotide sequences (colored boxes) flanking the cleavage site. A spacer of 5 or 7 bp is introduced between the *FokI* enzyme and the zinc finger to confer flexibility to the molecule and to allow the dimerization of the ZFNs. (c) Schematic representation of a transcription activator-like effector nucleases (TALEN) dimer. Like ZFNs each TALEN contains the catalytic domain of the *FokI* endonuclease flanked by modules that specifically recognize the sequence targeted for repair. Unlike ZFNs, however, each modular repeat binds to a specific base pair. Each color represents a module for each of the four nucleotide bases.

of the therapeutic applicability of nucleases for the treatment of DMD has been restricted to proof-of-concept studies which have been limited to the use of meganucleases (MGNs), also called homing endonucleases, zinc finger nucleases (ZFNs), and transcription activator-like effector nucleases (TALENs) (Figure 10.8).

MGNs are found in a large number of organisms such as bacteria, yeast, fungi, algae, and some plants. They are capable of recognizing unique DNA target sites of variable sizes, but usually greater than 12 base pairs (bp) in length. Tailor-made MGNs have been created either by modifying their specificity[82] or by fusing protein domains from different enzymes.[83] The major drawback of MGNs is represented by the limited number of sequences that can be targeted to date and by the difficulties encountered in ensuring proper recombination of heterodimers that form chimeric MGNs on the target DNA and which are necessary to induce activity.[83]

Unlike MGNs, ZFNs are easier to produce, but require the expression of two distinct nucleases each binding complementary regions of the genomic DNA (Figure 10.8(b)). They are generated through the fusion of the *FokI* domain with a DNA-binding domain that consists of anywhere between three to six zinc finger repeats. Each repeat recognizes a 3 bp DNA sequence allowing for the ZFNs to bind sequences comprised between 9 and 18 bp, a length that is sufficient to discriminate and identify unique sequences within the genomic DNA (Figure 10.8(b)). The activity of the nuclease requires the formation of dimers between two *FokI* domains and each domain needs to be separated by a sequence of 5–7 bp to allow proper formation of the *FokI* dimer. Similarly, TALENs are generated by fusing the transcription activator-like effectors (TALEs) DNA-binding domain with the *FokI* nuclease domain. The DNA-binding domains can recognize single DNA

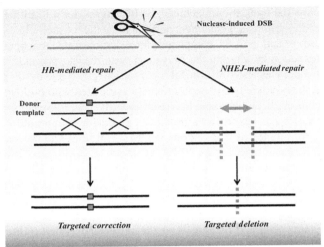

FIGURE 10.9 **Endonuclease-mediated genome editing.** The correction process is initiated by the association of the endonuclease with the region of the genomic DNA targeted for repair and the generation of a double strand break which triggers the activation of specific repair mechanisms. This usually results in deletions of one or more bases (indels). Correction of the genetic defects that leads to the restoration of normal, wild-type sequences occurs in the presence of a donor DNA template through the HR pathway (targeted correction). In this case, correction is achieved through the exchange of DNA sequences from the donor DNA to the genomic DNA targeted for repair. In the absence of donor DNA sequences, the break is repaired by the non-homologous end-joining (NHEJ) repair pathway by cleaving noncompatible overhang sequences and by joining the 3′ ends of the cleaved sequences (targeted deletion).

bases in a target site (Figure 10.8(c)), rendering TALENS more appealing as gene editing tools since assembly of the nuclease requires the simple combination of the four possible DNA-binding domains in the order specified by the sequence of the genomic DNA targeted for repair. Like ZFNs, a recognition sequence of 9–18 bp is sufficient to confer specificity toward their target, while a separation of 12–19 bp between *FokI* domains appears to be ideal to guarantee efficient dimerization of the catalytic domain of the nuclease.

The mechanisms of action of gene editing mediated by endonucleases are common to all nucleases and rely on the ability of MGNs, ZFNs, and TALENS to induce double-strand breaks (DSBs) at the DNA site targeted for repair (Figure 10.9). Once cleaved, the DNA can either be repaired through the activation of the NHEJ pathway or, in the presence of a donor DNA homologous to the region of the gene targeted for repair, through the activation of the HDR pathway. Activation of the NHEJ mechanism leads to the introduction of random micro-deletions (indels) at the site of the break. The efficiency of the repair process can vary widely, depending on the mechanism being recruited, the specific gene being targeted, and even the sequence within the same mutation targeted by the endonucleases, but it can reach frequencies of gene editing up to 20–30% in cells in culture.

Applications of nuclease-mediated gene editing for the treatment of DMD have, so far, been limited in scope and number and have focused primarily on demonstrating the feasibility of this approach to target muscle cells in culture.[84–87] MGNs and ZFNs have been used to introduce indels to restore the normal reading frame of a dog micro-dystrophin gene containing a frameshift mutation.[84,85] More recently, nuclease-mediated editing of dystrophin gene defects has been demonstrated in primary dermal fibroblasts isolated from a DMD patient harboring a deletion of exons 46 through 50 (Δ46-50) and has been used to induce targeted deletions of exon 51 to restore the coding reading frame of the dystrophin gene.[87] The feasibility of using TALENS to edit the genome was further confirmed in immortalized myoblast cell lines isolated from two DMD patients with a deletion of exons 48 through 50 (Δ48-50), treated by the same pair of TALENS used to correct the Δ46-50 fibroblasts.[87] Although it is still too early to determine the full applicability of the approach, these results are particularly promising because, like previously shown using chODNs,[79] they confirm the feasibility of using gene editing to permanently alter the intron/exon boundary of the dystrophin gene to restore its coding reading frame. As such, this approach could become a valid alternative to antisense-mediated exon-skipping strategies in the near future.

10.3.4 Suppression of Nonsense Mutations Mediated by RT Compounds

Among the approaches that have been pursued for the treatment of DMD, RT of premature stop codons is particularly appealing due to the potential of restoring full-length or near full-length dystrophin protein. These so-called nonsense mutations are usually single bp substitutions that generate a premature stop codon and that results in the absence of dystrophin expression or in a truncated, nonfunctional protein that is rapidly degraded.[88,89] Several studies have demonstrated the ability of aminoglycoside compounds to allow the translation machinery to RT stop codons, resulting in the expression of full-length protein.[89–95] Among aminoglycoside antibiotics, gentamicin appears to be the most efficient. Several elegant studies aimed at better characterizing the mechanism of action of RT compounds have demonstrated that gentamicin can substitute a different amino acid for the premature stop codon by altering the ribosomal proofreading process.[96,97] After the encouraging results obtained in animal models, gentamicin was tested in clinical trials for DMD[94,98] and cystic fibrosis,[99] but the results showed only limited effects. Furthermore, the ototoxicity and nephrotoxicity associated with prolonged administration of gentamicin hinders its long-term use, limiting its applicability to the treatment of DMD caused by nonsense mutations.

More recently, PTC Therapeutics has identified a nonaminoglycoside compound named ataluren (also known as translarna or PTC124) that can promote RT of stop codons in DMD cell lines and the *mdx* mouse.[100] Despite the apparent lack of specificity of the high-throughput screen (HTS) system used to identify the small molecule,[101,102] several subsequent studies have demonstrated the efficacy of the compound in inducing RT of premature stop codons at levels that are superior to those achieved using aminoglycosides.[89,103–105] A Phase IIb clinical trial of ataluren has shown that the drug is generally well tolerated,[106] but the clinical trial failed to meet its endpoint as administration of the drug for up to 54 weeks in DMD boys only showed limited improvement in the 6MWD test.[107] Although discouraging, a more detailed analysis of the results obtained from each of the subjects enrolled in the study and its correlation with the overall ambulatory state of the patient at the time of enrollment in the study, has evidenced some important factors that could have contributed to the overall limited effects observed at the end of the Phase IIb ataluren trial. A Phase III, multicenter, randomized, double-blind, placebo-controlled study is currently enrolling up to 220 DMD patients that will receive ataluren for up to 48 weeks of blinded treatment prior to the final analysis. Although the dosage used in the study is identical to the one used in the Phase II trial that showed the most improvement in motor function compared to placebo-treated controls, the criteria for subject eligibility are now more restrictive (http://clinicaltrials.gov).

More recently, two compounds, named RTC13 and RTC14, were identified by HTS and were found to have RT activity in human cells isolated from patients affected with ataxia telangiectasia and in *mdx* mice, both in vitro and in vivo.[89,95] Current studies are focused on further developing these compounds for the treatment of DMD and on further identifying and optimizing new small molecules that could potentially treat these disorders. Ultimately, only clinical trials will be able to determine the efficacy of these new compounds for the treatment of DMD and other genetic disorders caused by premature stop codons.

10.4 CELL-BASED THERAPIES FOR DMD

Cell transplantation for the treatment of DMD has gained attention during the past years due to the identification and better characterization of different cell types with stem cell-like properties capable of self-renewing and promoting muscle regeneration. The use of these cells is particularly appealing for diseases like DMD characterized by progressive loss of muscle mass since they could potentially be used in patients who have already progressed into late stages of the disease where the amount of muscle tissue left is not sufficient for the patient to benefit from gene augmentation or gene-correction strategies. Cell-based approaches generally employ cells expressing a functional dystrophin gene that are transplanted into muscle. Transplantation can use either autologous or allogeneic cells. Autologous transfer approaches utilize cells isolated from a patient ex vivo, corrected using gene replacement or gene editing strategies and reimplanted back into the patient, whereas allogeneic transfer uses cells obtained from an unaffected donor. Several cells of different origins have been tested for cell-based therapies with varying degrees of success (Figure 10.10) and clinical trials have been performed or are currently being tested in DMD patients. Nonetheless, cell-based therapies still remain in the early stages of development and several hurdles will have to be overcome before these approaches can be successfully implemented into clinical practice. Among those, is the need to demonstrate efficient distribution and homing of cells throughout the transplanted muscle following administration.

10.4.1 Cell-Based Therapies Using Muscle Progenitor or Muscle Stem Cells

First attempts using cells as potential therapy to restore myofibers in muscles were performed in the late 1980s and employed myoblasts, progenitor cells that are generated from the activation and differentiation of SCs. Their use was thought to be particularly advantageous due to the possibility of growing and expanding these cells in culture in large amounts, sufficient to satisfy clinical needs. The first evidence that myoblasts isolated from a healthy donor could restore dystrophin expression in muscle following transplantation came by studies conducted by Partridge et al in *mdx* mice that demonstrated that engrafted cells were able to fuse with preexisting or regenerating fibers.[108] Despite the encouraging findings obtained in animal models, clinical trials in DMD boys showed limited efficacy due primarily to poor survival of injected myoblasts and their low migratory ability beyond the injection site.[109–111]

Clinical trials using allogeneic myoblasts have been pivotal in understanding the potential of transplantation for the treatment of DMD and have prompted further studies aimed at determining the feasibility of restoring muscle function using

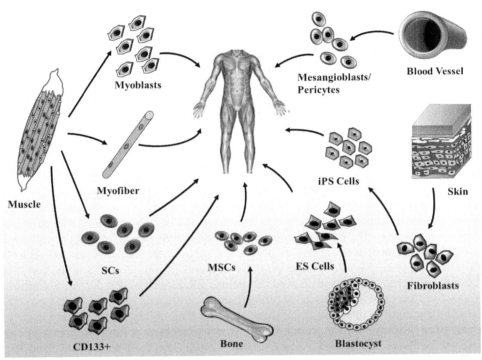

FIGURE 10.10   Cell-based therapies for Duchenne muscular dystrophy (DMD). Different populations tested to date as sources of transplantation for the treatment of DMD. These cells can be derived from muscles or obtained from tissues of different origins and then induced to commit to becoming muscle cells. Transplantation of myoblasts, single fibers, and satellite cells (SCs) have all shown to restore dystrophin expression following transplantation, although the rate of success differs depending on the type of cells used for the engraftment. Non muscle-derived stem cells include bone marrow (BM)-derived cells, embryonic stem (ES) cells that are derived from the inner cell mass of a blastocyst, induced pluripotent stem (iPS cells) derived from fibroblasts, and vessel-associated muscle-derived stem cells, such as mesoangioblasts or pericytes.

less differentiated cells capable of self-renewing following engraftment. The main advantage of this approach is the possibility of providing muscle with a continuous source of cells and is based on the hypothesis that muscle stem cells capable of surviving the transplantation procedure and homing into muscle could activate upon demand in response to degeneration or as a result of normal muscle turnover, thus limiting periodic administration of cells over prolonged periods of time. Single intact myofibers containing seven or fewer SCs were able to generate over 100 new myofibers after engraftment into muscles of *mdx* mice,[112] suggesting that SCs have regenerative capacities. Fluorescent activated cell sorting (FACS) has been used to purify SCs or subpopulations of SCs capable of efficiently contributing to fiber repair without affecting their ability to return to quiescence following activation. Recent studies have demonstrated the possibility of isolating specific subpopulations of SCs capable of restoring dystrophin expression, improving contractile function, and self-renewing following transplantation and engraftment, demonstrating that SCs are heterogeneous and that different SC populations have different regenerative capacities.[113–116]

A specific population of muscle-derived stem cells expressing the CD133 antigen has shown promise in clinical trials. CD133+stem cells are a subpopulation of circulating cells expressing the glycoprotein CD133. It has been shown that these cells can be isolated from the blood, as well as muscles of DMD patients and can be genetically modified ex vivo to express a functional dystrophin gene. Once injected in the circulatory system of *mdx* mice, transplanted cells restored dystrophin expression and improved function.[117] Phase I clinical trials using autologous transplantation of muscle-derived CD113+cells in eight boys with DMD gave promising results and no local or systemic side effects were observed in any of the treated patients, suggesting that cells may be safe to use in patients.[118] However, no additional studies have been conducted to further investigate the clinical applicability of this approach for the treatment of DMD.

10.4.2 Non Muscle-Derived Stem Cell Therapies

Over the past 10 years or so, several pivotal studies have demonstrated the existence of other cells with stem-like properties capable of homing into skeletal muscles and regenerating myofibers. Among those, stem cells derived from bone marrow (BM), mesenchymal stem cells (MSCs), induced pluripotent stem (iPS) cells, and mesoangioblasts are probably the most studied.

BM-derived stem cells systemically delivered into immunodeficient mice have been shown to migrate into muscle, differentiate into myogenic progenitors, and finally regenerate myofibers.[119] The results were followed by further observations that demonstrated that a side population of BM-derived stem cells, intravenously delivered into *mdx* mice, can partially restore dystrophin expression in the affected muscle.[120] Similar results have been obtained using BM-derived MSCs (also known as stromal cells) that have been reprogrammed in vitro to acquire the skeletal muscle lineage.[121,122] When transplanted into the muscles of rat and *mdx* mice, the cells were able to restore 20–30% of dystrophin-positive fibers. Interestingly, reprogrammed MSCs also contained SCs that contributed to muscle regeneration without additional transplantation of cells.[122]

Promising results have been achieved using embryonic stem (ES) cells, pluripotent stem cells derived from the inner cell mass of a blastocyst, or iPS cells (Figure 10.10). These cells can differentiate into all of the three primary germ layers (ectoderm, endoderm, and mesoderm). ES cells have been shown to replicate indefinitely, rendering these types of cells particularly valuable for therapeutic applications. However, differentiation of ES cells into skeletal muscle has proven to be challenging. Early studies involving transplantation of mouse and human differentiating ES cells were disappointing due to their low myogenic potential.[123,124] Unlike ES cells, iPS cells are pluripotent stem cells artificially derived from a nonpluripotent cell, similar to an adult somatic cell, by the induction of specific genes. The use of patient-derived iPS cells is appealing due to the possibility of obtaining and storing large quantities of cells to be used for autologous and allogeneic transplantation purposes. However, efficient conversion of iPS cells into myogenic precursors still appears to be out of reach, limiting the rapid advancement of this approach into clinical applications. Some success has been obtained in ES and iPS cells by conditionally expressing Pax7, a transcription factor that is expressed in SCs in the early stages of postnatal myogenesis and during adulthood. Darabi et al showed that human ES, as well as iPS-derived myogenic progenitors, can restore dystrophin-positive fibers and improve muscle strength upon transplantation in dystrophic muscle.[125]

Among the stem cell populations that have been investigated for their ability to reconstitute muscle and treat DMD, mesoangioblasts and pericytes have shown great promise. Mesoangioblasts are mesenchymal-like stem cells which are associated with the walls of large blood vessels (Figure 10.10). They can be isolated from the dorsal aorta of mouse embryos and can grow extensively in culture, generating virtually an unlimited source of cells for transplantation purposes.[126] Intra-arterial delivery of wild-type canine mesoangioblasts in a dog model of DMD resulted in up to 70% of dystrophin-positive fibers and consequently restored normal muscle function and muscle morphology.[127] Active studies conducted using human samples demonstrated the possibility of isolating cells expressing markers typical of mesoangioblasts from adult vessels, expanding the potential of cell-based therapies to autologous approaches.[128] These cells, termed pericytes, were reported to efficiently regenerate muscle after systemic delivery into immunodeficient *mdx* mice.[129] These results are particularly promising because they represent the first evidence that cell-based therapies can be used to target a large number of muscles following systemic delivery. The use of pericytes has now moved into phase I/II of clinical trials in DMD patients using allogeneic cells.

Altogether, these results clearly evidence the rapid progress made in the field of regenerative medicine and gene therapy. It is likely that the next few years will be characterized by an exponential growth in the number of approaches being pursued for DMD and other muscle disorders using cell-mediated regenerative approaches.

10.5 ALTERNATIVE STRATEGIES TO RESTORATION OF DYSTROPHIN EXPRESSION INTO MUSCLE

Concerns have been raised on the possibility that restoration of dystrophin expression alone may not be sufficient to guarantee clinical and/or long-lasting effects due to the possibility of immune responses toward dystrophin or the inability of the levels of protein being restored in muscle to be sufficient to confer functionality to myofibers. These concerns have prompted the search for therapeutic strategies that could be used in conjunction with or as an alternative to restoration of dystrophin expression in skeletal muscle of DMD patients. Upregulation of utrophin, a homolog of dystrophin, as well as downregulation of myostatin, a member of the TGFβ family of transcription factors known to regulate muscle growth, are among the therapies that are thought to have clinical value to treat the disease. Clinical trials for both strategies have been completed or are currently on their way and the results, although still preliminary, suggest that these strategies may have great therapeutic value for the treatment of the disease.

10.5.1 Increasing Levels of Utrophin as a Therapeutic Strategy for DMD

Utrophin is a component of the cytoskeleton that contains a C-terminus that interacts with dystroglycan, a central rod region consisting of a triple coiled-coil repeat and an actin-binding N-terminus domain. In normal adult muscle, utrophin is located at the neuromuscular and myotendinous junctions where it is important for normal membrane maintenance. However, in the fetus, utrophin expression localizes at the sarcolemma where dystrophin is absent and it disappears when dystrophin expression is activated, suggesting that utrophin substitutes for dystrophin in fetal and developing muscle fibers. Interestingly, utrophin is also upregulated at the sarcolemma of DMD patients and animal models of the disease, albeit at very low levels. This evidence suggests that utrophin might function as a surrogate for dystrophin in DMD muscles and that upregulating its expression into skeletal myofibers may obviate for the lack of the dystrophin protein. This hypothesis was supported by a study that demonstrated that transgenic mice overexpressing a truncated utrophin gene and crossed with *mdx* mice displayed a much milder phenotype than control *mdx* mice.[130] Similar results have been obtained following delivery into *mdx* and GRMD muscle of an Ad vector expressing a shortened version of utrophin capable of expressing critical components of the protein necessary to confer function once expressed in myofibers.[131,132]

In recent years, most of the efforts aimed at testing the feasibility of upregulating utrophin expression for the treatment of DMD have focused on identifying pharmacological approaches that could be used to increase the expression of endogenous utrophin, thus avoiding the need to deliver a new transgene into the skeletal muscle of patients. An HTS has led to the identification of a drug developed by Summit Corporation Plc named SMT-C1100 capable of increasing utrophin mRNA and protein expression up to two-fold in *mdx* mice.[133] SMT-C1100 entered Phase Ia trial in 2012 and the results of the study showed that the potential drug was well tolerated. Phase Ib trials are currently being conducted to determine the safety and tolerability of single and multiple oral doses of SMT-C1100 in DMD patients. Ultimately, the advancement of SMT-C1100 into the clinic will rely on the successful completion of efficacy studies in patients and on confirming the presence of utrophin upregulation in patients in response to active treatment.

10.5.2 Myostatin Blockers

Myostatin is a growth differentiation factor secreted by skeletal muscle that acts on muscle tissue by interacting with the activin type II receptor (ActRIIB). It is known that, during regeneration, myostatin controls muscle growth by inhibiting the activation of SCs and their differentiation. Downregulation of myostatin, achieved through genetic deletion or inhibition, results in a dramatic increase in muscle mass and in muscle regeneration. Therefore, blocking the activity of myostatin has been exploited as a therapeutic tool in treating muscle wasting diseases such as DMD. Several studies in animal models of DMD showed that myostatin inhibitors increase muscle mass but generally have variable effects on muscle function and histopathology.[134–137] Hypertrophy was observed in *mdx* mice after transgenic expression of the myostatin binding protein follistatin[137,138] and also after administration of neutralizing antibodies directed against myostatin.[139] Recently, it has been demonstrated that blocking binding of the myostatin ligand using a soluble form of ActRIIB increases muscle mass and muscle function in *mdx* mice[140] providing rationale for the use of this strategy to counteract muscle wasting in DMD.

So far, there are no myostatin-inhibiting drugs available on the market. Wyeth Pharmaceuticals developed a recombinant human antibody named Stamulumab (MYO-029), which was designed to bind and inhibit myostatin. Stamulumab underwent Phase I and II clinical trials in 2005 and 2006, respectively. However, the results of the trial were disappointing as administration of MYO-029 failed to show improvements in muscle strength or function in patients with BMD, limb-girdle, and fascioscapulohumeral muscular dystrophy.[141] Another myostatin blocker, called PF-06252616, is currently in Phase I clinical trials in healthy volunteers. Recently, a Phase II clinical trial of ACE-031, a soluble activin receptor Type IIB, showed an improvement in DMD boys, however, the trial was suspended, based on preliminary safety data that showed adverse effects such as reversible nosebleeds and telangiectasias, suggesting that further optimization may be required before this myostatin inhibitor can reach the clinic.

Although promising, limitations in the applicability of downregulating myostatin for the treatment of DMD do exist, as increasing muscle mass alone may not be sufficient to compensate for the loss of dystrophin. Approaches that can enhance regeneration while restoring the defective protein are likely to have important clinical application for the treatment of this disease. Some success has been achieved using transplantation of normal myoblasts downregulating myostatin expression, which was shown to increase the ability of transplanted cells to fuse into preexisting or nascent myofibers in *mdx* mice.[142] Along the same line, efforts have been undertaken to identify new genetic targets that could be used to increase regeneration of skeletal muscles after transplantation. Recently, an HTS of 16,000 individual small-interfering RNAs (siRNAs) identified cyclin D2 (CCND2) as a potential regulator of myoblast fusion. Inhibition of CCND2 expression in myoblasts prior to transplantation was shown to promote muscle regeneration and muscle repair in vivo.[143] Future studies are likely to be required before inhibition of CCND2 can be used as a therapeutic approach, but the results obtained are encouraging as they evidence the presence of a large number of genes that take an active role during the formation of muscle cells in culture and that, like myostatin, could be targeted through pharmacological or genetic approaches to counteract muscle regeneration into patients or to promote the ability of transplanted cells to engraft more efficiently.

10.6 CONCLUSION

The progress made during the past 10 years is a clear testimony of how the knowledge gained through basic science is finally bringing its fruits. The numerous clinical trials that have been performed and those that are currently being conducted demonstrate how the field of gene and molecular therapies for DMD is rapidly evolving toward the identification of effective treatments for the disease. This optimism should not be hampered by the apparent lack of success for some of the clinical testing. In fact, every attempt that is made and every trial that is conducted, even the one that does not meet the expected endpoint, represents a critical milestone in the development of a cure. The results obtained deepen our knowledge and our understanding of the parameters needed to bring a therapy from the bench to the bedside. Without question, these are exciting times for the field. The discoveries that have already been made and those that will come within the next few years are likely to make an impact on the lives of many patients affected by this disorder. The progress made could not have been achieved without the help of private agencies, government funds, and foundations that are working tirelessly toward one common goal. As we approach the critical stages of therapeutic development, it is becoming ever so clear that collaboration among different institutions is a key parameter to guarantee the success of each clinical approach currently being optimized. Ultimately, the identification of a cure for DMD and other muscle disorders may rely on the use of combinatory approaches which may be necessary to guarantee long-lasting effects in patients.

GLOSSARY

6MWD A method used to assess ambulatory capacity in patients and used as a primary endpoint in therapeutic trials. It measures the walking distance covered by a patient during the course of 6 min.

ActRIIB A receptor known to modulate TGFβ signaling by activating specific proteins that regulate transcription.

Base pairs (bp) Two nucleotide bases in complementary strands of DNA or DNA and RNA double-stranded molecules and consisting of a purine linked to a pyrimidine by hydrogen bonds.

cap The region within the genome of a virus responsible for encoding the VP1, VP2, and VP3 proteins.

cDNA DNA synthesized from an mRNA template often used to insert the gene coding sequence into vectors to induce its expression. The term is also used to refer to an mRNA sequence, expressed as DNA bases (guanine, cytosine, adenine and thymine) rather than RNA bases (guanine, cytosine, adenine and uracil).

Coding sequence The region of the gene that is retained in the mRNA.

Compound A substance containing two or more elements.

DNA A molecule made of twisted helical polymer chains that encodes the genetic instructions used by cells and organisms to function.

FDA Food and Drug Administration.

FokI The restriction enzyme found in a strain of bacteria capable of binding and cleaving DNA.

Frameshift mutation A mutation in a DNA sequence caused by insertions or deletions that alter the coding reading frame of the mRNA.

GRMD A dog model for DMD used for preclinical and clinical studies for the disease.

HTS A method used by researchers to screen thousands of biologically-active molecules simultaneously using miniaturized assays that mimic specific biological processes.

Indels A term used in biology to indicate the insertion or the deletion of bases in the DNA.

In-frame deletion A deletion in the coding region of a gene that does not alter the coding reading frame of the mRNA.

ITR Sequences of certain viruses capable of forming hairpins and required for certain critical functions of the virus such as replication, assembly of viral particles, and integration of the genetic material carried by the virus into the host genome.

Kilobases (kb) Unit of length equal to 1000 base pairs of DNA or RNA molecules.

LAGLIDADG A sequence of amino acids involved in the digestion of nucleic acid in the homing endonuclease. Each letter is a code that identifies a specific residue.

Luciferase An enzyme capable of catalyzing the release of photons of light from its substrate. The amount of light released can be accurately quantitated using standard enzymatic or luminescence techniques.

mdx **mice** A mouse model for DMD characterized by a nonsense mutation in exon 23 that creates a premature stop codon and results in the lack of dystrophin expression.

mRNA Molecules that specify the amino acid sequence of the protein products. They originated from the splicing of the pre-mRNA and convey genetic information from the nucleus to the cytoplasm.

Nonsense mutation A point mutation in a DNA sequence that results in a premature stop codon (UAA, UGA, UAG).

Nucleus The organelle of the cells that contains and stores all genetic information.

Oligoribonucleotides An oligonucleotide made of RNA bases.

PA site A sequence made of adenine bases that is added to the mRNA. It is believed to be critical for the export of the mRNA from the nucleus into the cytoplasm, its translation, and its stability.

Pre-mRNA mRNA synthesized from a DNA template in the nucleus containing sequences that are not translated into proteins. The removal of those sequences through splicing and processing results in the production of mRNA.

rep The region in the sequence of the viral genome encoding for the proteins necessary for replication.

siRNA Double-stranded RNA molecules capable of binding complementary sequences in the mRNA that transcribe a gene and of mediating their degradation thus preventing expression of the protein.

Stop codon A nucleotide triplet within the mRNA responsible for terminating its translation into a protein.

Virion Virus particles.

VP1, VP2, and VP3 Proteins encoded by certain viruses that are localized at the capsid, the structure that surrounds and protects the genetic material enclosed in the virus.

β-galactosidase Enzyme that catalyzes the hydrolysis of β-galactosides into simple sugars called monosaccharides, often used in molecular biology as a reporter marker to monitor gene expression.

Δ17-48 A cDNA lacking the region comprised between exon 17 and 48 of the dystrophin gene.

Δ46-50 A cDNA sequence of the dystrophin gene lacking exons 46 through 50.

Δ48-50 A cDNA sequence of the dystrophin gene lacking exons 48 through 50.

ΔH2-R19 A dystrophin protein lacking the amino acid sequence comprised between the hinge region 2 and the spectrin-like repeats 4 through 19.

ΔR4-R23 A dystrophin protein lacking the amino acid sequence comprised between the spectrin-like repeat 4 through 23.

ACKNOWLEDGMENTS

CB is supported by grants from the Muscular Dystrophy Association (MDA), The National Institutes of Health (NIH), and the Department of Defense (DoD), Congressionally Directed Medical Research Programs (CDMRP). The Authors would like to thank Thomas Gintjee and Fiona Bhondoekhan for their technical support during the preparation of this chapter. We apologize to the many colleagues whose work could not be cited due to constraints in the number of references allowed.

REFERENCES

1. Cox GA, Cole NM, Matsumura K, et al. Overexpression of dystrophin in transgenic mdx mice eliminates dystrophic symptoms without toxicity. *Nature* August 19, 1993;**364**:725–9.
2. Bertoni C, Jarrahian S, Wheeler TM, et al. Enhancement of plasmid-mediated gene therapy for muscular dystrophy by directed plasmid integration. *Proc Natl Acad Sci USA* 2006;**103**:419–24.
3. Kayali R, Bury F, Ballard M, Bertoni C. Site directed gene repair of the dystrophin gene mediated by PNA-ssODNs. *Hum Mol Genet* 2010;**19**:3266–81.
4. Hartigan-O'Connor D, Chamberlain JS. Developments in gene therapy for muscular dystrophy. *Microsc Res Tech* 2000;**48**:223–38.
5. Phelps SF, Hauser MA, Cole NM, et al. Expression of full-length and truncated dystrophin mini-genes in transgenic *mdx* mice. *Hum Mol Genet* 1995;**4**:1251–8.
6. Wang B, Li J, Xiao X. Adeno-associated virus vector carrying human minidystrophin genes effectively ameliorates muscular dystrophy in *mdx* mouse model. *Proc Natl Acad Sci USA* 2000;**97**:13714–9.
7. Harper SQ, Hauser MA, DelloRusso C, Duan D, Crawford RW, Phelps SF. Modular flexibility of dystrophin: implications for gene therapy of Duchenne muscular dystrophy. *Nat Med* 2002;**8**:253–61.
8. Lai CM, Lai YK, Rakoczy PE. Adenovirus and adeno-associated virus vectors. *DNA Cell Biol* 2002;**21**:895–913.
9. Ginsberg HS, Lundholm-Beauchamp U, Horswood RL, et al. Role of early region 3 (E3) in pathogenesis of adenovirus disease. *Proc Natl Acad Sci USA* 1989;**86**:3823–7.
10. Ragot T, Vincent N, Chafey P, et al. Efficient adenovirus-mediated transfer of a human minidystrophin gene to skeletal muscle of *mdx* mice. *Nature* 1993;**361**:647–50.
11. Vincent N, Ragot T, Gilgenkrantz H, et al. Long-term correction of mouse dystrophic degeneration by adenovirus-mediated transfer of a minidystrophin gene. *Nat Genet* 1993;**5**:130–4.
12. Deconinck N, Ragot T, Marechal G, Perricaudet M, Gillis JM. Functional protection of dystrophic mouse (*mdx*) muscles after adenovirus-mediated transfer of a dystrophin minigene. *Proc Natl Acad Sci USA* 1996;**93**:3570–4.
13. Yang L, Lochmuller H, Luo J, et al. Adenovirus-mediated dystrophin minigene transfer improves muscle strength in adult dystrophic (*MDX*) mice. *Gene Ther* 1998;**5**:369–79.
14. Kochanek S, Clemens PR, Mitani K, Chen HH, Chan S, Caskey CT. A new adenoviral vector: replacement of all viral coding sequences with 28 kb of DNA independently expressing both full-length dystrophin and beta-galactosidase. *Proc Natl Acad Sci USA* 1996;**93**:5731–6.
15. Kumar-Singh R, Chamberlain JS. Encapsidated adenovirus minichromosomes allow delivery and expression of a 14 kb dystrophin cDNA to muscle cells. *Hum Mol Genet* 1996;**5**:913–21.
16. Gilbert R, Nalbantoglu J, Howell JM, et al. Dystrophin expression in muscle following gene transfer with a fully deleted ("gutted") adenovirus is markedly improved by trans-acting adenoviral gene products. *Hum Gene Ther* 2001;**12**:1741–55.
17. DelloRusso C, Scott JM, Hartigan-O'Connor D, et al. Functional correction of adult *mdx* mouse muscle using gutted adenoviral vectors expressing full-length dystrophin. *Proc Natl Acad Sci USA* 2002;**99**:12979–84.
18. Gilchrist SC, Ontell MP, Kochanek S, Clemens PR. Immune response to full-length dystrophin delivered to Dmd muscle by a high-capacity adenoviral vector. *Mol Ther* 2002;**6**:359–68.

19. Clemens PR, Kochanek S, Sunada Y, et al. In vivo muscle gene transfer of full-length dystrophin with an adenoviral vector that lacks all viral genes. *Gene Ther* 1996;**3**:965–72.

20. Chen HH, Mack LM, Kelly R, Ontell M, Kochanek S, Clemens PR. Persistence in muscle of an adenoviral vector that lacks all viral genes. *Proc Natl Acad Sci USA* 1997;**94**:1645–50.

21. Chen HH, Mack LM, Choi SY, Ontell M, Kochanek S, Clemens PR. DNA from both high-capacity and first-generation adenoviral vectors remains intact in skeletal muscle. *Hum Gene Ther* 1999;**10**:365–73.

22. Dudley RW, Lu Y, Gilbert R, et al. Sustained improvement of muscle function one year after full-length dystrophin gene transfer into *mdx* mice by a gutted helper-dependent adenoviral vector. *Hum Gene Ther* 2004;**15**:145–56.

23. Jiang ZL, Reay D, Kreppel F, et al. Local high-capacity adenovirus-mediated mCTLA4Ig and mCD40Ig expression prolongs recombinant gene expression in skeletal muscle. *Mol Ther* 2001;**3**:892–900.

24. Yang EY, Kim HB, Shaaban AF, Milner R, Adzick NS, Flake AW. Persistent postnatal transgene expression in both muscle and liver after fetal injection of recombinant adenovirus. *J Pediatr Surg* 1999;**34**:766–72.

25. Bilbao R, Reay DP, Wu E, et al. Comparison of high-capacity and first-generation adenoviral vector gene delivery to murine muscle in utero. *Gene Ther* 2005;**12**:39–47.

26. Dunckley MG, Love DR, Davies KE, Walsh FS, Morris GE, Dickson G. Retroviral-mediated transfer of a dystrophin minigene into *mdx* mouse myoblasts in vitro. *FEBS Lett* 1992;**296**:128–34.

27. Salvatori G, Ferrari G, Mezzogiorno A, et al. Retroviral vector-mediated gene transfer into human primary myogenic cells leads to expression in muscle fibers in vivo. *Hum Gene Ther* 1993;**4**:713–23.

28. Dunckley MG, Wells DJ, Walsh FS, Dickson G. Direct retroviral-mediated transfer of a dystrophin minigene into *mdx* mouse muscle in vivo. *Hum Mol Genet* 1993;**2**:717–23.

29. Naldini L, Blomer U, Gallay P, et al. In vivo gene delivery and stable transduction of nondividing cells by a lentiviral vector. *Science* 1996;**272**:263–7.

30. Kafri T, Blomer U, Peterson DA, Gage FH, Verma IM. Sustained expression of genes delivered directly into liver and muscle by lentiviral vectors. *Nat Genet* 1997;**17**:314–7.

31. Li S, Kimura E, Fall BM, et al. Stable transduction of myogenic cells with lentiviral vectors expressing a minidystrophin. *Gene Ther* 2005;**12**:1099–108.

32. Kobinger GP, Louboutin JP, Barton ER, Sweeney HL, Wilson JM. Correction of the dystrophic phenotype by in vivo targeting of muscle progenitor cells. *Hum Gene Ther* 2003;**14**:1441–9.

33. MacKenzie TC, Kobinger GP, Kootstra NA, et al. Efficient transduction of liver and muscle after in utero injection of lentiviral vectors with different pseudotypes. *Mol Ther* 2002;**6**:349–58.

34. Kimura E, Li S, Gregorevic P, Fall BM, Chamberlain JS. Dystrophin delivery to muscles of mdx mice using lentiviral vectors leads to myogenic progenitor targeting and stable gene expression. *Mol Ther* 2010;**18**:206–13.

35. Xiao X, Li J, Samulski RJ. Efficient long-term gene transfer into muscle tissue of immunocompetent mice by adeno-associated virus vector. *J Virol* 1996;**70**:8098–108.

36. Konieczny P, Swiderski K, Chamberlain JS. Gene and cell-mediated therapies for muscular dystrophy. *Muscle Nerve* 2013;**47**:649–63.

37. Xiao X, Li J, Samulski RJ. Production of high-titer recombinant adeno-associated virus vectors in the absence of helper adenovirus. *J Virol* 1998;**72**:2224–32.

38. Blankinship MJ, Gregorevic P, Allen JM, et al. Efficient transduction of skeletal muscle using vectors based on adeno-associated virus serotype 6. *Mol Ther* 2004;**10**:671–8.

39. Gregorevic P, Blankinship MJ, Allen JM, et al. Systemic delivery of genes to striated muscles using adeno-associated viral vectors. *Nat Med* 2004;**10**:828–34.

40. Wang Z, Zhu T, Qiao C, et al. Adeno-associated virus serotype 8 efficiently delivers genes to muscle and heart. *Nat Biotechnol* 2005;**23**:321–8.

41. Inagaki K, Fuess S, Storm TA, et al. Robust systemic transduction with AAV9 vectors in mice: efficient global cardiac gene transfer superior to that of AAV8. *Mol Ther* 2006;**14**:45–53.

42. Rodino-Klapac LR, Janssen PM, Montgomery CL, et al. A translational approach for limb vascular delivery of the micro-dystrophin gene without high volume or high pressure for treatment of Duchenne muscular dystrophy. *J Transl Med* 2007;**5**:45.

43. Lai Y, Yue Y, Liu M, et al. Efficient in vivo gene expression by trans-splicing adeno-associated viral vectors. *Nat Biotechnol* 2005;**23**:1435–9.

44. Ghosh A, Yue Y, Duan D. Viral serotype and the transgene sequence influence overlapping adeno-associated viral (AAV) vector-mediated gene transfer in skeletal muscle. *J Gene Med* 2006;**8**:298–305.

45. Yuasa K, Yoshimura M, Urasawa N, et al. Injection of a recombinant AAV serotype 2 into canine skeletal muscles evokes strong immune responses against transgene products. *Gene Ther* 2007;**14**:1249–60.

46. Wang Z, Allen JM, Riddell SR, et al. Immunity to adeno-associated virus-mediated gene transfer in a random-bred canine model of Duchenne muscular dystrophy. *Hum Gene Ther* 2007;**18**:18–26.

47. Wang Z, Kuhr CS, Allen JM, et al. Sustained AAV-mediated dystrophin expression in a canine model of Duchenne muscular dystrophy with a brief course of immunosuppression. *Mol Ther* 2007;**15**:1160–6.

48. Mingozzi F, High KA. Immune responses to AAV in clinical trials. *Curr Gene Ther* 2007;**7**:316–24.

49. Bowles DE, McPhee SW, Li C, et al. Phase 1 gene therapy for Duchenne muscular dystrophy using a translational optimized AAV vector. *Mol Ther* 2012;**20**:443–55.

50. Wolff JA, Malone RW, Williams P, et al. Direct gene transfer into mouse muscle in vivo. *Science* 1990;**247**:1465–8.
51. Howell JM, Fletcher S, O'Hara A, Johnsen RD, Lloyd F, Kakulas BA. Direct dystrophin and reporter gene transfer into dog muscle in vivo. *Muscle Nerve* 1998;**21**:159–65.
52. Romero NB, Braun S, Benveniste O, et al. Phase I study of dystrophin plasmid-based gene therapy in Duchenne/Becker muscular dystrophy. *Hum Gene Ther* 2004;**15**:1065–76.
53. Zhang G, Budker V, Williams P, Subbotin V, Wolff JA. Efficient expression of naked dna delivered intraarterially to limb muscles of nonhuman primates. *Hum Gene Ther* 2001;**12**:427–38.
54. Liu F, Nishikawa M, Clemens PR, Huang L. Transfer of full-length Dmd to the diaphragm muscle of Dmd(*mdx/mdx*) mice through systemic administration of plasmid DNA. *Mol Ther* 2001;**4**:45–51.
55. Zhang G, Ludtke JJ, Thioudellet C, et al. Intraarterial delivery of naked plasmid DNA expressing full-length mouse dystrophin in the *mdx* mouse model of duchenne muscular dystrophy. *Hum Gene Ther* 2004;**15**:770–82.
56. Chen ZY, He CY, Ehrhardt A, Kay MA. Minicircle DNA vectors devoid of bacterial DNA result in persistent and high-level transgene expression in vivo. *Mol Ther* 2003;**8**:495–500.
57. Yant SR, Meuse L, Chiu W, Ivics Z, Izsvak Z, Kay MA. Somatic integration and long-term transgene expression in normal and haemophilic mice using a DNA transposon system. *Nat Genet* 2000;**25**:35–41.
58. Olivares EC, Hollis RP, Chalberg TW, Meuse L, Kay MA, Calos MP. Site-specific genomic integration produces therapeutic Factor IX levels in mice. *Nat Biotechnol* 2002;**20**:1124–8.
59. Takeshima Y, Nishio H, Sakamoto H, Nakamura H, Matsuo M. Modulation of in vitro splicing of the upstream intron by modifying an intra-exon sequence which is deleted from the dystrophin gene in dystrophin Kobe. *J Clin Invest* 1995;**95**:515–20.
60. Dunckley MG, Manoharan M, Villiet P, Eperon IC, Dickson G. Modification of splicing in the dystrophin gene in cultured *Mdx* muscle cells by antisense oligoribonucleotides. *Hum Mol Genet* 1998;**7**:1083–90.
61. Wilton SD, Lloyd F, Carville K, et al. Specific removal of the nonsense mutation from the *mdx* dystrophin mRNA using antisense oligonucleotides. *Neuromuscul Disord* 1999;**9**:330–8.
62. Lu QL, Mann CJ, Lou F, et al. Functional amounts of dystrophin produced by skipping the mutated exon in the *mdx* dystrophic mouse. *Nat Med* 2003;**9**:1009–14.
63. van Deutekom JC, Bremmer-Bout M, Janson AA, et al. Antisense-induced exon skipping restores dystrophin expression in DMD patient derived muscle cells. *Hum Mol Genet* 2001;**10**:1547–54.
64. van Deutekom JC, Janson AA, Ginjaar IB, et al. Local dystrophin restoration with antisense oligonucleotide PRO051. *N Engl J Med* 2007;**357**:2677–86.
65. Mann CJ, Honeyman K, Cheng AJ, et al. Antisense-induced exon skipping and synthesis of dystrophin in the *mdx* mouse. *Proc Natl Acad Sci USA* 2001;**98**:42–7.
66. Alter J, Lou F, Rabinowitz A, et al. Systemic delivery of morpholino oligonucleotide restores dystrophin expression bodywide and improves dystrophic pathology. *Nat Med* 2006;**12**:175–7.
67. Yin H, Lu Q, Wood M. Effective exon skipping and restoration of dystrophin expression by peptide nucleic acid antisense oligonucleotides in *mdx* mice. *Mol Ther* 2008;**16**:38–45.
68. Cirak S, rechavala-Gomeza V, Guglieri M, et al. Exon skipping and dystrophin restoration in patients with Duchenne muscular dystrophy after systemic phosphorodiamidate morpholino oligomer treatment: an open-label, phase 2, dose-escalation study. *Lancet* 2011;**378**:595–605.
69. Jearawiriyapaisarn N, Moulton HM, Buckley B, et al. Sustained dystrophin expression induced by peptide-conjugated morpholino oligomers in the muscles of *mdx* mice. *Mol Ther* 2008;**16**:1624–9.
70. Yin H, Moulton HM, Seow Y, et al. Cell-penetrating peptide-conjugated antisense oligonucleotides restore systemic muscle and cardiac dystrophin expression and function. *Hum Mol Genet* 2008;**17**:3909–18.
71. Wu B, Moulton HM, Iversen PL, et al. Effective rescue of dystrophin improves cardiac function in dystrophin-deficient mice by a modified morpholino oligomer. *Proc Natl Acad Sci USA* 2008;**105**:14814–9.
72. Kren BT, Parashar B, Bandyopadhyay P, Chowdhury NR, Chowdhury JR, Steer CJ. Correction of the UDP-glucuronosyltransferase gene defect in the gunn rat model of crigler-najjar syndrome type I with a chimeric oligonucleotide. *Proc Natl Acad Sci USA* 1999;**96**:10349–54.
73. Cole-Strauss A, Gamper H, Holloman WK, Munoz M, Cheng N, Kmiec EB. Targeted gene repair directed by the chimeric RNA/DNA oligonucleotide in a mammalian cell-free extract. *Nucleic Acids Res* 1999;**27**:1323–30.
74. Igoucheva O, Peritz AE, Levy D, Yoon K. A sequence-specific gene correction by an RNA-DNA oligonucleotide in mammalian cells characterized by transfection and nuclear extract using a lacZ shuttle system. *Gene Ther* 1999;**6**:1960–71.
75. Bertoni C. Clinical approaches in the treatment of Duchenne muscular dystrophy (DMD) using oligonucleotides. *Front Biosci* 2008;**13**:517–27.
76. Rando TA, Disatnik MH, Zhou LZ. Rescue of dystrophin expression in mdx mouse muscle by RNA/DNA oligonucleotides. *Proc Natl Acad Sci USA* 2000;**97**:5363–8.
77. Bertoni C, Rando TA. Dystrophin gene repair in mdx muscle precursor cells in vitro and in vivo mediated by RNA-DNA chimeric oligonucleotides. *Hum Gene Ther* 2002;**13**:707–18.
78. Bartlett RJ, Stockinger S, Denis MM, et al. In vivo targeted repair of a point mutation in the canine dystrophin gene by a chimeric RNA/DNA oligonucleotide. *Nat Biotechnol* 2000;**18**:615–22.
79. Bertoni C, Lau C, Rando TA. Restoration of dystrophin expression in *mdx* muscle cells by chimeraplast-mediated exon skipping. *Hum Mol Genet* 2003;**12**:1087–99.
80. Bertoni C, Morris GE, Rando TA. Strand bias in oligonucleotide-mediated dystrophin gene editing. *Hum Mol Genet* 2005;**14**:221–33.

81. Nik-Ahd F, Bertoni C. Ex vivo gene correction of the dystrophin gene in muscle stem cells using peptide nucleic acid single stranded oligodeoxynucleotides (PNA-ssODNs) induces stable expression of dystrophin in a mouse model for Duchenne muscular dystrophy. *Stem Cells* 2014;**32**:1817–30.

82. Seligman LM, Chisholm KM, Chevalier BS, et al. Mutations altering the cleavage specificity of a homing endonuclease. *Nucleic Acids Res* 2002;**30**:3870–9.

83. Arnould S, Chames P, Perez C, et al. Engineering of large numbers of highly specific homing endonucleases that induce recombination on novel DNA targets. *J Mol Biol* 2006;**355**:443–58.

84. Chapdelaine P, Pichavant C, Rousseau J, Paques F, Tremblay JP. Meganucleases can restore the reading frame of a mutated dystrophin. *Gene Ther* 2010;**17**:846–58.

85. Rousseau J, Chapdelaine P, Boisvert S, et al. Endonucleases: tools to correct the dystrophin gene. *J Gene Med* 2011;**13**:522–37.

86. Benabdallah BF, Duval A, Rousseau J, et al. Targeted gene addition of microdystrophin in mice skeletal muscle via human myoblast transplantation. *Mol Ther Nucleic Acids* 2013;**2**:e68.

87. Ousterout DG, Perez-Pinera P, Thakore PI, et al. Reading frame correction by targeted genome editing restores dystrophin expression in cells from Duchenne muscular dystrophy patients. *Mol Ther* 2013;**21**:1718–26.

88. Prior TW, Bartolo C, Pearl DK, et al. Spectrum of small mutations in the dystrophin coding region. *Am J Hum Genet* 1995;**57**:22–33.

89. Kayali R, Ku JM, Khitrov G, Jung ME, Prikhodko O, Bertoni C. Read-through compound 13 restores dystrophin expression and improves muscle function in the *mdx* mouse model for Duchenne muscular dystrophy. *Hum Mol Genet* 2012;**21**:4007–20.

90. Howard M, Frizzell RA, Bedwell DM. Aminoglycoside antibiotics restore CFTR function by overcoming premature stop mutations. *Nat Med* 1996;**2**:467–9.

91. Barton-Davis ER, Cordier L, Shoturma DI, Leland SE, Sweeney HL. Aminoglycoside antibiotics restore dystrophin function to skeletal muscles of *mdx* mice. *J Clin Invest* 1999;**104**:375–81.

92. Wagner KR, Hamed S, Hadley DW, et al. Gentamicin treatment of Duchenne and Becker muscular dystrophy due to nonsense mutations. *Ann Neurol* 2001;**49**:706–11.

93. Du M, Jones JR, Lanier J, et al. Aminoglycoside suppression of a premature stop mutation in a Cftr-/- mouse carrying a human CFTR-G542X transgene. *J Mol Med* 2002;**80**:595–604.

94. Politano L, Nigro G, Nigro V, et al. Gentamicin administration in Duchenne patients with premature stop codon. Preliminary results. *Acta Myol* 2003;**22**:15–21.

95. Du L, Damoiseaux R, Nahas S, et al. Nonaminoglycoside compounds induce readthrough of nonsense mutations. *J Exp Med* 2009;**206**:2285–97.

96. Manuvakhova M, Keeling K, Bedwell DM. Aminoglycoside antibiotics mediate context-dependent suppression of termination codons in a mammalian translation system. *RNA* 2000;**6**:1044–55.

97. Hobbie SN, Pfister P, Brull C, Westhof E, Bottger EC. Analysis of the contribution of individual substituents in 4,6-aminoglycoside-ribosome interaction. *Antimicrob Agents Chemother* 2005;**49**:5112–8.

98. Malik V, Rodino-Klapac LR, Viollet L, et al. Gentamicin-induced readthrough of stop codons in Duchenne muscular dystrophy. *Ann Neurol* 2010;**67**:771–80.

99. Wilschanski M, Yahav Y, Yaacov Y, et al. Gentamicin-induced correction of CFTR function in patients with cystic fibrosis and CFTR stop mutations. *N Engl J Med* 2003;**349**:1433–41.

100. Welch EM, Barton ER, Zhuo J, et al. PTC124 targets genetic disorders caused by nonsense mutations. *Nature* 2007;**447**:87–91.

101. Auld DS, Thorne N, Maguire WF, Inglese J. Mechanism of PTC124 activity in cell-based luciferase assays of nonsense codon suppression. *Proc Natl Acad Sci USA* 2009;**106**:3585–90.

102. McElroy SP, Nomura T, Torrie LS, et al. A lack of premature termination codon read-through efficacy of PTC124 (Ataluren) in a diverse array of reporter assays. *PLoS Biol* 2013;**11**:e1001593.

103. Goldmann T, Overlack N, Wolfrum U, Nagel-Wolfrum K. PTC124-mediated translational readthrough of a nonsense mutation causing Usher syndrome type 1C. *Hum Gene Ther* 2011;**22**:537–47.

104. Goldmann T, Overlack N, Moller F, et al. A comparative evaluation of NB30, NB54 and PTC124 in translational read-through efficacy for treatment of an USH1C nonsense mutation. *EMBO Mol Med* 2012;**4**:1186–99.

105. Yu H, Liu X, Huang J, Zhang Y, Hu R, Pu J. Comparison of read-through effects of aminoglycosides and PTC124 on rescuing nonsense mutations of HERG gene associated with long QT syndrome. *Int J Mol Med* 2014;**33**:729–35.

106. Hirawat S, Welch EM, Elfring GL, et al. Safety, tolerability, and pharmacokinetics of PTC124, a nonaminoglycoside nonsense mutation suppressor, following single- and multiple-dose administration to healthy male and female adult volunteers. *J Clin Pharmacol* 2007;**47**:430–44.

107. Finkel RS, Flanigan KM, Wong B, et al. Phase 2a study of ataluren-mediated dystrophin production in patients with nonsense mutation duchenne muscular dystrophy. *PLoS One* 2013;**8**:e81302.

108. Partridge TA, Morgan JE, Coulton GR, Hoffman EP, Kunkel LM. Conversion of mdx myofibres from dystrophin-negative to -positive by injection of normal myoblasts. *Nature* 1989;**337**:176–9.

109. Miller RG, Sharma KR, Pavlath GK, et al. Myoblast implantation in Duchenne muscular dystrophy: the San Francisco study. *Muscle Nerve* 1997;**20**:469–78.

110. Gussoni E, Blau HM, Kunkel LM. The fate of individual myoblasts after transplantation into muscles of DMD patients. *Nat Med* 1997;**3**:970–7.

111. Fan Y, Maley M, Beilharz M, Grounds M. Rapid death of injected myoblasts in myoblast transfer therapy. *Muscle Nerve* 1996;**19**:853–60.

112. Collins CA, Olsen I, Zammit PS, et al. Stem cell function, self-renewal, and behavioral heterogeneity of cells from the adult muscle satellite cell niche. *Cell* 2005;**122**:289–301.

113. Cerletti M, Jurga S, Witczak CA, et al. Highly efficient, functional engraftment of skeletal muscle stem cells in dystrophic muscles. *Cell* 2008;**134**: 37–47.
114. Sacco A, Doyonnas R, Kraft P, Vitorovic S, Blau HM. Self-renewal and expansion of single transplanted muscle stem cells. *Nature* 2008;**456**:502–6.
115. Rocheteau P, Gayraud-Morel B, Siegl-Cachedenier I, Blasco MA, Tajbakhsh S. A subpopulation of adult skeletal muscle stem cells retains all template DNA strands after cell division. *Cell* 2012;**148**:112–25.
116. Arpke RW, Darabi R, Mader TL, et al. A new immuno- dystrophin-deficient model, the NSG-*mdx* mouse, provides evidence for functional improvement following allogeneic satellite cell transplantation. *Stem Cells* 2013;**31**:1611–20.
117. Torrente Y, Belicchi M, Sampaolesi M, et al. Human circulating AC133(+) stem cells restore dystrophin expression and ameliorate function in dystrophic skeletal muscle. *J Clin Invest* 2004;**114**:182–95.
118. Torrente Y, Belicchi M, Marchesi C, et al. Autologous transplantation of muscle-derived CD133+ stem cells in Duchenne muscle patients. *Cell Transplant* 2007;**16**:563–77.
119. Ferrari G, Cusella-De AG, Coletta M, et al. Muscle regeneration by bone marrow-derived myogenic progenitors. *Science* 1998;**279**:1528–30.
120. Gussoni E, Soneoka Y, Strickland CD, et al. Dystrophin expression in the *mdx* mouse restored by stem cell transplantation. *Nature* 1999;**401**: 390–4.
121. De BC, Dell'Accio F, Vandenabeele F, Vermeesch JR, Raymackers JM, Luyten FP. Skeletal muscle repair by adult human mesenchymal stem cells from synovial membrane. *J Cell Biol* 2003;**160**:909–18.
122. Dezawa M, Ishikawa H, Itokazu Y, et al. Bone marrow stromal cells generate muscle cells and repair muscle degeneration. *Science* 2005;**309**:314–7.
123. Bhagavati S, Xu W. Isolation and enrichment of skeletal muscle progenitor cells from mouse bone marrow. *Biochem Biophys Res Commun* 2004;**318**:119–24.
124. Barberi T, Bradbury M, Dincer Z, Panagiotakos G, Socci ND, Studer L. Derivation of engraftable skeletal myoblasts from human embryonic stem cells. *Nat Med* 2007;**13**:642–8.
125. Darabi R, Arpke RW, Irion S, et al. Human ES- and iPS-derived myogenic progenitors restore DYSTROPHIN and improve contractility upon transplantation in dystrophic mice. *Cell Stem Cell* 2012;**10**:610–9.
126. De AL, Berghella L, Coletta M, et al. Skeletal myogenic progenitors originating from embryonic dorsal aorta coexpress endothelial and myogenic markers and contribute to postnatal muscle growth and regeneration. *J Cell Biol* 1999;**147**:869–78.
127. Sampaolesi M, Blot S, D'Antona G, et al. Mesoangioblast stem cells ameliorate muscle function in dystrophic dogs. *Nature* 2006;**444**:574–9.
128. Dellavalle A, Sampaolesi M, Tonlorenzi R, et al. Pericytes of human skeletal muscle are myogenic precursors distinct from satellite cells. *Nat Cell Biol* 2007;**9**:255–67.
129. Dellavalle A, Maroli G, Covarello D, et al. Pericytes resident in postnatal skeletal muscle differentiate into muscle fibres and generate satellite cells. *Nat Commun* 2011;**2**:499.
130. Tinsley JM, Potter AC, Phelps SR, Fisher R, Trickett JI, Davies KE. Amelioration of the dystrophic phenotype of *mdx* mice using a truncated utrophin transgene. *Nature* 1996;**384**:349–53.
131. Gilbert R, Nalbantoglu J, Petrof BJ, et al. Adenovirus-mediated utrophin gene transfer mitigates the dystrophic phenotype of *mdx* mouse muscles. *Hum Gene Ther* 1999;**10**:1299–310.
132. Cerletti M, Negri T, Cozzi F, et al. Dystrophic phenotype of canine X-linked muscular dystrophy is mitigated by adenovirus-mediated utrophin gene transfer. *Gene Ther* 2003;**10**:750–7.
133. Tinsley JM, Fairclough RJ, Storer R, et al. Daily treatment with SMTC1100, a novel small molecule utrophin upregulator, dramatically reduces the dystrophic symptoms in the *mdx* mouse. *PLoS One* 2011;**6**:e19189.
134. McPherron AC, Lee SJ. Double muscling in cattle due to mutations in the myostatin gene. *Proc Natl Acad Sci USA* 1997;**94**:12457–61.
135. McPherron AC, Lawler AM, Lee SJ. Regulation of skeletal muscle mass in mice by a new TGF-beta superfamily member. *Nature* 1997;**387**:83–90.
136. Lee SJ, McPherron AC. Myostatin and the control of skeletal muscle mass. *Curr Opin Genet Dev* 1999;**9**:604–7.
137. Wagner KR, McPherron AC, Winik N, Lee SJ. Loss of myostatin attenuates severity of muscular dystrophy in *mdx* mice. *Ann Neurol* 2002;**52**:832–6.
138. Nakatani M, Takehara Y, Sugino H, et al. Transgenic expression of a myostatin inhibitor derived from follistatin increases skeletal muscle mass and ameliorates dystrophic pathology in *mdx* mice. *FASEB J* 2008;**22**:477–87.
139. Bogdanovich S, Krag TO, Barton ER, et al. Functional improvement of dystrophic muscle by myostatin blockade. *Nature* 2002;**420**:418–21.
140. Pistilli EE, Bogdanovich S, Goncalves MD, et al. Targeting the activin type IIB receptor to improve muscle mass and function in the *mdx* mouse model of Duchenne muscular dystrophy. *Am J Pathol* 2011;**178**:1287–97.
141. Wagner KR, Fleckenstein JL, Amato AA, et al. A phase I/IItrial of MYO-029 in adult subjects with muscular dystrophy. *Ann Neurol* 2008;**63**: 561–71.
142. Benabdallah BF, Bouchentouf M, Rousseau J, et al. Inhibiting myostatin with follistatin improves the success of myoblast transplantation in dystrophic mice. *Cell Transplant* 2008;**17**:337–50.
143. Khanjyan MV, Yang J, Kayali R, Caldwell T, Bertoni C. A high-content, high-throughput siRNA screen identifies cyclin D2 as a potent regulator of muscle progenitor cell fusion and a target to enhance muscle regeneration. *Hum Mol Genet* 2013;**22**:3283–95.

Chapter 11

Gene Therapy for Retinal Disease

Michelle E. McClements[1,2,3] and Robert E. MacLaren[1,2,3]
¹Nuffield Laboratory of Ophthalmology, Department of Clinical Neurosciences, University of Oxford, Oxford, UK; ²NIHR Biomedical Research Centre, Oxford Eye Hospital, Oxford, UK; ³Moorfields Eye Hospital & NIHR Biomedical Research Centre for Opthalmology, London, UK

LIST OF ABBREVIATIONS

661W Murine photoreceptor-derived retinal cell line
AAV Adeno-associated virus
ABCA4 ATP-binding cassette, subfamily A, member 4
AIPL1 Aryl hydrocarbon receptor interacting protein like-1
ARPE19 Retinal pigment epithelia cell line
Cabp5 Calcium binding protein 5
CAG Cytomegalovirus early enhancer–promoter element fused with the chicken β-actin promoter and rabbit globin intron–exon fragment
CAR Cone arrestin
CBA Chicken β-actin
CD8+ T cells Cytotoxic T cells with CD8 surface protein
Chx10 Homeobox-containing transcription factor
CMV Cytomegalovirus
CNGB3 Cone photoreceptor cGMP-gated cation channel and the CBA Grm6 glutamate receptor, metabotropic 6
CNTF Ciliary neurotrophic factor
enhancer—S Short-wave (cone opsin)
ERG Electroretinography
F Phenylalanine residue
FGF-2 Fibroblast growth factor 2
FLT1/VEGF Fms-related tyrosine kinase 1/vascular endothelial growth factor
GCL Ganglion cell layer
GDNF Glial cell derived neurotrophic factor
GFP Green fluorescent protein
HEK293 Human embryonic kidney 293 cell line
hGRK1 Human rhodopsin kinase
INL Inner nuclear layer
IRDs Inherited retinal diseases
ITR Inverted terminal repeat
K296E Lysin 296 glutamic acid substitution
L Long-wave (cone opsin)
LCA Leber's congenital amaurosis
LV Lentiviral
M Medium-wave (cone opsin)
Mdx Mouse model for muscular dystrophy
Mertk c-mer proto-oncogene tyrosine kinase
MYO7A Myosin VIIA
OCT Optical coherence tomography
ONL Outer nuclear layer
P23H Proline 23 histidine substitution
PR2.1 L/M cone opsin promoter element, 2.1 kb
Prph2/rds Peripherin 2/retinal degeneration slow
rAAV Recombinant AAV
RCS Royal College of Surgeons

Translating Gene Therapy to the Clinic. http://dx.doi.org/10.1016/B978-0-12-800563-7.00011-7

173

Rd Retinal degeneration (mouse model)
REP1 Rab escort protein 1
RHO Rhodopsin
RP Retinitis pigmentosa
RPE Retinal pigment epithelium
RPE65 Retinal pigment epithelium-specific 65 kDa protein
ScAAV Self-complementary AAV
ssDNA Single-stranded DNA
UTR Untranslated region
VMD2 Bestrophin 1 (vitelliform macular dystrophy)
WT Wild-type
XIAP X-linked inhibitor of apoptosis
Y Tyrosine residue

11.1 INTRODUCTION TO THE RETINA AND INHERITED RETINAL DISEASES

The retina is a multilayered structure composed of various cell types (Figure 11.1). As light enters the eye it must pass through the various layers of neuronal cells (primarily ganglion cells and bipolar cells) before reaching the light-sensitive rod and cone photoreceptors at the back of the retina. The rod and cone photoreceptors comprise an outer segment, an inner segment, and a synaptic terminal (Figure 11.1). The outer segment contains discs, within the membranes of which the pigment that absorbs light is found and, via a G-protein cascade, the photopigment converts light into electrochemical signals that are transferred via the synaptic terminal through the bipolar cells, ganglion cells, and along the optic nerve to the brain. There are three types of cone photoreceptors, defined by the opsin type they express. Opsins bind with chromophore to form

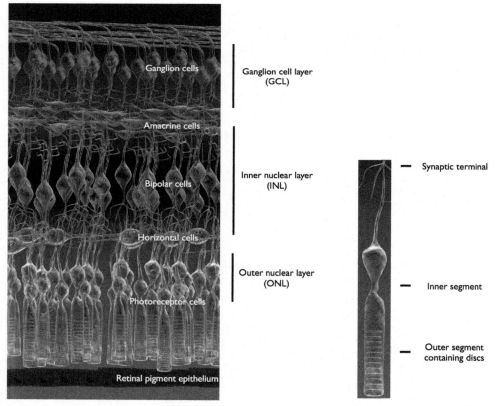

FIGURE 11.1 Basic vertebrate retinal structure and photoreceptor cell form. The retinal pigment epithelium (RPE) supports survival of the photoreceptors at the back of the retina. Photoreceptor cells form the outer retina and respond directly to light signals. Bipolar cells within the inner nuclear layer (INL) pass signals from the photoreceptor cells to the ganglion cells, which then travel through the optic nerve to the brain. Horizontal cells are interneurons in the INL that provide feedback to rod and cone photoreceptor cells. Amacrine cells perform a similar function by assessing bipolar cell signals to control ganglion cell responses. GCL, ganglion cell layer; ONL, outer nuclear layer.

the photopigments that sense different wavelengths of light—short (S, blue), medium (M, green), or long (L, red)—and each cone photoreceptor cell will express only one of these three opsin variants. Rod photoreceptors have a single photopigment type, rhodopsin, and are sensitive to low levels of light. This results in S, M, and L cone photoreceptors providing us with color vision and visual acuity in daylight (photopic vision), whereas rod photoreceptors allow for vision under dim light conditions (scotopic vision). Supporting the photoreceptors at the back of the retina is the retinal pigment epithelium (RPE), a single layer of cells essential for photoreceptor survival. Beyond the RPE is the vascular layer, the choroid.

The genetic cause of a particular inherited retinal disease (IRD) may result in the death of one cell type within the retina, such as with the cone dystrophies. Alternatively, more than one cell type may be lost; for example, retinitis pigmentosa (RP) begins with a primary loss of rod photoreceptor cells, but over time the cone photoreceptors also degenerate as their survival is dependent on rod survival. Mutations in genes specifically expressed in the RPE cells can lead to photoreceptor death prior to RPE degeneration because of the importance of RPE function in photoreceptor survival. The inner retinal layer does not tend to be the primary region of loss, and the majority of IRD cases result from mutations in genes specifically expressed in the photoreceptor or RPE cells. Over 200 genes have been linked to IRDs (http://www.sph.uth.tmc.edu/RetNet/) with a range of inheritance patterns exhibited. At least 30 genes alone have been associated with RP, the most common form of IRD. The cone and cone–rod dystrophies that primarily affect the cone photoreceptor cells are more rare, as are the macular disorders (dystrophies) in which degeneration is confined to the central retina (see Figure 11.2), often because the disease targets Bruch's membrane (part of the choroid).

As a target for gene therapy, the retina has many features that facilitate therapeutic intervention. Surgical access is relatively easy but complications may occur, particularly in patients with more advanced retinal degeneration (Rd). The success of the surgical procedure will have a major impact on the success of the gene therapy treatment itself. The route of administration can be adapted depending on the layer of the retina to be targeted: intravitreal injection is used to treat the inner retina, and subretinal injection into the subretinal space is used for targeting the outer retina and RPE. The blood–retina barrier separates the subretinal space from the blood supply and enhances the prospects of effective retinal gene therapy by protecting the retina from immune-mediated damage. After treatment, therapeutic outcomes can be safely evaluated using noninvasive methods such as electroretinography (ERG) and optical coherence tomography (OCT).

Preclinical studies in animal models of disease also benefit from the above features. Administration of gene therapy treatment is relatively uncomplicated, and many similar methods of assessing the therapeutic effect can be conducted in

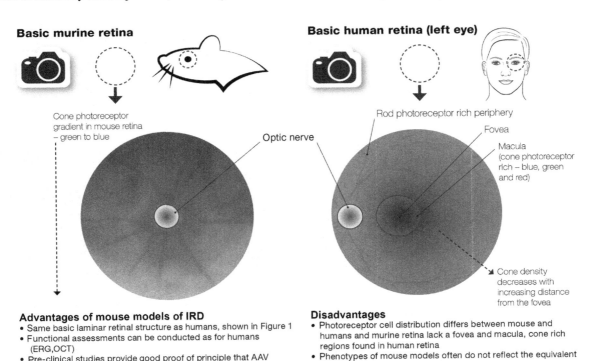

FIGURE 11.2 Basic representation of mouse and human retina as if taken by ophthalmoscopy, demonstrating general differences in photoreceptor cell distribution. ERG, electroretinography; IRD, inherited retinal disease; OCT, optical coherence tomography.

a similar way as for humans. There are numerous mouse models of the common forms of IRD available, both naturally occurring[1] and transgenic,[2] for studying the pathophysiology of IRDs and potential treatments. Gene therapy for retinal disease may not necessarily mean delivery of the correct copy of a gene; in some cases, it may be appropriate to deliver genes for the expression of antiapoptotic factors that restrict further cell loss. The timing of therapeutic intervention must also be considered. Gene replacement therapy may be attempted at an early stage of disease before there is complete loss of photoreceptors. Alternatively, at the terminal stages of disease when only the inner neuronal layers of the retina remain, an optogenetic approach may be considered. This strategy aims to encourage cells that are not inherently photosensitive to respond to light by delivery of a gene for a light-inducible pigment, such as channelrhodopsin-2.[3]

The physiology and progression of any given IRD must be considered when selecting a gene therapy treatment. It is also important to consider that, as the retina degenerates, transduction of the remaining retina may be less effective because of the structural changes that occur. In a mouse model of moderate RP, $Rpe65^{R91W/R91W}$, the loss of rod photoreceptors followed by cone photoreceptors is not as rapid as in other models (for example, the $rd10$ mouse model) and allows for investigation into how late into disease progression cone rescue can still be accomplished.[4] Given that cone photoreceptors are responsible for visual acuity and daylight vision, the ability to rescue and prevent their loss could have a critical impact on a patient's life. Knowing what form of gene therapy would be most appropriate, which cell types to target, and when to provide therapy intervention will increase the safety and efficacy of the gene therapy treatment.

11.2 CELL-SPECIFIC TARGETING WITHIN THE RETINA

Given that the majority of IRDs result from mutations in genes specifically expressed in the photoreceptor or RPE cells, the ideal gene replacement strategies would provide efficient transduction and restricted transgene expression to these cell types. Adeno-associated virus (AAV) is the current favored vector for retinal gene therapy, though lentiviral (LV) vectors are also being pursued. LV vectors are thought to be a less appropriate vessel for gene delivery to the various cell types of the retina because, as suggested by initial studies, their large size prevents diffusion from the subretinal space.[5,6] Although pseudotyped LV vector variants have had variable success at transducing photoreceptors *ex vivo*, doubts remain regarding the feasibility of LV vector use *in vivo*.[7] In contrast to this, recombinant AAV (rAAV) vectors have shown consistent success in being able to transduce all layers of the retina, particularly the RPE and photoreceptors (Table 11.1).[8] Intravitreal injections allow for targeting of ganglion cells, whereas subretinal injections of rAAV transduce the RPE, photoreceptors, and inner neuronal layers, depending on the serotype. Over 100 AAV serotypes have been identified, and pseudotyping of the variants has been a useful method for manipulating viral vector target specificity and transduction efficiency of cell types within the retina. The genome of AAV2 is commonly used to package transgenes in the capsid of a different variant, and intravitreal injections of rAAV2/2 and rAAV2/8 in C57BL/6 mice led to transduction of retinal ganglion cells.[9] Subretinal injections of rAAV2/1 and rAAV2/6 were shown to transduce only RPE cells,[10] whereas rAAV2/2, rAAV2/5, rAAV2/7, rAAV2/8, and rAAV2/9 were all shown to transduce RPE and photoreceptors in mice.[11] It was noted that rAAV2/8 and rAAV2/9 were particularly successful in transducing cone photoreceptors (75% compared to 35% by rAAV2/5). Although these studies are helpful as an indication of AAV serotype choice for *in vivo* testing of mouse models, it is important to remember that serotype tropism may differ between species. In nonhuman primates, rAAV2/2 has shown good transduction of rod photoreceptors and RPE,[12] with rAAV2/5 efficient in transducing rod photoreceptors but not cones.[13] However, rAAV2/5 was later shown to successfully target primate cones when used in combination with a cone-specific enhancer and promoter.[14]

The development of tyrosine-mutant serotypes has greatly enhanced transduction efficiencies of the rAAV vectors. Entry of AAV into a host cell triggers ubiquitin-mediated degradation by the proteasome, a process which affects the transduction efficiency of rAAV vectors.[15] The conversion of tyrosine residues to phenylalanine on the AAV capsid is thought to prevent ubiquitination of the virus within the host cell and subsequent degradation, thus enhancing transduction efficiency.[16] Targeting of the inner retinal layer, such as bipolar cells, has previously proven to be particularly difficult, but the tyrosine-mutant rAAV2/8 Y733F vector revealed reporter gene expression in bipolar cells of $rd10$ mice.[17] Other studies using tyrosine-mutant capsid forms have revealed significant improvement in transduction of all retinal layers, with intravitreal injections of rAAV2/2 Y444F showing transduction of the photoreceptor layer.[18] This suggests that AAV vectors can diffuse through all the layers of the retina, but require particular conditions in order to survive and transduce the various cell types. The single tyrosine mutants so far created do not appear to affect the cellular tropism of the serotype within the retina, but instead enhance their transduction potential. However, when multiple mutations were created in a single capsid, vectors exhibited different *in vivo* transduction properties. The sextuple mutant rAAV2/2 (Y252, 272, 444, 500, 704, 730F) exhibited transduction of all the retinal layers, including the ganglion cells, following subretinal injection.[19] Clearly, these tyrosine-mutant capsid variants have the potential to greatly enhance retinal gene therapy (assuming that they do not result

TABLE 11.1 Lentiviral (LV) and Adeno-Associated Virus (AAV) Vector Advantages and Disadvantages in the Development of Gene Therapy Treatments for Inherited Retinal Diseases

Vector	Primary Cell Targets Following Subretinal Injection	Advantages	Disadvantages	Promoter Options
Lentivirus (LV)	RPE	Large packaging capacity	Large size prevents diffusion through layers of the retina. Transduction limited to RPE. Greater potential for immune response.	*CMV, CBA, CAG, RPE65, VMD2*
Adeno-associated virus (AAV)	Photoreceptors and RPE	Uncomplicated *in vitro* testing of vectors (except for AAV2/5). Good evidence of transduction of all cell types within the retina of several animal models. Evidence of therapeutic expression levels from the delivered transgene. Little apparent toxicity in mouse models except at high concentrations (>E+10 genome particles per retina). Use in human clinical trials reveals generation of minimal immune response. Human clinical trials show positive signs of effective treatment following AAV delivery.	Limited packaging capacity. Unknown long-term effectiveness	
AAV2/2	Ganglion cells, Müller cells, photoreceptors, and RPE	Used in various human clinical trials for the treatment of retinal disease with minimal signs of toxicity or immune-mediated response. Encouraging signs of treatment effects.	Generally limited to targeting photoreceptors and RPE	*CMV, CBA, CAG, RHO, GRK1, VMD2*
AAV2/5	RPE and photoreceptors		Difficult to test vectors *in vitro*	*RPE65, GRK1*
AAV2/8	Photoreceptors and RPE	Particularly good photoreceptor tropism in mouse models and nonhuman primates. Used in human clinical trial for the treatment of hemophilia B.	Generally limited to targeting photoreceptors and RPE	*CMV, CBA, CAG, RHO, GRK1, CAR*
AAV2/2 Y444F	Inner and outer retina, RPE	Enhanced transduction of retinal cell layers compared to wild-type variant. Bipolar cell targeting enables optogenetic approaches.	Unknown downstream biological effects of the mutant capsid	*CMV, CBA, CAG, CHX10, mGLUR6*
AAV2/8 Y733F	Photoreceptors and RPE	Enhanced transduction of photoreceptors and RPE compared to wild-type variant in animal models.	Unknown downstream biological effects of the mutant capsid	*CBA, GRK1, CAR, VMD2*

CAG, cytomegalovirus early enhancer–promoter element fused with the chicken β-actin promoter and rabbit globin intron–exon fragment; *CAR*, cone arrestin; CBA, chicken β-actin; *CHX*10, homeobox-containing transcription factor; CMV, cytomegalovirus; *GRK*1, rhodopsin kinase; *mGLUR*6, metabotropic glutamate receptor 6; *RHO*, rhodopsin; RPE, retinal pigment epithelium; *RPE*65, retinal pigment epithelium-specific 65 kDa protein; *VMD2*, bestrophin 1 (vitelliform macular dystrophy).

in immune responses), and studies of their success in animal models of IRD will be discussed in another section of this chapter.

11.3 PROMOTER CHOICE FOR EXPRESSION IN SPECIFIC RETINAL CELL TARGETS

Use of ubiquitous promoters has enabled identification of the transduction patterns of various AAV vectors in the retina (as described above), but as human gene therapy trials are becoming more frequent, the development of vectors that specifically target the cell of interest will increase the safety and efficacy of the treatment. In retinal gene therapy animal studies, the commonly used ubiquitous promoters have been immediate-early cytomegalovirus (CMV) enhancer–promoter[20] and the CAG promoter, which combines the CMV enhancer with the chicken β-actin (CBA) promoter and rabbit globin intron–exon fragment.[21–23] Despite high expression levels generated from these promoters, the ideal scenario in a clinical application of retinal gene therapy will be to deliver therapeutic gene expression using a promoter specific to the targeted cell type. For RPE-specific expression, promoters from two genes involved in retinal disease have so far been tested. *RPE65* mutations cause both RP and Leber's congenital amaurosis (LCA). A fragment of the human *RPE65* promoter (−1556 to +23 bp) was used to induce RPE-specific expression of *RPE65* in the RPE cells of RPE65-deficient Briard dogs. This promoter was found to be only 10% as strong as the CMV promoter and ineffective in older animals.[24] This highlights a potential problem in the use of cell-specific promoters; that is, as the target cells are already in a state of disease, expression from the promoter may be downregulated, particularly in the later stages of disease. Despite this, a similar fragment of the human *RPE65* promoter was used to drive *RPE65* expression in a human LCA clinical trial[25] (discussed later in this chapter). A second potential RPE-specific promoter is that of *VMD2*, a gene that encodes bestrophin, mutations in which cause the macular dystrophy known as Best disease. This promoter appeared to drive only weak expression compared to CBA, cone arrestin (CAR), and mouse opsin promoters in *in vitro* transductions of retinal pigment epithelia cell line (ARPE19) and murine photoreceptor-derived retinal cell line (661W).[26] Concise reports of activity following rAAV administration to the retina are unavailable, but a 2005 review[27] reported that use of a −585 to +38 bp *VMD2*-promoter fragment in rAAV1 was able to limit green fluorescent protein (GFP) expression to the RPE but for a few GFP-positive photoreceptor cells.

In contrast, photoreceptor targeting in the murine retina has been successfully achieved using the rhodopsin and rhodopsin kinase promoters.[11] The combination of serotype and cell-specific promoter led to significantly higher photoreceptor cell expression in subretinally injected mice, though the onset of expression was reduced compared to the ubiquitous CMV and CBA promoters (3–4 weeks compared to 5–12 days). The human rhodopsin kinase (*hGRK1*) promoter is ideal for rAAV photoreceptor targeting because of its small size (292 bp) and effective targeting and activity in mice[28] (Figure 11.3) and nonhuman primate retina.[29] Numerous gene therapy studies using mouse models have been reported in which photoreceptor-specific expression was achieved using the *hGRK1* promoter with effective rescue of phenotype.[28,30–34] For some IRDs, such as the cone and cone–rod dystrophies, cone-specific photoreceptor expression may be desired. The human L and M cone opsin genes are located on the X chromosome and controlled by the same locus control region (LCR) but separate promoters. Various combinations of this LCR with L/M promoter sequence have been tested using rAAV2/5 in a canine model; the largest version at 2.1 kb (named PR2.1) was effective in generating L/M cone photoreceptor-specific gene expression.[35] However, not only is PR2.1 a very large promoter, and therefore limits the remaining space within the transgene, it also excludes expression in S cones. For these reasons, the *CAR* promoter may be a better option for cone-specific expression, as used in *CNGB3* (cone photoreceptor cGMP-gated cation channel β-subunit) treatment of the *Cngb3* knockout mouse model of achromatopsia.[36]

If significant degeneration of photoreceptors occurs before gene therapy intervention can be applied, then targeting of bipolar or ganglion cells for optogenetic approaches may be the desired treatment option. For bipolar cell targeting, a 200 bp fragment of the murine *Grm6* enhancer, 445 bp *Capb5* promoter, and 164 bp homeobox-containing transcription factor (*Chx10*) enhancer have been identified as potential elements to drive bipolar-specific gene expression.[37] For ganglion cell layer (GCL) targeting, only a fragment of the human connexin 36 promoter (2.8 kb) has revealed selective expression in foveal bipolar ganglion cells after intravitreal delivery rAAV2/2 to macaque eyes.[38]

Although use of a cell-specific promoter is desirable, often size can restrict the vector capacity and limit the therapeutic genes that can be delivered, as in the case of PR2.1. Expression levels may also vary between species, and so positive results achieved in animal models may not reflect what would be achieved in humans. Improvements in *ex vivo* retina survival[39] will enable further tropism testing of various serotypes and promoter combinations in human tissue using explants from routine retinectomy procedures (Figure 11.4); however, this procedure may alter the expression patterns within the cell types. The issue of promoter activity in the degenerating retina also needs to be considered as expression patterns in the targeted cell types may change during the pathophysiology of disease. This may explain, for instance, the poor results in older Briard dogs[24] and

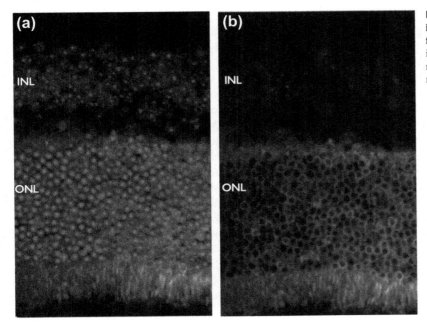

FIGURE 11.3   C57BL/6 retina after subretinal injection of the rAAV2/5 *hGRK*1 promoter-green fluorescent protein (GFP) vector. (a) Nuclei staining and GFP; (b) GFP expression only. *hGRK*1, human rhodopsin kinase; INL, inner nuclear layer; ONL, outer nuclear layer.

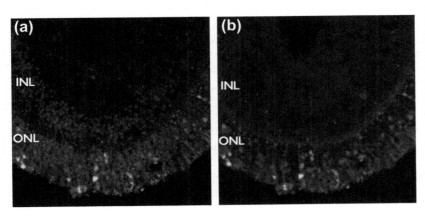

FIGURE 11.4   Human retinal explant transduced *ex vivo* with rAAV2/5 CAG promoter-green fluorescent protein (GFP). (a) Nuclei staining and GFP; (b) GFP expression only. CAG, cytomegalovirus early enhancer–promoter element fused with the chicken β-actin promoter and rabbit globin intron–exon fragment; INL, inner nuclear layer; ONL, outer nuclear layer. *Image provided by Dr S. De Silva.*

in the LCA clinical trial using the *hRPE*65 promoter,[25] as discussed later in this chapter. Clearly, the combination of tyrosine mutants, capsid serotype, and promoter has the potential to increase the safety and efficacy of retinal gene therapy treatments.

11.4 AAV TREATMENT OF AUTOSOMAL RECESSIVE MODELS OF RETINAL DISEASE

The majority of animal models of IRD exhibit null mutations, relevant to the study of autosomal recessive IRDs. Many studies have shown the feasibility of gene replacement therapy in such animal models. Peripherin 2 is a rod photoreceptor-specific structural protein, mutations in which lead to autosomal recessive RP or macular dystrophy. The retinal degeneration slow (*rds*) mouse is homozygous for a null mutation in the *rds/peripherin*-2 gene, completely lacking functional peripherin 2. Gene replacement therapy was attempted with rAAV2/2 carrying *Prph*2 driven by a rhodopsin promoter sequence.[40] Peripherin 2 was detected in rod cells and appeared to initiate morphological changes that led to an improvement in retinal function for up to 14 weeks posttreatment, beyond which improvements were not sustained. When treatment was given to younger animals there was a greater initial restoration of structure, but again the effect dropped over time.[41] Other models of autosomal recessive RP have shown more long-term success in gene replacement animal studies. *Rd*10 mice show rapid degeneration of rod photoreceptors because of a lack of rod cells and appeared to initiate morphological changes that led to an improvement in retinal function for up to 14 weeks posttreatment. *hPDE6sh* restored retinal function and preserved the retinal structure for up to 6 months.[42]

Rd in the Royal College of Surgeons (RCS) rat occurs because of the absence of c-mer proto-oncogene tyrosine kinase (Mertk), an RPE receptor tyrosine kinase required for phagocytosing shed photoreceptor outer segments. Mutations in

Mertk cause RP in humans. Initial rescue of the degeneration exhibited in the RCS rat model was demonstrated using a recombinant adenovirus vector,[43] an rAAV2/2 vector,[44] and more recently, rAAV2/8Y733F with expression of *hMERTK* driven by the RPE-specific *hVMD2* promoter. This latter study revealed rescue of RPE function and slowing of photoreceptor degeneration for at least 8 months after treatment.[45]

Other models of autosomal recessive IRD for which gene replacement therapy has shown promising recovery of function include various models of LCA, a severe form of childhood blindness. Aryl hydrocarbon receptor interacting protein like-1 (AIPL1) mutations in humans lead to LCA. AIPL1 is necessary for cone photoreceptor survival and several studies of gene replacement therapy have shown promising results using rAAV2/2, rAAV2/5, and rAAV2/8 vectors, with both ubiquitous and photoreceptor-specific promoters driving expression.[30,31] Another study used a self-complementary rAAV (scAAV) vector to deliver *hAIPL1*. The onset of expression after AAV infection is delayed by many factors, but the rate-limiting step is the synthesis of the complementary strand to the single-stranded DNA (ssDNA) transgene delivered by AAV.[46,47] Self-complementary AAV vectors have been developed to bypass this critical step by packaging a single-stranded transgene that contains both a forward and a reverse copy that can fold over on itself due to an altered ITR structure.[48] By design, this form of vector offers both an enhancement to retinal gene therapy and a downside: a reduction in packaging capacity, with two copies of the therapeutic transgene that need to fit within ~5 kb of DNA. If inclusion of a larger cell-specific promoter is also required, then scAAV vectors may not be the ideal choice for the gene therapy treatment of many IRDs. The benefits of scAAV vectors are likely to be greater for preclinical studies in mouse models than in human clinical trials. This is because the onset and speed of degeneration in the murine retina are often more rapid than in the equivalent human disease due to the shorter life span and so quicker onset of transgene expression *in vivo* will aid observed treatment effects. Several studies investigating scAAV transduction of the retina have been conducted in various animal models. Compared to the standard rAAV2/2 equivalent vector, scAAV initiated more rapid onset of reporter gene expression following subretinal delivery to mouse retina in both RPE and photoreceptor cells.[49] Comparisons of transduction from scAAV2/2, scAAV2/5, and scAAV2/8 compared to rAAV2/8 found consistently faster onset of reporter gene expression and higher levels of stable gene expression from the self-complementary vectors in adult murine retina.[50] On the basis of these data, scAAV vectors have been used in retinal gene therapy studies. The *Aipl1*[−/−] mouse shows rapid Rd and a severe phenotype, but subretinal delivery of an scAAV2/8Y733F vector in this model successfully delivered *mAipl1* during the active phase of degeneration, showing subsequent preservation of photoreceptors and indicating that a greater phenotypic effect was induced compared to the rAAV vector.[51] Another example of the use of scAAV vectors in the treatment of IRD was in the *rd*12 mouse model of LCA. Rescue of cone photoreceptors and halting of degeneration were shown in older mice after delivery of scAAV2/5-hRPE65, suggesting that improvements in the morphology and function of remaining cones can be made even in the later stages of disease.[52]

It is apparent that gene replacement studies using models of autosomal recessive IRD are showing consistent success in recovery of function and improved retina morphology. These positive effects are continuing over a relatively long time course and include successes when gene therapy intervention is provided during the later stages of disease. Combinations of serotypes and promoters to allow specific tropism and expression are adding to these positive results. The data are very encouraging for future prospects of human retinal gene therapy.

11.5 AAV TREATMENT OF AUTOSOMAL DOMINANT MODELS OF RETINAL DISEASE

The majority of animal models available for the study of IRD carry null mutations; yet, there are numerous IRDs that do not exhibit a complete lack of protein. Dominant mutations often yield mutated versions of a protein. If these impart a dominant-negative effect, then provision of the wild-type (WT) gene and subsequent protein should resolve the phenotype as in the gene replacement strategies above. Dominant-negative mutations tend to eliminate one or more functions of the encoded protein; for example, the common rhodopsin mutation proline 23 histidine substitution (P23H) is thought to preserve an ability of the opsin to oligomerize, which interferes with normal function.[53] However, if a toxic gain-of-function effect is generated, then the gene therapy approach will need to silence the mutant mRNA and provide a correct nonsilenced copy. Gain-of-function mutations can alter the protein to generate a novel activity; for example, the rhodopsin mutation K296E that leads to RP results in the photopigment constitutively activating rod transducin and the phototransduction pathway.[54] Clearly, it is important to determine the etiology and molecular pathology of autosomal dominant retinal diseases in order to determine the appropriate gene therapy approach. For rhodopsin alone, it is apparent that different mutations can trigger different responses. The main function of rod photoreceptors is to produce rhodopsin for initiating phototransduction, and around 85% of total protein in the outer segment is rhodopsin, which also makes it important for outer segment structure.[55] Outer segment proteins are synthesized in the inner segment of the photoreceptor and transported unidirectionally to the outer segments. Mutations that cause autosomal dominant RP can affect the stability, folding, and trafficking stages of this process. Because of the high production rate of rhodopsin, build up of mutants tends to have a toxic effect, and therefore, silencing of the mutant form is desired before supplying the WT form. AAV-delivered ribozymes have been used to destroy targeted mutant mRNA in an autosomal dominant RP rat model carrying the P23H rhodopsin mutation.[56,57] In this model, the vector was injected at different ages, including a later stage of degeneration when 40% of photoreceptor cells were missing. In all age groups, photoreceptor cell loss was delayed after injection of the AAV-ribozyme vector. AAV delivery of RNA interference has also been used to suppress rhodopsin, while codelivery of a codon-modified rhodopsin replacement gene resistant to suppression allowed for WT expression.[58,59] Whereas these approaches utilized codon-modified gene replacements, complete codon optimization of the transgene would not only be helpful in resistance to silencing, it may also enhance expression levels and therefore increase the chance and degree of physiological outcome.[60] *In vitro* studies in human embryonic kidney 293 (HEK293) cells have shown that codon optimized genes can produce a ninefold increase in protein level compared to the WT coding sequence.[61] Although no retinal codon optimized comparisons have currently been reported, a codon-optimized microdystrophin transgene showed significantly improved expression and physiological outcome in a mouse model for muscular dystrophy (*mdx*) mice after rAAV2/8 delivery.[62]

For the treatment of dominant-negative mutations, injections of an rAAV2/5-*RHO* vector without the use of a silencing mechanism into the *RHO*^{P23H} and *RHO*$^{+/+}$ mouse models revealed that expression of the WT rhodopsin slowed the rate of degeneration in the P23H mutant mice. However, in WT mice, expression from the vector resulted in damaging effects, suggesting that while expression of normal rhodopsin can suppress the P23H mutant effects, overproduction of rhodopsin can also be damaging.[63] In this scenario, use of a photoreceptor-specific promoter may be favored over a strong ubiquitous promoter because expression is more likely to be regulated, limiting the chance of overexpression.

Although more complex than gene replacement strategies, the cosuppression of mutant gene and provision of the WT replacement are showing promise via AAV delivery with many autosomal dominant IRD candidates for therapy.

11.6 AAV DELIVERY OF LARGE GENES TO THE RETINA

One issue in the reliance of rAAV vectors is that their maximum packaging size is limited to around 5 kb. To achieve greater packaging (up to 6 kb) and successful transduction with rAAV carrying such oversized genomes may require additional treatment with proteasome inhibitors.[64,65] However, many of the genes implicated in Rds exceed even this size, including ATP-binding cassette, subfamily A, member 4 (*ABCA4*), mutations in which are the most common cause of autosomal recessive IRD and inherited juvenile onset macular degeneration. ABCA4 is localized to the photoreceptor outer segment disc membranes and acts as a membrane transporter important for the recycling of the light-sensitive chromophore, a process necessary for phototransduction. A lack of ABCA4 function leads to build up of unwanted retinoids within the disc membranes. The RPE cells consume these discs as they become more distal due to the normal generation of new discs by photoreceptor cells. The presence of the unwanted retinoids in RPE cells leads to a toxic product, A2E, being generated, which disrupts correct RPE function and leads to both RPE and photoreceptor death.[66] Mutations in *ABCA4* can cause Stargardt disease, RP, cone–rod dystrophy, and bull's-eye maculopathy; therefore, a retinal gene therapy treatment for *ABCA4* would benefit a broad spectrum of patients. Until now, LV vectors have been favored for large gene delivery because of their greater packaging capacity, and LV vectors were able to successfully deliver *ABCA4* to an Abca4 knockout mouse model.[67] However, just 5–20% of photoreceptors were found to be transduced and these were limited to within the site of injection. Despite this limitation, the potential use of LV vectors is being explored with Oxford BioMedica currently testing the LV vector StarGen™ for the treatment of Stargardt disease.

Effective AAV transduction of the retina is widespread and extensively studied, leading to approaches that aim to deliver large transgenes by splitting a large AAV2-based transgene into two halves and packaging each half into a separate AAV capsid. The approach then relies on two vectors carrying different halves of the transgene transducing the same host cell, within which the two fragments can recombine. This is possible thanks to the retina being a relatively small, contained target area; such an approach for delivery of large genes may not be possible for nonretinal diseases in which AAV delivery is systemic. The joining of the two separate transgenes once inside the target cell can be achieved in different ways (Figure 11.5), with the most common being an overlapping vector approach[68–75] or a *trans*-splicing approach.[76–82] The overlapping dual vector approach has been successful in the recombination of a split *ABCA4* transgene *in vitro* and *in vivo*, with transductions leading to generation of the full-length protein.[72] However, a more recent study has suggested that the overlapping approach is not effective when targeting photoreceptors and that a hybrid *trans*-splicing/overlapping approach would be more effective.[83] The benefit of this hybrid approach is that it offers two opportunities for the opposite transgenes to recombine, either by homologous recombination or by ITR concatamerization. Each approach by itself relies on only one of these events occurring; therefore, by combining them, the chance of generating the desired large transgene from two separately packaged segments is supposedly increased. This seemed to be the case for both *ABCA4* and myosin VIIA (*MYO7A*) delivery to photoreceptor cells. *MYO7A* is another large gene and encodes myosin VIIA, mutations in which result in Usher syndrome, the most common form of deaf–blindness. Usher syndrome exhibits an RP phenotype of varying severity in addition to deafness. The poor performance of the overlapping approach in the above study[83] is in contrast to previous findings, which suggested that packaging of a

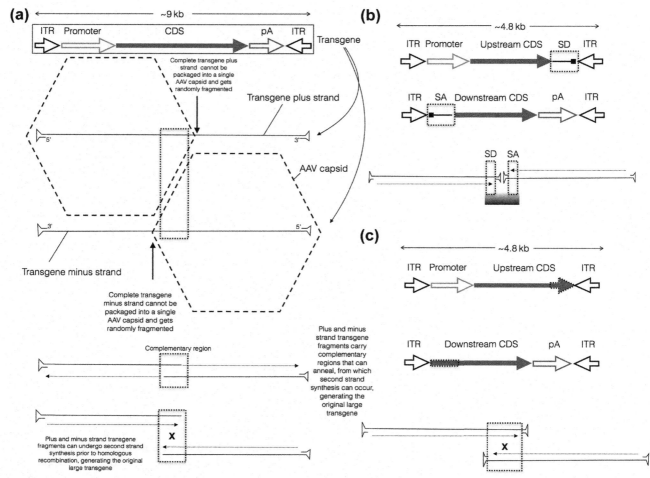

FIGURE 11.5 **Approaches for delivering large transgenes using adeno-associated virus (AAV) vectors.** (a) Generation of a single oversized transgene that is packaged as fragments with a single inverted terminal repeat (ITR) in different AAV capsids. Dual infection of vectors carrying opposite transgene fragments can lead to annealing of overlapping complementary transgenes prior to second-strand synthesis or recombination between two opposite overlapping transgenes after second-strand synthesis within the host cell, resulting in the regeneration of the complete large transgene; (b) dual vector approach in which the coding sequence (CDS) is split into defined fragments and two distinct transgenes are generated, one containing a splice donor (SD) and the other a splice acceptor (SA) site. Dual infection of both vectors leads to concatemerization between the ITRs, forming a large transgene from which the SD and SA sites are removed by splicing after transcription; (c) dual vector approach in which the CDS is split into two fragments and packaged into two distinct transgenes, each carrying a region of overlap within the CDS (highlighted by dotted lines), allowing for recombination between the two transgenes after dual infection. pA, poly(A) signal.

large *MYO7A* transgene using the overlapping approach in rAAV2/2 or rAAV2/5 led to successful transduction of photoreceptors and RPE cells and to recombination between opposite transgene fragments.[84] The solution to delivering large genes using AAV vectors to the retina is still a work in progress, but despite the current variable levels of expressed protein generated from these approaches, there are promising signs of success in improvement in the physiology of transduced retinas.[83,84]

Cotransduction efficiency will clearly be extremely important for this dual vector approach,[85] with correct serotype selection necessary to achieve optimal cotransduction.[79] The dual rAAV vector approach offers promise that effective gene therapy for IRDs will not be limited to smaller genes. Nanoparticles also offer potential for gene therapy treatment of large genes, with *ABCA4* delivery to knockout mice showing ABCA4 production 8 months after treatment.[86]

11.7 NEUROPROTECTION OF THE RETINA USING AAV

A further use of rAAV vectors is in the delivery of growth and antiapoptotic factors that can be used to prevent cell loss when Rd is advanced and the causative mutation unknown. Such vectors can also be used to enhance the effects of a codelivered therapeutic gene. Apoptosis of photoreceptor cells is common to most forms of IRD; therefore, preventing cell death could significantly slow degeneration and allow a greater window for gene therapy intervention. Minimizing rod photoreceptor death would also prolong cone survival, and AAV delivery of various neurotrophic factors has been shown to preserve photoreceptors.

rAAV2/2 subretinal delivery of glial cell line-derived neurotrophic factor (GDNF) to a rat model of autosomal dominant RP led to increased rod photoreceptor survival.[87] Another study of fibroblast growth factor 2 (FGF-2) had a preservation effect similar to that of photoreceptors, though improvements in ERG (as a measure of rod function) were minimal.[88] Other studies have confirmed the ability of GDNF to slow photoreceptor cell death, but similar testing of ciliary neurotrophic factor (CNTF) has shown an adverse effect on retinal function.[89] In *rds* mice, photoreceptor degeneration was shown to be delayed after subretinal and intravitreal delivery of AAV-*CNTF*, but too much CNTF appeared to have a detrimental effect on photoreceptor function in a dose-dependent manner.[90,91] In *rds* gene replacement therapy with rAAV2/2-*Prph*2, the beneficial effects of gene therapy were minimized by codelivery of CNTF.[92] These data suggest that low, regulated levels of CNTF may prove to be beneficial, but perhaps require the control of a weaker or inducible promoter to allow for consistent subtoxic but therapeutic levels of expression.

rAAV2/2 delivery of X-linked inhibitor of apoptosis (XIAP) has been shown to enhance survival of cultured human RPE cells after induced oxidative cell death.[93] Evidence of enhanced survival of transplanted rod precursors in a degenerating host retina due to XIAP has also been shown[94] and prolonged treatment effects of rAAV2/5-*PDE*62 gene therapy.[95]

These data suggest that antiapoptotic factors and growth factors have the potential to prevent further cell loss within the retina and also to enhance the effects of gene therapy. However, caution must be taken as the long-term effects and dose levels need to be better understood.

11.8 HUMAN AAV CLINICAL TRIALS FOR THE TREATMENT OF IRD

In 2008, data from three independent clinical trials revealed evidence of improved visual function after rAAV2/2 subretinal delivery of *hRPE*65 to LCA patients (2009 review[96]). RPE65 (retinal pigment epithelium-specific protein, 65 kDa) is expressed in the RPE and is essential in the visual pathway for its role in regenerating the active chromophore required for light absorption. Loss of function of RPE65 therefore leads to loss of vision. Although all three clinical trials were consistent in their use of rAAV2/2, the transgene structure differed in each: in one, a 1400 bp fragment of the *hRPE*65 promoter was used,[25] whereas the other two studies used the ubiquitous CAG promoter in combination with a modified Kozac consensus upstream of the start codon[97] or with a hybrid intron in the 5′UTR region upstream of the therapeutic gene.[98] The use of the less active *hRPE*65 promoter has been suggested to be the reason why less impressive vision changes were observed from that study compared to the others. Both studies using the CAG promoter revealed varied evidence of improved visual function. In one study, improvements in visual acuity were seen together with improved pupil responses,[97] whereas the other study reported a shift in fixation toward an extrafoveal region exposed to vector[99]—both objective assessments. Interpretation of mean vision outcomes in both of these studies was, however, impaired to some extent by surgical complications, such as macular hole formation. The other study with the *hRPE*65 promoter suggested anecdotal visual improvements in one of three patients, but the observed microperimetry gains did not match the area exposed to vector and should be interpreted with caution when retesting an amblyopic eye.[25] Similarly, while the maze navigation test showed apparent improvements for the treated eye, it also showed a similar improvement in the control eye, which raises questions about the reliability of the baseline measurements of this test. It was interesting to note in the preclinical testing of this vector that visual improvements were only seen in younger Briard dogs injected up to 11 months of age, but not at 30 months.[24] Hence, the lack of efficacy at later stages of the disease process in dogs and humans may be because the *hRPE*65 promoter is less active in RPE cells undergoing stress responses, where translation of cell-specific transcription factors would be inhibited.[96] Nevertheless, although evidence of vector efficacy is doubtful, this study did show safe detachment of the retina without surgical complications, such as macular hole formation. A follow-on study using the *hRPE*65 promoter at a higher dose is planned; however, many would argue that CAG/CBA is now the promoter of choice for clinical trials targeting the RPE, in view of the evidence of efficacy from a number of separate independent studies.[96]

One concern over AAV gene therapy is its ability to initiate an immune response. Neutralizing antibodies to the rAAV2/2 capsid temporarily increased in one patient,[97] with a marginal increase observed in another patient[98]; however, neither increase was significant. The findings from all three studies suggest that rAAV retinal gene therapy is safe and elicits a low immune response. Given that 30% of adults have neutralizing antibodies to AAV2 with others having antigen-specific memory CD8+ T cells,[100] use of other AAV serotypes may minimize further the chance of an immune response.[101] Treatment of the second eye in three patients has now been reported with no adverse effects after 6 months.[102]

More recent data from an ongoing clinical trial for the treatment of choroideremia have been published that show very encouraging signs of success.[103] Choroideremia is caused by mutations in the Rab escort protein 1 (REP1), resulting in a slow degeneration in which photoreceptors, RPE, and choroid (the vascular layer of the eye) are lost. These trials for the treatment of IRDs offer great encouragement for future trials and with recent developments in promoter and serotype choice, gene therapy treatment of IRDs has a promising future. There are several AAV gene therapy trials currently approved and ongoing (Table 11.2), including further studies to assess the safety and efficacy of *RPE*65 AAV2/2 subretinal delivery to LCA patients (NCT00481546, NCT00516477, NCT00643747, NCT01208389, NCT00749957, NCT00999609), *MERTK* AAV2/2 subretinal delivery to RP patients (NCT01482195), AAV2/2 *REP*1 subretinal injection in choroideremia patients

TABLE 11.2 Summary of Recent and Ongoing Clinical Trials for the Treatment of Inherited Retinal Diseases

Disease	Clinical Trial Reference	Therapeutic approach	Safety	Outcomes
LCA	Previous studies with data published in 2008	rAAV2/2 hRPE65, three independent studies, CAG or hRPE65 promoter	Surgical complication in one study (macular hole formation), safe retinal detachments in others. Temporary increase in neutralizing antibodies to rAAV2/2 capsid in one patient but not a significant increase. All three studies suggest rAAV retinal gene therapy is safe and elicits a minor immune response.	Varied evidence of improved visual function. Improved visual acuity and pupil responses. Shift in fixation toward extrafoveal region exposed to vector. Treatment of second eye revealed no adverse effects after 6 months.
	NCT00481546	rAAV2/2 CBSBpr.hRPE65	Phase 1 study to assess safety of subretinal administration of vector	Ocular and systemic toxicity to be assessed in LCA patients
	NCT00643747	rAAV2/2 hRPE65pr.hRPE65	Phase 1 dose-escalation study to assess safety and efficacy	To determine safety and efficacy of the treatment in LCA patients
	NCT00749957	rAAV2/2 CMV.CBApr.hRPE65	Phase 1/2 multiple site study to assess safety and efficacy of the vector	To evaluate the safety and efficacy of the vector in LCA patients
	NCT00516477	rAAV2/2 hRPE65v2	Phase 1 study to determine whether gene transfer will be safe and effective	To determine which vector is safe and well tolerated in LCA patients
	NCT01208389	rAAV2/2 hRPE65v2	Follow-on evaluation of the safety of readministration of the vector to the contralateral eye in progress	To discover whether readministration of the vector to the contralateral eye has any adverse effects in LCA patients previously injected
	NCT00999609	rAAV2/2 hRPE65v2	Safety determined from above studies	Phase 3 trial to determine if gene therapy intervention improves visual and retinal function in young LCA patients
Choroideremia	NCT01461213	rAAV2/2 CAGpr.hREP1	Phase 1 safety assessment completed. Safe retinal detachment, no surgical complications. Tolerance of low and high doses of vector.	Phase 1 showed good safety and tolerance of the vector. Initial findings indicate positive effects from the gene therapy, Phase 2 to begin to identify therapeutic benefits (by evidence of slowing of retinal degeneration compared to untreated eye)
RP	NCT01482195	rAAV2/2 hVMD2pr.hMERTK	Phase 1 study to assess safety of the vector	To determine safety and potential value of gene transfer in MERTK-associated retinal disease
AMD	NCT01024998	rAAV2/2 sFLT01(VEGF)	Phase 1 clinical study to examine safety and tolerability of vector delivered by intravitreal injection	To evaluate safety and tolerability of the vector in patients with neovascular age-related macular degeneration (AMD)
	NCT01494805	rAAV2/2 sFLT01(VEGF)	Phase 1/2 dose-escalation study to assess safety and efficacy of vector following subretinal injection	To determine the safety and efficacy of administration of the vector by subretinal injection in patients with exudative AMD

Continued

TABLE 11.2 Summary of Recent and Ongoing Clinical Trials for the Treatment of Inherited Retinal Diseases—cont'd

Disease	Clinical Trial Reference	Therapeutic approach	Safety	Outcomes
Stargardt disease	NCT01367444	StarGen™ LV *ABCA4*	Phase 1/2a dose-escalation study to assess the safety of StarGen	To examine the safety of the gene transfer agent StarGen when injected subretinally in patients with Stargardt disease
Usher 1B syndrome	NCT01505062	UshStat™ LV *MYO7A*	Phase 1/2a dose-escalation study to assess the safety of UshStat	To examine the safety of the gene transfer agent UshStat when injected subretinally in patients with Usher syndrome type 1B
General	NCT01543906	QLT091001, synthetic retinoid	Phase 1b study to assess the safety of oral QLT091001 in RP patients	To determine the safety of oral QLT091001 and evaluate any improvements in visual function in RP patients
	NCT01521793	QLT091001, synthetic retinoid	Phase 1 study to assess the safety of oral QLT091001 administration in LCA and RP patients previously treated	To determine the safety of additional courses of QLT091001 in LCA and RP patients
	NCT01014052	QLT091001, synthetic retinoid	Completed Phase 1 safety study of oral QLT091001 in LCA patients	To evaluate the safety of oral QLT091001 and assess for any improvements in visual function in LCA patients

AAV, adeno-associated virus; *ABCA4*, ATP-binding cassette, subfamily A, member 4; AMD, age-related macular degeneration; CAG, cytomegalovirus early enhancer–promoter element fused with the chicken β-actin promoter and rabbit globin intron–exon fragment; *CBSB*, cystathionine-beta-synthase b; CMV, cytomegalovirus; *FLT01/VEGF*, fms-related tyrosine kinase 1/vascular endothelial growth factor; LCA, Leber's congenital amaurosis; LV, lentiviral vector; *MYO7A*, myosin VIIA; RP, retinitis pigmentosa; *RPE65*, retinal pigment epithelium-specific 65 kDa protein; *VMD2*, bestrophin 1 (vitelliform macular dystrophy).

(NCT01461213), and AAV2/2 subretinal delivery of *FLT*1 (also known as vascular endothelial growth factor, VEGF) in patients with age-related macular degeneration (NCT01024998, NCT01494805). Oxford BioMedica has two LV-based Phase I/II studies ongoing: one to deliver *ABCA4* (with their transfer agent StarGen™) to patients with Stargardt macular degeneration (NCT01367444) and the other to deliver *MYO7A* to Usher 1B patients (using UshStat™, NCT01505062). Other clinical trials currently underway offering therapy for IRDs include a non-gene therapy-based treatment using the synthetic retinoid QLT091001 (NCT01543906, NCT01521793, NCT01014052). Many IRDs including LCA and RP have a reduced availability of the chromophore 11-*cis*-retinal, which is a consequence of the molecular changes that occur within retinal cells as a result of the underlying disease-causing mutation. These trials are attempting to determine if oral administration of a synthetic retinoid replacement can aid the symptoms of the IRDs.

11.9 SUMMARY

Gene therapy treatment of IRDs is an advancing field. Identification of AAV serotype tropism and enhancement of transduction efficiencies of the various retinal cell layers through generation of capsid mutants are allowing for targeted AAV treatment in various forms: growth factors and antiapoptotic factors to prevent cell loss and to enhance gene therapy in addition to gene replacement and suppression replacement therapies. All are showing success in animal models of retinal disease. Clinical trials of AAV gene therapy have shown that treatment is safe in humans and well tolerated, with improvements in vision only likely to improve further with advancements in the understanding of serotype and promoter selection.

REFERENCES

1. Baehr W, Frederick JM. Naturally occurring animal models with outer retina phenotypes. *Vision Res* 2009:2636–52.
2. Fletcher EL, Jobling AI, Vessey KA, Luu C, Guymer RH, Baird PN. Animal models of retinal disease. *Prog Mol Biol Transl Sci* 2011:211–86.
3. Ivanova E, Pan Z-H. Evaluation of the adeno-associated virus mediated long-term expression of channelrhodopsin-2 in the mouse retina. *Mol Vis* 2009:1680–9.

4. Kostic C, Crippa SV, Pignat V, Bemelmans A-P, Samardzija M, Grimm C, et al. Gene therapy regenerates protein expression in cone photoreceptors in Rpe65(R91W/R91W) mice. *PLoS One* 2011:e16588.

5. Auricchio A, Kobinger G, Anand V, Hildinger M, O'Connor E, Maguire AM, et al. Exchange of surface proteins impacts on viral vector cellular specificity and transduction characteristics: the retina as a model. *Hum Mol Genet* 2001:3075–81.

6. Grüter O, Kostic C, Crippa SV, Perez M-TR, Zografos L, Schorderet DF, et al. Lentiviral vector-mediated gene transfer in adult mouse photoreceptors is impaired by the presence of a physical barrier. *Gene Ther* 2005:942–7.

7. Lipinski DM, Barnard AR, Charbel Issa P, et al. Vesicular Stomatitis Virus Glycoprotein– and Venezuelan Equine Encephalitis Virus-Derived Glycoprotein–Pseudotyped Lentivirus Vectors Differentially Transduce Corneal Endothelium, Trabecular Meshwork, and Human Photoreceptors. *Hum Gene Ther* 2013;25(1):50–62.

8. Ali RR, Reichel MB, Thrasher AJ, Levinsky RJ, Kinnon C, Kanuga N, et al. Gene transfer into the mouse retina mediated by an adeno-associated viral vector. *Hum Mol Genet* 1996:591–4.

9. Lebherz C, Maguire A, Tang W, Bennett J, Wilson JM. Novel AAV serotypes for improved ocular gene transfer. *J Gene Med* 2008:375–82.

10. Yang GS, Schmidt M, Yan Z, Lindbloom JD, Harding TC, Donahue BA, et al. Virus-mediated transduction of murine retina with adeno-associated virus: effects of viral capsid and genome size. *J Virol* 2002:7651–60.

11. Allocca M, Mussolino C, Garcia-Hoyos M, Sanges D, Iodice C, Petrillo M, et al. Novel adeno-associated virus serotypes efficiently transduce murine photoreceptors. *J Virol* 2007:11372–80.

12. Bennett J, Maguire AM, Cideciyan AV, Schnell M, Glover E, Anand V, et al. Stable transgene expression in rod photoreceptors after recombinant adeno-associated virus-mediated gene transfer to monkey retina. *Proc Natl Acad Sci USA* 1999:9920–5.

13. Lotery AJ, Yang GS, Mullins RF, Russell SR, Schmidt M, Stone EM, et al. Adeno-associated virus type 5: transduction efficiency and cell-type specificity in the primate retina. *Hum Gene Ther* 2003:1663–71.

14. Mancuso K, Hendrickson AE, Connor TB, Mauck MC, Kinsella JJ, Hauswirth WW, et al. Recombinant adeno-associated virus targets passenger gene expression to cones in primate retina. *J Opt Soc Am A Opt Image Sci Vis* 2007:1411–6.

15. Yan Z, Zak R, Luxton GWG, Ritchie TC, Bantel-Schaal U, Engelhardt JF. Ubiquitination of both adeno-associated virus type 2 and 5 capsid proteins affects the transduction efficiency of recombinant vectors. *J Virol* 2002:2043–53.

16. Zhong L, Li B, Mah CS, Govindasamy L, Agbandje-McKenna M, Cooper M, et al. Next generation of adeno-associated virus 2 vectors: point mutations in tyrosines lead to high-efficiency transduction at lower doses. *Proc Natl Acad Sci USA* 2008:7827–32.

17. Doroudchi MM, Greenberg KP, Liu J, Silka KA, Boyden ES, Lockridge JA, et al. Virally delivered channelrhodopsin-2 safely and effectively restores visual function in multiple mouse models of blindness. *Mol Ther* 2011:1220–9.

18. Petrs-Silva H, Dinculescu A, Li Q, Min S-H, Chiodo V, Pang J-j, et al. High-efficiency transduction of the mouse retina by tyrosine-mutant AAV serotype vectors. *Mol Ther* 2009:463–71.

19. Petrs-Silva H, Dinculescu A, Li Q, Deng W-T, Pang J-j, Min S-H, et al. Novel properties of tyrosine-mutant AAV2 vectors in the mouse retina. *Mol Ther* 2011:293–301.

20. Grant CA, Ponnazhagan S, Wang XS, Srivastava A, Li T. Evaluation of recombinant adeno-associated virus as a gene transfer vector for the retina. *Curr Eye Res* 1997:949–56.

21. Sawicki JA, Morris RJ, Monks B, Sakai K, Miyazaki J. A composite CMV-IE enhancer/beta-actin promoter is ubiquitously expressed in mouse cutaneous epithelium. *Exp Cell Res* 1998:367–9.

22. Miyazaki J, Takaki S, Araki K, Tashiro F, Tominaga A, Takatsu K, et al. Expression vector system based on the chicken beta-actin promoter directs efficient production of interleukin-5. *Gene* 1989:269–77.

23. Niwa H, Yamamura K, Miyazaki J. Efficient selection for high-expression transfectants with a novel eukaryotic vector. *Gene* 1991:193–9.

24. Le Meur G, Stieger K, Smith AJ, Weber M, Deschamps JY, Nivard D, et al. Restoration of vision in RPE65-deficient Briard dogs using an AAV serotype 4 vector that specifically targets the retinal pigmented epithelium. *Gene Ther* 2007:292–303.

25. Bainbridge JWB, Smith AJ, Barker SS, Robbie S, Henderson R, Balaggan K, et al. Effect of gene therapy on visual function in Leber's congenital amaurosis. *N Engl J Med* 2008:2231–9.

26. Ryals RC, Boye SL, Dinculescu A, Hauswirth WW, Boye SE. Quantifying transduction efficiencies of unmodified and tyrosine capsid mutant AAV vectors in vitro using two ocular cell lines. *Mol Vis* 2011:1090–102.

27. Dinculescu A, Glushakova L, Min S-H, Hauswirth WW. Adeno-associated virus-vectored gene therapy for retinal disease. *Hum Gene Ther* 2005:649–63.

28. Khani SC, Pawlyk BS, Bulgakov OV, Kasperek E, Young JE, Adamian M, et al. AAV-mediated expression targeting of rod and cone photoreceptors with a human rhodopsin kinase promoter. *Invest Ophthalmol Vis Sci* 2007:3954–61.

29. Boye SE, Alexander JJ, Boye SL, Witherspoon CD, Sandefer K, Conlon T, et al. The human rhodopsin kinase promoter in an AAV5 vector confers rod and cone specific expression in the primate retina. *Hum Gene Ther* 2012:1101–15.

30. Sun X, Pawlyk B, Xu X, Liu X, Bulgakov OV, Adamian M, et al. Gene therapy with a promoter targeting both rods and cones rescues retinal degeneration caused by AIPL1 mutations. *Gene Ther* 2010:117–31.

31. Tan MH, Smith AJ, Pawlyk B, Xu X, Liu X, Bainbridge JB, et al. Gene therapy for retinitis pigmentosa and Leber congenital amaurosis caused by defects in AIPL1: effective rescue of mouse models of partial and complete Aipl1 deficiency using AAV2/2 and AAV2/8 vectors. *Hum Mol Genet* 2009:2099–114.

32. Boye SE, Boye SL, Pang J, Ryals R, Everhart D, Umino Y, et al. Functional and behavioural restoration of vision by gene therapy in the guanylate cyclase-1 (GC1) knockout mouse. *PLoS One* 2010:1–13.

33. Pawlyk BS, Bulgakov OV, Liu X, Xu X, Adamian M, Sun X, et al. Replacement gene therapy with a human RPGRIP1 sequence slows photoreceptor degeneration in a murine model of leber congenital amaurosis. *Hum Gene Ther* 2010:993–1004.

34. Petit L, Lhériteau E, Weber M, Le Meur G, Deschamps JY, Provost N, et al. Restoration of vision in the pde6β-deficient dog, a large animal model of rod-cone dystrophy. *Mol Ther* 2012;**20**:2019–30.

35. Komáromy AM, Alexander JJ, Cooper AE, Chiodo VA, Glushakova LG, Acland GM, et al. Targeting gene expression to cones with human cone opsin promoters in recombinant AAV. Gene Ther 2008:1049–55.

36. Carvalho LS, Xu J, Pearson RA, Smith AJ, Bainbridge JW, Morris LM, et al. Long-term and age-dependent restoration of visual function in a mouse model of CNGB3-associated achromatopsia following gene therapy. *Hum Mol Genet* 2011;**20**:3161–75.

37. Kim DS, Matsuda T, Cepko CL. A core paired-type and POU homeodomain-containing transcription factor program drives retinal bipolar cell gene expression. *J Neurosci* 2008:7748–64.

38. Yin L, Greenberg K, Hunter JJ, Dalkara D, Kolstad KD, Masella BD, et al. Intravitreal injection of AAV2 transduces macaque inner retina. *Invest Ophthalmol Vis Sci* 2011:2775–83.

39. Lipinski DM, Singh MS, MacLaren RE. Assessment of cone survival in response to CNTF, GDNF, and VEGF165b in a novel ex vivo model of end-stage retinitis pigmentosa. *Invest Ophthalmol Vis Sci* 2011:7340–6.

40. Schlichtenbrede FC, da Cruz L, Stephens C, Smith AJ, Georgiadis A, Thrasher AJ, et al. Long-term evaluation of retinal function in Prph2Rd2/Rd2 mice following AAV-mediated gene replacement therapy. *J Gene Med* 2003:757–64.

41. Sarra GM, Stephens C, de Alwis M, Bainbridge JW, Smith AJ, Thrasher AJ, et al. Gene replacement therapy in the retinal degeneration slow (rds) mouse: the effect on retinal degeneration following partial transduction of the retina. *Hum Mol Genet* 2001:2353–61.

42. Pang J-j, Dai X, Boye SE, Barone I, Boye SL, Mao S, et al. Long-term retinal function and structure rescue using capsid mutant AAV8 vector in the rd10 mouse, a model of recessive retinitis pigmentosa. *Mol Ther* 2011:234–42.

43. Vollrath D, Feng W, Duncan JL, Yasumura D, D'Cruz PM, Chappelow A, et al. Correction of the retinal dystrophy phenotype of the RCS rat by viral gene transfer of Mertk. *Proc Natl Acad Sci USA* 2001:12584–9.

44. Smith AJ, Schlichtenbrede FC, Tschernutter M, Bainbridge JW, Thrasher AJ, Ali RR. AAV-Mediated gene transfer slows photoreceptor loss in the RCS rat model of retinitis pigmentosa. *Mol Ther* 2003:188–95.

45. Deng W-T, Dinculescu A, Li Q, Boye SL, Li J, Gorbatyuk MS, et al. Tyrosine-mutant AAV8 delivery of human MERTK provides long-term retinal preservation in RCS rats. *Invest Ophthalmol Vis Sci* 2012:1895–904.

46. Fisher KJ, Gao GP, Weitzman MD, DeMatteo R, Burda JF, Wilson JM. Transduction with recombinant adeno-associated virus for gene therapy is limited by leading-strand synthesis. *J Virol* 1996:520–32.

47. Ferrari FK, Samulski T, Shenk T, Samulski RJ. Second-strand synthesis is a rate-limiting step for efficient transduction by recombinant adeno-associated virus vectors. *J Virol* 1996:3227–34.

48. McCarty DM, Fu H, Monahan PE, Toulson CE, Naik P, Samulski RJ. Adeno-associated virus terminal repeat (TR) mutant generates self-complementary vectors to overcome the rate-limiting step to transduction in vivo. *Gene Ther* 2003:2112–8.

49. Yokoi K, Kachi S, Zhang HS, Gregory PD, Spratt SK, Samulski RJ, et al. Ocular gene transfer with self-complementary AAV vectors. *Invest Ophthalmol Vis Sci* 2007:3324–8.

50. Natkunarajah M, Trittibach P, McIntosh J, Duran Y, Barker SE, Smith AJ, et al. Assessment of ocular transduction using single-stranded and self-complementary recombinant adeno-associated virus serotype 2/8. *Gene Ther* 2008:463–7.

51. Kirschman LT, Kolandaivelu S, Frederick JM, Dang L, Goldberg AFX, Baehr W, et al. The Leber congenital amaurosis protein, AIPL1, is needed for the viability and functioning of cone photoreceptor cells. *Hum Mol Genet* 2010:1076–87.

52. Li X, Li W, Dai X, Kong F, Zheng Q, Zhou X, et al. Gene therapy rescues cone structure and function in the 3-month-old rd12 mouse: a model for midcourse RPE65 leber congenital amaurosis. *Invest Ophthalmol Vis Sci* 2011:7–15.

53. Illing ME, Rajan RS, Bence NF, Kopito RR. A rhodopsin mutant linked to autosomal dominant retinitis pigmentosa is prone to aggregate and inter-acts with the ubiquitin proteasome system. *J Biol Chem* 2002:34150–60.

54. Robinson PR, Cohen GB, Zhukovsky EA, Oprian DD. Constitutively active mutants of rhodopsin. *Neuron* 1992:719–25.

55. Hargrave PA, McDowell JH. Rhodopsin and phototransduction: a model system for G protein-linked receptors. *FASEB J* 1992:2323–31.

56. Drenser KA, Timmers AM, Hauswirth WW, Lewin AS. Ribozyme-targeted destruction of RNA associated with autosomal-dominant retinitis pigmentosa. *Invest Ophthalmol Vis Sci* 1998:681–9.

57. Lewin AS, Drenser KA, Hauswirth WW, Nishikawa S, Yasumura D, Flannery JG, et al. Ribozyme rescue of photoreceptor cells in a transgenic rat model of autosomal dominant retinitis pigmentosa. *Nat Med* 1998:967–71.

58. O'Reilly M, Palfi A, Chadderton N, Millington-Ward S, Ader M, Cronin T, et al. RNA interference-mediated suppression and replacement of human rhodopsin in vivo. *Am J Hum Genet* 2007:127–35.

59. Millington-Ward S, Chadderton N, O'Reilly M, Palfi A, Goldmann T, Kilty C, et al. Suppression and replacement gene therapy for autosomal domi-nant disease in a murine model of dominant retinitis pigmentosa. *Mol Ther* 2011:642–9.

60. Fath S, Bauer AP, Liss M, Spriestersbach A, Maertens B, Hahn P, et al. Multiparameter RNA and codon optimization: a standardized tool to assess and enhance autologous mammalian gene expression. *PLoS One* 2011:e17596.

61. Padegimas L, Kowalczyk TH, Adams S, Gedeon CR, Oette SM, Dines K, et al. Optimization of hCFTR lung expression in mice using DNA nanopar-ticles. *Mol Ther* 2012:63–72.

62. Foster H, Sharp PS, Athanasopoulos T, Trollet C, Graham IR, Foster K, et al. Codon and mRNA sequence optimization of microdystro-phin transgenes improves expression and physiological outcome in dystrophic mdx mice following AAV2/8 gene transfer. *Mol Ther* 2008: 1825–32.

63. Mao H, James T, Schwein A, Shabashvili AE, Hauswirth WW, Gorbatyuk MS, et al. AAV delivery of wild-type rhodopsin preserves retinal function in a mouse model of autosomal dominant retinitis pigmentosa. *Hum Gene Ther* 2011:567–75.

64. Grieger JC, Samulski RJ. Packaging capacity of adeno-associated virus serotypes: impact of larger genomes on infectivity and postentry steps. *J Virol* 2005:9933–44.

65. Monahan PE, Lothrop CD, Sun J, Hirsch ML, Kafri T, Kantor B, et al. Proteasome inhibitors enhance gene delivery by AAV virus vectors expressing large genomes in hemophilia mouse and dog models: a strategy for broad clinical application. *Mol Ther* 2010:1907–16.

66. Cideciyan AV, Aleman TS, Swider M, Schwartz SB, Steinberg JD, Brucker AJ, et al. Mutations in ABCA4 result in accumulation of lipofuscin before slowing of the retinoid cycle: a reappraisal of the human disease sequence. *Hum Mol Genet* 2004:525–34.

67. Kong J, Kim S-R, Binley K, Pata I, Doi K, Mannik J, et al. Correction of the disease phenotype in the mouse model of Stargardt disease by lentiviral gene therapy. *Gene Ther* 2008:1311–20.

68. Duan D, Yue Y, Engelhardt JF. Expanding AAV packaging capacity with trans-splicing or overlapping vectors: a quantitative comparison. *Mol Ther* 2001:383–91.

69. Halbert CL, Allen JM, Miller AD. Efficient mouse airway transduction following recombination between AAV vectors carrying parts of a larger gene. *Nat Biotechnol* 2002:697–701.

70. Ghosh A, Yue Y, Duan D. Viral serotype and the transgene sequence influence overlapping adeno-associated viral (AAV) vector-mediated gene transfer in skeletal muscle. *J Gene Med* 2006:298–305.

71. Ghosh A, Yue Y, Lai Y, Duan D. A hybrid vector system expands adeno-associated viral vector packaging capacity in a transgene-independent manner. *Mol Ther* 2008:124–30.

72. Allocca M, Doria M, Petrillo M, Colella P, Garcia-Hoyos M, Gibbs D, et al. Serotype-dependent packaging of large genes in adeno-associated viral vectors results in effective gene delivery in mice. *J Clin Invest* 2008:1955–64.

73. Odom GL, Gregorevic P, Allen JM, Chamberlain JS. Gene therapy of mdx mice with large truncated dystrophins generated by recombination using rAAV6. *Mol Ther* 2011:36–45.

74. Ghosh A, Yue Y, Duan D. Efficient transgene reconstitution with hybrid dual AAV vectors carrying the minimized bridging sequences. *Hum Gene Ther* 2011:77–83.

75. Grose WE, Clark KR, Griffin D, Malik V, Shontz KM, Montgomery CL, et al. Homologous recombination mediates functional recovery of dysferlin deficiency following AAV5 gene transfer. *PLoS One* 2012:e39233.

76. Nakai H, Storm TA, Kay MA. Increasing the size of rAAV-mediated expression cassettes in vivo by intermolecular joining of two complementary vectors. *Nat Biotechnol* 2000:527–32.

77. Sun L, Li J, Xiao X. Overcoming adeno-associated virus vector size limitation through viral DNA heterodimerization. *Nat Med* 2000:599–602.

78. Yan Z, Zhang Y, Duan D, Engelhardt JF. Trans-splicing vectors expand the utility of adeno-associated virus for gene therapy. *Proc Natl Acad Sci USA* 2000:6716–21.

79. Reich SJ, Auricchio A, Hildinger M, Glover E, Maguire AM, Wilson JM, et al. Efficient trans-splicing in the retina expands the utility of adeno-associated virus as a vector for gene therapy. *Hum Gene Ther* 2003:37–44.

80. Lai Y, Yue Y, Liu M, Ghosh A, Engelhardt JF, Chamberlain JS, et al. Efficient in vivo gene expression by trans-splicing adeno-associated viral vectors. *Nat Biotechnol* 2005:1435–9.

81. Li J, Sun W, Wang B, Xiao X, Liu X-Q. Protein trans-splicing as a means for viral vector-mediated in vivo gene therapy. *Hum Gene Ther* 2008: 958–64.

82. Ghosh A, Yue Y, Shin J-H, Duan D. Systemic trans-splicing adeno-associated viral delivery efficiently transduces the heart of adult mdx mouse, a model for duchenne muscular dystrophy. *Hum Gene Ther* 2009:1319–28.

83. Trapani I, Colella P, Sommella A, Iodice C, Cesi G, De Simone S, et al. Effective delivery of large genes to the retina by dual AAV vectors. *EMBO Mol Med* 2014;**6**:194–211.

84. Lopes VS, Boye SE, Louie CM, Boye S, Dyka F, Chiodo V, et al. Retinal gene therapy with a large MYO7A cDNA using adeno-associated virus. *Gene Ther* 2013;**20**:824–33.

85. Palfi A, Chadderton N, McKee AG, Blanco-Fernandez A, Humphries P, Kenna PF, et al. Efficacy of co-delivery of dual AAV2/5 vectors in the murine retina and hippocampus. *Hum Gene Ther* 2012;**23**:847–58.

86. Han Z, Conley SM, Makkia RS, Cooper MJ, Naash MI. DNA nanoparticle-mediated ABCA4 delivery rescues Stargardt dystrophy in mice. *J Clin Invest* 2014;**122**:3221–6.

87. McGee Sanftner LH, Abel H, Hauswirth WW, Flannery JG. Glial cell line derived neurotrophic factor delays photoreceptor degeneration in a transgenic rat model of retinitis pigmentosa. *Mol Ther* 2001:622–9.

88. Lau D, McGee LH, Zhou S, Rendahl KG, Manning WC, Escobedo JA, et al. Retinal degeneration is slowed in transgenic rats by AAV-mediated delivery of FGF-2. *Invest Ophthalmol Vis Sci* 2000:3622–33.

89. Buch PK, MacLaren RE, DurRendahl KG, Manning WC, Escobedohlichtenbrede FC, et al. In contrast to AAV-mediated Cntf expression, AAV-mediated Gdnf expression enhances gene replacement therapy in rodent models of retinal degeneration. *Mol Ther* 2006:700–9.

90. Liang FQ, Aleman TS, Dejneka NS, Dudus L, Fisher KJ, Maguire AM, et al. Long-term protection of retinal structure but not function using RAAV.CNTF in animal models of retinitis pigmentosa. *Mol Ther* 2001:461–72.

91. Bok D, Yasumura D, Matthes MT, Ruiz A, Duncan JL, Chappelow AV, et al. Effects of adeno-associated virus-vectored ciliary neurotrophic factor on retinal structure and function in mice with a P216L rds/peripherin mutation. *Exp Eye Res* 2002:719–35.

92. Schlichtenbrede FC, Macneil A, Bainbridge JWB, Tschernutter M, Thrasher AJ, Smith AJ, et al. Intraocular gene delivery of ciliary neurotrophic factor results in significant loss of retinal function in normal mice and in the Prph2Rd2/Rd2 model of retinal degeneration. *Gene Ther* 2003:523–7.

93. Shan H, Ji D, Barnard AR, Lipinski DM, You Q, Lee EJ, et al. AAV-mediated gene transfer of human X-linked inhibitor of apoptosis protects against oxidative cell death in human RPE cells. *Invest Ophthalmol Vis Sci* 2011:9591–7.

94. Yao J, Feathers KL, Khanna H, Thompson D, Tsilfidis C, Hauswirth WW, et al. XIAP therapy increases survival of transplanted rod precursors in a degenerating host retina. *Invest Ophthalmol Vis Sci* 2011:1567–72.

95. Yao J, Jia L, Khan N, Zheng Q-D, Moncrief A, Hauswirth WW, et al. Caspase inhibition with XIAP as an adjunct to AAV vector gene-replacement therapy: improving efficacy and prolonging the treatment window. *PLoS One* 2012:e37197.

96. MacLaren RE. An analysis of retinal gene therapy clinical trials. *Curr Opin Mol Ther* 2009:540–6.

97. Maguire AM, Simonelli F, Pierce EA, Pugh EN, Mingozzi F, Bennicelli J, et al. Safety and efficacy of gene transfer for Leber's congenital amaurosis. *N Engl J Med* 2008:2240–8.

98. Hauswirth WW, Aleman TS, Kaushal S, Cideciyan AV, Schwartz SB, Wang L, et al. Treatment of leber congenital amaurosis due to RPE65 mutations by ocular subretinal injection of adeno-associated virus gene vector: short-term results of a phase I trial. *Hum Gene Ther* 2008:979–90.

99. Cideciyan AV, Aleman TS, Boye SL, Schwartz SB, Kaushal S, Roman AJ, et al. Human gene therapy for RPE65 isomerase deficiency activates the retinoid cycle of vision but with slow rod kinetics. *Proc Natl Acad Sci USA* 2008:15112–7.

100. Mingozzi F, High KA. Immune responses to AAV in clinical trials. *Curr Gene Ther* 2007:316–24.

101. Mingozzi F, Maus MV, Hui DJ, Sabatino DE, Murphy SL, Rasko JEJ, et al. CD8(+) T-cell responses to adeno-associated virus capsid in humans. *Nat Med* 2007:419–22.

102. Bennett J, Ashtari M, Wellman J, Marshall KA, Cyckowski LL, Chung DC, et al. AAV2 gene therapy readministration in three adults with congenital blindness. *Sci Transl Med* 2012:120ra15.

103. MacLaren RE, Groppe M, Barnard AR, et al. Retinal gene therapy in patients with choroideremia: initial findings from a phase 1/2 clinical trial. *Lancet* 2014;**383**:1129–37.

Chapter 12

Gene Therapy for Hemoglobinopathies: Progress and Challenges

Alisa Dong,[1] **Laura Breda**[1] **and Stefano Rivella**[1,2]

[1]*Department of Pediatrics, Division of Hematology-Oncology, Weill Cornell Medical College, New York, NY, USA;* [2]*Department of Cell and Development Biology, Weill Cornell Medical College, New York, NY, USA*

LIST OF ABBREVIATIONS

BM Bone marrow
GVHD Graft-versus-host disease
HbE Hemoglobin E
HMGA2 High mobility group AT-hook 2
HR Homologous recombination
HS Hypersensitive sites
HSC Hematopoietic stem cell
HSVtk Herpes simplex virus type 1 thymidine kinase
iCasp9 Inducible caspase 9
IDLV Integrase-defective lentiviral
IE Ineffective erythropoiesis
iPSCs Induced pluripotent stem cells
IS Integration sites
LCR Locus control region
PBMCs Peripheral blood mononuclear cells
RBC Red blood cell
shRNA Small hairpin RNA
UTR Untranslated region
VCN Vector copy number

12.1 WHY GENE THERAPY FOR HEMOGLOBINOPATHIES?

Hemoglobinopathies are conditions that result from defects in the genes that control the expression of the hemoglobin protein. Sickle cell disease (SCD), hemoglobin E (HbE), and the thalassemias are the most common hemoglobinopathies worldwide.[1] The extremely high frequency of hemoglobin disorders compared with other monogenic diseases reflects natural selection mediated by the relative resistance of carriers against *Plasmodium falciparum* malaria.[2] In SCD, structural defects are triggered by a single point mutation in the β-chain at position 6, leading to a change in the amino acid sequence from glutamic acid to valine.[3] Diminished production of α- or β-globins, the two subunits of the hemoglobin molecule, can lead to α- or β-thalassemia, respectively.[3–5] HbE is characterized by a single point mutation in the β-chain at position 26, associated with a change in the amino acid sequence from glutamic acid to lysine. The βE mutation also affects β-globin gene expression by creating an alternate splicing site in the mRNA at codons 25–27. Through this mechanism, there is a mild deficiency in normal β mRNA and production of small amounts of anomalous β mRNA.[6]

Palliative therapeutic options to correct patients' anemia are red blood cell (RBC) transfusion and iron chelation. Even in the less severe cases, anemia worsens with aging and can become symptomatic; therefore, blood transfusion becomes mandatory. In general, the level of chronic stable anemia determines the transfusion regimen. In addition to life-threatening anemia, patients may present with intrinsic and treatment-related features that exacerbate the pathology. For SCD patients, common complications are painful episodes, acute chest syndrome, and stroke; for thalassemia patients, common complications include hepatosplenomegaly, recurrent infections, and spontaneous fractures; for individuals who have HbE/β thalassemia, complications include heart failure, enlargement of the liver, and problems in the bones.[3,4,6] In all of these

Translating Gene Therapy to the Clinic. http://dx.doi.org/10.1016/B978-0-12-800563-7.00012-9

cases, transfusion-associated infections and organ damage are side effects of long-term treatment and unsatisfactory iron chelation.[3,4] Iron overload is also observed in patients who do not require blood transfusion for survival because of disorder-related ineffective erythropoiesis (IE). IE triggers increased iron absorption by reducing the expression of hepcidin, the hormone that controls iron absorption.[7,8] Although both transfusion and iron chelation treatments have remarkably improved over the years and thus improved quality of life, they do not provide a definitive cure, as they lack the ability to address the intrinsic genetic defect. To this end, hematopoietic stem cell (HSC) transplantation is the only presently available cure. Allogeneic blood marrow (BM) transplantation can cure hemoglobinopathies, but only a small proportion of patients have suitable donors. Myeloablative HSC transplantation carries a 5–10% mortality rate; in addition, graft-versus-host disease (GVHD) may limit the success of the procedure.[9,10] Given the limitations associated with allogeneic BM transplantation, gene therapy using a patients' own HSCs represents an alternative and potential cure, because it aims at the direct recovery of the hemoglobin protein function via globin-gene transfer. The development of gene therapy tools for hemoglobinopathies has been the object of research for the last few decades and has proved successful in mouse models in multiple studies. The following sections will provide insights about these studies and how gene therapy has paved its way to become the object of clinical investigations.

12.2 CHALLENGES TO HUMAN GENE THERAPY FOR HEMOGLOBINOPATHIES

Gene therapy offers a new approach to cure patients with hemoglobinopathies. As reported in the sections that follow, many studies based on animal models have shown that both SCD and β-thalassemia can be reversed using this alternative and novel approach. Although some studies utilized adenoviruses for gene delivery,[11] most studies focused on the use of onco-[12] and lentiviral vectors. The success of these vectors has led to consideration of their clinical applications for these diseases.[13–16] Researchers have also corrected abnormal globin gene function in induced pluripotent stem cells (iPSCs) generated from adult cells of β-thalassemia and SCD patients.[17–19] The fundamental conditions that must be met for safe and efficacious gene transfer in humans can be summarized by the following points[20–32]:

- highly efficient and stable transduction
- effective targeting of HSCs
- controlled transgene expression (erythroid-specific, stage-restricted, elevated, position-independent, and sustained over time)
- absent or low genomic toxicity
- correction of the phenotype in preclinical models and in transgenic mice.

Although clinical gene therapy studies for the hemoglobinopathies have been initiated, there are still many barriers to translating animal studies to the bedside. For example, the original studies in animal models of β-thalassemia utilized mice characterized by partial or complete deletion of the mouse β-globin genes. In contrast, β-thalassemia in humans is characterized by nearly 300 mutations, mostly associated with one or a limited number of nucleotides. This genetic heterogeneity adds a further level of complexity in the identification of the right conditions for each patient: conditions such as optimization of vector copy number, transgenic chimerism (the relative ratio between lentiviral-transduced and non-transduced HSCs), and the level of transgenic hemoglobin necessary to correct each patient's phenotype. Further concerns are the potential genome toxicity associated with the random integration of current gene transfer vectors and the limited homogeneity of gene transfer in HSCs (further elaboration provided in Sections 12.5–12.7).

12.3 PRECLINICAL STUDIES IN ANIMAL MODELS AND HUMAN CELLS

12.3.1 Oncoviral Vectors

Earlier gene therapy studies utilized recombinant oncoviruses as carriers to introduce a functional copy of the β-globin gene. These vectors have the ability to transfer the β-globin gene into murine HSCs without transferring any viral genes, but did not produce high and stable expression of the β-globin gene in mice.[33–36] Incorporation of core elements of the hypersensitive sites (HS) 2, 3, and 4 of the human β-globin locus control region (LCR) was integral to increased β-globin expression, and helped overcome the inability to reach higher hemoglobin levels. However, this incorporation failed to prevent position-effect variegation and to achieve therapeutic levels of β-globin expression.[37,38] Later work focused on blocking the activation of splicing sites within the LCR sequence of viral RNA, which caused rearrangements. These efforts were effective, but still did not satisfy the requirement of maintaining high β-globin gene expression. Additional studies introduced the α-locus HS-40 regulatory region in place of the large LCR[39,40]; alternative erythroid-specific promoters, such as ankyrin[41,42]; and mutant hereditary persistence of fetal hemoglobin (γ-globin) promoters.[43,44] Results from these studies failed to show their therapeutic relevance, but ultimately pioneered the way for successful lentiviral gene therapy.

12.3.2 Lentiviral Vectors

In the mid-1990s, the HIV-1-based lentiviral vector became available and offered a real alternative to oncoviral vectors as possible carriers of the globin genes.[45] As opposed to oncoviral vectors, lentiviral vectors have the ability to infect quiescent cells,[46] and can carry large transgene cassettes with introns and regulatory LCR elements with no or limited sequence rearrangement.[47–49] Incorporation of different promoters, enhancers, and other large regulatory elements from the β-globin locus in lentiviral vectors has been shown to be particularly suitable for achieving high and specific expression of α- and β-like genes in several mouse models for hemoglobinopathies. Using a lentiviral vector in which the α-globin gene and promoter were flanked by the HS 2, 3, and 4 regions of human LCR, Han and colleagues were able to rescue murine fetuses affected by α-thalassemia. Three months after birth, treated mice sustained 20% of α-globin transgenic expression compared to normal mice, although transgene expression started to decline at 7 months.[50] Several laboratories, including ours, have shown that lentiviral vectors carrying the human β- or γ-globin gene and its fundamental regulatory elements were able to cure and rescue animal models for β-thalassemia intermedia and major, respectively.[29–31,51–53] With a lentiviral vector with enhanced antisickling features, Pawliuk and colleagues corrected the phenotype of SAD and BERK SCD mouse models.[54] The BERK mouse model was also treated in a study in which Pestina and colleagues used a γ-globin-based lentiviral vector with a β-globin 3′UTR and enhancer.[55]

The study by Miccio and collaborators showed that genetically modified thalassemic erythroblasts undergo in vivo selection in mice, indicating that the corrected cells have a survival advantage over the thalassemic ones.[26] From a clinical point of view, this advantage could imply a lower myeloablative regimen for a patient, who could benefit even when full transgenic chimerism is not obtained after BM transplant. Much progress has also been made in the in vitro treatment of both human SCD and β-thalassemic cells with lentiviral vectors. Samakoglu and colleagues corrected human sickle cells by using a vector that combined a γ-globin gene with a small hairpin RNA (shRNA) targeting the sickle β-globin mRNA to, respectively, increase fetal hemoglobin expression and downregulate production of the sickle β-chains.[56]

Our laboratory and others utilized lentiviruses that incorporate insulators to maximize β-globin expression at the random integration site and protect the host genome from possible genotoxicity. Insulators can, in fact, shelter the transgenic cassette from the silencing effect of nonpermissive chromatin sites and simultaneously protect the genomic environment from the enhancer effect mediated by active regulatory elements (like the LCR) introduced with the vector. For example, the cHS4 insulator has a "core" that contains five footprints.[57] The footprints are involved in recruiting CTCF, an enhancer blocking protein; binding USF proteins to recruit histone-modification enzymes that make transcription-activation marks; and binding VEZF1, which prevents DNA methylation in the transcribed region. Arumugam[58] and colleagues compared vectors with the cHS4 insulator to those without, and consistently saw approximately double the β-globin expression with the insulator in vitro with MEL cells and in vivo with transduced and transplanted murine HSCs. Puthenveetil and colleagues utilized the 1.2-kb cHS4 insulator to rescue the phenotype of thalassemic CD34+ BM-derived cells.[32] While beneficial to expression, the 1.2-kb cHS4 insulator causes low viral titers. Thus, in 2009, Arumugam[59] identified a 400-bp extended core region of the cHS4 that still exhibits full insulator activity but does not have as severe an impact on titer.

The study by Wilber and others showed that fetal hemoglobin can be synthesized in human CD34+-derived cells after treatment with a lentiviral vector encoding the γ-globin gene, either in association with the 400-bp core of the cHS4 insulator or with a lentiviral vector carrying an shRNA targeting the γ-globin gene repressor protein BCL11A.[60] We recently showed[61] that using a 200-bp insulator, derived from the promoter of the ankyrin gene, could achieve significant amelioration of the thalassemic phenotype in mice and a high level of expression in both human thalassemic and SCD cells. Miccio and collaborators[62] instead used the HS2 enhancer of the GATA1 gene to achieve high β-globin gene expression in human cells from patients with β-thalassemia. A scheme of the most successful lentiviral-based gene therapy tools is provided in Figure 12.1.

12.3.3 Nonviral Vectors

Several studies have focused on nonviral vectors as a potentially safer method to genetically modify cells.[64] Their application is, however, challenged by relatively low gene transfer efficiency and the difficulty of maintaining long-term stable expression.[65] For the purpose of providing a stable genomic integration mechanism and long-term transgene expression, transposons have been developed.[66,67] The Sleeping Beauty transposon system has become increasingly useful in the field of gene therapy during the last decade, and the recent development of the SB100X transposase as well as its utility in

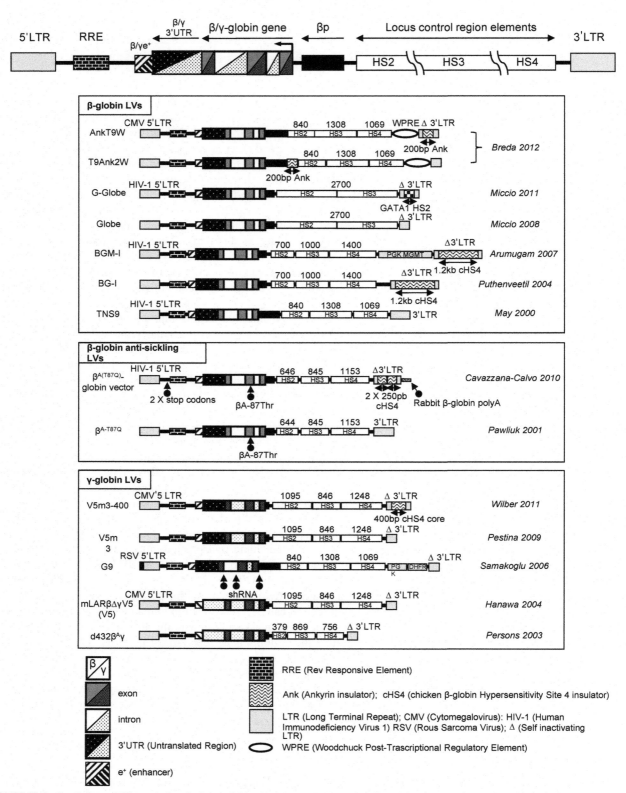

FIGURE 12.1 Lentiviruses (LV) expressing the β or γ-globin gene. A prototypical vector and its main components are represented on the top. Constructs are divided into three major categories: β-globin, β-globin antisickling, and γ-globin LVs. In each box they are represented in chronological order of publication, starting from the most recent. Names of vectors, first authors, and publication date are indicated. Hypersensitive sites (HS) of the LCR are indicated with the respective sizes. Specifics of each construct are described in the text. *Original figure from PMID: 23337292.*[63]

modifying primary CD34$^+$ cells serve as evidence of the system's continued improvement and potential clinical significance.[68,69] In a recent study, Sjeklocha and colleagues used the SB100X transposase system in combination with the IHK–β-globin gene and the cHS4 insulator to produce erythroid-specific expression of the β-globin gene integrated into primary CD34$^+$ cells, which differentiated into multiple lineages.[70] The results of this study support the potential use of an SB IHK–β-globin expression system for gene therapy of SCD in human clinical trials.

12.4 TARGETED REACTIVATION OF FETAL HEMOGLOBIN

The switch from fetal to adult Hb is critical to the pathogenesis of SCD and the β-thalassemias. Persistence of HbF in adults reduces the severity of both thalassemia and SCD traits. Three major loci (*Xmn1-HBG2* single nucleotide polymorphism, *HBS1L-MYB* intergenic region, and *BCL11A* encoding gene) have been identified as major regulatory regions of HbF. Other loci are expected to be involved,[71,72] but their contribution seems only marginal. Given that these loci are associated with putative binding sites for transcriptional repressors, studies have been initiated to characterize individuals carrying genetic variations associated with increased production of HbF, as well as to determine the correlation of this increased production to disease severity.[73,74] Short oligonucleotides have been employed to prevent the binding of repressive factors to their target sites. In one example, Xu and colleagues[75] used a double-stranded 22-bp decoy oligonucleotide containing the *Oct-1* consensus sequence to demonstrate its ability to bind *Oct-1* and cause significant increases in the level of the γ-globin mRNA when administered to K562 cells.

Some publications have focused on the direct manipulation of γ-globin gene transcription. To this end, both Gräslund[76] and Wilber[77] engineered artificial transcription factors to facilitate their interactions with specific DNA sequences and modulate endogenous gene expression within cells. Following this approach, several zinc finger-based transcriptional activators were designed to target the sites proximal to the −117 position of the γ-globin promoter. The results obtained by the two groups demonstrate that is possible to enhance HbF production in primary erythroblasts by using a zinc finger transcriptional activator designed to interact with the γ-globin promoter.

More recently, targeted reactivation of HbF induced by *BCL11A* repression proved to be successful in correcting the phenotype of transgenic knockout mice, which express exclusively human sickle hemoglobin.[78] One of the potential problems in targeting *BCL11A* is associated with the fact that this gene is not only expressed in erythroid cells but also in B cells. Therefore, activation of HbF will require targeting the expression specifically in erythroid cells to spare B cells and avoid undesired effects. Bauer and colleagues have identified a developmental stage-specific, lineage-restricted enhancer required in erythroid cells, but dispensable in B-lymphoid cells, for expression of *BCL11A*.[79] Therefore, this novel *BCL11A* enhancer represents a highly attractive target for therapeutic genome editing in HSCs and iPSCs.

Blobel and colleagues have newly developed an alternative strategy. It has been suggested that in adult erythroid cells, the LCR contacts the β-globin gene promoter by forming a loop that precludes interaction with other β-like globin genes, such as γ-globin.[80] Therefore, redirecting the LCR-complex to the γ-globin promoter could reactivate synthesis of γ-globin and silence synthesis of β-globin. Through this looping approach it may be possible to produce HbF in adult erythroid cells. Looping of the LCR complex to the γ-globin promoter was achieved by making a fusion protein.[81] This fusion protein brings together a zinc finger protein, which recognizes a specific sequence at the γ-globin promoter, to Ldb1, a protein associated with the LCR.[80,82] Ldb1 is a protein necessary for long-range chromatin interactions at the β-globin locus, triggering chromatin loop formation and transcription initiation. Introduction of this fusion protein into adult primary human erythroid cells strongly activated γ-globin expression with a concomitant reduction in β-globin transcription. Strikingly, γ-globin accounted for nearly 90% of total β-like globin transcription. Furthermore, fetal hemoglobin expression was nearly pan-cellular as determined by flow cytometry.[80–82]

Altogether, these studies offer new insights that may translate our understanding of HbF silencing into mechanism-based, improved therapy for the major hemoglobinopathies.

12.5 CLINICAL TRIALS FOR THE HEMOGLOBINOPATHIES

Three patients have been successfully treated in clinical trials for hemoglobinopathies in two independent trials. The first successful gene therapy trial for β-thalassemia was reported by Leboulch in 2010.[83] Using a βT87Q LentiGlobin vector, Leboulch and colleagues were able to achieve transfusion independence in one adult patient with severe βE/β0-thalassemia. The patient is a compound heterozygote (βE/β0), in which one allele (β0) is nonfunctioning and the other (βE) is a mutant allele whose mRNA may either be spliced correctly (producing a mutated βE-globin) or incorrectly (producing no β-globin).

The βT87Q LentiGlobin vector expresses a mutated β-globin distinguishable from transfused β-globin by an antisickling mutation at the 87th amino acid. This vector also contains 2×250-bp copies of the cHS4 insulator. Analysis of the patient's transduced cells revealed an intact vector, however, with the loss of one copy of the cHS4. Of the 24 chromosomal integration sites (IS) found, one of the sites, high mobility group AT-hook 2 (HMGA2), caused transcriptional activation of HMGA2 and conferred clonal dominance. Overexpression of HMGA2 can lead to a clonal growth advantage and is found in a number of benign and malignant tumors.[84] In the majority of these cases, overexpression of HMGA2 is associated with mutations affecting the 3′ untranslated region (UTR), which contains binding sites for the regulatory miRNA let-7.[84] Binding of let-7 miRNAs to the 3′UTR of HMGA2 negatively regulates HMGA2 mRNA and protein expression.[85] Transgenic mice carrying a 3′UTR-truncated HMGA2 expressed increased levels of HMGA2 protein in various tissues including hematopoietic cells, and exhibited an increased number of peripheral blood cells in all lineages, splenomegaly, and erythropoietin-independent erythroid colony formation. Furthermore, BM cells derived from these animals had a growth advantage over wild-type cells. Thus, overexpression of HMGA2 is associated with clonal expansion at the stem cell and progenitor levels.[84]

Cells from the clinical trial patient with an HGMA2 IS showed a loss of the 3′UTR of HGMA2, preventing the binding of the let-7 miRNAs to complementary sequences. Erythroid cells from this IS exhibited a dominant, myeloid-biased cell clone. The expression was reported to be erythroblast-specific, as HMGA2 mRNA was undetectable in granulocyte-monocytes. However, the clonal dominance of HMGA2 is represented in all populations in similar proportions (erythroblasts, granulocyte-monocyte and LTC-IC cells). Leboulch et al. hypothesize that this dominance is due to a transient expression of HMGA2 in a myeloid-restricted LT-HSC during β-LCR priming, before the β-LCR becomes restricted to the erythroid lineage. At the time of reporting, the patient had been transfusion-independent for two years, and showed stable Hb levels from 9 to 10 g/dL. Therapeutic Hb-βT87Q LentiGlobin, however, only accounted for 1/3 of total Hb, with endogenous HbE and HbF making up the rest. Without the additive effect of these endogenous hemoglobins, this first trial might not have been a success. This suggests that we not only need a predictive in vitro model with which to evaluate potential trial patients, but also better vectors that can achieve higher therapeutic hemoglobin expression.

Currently, the first United States phase I clinical trial has received FDA approval and is enrolling patients at Memorial Sloan Kettering Hospital in New York. The main goals of the trial, as briefly described in 2010,[86] are to assess insertional oncogenesis and replication-competent lentivirus safety and determine levels of engraftment and vector expression. The previously described TNS9 vector, which contains the cHS4 insulator,[30,51] will be used to induce transgenic expression of β-globin. Small and unpublished modifications have been made to this vector to increase titer, but the gene, promoter, enhancers, and LCR remain intact. In a pre-clinical study,[87] patient CD34+ hematopoietic progenitors were transduced and tested in vitro for BFU-Es and in vivo for engraftment in NOD-SCID mice. VCNs ranged from 0.2 to two copies/cell in different validations. In vitro experiments showed an improved β/α ratio, with three patients showing ratios of 0.36, 0.48, and 0.54. When normalizing for copy number, transgenic β-globin ranged from 73% to 100% of normal hemizygous β-globin output. Upon in vivo engraftment in immunodeficient mice, VCNs of 0.17–0.6 VC/cell persisted and the mean VCN of the engrafted cells ranged from 36% to 100% of the starting VCN, with an average of 69% retention upon engraftment.

Bluebird Bio, a company specializing in genetic and orphan diseases, has conducted a trial using a LentiGlobin BB305 T87Q virus, named the HGB-205 study.[88] The virus used is similar to the βT87Q LentiGlobin virus, but does not have the cHS4 insulator and is a third generation lentivirus with a CMV promoter in the 5′ LTR, as compared to the previous generation of lentivirus with the wild-type HIV 5′LTR. At the June 2014 European Hematology Association Meeting in Milan, Bluebird presented data on the first two patients treated in the trial, a 16-year-old female and an 18-year-old male who have been dependent on transfusion for most of their lives. These patients have now become transfusion-independent; thus far, one has lasted 3.5 months without transfusion and the other 6.5 months. After 2 months, one patient produced 4.2 g/dL of transgenic β-globin, and another patient was producing 6.6 g/dL of transgenic β-globin after 4.5 months. Vector copy numbers for the two patients were 1.5 and 2.1, a higher copy number than in the βT87Q LentiGlobin trial, where the copy number was 0.6 in CD34+ cells 1 week after transduction. Copy numbers are typically kept low to decrease the risk of insertional oncogenesis, and the clinical result of a higher copy number is unknown.[88]

12.5.1 Stem Cell Mobilization Methods

Li et al. in 1999 showed that β-thalassemia patients could undergo granulocyte colony-stimulating factor (G-CSF) mobilization.[89] In 2012, Yannaki and colleagues studied different mobilization methods in 23 patients with β-thalassemia.[90] Studying patients with and without splenectomy, they found that nonsplenectomized patients tolerated G-CSF, but splenectomized patients showed hyperleukocytosis and could not tolerate it. Poor tolerance could be overcome with a 1-month pretreatment with hydroxyurea and a 12–15 day interval between hydroxyurea withdrawal and G-CSF treatment. They additionally examined Plerixafor, which reversibly inhibits the CXCR4-SDF1 interaction with the BM microenvironment, to mobilize HSCs. Plerixafor proved successful for both splenectomized and nonsplenectomized patients. The New York study[87] has already successfully tested G-CSF mobilization of CD34+ cells. Five patients have been mobilized, with four completing mobilization, and a fifth unable to complete due to anxiety. Harvested cell counts ranged from 8 to 12×10^6 cells/kg in the four patients who completed mobilization. They had both splenectomized and nonsplenectomized patients, none of whom were pretreated with hydroxyurea. Plerixafor was not examined.[87]

Mobilization of sickle cell anemia patients still remains a clinical problem, as patients can show severe life-threatening side effects after G-CSF treatment. Fitzhugh and colleagues examined 11 cases of G-CSF mobilization of SCA patients, some of whom did not know that they had SCA until mobilization.[91] Of the 11 patients, 7 required hospitalization and 4 saw no side effects. In two of these cases, multiorgan failure was seen, with one case resulting in death, and the other resulting in multiple-month hospitalization. Attempts to reduce HbS percentage prior to undergoing mobilization still showed vaso-occulsive crisis. Fitzhugh and colleagues argue that the routine usage of G-CSF cannot be justified in a cost–benefit assessment, and thus should be avoided.[91]

12.5.2 Conditioning Regimen Intensity

The success of conditioning regimen intensity prior to transplant depends on a balance between toxicity and the amount of mixed transgenic chimerism (the amount of BM made up of transplanted cells that carry the vector) (Figure 12.2). Full myeloablation can, theoretically, result in a complete transgenic chimerism, where the BM is made entirely of transplanted cells, with a greater amount of therapeutic vector-transduced cells than in a partial myeloablation setting. However, full myeloablation is more toxic and puts the patient at greater risk, especially in the case of graft failure. A reduced-intensity conditioning regimen can be less toxic; however, it can also lead to a lower composition of therapeutic vector-transduced cells due to mixed transgenic chimerism. When a reduced-intensity conditioning regimen is used, β-globin expression from the vector must be high enough to give transduced cells a survival advantage compared to untransduced cells. In studies of traditional allogenic

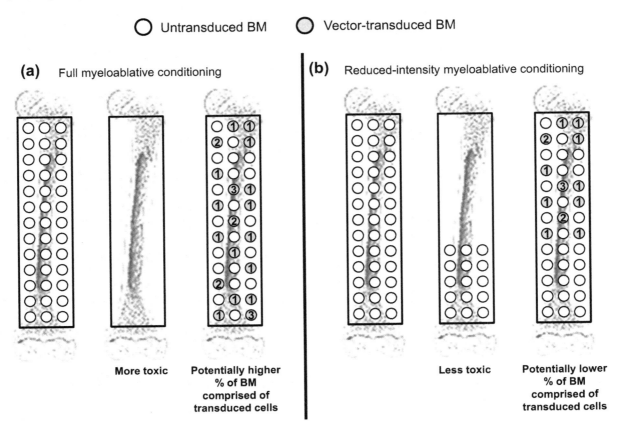

FIGURE 12.2 Schematic of full myeloablation versus reduced-intensity myeloablation. Left panel: BM before myeloablation. Middle panel: BM after myeloablation. Right panel: BM after transplant with transduced cells. Numbers indicate vector copy number, VCN. Transduced and untransduced cells are represented by gray and white circles, respectively. *Original figure from PMID: 23337292.*[63]

BM transplantation, patients with as low as 20% donor contribution were still able to achieve transfusion independence and normal hemoglobin levels; however, a high initial engraftment (>90% at 60 days posttransplant) is necessary if there is to be a reasonable chance of achieving stable mixed chimerism.[92] Lucarelli and colleagues identified three Pesaro risk classes for β-thalassemia patients based on previous iron chelation, hepato- and splenomegaly, and liver fibrosis.[93] Patients with less iron overload and liver damage fall into class one and class two. Patients with irregular iron chelation and more liver damage fall into class three. Full myeloablation for class one and class two patients involves obliteration of the bone marrow by 14–16 mg/kg busulfan, followed by 20 mg/kg of cyclophosphamide to suppress the immune system. An assessment of 886 β-thalassemia patients who received transplants from HLA-matched siblings or parents showed a 91 and 84% probability of thalassemia-free survival with this regimen for class one and class two patients, respectively.[94]

As reviewed by Bertaina et al.,[95] there has been a recent trend to lower intensity, nonmyeloablative conditioning regimens based on the following data: (1) lower morbidity and mortality is associated with these regimens; (2) patients not eligible for the traditional full myeloablative regimen have been safely transplanted with these regimens; and (3) mixed chimerism can be sustained thus leading to amelioration of disease in patients with allografts. Multiple groups have had success with reduced-intensity regimens with lower doses of busulfan, or by using alternatives such as thiotepa, treosulfan, fludarabine, busulfex, or antithmocyte globulin (as reviewed in Refs [95,96]). One report described partial myeloablative allogeneic transplants for 30 SCA patients aged 16 to 65.[97] Twenty-nine of these patients survived a median of 3.4 years at the time of reporting with no nonrelapse mortality, and 26 patients showed long-term stable donor engraftment without acute or chronic GVHD. Twenty-three of the 30 patients showed normalized hemoglobin levels, with ranges of 11.8–13.6 g/dL for women and 12.3–15.1 g/dL for men, demonstrating that sustained and therapeutic mixed chimerism can be achieved without full myeloablation.[97] This report and others, however, relate to allogeneic transplants with HLA-matched donors and not to autologous transplants done with vector-transduced cells. While evidence mixed chimerism with allogeneic transplants signals the potential success of mixed chimerism with autologous gene therapy transplants, this success still remains to be determined.

12.6 GENOME TOXICITY

Random integration of transgenes might be associated with genome toxicity and oncogenesis, a phenomenon already observed in trials for X-linked severe combined immunodeficiency, where leukemia developed as a result of aberrant gene activation from random integration.[98] Hargrove and colleagues investigated gene expression in genomic regions adjacent to the site of integration of lentiviral vectors containing enhancer elements of the β-globin LCR.[22] The analysis investigated gene expression changes following viral transduction in primary clonal murine β-thalassemia erythroid cells, where globin regulatory elements such as the LCR are highly active. The study revealed an overall incidence of perturbed expression in 28% of the transduced clones, with 11% of all genes contained within a 600-kilobase region surrounding the vector-insertion site demonstrating altered expression.

One method of preventing malignancy with lentiviruses is to have a failsafe way of getting rid of vector-transduced cells should they become malignant. In addition to the therapeutic gene, a suicide gene can also be transduced at the same time. If malignancy occurs, this suicide gene can be induced with drugs to cause apoptosis and ablate vector-transduced cells in the body. Two such suicide genes studied in the context of gene therapy are the herpes simplex virus type 1 thymidine kinase (HSVtk) and inducible caspase 9 (iCasp9).[99] HSVtk-transduced cells can be eliminated with the phosphorylation of acyclovir or ganciclovir by HSVtk. iCasp9 is expressed as a monomer but, upon addition of AP1903, dimerizes and causes apoptosis. In a study with T cells, iCasp9 affected immediate death, but HSVtk needed 3 days of treatment.[100]

Studies by Arumugam and colleagues utilized both γ-retroviral and lentiviral vectors containing LCR elements and their potential to transform primary hematopoietic cells.[101] In this study, lentiviral vectors exhibited approximately 200-fold lower transforming potential compared to the conventional γ-retroviral vectors. In an effort to shield cellular chromatin in the vicinity of the sites of chromosomal integration from enhancer effects brought upon by the vector, the cHS4 insulator was included,[102] showing that a further twofold reduction in transforming activity was observed when the LCR elements were flanked by this element. However, the failure to control genomic insertion still remains the main vulnerability of gene therapy.

12.7 PHENOTYPIC VARIABILITY AND GENE TRANSFER IN PATIENTS AFFECTED BY HEMOGLOBINOPATHIES

An additional concern associated with the attempt to cure hemoglobinopathies by gene transfer is the phenotypic variability of patients affected by hemoglobinopathies. The two major categories of the inherited hemoglobin disorders are structural hemoglobin variants and the thalassemia syndromes. The structural hemoglobin variants result primarily from amino acid substitutions in α- or β-globin chains.[103] The most widespread variants are associated with missense mutations, and the main structural changes are classified as HbS, HbC, and HbE.[103] In contrast, thalassemias are characterized by almost 300 mutations. An updated list of these mutations is accessible at the Globin Gene Server Website: http://globin.cse.psu.edu. Furthermore, the phenotype of the patients affected by hemoglobinopathies can be profoundly affected by the presence of modifiers. A major modifier of severity in both β-globin-associated hemoglobinopathies is the reduced or increased number of α-globin genes. Additionally, mutations and/or allelic variants associated with increased fetal hemoglobin expression can dramatically ameliorate these disorders.[104] In fact, patients with the same set of pathogenic globin mutations can have dramatically variable clinical courses based on modifier conditions.[105]

The vast majority of β-thalassemias are caused by point mutations within the gene or within its immediate flanking sequences, and are classified according to the mechanism by which they affect gene regulation: transcription, RNA processing, and mRNA translation.[103] Some transcription mutations involve either the conserved DNA sequences that form the β-globin promoter (−80T>A) or the stretch of 50 nucleotides in the 5′UTR. Generally, they result in a mild-to-minimal deficit of β-globin output and can be phenotypically *silent* carriers (Figure 12.3). An additional set of mutations introduces premature termination codons (PTCs) because of frameshift or nonsense mutations, and nearly all terminate within exon one or two. Mutations that result in PTCs are typically associated with minimal steady-state levels of β-mRNA in erythroid cells; this occurs because of rapid degradation of the mutant mRNA by nonsense-mediated mRNA decay (NMD) (Figure 12.3).[106–108] One of the most prevalent mutations in Western Mediterranean populations is the PTC at base 118 (C>T), which historically was named β0-39 because it introduces a PTC in codon 39 (Figure 12.3). In heterozygotes, no β-chain is produced from the mutant allele, but activity from the normal allele is sufficient to result in a typical asymptomatic phenotype.[104,109,110] Another subset of mutations abolishes or limits the use of the proper splice sites or creates novel aberrant cryptic splice sites. All these mutations lead to unspliced or improperly spliced products with varying severities of phenotype. When the mutations involve the invariant dinucleotides at the splice junctions (GT at 5′ and AG at 3′), normal splicing is completely abolished (IVS1-5; Figure 12.3). Other mutations within the splice consensus sequences only reduce the efficiency of normal splicing and produce phenotypes that range from mild to severe (IVS1-110 and IVS2-654; Figure 12.3). Furthermore, mutations that produce PTCs later in the β-globin sequence are not subjected to NMD and result in amounts of mutant β-mRNAs comparable to those from the normal allele (β0-121; Figure 12.3).[111] Heterozygotes for such late termination mutations tend to have a much more severe phenotype. Frequently, these aberrant mRNAs are stable, can be identified in patients, and can generate abnormal β-globin proteins leading to dominant forms of β-thalassemia (Figure 12.3).[112]

As mentioned previously, several preclinical studies showed that it is possible to rescue mice with β-thalassemia by lentiviral-mediated β-globin gene transfer. However, the original studies in animal models of hemoglobinopathies utilized mice characterized by only a handful of mutations[26,29–31,51–55] (Figure 12.1), while the complexity of the mutations and modifiers within the population of individuals affected by hemoglobinopathies is a source of great phenotypic variability. Therefore, some patients—based on their genetic profile and endogenous hemoglobin production—might be better candidates for lentiviral-based therapies than others. This is especially important in light of the intrinsic ability of the specific vectors used in the different clinical trials to make hemoglobin. In SCD, a further limitation of this curative approach through "gene addition" is the fact that the β-globin encoded by the vector must replace the mutant β-globin protein in the

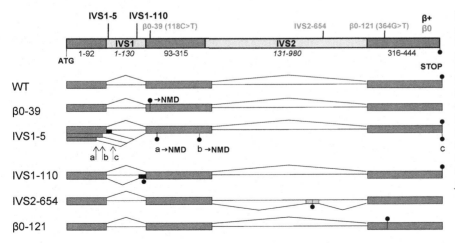

FIGURE 12.3 Structure of the human β-globin gene and the normal and alternative spliced forms for each mutant allele. Only the coding regions (dark boxes) and the introns (IVS, light boxes and lines) are indicated with their respective length. The size of the coding sequences plus introns is equal to 1424 bp. Circles indicate stop codons in the coding sequence. Aberrant mutant forms present PTCs due to point or frame-shift mutations. The effect of the mutation on β-globin synthesis is indicated as β0 or β+. Aberrant mRNA sensitive to nonsense-mediated RNA decay are indicated as NMD. *Original figure from PMID: 23337292.*[63]

adult hemoglobin tetramer without increasing the total amount of β-globin chains (normal + sickle) produced. Increase in the total amount of β-globin chains could potentially lead to a phenotype similar to that observed in α-thalassemia.

Thus, studies on the correlation between types of mutations in the β-globin gene and expression of globin from the unaffected gene have great potential to increase our understanding of gene expression in general and globin expression in thalassemia specifically. In addition, this knowledge may greatly improve the therapeutic options based on gene transfer, replacement, or gene correction. Based on these premises, we developed an in vitro protocol using CD34+ cells isolated from peripheral blood mononuclear cells (PBMCs) to predict if a patient has the potential to become transfusion-independent with a given vector. In β-thalassemia, mutant alleles that are associated with no β-globin synthesis are classified as β0, while those that allow some protein synthesis are designated β+. Thalassemic patients are therefore generally classified as β0/0, +/0 or +/+ based on the combination of these two alleles.[113] Using a wide range of samples from β0/0, β0/+, and β+/+ patients, we investigated the correlation between genotype, endogenous levels of hemoglobins, and VCN, with the phenotypic outcome.[61] We used a novel lentiviral vector, AnkT9W, which carries both the human β-globin gene and the erythroid-specific ankyrin 5′ HS barrier insulator.[114] This vector was able to maintain high yet stable levels of Hb synthesis in MEL cells and in β-thalassemic (th3/+) mice. Compared to β-thalassemic mice transplanted with bone marrow cells transduced with a vector that did not incorporate the ankyrin element (T9W), th3/+ animals treated with AnkT9W showed a greater improvement in Hb synthesis and hematological parameters.[61] Using 30 mL of blood, we developed a protocol for evaluating the correlation between the VCN of AnkT9W, the number of β-globin mRNA molecules, and the level of Hb synthesis in PBMC-derived human CD34+ and erythroid progenitor cells, following in vitro β-globin gene transfer and erythroid differentiation. Another goal of this study was to identify an in vitro protocol that could predict correction of both the thalassemic and SCD phenotype in a large pool of patient samples. In this regard, AnkT9W increased the synthesis of β-globin protein in most thalassemic and SCD specimens to levels comparable to those observed in carriers and control samples. However, after treatment, a subset of thalassemic samples, mostly β0/0, exhibited the lowest amount of total HbA (transgenic and endogenous). In contrast, β0/+ and β+/+ benefitted from the expression of the endogenous HbA and reached curative levels at lower VCNs.

Overall, these results indicate three major endpoints. First, β0/0 specimens, which exhibited the lowest level of β-globin chain synthesis, required higher amounts of vector to reach therapeutic levels of Hb, compared to specimens from β0/+ and β+/+ patients. For this reason, patients should be carefully investigated to determine if they would be good candidates for a clinical trial based on this approach, especially if the level of gene transfer achieved might not be sufficient. Second, when adding transgenic β-globin into SCD HSCs, there is a concern that after gene transfer, the total amount of β-chains (sickle + normal) might exceed the amount of α-chains, leading to an α-thalassemia-like phenotype. However, recent data from Khandros and colleagues indicate that erythroid cells are equipped with a series of protein quality control responses that can eliminate, to a certain extent, the excess of α-chains produced in β-thalassemia.[115] In fact, in SCD specimens, the treatment with AnkT9W was able to correct the phenotype by modifying the proportion of sickling versus functional Hb, without changing the overall Hb content, suggesting that a moderate excess of globin chains might be tolerated and eliminated efficiently by the erythroid cells.[61] Third, irrespective of the β-globin mutations, additional mechanisms potentially associated with genetic modifiers may contribute to the variability observed following gene transfer. Considering these points, we recommend that analysis of erythroid progenitors transduced with different amounts of lentiviral vectors could be extremely useful for testing the potential of each lentiviral construct prior to myeloablation and transplant.

12.8 FUTURE PERSPECTIVES

Lentiviral vectors are currently utilized in clinical trials; however, the fact that they integrate semirandomly throughout the genome is associated with several drawbacks. The expression of the transgene can be influenced by activating or repressing elements flanking the integration site, leading to expression variegation or silencing.[116] Some insertions may disrupt genes or perturb their transcription, altering cycle and survival of the transduced cell, and eventually leading to cancer.[117–119]

12.8.1 Safe Harbors and Homologous Recombination

Site-specific integration may limit the risks associated with lentiviral vectors and help achieve safe and uniform gene expression. Integrase-defective lentiviral (IDLV) or nonintegrating lentiviral vectors that carry a mutant integrase are under development to facilitate such a task.[120–122] Following reverse transcription, the DNA from these vectors does not integrate into the host genome, but forms episomal circles. The approach consists in generating two IDLV vectors. One vector carries the sequences to target a specific genomic region by homologous recombination (HR),[122] and the second IDLV vector expresses zinc finger nucleases that are target-specific endonucleases and induce site-specific DNA double-strand breaks, facilitating HR. With this approach, in principle, a curative sequence could be utilized to target specific mutant genes to achieve gene correction or for gene insertion into *safe harbors*.[122] Safe harbors are genomic regions in which a transgene, following integration, exhibits elevated and stable expression in different cell types or tissues, does not show position effect variegation, and does not disrupt coding or regulatory sequences due to the integration. Lombardo and colleagues identified one such safe harbor in the *AAVS1* genomic locus of human cells.[123] The same group also demonstrated that this approach can be utilized to generate genetically modified stem cells and human lymphocytes.[123,124]

12.8.2 Gene Transfer in iPS Cells

Somatic cells reprogrammed to iPSCs might also provide a possible new approach to treat β-thalassemia and SCD.[19] Currently, robust generation of iPSCs requires the introduction and integration of genes encoding the transcriptional factors OCT3/4, SOX2 with either KLF4 and c-MYC or NANOG and LIN28.[125–127] Recently, iPSCs with a single viral integration were generated by transduction with a single lentiviral vector (STEMCCA) expressing a stem cell cassette composed of the four transcription factors (OCT4, KLF4, SOX2, and c-MYC) and a combination of self-splicing peptides and internal ribosome entry site technology.[128,129] iPSCs generated in this manner display embryonic stem cell-like morphology, express embryonic stem cell markers, and exhibit in vivo pluripotency, as evidenced by their ability to form all three germ layers in teratoma assays and their robust contribution to mouse chimeras.

Chou and colleagues investigated epigenetic and gene expression signatures of multiple cell types such as fibroblasts and blood cells.[130] Their analysis suggested that newborn cord blood and adult PBMCs display unique signatures that are closer to iPSCs and human embryonic stem cells than age-matched fibroblasts, thus making PBMCs an easy and attractive cell choice for the generation of iPSCs. They generated iPSCs from unfractionated blood cells using two episomal vectors expressing the reprogramming factors and the SV40 Large T antigen. With this approach, they were able to generate between three and 12 iPSCs from 2×10^6 PBMCs. Ye and coworkers have shown that iPSCs can be generated from cells derived from skin fibroblasts, amniotic fluid, or chorionic villus sampling of patients with β-thalassemia, and these iPSCs can subsequently be differentiated into hematopoietic cells that synthesize hemoglobin.[19]

To minimize the oncogenic potential as a result of the presence of these transcriptional factors and random integrations, several precautions can be taken. Lentiviral vectors can be used to generate iPSCs and add therapeutic β-globin at the same time. Once the reprogramming has occurred, the genes encoding for the iPSC transcriptional factors can be excised from the cells, minimizing their role in potential future oncogenic events. Alternatively, iPSCs can be generated with episomes,[131] nonintegrating viruses, synthetic RNA, or proteins, as reviewed by Yanamaka's group.[132] Moreover, iPSCs can be analyzed to identify whether the β-globin gene integration events occurred in safe harbors within the genome.[133] Safe harbors can be identified based on insertion site. A safe harbor is typically considered to be more than 50–100 kb away from regions transcribing microRNAs, coding gene regions, and ultraconserved regions. Analysis of iPSCs infected with lentiviruses has demonstrated that ~10% of integrations occur in safe harbors. This analysis can include monitoring of the expression of genes in *cis* to the integration sites, to evaluate whether their regulation has been altered. Clones without insertional mutagenesis can be selected for high levels of transgene expression. However, this approach still relies heavily on cellular manipulation and random integrations. Furthermore, in SCD patients, this approach does not eliminate the issue of overproducing β-globin chains (therapeutic + sickle).

In another method, patient iPSCs could be corrected by HR techniques. This strategy has several potential advantages over conventional gene therapy because it avoids the risk of insertional mutagenesis by therapeutic vectors, and maintains β-globin expression by endogenous control elements rather than a constitutive promoter. The Kan and Townes laboratories utilized HR to correct mutations and generate iPSCs from, respectively, a β-thalassemic patient and sickle cell mice.[18,134] Using a humanized SCD mouse model, Jaenisch's laboratory demonstrated that mice can be rescued after transplantation with hematopoietic progenitors obtained in vitro from autologous iPSCs. This was achieved after correction of the human sickle hemoglobin allele by gene-specific targeting.[17] Further proof of principle that the β-globin gene can be repaired in human iPSCs was provided by two additional groups.[135,136] They achieved correction of the β-globin gene by HR in iPSCs derived from an SCD patient using β-globin zinc finger nucleases engineered to specifically stimulate HR at the β-globin locus. In 2008, Howden and colleagues used bacterial artificial chromosomes and components from adeno-associated virus to preferentially insert β-globin into the *AAVS1* in K562 cells.[137] Of the 36 insertion sites analyzed, only six of them (17%) occurred in *AAVS1*, and five out of those six were intact and functional.

Therefore, development of iPSCs technology combined with safe harbors or homologous recombination might represent a safer approach for the cure of hemoglobinopathies than traditional gene therapy. However, limitations with this approach are associated with the observations that red cell-derived iPSCs have limited ability to express adult Hb or home to the bone marrow. These are important problems that will need to be solved before the strategy can be utilized for therapeutic purposes.

12.9 CONCLUSION

Treatments for hemoglobinopathies have developed greatly over the years, with clinical trials now a possibility. The field has evolved from oncoretroviral vectors to third-generation lentiviral vectors to in situ genetic correction. Preclinical studies have examined many different types of genetic elements to increase transgene expression, increase viral titer, and indirectly increase β-globin expression. However, insertional oncogenesis, high and stable transgenic β-globin expression, and phenotypic variance across multiple patients are issues that still need to be addressed in preclinical and clinical settings. As our understanding of the β-globin locus, genetic regulation, and the hematopoietic system evolves, new treatments that can overcome some of these issues will also evolve.

ACKNOWLEDGMENT

This work is supported by the Children's Cancer and Blood Foundation and NIH grant NHLBI-R01HL102449-03 (to S. Rivella).

Competing Interests: S. Rivella is a consultant for Novartis, Biomarin, Bayer, and Isis Pharmaceuticals. S. Rivella holds equities in Merganser Biotech LLC. In addition, he is a coinventor for the patents US8058061 B2 C12N 20111115 and US7541179 B2C12N 20090602. The consulting work and intellectual property of S. Rivella did not affect in any way the design, conduct, or reporting of this research.

REFERENCES

1. Modell B, Darlison M. Global epidemiology of haemoglobin disorders and derived service indicators. *Bull World Health Organ* 2008;**86**:480–7.
2. Mohandas N, An X. Malaria and human red blood cells. *Med Microbiol Immunol* 2012;**201**(4):593–8.
3. Steinberg MH. *Sickle cell disease and other hemoglobinopathies. Goldman's cecil medicine.* 24 ed Elsevier; 2011. 1066–1075.
4. Cappellini MD. *The thalassemias. Goldman's cecil medicine.* 24 ed Elsevier; 2011. 1060–1066.
5. Ginzburg Y, Rivella S. Beta-thalassemia: a model for elucidating the dynamic regulation of ineffective erythropoiesis and iron metabolism. *Blood* 2011;**118**:4321–30.
6. Fucharoen S, Weatherall DJ. The hemoglobin E thalassemias. *Cold Spring Harb Perspect Med* 2012:2.
7. Gardenghi S, Marongiu MF, Ramos P, Guy E, Breda L, Chadburn A, et al. Ineffective erythropoiesis in beta-thalassemia is characterized by increased iron absorption mediated by down-regulation of hepcidin and up-regulation of ferroportin. *Blood* 2007;**109**:5027–35.
8. Gardenghi S, Ramos P, Marongiu MF, Melchiori L, Breda L, Guy E, et al. Hepcidin as a therapeutic tool to limit iron overload and improve anemia in beta-thalassemic mice. *J Clin Invest* 2010;**120**:4466–77.
9. La Nasa G, Giardini C, Argiolu F, Locatelli F, Arras M, De Stefano P, et al. Unrelated donor bone marrow transplantation for thalassemia: the effect of extended haplotypes. *Blood* 2002;**99**:4350–6.
10. Sodani P, Isgro A, Gaziev J, Polchi P, Paciaroni K, Marziali M, et al. Purified T-depleted, CD34+ peripheral blood and bone marrow cell transplantation from haploidentical mother to child with thalassemia. *Blood* 2010;**115**:1296–302.
11. Maina N, Zhong L, Li X, Zhao W, Han Z, Bischof D, et al. Optimization of recombinant adeno-associated viral vectors for human beta-globin gene transfer and transgene expression. *Hum Gene Ther* 2008;**19**:365–75.
12. Yi Y, Noh MJ, Lee KH. Current advances in retroviral gene therapy. *Curr Gene Ther* 2011;**11**:218–28.
13. Arumugam P, Malik P. Genetic therapy for beta-thalassemia: from the bench to the bedside. *Hematol Am Soc Hematol Educ Program* 2010;**2010**:445–50.
14. Sadelain M, Boulad F, Galanello R, Giardina P, Locatelli F, Maggio A, et al. Therapeutic options for patients with severe beta-thalassemia: the need for globin gene therapy. *Hum Gene Ther* 2007;**18**:1–9.
15. Sadelain M, Lisowski L, Samakoglu S, Rivella S, May C, Riviere I. Progress toward the genetic treatment of the beta-thalassemias. *Ann N Y Acad Sci* 2005;**1054**:78–91.
16. Breda L, Gambari R, Rivella S. Gene therapy in thalassemia and hemoglobinopathies. *Mediterr J Hematol Infect Dis* 2009;**1**(1).
17. Hanna J, Wernig M, Markoulaki S, Sun CW, Meissner A, Cassady JP, et al. Treatment of sickle cell anemia mouse model with iPS cells generated from autologous skin. *Science* 2007;**318**:1920–3.
18. Wu LC, Sun CW, Ryan TM, Pawlik KM, Ren J, Townes TM. Correction of sickle cell disease by homologous recombination in embryonic stem cells. *Blood* 2006;**108**:1183–8.
19. Ye L, Chang JC, Lin C, Sun X, Yu J, Kan YW. Induced pluripotent stem cells offer new approach to therapy in thalassemia and sickle cell anemia and option in prenatal diagnosis in genetic diseases. *Proc Natl Acad Sci USA* 2009;**106**:9826–30.
20. Case SS, Price MA, Jordan CT, Yu XJ, Wang L, Bauer G, et al. Stable transduction of quiescent CD34(+)CD38(−) human hematopoietic cells by HIV-1-based lentiviral vectors. *Proc Natl Acad Sci USA* 1999;**96**:2988–93.
21. Lisowski L, Sadelain M. Locus control region elements HS1 and HS4 enhance the therapeutic efficacy of globin gene transfer in beta-thalassemic mice. *Blood* 2007;**110**:4175–8.
22. Hargrove PW, Kepes S, Hanawa H, Obenauer JC, Pei D, Cheng C, et al. Globin lentiviral vector insertions can perturb the expression of endogenous genes in beta-thalassemic hematopoietic cells. *Mol Ther* 2008;**16**:525–33.
23. Zhou HS, Zhao N, Li L, Dong WJ, Wu XS, Hao DL, et al. Site-specific transfer of an intact beta-globin gene cluster through a new targeting vector. *Biochem Biophys Res Commun* 2007;**356**:32–7.

24. Yannaki E, Psatha N, Athanasiou E, Karponi G, Constantinou V, Papadopoulou A, et al. Mobilization of hematopoietic stem cells in a thalassemic mouse model: implications for human gene therapy of thalassemia. *Hum Gene Ther* 2010;**21**:299–310.

25. Kumar P, Woon-Khiong C. Optimization of lentiviral vectors generation for biomedical and clinical research purposes: contemporary trends in technology development and applications. *Curr Gene Ther* 2011;**11**:144–53.

26. Miccio A, Cesari R, Lotti F, Rossi C, Sanvito F, Ponzoni M, et al. In vivo selection of genetically modified erythroblastic progenitors leads to long-term correction of beta-thalassemia. *Proc Natl Acad Sci USA* 2008;**105**:10547–52.

27. Negre O, Fusil F, Colomb C, Roth S, Gillet-Legrand B, Henri A, et al. Correction of murine beta-thalassemia after minimal lentiviral gene transfer and homeostatic in vivo erythroid expansion. *Blood* 2011;**117**:5321–31.

28. Persons DA, Allay ER, Sawai N, Hargrove PW, Brent TP, Hanawa H, et al. Successful treatment of murine beta-thalassemia using in vivo selection of genetically modified, drug-resistant hematopoietic stem cells. *Blood* 2003;**102**:506–13.

29. Imren S, Payen E, Westerman KA, Pawliuk R, Fabry ME, Eaves CJ, et al. Permanent and panerythroid correction of murine beta thalassemia by multiple lentiviral integration in hematopoietic stem cells. *Proc Natl Acad Sci USA* 2002;**99**:14380–5.

30. May C, Rivella S, Callegari J, Heller G, Gaensler KM, Luzzatto L, et al. Therapeutic haemoglobin synthesis in beta-thalassaemic mice expressing lentivirus-encoded human beta-globin. *Nature* 2000;**406**:82–6.

31. May C, Rivella S, Chadburn A, Sadelain M. Successful treatment of murine beta-thalassemia intermedia by transfer of the human beta-globin gene. *Blood* 2002;**99**:1902–8.

32. Puthenveetil G, Scholes J, Carbonell D, Qureshi N, Xia P, Zeng L, et al. Successful correction of the human beta-thalassemia major phenotype using a lentiviral vector. *Blood* 2004;**104**:3445–53.

33. Bender MA, Gelinas RE, Miller AD. A majority of mice show long-term expression of a human beta-globin gene after retrovirus transfer into hematopoietic stem cells. *Mol Cell Biol* 1989;**9**:1426–34.

34. Dzierzak EA, Papayannopoulou T, Mulligan RC. Lineage-specific expression of a human beta-globin gene in murine bone marrow transplant recipients reconstituted with retrovirus-transduced stem cells. *Nature* 1988;**331**:35–41.

35. Karlsson S, Bodine DM, Perry L, Papayannopoulou T, Nienhuis AW. Expression of the human beta-globin gene following retroviral-mediated transfer into multipotential hematopoietic progenitors of mice. *Proc Natl Acad Sci USA* 1988;**85**:6062–6.

36. Lung HY, Meeus IS, Weinberg RS, Atweh GF. In vivo silencing of the human gamma-globin gene in murine erythroid cells following retroviral transduction. *Blood Cells Mol Dis* 2000;**26**:613–9.

37. Chang JC, Liu D, Kan YW. A 36-base-pair core sequence of locus control region enhances retrovirally transferred human beta-globin gene expression. *Proc Natl Acad Sci USA* 1992;**89**:3107–10.

38. Plavec I, Papayannopoulou T, Maury C, Meyer F. A human beta-globin gene fused to the human beta-globin locus control region is expressed at high levels in erythroid cells of mice engrafted with retrovirus-transduced hematopoietic stem cells. *Blood* 1993;**81**:1384–92.

39. Nishino T, Tubb J, Emery DW. Partial correction of murine beta-thalassemia with a gammaretrovirus vector for human gamma-globin. *Blood Cells Mol Dis* 2006;**37**:1–7.

40. Ren S, Wong BY, Li J, Luo XN, Wong PM, Atweh GF. Production of genetically stable high-titer retroviral vectors that carry a human gamma-globin gene under the control of the alpha-globin locus control region. *Blood* 1996;**87**:2518–24.

41. Sabatino DE, Wong C, Cline AP, Pyle L, Garrett LJ, Gallagher PG, et al. A minimal ankyrin promoter linked to a human gamma-globin gene demonstrates erythroid specific copy number dependent expression with minimal position or enhancer dependence in transgenic mice. *J Biol Chem* 2000;**275**:28549–54.

42. Sabatino DE, Seidel NE, Aviles-Mendoza GJ, Cline AP, Anderson SM, Gallagher PG, et al. Long-term expression of gamma-globin mRNA in mouse erythrocytes from retrovirus vectors containing the human gamma-globin gene fused to the ankyrin-1 promoter. *Proc Natl Acad Sci USA* 2000;**97**:13294–9.

43. Fragkos M, Anagnou NP, Tubb J, Emery DW. Use of the hereditary persistence of fetal hemoglobin 2 enhancer to increase the expression of oncoretrovirus vectors for human gamma-globin. *Gene Ther* 2005;**12**:1591–600.

44. Katsantoni EZ, Langeveld A, Wai AW, Drabek D, Grosveld F, Anagnou NP, et al. Persistent gamma-globin expression in adult transgenic mice is mediated by HPFH-2, HPFH-3, and HPFH-6 breakpoint sequences. *Blood* 2003;**102**:3412–9.

45. Naldini L, Blomer U, Gallay P, Ory D, Mulligan R, Gage FH, et al. In vivo gene delivery and stable transduction of nondividing cells by a lentiviral vector. *Science* 1996;**272**:263–7.

46. Lewis P, Hensel M, Emerman M. Human immunodeficiency virus infection of cells arrested in the cell cycle. *Embo J* 1992;**11**:3053–8.

47. Zufferey R, Dull T, Mandel RJ, Bukovsky A, Quiroz D, Naldini L, et al. Self-inactivating lentivirus vector for safe and efficient in vivo gene delivery. *J Virol* 1998;**72**:9873–80.

48. Miyoshi H, Blomer U, Takahashi M, Gage FH, Verma IM. Development of a self-inactivating lentivirus vector. *J Virol* 1998;**72**:8150–7.

49. Bukovsky AA, Song JP, Naldini L. Interaction of human immunodeficiency virus-derived vectors with wild-type virus in transduced cells. *J Virol* 1999;**73**:7087–92.

50. Han XD, Lin C, Chang J, Sadelain M, Kan YW. Fetal gene therapy of alpha-thalassemia in a mouse model. *Proc Natl Acad Sci USA* 2007;**104**:9007–11.

51. Rivella S, May C, Chadburn A, Riviere I, Sadelain M. A novel murine model of Cooley anemia and its rescue by lentiviral-mediated human beta-globin gene transfer. *Blood* 2003;**101**:2932–9.

52. Persons DA, Hargrove PW, Allay ER, Hanawa H, Nienhuis AW. The degree of phenotypic correction of murine beta -thalassemia intermedia following lentiviral-mediated transfer of a human gamma-globin gene is influenced by chromosomal position effects and vector copy number. *Blood* 2003;**101**:2175–83.

53. Hanawa H, Hargrove PW, Kepes S, Srivastava DK, Nienhuis AW, Persons DA. Extended beta-globin locus control region elements promote consistent therapeutic expression of a gamma-globin lentiviral vector in murine beta-thalassemia. *Blood* 2004;**104**:2281–90.

54. Pawliuk R, Westerman KA, Fabry ME, Payen E, Tighe R, Bouhassira EE, et al. Correction of sickle cell disease in transgenic mouse models by gene therapy. *Science* 2001;**294**:2368–71.

55. Pestina TI, Hargrove PW, Jay D, Gray JT, Boyd KM, Persons DA. Correction of murine sickle cell disease using gamma-globin lentiviral vectors to mediate high-level expression of fetal hemoglobin. *Mol Ther* 2009;**17**:245–52.

56. Samakoglu S, Lisowski L, Budak-Alpdogan T, Usachenko Y, Acuto S, Di Marzo R, et al. A genetic strategy to treat sickle cell anemia by coregulating globin transgene expression and RNA interference. *Nat Biotechnol* 2006;**24**:89–94.

57. Nienhuis AW, Persons DA. Development of gene therapy for thalassemia. *Cold Spring Harb Perspect Med* 2012:2.

58. Arumugam PI, Scholes J, Perelman N, Xia P, Yee JK, Malik P. Improved human beta-globin expression from self-inactivating lentiviral vectors carrying the chicken hypersensitive site-4 (cHS4) insulator element. *Mol Ther* 2007;**15**:1863–71.

59. Arumugam PI, Urbinati F, Velu CS, Higashimoto T, Grimes HL, Malik P. The 3' region of the chicken hypersensitive site-4 insulator has properties similar to its core and is required for full insulator activity. *PloS One* 2009;**4**:e6995.

60. Wilber A, Hargrove PW, Kim YS, Riberdy JM, Sankaran VG, Papanikolaou E, et al. Therapeutic levels of fetal hemoglobin in erythroid progeny of beta-thalassemic CD34+ cells after lentiviral vector-mediated gene transfer. *Blood* 2011;**117**:2817–26.

61. Breda L, Casu C, Gardenghi S, Bianchi N, Cartegni L, Narla M, et al. Therapeutic hemoglobin levels after gene transfer in beta-thalassemia mice and in hematopoietic cells of beta-thalassemia and sickle cells disease patients. *PLoS One* 2012;**7**:e32345.

62. Miccio A, Poletti V, Tiboni F, Rossi C, Antonelli A, Mavilio F, et al. The GATA1-HS2 enhancer allows persistent and position-independent expression of a beta-globin transgene. *PloS One* 2011;**6**:e27955.

63. Dong A, Rivella S, Breda L. Gene therapy for hemoglobinopathies: progress and challenges. *Transl Res* 2013;**161**:293–306.

64. Papapetrou EP, Zoumbos NC, Athanassiadou A. Genetic modification of hematopoietic stem cells with nonviral systems: past progress and future prospects. *Gene Ther* 2005;**12**(Suppl. 1):S118–30.

65. Jackson DA, Juranek S, Lipps HJ. Designing nonviral vectors for efficient gene transfer and long-term gene expression. *Mol Ther* 2006;**14**:613–26.

66. Hackett PB, Largaespada DA, Cooper LJ. A transposon and transposase system for human application. *Mol Ther* 2010;**18**:674–83.

67. Dalsgaard T, Moldt B, Sharma N, Wolf G, Schmitz A, Pedersen FS, et al. Shielding of sleeping beauty DNA transposon-delivered transgene cassettes by heterologous insulators in early embryonal cells. *Mol Ther* 2009;**17**:121–30.

68. Mates L, Chuah MK, Belay E, Jerchow B, Manoj N, Acosta-Sanchez A, et al. Molecular evolution of a novel hyperactive sleeping beauty transposase enables robust stable gene transfer in vertebrates. *Nat Genet* 2009;**41**:753–61.

69. Xue X, Huang X, Nodland SE, Mates L, Ma L, Izsvak Z, et al. Stable gene transfer and expression in cord blood-derived CD34+ hematopoietic stem and progenitor cells by a hyperactive sleeping beauty transposon system. *Blood* 2009;**114**:1319–30.

70. Sjeklocha LM, Park CW, Wong PY, Roney MJ, Belcher JD, Kaufman DS, et al. Erythroid-specific expression of beta-globin from sleeping beauty-transduced human hematopoietic progenitor cells. *PLoS One* 2011;**6**:e29110.

71. Forget BG. Progress in understanding the hemoglobin switch. *N Engl J Med* 2011;**365**:852–4.

72. Sankaran VG, Xu J, Byron R, Greisman HA, Fisher C, Weatherall DJ, et al. A functional element necessary for fetal hemoglobin silencing. *N Engl J Med* 2011;**365**:807–14.

73. Thein SL, Menzel S. Discovering the genetics underlying foetal haemoglobin production in adults. *Br J Haematol* 2009;**145**:455–67.

74. Galanello R. Recent advances in the molecular understanding of non-transfusion-dependent thalassemia. *Blood Rev* 2012;**26**(Suppl. 1):S7–11.

75. Xu XS, Hong X, Wang G. Induction of endogenous gamma-globin gene expression with decoy oligonucleotide targeting Oct-1 transcription factor consensus sequence. *J Hematol Oncol* 2009;**2**:15.

76. Graslund T, Li X, Magnenat L, Popkov M, Barbas 3rd CF. Exploring strategies for the design of artificial transcription factors: targeting sites proximal to known regulatory regions for the induction of gamma-globin expression and the treatment of sickle cell disease. *J Biol Chem* 2005;**280**:3707–14.

77. Wilber A, Tschulena U, Hargrove PW, Kim YS, Persons DA, Barbas 3rd CF, et al. A zinc-finger transcriptional activator designed to interact with the gamma-globin gene promoters enhances fetal hemoglobin production in primary human adult erythroblasts. *Blood* 2010;**115**:3033–41.

78. Xu J, Peng C, Sankaran VG, Shao Z, Esrick EB, Chong BG, et al. Correction of sickle cell disease in adult mice by interference with fetal hemoglobin silencing. *Science* 2011;**334**:993–6.

79. Bauer DE, Kamran SC, Lessard S, Xu J, Fujiwara Y, Lin C, et al. An erythroid enhancer of BCL11A subject to genetic variation determines fetal hemoglobin level. *Science* 2013;**342**:253–7.

80. Deng W, Lee J, Wang H, Miller J, Reik A, Gregory PD, et al. Controlling long-range genomic interactions at a native locus by targeted tethering of a looping factor. *Cell* 2012;**149**:1233–44.

81. Deng W, Rupon JW, Krivega I, Breda L, Motta I, Jahn KS, et al. Reactivation of developmentally silenced globin genes by forced chromatin looping. *Cell* 2014;**158**:849–60.

82. Rupon JW, Deng W, Wang H, Gregory PD, Reik AAD, et al. Using forced chromatin looping to overcome developmental silencing of embryonic and fetal β-type globin genes in adult erythroid cells. *Blood* 2013;**122**:433.

83. Cavazzana-Calvo M, Payen E, Negre O, Wang G, Hehir K, Fusil F, et al. Transfusion independence and HMGA2 activation after gene therapy of human beta-thalassaemia. *Nature* 2010;**467**:318–22.

84. Ikeda K, Mason PJ, Bessler M. 3'UTR-truncated Hmga2 cDNA causes MPN-like hematopoiesis by conferring a clonal growth advantage at the level of HSC in mice. *Blood* 2011;**117**:5860–9.

85. Young AR, Narita M. Oncogenic HMGA2: short or small? *Genes Dev* 2007;**21**:1005–9.

86. Sadelain M, Riviere I, Wang X, Boulad F, Prockop S, Giardina P, et al. Strategy for a multicenter phase I clinical trial to evaluate globin gene transfer in beta-thalassemia. *Ann N Y Acad Sci* 2010;**1202**:52–8.

87. Boulad F, Wang X, Qu J, Taylor C, Ferro L, Karponi G, et al. Safe mobilization of CD34⁺ cells in adults with beta-thalassemia and validation of effective globin gene transfer for clinical investigation. *Blood* 2014;**123**:1483–6.

88. Bluebird bio I. *Bluebird bio reports rapid transfusion independence in beta-thalassemia major patients treated with its LentiGlobin product candidate*; 2014.

89. Li K, Wong A, Li CK, Shing MM, Chik KW, Tsang KS, et al. Granulocyte colony-stimulating factor-mobilized peripheral blood stem cells in beta-thalassemia patients: kinetics of mobilization and composition of apheresis product. *Exp Hematol* 1999;**27**:526–32.

90. Yannaki E, Papayannopoulou T, Jonlin E, Zervou F, Karponi G, Xagorari A, et al. Hematopoietic stem cell mobilization for gene therapy of adult patients with severe beta-thalassemia: results of clinical trials using G-CSF or plerixafor in splenectomized and nonsplenectomized subjects. *Mol Ther* 2012;**20**:230–8.

91. Fitzhugh CD, Hsieh MM, Bolan CD, Saenz C, Tisdale JF. Granulocyte colony-stimulating factor (G-CSF) administration in individuals with sickle cell disease: time for a moratorium? *Cytotherapy* 2009;**11**:464–71.

92. Angelucci E. Hematopoietic stem cell transplantation in thalassemia. *Hematol Am Soc Hematol Educ Program* 2010;**2010**:456–62.

93. Lucarelli G, Galimberti M, Polchi P, Angelucci E, Baronciani D, Giardini C, et al. Bone marrow transplantation in patients with thalassemia. *N Engl J Med* 1990;**322**:417–21.

94. Lucarelli G, Andreani M, Angelucci E. The cure of thalassemia by bone marrow transplantation. *Blood Rev* 2002;**16**:81–5.

95. Bertaina A, Bernardo ME, Mastronuzzi A, La Nasa G, Locatelli F. The role of reduced intensity preparative regimens in patients with thalassemia given hematopoietic transplantation. *Ann N Y Acad Sci* 2010;**1202**:141–8.

96. Bernardo ME, Piras E, Vacca A, Giorgiani G, Zecca M, Bertaina A, et al. Allogeneic hematopoietic stem cell transplantation in thalassemia major: results of a reduced-toxicity conditioning regimen based on the use of treosulfan. *Blood* 2012;**120**:473–6.

97. Hsieh MM, Fitzhugh CD, Weitzel RP, Link ME, Coles WA, Zhao Xiongce, et al. Nonmyeloablative HLA-matched sibling allogeneic hematopoietic stem cell transplantation for severe sickle cell phenotype. *J Am Med Assoc* 2014;**312**:48–56.

98. Wang GP, Berry CC, Malani N, Leboulch P, Fischer A, Hacein-Bey-Abina S, et al. Dynamics of gene-modified progenitor cells analyzed by tracking retroviral integration sites in a human SCID-X1 gene therapy trial. *Blood* 2010;**115**:4356–66.

99. Sadelain M. Eliminating cells gone astray. *N Engl J Med* 2011;**365**:1735–7.

100. Marin V, Cribioli E, Philip B, Tettamanti S, Pizzitola I, Biondi A, et al. Comparison of different suicide-gene strategies for the safety improvement of genetically manipulated T cells. *Hum Gene Therapy Methods* 2012;**23**:376–86.

101. Arumugam PI, Higashimoto T, Urbinati F, Modlich U, Nestheide S, Xia P, et al. Genotoxic potential of lineage-specific lentivirus vectors carrying the beta-globin locus control region. *Mol Ther* 2009;**17**:1929–37.

102. Rivella S, Callegari JA, May C, Tan CW, Sadelain M. The cHS4 insulator increases the probability of retroviral expression at random chromosomal integration sites. *J Virol* 2000;**74**:4679–87.

103. Steinberg MH, Forget BG, Higgs DR, Nagel RL. *Disorders of hemoglobin: genetics, pathophysiology and clinical management*. Cambridge (UK): Cambridge University Press; 2001.

104. Sankaran VG, Lettre G, Orkin SH, Hirschhorn JN. Modifier genes in Mendelian disorders: the example of hemoglobin disorders. *Ann N Y Acad Sci* 2010;**1214**:47–56.

105. Weatherall DJ. Phenotype-genotype relationships in monogenic disease: lessons from the thalassaemias. *Nat Rev Genet* 2001;**2**:245–55.

106. Lim SK, Sigmund CD, Gross KW, Maquat LE. Nonsense codons in human beta-globin mRNA result in the production of mRNA degradation products. *Mol Cell Biol* 1992;**12**:1149–61.

107. Maquat LE, Kinniburgh AJ, Rachmilewitz EA, Ross J. Unstable beta-globin mRNA in mRNA-deficient beta o thalassemia. *Cell* 1981;**27**:543–53.

108. Gardner LB. Hypoxic inhibition of nonsense-mediated RNA decay regulates gene expression and the integrated stress response. *Mol Cell Biol* 2008;**28**:3729–41.

109. Weatherall DJ. The definition and epidemiology of non-transfusion-dependent thalassemia. *Blood Rev* 2012;**26**(Suppl. 1):S3–6.

110. Weatherall DJ, Williams TN, Allen SJ, O'Donnell A. The population genetics and dynamics of the thalassemias. *Hematol/oncol Clin N Am* 2010;**24**:1021–31.

111. Hall GW, Thein S. Nonsense codon mutations in the terminal exon of the beta-globin gene are not associated with a reduction in beta-mRNA accumulation: a mechanism for the phenotype of dominant beta-thalassemia. *Blood* 1994;**83**:2031–7.

112. Thein SL, Hesketh C, Taylor P, Temperley IJ, Hutchinson RM, Old JM, et al. Molecular basis for dominantly inherited inclusion body beta-thalassemia. *Proc Natl Acad Sci USA* 1990;**87**:3924–8.

113. Steinberg MH, Forget BG, Higgs DR, Nagel RL. *Molecular mechanism of β thalassemia; Bernard G. Forget*. Cambridge (UK): Cambridge University Press; 2001.

114. Gallagher PG, Steiner LA, Liem RI, Owen AN, Cline AP, Seidel NE, et al. Mutation of a barrier insulator in the human ankyrin-1 gene is associated with hereditary spherocytosis. *J Clin Invest* 2010;**120**:4453–65.

115. Khandros E, Thom CS, D'Souza J, Weiss MJ. Integrated protein quality-control pathways regulate free alpha-globin in murine beta-thalassemia. *Blood* 2012;**119**:5265–75.

116. Rivella S, Sadelain M. Genetic treatment of severe hemoglobinopathies: the combat against transgene variegation and transgene silencing. *Semin Hematol* 1998;**35**:112–25.

117. Nienhuis AW, Dunbar CE, Sorrentino BP. Genotoxicity of retroviral integration in hematopoietic cells. *Mol Ther* 2006;**13**:1031–49.

118. Bohne J, Cathomen T. Genotoxicity in gene therapy: an account of vector integration and designer nucleases. *Curr Opin Mol Ther* 2008;**10**:214–23.

119. Hacein-Bey-Abina S, Von Kalle C, Schmidt M, McCormack MP, Wulffraat N, Leboulch P, et al. LMO2-associated clonal T cell proliferation in two patients after gene therapy for SCID-X1. *Science* 2003;**302**:415–9.

120. Cornu TI, Cathomen T. Targeted genome modifications using integrase-deficient lentiviral vectors. *Mol Ther* 2007;**15**:2107–13.

121. Nightingale SJ, Hollis RP, Pepper KA, Petersen D, Yu XJ, Yang C, et al. Transient gene expression by nonintegrating lentiviral vectors. *Mol Ther* 2006;**13**:1121–32.

122. Lombardo A, Genovese P, Beausejour CM, Colleoni S, Lee YL, Kim KA, et al. Gene editing in human stem cells using zinc finger nucleases and integrase-defective lentiviral vector delivery. *Nat Biotechnol* 2007;**25**:1298–306.

123. Lombardo A, Cesana D, Genovese P, Di Stefano B, Provasi E, Colombo DF, et al. Site-specific integration and tailoring of cassette design for sustainable gene transfer. *Nat Methods* 2011;**8**:861–9.

124. Provasi E, Genovese P, Lombardo A, Magnani Z, Liu PQ, Reik A, et al. Editing T cell specificity towards leukemia by zinc finger nucleases and lentiviral gene transfer. *Nat Med* 2012;**18**:807–15.

125. Takahashi K, Yamanaka S. Induction of pluripotent stem cells from mouse embryonic and adult fibroblast cultures by defined factors. *Cell* 2006;**126**:663–76.

126. Okita K, Ichisaka T, Yamanaka S. Generation of germline-competent induced pluripotent stem cells. *Nature* 2007;**448**:313–7.

127. Wernig M, Meissner A, Foreman R, Brambrink T, Ku M, Hochedlinger K, et al. In vitro reprogramming of fibroblasts into a pluripotent ES-cell-like state. *Nature* 2007;**448**:318–24.

128. Sommer CA, Stadtfeld M, Murphy GJ, Hochedlinger K, Kotton DN, Mostoslavsky G. Induced pluripotent stem cell generation using a single lentiviral stem cell cassette. *Stem Cells* 2009;**27**:543–9.

129. Staerk J, Dawlaty MM, Gao Q, Maetzel D, Hanna J, Sommer CA, et al. Reprogramming of human peripheral blood cells to induced pluripotent stem cells. *Cell Stem Cell* 2010;**7**:20–4.

130. Chou BK, Mali P, Huang X, Ye Z, Dowey SN, Resar LM, et al. Efficient human iPS cell derivation by a non-integrating plasmid from blood cells with unique epigenetic and gene expression signatures. *Cell Res* 2011;**21**:518–29.

131. Okita K, Matsumura Y, Sato Y, Okada A, Morizane A, Okamoto S, et al. A more efficient method to generate integration-free human iPS cells. *Nat Methods* 2011;**8**:409–12.

132. Yamanaka S. Induced pluripotent stem cells: past, present, and future. *Cell Stem Cell* 2012;**10**:678–84.

133. Papapetrou EP, Lee G, Malani N, Setty M, Riviere I, Tirunagari LM, et al. Genomic safe harbors permit high beta-globin transgene expression in thalassemia induced pluripotent stem cells. *Nat Biotechnol* 2011;**29**:73–8.

134. Chang JC, Ye L, Kan YW. Correction of the sickle cell mutation in embryonic stem cells. *Proc Natl Acad Sci USA* 2006;**103**:1036–40.

135. Zou J, Mali P, Huang X, Dowey SN, Cheng L. Site-specific gene correction of a point mutation in human iPS cells derived from an adult patient with sickle cell disease. *Blood* 2011;**118**:4599–608.

136. Sebastiano V, Maeder ML, Angstman JF, Haddad B, Khayter C, Yeo DT, et al. In situ genetic correction of the sickle cell anemia mutation in human induced pluripotent stem cells using engineered zinc finger nucleases. *Stem Cells* 2011;**29**:1717–26.

137. Howden SE, Voullaire L, Wardan H, Williamson R, Vadolas J. Site-specific, rep-mediated integration of the intact beta-globin locus in the human erythroleukaemic cell line K562. *Gene Ther* 2008;**15**:1372–83.

Chapter 13

Hemophilia Gene Therapy

Christopher E. Walsh
Icahn School of Medicine at Mount Sinai, New York City, NY, USA

LIST OF ABBREVIATIONS

AAV Adeno-associated viral vector
FVIII Factor VIII
Hu FIX Human factor IX
ITR Inverted terminal repeats
MPS Mucopolysaccharidosis
rAAV Recombinant adeno-associated viral vector

13.1 INTRODUCTION

Genetic therapy for bleeding disorders has been achieved. Current and future work will focus on improving outcomes (coagulation factor levels) while limiting toxicities (viral mediated immune responses). The application of vector and other genetic technologies for widespread use is at hand via pharmaceutical corporations. This now signifies that gene transfer has matured as a viable therapeutic option for hemophilia patients.

13.2 HEMOPHILIA B GENE TRANSFER

Treatment of hemophilia B via gene therapy has always been considered the "low hanging fruit" of the inherited bleeding disorders. Patients with factor IX (FIX) deficiency manifest the disease with spontaneous joint and soft tissue bleeding. Severely affected patients (<1% FIX activity) may bleed on the average of 1–2 times a week if not treated, while mild patients (>5%) rarely bleed spontaneously. Prophylactic infusions of FIX are required 1–2 times a weeks to prevent bleeding. Converting a "severe" (<1%) to a "mild" phenotype (>5% activity) patient would effectively prevent spontaneous bleeding. Higher factor levels (approaching 50%) would effectively make the bleeding risk the same as that for a normal individual. Human factor IX (Hu FIX) cDNA is small (<1 kb) and requires no regulatory elements to guide its production. The only requirements are proper post-translational protein glycosylation and carboxylation that can be performed by most cell types. Several viral vectors have been tested with success in in vitro and animal models but were not effective in clinical trials. The recent demonstration of measurable FIX activity in six FIX patients has been described and is the first successful gene therapy trial in hemophilia patients.[1] This represents the culmination of decades of research as we described elsewhere.[2] Given the success of the recombinant adeno-associated viral vector (rAAV) in this successful trial, this review will focus on improvements made in rAAV production, capsid tropism, and packaging. Based on the success of this vector, several commercial groups are now supporting new hemophilia B gene therapy trials using AAV.

At the same time these gene therapy studies were published, a protein-based approach for treatment of hemophilia received a significant boost with the development of several long-acting FIX (LA-FIX). Current recombinant FIX has a half-life of ~18–24 h; new long-acting FIX has a half-life of 80–90 h. Alprolix (Biogen-Idec), the first in its class, was recently approved in Europe, Canada, and the United States for prophylaxis in FIX patients. The approvals were based on data supplied in a recent publication[3] where linking Hu FIX to a dimeric Fc fusion protein led to a fivefold extension of the FIX half-life. This new product now requires an intravenous infusion of factor once every 1 or 2 weeks rather than 1–2 times a week (as used currently) all while maintaining trough factor levels at 10–30%. In contrast, the successful gene therapy approach requires a single intravenous infusion of virus while maintaining blood levels of Hu FIX at 2–6%. The reduction of intravenous infusions using a gene transfer approach is a milestone. Yet despite great interest and education about gene therapy, patients are slow to enter gene therapy clinical trials. This reticence probably reflects an understandable caution about a new technology and concerns for short- and long-term side effects. It may also reflect the fact that patients may be willing to trade convenience (1 dose per year vs 52 doses per year) for higher factor levels. Thus, conventional protein therapy via new extended half-life Hu FIXs has raised the bar for FIX gene therapy to achieve and surpass.

13.2.1 Adeno-Associated Viral Vector

AAV as a viral vector has been reviewed many times;[4] however, the salient features of the virus are those of safety and simplicity. Wild-type AAV has not been associated with any significant human disease (although one could argue if in fact this has been studied thoroughly). At the core, this feature of safety is the driving force for the use of this vector. To date, there have been few studies demonstrating a pathological effect in humans related to a wild-type AAV infection; yet, ~30–40% of individuals have neutralizing AAV antibodies.[5,6] Development of liver tumors in a mucopolysaccharidosis (MPS) mouse model has been directly attributed to rAAV vector integration found within the tumor.[7,8] The use of AAV is also technically simple. The virus is so sparingly constructed that the wild-type virus requires another virus (adenovirus, herpesvirus) to supply the necessary genetic elements for its own replication. The vast majority of viral genomes do not integrate into host cell DNA.[8] Finally, the wild-type virus, when it does integrate, can preferentially integrate into specific areas of the genome termed AAVS1.[5] The AAV genome encodes ~4700 nucleotides consisting of one open reading frame that encodes for replication (Rep) and capsid (Cap) proteins. The virus possesses two palindromic ends or inverted terminal repeats; at least one is required for viral replication. Modification of the endonuclease binding site that allows for second-strand synthesis required for replication has been skillfully used to generate a double-stranded form that can be packaged by the virus at roughly half the wild-type size.[9,10] This double-stranded or self-complementary (scAAV) form when used as a vector should be more efficient for transgene expression as it bypasses the need for conversion of a single-to-double stranded form.[11]

Viral tropism is a well-established phenomenon based largely on the capacity of the virus to select particular cells for infection and replication. This is in part mediated via virus-specific receptors located on the cell membrane. AAV tropism has been reviewed[12–15] where different serotypes (capsids) can exhibit vast differences in the cell type infected. This can be achieved with small changes in the capsid structure.

Serotype infectivity in tissue culture cells may differ dramatically when used in vivo. Tropism does also differ in different animals (mouse, canine, primate) making preclinical assessment of vectors extremely difficult when deciding what serotype to use for human clinical trials. This simple virus can be easily modified in terms of its transgene (codon optimization, introduction of mutations/polymorphisms, scAAV transgene), transgene expression cassette (promoter/enhancer and poly(A) motifs), and most importantly, the capsid components that determine tropism; all these changes have been incorporated to successfully express a clinically relevant Hu FIX protein in a human being (Figure 13.1).

A major challenge that still faces this vector is its extremely poor infectivity; the number of vector particles (genomes) required to infect a cell is on the order of 10^{3-4} genomes/cell. This balance between rAAV low infectivity, which requires relatively massive numbers of genomes (~10^{12}/kg body weight), and induction of virus inflammatory and immune responses may have led to early unsuccessful gene therapy trials.[16,17]

13.2.2 AAV and Hemophilia B

The first successful rAAV Hu FIX trial incorporated changes such as the use of AAV8 serotype, codon optimization, and the use of an scAAV genome. Despite the vector changes, a dose of >10^{12} vg/kg still induced an immune response. Although investigators can mitigate the immune response with immunosuppressive agents, this may not be well received by a patient group who were savaged by two deadly viruses (hepatitis C and HIV).

Now that the field has matured to a point where the time and effort of a gene therapy trial may bear fruit, new challenges arise in terms of improving the vector infectivity efficiency as well as manufacturing and validation of recombinant virus for large scale clinical use.

13.2.3 Current AAV FIX Trials

At least three AAV trials are ongoing in the United States. All are FDA-approved Phase I/II trials performed using vector dose escalation. The ongoing trial performed at St Jude's Research Hospital and London University Hospital was successful.[1] Investigators used an AAV serotype 8 vector that contained a codon-optimized human FIX cDNA and an scAAV backbone. Six patients were initially reported receiving the vector in a dose-escalating fashion. At low dose (2×10^{10} vg/kg) circulating FIX was detected at 1–2% levels (studies were performed in patients with <1% FIX levels). However, at high vector dose (2×10^{12} vg/kg) ~10–12% FIX was measured before immune-mediated hepatitis developed 2 months after vector infusion. A coincident drop in

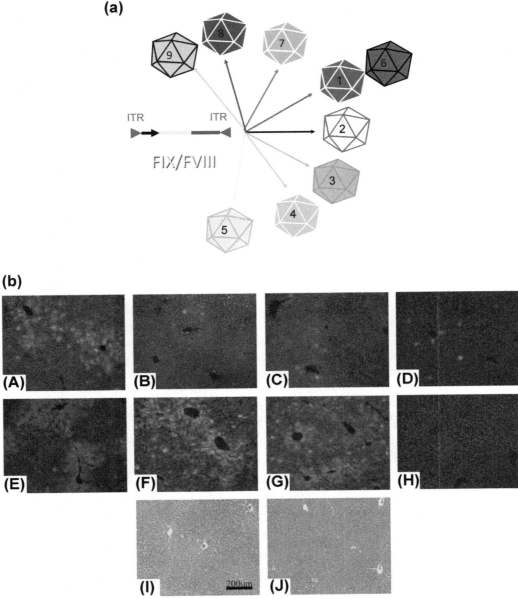

FIGURE 13.1 **Improvement in AAV vector technology.** Use and generation of novel serotypes (a) facilitate enhanced gene transfer (b) where staining for Hu FIX transgene expression is measured in mouse livers harvested months after intravenous injection (unpublished data). FIX expression in liver after infusion of AAV-hFIX serotype vectors, panel A: AAV 1-hFIX; panel B: AAV2-hFIX; panel C: AAV3-hFIX; panel D: AAV4-hFIX; panel E: AAV5-hFIX; panel F: AAV7-hFIX; panel G: AAV8-hFIX; panel H: control mouse liver; panel I: H/E staining of AAV8-hFIX liver; panel J: H/E staining of control mouse liver. AAV, adeno-associated viral vector; ITR, inverted terminal repeats.

FIX occurred with rising transaminase levels and was blunted by administration of corticosteroids. Patients maintained FIX levels of ~6% and have maintained these levels now 2–3 years later. Additional patients have received vector and this protocol continues.

Another trial initiated by the Children's Hospital of Philadelphia uses a conventional double-strand AAV8 vector and a robust promoter/enhancer element (www.clinicaltrials.gov, NCT 01620801). Patients have enrolled receiving a low dose vector (no data available).

The third trial builds on the success of the first trial (www.clinicaltrials.gov, NCT 01687608). The vector is nearly identical to the St Jude's vector (FIX codon-optimized, etc.), except that the human FIX cDNA has a gain-of-function mutation at amino acid position 338 that resides in the active site of the protein. The Padua mutation[18] was described in a child with FIX levels at 800–900% and diagnosed with thromboembolic disease. The mother was phenotypically normal with FIX of

200–300%. This mutation had previously been identified by mutational assessment of Hu FIX and when compared to native FIX has five- to eightfold greater activity.[19–21] This trial has enrolled patients; however, no data are available.

13.3 AAV AND HEMOPHILIA A

Successful gene therapy will likely revolutionize treatment for hemophilia A or factor VIII (FVIII) deficiency. Patients deficient in FVIII constitute 80% of all hemophilia patients.[22] The requirements for successful gene transfer include persistent expression of therapeutic levels of FVIII, lack of significant toxicity to the gene transfer vehicle (vector), and lack of host immune response to the normal factor VIII/IX and vector proteins. Gene therapy for FVIII poses several problems because of the large size of the FVIII gene.

13.3.1 Hemophilia A (Factor VIII Deficiency)

Similar to factor IX deficiency, factor VIII deficiency is a sex-linked recessive disorder manifested as frequent spontaneous intraarticular joint and soft tissue bleeding episodes.[23] The severity of the disease is directly related to the amount of functional circulating plasma levels of factor VIII. Patients with less than 1% of normal FVIII activity levels have phenotypically severe disease with frequent bleeding episodes that require treatment with either plasma-derived or recombinant factor products. Patients with mild disease maintain 5–30% of normal factor levels and typically have few spontaneous bleeding episodes, but are still at risk for trauma-induced bleeding. The majority of hemophilia A patients have <1% of FVIII and bleed spontaneously unless they receive treatment. Chronic intraarticular bleeding leads to arthropathy and eventual joint loss. Studies demonstrate that prophylactic factor infusion regimens have a dramatic effect on reducing the rate and severity of joint bleeding,[24–26] but prophylactic treatment is extremely expensive ($100–300,000 per year in the United States), requiring multiweek intravenous infusions. Currently, treatment with plasma-purified or recombinant protein at the time of bleeding is the standard of care. Thus, the goal for FVIII gene transfer is sustained long-term factor production at levels of >5%. This level would effectively convert severely affected patients to a milder phenotype. Treatment strategies that can produce constant levels approaching 50% would be considered curative. Extended half-life Hu FVIII is being tested; however, despite the different approaches used, all seem to be limited to 1.5× half-life extension. Unlike the Hu FIX extended products, these products still require one to two intravenous infusions per week.

13.3.2 Hemophilia A Gene Transfer Using AAV

FIX cDNA is ~0.9 kb in length and a perfect candidate gene to test in human subjects. However, the full-length FVIII cDNA (full-length FVIII is 9 kb) cannot fit within AAV. Experiments by Dong et al. demonstrated that the packaging of a cDNA in AAV was a function of size.[27] Efficient packaging of rAAV occurs when the transgene is close to 4.7 kb, the length of the wild-type AAV genome. Although larger transgenes can be packaged at >105% of wild-type size, the titer of rAAV is reported to fall by several logs. The FVIII B domain is encoded by a single ~3 kb exon and was found dispensable for FVIII procoagulation activity.[28] Chao et al.[29] demonstrated that by using a small viral promoter element, a B-domain-deleted FVIII (BDD FVIII, 4.4 kb) could be packaged in AAV. FVIII expression was modest. We later used small promoter cassettes that yield ~40–80% activity in FVIII KO mice (unpublished). Sarkar et al.[30,31] demonstrated long-term murine/canine FVIII gene expression in hemophilic mice and dogs, respectively, using transgene cassettes ~5.2–5.5 kb. Although there is a discrete packaging limitation of AAV with optimum packing near wild-type transgene size, larger than wild-type size vectors with deleted portions of the transgene can recombine yielding a full-length transgene.[32] Other groups have extended the size of BDD FVIII constructs using larger promoter/enhancer cassettes with reasonable FVIII levels; however, the viral dose used to obtain these results would preclude their use in human subjects (Table 13.1). Table 13.1 summarizes the results of several groups using AAV for FVIII gene transfer. Two approaches for rAAV FVIII gene transfer use either a single vector or two vectors, each carrying the heavy or light FVIII chain. To date, the amount of vector used in animal models, hemophilic mice and canines, would greatly exceed the amount of vector employed in three human subjects in each of which an immune response to the vector was observed.[1,17] It is important to note that canine FVIII cDNA has four- to eightfold higher expression compared with human FVIII; the reason is unknown.

13.3.3 Gene Repair/Gene Editing

Given the size limitation of the FVIII cDNA, several groups have attempted a gene correction approach rather than a gene insertion approach. We demonstrated the use of gene correction via RNA repair.[40,41] Alternatively, small base changes can

TABLE 13.1 AAV Vectors for Use in Hemophilia A (FVIII Deficiency)

AAV Vector	Dose	Animal Species Tested	FVIII Activity Level (%)	References
AAV2 canine FVIII	10^{10-11} vg/animal	NOD-scid mice	27	33
Dimerized AAV2 human BDD FVIII	10^{11} vg/animal	FVIII KO mice	2	34
AAV2 canine FVIII single vector	6×10^{12} vg/animal	FVIII KO dog	2.8	35
AAV2 canine FVIII, dual heavy and light chain	1×10^{10} vg/animal 3×10^{11} vg/animal 9×10^{11} vg/animal	FVIII KO mice	5 200 400	35
AAV8 canine FVIII	10^{11} vg/animal	FVIII KO mice	100	30
AAV8 canine FVIII AAV9 canine FVIII	1×10^{13} vg/kg 6×10^{12} vg/kg	FVIII KO dogs FVIII KO dogs	4.5 2.5	31
AAV8 dual vector; human BDD heavy and light chain	5×10^{11} vg/animal	FVIII KO mice	80	36
AAV8 canine FVIII + bortezomide	3×10^{10} vg/animal 10^{13} vg/kg	FVIII KO mice FVIII KO dogs	25–150+bor 0.5–3.8+bor	37
AAV8 canine FVIII single chain AAV8 canine FVIII heavy and light chain AAV9 canine FVIII heavy and light chain	4×10^{13} vg/kg 2×10^{13} vg/kg 1×10^{13} vg/vg 6×10^{12} vg/vg 1×10^{13} vg/vg 6×10^{12} vg/vg	FVIII KO dogs FVIII KO dogs FVIII KO dogs FVIII KO dogs FVIII KO dogs FVIII KO dogs	3–9 1–3 8 2 3–15 2	38
AAV8 huFVIII BDD (opt)	2×10^{12} vg/kg	FVIII KO mice Nonhuman primate	730 15	39

be affected by the use of zinc-finger proteins (ZFN),[42] TALENS,[40] or CRISPR[43,44] elements, all designed to cleave specific DNA sequences that are subsequently repaired with wild-type sequence.[45]

13.4 rAAV DOSE AND THE IMMUNE RESPONSE

It is a given that a human host given sufficient viral genomes (functional viral particles) will infect cells and produce a transgene product (in this case human FIX), which prevents spontaneous joint bleeding. It is a given that humans tend to develop clinical immune responses when the vg delivered intravenously is on the order of 10^{12}/kg (in an average 100 kg human that is equivalent to 10^{14} vg/5 L or 2×10^{10} vg/mL blood). In acute viral infections (hepatitis, HIV, etc.) patients with a high viral load would require a viral dose on the order of 10^{5-6} virions/mL. Thus, we are introducing artificially 10^{4-5} fold more virions in the rAAV gene therapy trials when compared to wild-type infectious viruses. It is no surprise that an immune response is elicited; more amazing is the fact that the response is relatively circumscribed and patients have no systemic effects. This all stems from the relative inefficiency of rAAV gene transduction. This again reinforces the original notion that AAV is intrinsically safer than other viruses, but that it is terribly inefficient at producing the desired protein product.

Experimental options abound with studies to block or blunt the development of rAAV neutralizing antibodies[46] to the transgene factor as well as block the immune response to the vector capsid proteins.[47] Manipulation of CpG motifs in the transgene cassette may also reduce immune responses to the infected cells.[48] Vector production and manufacturing will undoubtedly improve to ensure that those homogeneous functional recombinant virions are only produced. Most research lab-based AAV preparations are rarely tested for consistency/reliability and uniformity. That is no longer acceptable. The ratio of intact virions to nonfunctional or "empty capsids" may vary significantly.[12] Column purification with specific antibodies to purify functional particles is now required to further enrich for patient-quality vector. The removal of significant defective particles thus enriches in vivo infectivity while reducing the vector dose and potential for immune activation. A

10-fold enrichment using a vector with a 10-fold increase in activity should produce significant FIX levels. Conversely, the addition of empty virions may bind neutralizing rAAV antibodies, thus facilitating AAV infectivity.[47]

13.5 AAV-MEDIATED TRANSFER LASTS A LONG TIME

The majority of AAV when analyzed in animal models remains largely extrachromosomal.[13] The reasons for this are not well understood. This has been touted as another inherent safety feature—i.e., unlike retroviral vectors, nonintegrated genomes cannot cause tumors. However, a small percentage of rAAV vector DNA is incorporated within cellular genomes.[8] Vector genomes tend to associate at active areas in the chromatin near the promoter regions of coding genes.[49,50] To date, a paper has cited the relationship of tumor development in livers of MPS mice after delivery of rAAV virus.[7] Genomic analysis of patients who received rAAV vectors found DNA at muscle injection sites nearly a decade later.[51] Our own experience confirmed long-lasting FIX gene expression in hemophilic dogs (>8 years) (unpublished).

13.6 SUMMARY

Genetic correction of the hemophilias has been achieved in a few patients. Improvements in vector technology, vector production, and an understanding of host immune responses have facilitated positive outcomes in patients. This in turn has spurred interest in hemophilia gene therapy by investigators and the private sector. In addition to transfer of cDNAs, new methods for gene editing and repair may provide alternative methods for gene correction.

REFERENCES

1. Nathwani AC, Tuddenham EG, Rangarajan S, et al. Adenovirus-associated virus vector-mediated gene transfer in hemophilia B. *N Engl J Med* 2011;**365**:2357–65.
2. Walsh CE, Batt KM. Hemophilia clinical gene therapy: brief review. *Transl Res* 2013;**161**:307–12.
3. Powell JS, Pasi KJ, Ragni MV, et al. Phase 3 study of recombinant factor IX Fc fusion protein in hemophilia B. *N Engl J Med* 2013;**369**:2313–23.
4. Chuah MK, Evens H, VandenDriessche T. Gene therapy for hemophilia. *J Thromb Haemost* 2013;**11**(Suppl. 1):99–110.
5. Ward P, Walsh CE. Targeted integration of a rAAV vector into the AAVS1 region. *Virology* 2012;**433**:356–66.
6. Mimuro J, Mizukami H, Shima M, et al. The prevalence of neutralizing antibodies against adeno-associated virus capsids is reduced in young Japanese individuals. *J Med Virol* 2014;**86**(11):1990–97.
7. Donsante A, Miller DG, Li Y, et al. AAV vector integration sites in mouse hepatocellular carcinoma. *Science* 2007;**317**:477.
8. Russell DW. AAV vectors, insertional mutagenesis, and cancer. *Mol Ther* 2007;**15**:1740–3.
9. McCarty DM. Self-complementary AAV vectors; advances and applications. *Mol Ther* 2008;**16**:1648–56.
10. McCarty DM, Monahan PE, Samulski RJ. Self-complementary recombinant adeno-associated virus (scAAV) vectors promote efficient transduction independently of DNA synthesis. *Gene Ther* 2001;**8**:1248–54.
11. Ferrari FK, Samulski T, Shenk T, Samulski RJ. Second-strand synthesis is a rate-limiting step for efficient transduction by recombinant adeno-associated virus vectors. *J Virol* 1996;**70**:3227–34.
12. Asokan A, Schaffer DV, Samulski RJ. The AAV vector toolkit: poised at the clinical crossroads. *Mol Ther* 2012;**20**:699–708.
13. Grieger JC, Samulski RJ. Adeno-associated virus vectorology, manufacturing, and clinical applications. *Methods Enzymol* 2012;**507**:229–54.
14. Michelfelder S, Trepel M. Adeno-associated viral vectors and their redirection to cell-type specific receptors. *Adv Genet* 2009;**67**:29–60.
15. Van Vliet KM, Blouin V, Brument N, Agbandje-McKenna M, Snyder RO. The role of the adeno-associated virus capsid in gene transfer. *Methods Mol Biol* 2008;**437**:51–91.
16. Manno CS, Pierce GF, Arruda VR, et al. Successful transduction of liver in hemophilia by AAV-Factor IX and limitations imposed by the host immune response. *Nat Med* 2006;**12**:342–7.
17. Kay MA, Manno CS, Ragni MV, et al. Evidence for gene transfer and expression of factor IX in haemophilia B patients treated with an AAV vector. *Nat Genet* 2000;**24**:257–61.
18. Simioni P, Tormene D, Tognin G, et al. X-linked thrombophilia with a mutant factor IX (factor IX Padua). *N Engl J Med* 2009;**361**:1671–5.
19. Finn JD, Nichols TC, Svoronos N, et al. The efficacy and the risk of immunogenicity of FIX Padua (R338L) in hemophilia B dogs treated by AAV muscle gene therapy. *Blood* 2012;**120**:4521–3.
20. Kao CY, Yang SJ, Tao MH, Jeng YM, Yu IS, Lin SW. Incorporation of the factor IX Padua mutation into FIX-Triple improves clotting activity in vitro and in vivo. *Thromb Haemost* 2013;**110**:244–56.
21. Chang J, Jin J, Lollar P, et al. Changing residue 338 in human factor IX from arginine to alanine causes an increase in catalytic activity. *J Biol Chem* 1998;**273**:12089–94.
22. Hoyer LW. Hemophilia A. *N Engl J Med* 1994;**330**:38–47.
23. Kempton CL, Escobar MA, Roberts HR. Hemophilia care in the 21st century. *Clin Adv Hematol Oncol* 2004;**2**:733–40.
24. Carcao M. Changing paradigm of prophylaxis with longer acting factor concentrates. *Haemophilia* 2014;**20**(Suppl. 4):99–105.
25. Nilsson IM, Berntorp E, Lofqvist T, Pettersson H. Twenty-five years' experience of prophylactic treatment in severe haemophilia A and B. *J Intern Med* 1992;**232**:25–32.

26. Manco-Johnson MJ, Abshire TC, Shapiro AD, et al. Prophylaxis versus episodic treatment to prevent joint disease in boys with severe hemophilia. *N Engl J Med* 2007;**357**:535–44.
27. Dong JY, Fan PD, Frizzell RA. Quantitative analysis of the packaging capacity of recombinant adeno-associated virus. *Hum Gene Ther* 1996;**7**:2101–12.
28. Pittman DD, Alderman EM, Tomkinson KN, Wang JH, Giles AR, Kaufman RJ. Biochemical, immunological, and in vivo functional characterization of B-domain-deleted factor VIII. *Blood* 1993;**81**:2925–35.
29. Chao H, Mao L, Bruce AT, Walsh CE. Sustained expression of human factor VIII in mice using a parvovirus-based vector. *Blood* 2000;**95**:1594–9.
30. Sarkar R, Tetreault R, Gao G, et al. Total correction of hemophilia A mice with canine FVIII using an AAV 8 serotype. *Blood* 2004;**103**:1253–60.
31. Sarkar R, Mucci M, Addya S, et al. Long-term efficacy of adeno-associated virus serotypes 8 and 9 in hemophilia a dogs and mice. *Hum Gene Ther* 2006;**17**:427–39.
32. Dong B, Nakai H, Xiao W. Characterization of genome integrity for oversized recombinant AAV vector. *Mol Ther* 2010;**18**:87–92.
33. Chao H, Liu Y, Rabinowitz J, Li C, Samulski RJ, Walsh CE. Several log increase in therapeutic transgene delivery by distinct adeno-associated viral serotype vectors. *Mol Ther* 2000;**2**:619–23.
34. Chao H, Sun L, Bruce A, Xiao X, Walsh CE. Expression of human factor VIII by splicing between dimerized AAV vectors. *Mol Ther* 2002;**5**:716–22.
35. Scallan CD, Liu T, Parker AE, et al. Phenotypic correction of a mouse model of hemophilia A using AAV2 vectors encoding the heavy and light chains of FVIII. *Blood* 2003;**102**:3919–26.
36. Lai Y, Yue Y, Duan D. Evidence for the failure of adeno-associated virus serotype 5 to package a viral genome > or = 8.2 kb. *Mol Ther* 2010;**18**:75–9.
37. Monahan PE, Lothrop CD, Sun J, et al. Proteasome inhibitors enhance gene delivery by AAV virus vectors expressing large genomes in hemophilia mouse and dog models: a strategy for broad clinical application. *Mol Ther* 2010;**18**:1907–16.
38. Sabatino DE, Lange AM, Altynova ES, et al. Efficacy and safety of long-term prophylaxis in severe hemophilia A dogs following liver gene therapy using AAV vectors. *Mol Ther* 2011;**19**:442–9.
39. McIntosh J, Lenting PJ, Rosales C, et al. Therapeutic levels of FVIII following a single peripheral vein administration of rAAV vector encoding a novel human factor VIII variant. *Blood* 2013;**121**:3335–44.
40. Joung JK, Sander JD. TALENs: a widely applicable technology for targeted genome editing. *Nat Rev Mol Cell Biol* 2013;**14**:49–55.
41. Chao H, Mansfield SG, Bartel RC, et al. Phenotype correction of hemophilia A mice by spliceosome-mediated RNA trans-splicing. *Nat Med* 2003;**9**:1015–9.
42. Chou ST, Leng Q, Mixson AJ. Zinc finger nucleases: tailor-made for gene therapy. *Drugs Future* 2012;**37**:183–96.
43. Cai M, Yang Y. Targeted genome editing tools for disease modeling and gene therapy. *Curr Gene Ther* 2014;**14**:2–9.
44. Zhang F, Wen Y, Guo X. CRISPR/Cas9 for genome editing: progress, implications and challenges. *Hum Mol Genet* 2014.
45. Li H, Haurigot V, Doyon Y, et al. In vivo genome editing restores haemostasis in a mouse model of haemophilia. *Nature* 2011;**475**:217–21.
46. Scott DW. Inhibitors – cellular aspects and novel approaches for tolerance. *Haemophilia* 2014;**20**(Suppl. 4):80–6.
47. High KH, Nathwani A, Spencer T, Lillicrap D. Current status of haemophilia gene therapy. *Haemophilia* 2014;**20**(Suppl. 4):43–9.
48. Faust SM, Bell P, Cutler BJ, et al. CpG-depleted adeno-associated virus vectors evade immune detection. *J Clin Invest* 2013;**123**:2994–3001.
49. Nakai H, Montini E, Fuess S, Storm TA, Grompe M, Kay MA. AAV serotype 2 vectors preferentially integrate into active genes in mice. *Nat Genet* 2003;**34**:297–302.
50. Nakai H, Montini E, Fuess S, et al. Helper-independent and AAV-ITR-independent chromosomal integration of double-stranded linear DNA vectors in mice. *Mol Ther* 2003;**7**:101–11.
51. Buchlis G, Podsakoff GM, Radu A, et al. Factor IX expression in skeletal muscle of a severe hemophilia B patient 10 years after AAV-mediated gene transfer. *Blood* 2012;**119**:3038–41.

Chapter 14

Gene Transfer for Clinical Congestive Heart Failure

Tong Tang[1] and H. Kirk Hammond[2,3]

[1]*Zensun USA Inc., San Diego, CA, USA;* [2]*Department of Medicine, University of California, San Diego, La Jolla, CA, USA;* [3]*VA San Diego Healthcare System, San Diego, CA, USA*

LIST OF ABBREVIATIONS

AAV Adeno-associated virus
AC6 Adenylyl cyclase 6
α-MHC α-Myosin heavy chain
βAR β-Adrenergic receptor
βARKct Carboxyl-terminus of β-adrenergic receptor kinase
CBA Chicken β-actin
CHF Congestive heart failure
CMV Cytomegalovirus
EC coupling Excitation-contraction coupling
EF1α Elongation factor 1α
EMA European Medicines Agency
FDA Food and Drug Administration
GRK2 G-protein-coupled receptor kinase 2 (aka β-adrenergic receptor kinase)
IC Intracoronary
IGF-I Insulin growth factor I
IND Investigational new drug
IV Intravenous
LAD Left anterior descending
LCx Left circumflex
LV Left ventricular
MI Myocardial infarction
MLC-2v Ventricular myosin light chain 2
MMP Matrix metalloproteinase
mTOR Mammalian target of rapamycin
RCA Right coronary artery
RSV Rous sarcoma virus
scAAV Self-complementary adeno-associated virus
SDF-1 Stromal cell-derived factor-1
SERCA2a Sarco(endo)plasmic reticulum Ca^{2+}-ATPase 2a
SR Sarcoplasmic reticulum
ssAAV Single-stranded adeno-associated virus
TnT Troponin T

14.1 INTRODUCTION

Congestive heart failure (CHF) is a condition in which the heart cannot pump enough blood to meet the body's needs, or can do so only at abnormally high filling pressure. It is a major public health problem with unacceptably high morbidity, mortality, and economic cost. Despite advances in pharmacological, surgical, and device therapy, CHF has a 5-year mortality

Translating Gene Therapy to the Clinic. http://dx.doi.org/10.1016/B978-0-12-800563-7.00014-2

of 50%, a prognosis worse than most cancers.[1] In the United States, there are more than 300,000 deaths every year from CHF. The estimated number of CHF patients worldwide exceeds 23 million, and this number is projected to double by 2030.

This unmet medical need demands more effective treatment. Gene transfer has regained momentum in recent years.[2] There is no gene therapy drug on the market for CHF, but three clinical cardiac gene transfer trials for CHF are in progress. The focus of this chapter, which is based on a previous review,[3] is to discuss selection of vector, transgene, and delivery methods, and to summarize recent progress in cardiac gene transfer for clinical CHF.

14.2 GENERAL CONSIDERATIONS FOR CARDIAC GENE TRANSFER

Three factors must be considered for cardiac gene transfer: (1) a therapeutic transgene that corrects cardiac defects in CHF while having minimal side effects; (2) a suitable vector and promoter that will ensure long-term and sufficient transgene expression with little toxicity and immunogenicity; and (3) a delivery method that provides effective expression and yet is easy and safe to deploy. In certain circumstances, regulated or cardiac-specific expression may be desirable.

Careful selection of the therapeutic transgene is a prerequisite for successful cardiac gene transfer for CHF. Cardiac myocyte-directed transgene expression and gene deletion studies in mice have identified key regulators of cardiac function. Many of these regulators are deficient or dysfunctional in CHF. Cardiac gene transfer, by increasing expression of pivotal regulators of cardiac function, provides new potential treatments for CHF. In addition, cardiac gene transfer can also introduce dominant negative transgenes to counteract overly active regulators and reduce harmful cardiac effects. Understanding mechanisms by which specific transgenes improve function of the failing heart will not only expand therapeutic potential but also provide useful insight into cardiac function. Careful selection of vector and delivery method is critical to provide high yield transgene expression.

14.3 CANDIDATES FOR CHF GENE TRANSFER

CHF is associated with several cell-signaling pathways that are dysfunctional. Consequently, several potential therapeutic targets have been identified. We will summarize only those strategies that have proven to be effective in preclinical studies and have advanced or may soon advance to clinical CHF trials (Table 14.1). It is not surprising that many potential candidates influence β-adrenergic receptor (βAR) signaling and Ca^{2+} handling (Figure 14.1).

TABLE 14.1 Clinical Trials Using Cardiac Gene Transfer for Congestive Heart Failure

Transgene	SDF-1	SDF-1	SERCA2a	SERCA2a	AC6
Trial Identifier	NCT01082094	NCT01643590	NCT00454818	NCT01643330	NCT00787059
Phase	1	2	1 and 2	2	1/2
Vector	Plasmid	Plasmid	AAV1	AAV1	Adenovirus
Delivery Method	Endomyocardial injection	Endomyocardial injection	Intracoronary injection	Intracoronary injection	Intracoronary injection
Enrollment	17	90	39	200	56
Sample Size	5-mg dose: 4 15-mg dose: 6 30-mg dose: 7	Placebo: 30 15-mg dose: 30 30-mg dose: 30	Placebo: 14 Low dose: 8 Middle dose: 8 High dose: 9	Placebo: 100 Treatment: 100	Placebo: 12 Treatment: 42 (in 5 doses)
Design	Open label	Randomized Double-blind Placebo-controlled	Randomized Double-blind Placebo-controlled	Randomized Double-blind Placebo-controlled	Randomized Double-blind Placebo-controlled
Status	Completed	Recruiting	Completed	Recruiting	Recruiting
Outcome	Apparently safe Not powered for efficacy	N/A	Apparently safe Not powered for efficacy	Pending	Pending

AC6, adenylyl cyclase 6; SDF-1, stromal cell-derived factor-1; SERCA2a, sarco(endo)plasmic reticulum Ca^{2+}-ATPase 2a.

14.3.1 Angiogenesis and Cell Survival

14.3.1.1 SDF-1

Stromal cell-derived factor-1 (SDF-1), also known as chemokine CXCL12, activates STAT3 and Akt signaling pathways and protects the heart from ischemia reperfusion injury.[4,5] SDF-1 appears to recruit cardiac and bone marrow-derived stem cells and stimulate cardiac vasculogenesis and angiogenesis in a paracrine manner.[6–9] Recruited cardiac and bone marrow-derived stem cells, it is postulated, may differentiate into cardiac myocytes and integrate mechanically and electrically with existing cardiac myocytes.[10,11] Expression of SDF-1 is increased in left ventricle samples after myocardial infarction (MI).[12] However, transient increases of SDF-1 expression are not sufficient for tissue repair. Sustained SDF-1 expression is required to promote cell survival, angiogenesis, and stem cell mobilization and recruitment.

A phase 1 clinical trial (NCT01082094) showed no overt safety issues associated with gene transfer of plasmid DNA encoding human SDF-1.[13] A randomized double-blind placebo-controlled phase 2 clinical trial is underway to determine the efficacy of endomyocardial injection of plasmid SDF-1 in CHF (NCT01643590).

14.3.2 βAR Signaling

A hallmark of clinical CHF is impaired left ventricular (LV) βAR signaling.[14] The molecular basis for impaired βAR signaling includes decreased βAR density, βAR desensitization, uncoupling of βAR and Gαs, deficits in adenylyl cyclase (AC) expression, and subsequent defective cyclic adenosine 3′,5′-monophosphate (cAMP) production.[15] Clinical use of βAR antagonists (beta-blockers) reduces symptoms and prolongs life somewhat in CHF.[16] Correcting impaired βAR signaling safely has been a focus for CHF research for many years, and presents considerable challenges.

14.3.2.1 βARKct

G-protein-coupled receptor kinase 2 (GRK2) is a protein kinase that phosphorylates the βAR.[15] This phosphorylation promotes βAR binding to β-arrestin, which in turn promotes Gαs uncoupling and attenuation of βAR signaling. β-arrestin binding also leads to β$_1$AR internalization. The human failing heart is associated with increased expression and activity of GRK2.[17] In preclinical studies, deletion of GRK2 increases survival, attenuates LV remodeling, and reduces the extent of CHF after MI.[18]

Carboxyl-terminus of β-adrenergic receptor kinase (βARKct) was engineered to block GRK2 membrane translocation and activation. Expression of βARKct after virus-mediated gene transfer increases βAR density, cAMP production, and LV contractile function in MI-induced CHF in rats, rabbits, and pigs.[19–21] βARKct expression in cardiac myocytes from failing human hearts also increases cAMP production, cell shortening, and relaxation[22] and appears to work well with βAR antagonists.[21,23]

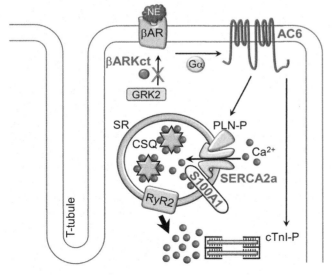

FIGURE 14.1 Schematic representation of β-adrenergic receptor (βAR) signaling and Ca^{2+} handling. Sarco(endo)plasmic reticulum Ca^{2+}-ATPase 2a (SERCA2a) and adenylyl cyclase 6 (AC6) gene transfer are being tested in clinical congestive heart failure (CHF) trials. Other gene transfer candidates for CHF include carboxyl-terminus of β-adrenergic receptor kinase (βARKct) and S100A1. CSQ, calsequestrin; cTnI-P, S23,24-phosphorylated cardiac troponin I; Gα, α subunit of GTP-binding protein; GRK2, G-protein-coupled receptor kinase 2; NE, norepinephrine; PLN-P, S16-phosphorylated phospholamban; RyR2, ryanodine receptor 2; SR, sarcoplasmic reticulum.

14.3.2.2 AC6

AC is the effector molecule that links βAR stimulation with cAMP production.[24–26] In the heart, it plays a pivotal role in LV contractile function and relaxation in response to βAR stimulation.[27–30] The failing heart is associated with decreased expression and activity of AC6, a major AC isoform in cardiac myocytes.[14,31–33] Cardiac-directed expression of AC6 in mice with genetically induced CHF increases LV function and prolongs life.[34,35] Associated with these beneficial effects in the failing heart are increased cAMP generating capacity in response to βAR stimulation, normalized protein kinase A activity, increased phospholamban phosphorylation, and increased sarcoplasmic reticulum (SR) Ca^{2+} uptake.[35,36] AC6 expression is not associated with increased heart rate or development of arrhythmias in the failing heart.[37] Increased LV AC6 expression is not associated with changes in contents of βAR, Gαs, or Gαi2.[27] Deletion of AC6 is associated with reduced SR Ca^{2+} uptake and decreased LV function in response to βAR stimulation.[38] In addition, AC6 expression normalizes prolonged action potential duration and attenuates ventricular arrhythmias.[39]

AC6 expression also has a pronounced favorable effect on cardiovascular function in CHF induced by MI. Although AC6 has no effect on infarct size, it prevents deleterious LV remodeling and reduces mortality in acute MI.[40] When AC6 expression is activated 5 weeks after MI, at which time severe CHF is evident, both LV systolic function and diastolic function are increased.[41] Intracoronary delivery of adenovirus encoding AC6 increases LV function in the failing pig heart.[42] AC6 expression increases LV Ca^{2+} handling and LV function in aged mice, suggesting a potential role in elderly subjects with CHF with preserved ejection fraction.[43] A phase 1/2 clinical trial is in progress to determine the safety and efficacy of gene transfer of adenovirus encoding AC6 (Ad5.AC6; NCT00787059).

14.3.3 Ca^{2+} Handling

Calcium plays a crucial role in controlling LV contraction and relaxation. During every heartbeat, Ca^{2+} is taken up and then released from SR. The failing heart is characterized by defective excitation-contraction (EC) coupling and dysfunctional SR Ca^{2+} uptake and release.[44]

14.3.3.1 SERCA2a

Sarco(endo)plasmic reticulum Ca^{2+}-ATPase 2a (SERCA2a) is the Ca^{2+} pump responsible for cardiac SR Ca^{2+} uptake. CHF is associated with abnormal SERCA2a expression and activity.[45,46] Cardiac-directed SERCA2a expression increases SR Ca^{2+} uptake and LV contractile function and relaxation.[47,48] Homogeneous SERCA2a deletion is lethal. Mice with deletion of only one SERCA2a allele show decreased SR Ca^{2+} uptake and are more prone to CHF after pressure overload.[49,50] These data suggest a role of SERCA2a in mediating Ca^{2+} handling and LV function.

SERCA2a gene transfer increases contractile function in vitro in cardiac myocytes isolated from CHF patients.[51] In pressure-overloaded hearts, expression of SERCA2a by adenovirus-mediated gene transfer increases contractile function and relaxation in rats.[52,53] SERCA2a gene transfer also increases LV function in aged hearts.[54] In volume-overloaded pigs, adeno-associated virus (AAV)-mediated SERCA2a gene transfer attenuated LV dysfunction and remodeling.[55] SERCA2a gene transfer shows similar beneficial effects in sheep with pacing-induced CHF.[56] A small phase 1/2 clinical trial found no apparent safety concerns associated with SERCA2a gene transfer (NCT00454818).[57] A larger clinical trial of SERCA2a gene transfer is in progress (NCT01643330).

14.3.3.2 S100A1

S100A1, a family member of EF-hand Ca^{2+}-binding proteins, is expressed in cardiac myocytes.[58] The subcellular locations of S100A1 include SR, mitochondria, and sarcomere. S100A1 can bind to RyR2, the SR Ca^{2+} release channel, and to SERCA2a, the SR Ca^{2+} uptake pump,[59,60] suggesting that S100A1 is not solely a Ca^{2+}-binding protein but also a regulator of Ca^{2+} homeostasis in cardiac myocytes. There is evidence that ischemic cardiomyopathy is associated with reduced S100A1 expression.[61]

Cardiac-directed S100A1 expression is associated with improved Ca^{2+} handling, decreased deleterious LV remodeling, and reduced mortality after MI in mice.[62] AAV-mediated gene transfer of S100A1 in CHF increases Ca^{2+} transients and LV function.[63] S100A1 gene transfer also improves Ca^{2+} handling and contractile function in vitro in cardiac myocytes isolated from human failing heart.[64] Recent data demonstrate that intracoronary delivery of AAV9 encoding S100A1 normalizes SR Ca^{2+} handling, attenuates LV remodeling, and increases contractile function.[19]

Although βARKct, AC6, SERCA2a, and S100A1 likely operate via effects on βAR and Ca^{2+} signaling, favorable cardiac effects may be mediated by additional mechanisms. For example, both βARKct and AC6 expression activate Akt—a kinase that promotes cell survival.[65,66] S100A1 inhibits 2-deoxyglucose and oxidative stress-induced apoptosis in neonatal cardiac myocytes in vitro.[67] SERCA2a and S100A1 appear to bind to endothelial cell nitric oxide synthase in endothelial cells, suggesting a role in modulating production of nitric oxide—a pivotal molecule for blood flow regulation.[68,69] Further exploration of the underlying mechanisms may help improve the efficacy of these CHF therapeutic candidates.

14.4 VECTORS AND METHODS FOR CARDIAC GENE TRANSFER

14.4.1 Plasmid Vectors

Plasmid vectors have been used in several angiogenesis gene transfer clinical trials. They have low immunogenicity and appear to be relatively safe. Injection of plasmid DNA encoding SDF-1 in rats 1 month after MI was associated with increased LV fractional shortening.[11] Virus vectors appear to be more effective than nonvirus vectors in increasing function of the failing heart in preclinical studies,[19,41,55,70,71] and are used preferentially in gene transfer for CHF.

14.4.2 Virus Vectors

Adenovirus and AAV are the most commonly used vectors for cardiac gene transfer. Adenovirus provides reasonable gene transfer efficiency,[72] particularly when used with mechanical or pharmaceutical adjuvants after intracoronary delivery.[73] Newer generation adenoviruses, including the so-called gutless adenovirus engineered by deleting all regions encoding virus proteins, may have lower immunogenicity than previous vectors.[72]

AAV, with an insert capacity <5 kb, provides potential long-term expression. Persistent transgene expression has been shown in rodents and larger animals years after a single injection of AAV.[74] Although recent clinical trials have found that some AAV serotypes incite immune responses after intramuscular injection,[75,76] other AAV vectors (AAVs 5, 6, 8 and 9) do not appear to have similar problems in nonhuman primates.[77] Previous exposure to AAV, with subsequent generation of neutralizing antibodies, impairs the effectiveness of AAV vectors in cardiac gene transfer.[57] Preexisting anti-AAV8 antibodies are present in 19% of human subjects. AAV1 and AAV2 have a 50–60% prevalence of neutralizing antibodies, which may limit their use in clinical trials.[78]

New vectors have been engineered to enable increased long-term expression. Self-complementary adeno-associated virus (scAAV) vectors may provide more rapid and perhaps higher transgene expression than their single-stranded adeno-associated virus (ssAAV) analogs.[79] Transgene expression using ssAAV vectors is delayed 4 weeks until the complementary DNA strand is synthesized. By encoding for the complementary DNA strand within the vector, scAAV (insert capacity 3.3 kb), enables transgene expression in 2 weeks.[79]

14.4.3 Promoters

The cytomegalovirus (CMV) promoter is widely used for cardiac gene transfer. It provides strong transgene expression in cardiac myocytes. However, the CMV promoter is susceptible to methylation and subsequent inactivation in the liver and skeletal muscle.[80,81] Rous sarcoma virus (RSV), chicken β-actin (CBA), and elongation factor 1α (EF1α) provide less robust transgene expression in cardiac myocytes, but are less susceptible to methylation. None of these promoters provides cardiac-specific transgene expression, though.

The α-myosin heavy chain (α-MHC) promoter is used for cardiac-directed transgene expression in transgenic mice.[82] However, it provides less robust transgene expression in virus vectors. More relevant, its size (~5.5 kb) is over the packaging capacity of AAV and prevents its use in AAV-mediated cardiac gene transfer. A 363-bp α-MHC minimal promoter has been identified, and a report showed that this minimal promoter directed transgene expression preferentially in the heart after AAV-mediated gene transfer in mice.[83] More studies are needed to test tissue specificity and expression level directed by this promoter. A 2.1-kb fragment of the ventricular myosin light chain 2 (MLC-2v) promoter may be an option.[84] A 418 bp fragment of the chicken cardiac troponin T (TnT) promoter provides 100-fold more expression in heart than liver after AAV-mediated gene transfer, although expression is lower than that provided by CMV or CBA promoters.[85]

14.4.4 Gene Delivery Methods

Delivery methods are often determined by vector selection. For example, intravascular delivery is not suitable for lentivirus and plasmid DNA because these vectors are unable to cross the capillary endothelium.[86] Direct intramyocardial injection appears to be the best delivery route for lentivirus and plasmid DNA for cardiac gene transfer. However, intramuscular injection provides transgene expression limited to the area adjacent to the needle tract. Therefore, injection at multiple sites is required when using lentivirus and plasmid DNA, as exemplified by cardiac gene transfer of plasmid DNA encoding SDF-1.[13]

For adenovirus and AAV, there are four effective delivery methods for cardiac gene transfer: direct intracoronary, indirect intracoronary (indirect IC), intramuscular (IM), and intravenous (IV) delivery.[87,88] Direct coronary delivery can be used in large animals, such as pigs.[42] In this procedure, a coronary catheter is advanced 1 cm into the ostium of the proximal portion of each of the three coronary arteries: left anterior descending (LAD), left circumflex (LCx), and right coronary artery (RCA). Virus vectors are infused into each coronary artery in proportion to the myocardial mass they perfuse. This procedure is also used in SERCA2a and AC6 clinical gene transfer trials with virus vectors (Table 14.1).

In mice, the coronary arteries are too small for direct IC delivery. Indirect IC delivery, instead, has been used to deliver virus vectors to the left ventricle in mice and rats. In this procedure, the aorta and pulmonary arteries are cross-clamped and virus vectors are delivered into the LV chamber. Continued LV contraction then forces the vector into the coronary arteries. Because vector exposure time promotes gene transfer, hypothermia is used to prolong dwell time but circumvent brain injury. IC infusion of pharmacological agents including histamine,[29] serotonin,[89] nitroprusside,[42,73] sildenafil,[90] and substance P[91] appear to increase adenovirus-mediated gene transfer efficiency. Indirect IC delivery provides greater cardiac transgene expression compared with IV delivery, regardless of AAV serotype.[92]

IM delivery of virus vectors into skeletal muscle has been used with success in a rat model of CHF.[93] In this study, a virus vector (AAV5) encoding a gene with paracrine activity (insulin growth factor (IGF)-1) was injected into skeletal muscle. Serum IGF-1, synthesized and secreted from the skeletal muscle, then could influence heart function without the need for cardiac gene transfer. Although IM delivery is appealing because of its simplicity and relative confinement to a single region, it has failed in clinical trials because of immune response.[75,76] It appears that newer generation AAV vectors (AAVs 5, 6, 8, and 9) do not have similar problems of immune response in primates,[77] but this has not been tested in clinical trials.

IV delivery of virus vectors (especially AAV vectors) has also been used with success in cardiac gene transfer in animals.[71,94,95] AAV and adenovirus vectors are widely disseminated after vascular delivery, and expression in liver and skeletal muscle, two large organs, provides the possibility of substantial generation of transgenes, and ideal circumstance if the goal is to increase serum levels of a peptide with paracrine activity.

14.4.5 Regulated Transgene Expression

Long-term expression vectors require, for safety in clinical trials, the ability to turn off transgene expression in the event that untoward effects develop. Regulated expression also enables the flexibility of intermittent rather than constant transgene expression. There are four regulated expression systems currently available: ecdysone, tamoxifen, tetracycline, and rapamycin.[96–99] The size of the ecdysone system requires a two-vector strategy, and tamoxifen presents difficult to resolve issues with toxicity. Tetracycline and rapamycin regulation systems (Table 14.2) are better suited for regulated transgene expression, but have only been tested in large animal models to date.[41,74,100–104]

14.4.5.1 Tetracycline-Regulated Expression

The tetracycline-regulation system has been extensively studied.[100,101,104] Unlike previous reverse tetracycline transactivator (rtTA) constructs, newer rtTA variants, such as rtTA2^S-M2, provide robust tetracycline-dependent expression with no

TABLE 14.2 Tetracycline versus Rapamycin Regulation

Feature	Tetracycline	Rapamycin
Activator	Doxycycline	AP22594
Basal Expression ("Leak")	Very low/none	None
Linear Dose-Response	Yes	Yes
Activator Side Effects	Low (avoid in pregnancy)	Immunosuppressant
Bacteria/Virus Proteins	Yes	No
Used in Clinical Trials	Not yet	Not yet

AP22594, oral rapamycin analog with 1/100th immune suppression versus rapamycin.

basal activity (i.e., no "leak") and 10-fold higher sensitivity to tetracycline (maximum transgene expression activation at 0.1 μg/mL).[28,101,104] A single daily dose of doxycycline of 10–20 mg may suffice for complete activation of transgene expression in human subjects.[105] Doses of 200 mg per day are well tolerated by patients using oral doxycycline chronically for acne and chronic infections.[105,106] Tetracyclines may attenuate matrix metalloproteinase (MMP) activity and affect LV remodeling when administered in the first few days after MI.[103] In clinical settings, tetracycline should be avoided in the acute phase of MI.

Immune responses to components of the rtTA system did not occur when AAV4.tet and AAV5.tet gene transfer (intraretinal) were used in nonhuman primates,[100,105] where tetracycline-dependent transgene expression persisted for the 2.5-year duration of the study. In our studies, inflammation was not observed in mouse hearts expressing high levels of rtTA[28,41] or in rats after AAV5-mediated regulated expression of IGF-I using the rtTA2^S-M2 regulation element.[93] Intramuscular delivery of AAV in nonhuman primates, unlike intraretinal or vascular delivery, does lead to attenuation of regulated expression, owing to immune responses to the bacterial and virus components of the transactivator fusion protein.[107] The rapamycin regulation system, which does not possess bacterial or virus proteins, and is not associated with provocation of the immune response,[74] may be a suitable alternative.

14.4.5.2 Rapamycin-Regulated Expression

In the rapamycin regulation system, transgene expression is triggered by nanomolar concentrations of rapamycin or a rapamycin analog, which is dose-dependent and reversible.[99] Rapamycin is used clinically to suppress immune response, forestalls deleterious effects of aging in mice,[102] and inhibits glioblastoma multiforme[108] by blocking the mammalian target of rapamycin (mTOR) signaling pathway.[109] The oral rapamycin analog AP22594, which activates transgene expression as effectively as rapamycin, exhibits minimal immune suppression, and does not inhibit mTOR.[74,109–111] Additional preclinical studies directly comparing tetracycline-regulated and rapamycin-regulated expression will be required before using these systems in clinical trials.

14.4.6 Alternative Methods for Cardiac Gene Transfer

Gene transfer of peptides with paracrine activities that may benefit the heart is an alternative to cardiac-targeted gene transfer and may be applicable for CHF and other cardiovascular diseases. A prerequisite for this approach is the selection of a transgene that has cardiac effects after being released to the circulation from a distant site. We have tested this concept using skeletal muscle injection of AAV5 encoding IGF-I under tetracycline regulation (AAV5.IGFI-tet).[93] In this study, AAV5.IGFI-tet was injected in the anterior tibialis muscle in rats with severe CHF induced by MI. Activation of IGF-I expression by addition of doxycycline to the drinking water increased serum IGF-I levels and improved function of the failing heart. This new approach enables transgene expression at a remote site and circumvents the problem of attaining high yield cardiac gene transfer. As mentioned previously, IV delivery may be superior to intramuscular delivery.

14.5 GENE TRANSFER CLINICAL TRIALS FOR CHF

Three candidate genes have been tested in early-phase clinical trials enrolling subjects with CHF (Table 14.1).

14.5.1 SDF-1

A phase 1 clinical trial (NCT01082094) was conducted to test the safety of plasmid DNA encoding human SDF-1.[13] In this open-label study, 17 CHF patients with ischemic-based heart failure received one of three escalating doses of SDF-1. SDF-1 was injected into the endocardium in peri-infarct zones using a needle catheter in a percutaneous intravascular approach. All 17 patients survived >28 days after SDF-1 injection, and 15 of 17 patients survived for 12 months. Endocardial injection of plasmid SDF-1 appeared to be safe in this small study. A randomized, double-blind, placebo-controlled phase 2 clinical trial is under way to determine the efficacy of SDF-1 endomyocardial injection in CHF (NCT01643590).

14.5.2 SERCA2a

A phase 1/2 clinical trial was conducted to determine the safety and efficacy of gene transfer of adenovirus encoding SERCA2a (SER; NCT00454818)[57] in subjects with New York Heart Association class III CHF. Phase 1 enrolled nine subjects in a limited open-label design. Subjects ($n=3$ per dose) received intracoronary AAV1 encoding SERCA2a (AAV1.SERCA2a) in doses of 1.4×10^{11}, 6×10^{11}, or 3×10^{12} DNase-resistant particles. Phase 2 was a randomized, double-blind, placebo-controlled study that enrolled 39 patients who received one of three doses of AAV1.SERCA2a (6×10^{11}, 3×10^{12}, or 1×10^{13} DNase-resistant particles) or placebo.

The results indicated that intracoronary administration of AAV1.SERCA2a appeared to be well tolerated. However, the trial was insufficiently powered to determine efficacy. A larger clinical trial of SERCA2a gene transfer is in progress (NCT01643330).

14.5.3 AC6

A phase 1/2 clinical trial is in progress to determine the safety and efficacy of gene transfer of adenovirus-5 encoding AC6 (Ad5.AC6; NCT00787059). In this randomized, double-blind, placebo-controlled study, a total of 56 patients will receive intracoronary delivery of Ad5.hAC6 or sucrose solution (placebo) in 3:1 randomization with dose escalation, starting at 3.2×10^9 to 3.2×10^{11} virus particles in five dose groups. Randomization is expected to be completed in 2014.

14.6 CONCLUSION

Three candidate genes are being tested in clinical trials enrolling subjects with CHF (Table 14.1), including two ongoing clinical trials using intracoronary delivery of AAV1.SERCA2a (NCT01643330) or Ad5.AC6 (NCT00787059). Another trial using endocardial injection of plasmid SDF-1 (NCT01643590) is also in progress. Optimization of virus vectors, regulated expression systems, gene delivery methods, and identification of new therapeutic candidates is likely to result in translation to clinical application in the coming decade.

GLOSSARY

β-Adrenergic receptor β-Adrenergic receptor is a G protein–coupled receptor with seven transmembrane domains that transduces extracellular catecholamine signals across cell membrane. Defective β-adrenergic receptor function is a hallmark of congestive heart failure.

Adenylyl cyclase Adenylyl cyclase is an enzyme that converts ATP to the second-messenger cAMP (3′,5′ cyclic AMP). Congestive heart failure is associated with defective β-adrenergic receptor signaling and decreased AC6 content in LV samples.

Congestive heart failure Congestive heart failure is a medical condition in which the heart cannot pump enough blood to meet the body's needs or can do so only at elevated filling pressure. The common known causes for congestive heart failure include myocardial infarction, hypertension, myocarditis, valvular heart disease, and congenital heart disease. A sizable proportion of patients with congestive heart failure do not have a known cause. Congestive heart failure is a major cause of morbidity and mortality.

SERCA2a SERCA2a is an enzyme that transports Ca^{2+} into intracellular sarcoplasmic reticulum in cardiac myocytes during diastole. Ca^{2+} released from sarcoplasmic reticulum through ryanodine receptor leads to myocyte contraction.

REFERENCES

1. Go AS, Mozaffarian D, Roger VL, Benjamin EJ, Berry JD, Borden WB, et al. Heart disease and stroke statistics–2013 update: a report from the American Heart Association. *Circulation* January 1, 2013;**127**(1):e6–245.
2. Miller N. Glybera and the future of gene therapy in the European Union. *Nat Rev Drug Discov* May 2012;**11**(5):419.
3. Tang T, Hammond HK. Gene transfer for congestive heart failure: update 2013. *Transl Res* April 2013;**161**(4):313–20.
4. Huang C, Gu H, Zhang W, Manukyan MC, Shou W, Wang M. SDF-1/CXCR4 mediates acute protection of cardiac function through myocardial STAT3 signaling following global ischemia/reperfusion injury. *Am J Physiol Heart Circ Physiol* October 2011;**301**(4):H1496–505.
5. Hu X, Dai S, Wu WJ, Tan W, Zhu X, Mu J, et al. Stromal cell derived factor-1 alpha confers protection against myocardial ischemia/reperfusion injury: role of the cardiac stromal cell derived factor-1 alpha CXCR4 axis. *Circulation* August 7, 2007;**116**(6):654–63.
6. Elmadbouh I, Haider H, Jiang S, Idris NM, Lu G, Ashraf M. Ex vivo delivered stromal cell-derived factor-1alpha promotes stem cell homing and induces angiomyogenesis in the infarcted myocardium. *J Mol Cell Cardiol* April 2007;**42**(4):792–803.
7. Hiasa K, Ishibashi M, Ohtani K, Inoue S, Zhao Q, Kitamoto S, et al. Gene transfer of stromal cell-derived factor-1alpha enhances ischemic vasculogenesis and angiogenesis via vascular endothelial growth factor/endothelial nitric oxide synthase-related pathway: next-generation chemokine therapy for therapeutic neovascularization. *Circulation* May 25, 2004;**109**(20):2454–61.
8. Saxena A, Fish JE, White MD, Yu S, Smyth JW, Shaw RM, et al. Stromal cell-derived factor-1alpha is cardioprotective after myocardial infarction. *Circulation* April 29, 2008;**117**(17):2224–31.
9. Abbott JD, Huang Y, Liu D, Hickey R, Krause DS, Giordano FJ. Stromal cell-derived factor-1alpha plays a critical role in stem cell recruitment to the heart after myocardial infarction but is not sufficient to induce homing in the absence of injury. *Circulation* November 23, 2004;**110**(21):3300–5.
10. Dong F, Harvey J, Finan A, Weber K, Agarwal U, Penn MS. Myocardial CXCR4 expression is required for mesenchymal stem cell mediated repair following acute myocardial infarction. *Circulation* July 17, 2012;**126**(3):314–24.

11. Sundararaman S, Miller TJ, Pastore JM, Kiedrowski M, Aras R, Penn MS. Plasmid-based transient human stromal cell-derived factor-1 gene transfer improves cardiac function in chronic heart failure. *Gene Ther* September 2011;**18**(9):867–73.

12. Askari AT, Unzek S, Popovic ZB, Goldman CK, Forudi F, Kiedrowski M, et al. Effect of stromal-cell-derived factor 1 on stem-cell homing and tissue regeneration in ischaemic cardiomyopathy. *Lancet* August 30, 2003;**362**(9385):697–703.

13. Penn MS, Mendelsohn FO, Schaer GL, Sherman W, Farr M, Pastore J, et al. An open-label dose escalation study to evaluate the safety of administration of nonviral stromal cell-derived factor-1 plasmid to treat symptomatic ischemic heart failure. *Circ Res* March 1, 2013;**112**(5):816–25.

14. Bristow MR, Ginsburg R, Minobe W, Cubicciotti RS, Sageman WS, Lurie K, et al. Decreased catecholamine sensitivity and beta-adrenergic-receptor density in failing human hearts. *N Engl J Med* July 22, 1982;**307**(4):205–11.

15. Rockman HA, Koch WJ, Lefkowitz RJ. Seven-transmembrane-spanning receptors and heart function. *Nature* January 10, 2002;**415**(6868):206–12.

16. Bristow MR. Treatment of chronic heart failure with beta-adrenergic receptor antagonists: a convergence of receptor pharmacology and clinical cardiology. *Circ Res* October 28, 2011;**109**(10):1176–94.

17. Ungerer M, Bohm M, Elce JS, Erdmann E, Lohse MJ. Altered expression of beta-adrenergic receptor kinase and beta 1-adrenergic receptors in the failing human heart. *Circulation* February 1993;**87**(2):454–63.

18. Raake PW, Vinge LE, Gao E, Boucher M, Rengo G, Chen X, et al. G protein-coupled receptor kinase 2 ablation in cardiac myocytes before or after myocardial infarction prevents heart failure. *Circ Res* August 15, 2008;**103**(4):413–22.

19. Raake PW, Schlegel P, Ksienzyk J, Reinkober J, Barthelmes J, Schinkel S, et al. AAV6.βARKct cardiac gene therapy ameliorates cardiac function and normalizes the catecholaminergic axis in a clinically relevant large animal heart failure model. *Eur Heart J* January 19, 2013;**34**(19):1437–47.

20. White DC, Hata JA, Shah AS, Glower DD, Lefkowitz RJ, Koch WJ. Preservation of myocardial beta-adrenergic receptor signaling delays the development of heart failure after myocardial infarction. *Proc Natl Acad Sci USA* May 9, 2000;**97**(10):5428–33.

21. Rengo G, Lymperopoulos A, Zincarelli C, Donniacuo M, Soltys S, Rabinowitz JE, et al. Myocardial adeno-associated virus serotype 6-betaARKct gene therapy improves cardiac function and normalizes the neurohormonal axis in chronic heart failure. *Circulation* January 6, 2009;**119**(1):89–98.

22. Williams ML, Hata JA, Schroder J, Rampersaud E, Petrofski J, Jakoi A, et al. Targeted beta-adrenergic receptor kinase (betaARK1) inhibition by gene transfer in failing human hearts. *Circulation* April 6, 2004;**109**(13):1590–3.

23. Harding VB, Jones LR, Lefkowitz RJ, Koch WJ, Rockman HA. Cardiac beta ARK1 inhibition prolongs survival and augments beta blocker therapy in a mouse model of severe heart failure. *Proc Natl Acad Sci USA* May 8, 2001;**98**(10):5809–14.

24. Sunahara RK, Dessauer CW, Gilman AG. Complexity and diversity of mammalian adenylyl cyclases. *Annu Rev Pharmacol Toxicol* 1996;**36**: 461–80.

25. Pierce KL, Premont RT, Lefkowitz RJ. Seven-transmembrane receptors. *Nat Rev Mol Cell Biol* 2002;**3**(9):639–50.

26. Marian AJ. β-adrenergic receptors signaling and heart failure in mice, rabbits and humans. *J Mol Cell Cardiol* July 2006;**41**(1):11–3.

27. Gao MH, Lai NC, Roth DM, Zhou J, Zhu J, Anzai T, et al. Adenylylcyclase increases responsiveness to catecholamine stimulation in transgenic mice. *Circulation* March 30, 1999;**99**(12):1618–22.

28. Gao MH, Bayat H, Roth DM, Yao Zhou J, Drumm J, Burhan J, et al. Controlled expression of cardiac-directed adenylylcyclase type VI provides increased contractile function. *Cardiovasc Res* November 2002;**56**(2):197–204.

29. Lai NC, Roth DM, Gao MH, Fine S, Head BP, Zhu J, et al. Intracoronary delivery of adenovirus encoding adenylyl cyclase VI increases left ventricular function and cAMP-generating capacity. *Circulation* November 7, 2000;**102**(19):2396–401.

30. Feldman AM. Adenylyl cyclase: a new target for heart failure therapeutics. *Circulation* 2002;**105**(16):1876–8.

31. Ping P, Anzai T, Gao M, Hammond HK. Adenylyl cyclase and G protein receptor kinase expression during development of heart failure. *Am J Physiol* August 1997;**273**(2 Pt 2):H707–17.

32. Ishikawa Y, Sorota S, Kiuchi K, Shannon RP, Komamura K, Katsushika S, et al. Downregulation of adenylylcyclase types V and VI mRNA levels in pacing-induced heart failure in dogs. *J Clin Invest* May 1994;**93**(5):2224–9.

33. Feldman MD, Copelas L, Gwathmey JK, Phillips P, Warren SE, Schoen FJ, et al. Deficient production of cyclic AMP: pharmacologic evidence of an important cause of contractile dysfunction in patients with end-stage heart failure. *Circulation* February 1987;**75**(2):331–9.

34. Roth DM, Bayat H, Drumm JD, Gao MH, Swaney JS, Ander A, et al. Adenylyl cyclase increases survival in cardiomyopathy. *Circulation* April 23, 2002;**105**(16):1989–94.

35. Roth DM, Gao MH, Lai NC, Drumm J, Dalton N, Zhou JY, et al. Cardiac-directed adenylyl cyclase expression improves heart function in murine cardiomyopathy. *Circulation* June 22, 1999;**99**(24):3099–102.

36. Tang T, Gao MH, Roth DM, Guo T, Hammond HK. Adenylyl cyclase type VI corrects cardiac sarcoplasmic reticulum calcium uptake defects in cardiomyopathy. *Am J Physiol Heart Circ Physiol* November 2004;**287**(5):H1906–12.

37. Roth DM, Drumm JD, Bhargava V, Swaney JS, Gao MH, Hammond HK. Cardiac-directed expression of adenylyl cyclase and heart rate regulation. *Basic Res Cardiol* November 2003;**98**(6):380–7.

38. Tang T, Gao MH, Lai NC, Firth AL, Takahashi T, Guo T, et al. Adenylyl cyclase type 6 deletion decreases left ventricular function via impaired calcium handling. *Circulation* January 1, 2008;**117**(1):61–9.

39. Timofeyev V, He Y, Tuteja D, Zhang Q, Roth DM, Hammond HK, et al. Cardiac-directed expression of adenylyl cyclase reverses electrical remodeling in cardiomyopathy. *J Mol Cell Cardiol* July 2006;**41**(1):170–81.

40. Takahashi T, Tang T, Lai NC, Roth DM, Rebolledo B, Saito M, et al. Increased cardiac adenylyl cyclase expression is associated with increased survival after myocardial infarction. *Circulation* August 1, 2006;**114**(5):388–96.

41. Lai NC, Tang T, Gao MH, Saito M, Takahashi T, Roth DM, et al. Activation of cardiac adenylyl cyclase expression increases function of the failing ischemic heart in mice. *J Am Coll Cardiol* April 15, 2008;**51**(15):1490–7.

42. Lai NC, Roth DM, Gao MH, Tang T, Dalton N, Lai YY, et al. Intracoronary adenovirus encoding adenylyl cyclase VI increases left ventricular function in heart failure. *Circulation* July 20, 2004;**110**(3):330–6.

43. Tang T, Hammond HK, Firth A, Yang Y, Gao MH, Yuan JX, et al. Adenylyl cyclase 6 improves calcium uptake and left ventricular function in aged hearts. *J Am Coll Cardiol* May 3, 2011;**57**(18):1846–55.

44. Bers DM. Altered cardiac myocyte Ca regulation in heart failure. *Physiology (Bethesda)* December 2006;**21**:380–7.

45. Gwathmey JK, Copelas L, MacKinnon R, Schoen FJ, Feldman MD, Grossman W, et al. Abnormal intracellular calcium handling in myocardium from patients with end-stage heart failure. *Circ Res* July 1987;**61**(1):70–6.

46. Hasenfuss G, Reinecke H, Studer R, Meyer M, Pieske B, Holtz J, et al. Relation between myocardial function and expression of sarcoplasmic reticulum Ca(2+)-ATPase in failing and nonfailing human myocardium. *Circ Res* September 1994;**75**(3):434–42.

47. He H, Giordano FJ, Hilal-Dandan R, Choi DJ, Rockman HA, McDonough PM, et al. Overexpression of the rat sarcoplasmic reticulum Ca^{2+} ATPase gene in the heart of transgenic mice accelerates calcium transients and cardiac relaxation. *J Clin Invest* July 15, 1997;**100**(2):380–9.

48. Baker DL, Hashimoto K, Grupp IL, Ji Y, Reed T, Loukianov E, et al. Targeted overexpression of the sarcoplasmic reticulum Ca^{2+}-ATPase increases cardiac contractility in transgenic mouse hearts. *Circ Res* December 14–28, 1998;**83**(12):1205–14.

49. Periasamy M, Reed TD, Liu LH, Ji Y, Loukianov E, Paul RJ, et al. Impaired cardiac performance in heterozygous mice with a null mutation in the sarco(endo)plasmic reticulum Ca^{2+}-ATPase isoform 2 (SERCA2) gene. *J Biol Chem* January 22, 1999;**274**(4):2556–62.

50. Schultz Jel J, Glascock BJ, Witt SA, Nieman ML, Nattamai KJ, Liu LH, et al. Accelerated onset of heart failure in mice during pressure overload with chronically decreased SERCA2 calcium pump activity. *Am J Physiol Heart Circ Physiol* March 2004;**286**(3):H1146–53.

51. del Monte F, Harding SE, Schmidt U, Matsui T, Kang ZB, Dec GW, et al. Restoration of contractile function in isolated cardiomyocytes from failing human hearts by gene transfer of SERCA2a. *Circulation* December 7, 1999;**100**(23):2308–11.

52. Miyamoto MI, del Monte F, Schmidt U, DiSalvo TS, Kang ZB, Matsui T, et al. Adenoviral gene transfer of SERCA2a improves left-ventricular function in aortic-banded rats in transition to heart failure. *Proc Natl Acad Sci USA* January 18, 2000;**97**(2):793–8.

53. del Monte F, Williams E, Lebeche D, Schmidt U, Rosenzweig A, Gwathmey JK, et al. Improvement in survival and cardiac metabolism after gene transfer of sarcoplasmic reticulum Ca(2+)-ATPase in a rat model of heart failure. *Circulation* September 18, 2001;**104**(12):1424–9.

54. Schmidt U, del Monte F, Miyamoto MI, Matsui T, Gwathmey JK, Rosenzweig A, et al. Restoration of diastolic function in senescent rat hearts through adenoviral gene transfer of sarcoplasmic reticulum Ca(2+)-ATPase. *Circulation* February 22, 2000;**101**(7):790–6.

55. Kawase Y, Ly HQ, Prunier F, Lebeche D, Shi Y, Jin H, et al. Reversal of cardiac dysfunction after long-term expression of SERCA2a by gene transfer in a pre-clinical model of heart failure. *J Am Coll Cardiol* March 18, 2008;**51**(11):1112–9.

56. Byrne MJ, Power JM, Preovolos A, Mariani JA, Hajjar RJ, Kaye DM. Recirculating cardiac delivery of AAV2/1SERCA2a improves myocardial function in an experimental model of heart failure in large animals. *Gene Ther* December 2008;**15**(23):1550–7.

57. Jessup M, Greenberg B, Mancini D, Cappola T, Pauly DF, Jaski B, et al. Calcium Upregulation by Percutaneous Administration of Gene Therapy in Cardiac Disease (CUPID): a phase 2 trial of intracoronary gene therapy of sarcoplasmic reticulum Ca^{2+}-ATPase in patients with advanced heart failure. *Circulation* July 19, 2011;**124**(3):304–13.

58. Most P, Bernotat J, Ehlermann P, Pleger ST, Reppel M, Borries M, et al. S100A1: a regulator of myocardial contractility. *Proc Natl Acad Sci USA* November 20, 2001;**98**(24):13889–94.

59. Kiewitz R, Acklin C, Schafer BW, Maco B, Uhrik B, Wuytack F, et al. Ca^{2+}-dependent interaction of S100A1 with the sarcoplasmic reticulum Ca^{2+}-ATPase2a and phospholamban in the human heart. *Biochem Biophys Res Commun* June 27, 2003;**306**(2):550–7.

60. Treves S, Scutari E, Robert M, Groh S, Ottolia M, Prestipino G, et al. Interaction of S100A1 with the Ca^{2+} release channel (ryanodine receptor) of skeletal muscle. *Biochemistry* September 23, 1997;**36**(38):11496–503.

61. Remppis A, Greten T, Schafer BW, Hunziker P, Erne P, Katus HA, et al. Altered expression of the Ca(2+)-binding protein S100A1 in human cardiomyopathy. *Biochim Biophys Acta* October 11, 1996;**1313**(3):253–7.

62. Most P, Remppis A, Pleger ST, Loffler E, Ehlermann P, Bernotat J, et al. Transgenic overexpression of the Ca^{2+}-binding protein S100A1 in the heart leads to increased in vivo myocardial contractile performance. *J Biol Chem* September 5, 2003;**278**(36):33809–17.

63. Pleger ST, Most P, Boucher M, Soltys S, Chuprun JK, Pleger W, et al. Stable myocardial-specific AAV6-S100A1 gene therapy results in chronic functional heart failure rescue. *Circulation* May 15, 2007;**115**(19):2506–15.

64. Brinks H, Rohde D, Voelkers M, Qiu G, Pleger ST, Herzog N, et al. S100A1 genetically targeted therapy reverses dysfunction of human failing cardiomyocytes. *J Am Coll Cardiol* August 23, 2011;**58**(9):966–73.

65. Brinks H, Boucher M, Gao E, Chuprun JK, Pesant S, Raake PW, et al. Level of G protein-coupled receptor kinase-2 determines myocardial ischemia/reperfusion injury via pro- and anti-apoptotic mechanisms. *Circ Res* October 29, 2010;**107**(9):1140–9.

66. Gao MH, Tang T, Guo T, Miyanohara A, Yajima T, Pestonjamasp K, et al. Adenylyl cyclase type VI increases Akt activity and phospholamban phosphorylation in cardiac myocytes. *J Biol Chem* November 28, 2008;**283**(48):33527–35.

67. Most P, Boerries M, Eicher C, Schweda C, Ehlermann P, Pleger ST, et al. Extracellular S100A1 protein inhibits apoptosis in ventricular cardiomyocytes via activation of the extracellular signal-regulated protein kinase 1/2 (ERK1/2). *J Biol Chem* November 28, 2003;**278**(48):48404–12.

68. Hulot JS, Senyei G, Hajjar RJ. Sarcoplasmic reticulum and calcium cycling targeting by gene therapy. *Gene Ther* June 2012;**19**(6):596–9.

69. Rohde D, Ritterhoff J, Voelkers M, Katus HA, Parker TG, Most P. S100A1: a multifaceted therapeutic target in cardiovascular disease. *J Cardiovasc Transl Res* October 2010;**3**(5):525–37.

70. Kaye DM, Preovolos A, Marshall T, Byrne M, Hoshijima M, Hajjar R, et al. Percutaneous cardiac recirculation-mediated gene transfer of an inhibitory phospholamban peptide reverses advanced heart failure in large animals. *J Am Coll Cardiol* July 17, 2007;**50**(3):253–60.

71. Pleger ST, Shan C, Ksienzyk J, Bekeredjian R, Boekstegers P, Hinkel R, et al. Cardiac AAV9-S100A1 gene therapy rescues post-ischemic heart failure in a preclinical large animal model. *Sci Transl Med* July 20, 2011;**3**(92):92ra64.

72. Alba R, Bosch A, Chillon M. Gutless adenovirus: last-generation adenovirus for gene therapy. *Gene Ther* October 2005;**12**(Suppl.):S18–27.

73. Roth DM, Lai NC, Gao MH, Fine S, McKirnan MD, Roth DA, et al. Nitroprusside increases gene transfer associated with intracoronary delivery of adenovirus. *Hum Gene Ther* October 2004;**15**(10):989–94.

74. Rivera VM, Gao GP, Grant RL, Schnell MA, Zoltick PW, Rozamus LW, et al. Long-term pharmacologically regulated expression of erythropoietin in primates following AAV-mediated gene transfer. *Blood* February 15, 2005;**105**(4):1424–30.

75. Mingozzi F, Meulenberg JJ, Hui DJ, Basner-Tschakarjan E, Hasbrouck NC, Edmonson SA, et al. AAV-1-mediated gene transfer to skeletal muscle in humans results in dose-dependent activation of capsid-specific T cells. *Blood* September 3, 2009;**114**(10):2077–86.

76. Manno CS, Pierce GF, Arruda VR, Glader B, Ragni M, Rasko JJ, et al. Successful transduction of liver in hemophilia by AAV-Factor IX and limitations imposed by the host immune response. *Nat Med* March 2006;**12**(3):342–7.

77. Hildinger M, Auricchio A, Gao G, Wang L, Chirmule N, Wilson JM. Hybrid vectors based on adeno-associated virus serotypes 2 and 5 for muscle-directed gene transfer. *J Virol* July 2001;**75**(13):6199–203.

78. Boutin S, Monteilhet V, Veron P, Leborgne C, Benveniste O, Montus MF, et al. Prevalence of serum IgG and neutralizing factors against adeno-associated virus (AAV) types 1, 2, 5, 6, 8, and 9 in the healthy population: implications for gene therapy using AAV vectors. *Hum Gene Ther* June 2010;**21**(6):704–12.

79. Wang Z, Ma HI, Li J, Sun L, Zhang J, Xiao X. Rapid and highly efficient transduction by double-stranded adeno-associated virus vectors in vitro and in vivo. *Gene Ther* December 2003;**10**(26):2105–11.

80. Everett RS, Evans HK, Hodges BL, Ding EY, Serra DM, Amalfitano A. Strain-specific rate of shutdown of CMV enhancer activity in murine liver confirmed by use of persistent [E1(−), E2b(−)] adenoviral vectors. *Virology* July 20, 2004;**325**(1):96–105.

81. Brooks AR, Harkins RN, Wang P, Qian HS, Liu P, Rubanyi GM. Transcriptional silencing is associated with extensive methylation of the CMV promoter following adenoviral gene delivery to muscle. *J Gene Med* April 2004;**6**(4):395–404.

82. Subramaniam A, Jones WK, Gulick J, Wert S, Neumann J, Robbins J. Tissue-specific regulation of the alpha-myosin heavy chain gene promoter in transgenic mice. *J Biol Chem* December 25, 1991;**266**(36):24613–20.

83. Aikawa R, Huggins GS, Snyder RO. Cardiomyocyte-specific gene expression following recombinant adeno-associated viral vector transduction. *J Biol Chem* May 24, 2002;**277**(21):18979–85.

84. Griscelli F, Gilardi-Hebenstreit P, Hanania N, Franz WM, Opolon P, Perricaudet M, et al. Heart-specific targeting of beta-galactosidase by the ventricle-specific cardiac myosin light chain 2 promoter using adenovirus vectors. *Hum Gene Ther* September 1, 1998;**9**(13):1919–28.

85. Prasad KM, Xu Y, Yang Z, Acton ST, French BA. Robust cardiomyocyte-specific gene expression following systemic injection of AAV: in vivo gene delivery follows a Poisson distribution. *Gene Ther* January 2011;**18**(1):43–52.

86. Fleury S, Simeoni E, Zuppinger C, Deglon N, von Segesser LK, Kappenberger L, et al. Multiply attenuated, self-inactivating lentiviral vectors efficiently deliver and express genes for extended periods of time in adult rat cardiomyocytes in vivo. *Circulation* May 13, 2003;**107**(18):2375–82.

87. Tilemann L, Ishikawa K, Weber T, Hajjar RJ. Gene therapy for heart failure. *Circ Res* March 2, 2012;**110**(5):777–93.

88. Katz MG, Swain JD, Tomasulo CE, Sumaroka M, Fargnoli A, Bridges CR. Current strategies for myocardial gene delivery. *J Mol Cell Cardiol* May 2011;**50**(5):766–76.

89. Roth DM, Lai NC, Gao MH, Drumm JD, Jimenez J, Feramisco JR, et al. Indirect intracoronary delivery of adenovirus encoding adenylyl cyclase increases left ventricular contractile function in mice. *Am J Physiol Heart Circ Physiol* July 2004;**287**(1):H172–7.

90. Donahue JK, Heldman AW, Fraser H, McDonald AD, Miller JM, Rade JJ, et al. Focal modification of electrical conduction in the heart by viral gene transfer. *Nat Med* December 2000;**6**(12):1395–8.

91. Iwatate M, Gu Y, Dieterle T, Iwanaga Y, Peterson KL, Hoshijima M, et al. In vivo high-efficiency transcoronary gene delivery and Cre-LoxP gene switching in the adult mouse heart. *Gene Ther* October 2003;**10**(21):1814–20.

92. Fang H, Lai NC, Gao MH, Miyanohara A, Roth DM, Tang T, et al. Comparison of adeno-associated virus serotypes and delivery methods for cardiac gene transfer. *Hum Gene Ther Methods* August 2012;**23**(4):234–41.

93. Lai NC, Tang T, Gao MH, Saito M, Miyanohara A, Hammond HK. Improved function of the failing rat heart by regulated expression of insulin-like growth factor I via intramuscular gene transfer. *Hum Gene Ther* January 12, 2012;**23**(3):255–61.

94. Inagaki K, Fuess S, Storm TA, Gibson GA, McTiernan CF, Kay MA, et al. Robust systemic transduction with AAV9 vectors in mice: efficient global cardiac gene transfer superior to that of AAV8. *Mol Ther* July 2006;**14**(1):45–53.

95. Raake PW, Tscheschner H, Reinkober J, Ritterhoff J, Katus HA, Koch WJ, et al. Gene therapy targets in heart failure: the path to translation. *Clin Pharmacol Ther* October 2011;**90**(4):542–53.

96. Hoppe UC, Marban E, Johns DC. Adenovirus-mediated inducible gene expression in vivo by a hybrid ecdysone receptor. *Mol Ther* February 2000;**1**(2):159–64.

97. Sipo I, Wang X, Hurtado Pico A, Suckau L, Weger S, Poller W, et al. Tamoxifen-regulated adenoviral E1A chimeras for the control of tumor selective oncolytic adenovirus replication in vitro and in vivo. *Gene Ther* January 2006;**13**(2):173–86.

98. Goverdhana S, Puntel M, Xiong W, Zirger JM, Barcia C, Curtin JF, et al. Regulatable gene expression systems for gene therapy applications: progress and future challenges. *Mol Ther* August 2005;**12**(2):189–211.

99. Rivera VM, Clackson T, Natesan S, Pollock R, Amara JF, Keenan T, et al. A humanized system for pharmacologic control of gene expression. *Nat Med* September 1996;**2**(9):1028–32.

100. Stieger K, Le Meur G, Lasne F, Weber M, Deschamps JY, Nivard D, et al. Long-term doxycycline-regulated transgene expression in the retina of nonhuman primates following subretinal injection of recombinant AAV vectors. *Mol Ther* May 2006;**13**(5):967–75.

101. Stieger K, Belbellaa B, Le Guiner C, Moullier P, Rolling F. In vivo gene regulation using tetracycline-regulatable systems. *Adv Drug Deliv Rev* July 2, 2009;**61**(7–8):527–41.

102. Harrison DE, Strong R, Sharp ZD, Nelson JF, Astle CM, Flurkey K, et al. Rapamycin fed late in life extends lifespan in genetically heterogeneous mice. *Nature* July 16, 2009;**460**(7253):392–5.

103. Villarreal FJ, Griffin M, Omens J, Dillmann W, Nguyen J, Covell J. Early short-term treatment with doxycycline modulates postinfarction left ventricular remodeling. *Circulation* 2003;**108**(12):1487–92.

104. Stieger K, Mendes-Madeira A, Meur GL, Weber M, Deschamps JY, Nivard D, et al. Oral administration of doxycycline allows tight control of transgene expression: a key step towards gene therapy of retinal diseases. *Gene Ther* December 2007;**14**(23):1668–73.

105. Rolain JM, Mallet MN, Raoult D. Correlation between serum doxycycline concentrations and serologic evolution in patients with Coxiella burnetii endocarditis. *J Infect Dis* November 1, 2003;**188**(9):1322–5.

106. Berman B, Perez OA, Zell D. Update on rosacea and anti-inflammatory-dose doxycycline. *Drugs Today (Barc)* January 2007;**43**(1):27–34.

107. Herzog RW, Hagstrom JN, Kung SH, Tai SJ, Wilson JM, Fisher KJ, et al. Stable gene transfer and expression of human blood coagulation factor IX after intramuscular injection of recombinant adeno-associated virus. *Proc Natl Acad Sci USA* May 27, 1997;**94**(11):5804–9.

108. Minniti G, Muni R, Lanzetta G, Marchetti P, Enrici RM. Chemotherapy for glioblastoma: current treatment and future perspectives for cytotoxic and targeted agents. *Anticancer Res* December 2009;**29**(12):5171–84.

109. Zoncu R, Efeyan A, Sabatini DM. mTOR: from growth signal integration to cancer, diabetes and ageing. *Nat Rev Mol Cell Biol* January 2011;**12**(1):21–35.

110. Yang W, Digits CA, Hatada M, Narula S, Rozamus LW, Huestis CM, et al. Selective epimerization of rapamycin via a retroaldol/aldol mechanism mediated by titanium tetraisopropoxide. *Org Lett* December 16, 1999;**1**(12):2033–5.

111. Abraham RT, Wiederrecht GJ. Immunopharmacology of rapamycin. *Annu Rev Immunol* 1996;**14**:483–510.

Chapter 15

Gene Therapy for the Prevention of Vein Graft Disease

Sarah B. Mueller[1,2] and Christopher D. Kontos[1,2,3]

[1]*Medical Scientist Training Program, Duke University School of Medicine;* [2]*Department of Pharmacology and Cancer Biology;* [3] *Department of Medicine, Duke University Medical Center, Durham, NC, USA*

LIST OF ABBREVIATIONS

Ad Adenovirus
AAV Adeno-associated virus
CABG Coronary artery bypass grafting
CCR2 C–C chemokine receptor-2
CNP C-type natriuretic peptide
COX-1 Cyclooxygenase-1
EC Endothelial cell
ECM Extracellular matrix
eNOS Endothelial nitric oxide synthase
HDAd Helper-dependent adenoviruses
HVJ Hemagglutinating virus of Japan
ICAM Intercellular adhesion molecule
IH Intimal hyperplasia
I/M Intima to media
IVC Inferior vena cava
MCP-1 Monocyte chemoattractant protein-1
mTOR Mammalian target of rapamycin
NFκB Nuclear factor κB
NO Nitric oxide
O_2^- Superoxide
ODN Oligodeoxynucleotide
PCNA Proliferating cell nuclear antigen
PGI_2 Prostaglandin I_2
PI3K Phosphoinositide 3-kinase
PREVENT *Pr*oject of *Ex* Vivo *Ve*in Graft *En*gineering via *T*ransfection
PTEN Phosphatase and tensin homology deleted on chromosome 10
ROS Reactive oxygen species
shRNA Short hairpin RNA
siRNA Small interfering RNA
SMC Smooth muscle cell
SOD Superoxide dismutase
SVG Saphenous vein graft
TF Tissue factor
TIMP Tissue inhibitor of metalloproteinase
TM Thrombomodulin
tPA Tissue plasminogen activator
VCAM Vascular cell adhesion molecule
VGD Vein graft disease
VSMC Vascular smooth muscle cell

Translating Gene Therapy to the Clinic. http://dx.doi.org/10.1016/B978-0-12-800563-7.00015-4

15.1 INTRODUCTION TO VEIN GRAFT DISEASE

Atherosclerosis resulting in obstruction of the coronary and peripheral arteries is a major cause of morbidity and mortality in the United States and other developed countries,[1] and the global burden of atherosclerotic cardiovascular disease is expected to grow rapidly in the coming decades. In the United States alone, one person has a coronary event due to coronary heart disease approximately every 25 seconds, and one person per minute will die as a result. Peripheral artery disease (PAD) is now recognized to be nearly as common as coronary artery disease (CAD), with over 8 million affected individuals in the United States.[1] Severe lower limb ischemia resulting from PAD can lead to tissue loss, such as skin ulceration and gangrene, which frequently necessitates amputation of the ischemic limb.

Several approaches, including pharmacological therapies and aggressive lifestyle modifications, can help control symptoms, slow disease progression, and stabilize chronic disease; however, these noninvasive interventions tend to be less effective in PAD than in CAD (reviewed in Refs 2,3). In both CAD and PAD, invasive revascularization is required in acute cases and to relieve severe symptoms that are not responsive to medical management. Less invasive percutaneous and endovascular approaches have become increasingly popular as a result of scientific and technological advances; however, bypass grafting remains the method of choice for patients in whom less invasive procedures have failed and in patients with multivessel CAD or complex lower extremity PAD. In the United States, over 400,000 patients undergo coronary artery bypass grafting (CABG) and over 40,000 patients undergo peripheral artery bypass surgery annually.[1,4] Arterial conduits are preferred whenever possible for both coronary and peripheral bypass grafts because of their superior long-term patency and improved event-free survival in patients with arterial grafts.[5,6] However, arterial conduits are of limited availability, and venous conduits, primarily the greater saphenous vein, are still the most commonly used vascular grafts.

Unfortunately, vein grafts are susceptible to failure as a result of a process referred to as vein graft disease (VGD). Although vein grafts can become occluded early after engraftment because of thrombosis, VGD is typically defined as progressive graft stenosis as a result of intimal hyperplasia (IH) and accelerated atherosclerosis. Collectively, these factors lead to eventual obstruction of blood flow in the vein graft, tissue ischemia, and the need for repeat revascularization procedures. Medical therapies, including antiplatelet agents and aggressive lipid-lowering therapy, have proven successful in reducing saphenous vein graft (SVG) atherosclerosis, the need for revascularization procedures, and the incidence of major adverse cardiovascular events in patients with vein grafts.[7] However, overall vein graft failure rates have remained fairly constant since the introduction of coronary artery bypass grafting in 1968 despite appreciable technical improvements since then.[8-10] One large study has estimated the rate of angiographic occlusion of coronary SVGs to be 15–20% at 1 year, approximately 25% at 5 years, and over 50% at 15 years or longer.[8] In the peripheral vasculature, failure rates as high as 50% within 5 years have been reported.[11]

Repeat CABG in patients with failed vein grafts is as effective as the initial procedure in terms of preventing subsequent cardiac events; however, it is costlier and riskier.[12] Repeat revascularization is also effective in the case of failed peripheral vein grafts and can extend the period between the diagnosis of severe limb ischemia and amputation.[13] However, the limited availability of conduits for repeat revascularization results in a 20–30% amputation rate within 1 year of bypass grafting.[14,15] Therefore, alternative strategies are being pursued to maintain long-term vein graft patency and prolong symptom-free survival after bypass grafting.

One promising alternative strategy to address this problem involves genetic reprogramming of vascular cells within the vein graft ex vivo prior to grafting into the arterial circulation. Vein grafts are ideal candidates for gene therapy for several reasons. First, the time between harvesting the vein and surgical engraftment into the arterial circulation provides a natural window for gene delivery. Second, VGD begins with the surgical dissection and removal of the vein graft from its native location; therefore, intraoperative gene therapy with rapidly expressing vectors allows molecular reprogramming to occur almost simultaneously with the initial injury that results in VGD (Figure 15.1). This strategy also provides the opportunity to pretreat the vein graft to protect the graft from injury or at least mitigate the ill effects of engraftment into the high pressure arterial circulation. Third, gene delivery to vein grafts can be performed ex vivo, which substantially diminishes many of the concerns inherent in systemic gene therapy.

This chapter will provide an overview of the pathophysiology of VGD with an emphasis on the temporal progression of disease and the role of different cell types in VGD development. The pros and cons of different gene delivery strategies will be discussed, with an emphasis on choice of vector and delivery strategy based on a thorough understanding of the role of the gene of interest in VGD pathogenesis. We will then briefly review gene targets used in prior preclinical studies and highlight promising and unexpected results. The PREVENT trials, the only clinical trials of gene therapy for VGD to date, will be discussed with a focus on potential reasons for the failure of these trials. Finally, we will summarize important considerations for designing future studies with the goal of translating the wealth of basic research in animal models of VGD into clinically meaningful benefits to patients.

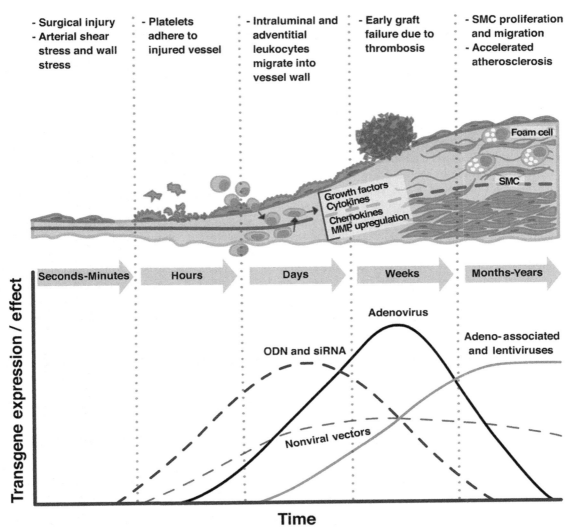

FIGURE 15.1 **Schematic representation of the phases of vein graft injury and timeline of different gene therapy effects.** Surgical harvesting induces injury of the vein graft endothelium, which is compounded by arterial shear force and wall stress after anastomosis. Rapid endothelial damage results in exposure of tissue factor and upregulation of adhesion molecules, increasing the risk of thrombosis and leading to adhesion and migration of inflammatory leukocytes into the vessel media. Upregulation of growth factors, cytokines, and matrix metalloproteinases (MMP) in the vein graft wall leads to smooth muscle cell (SMC) migration into the neointima, SMC proliferation, and resultant intimal hyperplasia and increased susceptibility to development of atherosclerotic plaques (depicted above as lipid-laden macrophages, or "foam cells," within the neointimal lesion). Onset and magnitude of transgene expression or effect varies depending on the gene delivery modality. Oligodeoxynucleotide (ODN) decoys and siRNA, which do not require gene expression, can act almost immediately; adenoviruses express high levels of transgene for brief periods; and adeno-associated viruses express moderate levels of transgene for extended periods.

15.2 PATHOPHYSIOLOGY OF VEIN GRAFT DISEASE

The gross and histopathological changes associated with each stage of VGD progression in humans are well characterized. However, the molecular mechanisms that underlie VGD progression are complex and incompletely understood. A comprehensive review of VGD pathogenesis is beyond the scope of this chapter, and for more detailed information the reader is referred to an excellent review on the topic by Mitra et al.[16]

Although the vasculature was historically perceived as a network of passive conduits, it is actually a complex organ system composed of heterogeneous blood vessels that are themselves complex multicellular tissues. Viewed simply, vessels larger than capillaries are composed of three layers: the intima, media, and adventitia. The intima refers to the innermost lining of the blood vessel and, in mature vessels, is composed of a continuous monolayer of quiescent endothelial cells (ECs) and the extracellular matrix (ECM) that they secrete. The media lies between the intima and adventitia and, in the saphenous vein, consists primarily of contractile smooth muscle cells (SMCs) surrounded by ECM. The adventitia is the outermost layer of the vessel and contains connective tissue, nerves, and the smaller vessels that supply the vessel wall with oxygen and nutrients (the vasa vasorum).

In large veins and arteries, these layers are separated by the internal and external elastic laminae. The composition, thickness, and function of these layers depend considerably on vessel size, anatomic location, and arterial or venous identity of the vessel. Moreover, cells of a single type in one vessel may have very different baseline characteristics, gene expression profiles, and responses to stimuli than cells of that type in another vascular bed (reviewed in Refs 17,18). The significant differences between arteries and veins likely contribute to the pathogenesis of VGD, and the heterogeneity between similar vessel types in different anatomic locations (for example, coronary arteries vs lower extremity arteries) may be important in the design and implementation of gene therapy approaches.

The pathogenesis of vein graft disease begins as soon as the vein is surgically harvested (Figure 15.2). Removal of the vein from the circulation cuts off blood supply to the lumen and vasa vasorum, resulting in ischemia followed by cellular hypoxia and increased susceptibility to ischemia-reperfusion injury after engraftment. Mechanical injury due to physical manipulation of the vein during harvesting contributes as well, and a "no-touch" technique for vein graft harvesting has been shown to improve patency rates (reviewed in Ref. 19). Once the vein is grafted into the high-pressure arterial circulation, increased wall stretch, high shear stress, and turbulent flow around residual venous valves result in additional injury to the vessel wall.[20] Disruption of the EC monolayer due to mechanical injury and EC death can lead to platelet adhesion, activation, and aggregation, activation of the coagulation cascade, and thrombosis.[21] The postoperative thrombogenicity of the injured vein can lead to acute thrombotic occlusion of the graft, which is the primary mechanism of vein graft failure within the first month.

After engraftment, vein grafts undergo a process referred to as "arterialization," in which the vessel adapts to the high pressure, high shear stress, and pulsatile flow of the arterial circulation. Vein graft arterialization is similar in concept to the thickening of renal and pulmonary veins in the setting of chronic hypertension. Arterialization is thought to be necessary for the vein to function in the arterial circulation,[22] and the ability of veins to arterialize is necessary for the maturation of arteriovenous fistulas used for dialysis access. However, this process also destroys the normal architecture of the vein, disturbing vascular homeostasis and potentially contributing to the pathogenesis of VGD.

The signaling pathways activated by loss of the protective EC monolayer and by arterial biomechanical forces converge to promote inflammation and vascular SMC proliferation. The secretion of growth factors and hormones by injured graft cells results in recruitment of graft-extrinsic inflammatory cells, which in turn secrete factors that perpetuate the injury response. Although numerous cell types are affected by this proliferative, proinflammatory microenvironment, the SMC response is the best understood. In the first 24h postoperatively, vascular SMCs undergo a well-characterized phenotypic transition from the quiescent, contractile cells responsible for normal regulation of vascular tone to proliferative, motile cells that migrate into the intima and upregulate expression of ECM genes (reviewed in Ref. 23). At the end of the first postoperative week, the new intima, or "neo-intima," is composed of SMCs and the ECM components they secrete. Thickening of the intima because of SMC infiltration, proliferation, and ECM deposition is known as intimal hyperplasia (IH) and is a histopathological hallmark of VGD.

IH is often cited as the primary cause of vein graft failure 1 month to 1 year postoperatively. Except in the case of occlusive perianastomotic IH, it is unclear whether luminal narrowing due to IH contributes directly to morbidity and mortality (reviewed in Ref. 24). However, the neointima is known to form a nidus that leads to accelerated atherosclerosis, which is

FIGURE 15.2 Vector delivery strategy to prevent vein graft disease. Prior to surgical harvesting of the vein graft, both the endothelium and underlying SMCs are in a quiescent state. Injury occurs within minutes as a result of ischemia and mechanical trauma and is exacerbated upon anastomosis into the arterial circulation. Transgene delivery is performed ex vivo and occurs rapidly and with a low risk of systemic transgene toxicity. Endothelial injury, thrombosis, inflammation, and SMC activation have been targeted to prevent vein graft intimal hyperplasia.

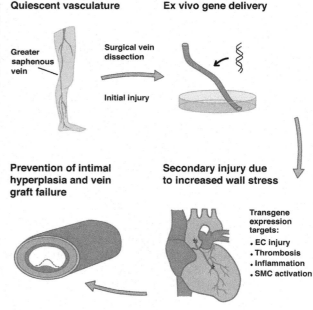

responsible for the majority of late coronary graft failures.[25] Additionally, the atherosclerotic plaques that develop in vein grafts tend to display many of the hallmarks of unstable plaques, such as lack of a fibrous cap, reduced calcium deposition, and high lipid content.[26] Moreover, vein graft atherosclerosis appears to be more resistant to medical management than atherosclerosis in native coronary arteries, necessitating alternative approaches such as gene therapy (reviewed in Ref. 27).

15.3 GENE DELIVERY STRATEGIES

Based on the pathogenic mechanisms that cause VGD, numerous potential gene therapy targets exist and many have been tested in preclinical models. These gene targets will be discussed in detail below. First, however, this section will address the specific mechanisms used to deliver therapeutic genes to the vein graft. As described, the pathogenesis of VGD is complex, depends on numerous cell types, and varies significantly over time. The optimal gene delivery strategy would target the appropriate cell type(s), induce expression during the relevant time period, and result in an adequate expression level to achieve the desired biological effect. This section reviews the features of existing gene delivery strategies as they relate to the pathophysiology of VGD and highlights important considerations when using these strategies. Important features of each vector and gene delivery strategy are summarized in Table 15.1.

TABLE 15.1 Comparison of Gene Delivery Strategies

Vector	Cell Tropism	Packaging Capacity	Transgene Expression	Delivery Strategies	Production Concerns	Limitations
Naked DNA or RNA	Nonspecific	N/A	Onset in minutes to hours Duration intermediate, from days to months Magnitude of expression low	Pressure-mediated transfection,[28–31] lipofection,[32,33] nanoparticles, biopolymer gels	Very easy to produce on large scale	Potential damage to vein grafts by pressure-mediated transfection and lipofection; potential nonspecific toxicity at high concentrations
Adenovirus	Transduces most cell types including dividing and quiescent cells	Large (up to around 10 kb)	Onset in hours to days Duration intermediate, from days to weeks Magnitude of expression high	Intra- and/or extraluminal incubation with[34] or without[35] distension	Easily produced in very high titers for in vivo studies	Host inflammatory response may reduce efficiency and contribute to pathological inflammation Patient death in clinical trial of systemic adenovirus delivery, may or may not be of concern for ex vivo delivery
Adeno-associated virus (AAV)	Transduces a variety of cell types including dividing and quiescent cells Different serotypes of recombinant AAV can be used to selectively target specific cell types	Small (2–5 kb depending on the vector)	Onset in days to weeks Duration long, from months to years Magnitude of expression intermediate to high depending on target tissue and AAV serotype	Intra- and/or extraluminal incubation with or without distension	Difficult to produce in high titers	Likely very low risk of insertional mutagenesis with recombinant AAV Potential for host inflammatory response, but substantially less than adenovirus Presence of circulating antibodies to some serotypes may limit efficiency
Lentivirus	Transduces a variety of cell types including dividing and quiescent cells	Generally <10 kb, but titer significantly decreases with insert size	Onset in days to weeks Duration long, from months to years Magnitude of expression intermediate	Intra- and/or extraluminal incubation with or without distension, biopolymer gels[36]	Easy to generate, but difficult to produce in high titers	Risk of insertional mutagenesis

15.3.1 Nonviral Vectors

Nonviral vectors are attractive options for gene therapy because they are considered safer than viral vectors and are generally easier to prepare. However, they are associated with reduced expression (both in magnitude and duration) compared to viral vectors. In the following section, we will review nonviral vectors including plasmid DNA, oligodeoxynucleotides (ODN), DNAzymes, small interfering RNA (siRNA), and ribozymes.

DNA- and RNA-based vectors have several advantages for use in the prevention of VGD. First, they have no inherent tropism and can therefore be optimized to transfect both ECs and SMCs; however, transfection efficiency tends to be low. Second, the onset of transgene expression using these vectors can begin within minutes of transfection and has been shown to last as long as 8 weeks in animal models.[37] This makes these vectors ideal for use with gene targets that play a role in the earliest stages of endothelial injury and supports their use with transgenes thought to be important during the critical period of remodeling within the first month after engraftment. Third, ex vivo delivery of ODNs was shown to be safe in a clinical study of gene therapy for VGD.[38]

There are numerous plasmid DNA vectors that allow expression of beneficial transgenes in mammalian cells. The basic components required for protein expression from a plasmid are a promoter, the cDNA of the gene of interest, and a polyadenylation sequence downstream of the transgene to ensure proper handling of the RNA transcript. The promoter sequence can be varied to achieve specific desired effects. Viral promoters, such as those from cytomegalovirus (CMV) or simian virus 40 (SV40), are the most commonly used as they result in constitutive, high-level expression. Tissue-specific promoters, such as the Tie2 promoter (which is mostly active in ECs) or the SM22α promoter (which is active in vascular SMCs), can be used to limit expression to the desired cell type given that nonviral vectors have no inherent tropism. Plasmids can also be designed to express short hairpin RNA (shRNA) to reduce the expression of genes that contribute to VGD progression.

The remaining nonviral vectors—siRNA, ribozymes, and DNAzymes—are oligonucleotides designed to knock down gene expression by degrading messenger RNA (mRNA) and thus preventing translation. siRNA is commonly used in many research laboratories, and a significant amount of preclinical data has shown that siRNAs can effectively knock down gene expression in tissues in vivo as well as in cultured cells in vitro. DNAzymes and ribozymes are catalytic oligonucleotides that can be designed to target mRNA sequences with high specificity (reviewed in Ref. 39). These vectors are relatively new in the field of gene therapy for VGD; however, DNAzymes have recently been used successfully in a preclinical model of VGD and been shown to be safe in clinical trials for other conditions.[40]

ODNs are typically composed of short double-stranded DNA sequences that mimic transcription factor binding sites, thereby acting as "decoys" to block the binding of transcription factors involved in regulating cellular proliferation and other relevant processes. This represents a unique strategy because it does not target a single transcript for over- or under-expression but rather targets the expression of a panel of genes regulated by a given transcription factor. The PREVENT trials utilized an ODN approach with edifoligide, an ODN designed to act as a decoy for the proliferative transcription factor E2F1.[38]

15.3.2 Viral Vectors

Viral vectors commonly used in preclinical VGD models include adenovirus (Ad), adeno-associated virus (AAV), and lentivirus. Each vector has distinct cell tropism, transgene expression profiles, and safety concerns that impact its potential for use in human VGD (Table 15.1). Similar to the plasmid vectors described earlier, viral vectors can be designed to achieve transgene overexpression or gene expression knock down by incorporating specific shRNAs.

15.3.2.1 Adenovirus

Adenoviruses (Ad) are nonenveloped double-stranded DNA viruses with large packaging capacities that can be produced in high titers. They are able to infect both dividing and quiescent cells and can efficiently transduce a wide variety of cell types, including ECs and SMCs. This is particularly beneficial in the case of VGD given that cells in the graft are quiescent at the time of tissue harvest but rapidly begin to proliferate after engraftment. The onset of gene expression from Ad vectors is rapid (sometimes within hours), but tends to peak by day 7 and dissipate by day 28, potentially making them less useful in treating later stages of disease progression (such as the development of atherosclerosis).

There are several potential reasons for this effect. First, Ad vectors do not incorporate into the genome, which obviates the risk of insertional mutagenesis. However, when SMCs proliferate within the neointima, new, nontransduced cells gradually replace the cells expressing the transgene. Second, approximately 60% of the human population is seropositive for antibodies against adenovirus, and cells expressing adenoviral proteins are aggressively eliminated by cytotoxic T cells. This inflammatory response selectively removes cells expressing the transgene and may perpetuate endothelial injury and exacerbate IH development.[41] Third-generation or helper-dependent adenoviruses (HDAd) do not express viral genes and are therefore less immunogenic.[42] Compared to first-generation Ad, HDAd vectors appear to induce significantly less inflammation and IH in rabbit carotid injury models; however, HDAd has not been tested in VGD models to our knowledge.[43]

Additionally, there are legitimate safety concerns regarding the use of adenovirus in human patients stemming from a patient death during a clinical trial of gene therapy using direct intrahepatic adenovirus administration.[44] Because vein grafts are treated ex vivo and can be flushed with saline prior to engraftment to remove excess unincorporated virus, this concern may be less applicable but is still worthy of careful consideration.

15.3.2.2 Adeno-Associated Virus

Recombinant adeno-associated viruses (AAVs) are single-stranded DNA viruses of the parvovirus family that do not integrate into the genome but nonetheless effect long-term gene expression (months to years). AAV are able to infect quiescent cells, and modifications to the AAV capsid have dramatically enhanced the ability of AAV vectors to transduce ECs and SMCs despite their lack of native tropism for these cell types.[45,46] The long duration of expression from AAV vectors may make them particularly useful in suppressing SMC proliferation and, potentially, in preventing the development of atherosclerosis later in the disease process. Recombinant AAVs are deleted of all viral genes and therefore have substantially lower immunogenicity than adenoviruses, and they have shown an excellent safety profile in a large number of preclinical and clinical trials for other conditions.[47,48]

However, there are several potential limiting factors in using AAV vectors for VGD. The first is that transgene expression is delayed as a result of the time required to replicate the single-stranded viral DNA. This can be circumvented somewhat by the use of self-complementary AAV vectors, which enable the genome to form double-stranded DNA through folding rather than replication (reviewed in Ref. 49); however, these vectors have not yet been tested in the vasculature. An additional minor concern is that it is possible that recombinant AAV could incorporate into the host genome. Wild-type AAV is known to incorporate into a specific location on human chromosome 19, although it is generally felt that this potential is lost in recombinant vectors. Thus, although the risk of insertional mutagenesis is low, it remains a consideration. Finally, AAV vectors have a small packaging capacity and are difficult to produce in the high titers needed for many in vivo applications.

15.3.2.3 Lentivirus

Lentiviruses are a genus within the retrovirus family but, unlike most retroviruses, they are able to infect nondividing cells. Recombinant lentiviruses are able to efficiently transduce ECs and SMCs and are far less immunogenic than adenoviruses.[50] Like retroviruses, lentiviral vectors incorporate into the host genome, resulting in transgene expression over months to years. However, this does raise safety concerns regarding the possibility of insertional mutagenesis. Nonintegrating vectors have been developed, but the safety and transduction efficiency of these modified vectors in the vasculature is unclear.[51]

15.3.3 Vector Delivery Strategies

Delivery of gene transfer vectors to vein grafts can be accomplished in a variety of ways. They can be applied to the outer surface of the vein, intraluminally, or both, as well as with or without pressurization (Table 15.1).

Viral vectors have typically been delivered using an intraluminal method. After the vein is harvested, a solution containing the viral vector at the desired titer is infused into the vein lumen and the ends are clamped. The vein is then incubated in a buffer solution until the bypass site has been prepared; a typical incubation time is 20–30 min at room temperature, although other conditions have been explored in preclinical studies. Intraluminal delivery is typically performed either without distension or with a distension pressure of approximately 10 mmHg. Prior to engraftment, the vein is flushed gently with heparinized saline, thereby removing the majority of unincorporated viral particles.[34] There is some concern that intraluminal strategies, particularly with distending pressure, may damage the intima of the vein graft further, hastening the development of IH.[52] DNA- and RNA-based vectors can be introduced intraluminally as well; however, transfection efficiency tends to be low without the aid of transfection reagents, such as liposomes or nanoparticles.

Nondistending pressure-mediated gene transfer refers to a technique in which the vector solution is introduced into the lumen and also bathes the outside of the vein. The bath containing the vein graft is incubated in a chamber pressurized to

300 mmHg, theoretically facilitating pressure-mediated transfection without damaging the vein by distension. This method is typically used for DNA- and RNA-based vectors and can achieve efficient transfection without the use of additional transfection reagents. Although this method avoids distending pressure, it is possible that the pressurization technique may damage the graft. This method has been used extensively in preclinical trials and was used to deliver the E2F decoy ODN edifoligide in the PREVENT trials, the only clinical trials to date of gene therapy for VGD.[9,38,53,54] The PREVENT trials are discussed more extensively below.

Extraluminal gene delivery is generally less common, as it would seem more efficient to target ECs and SMCs via the graft lumen. However, several preclinical studies targeting SMC proliferation have successfully employed an extraluminal approach.[55–57] In these studies, a gel (e.g., poloxamer) containing the DNA or RNA vector together with a transfection reagent was applied to the adventitia of the graft. The transfection complex then diffuses into the media and intima, promoting transgene expression throughout the vessel wall. This delivery modality appears to be safe in preclinical studies, and the application of a poloxamer gel alone does not exacerbate IH, unlike gene delivery methods that require intraluminal distending pressure.[57]

15.4 ANIMAL MODELS OF VEIN GRAFT DISEASE

A brief introduction to the common animal models of VGD is necessary to understand and interpret the wealth of preclinical data generated in animals. Animal VGD experiments are performed in mammals ranging in size from mice to pigs. Typically, models in smaller animals (e.g., mice, rats, and rabbits) utilize interposition grafts in which a segment of vein is anastomosed "end-to-end" to the artery, replacing a segment of the artery entirely. These grafts would be expected to have distinct hemodynamic characteristics from coronary and peripheral bypass grafts, which are typically anastomosed end-to-side, thereby branching off of the aorta or a proximal peripheral artery to bypass the atherosclerotic occlusion. Moreover, these models often use very different vessels than those in human bypass surgeries. For example, jugular vein–carotid artery, inferior vena cava (IVC)–carotid artery, and IVC–aorta interposition grafts are the most commonly used models in rabbits and mice. However, larger animals (e.g., dogs, sheep, pigs, and nonhuman primates) have been used in bypass models that more closely resemble CABG and peripheral bypass grafting in humans.

One major flaw in both small and large animal models is the lack of relevant comorbid diseases, such as hypercholesterolemia, hypertension, and diabetes mellitus, which are seen routinely in patients undergoing bypass operations. Furthermore, even when these conditions can be induced in animal models, the pathophysiology and disease progression may be very different from human disease. For example, development of atherosclerosis in pigs is similar to humans in terms of induction by diet, distribution of plaques, and morphology of lesions. In contrast, most rodent strains and dogs are not susceptible to coronary atherosclerosis, and even transgenic mouse models of atherosclerosis result in lesions that are very different in distribution and morphology than those seen in human disease. There is also significant variability between humans and animal models in terms of the expression and distribution of certain genes. For example, humans, dogs, and nonhuman primates express the angiotensin II-forming enzyme chymase, while rodents express only angiotensin-converting enzyme (ACE). Therefore, studies targeting the renin–angiotensin pathway might have very different results depending on the choice of animal model.

Finally, the effect of a given treatment on VGD in an animal model is typically measured by quantifying IH. As mentioned earlier, clinically relevant endpoints (such as graft occlusion, acute coronary events, and death) typically do not occur in animal models of VGD or are not assessed, and the relationship between IH and these endpoints is unclear. Several studies have recorded additional metrics, such as macrophage infiltration into the graft wall, the proportion of proliferating cells, and the number of SMCs. However, few studies have measured functional readouts, such as thrombogenicity and the ability to respond to physiological signals like acetylcholine.

No animal model will perfectly mimic human VGD in terms of pathophysiology or molecular mechanisms. However, it is important to carefully consider all aspects of a potential model in order to choose the one that best approximates human disease with respect to the specific pathway of interest. For a more detailed review of these models and important technical considerations, the reader is referred to reviews in Refs 58–60.

15.5 GENE TARGETS AND PRECLINICAL STUDIES

This section provides a broad overview of gene therapy approaches that have been tested in preclinical models of VGD, which are summarized in Table 15.2. For a more comprehensive discussion of gene therapy targets that were tested in preclinical models prior to 2013, the reader is referred to the review in Ref. 73.

TABLE 15.2 Summary of Preclinical and Clinical Studies of Gene Delivery Modalities for the Prevention of Vein Graft Disease

Target	Approach	Vector	Model	Delivery Modality	Summary of Results	References
Preclinical Studies						
EC health	EC-SOD over-expression (in combination with 35K and TIMP-1)	Adenovirus	Rabbit jugular–carotid artery interposition graft	Intraluminal with distension	Reduced I/M ratio at 2 weeks in combination with 35K and at 2 and 4 weeks with TIMP-1 Reduced macrophage infiltration and BrDU staining in combination with 35K at 4 weeks Slight reduction in superoxide anion production at 2 weeks with EC-SOD alone	61
	eNOS over-expression	HVJ-liposome	Canine femoral artery bypass	Intraluminal distending pressure-mediated transfection	>50% reduction in intimal thickness at 4 weeks	32
	eNOS over-expression	HVJ-liposome	Hypercholesterolemic rabbit jugular vein–carotid artery interposition graft	Intraluminal distending pressure-mediated transfection	30% reduction in intimal thickness at 4 weeks Reduction in neointimal macrophages at 4 weeks, decreased PCNA positive cells at 2 weeks	33
	COX-1 (PGI_2) over-expression	Adenovirus	Hypercholesterolemic rabbit jugular vein–carotid artery interposition graft	Intraluminal without distension	No difference in IH Increased luminal area, blood flow, and PGI_2 at 4 weeks	35
	CNP over-expression	Adenovirus	Rabbit jugular–carotid artery interposition graft	Intraluminal with distension	Increased reendothelialization, decreased thrombus formation, and 60% reduction in intimal thickness at 2 and 4 weeks	62
Thrombosis	Thrombomodulin over expression	Erroneous vector	Rabbit jugular–carotid artery interposition graft	Intraluminal with distension	No difference in IH Reduced thrombin generation at 7 days	63
	tPA over-expression	Adenovirus	Porcine saphenous–carotid artery interposition graft	Intraluminal without distension	Reduction in flow-restricting thrombi at 1 day	64

Continued

TABLE 15.2 Summary of Preclinical and Clinical Studies of Gene Delivery Modalities for the Prevention of Vein Graft Disease—cont'd

Target	Approach	Vector	Model	Delivery Modality	Summary of Results	References
Inflamma-tion	MCP-1 inhibition (7ND-MCP-1)	Naked plasmid	Hypercholesterolemic mouse donor IVC–carotid artery interposition graft	Systemic expression (IM injection and electroporation)	51% decrease in IH at 4 weeks No change in macrophage infiltration	65
	MCP-1 inhibition (7ND-MCP-1)	Adenovirus	Canine jugular–carotid artery interposition graft	Submersion in adenoviral solution without pressurization	Decreased IH and lumen stenosis at 4 weeks Decreased PCNA positive cells and macrophages at 4 weeks	66
	MCP-1 inhibition (CCR2 silencing)	Lentiviral shRNA	Hypercholesterolemic mouse donor IVC–carotid artery interposition graft	Adventitial application of bioprotein gel	38% reduction in IH at 4 weeks	36
	CC-Chemokine inhibition (35K)	Adenovirus	Hypercholesterolemic mouse donor IVC–carotid artery interposition graft	Systemic expression (IV injection 2 days pre-op)	Reduced wall thickness, Ki67 positive cells, SMA positive cells, and macrophage infiltration at 2 and 4 weeks	67
	CC-Chemokine inhibition (35K)	Adenovirus	Rabbit jugular–carotid artery interposition graft	Intraluminal with distension	Reduced I/M ratio at 2 weeks Reduced macrophage infiltration and BrdU incorporation at 4 weeks	68
	NFκB inhibition	ODN	Canine aortocoronary bypass	Nondistending pressure-mediated transfection	Decreased intimal area and PCNA positive cells at 4 weeks	69
	NFκB inhibition	ODN	Hypercholesterolemic rabbit jugular vein–carotid artery interposition graft	Nondistended; pressure-mediated transfection	Decrease in IH at 4 weeks Improved endothelium-mediated vasorelaxation	53
VSMC	PTEN over-expression	Adenovirus	Canine aortocoronary bypass	Intraluminal pressure-mediated delivery	Reduced intimal area and I/M ratio at 3 months	34
	TIMP-3 over-expression	Adenovirus	Porcine saphenous–carotid artery interposition graft	Intraluminal delivery without distention	58% reduction in IH at 28 days	70
	TIMP-3 over-expression	Adenovirus	Porcine saphenous–carotid artery interposition graft	Intraluminal delivery without distention	Reduced IH at 3 months	71
	E2F inhibition	ODN	Hypercholesterolemic rabbit jugular vein–carotid artery interposition graft	Nondistended; pressure-mediated transfection	Decrease in IH at 6 months	28–31

TABLE 15.2 Summary of Preclinical and Clinical Studies of Gene Delivery Modalities for the Prevention of Vein Graft Disease—cont'd

Target	Approach	Vector	Model	Delivery Modality	Summary of Results	References
Clinical Studies						
PREVENT I (phase I)	E2F inhibition	ODN	Human infrainguinal arterial bypass with autologous vein	Nondistended; pressure-mediated transfection	Decrease graft occlusions, revisions, or critical stenoses at 12 months	38
PREVENT II (phase IIb)	E2F inhibition	ODN	Human coronary artery bypass grafting with autologous vein	Nondistended; pressure-mediated transfection	30% relative reduction in critical stenosis at 12 months	72
PREVENT III (phase III)	E2F inhibition	ODN	Human infrainguinal arterial bypass with autologous vein	Nondistended; pressure-mediated transfection	No significant difference in primary graft patency or limb salvage at 12 months	14
PREVENT IV (phase III)	E2F inhibition	ODN	Human coronary artery bypass grafting with autologous vein	Nondistended; pressure-mediated transfection	No significant difference in critical graft stenosis at 12 months	9

Abbreviations: CCR2, C–C chemokine receptor-2; CNP, C-type natriuretic peptide; COX-1, cyclooxygenase-1; EC, endothelial cell; EC-SOD, extracellular superoxide dismutase; eNOS, endothelial nitric oxide synthase; HVJ, hemagglutinating virus of Japan; IH, intimal hyperplasia; I/M, intima to media; IVC, inferior vena cava; MCP-1, monocyte chemoattractant protein-1; NFκB, nuclear factor κB; ODN, oligodeoxynucleotide; PCNA, proliferating cell nuclear antigen; PD-ECGF, platelet-derived endothelial cell growth factor; PGI$_2$, prostaglandin I$_2$; PREVENT, *Project of Ex Vivo Vein Graft Engineering via Transfection*; PTEN, phosphatase and tensin homology deleted on chromosome 10; SMA, smooth muscle actin; TIMP, tissue inhibitor of metalloproteinase; tPA, tissue plasminogen activator; VSMC, vascular smooth muscle cell.

15.5.1 Endothelial Injury and Reendothelialization

Normal vascular function is maintained largely by the quiescent ECs that line mature vessels (reviewed in Ref. 74). During the process of vein graft harvesting and arterial engraftment, the endothelium within the graft is injured by mechanical manipulation, ischemia, and the altered biomechanical forces of the arterial circulation (Figure 15.3).

Endothelial cell denudation and death because of these insults result in loss of the physical barrier between the blood and media of the vessel as well as loss of many vasoprotective substances that are normally produced by healthy ECs. Gene therapy approaches to combat these effects aim to promote EC survival, increase the production of vasoprotective factors, and enhance reendothelialization of the injured vessel, thereby preventing subsequent development of VGD.

The primary strategies for promoting EC survival have focused on ameliorating oxidative damage from reactive oxygen species (ROS) generated when the graft is reperfused after a period of ischemia. The principal ROS studied in this context is O_2^-, which is degraded by superoxide dismutases (SOD). Compared to SVGs, arterial grafts have greater SOD activity at the time of harvest, which may contribute to their lower failure rates after engraftment.[75] In a rabbit model, intraoperative delivery of an adenovirus encoding extracellular SOD reduced macrophage accumulation when used alone and synergistically reduced the intima-to-media (I/M) ratio and proliferation index in the graft wall when used in combination with other gene therapies.[61] Gene therapies with other antioxidant enzymes including catalase, heme oxygenase-1, and glutathione peroxidase have been used successfully to reduce atherosclerosis and arterial neointimal hyperplasia, but they have not been tested in models of VGD.[76,77]

Among the vasoprotective factors that have been targeted using gene therapy approaches are nitric oxide (NO) and prostacyclin (PGI$_2$). NO is synthesized primarily by the constitutively expressed enzyme endothelial nitric oxide synthase (eNOS) and is critical for normal vascular function. NO induces vasodilation, prevents SMC proliferation, and inhibits adhesion of platelets and leukocytes to the endothelium.[78,79] Impaired release of NO from the damaged vein graft endothelium is therefore thought to contribute to the development of VGD. Although ex vivo studies have provided a strong rationale for enhancing NO production by overexpressing eNOS, the results of in vivo testing have been modest. Intraluminal transfection of bovine eNOS cDNA encapsulated in hemagglutinating virus of Japan (HVJ) liposomes reduced intimal

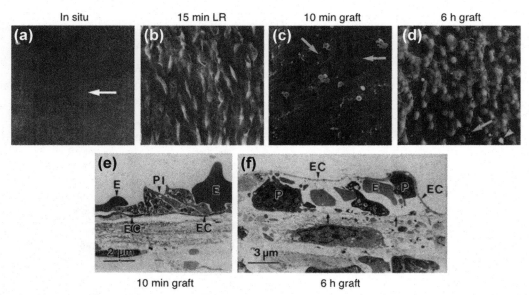

FIGURE 15.3 **Early effects of engraftment on endothelial injury in saphenous vein grafts.** Scanning electron micrographs of rabbit veins: (a) fixed in situ before harvesting (arrow, intact EC junctions); (b) after harvest and incubation for 15 min in lactated Ringer's solution (LR), demonstrating disruption of the EC monolayer; (c) 10 min after engraftment into the arterial circulation, demonstrating adherent platelets and red blood cells (arrows, bulging EC nuclei); and (d) 6 h after engraftment, showing a thin but intact endothelium with numerous cells in the subendothelial space (arrow, adherent platelets; arrowhead, adherent polymorphonuclear leukocyte (PMN)). Transition electron microscopy (TEM) demonstrates thinned ECs with adherent platelets (Pl) and erythrocytes (E) 10 min after engraftment (e). (f) Six hours after anastomosis TEM demonstrates persistent EC thinning with erythrocytes (E) and PMNs (P) in the subendothelial space (arrows, basal lamina). Original magnification: (a)–(d) ×600; (e) ×5000; (f) ×3400. *Adapted with permission from Ref. 20.*

thickness at 4 weeks postgrafting in both canine and hypercholesterolemic rabbit models.[32,33] However, the reduction in intimal thickness was modest (approximately 30% in the rabbit model and 50% in the dog model), potentially because of damage to the vein graft resulting from the distending pressure utilized for transfection.[52]

PGI$_2$ is also constitutively produced by healthy endothelium in a series of enzymatic reactions that are rate limited by the enzyme cyclooxygenase-1 (COX-1). PGI$_2$ is a potent vasodilator that also inhibits platelet aggregation, leukocyte adhesion, and SMC proliferation in vitro.[80] COX-1 catalytic activity results in the enzyme's own rapid degradation; therefore, its activity is primarily controlled by its expression level.[81] Treatment of rabbit vein grafts with an adenovirus encoding human COX-1 resulted in improved PGI$_2$ production and blood flow through the graft; however, it had no effect on intimal thickness.[35] These results highlight that, as in humans, the degree of IH does not always correlate with graft function (i.e., perfusion) in animal models.

An additional strategy is to promote reendothelialization or endothelial "healing" in the graft to more quickly regain normal protective endothelial function. Data from animal models indicate that higher rates of reendothelialization of injured areas correlate with reduced IH in vein grafts.[82] There are few studies explicitly testing gene therapy to promote reendothelialization in the context of VGD. However, one study utilizing the C-type natriuretic peptide (CNP) gene represents a novel approach that targets several aspects of VGD simultaneously. CNP is produced by the endothelium and influences vascular tone and physiology in a variety of ways, including promoting vasorelaxation, inhibiting SMC proliferation, and potentially suppressing certain prothrombotic genes (reviewed in Ref. 83). In a rabbit vein graft model, adenoviral expression of CNP reduced mural thrombus formation, decreased neointimal thickness, and improved endothelial coverage of the graft lumen at weeks two and four postoperatively.[62] There is also a significant body of literature describing attempts to enhance endothelialization of vein grafts and other vascular conduits by administering genetically engineered exogenous cells (e.g., progenitor cells); however, these studies are beyond the scope of this chapter.

15.5.2 Thrombosis

In addition to the loss of endothelial vasoprotective substances, disruption of the endothelium exposes the subendothelial basement membrane to platelets, initiating the coagulation cascade. Moreover, SMCs in the saphenous vein media, which may also be exposed as a result of endothelial denudation, express prothrombotic genes such as tissue factor (TF) and tissue-type plasminogen activator (tPA).[84] Even if the graft does not occlude postoperatively because of thrombosis, this early activation of the coagulation cascade contributes to the development of IH via the release of factors that promote SMC proliferation and migration.[85–87] Anticoagulation does not eliminate the risk of early graft thrombosis, and systemic anticoagulation is associated with bleeding complications, which can be particularly dangerous postoperatively. Gene therapy of the graft ex vivo to prevent thrombosis could therefore circumvent these complications.

The two primary gene targets that have been assessed in VGD models are the anticoagulant proteins thrombomodulin (TM) and tPA. TM binds to the procoagulant protein thrombin, inactivating it and activating the antithrombotic protein C pathway. The expression of TM in vein grafts decreases dramatically upon exposure to the arterial circulation, so gene therapy approaches have aimed to replace this activity by overexpressing TM.[88] The success of TM gene therapy in preclinical studies varies depending on the model and primary endpoint. In a rat model, liposomal transfection of TM improved thromboresistance in the graft; however, adenoviral delivery of human TM had no effect on IH in a rabbit model.[63,88] Similarly, the effects of tPA gene therapy on the development of IH are unclear. Like TM, tPA expression is decreased after vein engraftment, impairing lysis of thrombi that form in the graft. In a rabbit VGD model, Ad-tPA reduced IH at 4 weeks, but in a rabbit arterial injury model, Ad-tPA treatment promoted IH.[64,89] These conflicting results are likely due in part to differences in the pathogenesis of IH in arteries and veins, and they highlight the importance of using disease-specific models when testing gene therapy approaches.

15.5.3 Inflammation

Given that most graft failures after the first year are due to atherosclerosis and its sequelae, targeting inflammation in the graft wall may yield important and clinically relevant results. Platelet activation at the sites of endothelial denudation initiates an inflammatory cascade that promotes the development of IH and the accelerated atherosclerosis that is characteristic of vein grafts. The molecular mechanisms of EC interaction with leukocytes are reviewed in detail elsewhere;[90] however, a brief overview is included here to provide context for the gene therapy targets discussed below. First, leukocytes rolling on the luminal surface of the graft interact with P-selectin on adherent platelets via P-selectin glycoprotein ligand-1 (PSGL-1) on the leukocyte surface. Binding of PSGL-1 to P-selectin activates leukocytes to migrate into the vessel wall via the engagement of leukocyte transmembrane proteins with the EC receptors ICAM-1 and VCAM-1. The inflammatory response is perpetuated by cytokines and growth factors secreted by activated leukocytes as well as by graft-intrinsic cells such as ECs and SMCs. A number of these secreted factors not only promote inflammation but also induce EC apoptosis and promote SMC proliferation, further contributing to the pathogenesis of VGD.

One important factor that has been targeted in several gene therapy approaches is monocyte chemoattractant protein-1 (MCP-1), also known as CCL2. MCP-1 is a well-studied chemokine that is strongly linked to atherosclerosis in humans and experimental models (reviewed in Ref. 91). It is secreted by ECs, SMCs, fibroblasts, and leukocytes and potently recruits monocytes and macrophages, which eventually form the atherosclerotic plaque. In rat models of VGD, MCP-1 is upregulated within hours after engraftment, and expression persists for at least 8 weeks, indicating that targeting this gene in the early and intermediate stages of VGD may be useful.[92] Because overexpression of MCP-1 would likely exacerbate atherosclerosis and plaque instability, approaches to target MCP-1 differ from other studies described above, which have focused primarily on gene overexpression. One strategy targeted MCP-1's effects by using lentivirus to deliver shRNAs targeting the MCP-1 receptor, CCR2. In an ApoE3 Leiden mouse model, this approach decreased macrophage recruitment as well as SMC proliferation and graft wall thickness.[36] Another innovative approach utilized 7ND-MCP-1, which is a dominant negative N-terminal deletion mutant of MCP-1 that binds to CCR2, thereby preventing it from binding wild-type MCP-1.[93] Transfection of a 7ND-MCP-1-encoding plasmid reduced IH in the ApoE3 Leiden mouse model and in human saphenous vein segments in vitro.[65] In a canine model, adenoviral expression of 7ND-MCP-1 also reduced IH without impairing reendothelialization.[66]

Another innovative approach to preventing inflammation in vein grafts is to target NFκB, a transcription factor that is downstream of several inflammatory stimuli and cell stressors and controls the expression of numerous important chemokines and receptors. NFκB is an important factor in a number of inflammatory conditions, including atherosclerosis (reviewed in Ref. 94). Because NFκB is a transcription factor, it has most often been targeted using the decoy ODN approach. In hypercholesterolemic rabbit models, administration of a decoy ODN only modestly reduced the I/M ratio but dramatically reduced macrophage infiltration into the graft wall, indicating that this approach may be more effective at targeting atherosclerosis than IH.[53] NFκB decoy-treated grafts also displayed improved NO-mediated vasodilation induced by acetylcholine, indicating that inhibition of NFκB may preserve the function of surviving ECs. A study using the same decoy ODN in a canine model demonstrated that neointimal area and SMC proliferation were reduced in the treatment group.[69] Markers of atherosclerosis were not assessed because the canine model does not develop atherosclerosis spontaneously.

All of the approaches discussed above have utilized wild-type or mutant mammalian genes. The following strategy is novel in that it utilized a viral protein, vaccinia virus protein 35K, which binds to CC chemokines, such as MCP-1, blocking their interactions with chemokine receptors and partially suppressing the inflammatory response to infection.[95] Ad35K administered in an ApoE[−/−] mouse VGD model reduced vein graft wall thickness, SMC proliferation, and macrophage infiltration.[67] The results of this study were most dramatic at 2 weeks postengraftment; however, the treatment and control groups began to converge at 4 weeks, possibly because of the use of an adenoviral vector and Ad-induced inflammation. Similar results were reported in a rabbit vein graft model, in which intraluminal infection of the graft with Ad35K resulted in modest decreases in I/M ratio but dramatic decreases in macrophage infiltration.[68] A membrane-targeted mutant of 35K, m35K, was more effective than 35K in inhibiting CC chemokines in murine peritonitis and hepatitis models but has not yet been tested in models of VGD.[96]

15.5.4 Smooth Muscle Cell Proliferation and Migration

SMCs are the predominant cells within the neointimal lesion from 1 month to 1 year after placement of the vein graft. Although the contribution of SMC proliferation specifically to the clinical manifestations of VGD is unclear, the neointima is the setting for accelerated atherosclerosis. The goal of gene therapy targeting SMCs is to inhibit proliferation and restore the contractile phenotype to allow normal regulation of vascular tone. Although there have been numerous studies of how specific mitogens affect SMC proliferation, the strategies described below focus on points of convergence of the many proliferative pathways that are activated in SMCs within the graft wall.

In mature vessels, integrin-mediated contacts with the ECM inhibit SMC proliferation and migration. When SMCs transition from a quiescent phenotype to a proliferative phenotype, they express matrix-degrading metalloproteinases (MMPs), which digest the ECM, allowing them to proliferate and migrate into the neointima. The activity of these MMPs can be targeted by overexpressing their endogenous inhibitors. Tissue inhibitor of metalloproteinase (TIMP)-3 has been used in this context to significantly reduce short-term (28-day) and long-term (3-month) IH in a porcine vein graft model.[70,71]

Our group has focused on the phosphoinositide 3-kinase (PI3K) pathway, specifically the lipid phosphatase and tumor suppressor phosphatase and tensin (PTEN) homolog, an endogenous negative regulator of this pathway. PI3K is a point of convergence of numerous proliferative stimuli induced by both growth factor receptors and G protein-coupled receptors (GPCRs). After activation, PI3K induces a signaling cascade that activates multiple downstream effectors, such as mammalian target of rapamycin (mTOR), that promote SMC growth, proliferation, and migration. PTEN indirectly antagonizes PI3K's catalytic function by hydrolyzing the 3-phosphoinositide lipid products of PI3K, but PTEN expression is reduced in SVGs after engraftment. AdPTEN delivery to vein grafts in a canine CABG model significantly inhibited IH up to 90 days postoperatively.[34] Notably, the PTEN transgene could be detected by PCR only in the treated vein graft but not in any of a number of other tissues, demonstrating that ex vivo delivery of recombinant adenoviruses such as AdPTEN to vein grafts poses a low risk of systemic infection.

15.6 THE PREVENT TRIALS

As discussed above, numerous animal studies have demonstrated the preclinical efficacy of a variety of gene therapy approaches to treat vein graft disease. However, the E2F decoy oligonucleotide edifoligide is the only strategy that has been evaluated in clinical trials. E2F family transcription factors bind directly to DNA and modulate the expression of numerous genes associated with the regulation of cell cycle progression. The family is composed of eight members, with isoforms 1–3 associated with transcriptional activation of proliferative genes and isoforms 4–8 associated with repression of these genes.[97] Edifoligide is a 14-base pair double-stranded ODN designed to mimic the conserved binding site found in E2F1–3 target genes. After delivery into cells, edifoligide acts as a decoy, sequestering E2F from genomic DNA and thereby preventing expression of its target genes.[37] At the time the E2F decoy strategy was first described, E2F was thought to play a role exclusively in promoting cell cycle progression and cellular proliferation. Because IH is characterized largely by increased proliferation of SMCs, it was hypothesized that edifoligide would limit IH by preventing the proliferative response of SMCs. Indeed, promising preclinical studies in rabbit models and cultured human vein segments demonstrated that edifoligide not only inhibited SMC proliferation but also significantly reduced IH, preserved endothelial function, and limited the development of atherosclerosis.[28–31] These exciting data led to the development of a series of clinical trials named the *P*roject of *E*x *V*ivo *V*ein Graft *E*ngineering via *T*ransfection (PREVENT). PREVENT I was a phase I trial that demonstrated the safety and biologic efficacy of edifoligide decoy ODN delivery in 41 patients undergoing lower extremity bypass surgery.[38] PREVENT II was a phase IIb trial of 200 patients undergoing coronary artery bypass grafting in which the treatment group demonstrated a 30% relative reduction in vein graft critical stenosis by angiography and a 30% reduction in total vein graft wall volume by intravascular ultrasound.[72] On the basis of these results, the E2F decoy strategy was investigated in two separate phase III trials of vein graft disease: PREVENT III in lower extremity bypass surgery and PREVENT IV in coronary artery bypass grafting.

Unfortunately, neither study demonstrated a significant difference in the primary endpoints of primary graft patency and critical graft stenosis at 1 year.[9,14] Further analysis of PREVENT IV data indicated that angiographically determined vein graft failure within the first year was associated with an increased risk of repeat revascularization at 4 years following angiography. However, this angiographic vein graft failure was not associated with an increased risk of myocardial infarction or death within the follow-up period.[98] These results highlight the need for a better understanding of how the pathophysiology of VGD, particularly the development of IH, correlates with clinically relevant endpoints.

The reasons underlying the lack of clinical success of edifoligide are not entirely clear; however, there are several factors that may have contributed. As noted, edifoligide was designed to bind to the activating E2F isoforms but may have inadvertently bound inhibitory E2F isoforms due to their highly conserved DNA-binding domains.[37] Even traditionally activating isoforms can act as context-dependent transcriptional repressors, potentially counteracting the intended antiproliferative effects of edifoligide.[99]

Another surprising result from the PREVENT IV trial was the higher-than-expected incidence of vein graft failure demonstrated by angiography. Over 45% of patients in both treatment groups met this criterion at 1 year, which is significantly higher than the historical rate of approximately 15%.[8] Interestingly, the study lacked a "standard of care" control group, meaning that vein grafts in both the treatment and placebo groups were exposed to 310 mmHg of nondistending pressure during gene delivery. It is possible that the application of pressure and/or incubation in transfection solution exacerbated endothelial injury, highlighting the importance of understanding the effects of specific gene delivery strategies independent of the effects of a given target gene.[9]

Finally, there is accumulating evidence that graft-extrinsic cells play a large role in formation of the neointimal lesion.[100] The treatment strategies in the PREVENT trials and in many of the preclinical trials discussed above primarily target graft-intrinsic cells. Unfortunately, systemic gene therapy to target graft-extrinsic cells obviates many of the potential benefits of ex vivo gene therapy delivered to vein grafts intraoperatively, and such approaches are beyond the scope of this review.

15.7 ADDITIONAL CONSIDERATIONS FOR TRANSLATION

From a detailed review of existing studies of gene therapy for vein graft disease, it becomes clear that in order to properly assess the promise of preclinical studies, endpoints in animal models must correlate with clinically relevant endpoints in patients. Many published preclinical studies have used IH as the primary endpoint, but this may not be clinically meaningful, as it remains unclear how IH correlates with vein graft failure, as assessed in PREVENT IV. However, there is no universal, consistent definition of vein graft failure (reviewed in Ref. 24). Vein graft stenosis, as measured by angiography or other imaging modalities, is another frequently used metric; however, there is no strict cutoff for the degree of stenosis to define graft failure. Furthermore, it is also unclear exactly how angiographic determinations of graft failure relate to clinical outcomes. For example, analysis of PREVENT IV demonstrates that critical graft stenosis is associated with an increased need for repeat vascularization but not an increased risk of death and/or myocardial infarction.[98] Other readouts such as graft atherosclerosis, measures of systemic inflammation, and plaque stability may be more meaningful, particularly in late graft failure, but many of these measures remain poorly characterized even in a nonvein graft setting. Focusing exclusively on IH in animal models may exclude promising therapies that may have modest effects on IH but dramatic effects on other aspects of VGD (for example, macrophage recruitment). Moreover, many of the studies reviewed here demonstrate that increased graft wall thickness does not necessarily reflect poor graft function (Table 15.3).

Finally, even if gene therapy for VGD is successfully applied in the clinic, it is unlikely that targeting a single pathway will be successful in all patients as a result of genetic variation. Moreover, targeting different pathways and cell types simultaneously may be synergistic, as in the case of SOD, TIMP-1, and 35K combination gene therapy discussed above.[61,68] Similarly, combining gene therapy with other approaches may yield better outcomes than either treatment alone. Potential examples include genetic modification of progenitor cells used to seed bioengineered grafts, genetic modification of grafts to promote progenitor cell homing, the use of sheaths to provide structural support for vein grafts treated with gene therapy, and the use of biological scaffolds for long-term delivery of gene therapy to engrafted veins.

TABLE 15.3 Summary of Clinically Relevant Endpoints and Challenges Translating Preclinical Endpoints to Human Disease

Clinical Endpoint from the PREVENT Trials	Related Preclinical Endpoints	Challenges in Translation
Need for graft revision	N/A	
Need for repeat revascularization	N/A	
Major amputation (peripheral bypass); myocardial infarction (coronary bypass)	Canine femoral artery bypass model; did not report on limb function, incidence of necrosis, etc., within the 4-week study period[32]	Commonly used animal models (e.g., jugular vein–carotid artery interposition model grafting) do not allow for monitoring of these clinically relevant sequelae of VGD Time course of preclinical studies is often insufficient to reach these endpoints Late adverse events (e.g., myocardial infarction due to atherosclerotic plaque rupture) are not common or do not occur in animal models
Evidence of ≥70% stenosis (peripheral artery) or ≥75% stenosis (coronary artery) by ultrasound and/or angiography	Coronary arteriography on postoperative days 30 and 90 in a canine aortocoronary bypass model[34] Laser Doppler blood flow monitoring at sacrifice in a rabbit jugular vein–carotid artery interposition graft[35] Blood flow monitoring in the femoral artery at sacrifice in a canine femoral artery bypass model[32]	Although associated with increased need for revascularization, unclear how they relate to important clinical outcomes (e.g., amputation, myocardial infarction, death)
Additional Clinically Relevant Endpoints	**Related Preclinical Endpoints**	**Challenges in Translation**
Thrombosis	Histological assessment of thrombus formation in excised grafts in a rabbit jugular vein–carotid artery interposition model[62] Assessment of flow-restricting thrombi by ultrasound detection of cyclic flow reductions in a porcine saphenous vein–carotid artery interposition model[64]	Significant differences in expression of components of the thrombotic and thrombolytic pathways between humans and animal models (e.g., dogs)
Graft atherosclerosis; unstable plaque	Use of hypercholesterolemic rabbits[28–31,33,35,53] and ApoE knockout mice[36,65,67] Quantification of macrophages in the neointimal lesion[33,61,65–68]	Commonly used animal models do not develop spontaneous atherosclerosis (e.g., dogs) Hypercholesterolemic animal models do not typically develop unstable plaque; differences in expression of relevant inflammatory ligands/receptors between humans and animal models (e.g., mice) Lack of established biomarkers of clinical plaque instability preclude study in preclinical models

15.8 CONCLUSIONS

Cardiovascular disease is the leading cause of morbidity and mortality in the developed world, and surgical revascularization is likely to remain an important therapy for patients with arterial occlusive disease. Although there have been significant advances in arterial bypass grafting and in the field of synthetic grafts, autologous veins will likely remain the most used bypass conduit for some time to come, especially for complex patients with multivessel disease. The population of patients requiring bypass grafting is aging and becoming more complex; therefore, it is critical that we efficiently translate basic discoveries in VGD pathophysiology into the clinic. Although the PREVENT trials did not yield a viable therapeutic, they demonstrated that gene therapy for VGD is feasible and safe. The development of more sensitive and clinically meaningful endpoints, combined with a better understanding of how the molecular mechanisms of VGD in animal models relate to clinical disease will allow for more rapid translation of gene therapy strategies for vein graft disease to the clinic.

REFERENCES

1. Roger VL, Go AS, Lloyd-Jones DM, Benjamin EJ, Berry JD, Borden WB, et al. Executive summary: heart disease and stroke statistics–2012 update: a report from the American Heart Association. *Circulation* 2012;**125**(1):188–97.

2. Pellicori P, Costanzo P, Joseph AC, Hoye A, Atkin SL, Cleland JG. Medical management of stable coronary atherosclerosis. *Curr Atheroscler Rep* 2013;**15**(4):313.

3. Tattersall MC, Johnson HM, Mason PJ. Contemporary and optimal medical management of peripheral arterial disease. *Surg Clin North Am* 2013;**93**(4):761–78. vii.

4. Goodney PP, Beck AW, Nagle J, Welch HG, Zwolak RM. National trends in lower extremity bypass surgery, endovascular interventions, and major amputations. *J Vasc Surg* 2009;**50**(1):54–60.

5. Loop FD, Lytle BW, Cosgrove DM, Stewart RW, Goormastic M, Williams GW, et al. Influence of the internal-mammary-artery graft on 10-year survival and other cardiac events. *N Engl J Med* 1986;**314**(1):1–6.

6. Silverberg D, Sheick-Yousif B, Yakubovitch D, Halak M, Schneiderman J. The deep femoral artery, a readily available inflow vessel for lower limb revascularization: a single-center experience. *Vascular* 2013;**21**(2):75–8.

7. The effect of aggressive lowering of low-density lipoprotein cholesterol levels and low-dose anticoagulation on obstructive changes in saphenous-vein coronary-artery bypass grafts. The Post Coronary Artery Bypass Graft Trial Investigators. *N Engl J Med* 1997;**336**(3):153–62.

8. Fitzgibbon GM, Kafka HP, Leach AJ, Keon WJ, Hooper GD, Burton JR. Coronary bypass graft fate and patient outcome: angiographic follow-up of 5,065 grafts related to survival and reoperation in 1,388 patients during 25 years. *J Am Coll Cardiol* 1996;**28**(3):616–26.

9. Alexander JH, Hafley G, Harrington RA, Peterson ED, Ferguson Jr TB, Lorenz TJ, et al. Efficacy and safety of edifoligide, an E2F transcription factor decoy, for prevention of vein graft failure following coronary artery bypass graft surgery: PREVENT IV: a randomized controlled trial. *JAMA* 2005;**294**(19):2446–54.

10. Favaloro RG. Saphenous vein autograft replacement of severe segmental coronary artery occlusion: operative technique. *Ann Thorac Surg* 1968;**5**(4):334–9.

11. Conte MS, Lorenz TJ, Bandyk DF, Clowes AW, Moneta GL, Seely BL. Design and rationale of the PREVENT III clinical trial: edifoligide for the prevention of infrainguinal vein graft failure. *Vasc Endovasc Surg* 2005;**39**(1):15–23.

12. Foster ED, Fisher LD, Kaiser GC, Myers WO. Comparison of operative mortality and morbidity for initial and repeat coronary artery bypass grafting: the Coronary Artery Surgery Study (CASS) registry experience. *Ann Thorac Surg* 1984;**38**(6):563–70.

13. Lipsitz EC, Veith FJ, Cayne NS, Harvey J, Rhee SJ. Repetitive bypass and revisions with extensions for limb salvage after multiple previous failures. *Vascular* 2013;**21**(2):63–8.

14. Conte MS, Bandyk DF, Clowes AW, Moneta GL, Seely L, Lorenz TJ, et al. Results of PREVENT III: a multicenter, randomized trial of edifoligide for the prevention of vein graft failure in lower extremity bypass surgery. *J Vasc Surg* 2006;**43**(4):742–51. discussion 51.

15. Bradbury AW, Adam DJ, Bell J, Forbes JF, Fowkes FG, Gillespie I, et al. Bypass versus Angioplasty in Severe Ischaemia of the Leg (BASIL) trial: an intention-to-treat analysis of amputation-free and overall survival in patients randomized to a bypass surgery-first or a balloon angioplasty-first revascularization strategy. *J Vasc Surg* 2010;**51**(5 Suppl.):5S–17S.

16. Mitra AK, Gangahar DM, Agrawal DK. Cellular, molecular and immunological mechanisms in the pathophysiology of vein graft intimal hyperplasia. *Immunol Cell Biol* 2006;**84**(2):115–24.

17. Aird WC. Endothelial cell heterogeneity. *Cold Spring Harb Perspect Med* 2012;**2**(1):a006429.

18. Rensen SS, Doevendans PA, van Eys GJ. Regulation and characteristics of vascular smooth muscle cell phenotypic diversity. *Neth Heart J* 2007;**15**(3):100–8.

19. Dashwood MR, Tsui JC. 'No-touch' saphenous vein harvesting improves graft performance in patients undergoing coronary artery bypass surgery: a journey from bedside to bench. *Vasc Pharmacol* 2013;**58**(3):240–50.

20. Davies MG, Klyachkin ML, Dalen H, Massey MF, Svendsen E, Hagen PO. The integrity of experimental vein graft endothelium–implications on the etiology of early graft failure. *Eur J Vasc Surg* 1993;**7**(2):156–65.

21. Weiss DR, Juchem G, Kemkes BM, Gansera B, Nees S. Extensive deendothelialization and thrombogenicity in routinely prepared vein grafts for coronary bypass operations: facts and remedy. *Int J Clin Exp Med* 2009;**2**(2):95–113.

22. Gasper WJ, Owens CD, Kim JM, Hills N, Belkin M, Creager MA, et al. Thirty-day vein remodeling is predictive of midterm graft patency after lower extremity bypass. *J Vasc Surg* 2013;**57**(1):9–18.

23. Schwartz SM. Smooth muscle migration in atherosclerosis and restenosis. *J Clin Invest* 1997;**100**(11 Suppl.):S87–9.

24. Harskamp RE, Williams JB, Hill RC, de Winter RJ, Alexander JH, Lopes RD. Saphenous vein graft failure and clinical outcomes: toward a surrogate end point in patients following coronary artery bypass surgery? *Am Heart J* 2013;**165**(5):639–43.

25. Bulkley BH, Hutchins GM. Accelerated "atherosclerosis". A morphologic study of 97 saphenous vein coronary artery bypass grafts. *Circulation* 1977;**55**(1):163–9.

26. Jim MH, Hau WK, Ko RL, Siu CW, Ho HH, Yiu KH, et al. Virtual histology by intravascular ultrasound study on degenerative aortocoronary saphenous vein grafts. *Heart Vessels* 2010;**25**(3):175–81.

27. Harskamp RE, Lopes RD, Baisden CE, de Winter RJ, Alexander JH. Saphenous vein graft failure after coronary artery bypass surgery: pathophysiology, management, and future directions. *Ann Surg* 2013;**257**(5):824–33.

28. Ehsan A, Mann MJ, Dell'Acqua G, Dzau VJ. Long-term stabilization of vein graft wall architecture and prolonged resistance to experimental atherosclerosis after E2F decoy oligonucleotide gene therapy. *J Thorac Cardiovasc Surg* 2001;**121**(4):714–22.

29. Ehsan A, Mann MJ, Dell'Acqua G, Tamura K, Braun-Dullaeus R, Dzau VJ. Endothelial healing in vein grafts: proliferative burst unimpaired by genetic therapy of neointimal disease. *Circulation* 2002;**105**(14):1686–92.

30. Mann MJ, Gibbons GH, Tsao PS, von der Leyen HE, Cooke JP, Buitrago R, et al. Cell cycle inhibition preserves endothelial function in genetically engineered rabbit vein grafts. *J Clin Invest* 1997;**99**(6):1295–301.

31. Mann MJ, Kernoff R, Dzau VJ. Vein graft gene therapy using E2F decoy oligonucleotides: target gene inhibition in human veins and long term resistance to atherosclerosis in rabbits. *Surg Forum* 1997;**48**:242–4.

32. Matsumoto T, Komori K, Yonemitsu Y, Morishita R, Sueishi K, Kaneda Y, et al. Hemagglutinating virus of Japan-liposome-mediated gene transfer of endothelial cell nitric oxide synthase inhibits intimal hyperplasia of canine vein grafts under conditions of poor runoff. *J Vasc Surg* 1998;**27**(1):135–44.

33. Ohta S, Komori K, Yonemitsu Y, Onohara T, Matsumoto T, Sugimachi K. Intraluminal gene transfer of endothelial cell-nitric oxide synthase suppresses intimal hyperplasia of vein grafts in cholesterol-fed rabbit: a limited biological effect as a result of the loss of medial smooth muscle cells. *Surgery* 2002;**131**(6):644–53.

34. Hata JA, Petrofski JA, Schroder JN, Williams ML, Timberlake SH, Pippen A, et al. Modulation of phosphatidylinositol 3-kinase signaling reduces intimal hyperplasia in aortocoronary saphenous vein grafts. *J Thorac Cardiovasc Surg* 2005;**129**(6):1405–13.

35. Eichstaedt HC, Liu Q, Chen Z, Bobustuc GC, Terry T, Willerson JT, et al. Gene transfer of COX-1 improves lumen size and blood flow in carotid bypass grafts. *J Surg Res* 2010;**161**(1):162–7.

36. Eefting D, Bot I, de Vries MR, Schepers A, van Bockel JH, Van Berkel TJ, et al. Local lentiviral short hairpin RNA silencing of CCR2 inhibits vein graft thickening in hypercholesterolemic apolipoprotein E3-Leiden mice. *J Vasc Surg* 2009;**50**(1):152–60.

37. Morishita R, Gibbons GH, Horiuchi M, Ellison KE, Nakama M, Zhang L, et al. A gene therapy strategy using a transcription factor decoy of the E2F binding site inhibits smooth muscle proliferation in vivo. *Proc Natl Acad Sci USA* 1995;**92**(13):5855–9.

38. Mann MJ, Whittemore AD, Donaldson MC, Belkin M, Conte MS, Polak JF, et al. Ex-vivo gene therapy of human vascular bypass grafts with E2F decoy: the PREVENT single-centre, randomised, controlled trial. *Lancet* 1999;**354**(9189):1493–8.

39. Cairns MJ, Saravolac EG, Sun LQ. Catalytic DNA: a novel tool for gene suppression. *Curr Drug Targets* 2002;**3**(3):269–79.

40. Li Y, Bhindi R, Deng ZJ, Morton SW, Hammond PT, Khachigian LM. Inhibition of vein graft stenosis with a c-jun targeting DNAzyme in a cationic liposomal formulation containing 1,2-dioleoyl-3-trimethylammonium propane (DOTAP)/1,2-dioleoyl-sn-glycero-3-phosphoethanolamine (DOPE). *Int J Cardiol* 2013;**168**(4):3659–64.

41. Newman KD, Dunn PF, Owens JW, Schulick AH, Virmani R, Sukhova G, et al. Adenovirus-mediated gene transfer into normal rabbit arteries results in prolonged vascular cell activation, inflammation, and neointimal hyperplasia. *J Clin Invest* 1995;**96**(6):2955–65.

42. Parks RJ, Chen L, Anton M, Sankar U, Rudnicki MA, Graham FL. A helper-dependent adenovirus vector system: removal of helper virus by Cre-mediated excision of the viral packaging signal. *Proc Natl Acad Sci USA* 1996;**93**(24):13565–70.

43. Wen S, Graf S, Massey PG, Dichek DA. Improved vascular gene transfer with a helper-dependent adenoviral vector. *Circulation* 2004;**110**(11):1484–91.

44. Lehrman S. Virus treatment questioned after gene therapy death. *Nature* 1999;**401**(6753):517–8.

45. Work LM, Nicklin SA, Brain NJ, Dishart KL, Von Seggern DJ, Hallek M, et al. Development of efficient viral vectors selective for vascular smooth muscle cells. *Mol Ther* 2004;**9**(2):198–208.

46. Nicklin SA, Buening H, Dishart KL, de Alwis M, Girod A, Hacker U, et al. Efficient and selective AAV2-mediated gene transfer directed to human vascular endothelial cells. *Mol Ther* 2001;**4**(3):174–81.

47. Kaplitt MG, Feigin A, Tang C, Fitzsimons HL, Mattis P, Lawlor PA, et al. Safety and tolerability of gene therapy with an adeno-associated virus (AAV) borne GAD gene for Parkinson's disease: an open label, phase I trial. *Lancet* 2007;**369**(9579):2097–105.

48. Pankajakshan D, Makinde TO, Gaurav R, Del Core M, Hatzoudis G, Pipinos I, et al. Successful transfection of genes using AAV-2/9 vector in swine coronary and peripheral arteries. *J Surg Res* 2012;**175**(1):169–75.

49. McCarty DM. Self-complementary AAV vectors; advances and applications. *Mol Ther* 2008;**16**(10):1648–56.

50. Cefai D, Simeoni E, Ludunge KM, Driscoll R, von Segesser LK, Kappenberger L, et al. Multiply attenuated, self-inactivating lentiviral vectors efficiently transduce human coronary artery cells in vitro and rat arteries in vivo. *J Mol Cell Cardiol* 2005;**38**(2):333–44.

51. Philippe S, Sarkis C, Barkats M, Mammeri H, Ladroue C, Petit C, et al. Lentiviral vectors with a defective integrase allow efficient and sustained transgene expression in vitro and in vivo. *Proc Natl Acad Sci USA* 2006;**103**(47):17684–9.

52. Khaleel MS, Dorheim TA, Duryee MJ, Durbin Jr HE, Bussey WD, Garvin RP, et al. High-pressure distention of the saphenous vein during preparation results in increased markers of inflammation: a potential mechanism for graft failure. *Ann Thorac Surg* 2012;**93**(2):552–8.

53. Miyake T, Aoki M, Shiraya S, Tanemoto K, Ogihara T, Kaneda Y, et al. Inhibitory effects of NFκB decoy oligodeoxynucleotides on neointimal hyperplasia in a rabbit vein graft model. *J Mol Cell Cardiol* 2006;**41**(3):431–40.

54. Mann MJ, Gibbons GH, Hutchinson H, Poston RS, Hoyt EG, Robbins RC, et al. Pressure-mediated oligonucleotide transfection of rat and human cardiovascular tissues. *Proc Natl Acad Sci USA* 1999;**96**(11):6411–6.

55. Handa M, Li W, Morioka K, Takamori A, Yamada N, Ihaya A. Adventitial delivery of platelet-derived endothelial cell growth factor gene prevented intimal hyperplasia of vein graft. *J Vasc Surg* 2008;**48**(6):1566–74.

56. Sun J, Zheng J, Ling KH, Zhao K, Xie Z, Li B, et al. Preventing intimal thickening of vein grafts in vein artery bypass using STAT-3 siRNA. *J Transl Med* 2012;**10**(1):2.

57. Wang J, Liu K, Shen L, Wu H, Jing H. Small interfering RNA to c-myc inhibits vein graft restenosis in a rat vein graft model. *J Surg Res* 2011;**169**(1):e85–91.

58. Thomas AC. Animal models for studying vein graft failure and therapeutic interventions. *Curr Opin Pharmacol* 2012;**12**(2):121–6.

59. Schachner T, Laufer G, Bonatti J. In vivo (animal) models of vein graft disease. *Eur J Cardiothorac Surg* 2006;**30**(3):451–63.

60. Shofti R, Zaretzki A, Cohen E, Engel A, Bar-El Y. The sheep as a model for coronary artery bypass surgery. *Lab Anim* 2004;**38**(2):149–57.
61. Turunen P, Puhakka HL, Heikura T, Romppanen E, Inkala M, Leppanen O, et al. Extracellular superoxide dismutase with vaccinia virus anti-inflammatory protein 35K or tissue inhibitor of metalloproteinase-1: combination gene therapy in the treatment of vein graft stenosis in rabbits. *Hum Gene Ther* 2006;**17**(4):405–14.
62. Ohno N, Itoh H, Ikeda T, Ueyama K, Yamahara K, Doi K, et al. Accelerated reendothelialization with suppressed thrombogenic property and neointimal hyperplasia of rabbit jugular vein grafts by adenovirus-mediated gene transfer of C-type natriuretic peptide. *Circulation* 2002;**105**(14):1623–6.
63. Tabuchi N, Shichiri M, Shibamiya A, Koyama T, Nakazawa F, Chung J, et al. Non-viral in vivo thrombomodulin gene transfer prevents early loss of thromboresistance of grafted veins. *Eur J Cardiothorac Surg* 2004;**26**(5):995–1001.
64. Thomas AC, Wyatt MJ, Newby AC. Reduction of early vein graft thrombosis by tissue plasminogen activator gene transfer. *Thromb Haemost* 2009;**102**(1):145–52.
65. Schepers A, Eefting D, Bonta PI, Grimbergen JM, de Vries MR, van Weel V, et al. Anti-MCP-1 gene therapy inhibits vascular smooth muscle cells proliferation and attenuates vein graft thickening both in vitro and in vivo. *Arterioscler Thromb Vasc Biol* 2006;**26**(9):2063–9.
66. Tatewaki H, Egashira K, Kimura S, Nishida T, Morita S, Tominaga R. Blockade of monocyte chemoattractant protein-1 by adenoviral gene transfer inhibits experimental vein graft neointimal formation. *J Vasc Surg* 2007;**45**(6):1236–43.
67. Ali ZA, Bursill CA, Hu Y, Choudhury RP, Xu Q, Greaves DR, et al. Gene transfer of a broad spectrum CC-chemokine inhibitor reduces vein graft atherosclerosis in apolipoprotein E-knockout mice. *Circulation* 2005;**112**(9 Suppl.):I235–41.
68. Puhakka HL, Turunen P, Gruchala M, Bursill C, Heikura T, Vajanto I, et al. Effects of vaccinia virus anti-inflammatory protein 35K and TIMP-1 gene transfers on vein graft stenosis in rabbits. *In Vivo* 2005;**19**(3):515–21.
69. Shintani T, Sawa Y, Takahashi T, Matsumiya G, Matsuura N, Miyamoto Y, et al. Intraoperative transfection of vein grafts with the NFkappaB decoy in a canine aortocoronary bypass model: a strategy to attenuate intimal hyperplasia. *Ann Thorac Surg* 2002;**74**(4):1132–7. discussion 7–8.
70. George SJ, Lloyd CT, Angelini GD, Newby AC, Baker AH. Inhibition of late vein graft neointima formation in human and porcine models by adenovirus-mediated overexpression of tissue inhibitor of metalloproteinase-3. *Circulation* 2000;**101**(3):296–304.
71. George SJ, Wan S, Hu J, MacDonald R, Johnson JL, Baker AH. Sustained reduction of vein graft neointima formation by ex vivo TIMP-3 gene therapy. *Circulation* 2011;**124**(11 Suppl. 1):S135–42.
72. Late-breaking clinical trial abstracts. *Circulation* 2001;**104**(25):1b–4b.
73. Southerland KW, Frazier SB, Bowles DE, Milano CA, Kontos CD. Gene therapy for the prevention of vein graft disease. *Transl Res* 2013;**161**(4): 321–38.
74. Murakami M. Signaling required for blood vessel maintenance: molecular basis and pathological manifestations. *Int J Vasc Med* 2012;**2012**:293641.
75. Schmalfuss CM, Chen LY, Bott JN, Staples ED, Mehta JL. Superoxide anion generation, superoxide dismutase activity, and nitric oxide release in human internal mammary artery and saphenous vein segments. *J Cardiovasc Pharmacol Ther* 1999;**4**(4):249–57.
76. Van-Assche T, Huygelen V, Crabtree MJ, Antoniades C. Gene therapy targeting inflammation in atherosclerosis. *Curr Pharm Des* 2011;**17**(37):4210–23.
77. Durante W. Targeting heme oxygenase-1 in vascular disease. *Curr Drug Targets* 2010;**11**(12):1504–16.
78. Lamas S, Marsden PA, Li GK, Tempst P, Michel T. Endothelial nitric oxide synthase: molecular cloning and characterization of a distinct constitutive enzyme isoform. *Proc Natl Acad Sci USA* 1992;**89**(14):6348–52.
79. Gkaliagkousi E, Ferro A. Nitric oxide signalling in the regulation of cardiovascular and platelet function. *Front Biosci* 2011;**16**:1873–97.
80. Hara S, Morishita R, Tone Y, Yokoyama C, Inoue H, Kaneda Y, et al. Overexpression of prostacyclin synthase inhibits growth of vascular smooth muscle cells. *Biochem Biophys Res Commun* 1995;**216**(3):862–7.
81. Wu KK. Regulation of prostaglandin H synthase-1 gene expression. *Adv Exp Med Biol* 1997;**400A**:121–6.
82. Shiokawa Y, Rahman MF, Ishii Y, Sueishi K. The rate of re-endothelialization correlates inversely with the degree of the following intimal thickening in vein grafts. Electron microscopic and immunohistochemical studies. *Virchows Arch A Pathol Anat Histopathol* 1989;**415**(3):225–35.
83. Rose RA, Giles WR. Natriuretic peptide C receptor signalling in the heart and vasculature. *J Physiol* 2008;**586**(2):353–66.
84. Payeli SK, Latini R, Gebhard C, Patrignani A, Wagner U, Luscher TF, et al. Prothrombotic gene expression profile in vascular smooth muscle cells of human saphenous vein, but not internal mammary artery. *Arterioscler Thromb Vasc Biol* 2008;**28**(4):705–10.
85. Pakala R. Coagulation factor Xa synergistically interacts with serotonin in inducing vascular smooth muscle cell proliferation. *Cardiovasc Radiat Med* 2003;**4**(2):69–76.
86. Bretschneider E, Braun M, Fischer A, Wittpoth M, Glusa E, Schror K. Factor Xa acts as a PDGF-independent mitogen in human vascular smooth muscle cells. *Thromb Haemost* 2000;**84**(3):499–505.
87. Pyo RT, Sato Y, Mackman N, Taubman MB. Mice deficient in tissue factor demonstrate attenuated intimal hyperplasia in response to vascular injury and decreased smooth muscle cell migration. *Thromb Haemost* 2004;**92**(3):451–8.
88. Kim AY, Walinsky PL, Kolodgie FD, Bian C, Sperry JL, Deming CB, et al. Early loss of thrombomodulin expression impairs vein graft thromboresistance: implications for vein graft failure. *Circ Res* 2002;**90**(2):205–12.
89. Hilfiker PR, Waugh JM, Li-Hawkins JJ, Kuo MD, Yuksel E, Geske RS, et al. Enhancement of neointima formation with tissue-type plasminogen activator. *J Vasc Surg* 2001;**33**(4):821–8.
90. Vestweber D. Adhesion and signaling molecules controlling the transmigration of leukocytes through endothelium. *Immunol Rev* 2007;**218**:178–96.
91. Daly C, Rollins BJ. Monocyte chemoattractant protein-1 (CCL2) in inflammatory disease and adaptive immunity: therapeutic opportunities and controversies. *Microcirculation* 2003;**10**(3–4):247–57.
92. Stark VK, Hoch JR, Warner TF, Hullett DA. Monocyte chemotactic protein-1 expression is associated with the development of vein graft intimal hyperplasia. *Arterioscler Thromb Vasc Biol* 1997;**17**(8):1614–21.

93. Zhang YJ, Rutledge BJ, Rollins BJ. Structure/activity analysis of human monocyte chemoattractant protein-1 (MCP-1) by mutagenesis. Identification of a mutated protein that inhibits MCP-1-mediated monocyte chemotaxis. *J Biol Chem* 1994;**269**(22):15918–24.

94. Pamukcu B, Lip GY, Shantsila E. The nuclear factor – kappa B pathway in atherosclerosis: a potential therapeutic target for atherothrombotic vascular disease. *Thromb Res* 2011;**128**(2):117–23.

95. Bursill CA, Cai S, Channon KM, Greaves DR. Adenoviral-mediated delivery of a viral chemokine binding protein blocks CC-chemokine activity in vitro and in vivo. *Immunobiology* 2003;**207**(3):187–96.

96. Bursill CA, Cash JL, Channon KM, Greaves DR. Membrane-bound CC chemokine inhibitor 35K provides localized inhibition of CC chemokine activity in vitro and in vivo. *J Immunol* 2006;**177**(8):5567–73.

97. Attwooll C, Lazzerini Denchi E, Helin K. The E2F family: specific functions and overlapping interests. *EMBO J* 2004;**23**(24):4709–16.

98. Lopes RD, Mehta RH, Hafley GE, Williams JB, Mack MJ, Peterson ED, et al. Relationship between vein graft failure and subsequent clinical outcomes after coronary artery bypass surgery. *Circulation* 2012;**125**(6):749–56.

99. Chong JL, Wenzel PL, Saenz-Robles MT, Nair V, Ferrey A, Hagan JP, et al. E2f1-3 switch from activators in progenitor cells to repressors in differentiating cells. *Nature* 2009;**462**(7275):930–4.

100. Zhang L, Freedman NJ, Brian L, Peppel K. Graft-extrinsic cells predominate in vein graft arterialization. *Arterioscler Thromb Vasc Biol* 2004;**24**(3):470–6.

Chapter 16

Gene Therapy in Cystic Fibrosis

Michelle Prickett and Manu Jain
Division of Pulmonary and Critical Care Medicine, Department of Medicine, Feinberg School of Medicine, Northwestern University, Chicago, IL, USA

LIST OF ABBREVIATIONS

AAV2 Adeno-associated vector 2
BAL Bronchoalveolar lavage
CF Cystic fibrosis
CFTR Cystic fibrosis transmembrane regulator
DC-Chol 3b-[N-(N9,N9-dimethylaminoethane)-carbamoyl]-cholesterol
DOPE 1,2-Dioleoyl-sn-glycerol-phosphoethanolamine
DOTAP 1,2-Dioleoyl-3-trimethylammoniopropane
ENaC Epithelial sodium channel
HTS High-throughput screening
IRT Immunoreactive trypsinogen
LCI Lung clearance index
mRNA Messenger RNA
NPD Nasal potential difference
pDNA Plasmid DNA
PS-CF Pancreatic sufficient CF
PTC Premature termination codons

16.1 A BRIEF HISTORY OF CYSTIC FIBROSIS GENETICS

Recognition of characteristic cystic fibrosis (CF) symptoms and their relationship to mortality has existed for centuries. References in medieval folklore spoke of premature death for an infant that tastes "salty."[1] Alonso y de los Ruyzes de Fonteca, professor of medicine at Henares in Spain, wrote that it was known that fingers tasted salty after rubbing the forehead of a bewitched child.[2] Subsequently, various other reports associated defective pancreatic function with steatorrhea and meconium ileus.[3] CF as a distinct entity was first recognized in the 1930s when the characteristic cystic changes in the pancreatic ducts of children with malnutrition were defined.[4] These early descriptions were vital to establishing a phenotypic basis for disease notable for malabsorption, failure to thrive, and pulmonary infection.[5] It was later recognized that these changes are acquired through autosomal recessive distribution, identifying a genetic basis to the disease.

16.1.1 Diagnostic Testing for CF is the First Step

In 1953, di Sant'Agnese recognized that excessive salt loss in sweat was the explanation for heat prostration in several CF patients,[6] which led to the development of sweat chloride testing,[7] the most common and widely available test to make a CF diagnosis. The advent of sweat chloride testing allowed clinicians and researchers to use a standardized test to confirm, or refute, the diagnosis in patients at early stages of disease. For clinicians, this relatively simple diagnostic test helped to identify a large population of patients before development of symptoms, which could facilitate early intervention. For researchers, it also provided a method to ensure inclusion of patients with uniform criteria for clinical trials and an outcome measure for interventions targeting the genetic defect.

Translating Gene Therapy to the Clinic. http://dx.doi.org/10.1016/B978-0-12-800563-7.00016-6

16.1.2 Beyond the Sweat Test

During the past 10 years, newborn screening for CF has been implemented in the United States and many other countries worldwide. Methods to screen for a CF diagnosis have been developed to identify at-risk infants in early disease or in atypical cases.[8] Once screening identifies an at-risk individual, confirming a diagnosis is important to ensure that appropriate therapy can be started before the onset of symptoms. Newborn screening for CF is now standard in all 50 states and measures immunoreactive trypsinogen, a chemical secreted by the pancreas, in blood samples collected within the first few days of life. If immunoreactive trypsinogen is elevated, then a confirmatory algorithm is implemented. Though the confirmatory algorithm differs from state to state, ultimately all algorithms lead to a sweat test to confirm the diagnosis. For atypical cases or in the research setting, nasal potential difference (NPD) testing can also be used. This is a more sensitive method for evaluating chloride and sodium movement across the nasal epithelium. Although only available through select centers, this testing is highly specific for a CF diagnosis and provides a technique to measure treatment responses to therapeutic interventions targeting the basic genetic defect.

16.1.3 Genetics Opens the Door to New Treatments

The next hallmark in the history of CF was in 1989 when the causative gene was discovered. After years of research and early work describing the presence of abnormal sodium chloride transport in the respiratory epithelium of CF patients,[9,10] the gene responsible was uncovered. The gene for the *cystic fibrosis transmembrane regulator* (CFTR) protein was reported simultaneously by three groups of investigators,[11–13] which spawned the era of novel genetic strategies for both CF diagnostics and therapeutics. Since that time, more than 1900 mutations have been reported[14]; however, only a small number of these mutations (Table 16.1) account for the vast majority of both typical and atypical CF phenotypes.[15] With respect to genetic treatment approaches, gene therapy is relatively unique in that it offers the potential of mutation-independent therapy by replacing the mutated gene with a normal copy. However, the relatively small number of mutations that account for most cases of CF also affords the opportunity to target a small number of mutated proteins with a protein modifier approach. Both of these modalities have recently been reviewed but will be expanded upon in the following sections.[16]

16.2 CFTR MUTATIONS

As the number of genetic mutations in CFTR continues to expand, it has become increasingly important to categorize mutations based on the protein defect caused by specific mutations. There have been five functional classes of CFTR mutations defined (Table 16.2), each affecting a different stage of protein synthesis or function.[17] Mutations belonging to classes I–III have little or no functional CFTR production and typically lead to more severe or classic CF phenotypes. Conversely, mutations from classes IV and V have some residual CFTR function and are more commonly seen with atypical or milder disease.

16.2.1 Class I, II, and III Mutations

Class I mutations produce biosynthetic defects, including premature termination codons (PTCs) and frame shift mutations that result in low levels of truncated and/or dysfunctional CFTR proteins. The most common mutation to cause CF is F508del, a class II mutation, which allows proteins to be translated through messenger RNA (mRNA) reading frames but causes protein misfolding. Thus, the mutant protein is unable to progress through its normal maturation and is almost immediately degraded with very little protein reaching the epithelial cell surface. Proteins from class III mutations reach the plasma membrane, but have gating defects that limit channel opening (i.e., G551D).

16.2.2 Class IV and V Mutations

Class IV and V mutations allow CFTR protein to reach the cell surface, but these proteins have defects with limited channel conductance or reduced quantity because of abnormal mRNA splicing, respectively. Individual mutations may express more than one type of abnormality, which makes the idea of combined therapies important. An example of this is F508del; in addition to causing impaired folding, this mutation has been shown to have decreased channel conductance and reduced membrane half-life.[18,19]

16.3 CF GENE THERAPY CHALLENGES

Since the identification of the mutant CFTR gene as the cause of CF, there has been a significant effort to harness gene therapy to correct the mutation at a cellular level. Traditional gene therapy would aim to place a copy of normal CFTR

TABLE 16.1 Most Common Mutations in CFTR

Rank	Mutation Name	# Alleles in CFTR2	Frequency in CFTR2	Mutation Resulting Disease
1	F508del	49,740	0.70277	CF-causing
2	G542X	1856	0.02622	CF-causing
3	G551D	1427	0.02016	CF-causing
4	N1303K	1242	0.01755	CF-causing
5	W1282X	1056	0.01492	CF-causing
6	621+1G->T	817	0.01154	CF-causing
7	R117H	808	0.01142	Varying
8	R553X	645	0.00911	CF-causing
9	1717-1G->A	635	0.00897	CF-causing
10	2789+5G->A	538	0.0076	CF-causing
11	Protein	524	0.0074	CF-causing
12	R1162X	346	0.00489	CF-causing
13	I507del	319	0.00451	CF-causing
14	G85E	316	0.00446	CF-causing
15	2183AA->G	292	0.00413	CF-causing
16	CFTRdele2,3	277	0.00391	CF-causing
17	3120+1G->A	266	0.00376	CF-causing
18	3659delC	248	0.0035	CF-causing
19	1898+1G->A	245	0.00346	CF-causing
20	R347P	234	0.00331	CF-causing

The top 20 CFTR mutations represent more than 85% of the 1900 mutations identified to date. Each patient will have either the same mutation (homozygote) or two different mutations (heterozygote). The most common mutation is dF508 for which 50% of US cases have two copies and 40% will have one copy. Table modified from the CFTR2 Database from 70,777 alleles submitted from US and other worldwide registries (Reference: The Clinical and Functional Translation of CFTR (CFTR2). Available at http://cftr2.org.)

TABLE 16.2 Nomenclature of CFTR Mutations

Class	Classification	Common Mutations	Worldwide Frequency (%)	Predominant Phenotype	CFTR Protein Outcome
I	Defective synthesis	W1282X, G542X	10	Severe	No protein
II	Defective maturation	F508del, N1303K	70	Severe	Intracellular degradation
III	Blocked regulation	G551D	3–4	Severe	Gating defect
IV	Decreased conductance	R117H R347P	<2	Mild	Decreased Cl movement
V	Decreased abundance	A455E, 3849+10kb°C→T	<1	Variable	Reduced wild-type CFTR

Shown in the table are five classes of CFTR mutations, an example of each class and associated genetic and biologic changes. Class I, II, and III mutations are typically associated with more severe disease, whereas classes IV and V are commonly milder phenotypes.

into the patients' cells. This approach is attractive because it could potentially be curative for all CF patients regardless of genotype. Despite the theoretical allure of gene therapy, the reality of bringing it to the clinical setting has proven a significant challenge.[20–25]

16.3.1 CFTR Cell Expression

Two important unanswered questions regarding gene therapy in CF are which cells need to express CFTR and what level of expression is needed to reverse the CF phenotype. Because lung disease is responsible for 90% of mortality in CF, most investigators have focused gene therapy efforts on the airways. It has been reported that CFTR is normally expressed in airway epithelial cells and cells of the submucosal glands. Expression of CFTR is also tissue specific, with a nearly universal absence of CFTR in sweat glands from F508del patients and higher levels detected in ductal cells of the respiratory and intestinal mucosa of these same patients.[26] With respect to the amount of CFTR expression needed to modulate the CF phenotype, some insight is provided by examining relationships between residual CFTR function and associated clinical phenotypes in CF patients (Figure 16.1). Obligate heterozygotes (i.e., parents of CF patients) have half the normal amount of CFTR, similar in vivo CFTR function (sweat chloride, NPD), and essentially an indistinguishable phenotype as people with two normal CFTR genes. At the other end of the spectrum, CF patients with two severe CFTR mutations have little if any detectable CFTR function, with sweat chloride values near 100 mmol. Increasing levels of CFTR function are found in CF patients who are pancreatic-sufficient CF and in non-CF patients with congenital bilateral absence of the vas deferens and other disorders (e.g., chronic idiopathic pancreatitis).[27] Using these benchmarks, one might predict that moving CFTR function to 50% of wild-type would lead to a normal phenotype and to 30% of normal to a significant improvement in clinical outcomes.

16.3.2 Physical and Immunologic Barriers to CF Gene Therapy

The normal clearance and protective mechanisms of the lung pose significant barriers to the successful implementation of gene therapy in normal lungs, which are likely to be magnified in CF patients. The most important hurdles for gene therapy relate to technical difficulties created by the mucus barrier and the innate and adaptive immune systems in the lung itself. Most gene therapy trials have applied the gene transfer agents directly to the sinopulmonary epithelium via aerosols. Therefore, the first step in transfection with functioning CFTR via any vector is to physically reach the target host cell. From a structural perspective, the lung is well equipped to defend against particle uptake by a mucus layer serving as a primary barrier. The clinical course of CF is characterized by a progressive increase in sputum and airway obstruction over time. Thus, the physical barrier is intensified in patients with CF who have a thicker, more viscous mucus layer that can impede the uptake of the vector and prevent its delivery to distal airways. There are data to suggest that the most commonly used vectors for gene therapy are unable to penetrate CF sputum.[28] Thus, current gene therapy approaches may only be feasible in early-stage disease before the development of significant sputum accumulation or airway changes.

FIGURE 16.1 Residual CFTR function and clinical phenotype. The severity of the clinical phenotype is directly correlated with the amount of CFTR expressed in tissue. Fifty percent of wild-type CFTR expression is associated with a normal phenotype as is the case with obligate carriers such as parents of children with CF. CF-PI = cystic fibrosis with pancreatic insufficiency, CF-PS = cystic fibrosis with pancreatic sufficiency, CBAVD = congenital bilateral absence of the vas deferens.

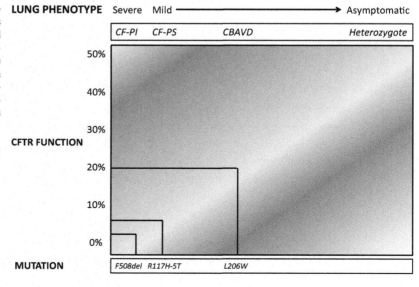

Once the epithelial layer has been traversed, there are still immune barriers to overcome. These include innate immunity by routine digestion through macrophages and acquired humoral immunity, which is most significant for viral vectors (see the following section). Alveolar macrophages have a multifaceted role as the first line of cellular host defense to invading pathogens.[29] Alveolar macrophages can directly damage invading particles by releasing such substances as lysozyme and proteases and by recruiting neutrophils, lymphocytes, and dendritic cells via expression of cytokines and chemokines.[30] Experiments with adenoviral vectors in immunocompetent and immunodeficient mice have shown that nearly 70% of the vector is lost in the first 24h, a loss that was reduced in the absence of macrophages.[31] In addition, the lung is highly immunogenic, and if pathogens can evade the primary macrophage-mediated immune barrier, they are still subject to removal by humoral or cellular immune systems as a second protective step. Furthermore, humoral immune responses induced by helper T cells targeting the viral vector reduce the transfection potential with each subsequent administration of a viral vector.[32,33] In addition, pretreatment with cyclophosphamide has been demonstrated to improve transfection efficiency.[34] These issues are important considerations with respect to safe use of immunosuppression in CF with its associated chronic airway infection.

16.3.3 Safety Concerns and Duration of Expression in CF

In addition to the technical concerns outlined previously, there are also important safety and efficacy issues that need to be considered. Generally, gene therapy runs the risk of an exuberant, potentially fatal immune response, and the risk of insertional mutagenesis. The risk of severe cytokine response is greatest in the setting of systemic administration; however, it is still present with topical administration of a vector. Immunological response to the vector is a well-established concern in gene therapy, with documented cases of death in other disease states directly related to transfection.[35] Although the risk of a systemic inflammatory response is present in CF, this is mitigated by the topical instillation of vector on the airways in all trials to date. Direct application of vector on the airways, however, leads to a transient duration of action and often requires repeat administration of vector to maintain effect. The repeat administration of the therapeutic gene increases the risk of a heightened immune response of the host toward the gene therapy vector and transgene.

Even if the vector causes no systemic reaction and does not require repeat administration, there is still the risk related to the location of the inserted gene and the subsequent host response. Vector integration into the host genome is typically not controlled with gene therapy and, therefore, may occur at any place within the genome. If insertion takes place within existing genes, it may alter their function and lead to new problems. This phenomenon has been well documented in gene therapy trials for X-linked severe combined immunodeficiency.[36,37] In this instance, the primary disease was essentially cured by gene therapy; however, a few years after the trial, two patients subsequently developed leukemia related to interruption of an oncogene by the inserted vector. Further reports have shown that integrating vectors used in gene therapy have a predilection for regulatory regions of transcriptionally active genes.[38] Other potential mechanisms of altered host response include gene silencing, horizontal or vertical transmission of therapeutic DNA, and immunotoxicity by the host cell to the vector.

16.4 CF GENE THERAPY IN CLINICAL TRIALS

Despite the theoretical and technical challenges outlined, many investigators have invested significant resources in an effort to develop gene therapy as a mutation-independent cure in CF. There has been a great deal of preclinical research and early-phase clinical trials investigating viable approaches for gene therapy. These studies have included testing of viral and nonviral vectors administered locally and systemically in in vitro and in vivo animal models. Many of these vectors have also been tested in CF patients as part of phase 1 and phase 2 studies (Tables 16.3 and 16.4). This chapter will review the notable articles in the field, with particular emphasis on the hurdles and limitations of gene therapy in CF and assessment of a possible path in the future.

16.4.1 Gene Therapy: Early Studies with Adenovirus Vectors

Adenoviruses, a well-known cause of the common cold, have a natural propensity for infecting sinopulmonary epithelial cells. Adenoviruses are primarily taken up by host cells through the coxsackie-adenovirus receptor in the apical surface of airway epithelial cells. These receptors are present in many species but notably scarce on the apical surface in the majority of human airway cells.[64] This limits their uptake in humans compared with animal models. Because of its low transfection efficiency and the immunologic response it induces, this vector has shown poor clinical efficacy; thus, there has been limited enthusiasm for its continued development. Later generations of these nonenveloped, double-stranded DNA viruses have been modified to serve as gene vectors through intricate removal of all viral gene DNA, forming so-called "gutless" viral vectors. When coupled with target DNA, such as CFTR, they remain in the cytoplasm of the cell without integrating into the host cell genome.

TABLE 16.3 Viral Vector Gene Therapy Clinical Trials

Reference	Vector	Route(s)	n	Safety Concerns	Results
Zabner et al.[39]	Ad	Nose	3	No	NPD: decreased baseline toward normal values CFTR mRNA and protein: not detected
Crystal et al.[40]	Ad	Nose–lung	4	Yes, transient inflammation at highest dose	NPD: variable CFTR mRNA: detectable in nose but not lung
Knowles et al.[41]	Ad	Nose	12	Local inflammation at highest dose	NPD: unchanged CFTR mRNA: detectable mostly at higher doses
Hay et al.[42]	Ad	Nose	9	No	NPD: partial correction
Zabner et al.[43]	Ad	Nose	6	No	NPD: mild normalization
Bellon et al.[44]	Ad	Nose–lung	6	No	CFTR mRNA and protein: some detected nose > lung
Harvey et al.[45]	Ad	Lung	14	No	CFTR mRNA: partially positive but reduced in subsequent administration
Zuckerman et al.[46]	Ad	Lung	11	Flu-like symptoms	Vector DNA: detectable
Joseph et al.[48] Perricone et al.[47]	Ad	Lung	36	Transient fevers, myalgias	CFTR mRNA and DNA: transient detection noted
Wagner et al.[49]	AAV2	Nose	10	No	NPD: partial correction CFTR mRNA: none detected
Aitkin et al.[50]	AAV2	Lung	12	Possible	mRNA: none detected
Wagner et al.[51]	AAV2	Nose	23	No	NPD: no change No change in histology of IL-8
Flotte et al.[52]	AAV2	Nose–lung	25	Well tolerated, possible side effects reported	NPD: unchanged
Moss et al.[53]	AAV2	Lung	37	No	FEV_1: improvement trend CFTR mRNA: undetectable
Moss et al.[54]	AAV2	Lung	102	No	FEV_1, sputum markers, antibiotic days: no improvement

Table of clinical trials using viral vectors in patients with cystic fibrosis. Viral vectors have been the best studied in clinical trials in CF to date. Ad, adenovirus; AAV, adeno-associated virus; CFTR, cystic fibrosis transmembrane conductance regulator; FEV_1, forced expiratory volume in 1 s; NPD, nasal potential difference.

Adenovirus coupled with CFTR DNA was the first viral vector used in clinical trials only a few years after the discovery of CFTR. In 1993, Zabner et al. published the first study of three patients in which wild-type CFTR was instilled nasally using an adenovirus vector.[39] They found no evidence of CFTR expression, but did detect a slight decrease in NPD toward normal values. This study demonstrated proof of principle that gene therapy was well tolerated and could lead to detectable physiologic changes in CF patients. However, the study was limited because there was only a single administration of the CFTR carrying viral vector in a very small number of patients and no assessment of long-term sustainability.

The initial study by Zabner et al. was followed by nine additional studies with adenoviruses in the subsequent 8 years.[40–48] Most of these studies assessed the feasibility of a single administration of adenovirus gene transfer therapy in up to 101 patients. They tested a variety of methods including nasal instillation, lung nebulization, and endobronchial instillation. There were, however,

TABLE 16.4 Nonviral Vectors Gene Therapy Clinical Trials

Reference	Vector	Route(s)	n	Safety Concerns	Results
Caplen et al.[55]	DC-Chol/DOPE	Nose	15	No	NPD: partial correction CFTR mRNA: not detected
Gill et al.[56]	DC-Chol/DOPE	Nose	12	Possible	NPD: partial correction
Hyde et al.[57]	DC-Chol/DOPE	Nose	11	No	NPD: partial correction CFTR mRNA: detected in some
Porteous et al.[58]	DOTAP	Nose	16	No	NPD: partial correction CFTR mRNA: detected in some
Noone et al.[59]	EDMPC	Nose	11	No	NPD: unchanged CFTR mRNA: undetectable
Zabner et al.[60]	GL67 vs naked pDNA	Nose	12	No	NPD: significant response with both vectors CFTR mRNA: non detected
Alton et al.[61]	GL67	Nose–lung	16	Flu-like symptoms	NPD: significant correction CFTR mRNA: undetectable CFTR DNA: detectable
Ruiz et al.[62]	GL67	Lung	8	Flu-like symptoms	CFTR mRNA: partially detectable
Konstan et al.[63]	DNA nanoparticles	Nose	12	No	NPD: partial correction CFTR DNA: detectable in treatment and some controls

Table of clinical trials using nonviral vectors in patients with cystic fibrosis. Nonviral vectors offer advantages to viral vectors with less severe immune response and better tolerance to repeated administrations.
CFTR, cystic fibrosis transmembrane conductance regulator; DC-Chol, 3b-[N-(N9,N9-dimethylaminoethane)-carbamoyl]-cholesterol; DOPE, 1,2-dioleoyl-sn-glycerol-phosphoethanolamine; DOTAP, 1,2-dioleoyl-3-trimethylammoniopropane; NPD, nasal potential difference; pDNA, plasmid DNA.

two studies[43,45] in which repeated dosing strategies were used. Collectively, these studies demonstrated relative safety of repeated administration, with few side effects that primarily included self-limited flu-like symptoms. These studies also highlighted some of the difficulties with an adenovirus-based approach in the CF lung. The problems included the following: induction of inflammatory responses in host tissues toward the virus; poor penetration in mucus-laden tissue of the lower lung compared with nasal epithelium; and transient, low-level CFTR expression. Attempts to achieve a sustained response by repeat administration only increased viral antibody responses in the host without an increase in CFTR expression.

16.4.2 Adenovirus-Associated Viral Vectors

The failure of an adenovirus-based approach led to a search for alternative viral vectors in which the virus induces less of an immune response. Of these alternatives, adeno-associated viruses (AAV) were the next logical step. These single-strand linear

DNA viruses are similar to adenoviruses but are not pathologic to humans and have a longer duration of exogenous gene expression. In late 1999 to the early 2000s, AAV2 was used in multiple clinical trials[49–53] involving a total of 209 patients and making it the most well-studied gene therapy vector in CF patients to date. The first of these were phase 1 safety studies testing single inoculations in patients with CF. Significant findings from these studies were that the AAV2 vectors were generally well tolerated with variable detection of transfected CFTR expression and unclear clinical implications.

Only two of the phase 1 AAV-CFTR trials led to phase 2 studies in which repeat administration of AAV2-CFTR was evaluated.[53,54] The first study published in 2004 evaluated 37 CF patients who received either three doses of AAV2-CFTR or placebo spaced 30 days apart. After the final instillation, vector DNA was detectable in all samples tested; however, CFTR mRNA was not detected. A trend in forced expiratory volume in 1 s (FEV_1) improvement was noted in the treatment arm compared with placebo, although the differences were not statistically significant. In addition, evidence of AAV2-mediated antibodies in treated patients suggested limitations on its use. This was followed by a phase 2B study[54] in which 102 patients were randomized to treatment with either nebulized AAV2-CFTR or placebo administered in two doses given 30 days apart. The safety profile was similar in both groups, though there was again a significant humoral immune response with 100% of the treatment group having a fourfold increase in AAV2 antibody titers compared to only 2% in the placebo group. There were no significant improvements in endpoints such as FEV_1, sputum inflammatory biomarkers, or days of antibiotic use. Given the need for repeat administration, absence of improvement in clinical endpoints, and viral-associated antibody response, evaluation of alternative viruses as vector choices has been pursued.

16.4.3 Additional Viral Vectors

Other viruses that have been explored as CF gene therapy vectors include simian vacuolating virus 40 (SV40), RNA viruses, and lentivirus. Simian vacuolating virus is a circular DNA virus with the ability to integrate into the host cell genome. There was initial optimism that SV40 may be a better alternative to adenoviruses; however, it has had limited utility because of its relatively smaller size, which has made the packaging of the large CFTR transcript challenging. In in vivo experiments with CF mouse models, transfection with rSV40CFTR demonstrated a modest improvement in chloride channel function at the apical surface compared with empty vector.[65] This study also examined CFTR null mice transfected with rSV40/CFTR after *Pseudomonas aeruginosa* challenge and reported improvements in weight and inflammatory markers such as bronchoalveolar lavage fluid neutrophil counts, interleukin (IL)-1β, and IL-8. However, the in vivo responses in mice were not expanded to humans.

RNA viruses have negative strands and remain in the cytoplasm without integration into the host cell genome. Examples include parainfluenza virus, respiratory syncytial virus, and Sendai virus (SeV). Of these, the only vector that has been used in animal models is SeV. In a study by Griesenbach et al., mice were transfected through nasal instillation of an F-protein-deleted SeV viral vector (ΔF/SeV) coupled to a luciferase reporter gene with either a single or repeat administrations.[66] The results demonstrated a robust response to initial administration at 2 days after instillation; however, there was a 60% decline in target gene expression when empty vectors were administered for two doses before luciferase instillation. This suggests a vector-related immune response that modulates target gene expression. The lower response to repeat dosing was similar to response levels seen with nonviral vectors. With its very short time (days) for gene expression and difficulty with repeat administration because of neutralizing antibody responses,[66] translation of SeV viruses to human trials would be challenging.

Lentiviruses are viruses that are currently being evaluated most actively as gene therapy vectors in CF. They are linear RNA viruses that have the capability of integrating into the host genome and offer the potential of sustained CFTR expression. There are, however, still limitations to overcome with this virus because it has relatively poor uptake in airway epithelium. For this reason, adjunct use of tight junction modulators such as lysophosphatidylcholine has been used to facilitate transfer into host progenitor cells in the deeper epithelium.[67,68] Nasal expression of CFTR in CF knockout mice has resulted in partial correction of chloride transport, which was sustained for up to 12 months after a single administration of modified lentivirus.[69] Subsequent studies have shown persistent luciferase reporter expression up to 18 months after transduction of Sendai virus-F/HN pseudotyped lentivirus after lysophosphatidylcholine pretreatment in CF mice.[70] Follow-up work by this same group recently reported that, in CF mice, a single dose with this regimen showed sustained luciferase activity for the lifetime of the mouse. CFTR expression was increased with repeated monthly inoculation and sustained tolerance was also demonstrated.[71] For these reasons, lentiviruses are the current vector of choice for ongoing viral-based studies in CF, although no human trials are currently under way.

16.4.4 Nonviral Vectors in CF Gene Therapy

An alternative to viruses are nonviral vectors that typically comprise plasmids or carrier molecules made of cations or lipids that can deliver target DNA into a cell. The potential advantage of nonviral vectors compared with viral vectors is a decreased immunomodulatory response and better tolerance to repeat administration. One of the theoretical disadvantages of nonviral vectors is lower efficiency of transfection, owing in part to the absence of functional proteins that aid with intracellular transport and integration. It has been demonstrated in a small number of CF patients that repeated nasal administration of DNA liposome complex, DC-Chol/DOPE, led to detectable CFTR expression and no evidence of an immune response. However, the transfection rates were low and there was minimal change in chloride transport as measured by NPD.[57]

Additional data come from work performed in the 1990s when there was significant enthusiasm for nonviral vectors. There were nine separate phase 1 clinical trials examining five different nonviral vectors, each tested in a small number of CF patients. The studies using nasal administration showed reasonable tolerability; however, in the two studies in which lower respiratory tract administration was performed,[62,72] the majority of patients were found to have significant inflammatory responses. Both of these studies used the nonviral vector GL67A coupled to plasmid CFTR, and work has been ongoing to modify the vector to improve tolerability and efficacy. A clinical trial in 35 CF patients was recently completed in which a single nebulized treatment of GL67A (pGM169/GL67A) was administered at variable doses. The investigators found evidence of CFTR mRNA in bronchial lining cells and improved NPD in the majority of patients. This was most pronounced with the mid-level dose, which had a favorable safety profile.[73] A randomized, double blind, multidose, placebo-controlled clinical trial is currently under way in which CF patients will undergo repeat instillations every month for 1 year. This study design was recently published; however, the results of the trial are likely several years off.[74]

16.4.5 Stem Cells to Restore CFTR Function

There has also been significant interest in the use of stem cells as a therapeutic option for restoring functional CFTR in patients with CF. This includes the use of early progenitor or bone marrow-derived cells in which CFTR function has been restored and administered to the host intratracheally or systemically. This field is only beginning to emerge and its merits and potential limitations have recently been reviewed by several authors.[23,75] Mice lacking CFTR have been tested with the use of autologous bone marrow cells that have had CFTR function restored in vitro. Some of these studies have used direct instillation of early stem cells similar to methods used in viral vector transfection as described previously. However, murine studies have also used systemic administration of early progenitor cells, in the absence of irradiation or immunosuppression, with salutary effects on P. aeruginosa lung infection.[76] This approach shows promise with several early animal models demonstrating engraftment of donor cells. However, the efficiency of engraftment has been low raising the question of whether clinically significant correction is possible in humans with this approach.

16.5 MUTANT PROTEIN REPAIR

After its discovery as the cause of CF, CFTR was identified as a member of the traffic ATPase family and a chloride channel.[11,12,77] Subsequently, it was shown to transport other anions (e.g., thiocyanate, bicarbonate) and to regulate ion transporters such as the epithelial sodium channel and alternative chloride channels.[74-81] This work provided a new perspective for relationships between CFTR mutations and the molecular mechanism of their associated dysfunction. The breakdown of mutations into five classes based on mechanism of protein dysfunction provides a useful approach to therapy because it allows the potential of designing interventions targeting mutant CFTR function based on mutation class. Using support from the CF Foundation, multiple pharmaceutical companies have undertaken separate high-throughput screening efforts that address three different CFTR mutation targets, including (1) suppression of premature termination codons (PTC), (2) "potentiation" of mutant CFTR at the plasma membrane (e.g., G551D CFTR), and (3) "correction" of CFTR trafficking defects (F508del CFTR). Before and in parallel to these efforts, proof of concept studies in independent laboratories demonstrated that mutant CFTR function could be improved by a variety of small molecules in preclinical model systems and in vivo.[19,82-85]

16.5.1 CFTR Potentiators

Potentiators are CFTR modulators that increase the time CFTR spends in the open channel configuration and thereby lead to increased chloride transport. Ivacaftor, the first potentiator to gain Food and Drug Administration approval, is indicated for CF patients with the G551D mutation.[82] Though it is the second most common cause of CF, G551D is present in only about 4% of CF patients.[14] Ivacaftor has been studied in one phase 2[86] and two pivotal phase 3 trials.[87,88] In all three studies, ivacaftor treatment led to rapid, dramatic, and sustained improvements in FEV_1 and biomarkers of CFTR function (e.g., sweat chloride values below the 60 mmol CF diagnostic threshold). In the larger phase 3 trial in adults, ivacaftor-treated patients demonstrated more than a 10% absolute improvement in FEV$_1$, which was highly significant. There were also clinically significant improvements in weight, quality of life measures, and pulmonary exacerbation frequency. Similar findings were seen in the phase 3 trial of patients aged 6–11 years.[88] The responses in younger CF patients were particularly noteworthy, confirming the suspicion of "silent" disease in seemingly normal CF patients, and providing support to the notion that early disease is reversible (and potentially preventable). In a more recent phase 2 trial of patients with milder lung disease (FEV$_1$ > 90%), ivacaftor treatment reduced ventilation inhomogeneity, as assessed by the lung clearance index, and improved sweat chloride and FEV$_1$.[89] It remains to be seen whether patients with other gating mutations (class III) or conductance mutations (class IV) can benefit from ivacaftor, but preclinical studies indicate reduced chloride transport caused by other CFTR-gating mutations can be increased by ivacaftor.[90] Clinical trials have been completed in both categories of patients with results pending. Based on submitted though unpublished data, the Food and Drug Administration approved ivacaftor for seven additional low-prevalent mutations.

16.5.2 PTC Mutation Suppressors

PTCs result when single base-pair substitutions create an erroneous stop codon within the open reading frame of a gene. Suppressors of PTCs, such as aminoglycoside antibiotics, are able to bind eukaryotic ribosomes and cause the insertion of a near cognate amino-acyl transfer RNA into the ribosomal A site.[91] This process can allow the ribosome to "readthrough" the PTC and produce some full-length protein and has been extensively tested in proof-of-concept studies using aminoglycosides to suppress PTCs (gentamicin, amikacin, geneticin).[91] There has been demonstrated efficacy in in vitro studies, in animal models of CF and muscular dystrophy, and in small numbers of CF patients. Currently, there is one oral compound, ataluren (PTC Therapeutics), in clinical trials to treat CF caused by PTCs. Ataluren was studied in three phase 2 randomized, dose-ascending, open-label trials in CF.[83,92,93] Each study demonstrated short-term tolerability of ataluren, and two studies demonstrated improvements in CFTR function (across a number of PTC mutations) as measured by NPD. One study also demonstrated improvements in CFTR localization to the nasal cell membrane, whereas another demonstrated improvement in cough over 3 months.[83,92] The third study failed to demonstrate improvements in NPD, and all three studies were limited by small numbers and absence of placebo groups.[94] It was then studied in a large phase 3 randomized, 48-week, double-blind, placebo-controlled trial designed to test safety, efficacy, and tolerability. The results have only been reported in abstracts to date.[95,96] The phase 3 study consisted of 232 patients and showed that ataluren was associated with a trend toward slower loss of FEV$_1$ (−2.5% ataluren vs. −5.5% placebo) and fewer pulmonary exacerbations (23% decrease compared with placebo).[97] The primary significant drug-related toxicity was to the kidney.

16.5.3 CFTR Correctors

The F508del mutation disrupts folding of nascent CFTR, causing retention in the endoplasmic reticulum and subsequent proteasomal degradation.[81] The result is minimal protein escaping intracellular degradation. CFTR correctors induce more F508del CFTR protein delivery to the plasma membrane, and Vertex Pharmaceuticals has developed two F508del correctors that have advanced to clinical trials. One F508del CFTR corrector (VX-809, Lumacaftor) has been assessed in a phase 2, multicenter, randomized, double-blind, placebo-controlled trial in adults homozygous for F508del CFTR. Eighty-nine subjects were enrolled to evaluate safety and efficacy of 28 days of VX-809 treatment compared with placebo across four doses.[98] Although VX-809 was found to be safe for patients, there was no impact on any clinically significant outcome. There was, however, a dose-dependent reduction in sweat chloride, demonstrating that F508del CFTR was a viable drug target and laying the foundation for combinational trials with the CFTR potentiator, ivacaftor. Indeed, in vitro results of VX-809 combined with ivacaftor have reported increased F508del CFTR activity relative to VX-809 alone (as assessed by chloride currents in primary human bronchial epithelial cells).[99] In a phase 2 study of 82 F508del CF patients, using VX-809 in combination with ivacaftor was associated with an approximately 10 mmol/L reduction in sweat chloride compared with placebo. In the subgroup of patients who received the largest dose of VX809, FEV$_1$ increased by 6.62% more than with placebo.[100] A phase 3 study is ongoing and should have results reported in mid- to late 2014. In addition, a second F508del CFTR corrector (VX-661) has also entered clinical trials in F508del CFTR homozygous adults, which will assess treatment with VX-661 alone and in combination with ivacaftor in a dual-dose ascending study format (NCT01531673).

One unanswered question in regard to mutation-specific therapy in CF is whether both mutations need to be targeted in patients who are heterozygous for the CFTR genotype. Data from the ivacaftor in G551D study, in which >99% of patients were heterozygotes,[87] suggest that only one mutation needs to be targeted if the intervention is sufficiently potent. It is likely, however, that for less potent interventions, both mutations may need to be targeted to obtain a robust clinical effect.

16.6 CONCLUSION

The discovery that mutant CFTR is the cause of CF engendered hope that gene therapy would lead to a disease cure. There have been substantial gains in the diagnosis and understanding of the basic cellular mechanisms to better characterize the disease and assess treatment response. In addition, the breadth of potential gene mutations and subsequent categorization into classes has helped to shape the working model of the disease at the molecular level. Replacement of the mutated gene with a functional copy has been the hope since the discovery of CFTR more than 20 years ago and still remains elusive. Much has been learned in the intervening period with respect to hurdles and limitations of gene therapy. Some of the largest obstacles have been the inability to transfect the target cells with an intact copy of the gene because of physical and immunologic barriers. Clinical trials have been conducted using several viral vectors of which the most common have been adeno- and adeno-associated viruses with limited success. Even when vector uptake is possible, repeated administration has been required with each administration associated with a diminished clinical response. Current gene therapy research is targeting use of lentiviruses and nonviral vectors. Additional research using bone marrow-derived stem cells and early progenitor cells also provides hope for gene replacement, though this has not been expanded to human trials to date. Newer approaches targeting repair of mutant CFTR proteins are a promising alternative to gene therapy. One small molecule potentiator has been approved for use in several class III mutations, whereas several other compounds targeting additional genotypes are being tested in ongoing clinical trials. Although gene replacement has been sought for many years, the most successful approaches have used single and combination small molecule compounds to correct and potentiate protein function with moderate success and improved clinical outcomes. Work continues on all fronts to develop a long-term cure that will likely require a combination of strategies currently discussed.

GLOSSARY

Obligate heterozygotes Possessing one mutated gene and one functioning gene (i.e., CF carriers or parents of offspring with CF)
Premature termination codons DNA mutations resulting in stop mutations and truncated mRNA
Potentiators Molecule that improves channel function

REFERENCES

1. Busch R. On the history of cystic fibrosis. *Acta Univ Carol Med* 1990;**36**(1–4):13–5.
2. Alonso Y De Los Ruyzes De Fonteca J. Diez Previlegios para Mgeres Prenadas. *Alcala Hen* 1606;**212**.
3. Garrod AE, Hurtley WH. Congenital family steatorrhea. *Quart J Med* 1912;**6**:242–58.
4. Andersen DH. Cystic fibrosis of the pancreas and its relation to celiac disease. *Am J Dis Child* 1938;**56**:344–99.
5. Andersen DH, Hodges RG. Celiac syndrome; genetics of cystic fibrosis of the pancreas, with a consideration of etiology. *Am J Dis Child* July 1946;**72**:62–80.
6. Di Sant'Agnese PA, Darling RC, Perera GA, Shea E. Abnormal electrolyte composition of sweat in cystic fibrosis of the pancreas; clinical significance and relationship to the disease. *Pediatrics* November 1953;**12**(5):549–63.
7. Gibson LE, Cooke RE. A test for concentration of electrolytes in sweat in cystic fibrosis of the pancreas utilizing pilocarpine by iontophoresis. *Pediatrics* March 1959;**23**(3):545–9.
8. Farrell PM, Rosenstein BJ, White TB, et al. Guidelines for diagnosis of cystic fibrosis in newborns through older adults: cystic Fibrosis Foundation consensus report. *J Pediatr* August 2008;**153**(2):S4–14.
9. Boucher RC, Stutts MJ, Knowles MR, Cantley L, Gatzy JT. Na+ transport in cystic fibrosis respiratory epithelia. Abnormal basal rate and response to adenylate cyclase activation. *J Clin Invest* November 1986;**78**(5):1245–52.
10. Knowles MR, Stutts MJ, Spock A, Fischer N, Gatzy JT, Boucher RC. Abnormal ion permeation through cystic fibrosis respiratory epithelium. *Science* September 9, 1983;**221**(4615):1067–70.
11. Kerem B, Rommens JM, Buchanan JA, et al. Identification of the cystic fibrosis gene: genetic analysis. *Science* September 8, 1989;**245**(4922):1073–80.
12. Riordan JR, Rommens JM, Kerem B, et al. Identification of the cystic fibrosis gene: cloning and characterization of complementary DNA. *Science* September 8, 1989;**245**(4922):1066–73.
13. Rommens JM, Iannuzzi MC, Kerem B, et al. Identification of the cystic fibrosis gene: chromosome walking and jumping. *Science* September 8, 1989;**245**(4922):1059–65.
14. Cystic Fibrosis Mutation Database. *Cystic Fibrosis Centre at the Hospital for Sick Children in Toronto*; 1989 [accessed 09.04.12].
15. Zielenski J, Tsui LC. Cystic fibrosis: genotypic and phenotypic variations. *Annu Rev Genet* 1995;**29**:777–807.
16. Prickett M, Jain M. Gene therapy in cystic fibrosis. *Transl Res* April 2013;**161**(4):255–64.
17. Welsh MJ, Smith AE. Molecular mechanisms of CFTR chloride channel dysfunction in cystic fibrosis. *Cell* July 2, 1993;**73**(7):1251–4.
18. Varga K, Goldstein RF, Jurkuvenaite A, et al. Enhanced cell-surface stability of rescued DeltaF508 cystic fibrosis transmembrane conductance regulator (CFTR) by pharmacological chaperones. *Biochem J* March 15, 2008;**410**(3):555–64.

19. Hwang TC, Wang F, Yang IC, Reenstra WW. Genistein potentiates wild-type and delta F508-CFTR channel activity. *Am J Physiol* September 1997;**273**(3 Pt 1):C988–98.

20. McLachlan G, Davidson H, Holder E, et al. Pre-clinical evaluation of three non-viral gene transfer agents for cystic fibrosis after aerosol delivery to the ovine lung. *Gene Ther* October 2011;**18**(10):996–1005.

21. Griesenbach U, Alton EW. Cystic fibrosis gene therapy: successes, failures and hopes for the future. *Expert Rev Respir Med* August 2009;**3**(4):363–71.

22. Griesenbach U, Alton EW. Current status and future directions of gene and cell therapy for cystic fibrosis. *BioDrugs* April 1, 2011;**25**(2):77–88.

23. Griesenbach U, Alton EW. Progress in gene and cell therapy for cystic fibrosis lung disease. *Curr Pharm Des* 2012;**18**(5):642–62.

24. Griesenbach U, Geddes DM, Alton EW. Gene therapy progress and prospects: cystic fibrosis. *Gene Ther* July 2006;**13**(14):1061–7.

25. Davies JC, Alton EW. Gene therapy for cystic fibrosis. *Proc Am Thorac Soc* November 2010;**7**(6):408–14.

26. Kalin N, Claass A, Sommer M, Puchelle E, Tummler B. DeltaF508 CFTR protein expression in tissues from patients with cystic fibrosis. *J Clin Invest* May 15, 1999;**103**(10):1379–89.

27. Wilschanski M, Dupuis A, Ellis L, et al. Mutations in the cystic fibrosis transmembrane regulator gene and in vivo transepithelial potentials. *Am J Respir Crit Care Med* October 1, 2006;**174**(7):787–94.

28. Hida K, Lai SK, Suk JS, Won SY, Boyle MP, Hanes J. Common gene therapy viral vectors do not efficiently penetrate sputum from cystic fibrosis patients. *PLoS One* 2011;**6**(5):e19919.

29. Sibille Y, Reynolds HY. Macrophages and polymorphonuclear neutrophils in lung defense and injury. *Am Rev Respir Dis* February 1990;**141**(2): 471–501.

30. Haslett C. Granulocyte apoptosis and its role in the resolution and control of lung inflammation. *Am J Respir Crit Care Med* November 1999; **160**(5 Pt 2):S5–11.

31. Worgall S, Leopold PL, Wolff G, Ferris B, Van Roijen N, Crystal RG. Role of alveolar macrophages in rapid elimination of adenovirus vectors administered to the epithelial surface of the respiratory tract. *Hum Gene Ther* September 20, 1997;**8**(14):1675–84.

32. Yang Y, Li Q, Ertl HC, Wilson JM. Cellular and humoral immune responses to viral antigens create barriers to lung-directed gene therapy with recombinant adenoviruses. *J Virol* April 1995;**69**(4):2004–15.

33. Mingozzi F, Maus MV, Hui DJ, et al. CD8(+) T-cell responses to adeno-associated virus capsid in humans. *Nat Med* April 2007;**13**(4):419–22.

34. Cao H, Yang T, Li XF, et al. Readministration of helper-dependent adenoviral vectors to mouse airway mediated via transient immunosuppression. *Gene Ther* February 2011;**18**(2):173–81.

35. Raper SE, Chirmule N, Lee FS, et al. Fatal systemic inflammatory response syndrome in a ornithine transcarbamylase deficient patient following adenoviral gene transfer. *Mol Genet Metab* September-October 2003;**80**(1–2):148–58.

36. Hacein-Bey-Abina S, Von Kalle C, Schmidt M, et al. LMO2-associated clonal T cell proliferation in two patients after gene therapy for SCID-X1. *Science* October 17, 2003;**302**(5644):415–9.

37. Hacein-Bey-Abina S, von Kalle C, Schmidt M, et al. A serious adverse event after successful gene therapy for X-linked severe combined immuno-deficiency. *N Engl J Med* January 16, 2003;**348**(3):255–6.

38. Bushman F, Lewinski M, Ciuffi A, et al. Genome-wide analysis of retroviral DNA integration. *Nat Rev Microbiol* November 2005;**3**(11):848–58.

39. Zabner J, Couture LA, Gregory RJ, Graham SM, Smith AE, Welsh MJ. Adenovirus-mediated gene transfer transiently corrects the chloride transport defect in nasal epithelia of patients with cystic fibrosis. *Cell* October 22, 1993;**75**(2):207–16.

40. Crystal RG, McElvaney NG, Rosenfeld MA, et al. Administration of an adenovirus containing the human CFTR cDNA to the respiratory tract of individuals with cystic fibrosis. *Nat Genet* September 1994;**8**(1):42–51.

41. Knowles MR, Hohneker KW, Zhou Z, et al. A controlled study of adenoviral-vector-mediated gene transfer in the nasal epithelium of patients with cystic fibrosis. *N Engl J Med* September 28, 1995;**333**(13):823–31.

42. Hay JG, McElvaney NG, Herena J, Crystal RG. Modification of nasal epithelial potential differences of individuals with cystic fibrosis consequent to local administration of a normal CFTR cDNA adenovirus gene transfer vector. *Hum Gene Ther* November 1995;**6**(11):1487–96.

43. Zabner J, Ramsey BW, Meeker DP, et al. Repeat administration of an adenovirus vector encoding cystic fibrosis transmembrane conductance regula-tor to the nasal epithelium of patients with cystic fibrosis. *J Clin Invest* March 15, 1996;**97**(6):1504–11.

44. Bellon G, Michel-Calemard L, Thouvenot D, et al. Aerosol administration of a recombinant adenovirus expressing CFTR to cystic fibrosis patients: a phase I clinical trial. *Hum Gene Ther* January 1, 1997;**8**(1):15–25.

45. Harvey BG, Leopold PL, Hackett NR, et al. Airway epithelial CFTR mRNA expression in cystic fibrosis patients after repetitive administration of a recombinant adenovirus. *J Clin Invest* November 1999;**104**(9):1245–55.

46. Zuckerman JB, Robinson CB, McCoy KS, et al. A phase I study of adenovirus-mediated transfer of the human cystic fibrosis transmembrane con-ductance regulator gene to a lung segment of individuals with cystic fibrosis. *Hum Gene Ther* December 10, 1999;**10**(18):2973–85.

47. Perricone MA, Morris JE, Pavelka K, et al. Aerosol and lobar administration of a recombinant adenovirus to individuals with cystic fibrosis. II. Transfection efficiency in airway epithelium. *Hum Gene Ther* July 20, 2001;**12**(11):1383–94.

48. Joseph PM, O'Sullivan BP, Lapey A, et al. Aerosol and lobar administration of a recombinant adenovirus to individuals with cystic fibrosis. I. Methods, safety, and clinical implications. *Hum Gene Ther* July 20, 2001;**12**(11):1369–82.

49. Wagner JA, Messner AH, Moran ML, et al. Safety and biological efficacy of an adeno-associated virus vector-cystic fibrosis transmembrane regula-tor (AAV-CFTR) in the cystic fibrosis maxillary sinus. *Laryngoscope* February 1999;**109**(2 Pt 1):266–74.

50. Aitken ML, Moss RB, Waltz DA, et al. A phase I study of aerosolized administration of tgAAVCF to cystic fibrosis subjects with mild lung disease. *Hum Gene Ther* October 10, 2001;**12**(15):1907–16.

51. Wagner JA, Nepomuceno IB, Messner AH, et al. A phase II, double-blind, randomized, placebo-controlled clinical trial of tgAAVCF using maxillary sinus delivery in patients with cystic fibrosis with antrostomies. *Hum Gene Ther* July 20, 2002;**13**(11):1349–59.

52. Flotte TR, Zeitlin PL, Reynolds TC, et al. Phase I trial of intranasal and endobronchial administration of a recombinant adeno-associated virus serotype 2 (rAAV2)-CFTR vector in adult cystic fibrosis patients: a two-part clinical study. *Hum Gene Ther* July 20, 2003;**14**(11):1079–88.

53. Moss RB, Rodman D, Spencer LT, et al. Repeated adeno-associated virus serotype 2 aerosol-mediated cystic fibrosis transmembrane regulator gene transfer to the lungs of patients with cystic fibrosis: a multicenter, double-blind, placebo-controlled trial. *Chest* February 2004;**125**(2):509–21.

54. Moss RB, Milla C, Colombo J, et al. Repeated aerosolized AAV-CFTR for treatment of cystic fibrosis: a randomized placebo-controlled phase 2B trial. *Hum Gene Ther* August 2007;**18**(8):726–32.

55. Caplen NJ, Alton EW, Middleton PG, et al. Liposome-mediated CFTR gene transfer to the nasal epithelium of patients with cystic fibrosis. *Nat Med* Jan 1995;**1**(1):39–46.

56. Gill DR, Southern KW, Mofford KA, et al. A placebo-controlled study of liposome-mediated gene transfer to the nasal epithelium of patients with cystic fibrosis. *Gene Ther* Mar 1997;**4**(3):199–209.

57. Hyde SC, Southern KW, Gileadi U, et al. Repeat administration of DNA/liposomes to the nasal epithelium of patients with cystic fibrosis. *Gene Ther* July 2000;**7**(13):1156–65.

58. Porteous DJ, Dorin JR, McLachlan G, et al. Evidence for safety and efficacy of DOTAP cationic liposome mediated CFTR gene transfer to the nasal epithelium of patients with cystic fibrosis. *Gene Ther* Mar 1997;**4**(3):210–8.

59. Noone PG, Hohneker KW, Zhou Z, et al. Safety and biological efficacy of a lipid-CFTR complex for gene transfer in the nasal epithelium of adult patients with cystic fibrosis. *Mol Ther* Jan 2000;**1**(1):105–14.

60. Zabner J, Cheng SH, Meeker D, et al. Comparison of DNA-lipid complexes and DNA alone for gene transfer to cystic fibrosis airway epithelia in vivo. *J Clin Invest* Sep 15, 1997;**100**(6):1529–37.

61. Alton EW, Stern M, Farley R, et al. Cationic lipid-mediated CFTR gene transfer to the lungs and nose of patients with cystic fibrosis: a double-blind placebo-controlled trial. *Lancet* Mar 20, 1999;**353**(9157):947–54.

62. Ruiz FE, Clancy JP, Perricone MA, et al. A clinical inflammatory syndrome attributable to aerosolized lipid-DNA administration in cystic fibrosis. *Hum Gene Ther* May 1, 2001;**12**(7):751–61.

63. Konstan MW, Davis PB, Wagener JS, et al. Compacted DNA nanoparticles administered to the nasal mucosa of cystic fibrosis subjects are safe and demonstrate partial to complete cystic fibrosis transmembrane regulator reconstitution. *Hum Gene Ther* Dec 2004;**15**(12):1255–69.

64. Walters RW, Grunst T, Bergelson JM, Finberg RW, Welsh MJ, Zabner J. Basolateral localization of fiber receptors limits adenovirus infection from the apical surface of airway epithelia. *J Biol Chem* April 9, 1999;**274**(15):10219–26.

65. Mueller C, Strayer MS, Sirninger J, et al. In vitro and in vivo functional characterization of gutless recombinant SV40-derived CFTR vectors. *Gene Ther* February 2010;**17**(2):227–37.

66. Griesenbach U, McLachlan G, Owaki T, et al. Validation of recombinant Sendai virus in a non-natural host model. *Gene Ther* February 2011;**18**(2):182–8.

67. Cmielewski P, Anson DS, Parsons DW. Lysophosphatidylcholine as an adjuvant for lentiviral vector mediated gene transfer to airway epithelium: effect of acyl chain length. *Respir Res* 2010;**11**:84.

68. Kremer KL, Dunning KR, Parsons DW, Anson DS. Gene delivery to airway epithelial cells in vivo: a direct comparison of apical and basolateral transduction strategies using pseudotyped lentivirus vectors. *J Gene Med* May 2007;**9**(5):362–8.

69. Stocker AG, Kremer KL, Koldej R, Miller DS, Anson DS, Parsons DW. Single-dose lentiviral gene transfer for lifetime airway gene expression. *J Gene Med* October 2009;**11**(10):861–7.

70. Mitomo K, Griesenbach U, Inoue M, et al. Toward gene therapy for cystic fibrosis using a lentivirus pseudotyped with Sendai virus envelopes. *Mol Ther* June 2010;**18**(6):1173–82.

71. Griesenbach U, Inoue M, Meng C, et al. Assessment of F/HN-pseudotyped lentivirus as a clinically relevant vector for lung gene therapy. *Am J Respir Crit Care Med* 2012;**186**(9):846–56.

72. Conese M. Cystic fibrosis and the innate immune system: therapeutic implications. *Endocr Metab Immune Disord Drug Targets* March 2011;**11**(1):8–22.

73. Davies JC, Davies G, Gill D, et al. Safety and expression of a single dose of lipid-mediated CFTR gene therapy to the upper and lower airways of patients with CF. *Pediatr Pulmonol* 2011;**46**(Suppl. 2011):S34.

74. Alton EW, Boyd AC, Cheng SH, et al. A randomised, double-blind, placebo-controlled phase IIB clinical trial of repeated application of gene therapy in patients with cystic fibrosis. *Thorax* November 2013;**68**(11):1075–7.

75. Conese M, Ascenzioni F, Boyd AC, et al. Gene and cell therapy for cystic fibrosis: from bench to bedside. *J Cyst Fibros* June 2011;**2**(10 Suppl.):S114–28.

76. Bonfield TL, Lennon D, Ghosk SK, DiMarino AM, Weinberg A, Caplan AI. Cell based therapy aides in infection and inflammation resolution in the murine model of cystic fibrosis lung disease. *Stem Cell Discovery* 2013;**3**(2):139–53.

77. Anderson MP, Gregory RJ, Thompson S, et al. Demonstration that CFTR is a chloride channel by alteration of its anion selectivity. *Science* July 12, 1991;**253**(5016):202–5.

78. Tang L, Fatehi M, Linsdell P. Mechanism of direct bicarbonate transport by the CFTR anion channel. *J Cyst Fibros* March 2009;**8**(2):115–21.

79. Moskwa P, Lorentzen D, Excoffon KJ, et al. A novel host defense system of airways is defective in cystic fibrosis. *Am J Respir Crit Care Med* January 15, 2007;**175**(2):174–83.

80. Donaldson SH, Boucher RC. Sodium channels and cystic fibrosis. *Chest* November 2007;**132**(5):1631–6.

81. Rowe SM, Miller S, Sorscher EJ. Cystic fibrosis. *N Engl J Med* May 12, 2005;**352**(19):1992–2001.

82. Van Goor F, Hadida S, Grootenhuis PD, et al. Rescue of CF airway epithelial cell function in vitro by a CFTR potentiator, VX-770. *Proc Natl Acad Sci USA* November 3, 2009;**106**(44):18825–30.

83. Sermet-Gaudelus I, Boeck KD, Casimir GJ, et al. Ataluren (PTC124) induces cystic fibrosis transmembrane conductance regulator protein expression and activity in children with nonsense mutation cystic fibrosis. *Am J Respir Crit Care Med* November 15, 2010;**182**(10):1262–72.

84. Wilschanski M, Yahav Y, Yaacov Y, et al. Gentamicin-induced correction of CFTR function in patients with cystic fibrosis and CFTR stop mutations. *N Engl J Med* October 9, 2003;**349**(15):1433–41.

85. Howard M, Frizzell RA, Bedwell DM. Aminoglycoside antibiotics restore CFTR function by overcoming premature stop mutations. *Nat Med* April 1996;**2**(4):467–9.

86. Accurso FJ, Rowe SM, Clancy JP, et al. Effect of VX-770 in persons with cystic fibrosis and the G551D-CFTR mutation. *N Engl J Med* November 18, 2010;**363**(21):1991–2003.

87. Ramsey BW, Davies J, McElvaney NG, et al. A CFTR potentiator in patients with cystic fibrosis and the G551D mutation. *N Engl J Med* November 3, 2011;**365**(18):1663–72.

88. Davies JC, Wainwright CE, Canny GJ, et al. Efficacy and safety of ivacaftor in patients aged 6 to 11 years with cystic fibrosis with a G551D mutation. *Am J Respir Crit Care Med* June 1, 2013;**187**(11):1219–25.

89. Davies JC, Sheridan H, Lee P, Song T, Stone A, Ratjen F. Lung clearance Index to evaluate the effect of ivacaftor on lung function in subjects with CF who have the G551D-CFTR mutation and mild lung disease. *Pediatr Pulmonol* 2012;**35**(Suppl.):A249.

90. Yu H, Burton B, Huang CJ, et al. Ivacaftor potentiation of multiple CFTR channels with gating mutations. *J Cyst Fibros* January 30, 2012.

91. Rowe SM, Clancy JP. Pharmaceuticals targeting nonsense mutations in genetic diseases: progress in development. *BioDrugs* 2009;**23**(3):165–74.

92. Kerem E, Hirawat S, Armoni S, et al. Effectiveness of PTC124 treatment of cystic fibrosis caused by nonsense mutations: a prospective phase II trial. *Lancet* August 30, 2008;**372**(9640):719–27.

93. Wilschanski M, Miller LL, Shoseyov D, et al. Chronic ataluren (PTC124) treatment of nonsense mutation cystic fibrosis. *Eur Respir J* July 2011;**38**(1):59–69.

94. Clancy JP, Konstan MW, Rowe SM, Accurso F, Zeitlin P, Hirawat S. A Phase II Study of PTC124 in CF patients harboring premature stop mutations. *Pediatr Pulmonol* 2006;**41**(Suppl.):A269.

95. Konstan MW, Accurso F, De Boeck K, Rowe S. Results of the phase 3 study of ataluren in nonsense mutation cystic fibrosis (nmCF). *J Cyst Fibros* 2012;**11**.

96. Kerem E, Wilschanski M, Sermet-Gaudelus I, et al. Interim reults of the phase 3 open-label study of ataluren in nonsense cystic fibrosis (nmCF). *J Cyst Fibros* 2013;**12**(Suppl. 1):S15.

97. Rowe S, Sermet-Gaudelus I, Konstan MW, et al. Results of the phase 3 study of ataluren in nonsense mutation cystic fibrosis (NMCF). *Pediatr Pulmonol* 2012;(Suppl. 35):A193.

98. Clancy JP, Rowe SM, Accurso FJ, et al. Results of a phase IIa study of VX-809, an investigational CFTR corrector compound, in subjects with cystic fibrosis homozygous for the F508del-CFTR mutation. *Thorax* January 2012;**67**(1):12–8.

99. Van Goor F, Hadida S, Grootenhuis PD, et al. Correction of the F508del-CFTR protein processing defect in vitro by the investigational drug VX-809. *Proc Natl Acad Sci USA* November 15, 2011;**108**(46):18843–8.

100. Boyle MP, Bell S, Konstan MW, et al. The investigational CFTR corrector VX-809 (Lumacaftor) Co-administered with the oral potentiator ivacaftor improved CFTR and lung function in F508Del homozygous patients: phase II study results. *Pediatr Pulmonol* 2012;(Suppl. 35):A260.

Chapter 17

Genetic Engineering of Oncolytic Viruses for Cancer Therapy

Manish R. Patel and Robert A. Kratzke

Department of Medicine, Division of Hematology, Oncology, and Transplantation, University of Minnesota Medical School, Minneapolis, USA

LIST OF ABBREVIATIONS

BOEC Blood outgrowth endothelial cells
CRAD Conditionally-replicating adenovirus
CEA Carcinoembryonic antigen
eIF4F Eukaryotic initiation factor 4 F
GMCSF Granulocyte-macrophage colony stimulating factor
HSV Herpes simplex virus
HSP70 Heat shock protein 70
IFN Interferon
MV Measles virus
MSC Mesenchymal stem cells
PKR Double-stranded RNA kinase
RR Ribonucleotide reductase
RT3D Reovirus Type 3 Dearing
Rb Retinoblastoma
SCCHN Squamous cell carcinoma of the head and neck
TK Thymidine kinase
VV Vaccinia virus

17.1 INTRODUCTION

Oncolytic viruses are a group of viruses that selectively infect and kill cancer cells while sparing normal tissues. The idea for using these viruses as a cancer therapy is not a new one, and is based on patients demonstrating tumor regression after a natural viral infection. Experimental uses of impure viral products and even infected body fluids have been given to patients in the twentieth century as a mode of therapy for cancer patients, often with disastrous consequences. However, serologic techniques for purifying viruses and titering viruses has allowed for safer and more precise use of viruses for therapy between 2000 and 2010. Furthermore, advances in molecular engineering have allowed manipulation of the viral genome to both make them safer by deleting viral genes involved in pathogenesis and by insertion of novel transgenes to enhance antitumor activity (Figure 17.1). As a result, several viruses have now been tested in clinical trials with varying degrees of success (Table 17.1). These trials have led to the licensing in China of an adenoviral vector, H101 (commercially sold as Oncorine), and to late phase clinical trials of recombinant herpes virus and reovirus vectors.[1-3] The aim of this chapter is to provide a primer for the use of oncolytic viruses for cancer therapy focusing primarily on current viruses in clinical use and the therapeutic approaches that are being used to improve oncolytic viruses. The list of viruses discussed here is not meant to be exhaustive and there are several other vectors in development. This chapter is an update and extension of a review article that was published in Translational Research.[4]

17.2 CONDITIONALLY REPLICATING ADENOVIRUSES (CRADs)

Adenoviral vectors were one of the first viruses recognized with oncolytic activity and one of the first viral-based cancer therapies used in modern clinical testing. The wild-type adenovirus was used as a therapy for patients with cervical cancer in the 1950s both by intratumoral and intravenous injections. Although the treatment was found to be safe, it was only

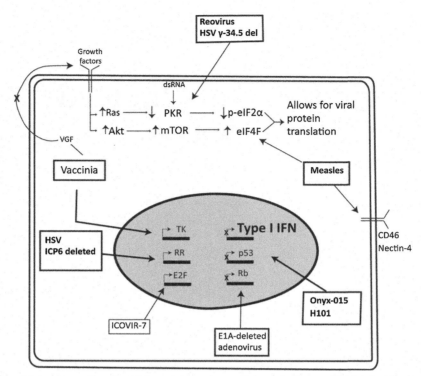

FIGURE 17.1 Mechanisms of tumor tropism of various oncolytic viruses. Schematic diagram highlighting the mechanisms of tumor tropism employed by several oncolytic viruses in clinical development. Defective antiviral interferon (IFN) responses in cancers are a common mechanism for tumor specific viral replication of most oncolytic viruses. DNA viruses have been engineered to rely upon tumor expression of ribonucleotide reductase (RR) and thymidine kinase (TK) to retain tumor specificity of viral replication. Additionally, several of the adenovirus vectors are engineered to replicate in p53 and Rb defective cancer cells. Inhibition of protein translation is a critical checkpoint to limit viral replication. Efficient reovirus and HSV replication is enabled by Ras-mediated inhibition of protein kinase of double-stranded RNA (PKR) and enhanced cap-dependent translation, which results in constitutive translation in cancer cells. Vaccinia viruses (VV) make vaccinia growth factor (VGF), which stimulates similar signaling pathways and increases the pathogenesis of this virus. Deletion of VGF can also result in enhanced tumor specificity as growth factor pathways that are active in cancer cells result in efficient replication. MV utilizes CD46 and Nectin4 expressed on many tumor types as its epithelial receptor and accounts for some of its tumor tropism. In lung cancer and mesothelioma, MV replication is partly dependent upon aberrant activation of protein translation.

modestly effective, and as a result fell out of favor until the 1990s. During this time, nonreplicating adenoviral vectors were undergoing testing for cancer gene therapy; however, the efficacy was quite poor due to the limited number of cells that could be transduced with a nonreplicating virus. Therefore, the concept of using replicating vectors resurfaced as a way to increase the transduction of gene therapy to a larger percentage of a tumor mass by productive infection.

Adenovirus (Ad) is a 36 kb double-stranded DNA virus that is encapsulated by an icosahedral capsid. The capsid structure is composed of several proteins, perhaps the most important of which are the fiber protein and penton base protein (Figure 17.2). The fiber protein projects out from the icosahedral capsid and is a heterotrimer. There are two major domains of the fiber protein consisting of the shaft and knob region. The knob region forms the attachment site for the virus on the target cell. The majority of oncolytic adenoviruses are of serotype 5. The knob region of Ad5 binds specifically to the coxsackie and adenovirus receptor (CAR) on the surface of the target cell, whereas the knob region of other serotypes has specificity for alternate receptors. The penton protein is located at the base of each fiber and is composed of an Arg-Gly-Asp (RGD) motif, which serves to facilitate internalization of the virus after binding to the target cell. Once a virus has attached to the cell, the RGD motif interacts with integrins on the cell surface, which then allows uptake via clathrin-coated vesicles. Subsequent acidification of the vesicle then results in release of the genome, which then traffics to the nucleus for replication. The Ad genome consists of four early genes (E1 through E4) that encode transcriptional regulators and five late genes (L1 through L5) that encode structural proteins. The E1A gene product is required for viral replication, and E1B facilitates infection of normal healthy tissue. Thus, for engineering oncolytic adenoviruses, approaches to improve therapeutic targeting to cancer cells encompass modifications to the early genes, fiber protein, and RGD motif. In addition, oncolytic adenoviral vectors have been "armed" with insertion of transgenes aimed at enhancing the therapeutic effect either by causing expression of a suicide gene or by stimulating an immune response to infected tumor cells. These modifications are discussed in the subsequent passages.

TABLE 17.1 Clinically Tested Oncolytic Viruses

Virus	Cancer Type	Modification	Response	Reference
Adenovirus				
ONYX-015 H101	SCCHN Glioma Ovarian cancer	E1B-55k deleted	PR	1–5
CGTG-401	Solid tumors	hTERT promoter CD40L expression	SD	6
Ad5/3-D24-GMCSF	Solid tumors	Ad3 fiber E1A-d24 mutation (Rb selective) GMCSF	SD	7
Ad5-D24-GMCSF	Solid tumors	E1A-d24 mutation GMCSF	2 CR 1 PR	8
Ad3-hTERT-E1A	Solid tumors	E1A-deleted hTERT promoter Ad3 fiber	SD	9
ICOVIR-7	Solid tumors	E2F promoter controlling E1A-d24 mutation RGD-4C modification	1 PR	10
H103	Solid tumors	HSP70 expression	2 PR	11
CV706 (CG7060)	Prostate cancer	PSA-selective	SD	
CV787 (CG7870)	Prostate cancer	PSA-selective Expresses E3 region	SD	
INGN 241	Solid tumors	Expresses mda-7	2 PR	
Ad5-CD/TKrep	Prostate cancer	Expresses cytosine deaminase and thymidine kinase		
Herpes Virus				
Oncovex-GMCSF	Breast, SCCHN, melanoma IT Liver tumors	ICP6 deleted α47 deleted γ-34.5 deleted GMCSF expression	8 CR 13 PR	12–14
NV1020	Liver tumors IA	ICP0, ICP4 deleted γ-34.5 deleted LAT TK insertion	1 PR	15
G207	Glioma IT	ICP6 deleted γ-34.5 deleted	8 PR	16,17
Vaccinia Virus				
GL-ONC1	Solid tumors	GFP expression	SD	
JX-594	Liver tumors Solid tumors IV	TK deleted GMCSF expression	2 PR	18–20
Reovirus Type 3 Dearing	Solid tumors IV SCCHN IT Colon cancer IV Combination with chemotherapy	N/A	SD 3 PR	21,22

Continued

TABLE 17.1 Clinically Tested Oncolytic Viruses—cont'd

Virus	Cancer Type	Modification	Response	Reference
Measles Virus				
MV-CEA	Ovarian Ca IP	CEA expression	SD	23
MV-NIS	Ovarian Ca IP Myeloma IV Glioma IT Mesothelioma IPl	Sodium-iodide symporter expression	SD	Unpublished
Vesicular Stomatits Virus				
VSV-IFN-β	Liver tumors	IFN-β		Unpublished

IT: Intratumoral, IV: Intravenous, IP: Intraperitoneal, IPl: Intrapleural, IA: Intra-arterial. SCCHN: head and neck cancers. Responses are according to RECIST criteria: CR=complete response, PR=partial response, SD=stable disease.

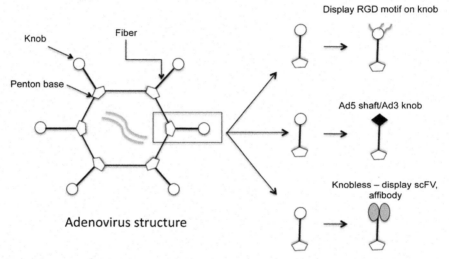

FIGURE 17.2 Adenovirus structural modifications. The adenoviral particle consists of an icosahedral capsid with a penton base at each vertex. A fiber structure projects from the penton base with a knob structure attached at the end. The knob/fiber region is critical for cell attachment and determines binding specificity. Serotype 5 Ads bind to coxsackie and adenovirus receptors (CAR), which are not highly expressed on most tumors. Therefore, modifications to the knob region have been generated to enhance tumor specificity. An arginine-glycine-aspartic acid (RGD) motif can be added to the C-terminus of the knob structure, which mediates binding to cellular integrins (top). Ad3 serotype knobs have been incorporated with the Ad5 fiber; this allows the chimeric virus to utilize both CAR and non-CAR receptors to attach to cells (middle). Knob-less Ads have been generated with a single-chain variable fragment (scFV) or an affibody display that directs attachment to its target receptor (bottom).

17.2.1 Modifications to Improve Tumor Tropism

The initial adenoviral vector used in clinical trials employed recombinant engineering to delete the E1B-55K gene and thereby make adenoviral replication conditional. This approach relies on one of the major cellular defenses against adenoviral infection: activation of cell cycle checkpoint proteins, such as p53, to prevent viral replication. Virulent strains of adenovirus have evolved to express the E1B-55K protein to bind and inactivate p53, allowing for progression through the cell cycle and subsequent viral replication. Because p53 is inactivated or mutated in a large majority of tumor types, the deletion of E1B-55K attenuates the virus for normal cells with intact p53, but allows viral replication in cancer cells lacking functional p53. ONYX-015 is an adenoviral vector that was designed to be replicated conditionally in p53-negative cells. The deletion of E1B-55K resulted in a markedly attenuated virus for normal cells, while retaining replicative capacity in p53-null cancer cells. However, E1B-55K deletion was observed to reduce viral replication compared to the wild-type virus in many models both in vitro and in vivo. In addition to its role as a p53 inhibitor, E1B-55K is involved in mRNA transport functions. Deletion of E1B-55K may have reduced the efficacy of ONYX-015. There are now adenoviral vectors that have been constructed with a more precise mutation in E1B-55K that eliminates the p53-binding site, but maintains its function in mRNA transport. This viral vector is less attenuated compared to wild-type adenovirus.

Similar to ONYX-015, the delta-24 adenovirus (Ad-D24) was engineered to encode an adenoviral E1A protein that lacks the retinoblastoma (Rb) binding site. The Rb pathway is a major inhibitor of cell cycle progression and its inactivation is important for DNA synthesis required for viral replication and the uncontrolled growth of many cancers. Ad-D24 is markedly attenuated in cells with wild-type Rb, however, this modification does not affect viral replication in cancer cells with an inactivated Rb pathway.

An alternate approach to control viral replication in tumor cells is to design vectors in which genes essential for viral replication are under the control of tissue-specific promoters that restrict viral replication to cancer cells. Several viral constructs have been made that incorporate this strategy, including human telomerase (hTERT), prostate-specific antigen (PSA), survivin, alpha-fetoprotein (AFP), carcinoembryonic antigen (CEA), and others as activators for tissue specificity. While these have all been demonstrated to improve tumor selectivity, only the PSA- and hTERT-targeted viruses have gone on to clinical testing in patients (DeWeese, 2001 #949[24]) (Yu, 1999 #950[25]) (Nemunaitis, 2010 #951[26]) (see below for further details).

17.2.2 Improving Transduction of Adenoviral Vectors

The vast majority of adenoviral vectors used as oncolytic viruses are serotype 5, which utilize the coxsackie and adenovirus receptor (CAR) for entry into the cell. Many tumor types do not express the CAR receptor, and importantly, this receptor is expressed highly in liver tissue resulting in potential liver toxicity. Thus, fiber modifications have been made that allow binding of adenovirus to tumor cells in a CAR-independent manner as well as limit off-target transduction of normal liver tissue (Figure 17.2). Initial attempts modified the fiber protein, which is responsible for cell binding, to the serotype 3 fiber (Ad3), which is superior to wild-type Ad5 in transduction of many tumor types including melanoma, ovarian cancer, and others.[5,27] Interestingly, a chimeric Ad5/3 vector that incorporates a truncated fiber shaft of Ad5 with Ad3 knob structure results in enhanced infection compared to wild-type Ad5 and has the advantage of infecting in a CAR-dependent and CAR-independent manner, broadening its specificity to tumor cells that do not express CAR.[28] Further modifications could be made to the knob region of the fiber protein to alter the targeting of adenoviral vectors. Introduction of an RGD motif in the knob region results in targeting directly to integrins. This particular modification is now being tested in several clinical evaluations of replicating Ads.

Another approach to retarget adenoviral vectors has been to use adapters to eliminate the binding of the knob region to CAR and replace that with a tumor-specific ligand or single chain variable fragment (scFv) of an antibody. The adapter approach uses an external adapter that binds to the knob region on one end and on the other binds to a tumor-specific receptor. These adapters are just that and are not permanent modifications to the virus, but are admixed with virus prior to treatment. This has been accomplished using chemical conjugation of a Fab fragment specific for the knob region on one end and with a ligand such as fibroblast growth factor 2 (FGF2). More recently, bispecific adapters have been genetically fused to create a single fusion protein, which simplifies the manufacture of these molecules.[29] This approach depends upon an adapter binding to the knob region to eliminate binding to cellular CAR. An alternate approach is to genetically alter the virus such that instead of displaying the knob region, the adenovirus is genetically constructed to display a peptide ligand, Affibody, or scFv.[6–8] Probably the best example is a knob-less adenoviral vector expressing an Affibody against Her2/neu that effectively retargets adenovirus to infect Her2-positive cells.[9]

17.2.3 "Armed" Adenoviral Vectors

One of the conceptual approaches to viral cancer gene therapy has been to introduce a gene into cancer cells that results in cell death. The difficulty with this approach is that the transduction efficiencies are so low that the number of infected cells is fairly small, and the cells that do not receive the gene of interest are not affected. The use of replicating viruses can overcome some of these limitations because viral replication can amplify the number of cells that are infected. Moreover, the introduction of a suicide gene can induce a bystander effect in which surrounding cells can be exposed to the lethal gene product, causing cell death. These so-called "armed" adenoviral vectors have been employed in several models and have now undergone early clinical testing. Multiple armed adenoviral vectors have been constructed with numerous therapeutic transgenes; these have exhibited preclinical efficacy compared to parent vectors. One approach to suicide gene therapy is to enhance apoptosis by introduction of a proapoptotic protein such as tumor necrosis factor-related apoptosis inducing ligand (TRAIL)[10,30] or socs3.[11] Another approach is to introduce a tumor suppressor such as mda-7.[31] Alternately, adenoviral vectors have been armed with a transgene that enhances sensitivity to combination therapy with chemotherapeutic agents. Introduction of cytosine deaminase using adenoviral vectors enhances the conversion of 5-fluorocytosine into the cytotoxic drug, 5-fluorouracil, and has been shown to increase cell death in vitro and improve tumor control in animal models of colon and breast cancers.[16,32] Despite promising preclinical data, only Ad-mda-7 (INGN 241) and Ad-cytosine deaminase have gone on to clinical testing in patients.

Perhaps the most promising approach to modification of oncolytic viruses in general has been the introduction of armed viruses to enhance immune-mediated killing of cancer cells. Several adenoviral vectors have been constructed with this goal in mind and have shown the greatest promise so far. The introduced transgene is designed to recruit immune effector cells such as cytotoxic T cells, NK cells, or dendritic cells to enhance therapy and ultimately to create a long-term adaptive immune response against tumors. The introduction of granulocyte-macrophage colony-stimulating factors (GM-CSF) into adenoviral vectors has been particularly promising. These vectors have been seen to eradicate tumors in several preclinical models and have shown recruitment of immune mediators into the tumor microenvironment.[15,17] Expression of CD40L by oncolytic adenoviral vectors also results in enhanced immune response through activation of a T-helper type 1 (Th1) response, leading to activation of cytotoxic T cells. This approach has been effective in immune competent mouse models of bladder cancer and melanoma. Furthermore, this improvement was dependent on the presence of T cells demonstrating the immunotherapeutic effect.[12] The expression of interleukin-12 (IL-12) to enhance the antitumor activity of NK cells and T cells has been shown to enhance therapy in an immune-competent Syrian hamster model of pancreas cancer. This is significant because Syrian hamsters are possibly the only small animal model that is permissive for adenovirus replication.[33] The above viruses have now entered clinical testing with promising early results (see below). Numerous additional immunotherapeutic approaches have been tried and are in various stages of preclinical development.

17.2.4 Clinical Observations

In clinical testing, ONYX-015 was shown to be safe at doses up to 10^{11} plaque forming units (pfu); however, it demonstrated limited activity by itself.[13,34] Tumor response was seen after intratumoral injection in patients with accessible head and neck cancer, which led to trials using ONYX-015 in combination with chemotherapy.[14] A phase II study suggested improved outcomes compared to what was expected with chemotherapy alone and tumor biopsies showed evidence of tumor-specific viral replication.[18] A Chinese company used an identical virus (H101) in a randomized phase III trial in 160 patients with nasopharyngeal carcinoma in combination with platinum and 5-fluorouracil.[19] The addition of H101 to chemotherapy resulted in a more than doubling of the response rate. On the basis of these results, Chinese authorities licensed H101. The results obtained in China have not been duplicated and further clinical development of ONYX-015 was abandoned at the time, in part because of the low objective response rates of 15%. Since then, numerous iterations of conditionally replicating adenoviruses (CRADs) have been developed and tested in preclinical models and early clinical trials incorporating one or more genetic modifications. These modifications attempt to enhance tumor specificity of viral replication either through targeting viral entry to tumor-specific receptors or by engineering viral gene expression under the control of tumor-specific promoters.[20,35] For example, 20 men with biochemical local recurrence of prostate cancer were treated with CV706, a PSA-selective adenoviral vector. Treatment was safe with no dose-limiting toxicity by intratumoral injection. PSA decreases of up to 73% were seen in the highest dose cohorts.[24] CG7870 was similarly targeted to PSA expressing cancer cells but also expressed the adenoviral E3 region, which was thought to play a role in evading the immune response.[25] CG7870 was given to 23 patients with metastatic hormone refractory prostate cancer through intravenous injection. Though stable disease was the best response, there were PSA responses seen at the higher dose cohorts. Telomelysin is an hTERT targeted adenoviral vector that has been tested clinically in a phase I study by intratumoral injection. One patient with melanoma had a 56% reduction in the size of a cutaneous lesion.[26] Additionally, engineering of adenovirus to express genes designed to augment antitumor immunity has now been tested.[21,36] One in particular, Ad5-D24-GMCSF, has been engineered to express granulocyte-monocyte colony stimulating factor (GM-CSF). In 21 heavily pretreated patients with advanced cancer, two complete responses and one partial response were observed. Furthermore, evidence of antitumor activity at sites distant to the injection site and induction of tumor-specific immunity were observed.[17,21] Several others have now advanced to phase I clinical trials. These trials have demonstrated safety and feasibility for adenoviral-based virotherapy and have shown some evidence of activity.[17,21,22,37–39]

17.3 HERPES SIMPLEX VIRUS (HSV)

HSV is a double-stranded DNA virus and a well-known human pathogen. The genome of HSV is large consisting of >150 kbp and encoding over 84 polypeptides, several of which are essential for viral replication. The HSV-1 genome sits inside a capsid structure that is surrounded by a fibrous constituent called the integument. The entire virus is encapsulated within a glycolipid envelope structure that has glycoproteins attached to the surface. Four glycoproteins (GA-D) mediate attachment to the target cell through heparan sulfate (HS). Subsequently, the virus envelope binds to specific cellular receptors leading to fusion of the viral envelope with the cellular membrane. The capsid then undergoes transport to the nucleus where the genome is released. Immediate-early genes are the first to be expressed and serve to initiate transcription of both early and late genes. Early genes encode for infected cell proteins (ICP) that include viral ribonucleotide reductase (ICP6), thymidine kinase (TK), and DNA polymerase (UL30), which initiate DNA synthesis. Late genes include γ-genes, which are structural proteins that are involved in capsid assembly. In addition, several late genes are virulence genes that inhibit cellular defenses against viral replication and spread. HSV-1 infections in humans often can be latent related to transcription of latency-associated transcript (LAT) proteins. Though the mechanism of latent infection is not clear, these transcripts allow the viral genome to remain in the nucleus and often can result in recurrent infections throughout a person's lifetime.

17.3.1 Attenuation of HSV-1

As a human pathogen, HSV-1 requires attenuation of its virulence genes. Many oncolytic HSV-1 vectors have had the LAT genes deleted to prevent latent infections. Other nonessential genes that predominantly mediate virulence but are not strictly required for replication have also been an area of major focus. The viral γ-34.5 gene is thought to be the principal neurovirulence gene as deletion of this gene results in abrogation of encephalitis in animal models. Expression of the γ-34.5 genes is required to inhibit the double-stranded RNA kinase (PKR) antiviral response and to inhibit the shut off of host protein synthesis.[40] This is one of the chief innate cellular defenses against viral infection. As many cancer cells already have attenuated PKR responses because of activation of upstream signaling, deletion of this gene does limit viral replication within tumor tissue. Virtually all of the HSV vectors for therapeutic use have had this critical neurovirulence factor deleted. Additionally, pathogenic strains of HSV express the ICP6 gene, which encodes for viral ribonucleotide reductase (RR), which is required for DNA synthesis. The deletion of this gene from the viral genome results in restriction of viral replication within cancer cells that express homologs of RR at high levels.[41] Several vectors have been developed on the platform of deletion of γ-34.5. G207 is deleted for γ-34.5 and ICP6, while NV1020 is deleted for γ-34.5, ICP0, ICP4, and LAT genes. As both of these vectors have now been tested in clinical trials (discussed below), they serve as the backbone template from which further modifications have been made.

17.3.2 Enhancing Antitumor Immune Responses with HSV Vectors

Early experiences with HSV vectors showed that they were strongly immunogenic and could induce antitumor immune effects. Thus, several attempts have been made to augment the immune response to boost antitumor activity and potentially improve durability of responses by creating a memory T cell response. The α47 gene of HSV functions to inhibit the transporter protein associated with antigen presentation (TAP) to prevent immune recognition of virally infected cells. Deletion of the α47 gene from the G207 virus (G47Δ) resulted in enhanced MHC-I expression on tumor cells and enhanced recognition of virally infected cells. This gene deletion also resulted in enhanced viral replication due to the juxtaposition of the US11 gene to the α47 promoter, leading to early expression of the US11 gene.[42] Through unclear mechanisms, this resulted in protein synthesis inhibition necessary for attenuating viral replication. Nonetheless, in animal models, G47Δ was as safe as its parent vector G207 (γ-34.5 mutant).

Similar to adenoviral vectors, several HSV vectors have been armed with therapeutic transgenes aimed at improving the tumor kill above and beyond the effects of viral replication. Talimogene laherparepvec (formerly known as Oncovex-GMCSF) is an armed HSV vector with a GM-CSF transgene inserted in the deleted γ-34.5 locus. Viral expression of GM-CSF results in enhanced efficacy compared to the parent virus in immune-competent animal models, and mice that were cleared of tumors displayed evidence of antitumor immune responses as well as resistance to further rechallenge with syngeneic tumor cells. Thus, this showed proof of principle that talimogene was able to not only induce tumor regression, but also lead to antitumor immunity. As a result, this virus has been taken to clinical testing and has nearly completed a phase III registration trial (discussed below). Similarly, armed HSV vectors have been constructed on the backbone of G207 or NV1020 virus with the transgene inserted within loci of deleted genes. Several vectors have been engineered to express numerous immune cytokines including IL-12 (NV1042), IL-4, and CD40 ligand. While it is not clear that these vectors are more efficacious than talimogene, they each have shown some promise in preclinical studies.[42]

17.3.3 Other Armed HSV Vectors

Transgenes that have been inserted into HSV vectors include antiangiogenic genes such as thrombospondin-1 or platelet factor 4.[43,44] These vectors demonstrate enhanced antitumor efficacy compared to the parent virus. Alternatively, HSV vectors have been armed with a gene designed to enhance chemosensitivity of infected tumor cells. Expression of CYP2B1, a cytochrome P450 enzyme needed for conversion of cyclophosphamide to its active metabolite, and intestinal carboxylesterase (needed for activation of irinotecan), dubbed MGH2, resulted in enhanced activity in a glioma model when combined with the two chemotherapy agents. The combination of MGH2 with irinotecan was lethal to mice at viral doses greater than 10^6 pfus. However, below lethal doses, all mice were able to survive the treatment without significant toxicity.[45] At this time, clinical trials are being planned utilizing this combination.

17.3.4 Clinical Observations

G207 is an HSV vector with the γ-34.5 and ICP6 genes deleted. Twenty-one patients with recurrent gliomas were given intratumoral injections of G207 in a phase I dose-escalation trial.[46] Overall, the treatment was safe and no cases of encephalitis were reported. By imaging, eight patients responded to treatment. HSV DNA was demonstrated in two out of six post-inoculation biopsy specimens suggesting virus-mediated cell death. G207 has completed a phase Ib trial of intratumoral treatment followed by inoculation of the tumor bed after resection of target lesions.[47] Once again, the treatment was safe; however, one patient seemingly had inadvertent inoculation of the cerebral ventricle resulting in detectable HSV replication in the cerebrospinal fluid. None of the patients on this trial had a response at tumors distant to the injected tumor. Injected tumors were resected and evidence of viral replication was seen in only 2/6 patients treated.

NV1020 is a recombinant HSV virus in which the γ-34.5, ICP0, ICP4, and LAT genes have been deleted. It also has an insertion of an exogenous copy of the HSV-1 thymidine kinase (TK) gene making it highly sensitive to the antiviral drug, acyclovir. This virus was tested in a phase I trial of intrahepatic arterial infusion to patients with colorectal cancer with liver metastasis.[48] Twelve patients were treated with a single dose of NV1020 at four separate dose levels. This treatment was followed by placement of an intra-arterial pump to deliver chemotherapy locoregionally. Although all patients developed transient fever, the treatment was safe and no patient developed systemic viral infection, encephalitis, or other major toxicity. There were two patients treated on the study who had an objective tumor response; however, because of the concomitant delivery of regional chemotherapy, it is not clear that NV1020 played a substantial role. A subsequent phase I/II study was done with NV1020 given into the hepatic artery weekly for four doses followed by two cycles of systemic chemotherapy chosen by the investigator.[49] Similar toxicity was seen and no dose-limiting toxicity was found. Of the 22 patients treated at the highest dose, 1×10^8 pfu, there was one objective responder and 14 patients with stable disease.

Similar to the case with adenovirus, the insertion of a granulocyte-monocyte colony stimulating factor (GMCSF) expression cassette has been used with some success. OncovexGM-CSF® is an HSV virus with deletions of ICP6 and γ-34.5 and insertion of GMCSF. A two-part phase I study was conducted using intralesional injection for patients with accessible cutaneous melanoma, breast cancer, head and neck cancer, and GI cancer.[50] The first part was a dose-escalation design in which four patients were treated per dose cohort. In the second part, each patient was administered with three injections and the dose was determined based on their pretreatment HSV serology status. Treatment-related toxicity was mild, and all of the toxicity was grade one and two. No complete or partial responses were seen in this study; however, tumor biopsies obtained after treatment revealed evidence of tumor necrosis in 14 of 19 biopsies. Furthermore, several patients experienced flattening of injected lesions, expression of GMCSF, and viral replication at the tumor site. A phase II trial in cutaneous melanoma further employed OncovexGM-CSF at doses of 10^8/mL by direct intralesional injection every 2 weeks for up to 24 doses to 50 patients.[51,52] There were 13 objective responses and eight complete responses in this trial with examples of distant visceral disease responding to intratumoral injection at cutaneous sites. Two cases of autoimmune vitiligo developed after treatment in responding patients. The median overall survival was 16 months, and one-year survival for patients where an initial response was 93%. Evidence of a melanoma-specific cytotoxic T cell response was demonstrated.[51] A phase III randomized study has now been completed and achieved its primary endpoint of durable response rate. However, overall survival data are not yet mature and final results are expected soon. If OS data are positive, it is very likely that Oncovex (now called talimogene laherparepvec) would likely be the first oncolytic virus therapy licensed in the United States.[53]

The same virus has been tested in combination with chemotherapy and radiation therapy for head and neck cancers. The virus was injected four times during the course of concurrent cisplatin-based chemoradiotherapy in a dose-escalation design.[54] Pathologic complete remission was obtained in 93% of patients and evidence of viral replication was detected in tumors at the time of planned neck dissection. While there were only 17 patients treated on this trial, the rate of complete pathologic response was higher than historical rates of anywhere between 60 and 75%. As a result, a large multicenter randomized phase III trial has begun.

17.4 VACCINIA VIRUS (VV)

VV is a double-stranded DNA virus that has found prominent use in medicine as the vaccine for smallpox. Despite the fact that VV is a DNA virus, it is somewhat unique in that viral replication occurs entirely in the cytoplasm in viral factories. The VV genome is 200 kb in length and encodes approximately 200 genes. One of the unique aspects of VV is that it exists within the host in several forms (Figure 17.3). The mature virion (MV) is found within the cytoplasm and is released only upon cell lysis. It is a biconcave particle surrounded by a membrane. There is also a wrapped virion (WV) that is wrapped in the Golgi body with an additional membrane layer. This form is also exclusively found within the cell cytoplasm.

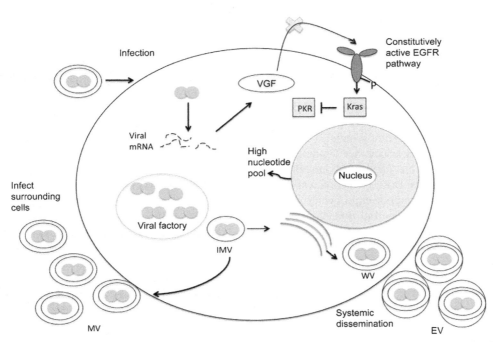

FIGURE 17.3   Oncolytic vaccinia virus life cycle in cancer cells. Viral attachment leads to uptake of the viral particle and uncoating of the envelope. Early viral mRNA encode for VGF, which is dispensable in cancer cells that possess constitutive activation of the EGFR/Ras pathway. Furthermore, viral TK is not necessary in cancer cells because of the presence of a high nucleotide pool. Viral replication occurs entirely within viral factories in the cytosol. Some mature virions (MV) emerge from viral factories and are released from the cell, whereas other immature virions (IMV) are transported to Golgi apparatus and acquire another envelope, termed wrapped virion (WV). This then fuses with the host cell membrane to form the extracellular virion (EV), which allows VV to disseminate systemically.

The extracellular virion (EV) results from the transport of WV to the plasma membrane. Through fusion with the plasma membrane, the EV acquires an additional outer lipid layer. The EV is found outside the cell and is thought to play an important role in systemic dissemination of VV throughout the host. The EV, by virtue of being encapsulated by the host cell membrane, also allows VV to avoid recognition by complement or neutralizing antibodies. To initiate infection, the MV is able to fuse with the host plasma membrane and release its core into the cytosol. The viral core contains all the necessary enzymes for synthesis of early mRNA including an RNA polymerase, a viral early transcription factor (VETF), an mRNA capping enzyme, as well as a poly(A) polymerase. The viral early genes encode genes for DNA replication, and therefore, viral replication occurs within the cytoplasm at sites termed "viral factories." An endoplasmic reticulum-derived membrane surrounds these factories within the cytoplasm. While detailed mechanisms of the structure of the vaccinia virus remain partially unknown, the first signs of replication are the formation of crescents that expand to become enclosed circles of lipid membrane termed the immature virion (IMV). Around the same time, the IMV acquires a nucleoid body, which contains the viral genome. Thereafter, proteolytic cleavage occurs and MV is discharged from the viral factories. These MVs are fully infectious; however, a portion of these undergo wrapping in the Golgi body to form WV, and these can be extruded by fusion with plasma membrane to become an EV. The detailed mechanism of how these steps occur is beyond the scope of this chapter, but it is a highly orchestrated and complex process.

VV has many features that make it an attractive vector for oncolytic therapy. VV expresses a viral thymidine kinase (TK) gene, which is required for DNA replication in quiescent cells. However, viral TK is dispensable in rapidly proliferating cancer cells with a high pool of available nucleotides. One of the VV virulence factors is the vaccinia growth factor (VGF), which is a ligand for the epidermal growth factor receptors (EGFRs). The production of VGF stimulates cell signaling pathways that prevent host shut-off of protein synthesis. In many tumor types, VGF is unnecessary because of the constitutive activity of these signaling pathways. Therefore, engineering of VV with deleted TK and VGF genes has resulted in selective VV replication within tumors. Perhaps the most important feature of VV is the ability to encapsulate itself within a host cell-derived envelope. As a result, this virus is able to traffic within the blood stream allowing for distribution to metastatic tumor sites and avoid neutralization by complement or the humoral immune response. Furthermore, the genome of VV is relatively flexible and can accommodate insertion of large transgenes to complement antitumor activity.

17.4.1 Clinical Observations

GL-ONC1 is an attenuated vaccinia virus in which the viral F14.5, TK, and hemagglutinin genes are replaced with Renilla luciferase/green fluorescence protein (RLuc/GFP) fusion gene, LacZ, and β-glucoronidase genes, respectively. A phase I dose escalation trial of intravenous injections of GL-ONC1 in patients with solid tumors was presented at the annual American Society of Clinical Oncology meeting in 2012.[55] One patient had a dose-limiting toxicity of transient AST elevation at the 1×10^9 pfu dose. The best response was stable disease in nine of 27 patients treated. Four of these had durable disease stability >24 weeks. In one patient with head and neck cancer, a posttreatment biopsy documented vaccinia virus in the primary tumor by IHC staining, thus showing proof of principle of the ability of vaccinia virus to traffic to tumor sites.

JX-594 (now called Pexa-vec) is another TK-deleted vaccinia virus with a GM-CSF insertion cassette. It has been tested in three phase I clinical trials thus far.[23,56,57] Fourteen patients were treated with JX-594 by intratumoral injection into primary or metastatic liver tumors. Dose escalation was done up to 3×10^9 pfu in a standard 3×3 design. Treatment was given every three weeks for up to a total of eight cycles in responding patients. The maximally tolerated dose was 1×10^9 pfu. The dose-limiting toxicity was hyperbilirubinemia resulting from tumor swelling and biliary obstruction occurring at the highest dose level. As in previous examples, toxicity was mainly constitutional symptoms and these were transient. Only patients treated at high doses had detectable levels of GM-CSF in the blood. Remarkably, viral replication was observed at distant tumor sites, suggesting systemic spread to metastatic tumor sites. For example, one patient had viral DNA detected in malignant pleural effusion and ascites. A second patient had histologically identified replicating virus in a neck lymph node that had a 57% decrease in diameter of the lesion.[23] A dose-finding randomized phase II trial using JX-594 was recently completed in patients with hepatocellular carcinomas delivered intratumorally using image guidance. Similar antitumor activity was seen in both dose cohorts (10^8 or 10^9 pfu) with antitumor responses at both local and distant tumor sites seen in 15% by RECIST criteria but 62% by Choi criteria. Furthermore, there was evidence of antitumor humoral immune responses and activation of cellular immunity in these patients. The treatment was safe, and there were no dose-limiting toxicities; however, the higher dose cohorts did have significantly more anorexia (30% vs 0) than the low-dose group. Perhaps the most striking result was the overall survival, which was significantly better for the high-dose group. In multivariate analysis, the only parameters that were associated with overall survival were high dose (14.1 months vs 6.7 months, $p = 0.02$) and viral levels in the blood, suggesting that viral dissemination was an important factor in determining outcome.[58] A second phase I trial employed intravenously delivered JX-594 to patients with advanced cancers.[57] Again, a dose-escalation design was used. One patient with mesothelioma had a partial response to treatment. Twelve of 23 patients had stable disease. Importantly, this study demonstrated that intravenous virus delivery to tumors was achievable in the presence of neutralizing antibodies and was dose dependent. Greater than 85% of biopsies obtained posttreatment showed evidence of viral replication at the tumor sites using doses above 10^9 pfu. Though these results are exciting, a recent press release by the manufacturer revealed that a phase IIb randomized trial compared to best supportive care failed to meet its overall survival endpoint. Though randomized phase III trials are being planned, these results have somewhat dampened enthusiasm for this virus. On the basis of prior data indicating that JX-594 might be a vascular disrupting agent, it is now being tested in combination with the angiogenesis inhibitor, sorafenib, for patients with hepatocellular carcinoma.[59,60] Preclinical and pilot clinical data have now been published highlighting the potential of this approach.[60]

17.5 REOVIRUS TYPE 3 DEARING (RT3D OR REOLYSIN®)

RT3D is a nonenveloped double-stranded RNA virus that is ubiquitous in the environment. It is a common isolate from the respiratory tract of people; however, it is not known to cause any human disease. The genome of reovirus consists of 10 discrete segments that encode for structural proteins and replication machinery. The viral structure consists of the inner capsid, which contains the genome, and five associated proteins, which function as the transcription unit. There is also an outer capsid that is composed of proteins that function to allow binding to the target cell and invasion. The virus binds to sialic acid or junction adhesion molecule on the cell surface and is internalized by endocytosis into a lysosome where it requires an acidic pH to activate proteolytic enzymes. After proteolysis of the outer capsid, the inner capsid then binds to the endosomal membrane, which activates its RNA-dependent RNA polymerase resulting in primary RNA transcription. Within the inner capsid, early translation occurs as well. The resulting RNA and protein then become associated into RNA assortment complexes followed by secondary RNA transcription and translation. Finally, these complexes associate into viral factories and final virion assembly takes place leading to cell death and release of progeny virions.[61] It is important to note that the process is dependent upon host cell ribonucleotides as well as host translational machinery. As a result of its double-stranded RNA, reovirus is a potent activator of antiviral mechanisms including the antiviral PKR, which recognizes double-stranded RNA leading to autophosphorylation and host translational shut off. This is probably one reason that reovirus is nonpathogenic to humans. The Ras signaling pathway is active in the majority of human cancers and results in inhibition of PKR antiviral mechanisms, thus resulting in efficient viral replication within cancer cells. Especially considering that RT3D is not a human pathogen, there has been a great deal of interest in human trials with this virus.[62] Notably, because its genome is divided into 10 distinct segments and it has a highly rigid outer capsid, reovirus is fairly difficult to manipulate even with advanced bioengineering. Thus, while a few recombinant reoviral vectors have been successfully generated,[63] they have not yet been used clinically and much of the preclinical and clinical investigation has been done with the natural virus.

Despite the engineering constraints, RT3D has shown remarkable activity in vivo against a variety of tumor types, including tumors harboring mutant *Kras* and those not. It should be pointed out that, despite lack of *Kras* mutations, the majority of tumor types have activation of Ras pathways by upstream or downstream signaling.[64] Complete tumor regressions have been observed in glioma models and in immune competent murine tumor models. In models of melanoma, RT3D has also been demonstrated to be a potent activator of the host immune system and this may augment therapy. Furthermore, preclinical work has shown that RT3D can work synergistically with both chemotherapy and radiotherapy. This has been repeatedly shown in numerous tumor models including non-small cell lung cancer, head and neck cancer, prostate, and melanoma.[65–68] These results have driven clinical trials that have been done using combination therapies (discussed below).

17.5.1 Clinical Observations

The initial phase I clinical trial in patients with advanced malignancies administered the virus directly into accessible tumors.[69] The maximally tolerated dose was not reached at the highest dose level of 10^{10} pfu. Three of 19 tumors injected showed an objective response and 37% of patients had stable disease or better. Toxicities were mild with only mild flu-like symptoms and local transient erythema at injection sites. Subsequently, a phase I study of reovirus by intravenous injection was completed for patients with solid tumors.[70] Most of the patients treated on this study received reovirus on days one to five every four weeks with a dose-escalation up to 3×10^{10} pfu. While no objective responses by RECIST criteria were seen in this study, several patients had a slight decrease in tumor size and one patient had a 20% reduction in tumor burden. Toxicity was mild with two patients at the highest dose level experiencing grade three neutropenia. Most patients did have grade one to two fever and headache, but no dose-limiting toxicity was seen. In patients who had a posttreatment biopsy, replicating reovirus was detected. Neutralizing antibody titers against reovirus increased in every patient treated.

As a result of the excellent safety profile, reovirus has now been tested in three phase I clinical trials in combination with systemic chemotherapy.[71–73] Overall, the toxicity of combined therapy was predictable across trials. The combination with gemcitabine resulted in toxicity when reovirus was administered for five consecutive days. Therefore, the majority of patients on that trial were treated with one injection per three-week cycle. Interestingly, concomitant gemcitabine treatment had an impact on the neutralizing antibody response in which the fold increase from baseline and the peak levels were less than that seen on previous clinical trials. This was not observed when reovirus was combined with docetaxel. The response to reolysin may be independent of the presence of neutralizing antibodies. Evidence from an interesting translational study in patients with metastatic colorectal cancer undergoing planned surgery showed that host leukocytes could take up the virus in the circulation and carry the virus to tumor tissue perhaps shielding it from neutralization.[74] Combination reovirus with carboplatin and taxol resulted in delay to development of peak neutralizing antibodies, but all patients did develop a robust increase in the antibody response from the baseline. The carboplatin/taxol study was weighted to include patients with metastatic squamous cancer of the head and neck (SCCHN). In this heavily pretreated group of patients who had received prior platinum therapy, nine out of 14 patients had a response to treatment with one complete response. Certainly, some of this could be attributed to chemotherapy; however, objective responses to chemotherapy in SCCHN are historically 10% or less. Therefore, based on these promising data, a phase III randomized trial for patients with metastatic or locally recurrent SCCHN has been undertaken. While this study has not been published, the company, Oncolytics Biotech, issued a press release indicating that there was a signal of a progression-free survival advantage in an unplanned subset of patients with loco-regional recurrence.[75] There has been no further guidance regarding statistical analysis or the primary endpoint of overall survival. Nonetheless, there are 16 planned or ongoing clinical trials utilizing reovirus alone or in combination with chemotherapies.

17.6 VACCINE STRAINS OF MEASLES VIRUS (vMV)

Measles virus is a single-stranded negative sense RNA virus and the wild-type (wtMV) virus is a well-known pathogen. A number of anecdotal case reports described spontaneous regressions and/or sustained complete remissions following a natural wtMV infection in patients with hematological malignancies.[76–79] These observations and the generation of an attenuated vMV with an excellent safety profile triggered scientific interest in the therapeutic use of modified vMV for cancer therapy.[80] vMV are attenuated strains that were derived through serial passages of wtMV in the laboratory. These strains have evolved to utilize a different cellular receptor than the wild-type virus. While the wtMV utilizes the SLAM receptor (CD150) targeting viral infection of T and B cells and marked immune suppression of infected patients, the vMVs utilize CD46 expressed on epithelial cells and overexpressed on many tumors.[80] Recently, nectin-4 has also been identified as an epithelial receptor for vMV viral entry.[81,82] Nectin-4, which is normally expressed in fetal tissue, is also overexpressed in many tumor types.[83–85] The importance of nectin-4 is not entirely clear but appears to be context-dependent. While wtMV is able to inhibit interferon signaling, vMV is not able to, thereby conferring tumor tropism on interferon defective cancer cells.

MV is a member of the *paramyxoviridae* family of viruses. The structure of the virus is similar to other members of the family. The viral particle is enveloped by a lipid bilayer consisting of glycoproteins. The 15.5 kb genome is associated with ribonucleoprotein primarily consisting of the negative sense RNA and nucleocapsid protein (N) and the L protein which performs all the functions of RNA synthesis and posttranscriptional processing of viral mRNA including 5′ capping and addition of a polyA tail. The envelope contains viral H and fusion (F) proteins that mediate attachment and fusion to target cells, respectively. Expression of F protein on infected cells also mediates fusion of cells to adjacent cells resulting in large multinucleated cells, termed syncytia. Cell fusion results in cellular necrosis and also likely contributes to the spread of viral infection.

vMV is highly immunogenic and as a result has been very successful in preventing illness due to wtMV infections. Several lines of evidence suggest that the immune response to MV therapy likely plays a role in antitumor efficacy. vMV infection results in a large burst of interferon α production likely related to activation of toll-like receptors on dendritic cells.[86] As a result, MV infection of mesothelioma cells can activate plasmacytoid dendritic cells from human sera and cross-present tumor antigens to mesothelioma-specific CD8+ T cells, leading to the potential for adaptive antitumor immunity.[87] While activation of pDCs in the tumor microenvironment plays a role in developing an antitumor immune response, it is also likely that interferon also has an antitumor effect. A recombinant MV engineered to express mouse interferon β given to immune deficient mice bearing mesothelioma xenografts showed infiltration of macrophages to the tumor and prevention of neo angiogenesis in the tumor.[88] Thus, in a model devoid of T cells, innate immune and antiangiogenic effects of interferon likely play an independent role in MV efficacy.

Preclinical models of a variety of human malignancies, including MPM, demonstrate the efficacy of measles.[89–96] Since murine tissue is not within the host range of vMV, these models do not reflect the potential for decreased activity in the setting of immune competent models or human tumors (neutralizing antibodies and memory T lymphocytes). Indeed, in passively immunized mice, vMV activity is very much curtailed particularly when delivered via the intravenous route. Since 95% of the Western population has preexisting neutralizing antibodies, most animal models fail to account for this. As a result, much of the development of MV as a cancer therapy has been aimed at tumors in which intra/peritumoral injection would be feasible. Preclinical work has delineated a potential therapeutic benefit of lymphodepletion using cyclophosphamide to enhance MV replication when delivered systemically. Studies in squirrel monkeys and engineered mice that can be infected with MV have demonstrated enhanced viral titer when vMV is given concomitantly with cyclophosphamide.[97] Thus, this strategy has been taken forward into clinical trials of patients with myeloma, a disease in which the humoral immune response is already somewhat suppressed.

One advantage of vMV as a platform for cancer therapy is the relative flexibility with which it can be manipulated. Numerous recombinant vectors have been generated including retargeted MV as well as armed MV expressing enzymes to allow concurrent chemotherapy similar to approaches used with other viruses.[98–104] Somewhat unique to oncolytic virus research is the aim to create a recombinant virus with transgenes to allow noninvasive monitoring of viral replication and imaging. Thus, MV carrying insertion of sequences encoding for carcinoembryonic antigen (CEA) has been generated to allow for detection of viral transgene by simple ELISA assay. MV-CEA was used in the first clinical trial using vMV in patients with ovarian cancer. Subsequently, MV encoding for the sodium iodine transporter (MV-NIS) has been generated that expresses the NIS protein on the surface of infected cells allowing anatomical localization of MV replication noninvasively using radioiodine single positron emission computed tomography (SPECT) scans. This modification also will make possible the adjunct use of radioablative iodine (I^{131}) to kill infected tumor cells that express NIS allowing uptake of this radiotoxin.[105] MV-NIS is now being used in ongoing phase I studies in myeloma and in mesothelioma.

17.6.1 Clinical Observations

The results of the first clinical trial employing intraperitoneal delivery of a modified vMV carrying the human CEA gene (MV-CEA) in patients with refractory ovarian cancer were published.[106] Despite suboptimal viral replication as suggested by undetectable serum and peritoneal CEA levels in 17 of 21 patients, the majority of the patients (14 of 21) had prolonged periods of disease stability. The median overall survival was 12 months (range 1.3–55.8 months), which compares very favorably with the expected median survival of 6 months for this patient population.[106–108] Though there were no objective responses, there were some patients who had a biochemical response with decreases in ovarian tumor marker, CA-125. Furthermore, there was an indication that there might have been an antitumor immune response. This potential immune mechanism of MV therapy is supported by in vitro observations demonstrating that dendritic cells (DC) cocultured with MV infected mesothelioma cells acquire a mature phenotype and are capable of expanding tumor specific CD8+ T cells from PBMC.[109,110]

MV that expresses the sodium-iodide symporter (MV-NIS) has been generated to allow noninvasive monitoring of viral replication. Infected cells can take up radioiodine and thus can be imaged using SPECT scans. MV-NIS is currently being examined in two additional phase I clinical trials. MV-NIS will be administered intravenously and into the pleural space of patients with myeloma and mesothelioma, respectively. To inhibit the neutralizing antibody response during intravenous administration, treatment will be given with and without cyclophosphamide to see if immune suppression might enhance viral replication as was seen in primate studies.[97] Both trials will utilize radioiodine uptake as a noninvasive method of monitoring the anatomical sites of viral replication.

17.7 VESICULAR STOMATITIS VIRUS (VSV)

VSV is another single-stranded RNA virus of the *rhabdoviridae* family, which includes the rabies virus. Because it is closely related to rabies in structure and biology, yet less pathogenic, it has been studied as a proxy to understanding rabies biology. In the wild, VSV is a known cause of outbreaks in animal livestock. It causes oral ulcers in cattle and horses, which is typically a self-limiting illness. There have been documented exposures in people, particularly among livestock workers. Several cases of documented infection have been described from Central America where VSV can be carried by sandflies. By and large, the infection in humans is self-limiting without any sequelae; however, there has been one case reported in the literature of an encephalitis-like syndrome in which VSV was isolated from a throat culture but not from cerebrospinal fluid.[111] Several series have described the accidental infection of laboratory workers in which all of them suffered from a febrile illness that was self-limited and without clinical sequelae.[112,113] VSV has a broad host range and can infect mice, dogs, and other vertebrates, making preclinical models more pertinent to clinical translation. However, it certainly seems to cause less severe disease in humans than other species.

VSV structure is a bullet-shaped virus. The single-stranded negative sense RNA genome is associated with nucleoprotein (N), phosphoprotein (P), and RNA-dependent polymerase (L). The genome is arranged in a helical fashion and can accommodate insertion of one or more transgenes. A lipid-glycoprotein envelope surrounds the nucleoprotein core. This is composed of glycoprotein (G) protein, which serves to attach to target cells, and the matrix (M) protein that not only is a structural protein, but also likely plays a role in pathogenesis by inhibiting innate interferon responses. VSV life cycle is extremely rapid and progeny virion can be detected as early as six hours after infection. VSV replication can be thwarted effectively by the production of type I interferon such that normal cells with intact innate interferon signaling pathways quickly inhibit the spread of infection. However, VSV spreads and replicates very efficiently in tumor cells lacking interferon responses.[114] As mentioned above, VSV is readily manipulated through reverse genetics techniques similar to other RNA viruses without significantly altering its infectious capacity. For clinical translation, fears of neurovirulence have led to the development of attenuated recombinant viruses that express interferon-β. The reason for expressing this transgene is to limit viral spread in normal cells while allowing for replication within tumor cells lacking interferon responsiveness. In mice, VSV-IFNβ protects mice from lethal neurotoxicity up to three logs of infectious dose higher than parent VSV. Despite this attenuation, VSV-IFNβ retains similar antitumor activity against several nude mouse xenograft models including mesothelioma, myeloma, and liver cancers.[115,116] In immune competent mice, VSV-IFNβ has been observed to be more efficacious than the immune deficient model.[117,118] At least in the mesothelioma model, this effect was dependent upon the presence of CD8+ T cells suggesting that enhanced efficacy was a result of raising an antitumor immune response.

As a result of the above, the first phase I trial of VSV encoding human interferon-β is now underway for treatment of hepatocellular carcinomas. The VSV will be delivered by hepatic artery infusion with a careful dose escalation design to evaluate toxicity. Though still very early, the results of this trial will guide future development of VSV-IFNβ for clinical translation and could be applied to multiple tumor types.

17.8 CHALLENGES TO ONCOLYTIC VIROTHERAPY

The early clinical experience with oncolytic viruses has largely been very promising. Fortunately, many patients have been treated with oncolytic viruses and toxicity has been extraordinarily minor. There are only scant reports of grade four toxicities so far. As a result several clinical trials are now open for enrollment or in the planning phase using novel viruses or novel combination therapies.

Though there have been glimmers of success, the results thus far fall well short of the dramatic preclinical results in animal models. This is no doubt due, in part, to differences in viral infection in animals and humans. Pertinent preclinical studies of vMV are difficult, as there are no immune-competent animal tumor models that suitably model MV infection in humans. Adenovirus suffers from the same problem. As a result most of the preclinical work is done in immune-deficient xenograft models, which cannot approximate either the negative or potentially positive effect of an intact immune system. On the contrary, vesicular stomatitis virus, which has broad tumor tropism and antitumor activity, has substantial toxicity in mice, but is not known to be pathogenic to humans. Ultimately, even if there are suitable immune competent models, the differences between murine and human immune systems are significant, and it is not always clear that data from mouse studies will yield meaningful responses in humans. Thus, reproducing preclinical data in clinical trials has remained a challenge. Now that trials are ongoing in humans, it will be imperative to learn about the complex biology behind the antitumor activity in humans. Carefully designed translational studies of early phase trials have already shed some light on mechanistic aspects of oncolytic viruses in humans.[57,74]

Clearly, the other major challenge to successful implementation of viral therapy is the neutralizing antibody response. For many of the above-mentioned viruses, most people have been previously exposed to the virus through previous vaccination or infection. Therefore, circulating antibodies can limit the potential of the virus ever reaching the tumor site. Even if a person is naïve to the virus, neutralizing antibodies inevitably develop. Therefore, there is only a narrow window to achieve oncolytic activity. Faced with that problem, there are potentially two approaches for success: immune suppression or evasion. Immune suppression with cyclophosphamide or gemcitabine has been used to suppress the neutralizing antibody response with success in preclinical models.[119–122] The use of gemcitabine was associated with increased toxicity in combination with reovirus.[72] B cell depletion is a potential strategy for overcoming the neutralizing antibody response and has been used in animal studies.[123] Highlighting this point, one study employed several immune suppression tactics similar to that used in organ transplantation with glucocorticoids, mycophenolate mofetil, and tacrolimus in combination with poxvirus infection in preimmunized mice bearing murine tumors. In this setting, immune suppression was insufficient to allow recovery of replicating virus from tumors, and the virus was only recovered from transgenic B cell knockout mice.[123] The anti-CD20 monoclonal antibody rituximab could be employed to inhibit humoral immunity, but thus far has not been successfully applied to oncolytic virus therapy. Rituximab has been shown to decrease neutralizing antibody titers to adeno-associated virus in patients with rheumatoid arthritis and allowed for enhanced gene therapy in a primate model of hemophilia B.[124,125]

Another intriguing method to overcome the neutralizing antibody response is to "hide" the virus from exposure to antibodies. For the most part, this approach relies upon use of an immunologically inert cell carrier that is infected ex vivo followed by administration to the mouse. During the time that the virus is within the cell and replicating, it remains hidden from the antibody response, allowing the cell suitable time to traffic to tumor sites and deliver its payload through lytic infection of the carrier cell. The efficacy of this approach has been demonstrated in several preclinical models using a variety of cell types.[126–132] While tumor-specific T cells have been successful in mouse models, clinical translation is potentially challenging as obtaining these cells in sufficient numbers remains a challenge.[133] Mesenchymal stem cells (MSC) and blood outgrowth endothelial cells (BOEC) have been used successfully in preclinical models and have the advantage of being easy to obtain and culture from a variety of donors including the patient themselves.[131,132,134–138] The groundwork has been laid for the first clinical trial utilizing MSCs to carry vMV for intraperitoneal delivery for patients with ovarian cancer.[139]

17.9 CONCLUSIONS AND FUTURE DIRECTIONS

Early clinical trials have demonstrated the proof of principle that replicating viruses have antitumor efficacy in humans. Cautious early trials have proven that the vast majority of oncolytic viruses are safe. Though there have been instances of efficacy, the results so far have fallen short of what was expected after more dramatic results in preclinical models. An intriguing finding from early trials is the suggestion that oncolytic virotherapy might be an important vehicle for immunotherapy. This has now been repeatedly demonstrated with numerous viruses. Just as live virus vaccines are somewhat more immunogenic than killed vaccines, data suggest that the awakening of the antitumor immune response may be an important mechanism by which tumor control is achieved. Data from preclinical studies show that NK cells and CD8+ T cells are required for effective tumor control in immune-competent animals.[116,140] As mentioned above, the expression of GM-CSF has enhanced the therapy of VV and HSV therapy. MV has been shown to activate tumor-specific CD8+ T cells through maturation of plasmacytoid dendritic cells in mesothelioma.[87] MV and VSV engineered to produce interferon-β (IFNβ) can augment innate and adaptive immunity, respectively, in mesothelioma models and, as a result, augment virotherapy.[88,116] Results from clinical trials suggest that an antitumor immune response plays a role in MV efficacy. Recent immunotherapy successes in melanoma and lung cancer targeting immune checkpoints have spawned attempts at exploiting the immune system to augment virotherapy.[140,141] The downside, of course, remains antiviral immune responses which can thwart the spread and replication of the therapeutic agent. Thus, future work will need to elucidate ways to tip the balance of immunity towards antitumor activity and away from antiviral activity; this approach will be critical for improving therapy. Given the limitations of animal models described above, it will be critical to design early phase trials with strong correlative science so that we can learn detailed information about the biology of these agents in humans. Fortunately, ongoing clinical trials will provide a wealth of information in that regard.

ACKNOWLEDGMENTS

The authors would like to thank Blake Jacobson for his critical review of the manuscript.

REFERENCES

1. Eager RM, Nemunaitis J. Clinical development directions in oncolytic viral therapy. *Cancer Gene Ther* May 2011;**18**(5):305–17.
2. Hammill AM, Conner J, Cripe TP. Oncolytic virotherapy reaches adolescence. *Pediatr Blood Cancer* December 15, 2010;**55**(7):1253–63.
3. Breitbach CJ, Reid T, Burke J, Bell JC, Kirn DH. Navigating the clinical development landscape for oncolytic viruses and other cancer therapeutics: no shortcuts on the road to approval. *Cytokine Growth Factor Rev* April–June 2010;**21**(2–3):85–9.
4. Patel MR, Kratzke RA. Oncolytic virus therapy for cancer: the first wave of translational clinical trials. *Transl Res* April 2013;**161**(4):355–64.
5. Kawakami Y, Li H, Lam JT, Krasnykh V, Curiel DT, Blackwell JL. Substitution of the adenovirus serotype 5 knob with a serotype 3 knob enhances multiple steps in virus replication. *Cancer Res* March 15, 2003;**63**(6):1262–9.
6. Myhre S, Henning P, Friedman M, Stahl S, Lindholm L, Magnusson MK. Re-targeted adenovirus vectors with dual specificity; binding specificities conferred by two different Affibody molecules in the fiber. *Gene Ther* February 2009;**16**(2):252–61.
7. Glasgow JN, Mikheeva G, Krasnykh V, Curiel DT. A strategy for adenovirus vector targeting with a secreted single chain antibody. *PLoS One* 2009;**4**(12):e8355.
8. Gu DL, Gonzalez AM, Printz MA, Doukas J, Ying W, D'Andrea M, et al. Fibroblast growth factor 2 retargeted adenovirus has redirected cellular tropism: evidence for reduced toxicity and enhanced antitumor activity in mice. *Cancer Res* June 1, 1999;**59**(11):2608–14.
9. Magnusson MK, Kraaij R, Leadley RM, De Ridder CM, van Weerden WM, Van Schie KA, et al. A transductionally retargeted adenoviral vector for virotherapy of Her2/neu-expressing prostate cancer. *Hum Gene Ther* January 2012;**23**(1):70–82.
10. Wohlfahrt ME, Beard BC, Lieber A, Kiem HP. A capsid-modified, conditionally replicating oncolytic adenovirus vector expressing TRAIL leads to enhanced cancer cell killing in human glioblastoma models. *Cancer Res* September 15, 2007;**67**(18):8783–90.
11. Cui Q, Jiang W, Wang Y, Lv C, Luo J, Zhang W, et al. Transfer of suppressor of cytokine signaling 3 by an oncolytic adenovirus induces potential antitumor activities in hepatocellular carcinoma. *Hepatology* January 2008;**47**(1):105–12.
12. Diaconu I, Cerullo V, Hirvinen ML, Escutenaire S, Ugolini M, Pesonen SK, et al. Immune response is an important aspect of the antitumor effect produced by a CD40L-encoding oncolytic adenovirus. *Cancer Res* May 1, 2012;**72**(9):2327–38.
13. Vasey PA, Shulman LN, Campos S, Davis J, Gore M, Johnston S, et al. Phase I trial of intraperitoneal injection of the E1B-55-kd-gene-deleted adenovirus ONYX-015 (dl1520) given on days 1 through 5 every 3 weeks in patients with recurrent/refractory epithelial ovarian cancer. *J Clin Oncol* March 15, 2002;**20**(6):1562–9.
14. Nemunaitis J, Khuri F, Ganly I, Arseneau J, Posner M, Vokes E, et al. Phase II trial of intratumoral administration of ONYX-015, a replication-selective adenovirus, in patients with refractory head and neck cancer. *J Clin Oncol* January 15, 2001;**19**(2):289–98.
15. Ramesh N, Ge Y, Ennist DL, Zhu M, Mina M, Ganesh S, et al. CG0070, a conditionally replicating granulocyte macrophage colony-stimulating factor–armed oncolytic adenovirus for the treatment of bladder cancer. *Clin Cancer Res* January 1, 2006;**12**(1):305–13.
16. Liu Y, Ye T, Maynard J, Akbulut H, Deisseroth A. Engineering conditionally replication-competent adenoviral vectors carrying the cytosine deaminase gene increases the infectivity and therapeutic effect for breast cancer gene therapy. *Cancer Gene Ther* April 2006;**13**(4):346–56.
17. Cerullo V, Pesonen S, Diaconu I, Escutenaire S, Arstila PT, Ugolini M, et al. Oncolytic adenovirus coding for granulocyte macrophage colony-stimulating factor induces antitumoral immunity in cancer patients. *Cancer Res* June 1, 2010;**70**(11):4297–309.
18. Khuri FR, Nemunaitis J, Ganly I, Arseneau J, Tannock IF, Romel L, et al. a controlled trial of intratumoral ONYX-015, a selectively-replicating adenovirus, in combination with cisplatin and 5-fluorouracil in patients with recurrent head and neck cancer. *Nat Med* August 2000;**6**(8):879–85.
19. Xia ZJ, Chang JH, Zhang L, Jiang WQ, Guan ZZ, Liu JW, et al. Phase III randomized clinical trial of intratumoral injection of E1B gene-deleted adenovirus (H101) combined with cisplatin-based chemotherapy in treating squamous cell cancer of head and neck or esophagus. *Ai Zheng* December 2004;**23**(12):1666–70.
20. Douglas JT, Rogers BE, Rosenfeld ME, Michael SI, Feng M, Curiel DT. Targeted gene delivery by tropism-modified adenoviral vectors. *Nat Biotechnol* November 1996;**14**(11):1574–8.
21. Koski A, Kangasniemi L, Escutenaire S, Pesonen S, Cerullo V, Diaconu I, et al. Treatment of cancer patients with a serotype 5/3 chimeric oncolytic adenovirus expressing GMCSF. *Mol Ther* October 2010;**18**(10):1874–84.
22. Hemminki O, Diaconu I, Cerullo V, Pesonen SK, Kanerva A, Joensuu T, et al. Ad3-hTERT-E1A, a fully serotype 3 oncolytic adenovirus, in patients with chemotherapy refractory cancer. *Mol Ther* September 2012;**20**(9):1821–30.
23. Park BH, Hwang T, Liu TC, Sze DY, Kim JS, Kwon HC, et al. Use of a targeted oncolytic poxvirus, JX-594, in patients with refractory primary or metastatic liver cancer: a phase I trial. *Lancet Oncol* June 2008;**9**(6):533–42.

24. DeWeese TL, van der Poel H, et al. A phase I trial of CV706, a replication-competent, PSA selective oncolytic adenovirus, for the treatment of locally recurrent prostate cancer following radiation therapy. *Cancer Res* 2001;**61**(20):7464–72.

25. Yu DC, Chen Y, et al. The addition of adenovirus type 5 region E3 enables calydon virus 787 to eliminate distant prostate tumor xenografts. *Cancer Res* 1999;**59**(17):4200–3.

26. Nemunatis M, Tong AW, et al. A phase I study of telomerase-specific replication competent oncolytic adenovirus (telomelysin) for various solid tumors. *Mol Ther* 2010;**18**(2):429–34.

27. Takayama K, Reynolds PN, Short JJ, Kawakami Y, Adachi Y, Glasgow JN, et al. A mosaic adenovirus possessing serotype Ad5 and serotype Ad3 knobs exhibits expanded tropism. *Virology* May 10, 2003;**309**(2):282–93.

28. Rivera AA, Davydova J, Schierer S, Wang M, Krasnykh V, Yamamoto M, et al. Combining high selectivity of replication with fiber chimerism for effective adenoviral oncolysis of CAR-negative melanoma cells. *Gene Ther* December 2004;**11**(23):1694–702.

29. Dreier B, Honegger A, Hess C, Nagy-Davidescu G, Mittl PR, Grutter MG, et al. Development of a generic adenovirus delivery system based on structure-guided design of bispecific trimeric DARPin adapters. *Proc Natl Acad Sci USA* March 5, 2013;**110**(10):E869–77.

30. Sova P, Ren XW, Ni S, Bernt KM, Mi J, Kiviat N, et al. A tumor-targeted and conditionally replicating oncolytic adenovirus vector expressing TRAIL for treatment of liver metastases. *Mol Ther* April 2004;**9**(4):496–509.

31. Luo J, Xia Q, Zhang R, Lv C, Zhang W, Wang Y, et al. Treatment of cancer with a novel dual-targeted conditionally replicative adenovirus armed with mda-7/IL-24 gene. *Clin Cancer Res* April 15, 2008;**14**(8):2450–7.

32. Akbulut H, Zhang L, Tang Y, Deisseroth A. Cytotoxic effect of replication-competent adenoviral vectors carrying L-plastin promoter regulated E1A and cytosine deaminase genes in cancers of the breast, ovary and colon. *Cancer Gene Ther* May 2003;**10**(5):388–95.

33. Bortolanza S, Bunuales M, Otano I, Gonzalez-Aseguinolaza G, Ortiz-de-Solorzano C, Perez D, et al. Treatment of pancreatic cancer with an oncolytic adenovirus expressing interleukin-12 in Syrian hamsters. *Mol Ther* April 2009;**17**(4):614–22.

34. Chiocca EA, Abbed KM, Tatter S, Louis DN, Hochberg FH, Barker F, et al. A phase I open-label, dose-escalation, multi-institutional trial of injection with an E1B-Attenuated adenovirus, ONYX-015, into the peritumoral region of recurrent malignant gliomas, in the adjuvant setting. *Mol Ther* November 2004;**10**(5):958–66.

35. Rodriguez R, Schuur ER, Lim HY, Henderson GA, Simons JW, Henderson DR. Prostate attenuated replication competent adenovirus (ARCA) CN706: a selective cytotoxic for prostate-specific antigen-positive prostate cancer cells. *Cancer Res* July 1, 1997;**57**(13):2559–63.

36. Pesonen S, Diaconu I, Kangasniemi L, Ranki T, Kanerva A, Pesonen SK, et al. Oncolytic immunotherapy of advanced solid tumors with a CD40L-expressing replicating adenovirus: assessment of safety and immunologic responses in patients. *Cancer Res* April 1, 2012;**72**(7):1621–31.

37. Kimball KJ, Preuss MA, Barnes MN, Wang M, Siegal GP, Wan W, et al. A phase I study of a tropism-modified conditionally replicative adenovirus for recurrent malignant gynecologic diseases. *Clin Cancer Res* November 1, 2010;**16**(21):5277–87.

38. Nokisalmi P, Pesonen S, Escutenaire S, Sarkioja M, Raki M, Cerullo V, et al. Oncolytic adenovirus ICOVIR-7 in patients with advanced and refractory solid tumors. *Clin Cancer Res* June 1, 2010;**16**(11):3035–43.

39. Li JL, Liu HL, Zhang XR, Xu JP, Hu WK, Liang M, et al. A phase I trial of intratumoral administration of recombinant oncolytic adenovirus overexpressing HSP70 in advanced solid tumor patients. *Gene Ther* March 2009;**16**(3):376–82.

40. Smith KD, Mezhir JJ, Bickenbach K, Veerapong J, Charron J, Posner MC, et al. Activated MEK suppresses activation of PKR and enables efficient replication and in vivo oncolysis by Deltagamma(1)34.5 mutants of herpes simplex virus 1. *J Virol* February 2006;**80**(3):1110–20.

41. Aghi M, Visted T, Depinho RA, Chiocca EA. Oncolytic herpes virus with defective ICP6 specifically replicates in quiescent cells with homozygous genetic mutations in p16. *Oncogene* July 10, 2008;**27**(30):4249–54.

42. Liu S, Dai M, You L, Zhao Y. Advance in herpes simplex viruses for cancer therapy. *Sci China Life Sci* April 2013;**56**(4):298–305.

43. Liu TC, Zhang T, Fukuhara H, Kuroda T, Todo T, Martuza RL, et al. Oncolytic HSV armed with platelet factor 4, an antiangiogenic agent, shows enhanced efficacy. *Mol Ther* December 2006;**14**(6):789–97.

44. Tsuji T, Nakamori M, Iwahashi M, Nakamura M, Ojima T, Iida T, et al. An armed oncolytic herpes simplex virus expressing thrombospondin-1 has an enhanced in vivo antitumor effect against human gastric cancer. *Int J Cancer* January 15, 2013;**132**(2):485–94.

45. Tyminski E, Leroy S, Terada K, Finkelstein DM, Hyatt JL, Danks MK, et al. Brain tumor oncolysis with replication-conditional herpes simplex virus type 1 expressing the prodrug-activating genes, CYP2B1 and secreted human intestinal carboxylesterase, in combination with cyclophosphamide and irinotecan. *Cancer Res* August 1, 2005;**65**(15):6850–7.

46. Markert JM, Medlock MD, Rabkin SD, Gillespie GY, Todo T, Hunter WD, et al. Conditionally replicating herpes simplex virus mutant, G207 for the treatment of malignant glioma: results of a phase I trial. *Gene Ther* May 2000;**7**(10):867–74.

47. Markert JM, Liechty PG, Wang W, Gaston S, Braz E, Karrasch M, et al. Phase Ib trial of mutant herpes simplex virus G207 inoculated pre-and post-tumor resection for recurrent GBM. *Mol Ther* January 2009;**17**(1):199–207.

48. Kemeny N, Brown K, Covey A, Kim T, Bhargava A, Brody L, et al. Phase I, open-label, dose-escalating study of a genetically engineered herpes simplex virus, NV1020, in subjects with metastatic colorectal carcinoma to the liver. *Hum Gene Ther* December 2006;**17**(12):1214–24.

49. Geevarghese SK, Geller DA, de Haan HA, Horer M, Knoll AE, Mescheder A, et al. Phase I/II study of oncolytic herpes simplex virus NV1020 in patients with extensively pretreated refractory colorectal cancer metastatic to the liver. *Hum Gene Ther* September 2010;**21**(9):1119–28.

50. Hu JC, Coffin RS, Davis CJ, Graham NJ, Groves N, Guest PJ, et al. A phase I study of OncoVEXGM-CSF, a second-generation oncolytic herpes simplex virus expressing granulocyte macrophage colony-stimulating factor. *Clin Cancer Res* November 15, 2006;**12**(22):6737–47.

51. Kaufman HL, Kim DW, DeRaffele G, Mitcham J, Coffin RS, Kim-Schulze S. Local and distant immunity induced by intralesional vaccination with an oncolytic herpes virus encoding GM-CSF in patients with stage IIIc and IV melanoma. *Ann Surg Oncol* March 2010;**17**(3):718–30.

52. Senzer NN, Kaufman HL, Amatruda T, Nemunaitis M, Reid T, Daniels G, et al. Phase II clinical trial of a granulocyte-macrophage colony-stimulating factor-encoding, second-generation oncolytic herpesvirus in patients with unresectable metastatic melanoma. *J Clin Oncol* December 1, 2009;**27**(34):5763–71.

53. Kaufman HL, Bines SD. OPTIM trial: a phase III trial of an oncolytic herpes virus encoding GM-CSF for unresectable stage III or IV melanoma. *Future Oncol* June 2010;**6**(6):941–9.

54. Harrington KJ, Hingorani M, Tanay MA, Hickey J, Bhide SA, Clarke PM, et al. Phase I/II study of oncolytic HSV GM-CSF in combination with radiotherapy and cisplatin in untreated stage III/IV squamous cell cancer of the head and neck. *Clin Cancer Res* August 1, 2010;**16**(15):4005–15.

55. Corral JJ, Young AM, Joaquin M, Yap TA, Denholm KA, Shah KJ, et al. Phase I clinical trial of a genetically modified and oncolytic vaccinia virus GL-ONC1 with green fluorescent protein imaging. American Society of Clinical Oncology. *J Clin Oncol* 2012;**30**:(suppl)2530.

56. Hwang TH, Moon A, Burke J, Ribas A, Stephenson J, Breitbach CJ, et al. A mechanistic proof-of-concept clinical trial with JX-594, a targeted multi-mechanistic oncolytic poxvirus, in patients with metastatic melanoma. *Mol Ther* October 2011;**19**(10):1913–22.

57. Breitbach CJ, Burke J, Jonker D, Stephenson J, Haas AR, Chow LQ, et al. Intravenous delivery of a multi-mechanistic cancer-targeted oncolytic poxvirus in humans. *Nature* September 1, 2011;**477**(7362):99–102.

58. Heo J, Reid T, Ruo L, Breitbach CJ, Rose S, Bloomston M, et al. Randomized dose-finding clinical trial of oncolytic immunotherapeutic vaccinia JX-594 in liver cancer. *Nat Med* March 2013;**19**(3):329–36.

59. Liu TC, Hwang T, Park BH, Bell J, Kirn DH. The targeted oncolytic poxvirus JX-594 demonstrates antitumoral, antivascular, and anti-HBV activities in patients with hepatocellular carcinoma. *Mol Ther* September 2008;**16**(9):1637–42.

60. Heo J, Breitbach CJ, Moon A, Kim CW, Patt R, Kim MK, et al. Sequential therapy with JX-594, a targeted oncolytic poxvirus, followed by sorafenib in hepatocellular carcinoma: preclinical and clinical demonstration of combination efficacy. *Mol Ther* June 2011;**19**(6):1170–9.

61. Chandran K, Nibert ML. Animal cell invasion by a large nonenveloped virus: reovirus delivers the goods. *Trends Microbiol* August 2003;**11**(8):374–82.

62. Maitra R, Ghalib MH, Goel S. Reovirus: a targeted therapeutic–progress and potential. Molecular cancer research. *MCR* December 2012;**10**(12):1514–25.

63. Boehme KW, Ikizler M, Kobayashi T, Dermody TS. Reverse genetics for mammalian reovirus. *Methods* October 2011;**55**(2):109–13.

64. Patel MR, Jacobson BA, De A, Frizelle SP, Janne P, Thumma SC, et al. Ras pathway activation in malignant mesothelioma. *J Thorac Oncol* September 2007;**2**(9):789–95.

65. Heinemann L, Simpson GR, Boxall A, Kottke T, Relph KL, Vile R, et al. Synergistic effects of oncolytic reovirus and docetaxel chemotherapy in prostate cancer. *BMC Cancer* 2011;**11**(221).

66. Pandha HS, Heinemann L, Simpson GR, Melcher A, Prestwich R, Errington F, et al. Synergistic effects of oncolytic reovirus and cisplatin chemotherapy in murine malignant melanoma. *Clin Cancer Res* October 1, 2009;**15**(19):6158–66.

67. Roulstone V, Twigger K, Zaidi S, Pencavel T, Kyula JN, White C, et al. Synergistic cytotoxicity of oncolytic reovirus in combination with cisplatin-paclitaxel doublet chemotherapy. *Gene Ther* May 2013;**20**(5):521–8.

68. Sei S, Mussio JK, Yang QE, Nagashima K, Parchment RE, Coffey MC, et al. Synergistic antitumor activity of oncolytic reovirus and chemotherapeutic agents in non-small cell lung cancer cells. *Mol Cancer* 2009;**8**:47.

69. Morris DG, Feng X, Difrancesco LM, Fonseca K, Forsyth PA, Paterson AH, et al. REO-001: a phase I trial of percutaneous intralesional administration of reovirus type 3 dearing (Reolysin(R)) in patients with advanced solid tumors. *Invest New Drugs* June 2013;**31**(3):696–706.

70. Vidal L, Pandha HS, Yap TA, White CL, Twigger K, Vile RG, et al. A phase I study of intravenous oncolytic reovirus type 3 Dearing in patients with advanced cancer. *Clin Cancer Res* November 1, 2008;**14**(21):7127–37.

71. Karapanagiotou EM, Roulstone V, Twigger K, Ball M, Tanay M, Nutting C, et al. Phase I/II trial of carboplatin and paclitaxel chemotherapy in combination with intravenous oncolytic reovirus in patients with advanced malignancies. *Clin Cancer Res* April 1, 2012;**18**(7):2080–9.

72. Lolkema MP, Arkenau HT, Harrington K, Roxburgh P, Morrison R, Roulstone V, et al. A phase I study of the combination of intravenous reovirus type 3 Dearing and gemcitabine in patients with advanced cancer. *Clin Cancer Res* February 1, 2011;**17**(3):581–8.

73. Comins C, Spicer J, Protheroe A, Roulstone V, Twigger K, White CM, et al. REO-10: a phase I study of intravenous reovirus and docetaxel in patients with advanced cancer. *Clin Cancer Res* November 15, 2010;**16**(22):5564–72.

74. Adair RA, Roulstone V, Scott KJ, Morgan R, Nuovo GJ, Fuller M, et al. Cell carriage, delivery, and selective replication of an oncolytic virus in tumor in patients. *Sci Transl Med* June 13, 2012;**4**(138):138ra77.

75. (http://www.oncolyticsbiotech.com/English/news/press-release-details/2013/Oncolytics-Biotech-Inc-Announces-Positive-Top-Line-Data-from-REO-018-Randomized-Study-of-REOLYSIN-in-Head-and-Neck-Cancers). 2013 cited; Available from: (http://www.oncolyticsbiotech.com/English/news/press-release-details/2013/Oncolytics-Biotech-Inc-Announces-Positive-Top-Line-Data-from-REO-018-Randomized-Study-of-REOLYSIN-in-Head-and-Neck-Cancers).

76. Taqi AM, Abdurrahman MB, Yakubu AM, Fleming AF. Regression of Hodgkin's disease after measles. *Lancet* May 16, 1981;**1**(8229):1112.

77. Gross S. Measles and leukaemia. *Lancet* February 20, 1971;**1**(7695):397–8.

78. Pasquinucci G. Possible effect of measles on leukaemia. *Lancet* January 16, 1971;**1**(7690):136.

79. Zygiert Z. Hodgkin's disease: remissions after measles. *Lancet* March 20, 1971;**1**(7699):593.

80. Griffin D, Bellini W. Measle virus. In: Knipe D, Howley P, editors. *Fields Virology*. Philadelphia: Lippencott Williams & Wilkins; 2001. pp. 1401–41.

81. Noyce RS, Richardson CD. Nectin 4 is the epithelial cell receptor for measles virus. *Trends Microbiol* September 2012;**20**(9):429–39.

82. Muhlebach MD, Mateo M, Sinn PL, Prufer S, Uhlig KM, Leonard VH, et al. Adherens junction protein nectin-4 is the epithelial receptor for measles virus. *Nature* December 22, 2011;**480**(7378):530–3.

83. Fabre-Lafay S, Monville F, Garrido-Urbani S, Berruyer-Pouyet C, Ginestier C, Reymond N, et al. Nectin-4 is a new histological and serological tumor associated marker for breast cancer. *BMC Cancer* 2007;**7**:73.

84. Takano A, Ishikawa N, Nishino R, Masuda K, Yasui W, Inai K, et al. Identification of nectin-4 oncoprotein as a diagnostic and therapeutic target for lung cancer. *Cancer Res* August 15, 2009;**69**(16):6694–703.

85. Derycke MS, Pambuccian SE, Gilks CB, Kalloger SE, Ghidouche A, Lopez M, et al. Nectin 4 overexpression in ovarian cancer tissues and serum: potential role as a serum biomarker. *Am J Clin Pathol* November 2010;**134**(5):835–45.

86. Kim D, Niewiesk S. Synergistic induction of interferon alpha through TLR-3 and TLR-9 agonists stimulates immune responses against measles virus in neonatal cotton rats. *Vaccine* January 3, 2014;**32**(2):265–70.

87. Gauvrit A, Brandler S, Sapede-Peroz C, Boisgerault N, Tangy F, Gregoire M. Measles virus induces oncolysis of mesothelioma cells and allows dendritic cells to cross-prime tumor-specific CD8 response. *Cancer Res* June 15, 2008;**68**(12):4882–92.

88. Li H, Peng KW, Dingli D, Kratzke RA, Russell SJ. Oncolytic measles viruses encoding interferon beta and the thyroidal sodium iodide symporter gene for mesothelioma virotherapy. *Cancer Gene Ther* August 2010;**17**(8):550–8.

89. Dingli D, Peng KW, Harvey ME, Greipp PR, O'Connor MK, Cattaneo R, et al. Image-guided radiovirotherapy for multiple myeloma using a recombinant measles virus expressing the thyroidal sodium iodide symporter. *Blood* 2004;**103**(5):1641–6.

90. Grote D, Russell SJ, Cornu TI, Cattaneo R, Vile R, Poland GA, et al. Live attenuated measles virus induces regression of human lymphoma xenografts in immunodeficient mice. *Blood* 2001;**97**(12):3746–54.

91. Hasegawa K, Pham L, O'Connor MK, Federspiel MJ, Russell SJ, Peng KW. Dual therapy of ovarian cancer using measles viruses expressing carcinoembryonic antigen and sodium iodide symporter. *Clin Cancer Res* 2006;**12**(6):1868–75.

92. Li H, Peng KW, Dingli D, Kratzke RA, Russell SJ. Oncolytic measles viruses encoding interferon beta and the thyroidal sodium iodide symporter gene for mesothelioma virotherapy. *Cancer Gene Ther* 2010;**17**:550–8.

93. McDonald CJ, Erlichman C, Ingle JN, Rosales GA, Allen C, Greiner SM, et al. A measles virus vaccine strain derivative as a novel oncolytic agent against breast cancer. *Breast Cancer Res Treat* 2006;**99**(2):177–84.

94. Msaouel P, Iankov ID, Allen C, Morris JC, von Messling V, Cattaneo R, et al. Engineered measles virus as a novel oncolytic therapy against prostate cancer. *Prostate* January 1, 2009;**69**(1):82–91.

95. Peng KW, TenEyck CJ, Galanis E, Kalli KR, Hartmann LC, Russell SJ. Intraperitoneal therapy of ovarian cancer using an engineered measles virus. *Cancer Res* 2002;**62**(16):4656–62.

96. Phuong LK, Allen C, Peng KW, Giannini C, Greiner S, TenEyck CJ, et al. Use of a vaccine strain of measles virus genetically engineered to produce carcinoembryonic antigen as a novel therapeutic agent against glioblastoma multiforme. *Cancer Res* 2003;**63**(10):2462–9.

97. Myers RM, Greiner SM, Harvey ME, Griesmann G, Kuffel MJ, Buhrow SA, et al. Preclinical pharmacology and toxicology of intravenous MV-NIS, an oncolytic measles virus administered with or without cyclophosphamide. *Clin Pharmacol Ther* December 2007;**82**(6):700–10.

98. Bach P, Abel T, Hoffmann C, Gal Z, Braun G, Voelker I, et al. Specific elimination of CD133+ tumor cells with targeted oncolytic measles virus. *Cancer Res* January 15, 2013;**73**(2):865–74.

99. Nakamura T, Peng KW, Harvey M, Greiner S, Lorimer IA, James CD, et al. Rescue and propagation of fully retargeted oncolytic measles viruses. *Nat Biotechnol* February 2005;**23**(2):209–14.

100. Allen C, Vongpunsawad S, Nakamura T, James CD, Schroeder M, Cattaneo R, et al. Retargeted oncolytic measles strains entering via the EGFRvIII receptor maintain significant antitumor activity against gliomas with increased tumor specificity. *Cancer Res* December 15, 2006;**66**(24):11840–50.

101. Paraskevakou G, Allen C, Nakamura T, Zollman P, James CD, Peng KW, et al. Epidermal growth factor receptor (EGFR)-retargeted measles virus strains effectively target EGFR- or EGFRvIII expressing gliomas. *Mol Ther* April 2007;**15**(4):677–86.

102. Allen C, Paraskevakou G, Iankov I, Giannini C, Schroeder M, Sarkaria J, et al. Interleukin-13 displaying retargeted oncolytic measles virus strains have significant activity against gliomas with improved specificity. *Mol Ther* September 2008;**16**(9):1556–64.

103. Jing Y, Tong C, Zhang J, Nakamura T, Iankov I, Russell SJ, et al. Tumor and vascular targeting of a novel oncolytic measles virus retargeted against the urokinase receptor. *Cancer Res* February 15, 2009;**69**(4):1459–68.

104. Liu C, Hasegawa K, Russell SJ, Sadelain M, Peng KW. Prostate-specific membrane antigen retargeted measles virotherapy for the treatment of prostate cancer. *Prostate* July 1, 2009;**69**(10):1128–41.

105. Reddi HV, Madde P, McDonough SJ, Trujillo MA, Morris 3rd JC, Myers RM, et al. Preclinical efficacy of the oncolytic measles virus expressing the sodium iodide symporter in iodine non-avid anaplastic thyroid cancer: a novel therapeutic agent allowing noninvasive imaging and radioiodine therapy. *Cancer Gene Ther* September 2012;**19**(9):659–65.

106. Galanis E, Hartmann LC, Cliby WA, Long HJ, Peethambaram PP, Barrette BA, et al. Phase I trial of intraperitoneal administration of an oncolytic measles virus strain engineered to express carcinoembryonic antigen for recurrent ovarian cancer. *Cancer Res* February 1, 2010;**70**(3):875–82.

107. Markman M, Webster K, Zanotti K, Peterson G, Kulp B, Belinson J. Phase 2 trial of carboplatin plus tamoxifen in platinum-resistant ovarian cancer and primary carcinoma of the peritoneum. *Gynecol Oncol* August 2004;**94**(2):404–8.

108. Markman M, Webster K, Zanotti K, Peterson G, Kulp B, Belinson J. Survival following the documentation of platinum and taxane resistance in ovarian cancer: a single institution experience involving multiple phase 2 clinical trials. *Gynecol Oncol* June 2004;**93**(3):699–701.

109. Gauvrit A, Brandler S, Sapede-Peroz C, Boisgerault N, Tangy F, Gregoire M. Measles virus induces oncolysis of mesothelioma cells and allows dendritic cells to cross-prime tumor-specific CD8 response. *Cancer Res* 2008;**68**(12):4882–92.

110. Donnelly OG, Errington-Mais F, Steele L, Hadac E, Jennings V, Scott K, et al. Measles virus causes immunogenic cell death in human melanoma. *Gene Ther* 2013;**20**(1):7–1.

111. Quiroz E, Moreno N, Peralta PH, Tesh RB. A human case of encephalitis associated with vesicular stomatitis virus (Indiana serotype) infection. *Am J Trop Med Hygiene* September 1988;**39**(3):312–4.

112. Johnson KM, Vogel JE, Peralta PH. Clinical and serological response to laboratory-acquired human infection by Indiana type vesicular stomatitis virus (VSV). *Am J Trop Med Hygiene* March 1966;**15**(2):244–6.

113. Patterson WC, Mott LO, Jenney EW. A study of vesicular stomatitis in man. *J Am Veterinary Med Assoc* July 1, 1958;**133**(1):57–62.

114. Barber GN. Vesicular stomatitis virus as an oncolytic vector. *Viral Immunol* 2004;**17**(4):516–27.

115. Jenks N, Myers R, Greiner SM, Thompson J, Mader EK, Greenslade A, et al. Safety studies on intrahepatic or intratumoral injection of oncolytic vesicular stomatitis virus expressing interferon-beta in rodents and nonhuman primates. *Hum Gene Ther* April 2010;**21**(4):451–62.

116. Willmon CL, Saloura V, Fridlender ZG, Wongthida P, Diaz RM, Thompson J, et al. Expression of IFN-beta enhances both efficacy and safety of oncolytic vesicular stomatitis virus for therapy of mesothelioma. *Cancer Res* October 1, 2009;**69**(19):7713–20.

117. Naik S, Nace R, Barber GN, Russell SJ. Potent systemic therapy of multiple myeloma utilizing oncolytic vesicular stomatitis virus coding for interferon-beta. *Cancer Gene Ther* July 2012;**19**(7):443–50.

118. Saloura V, Wang LC, Fridlender ZG, Sun J, Cheng G, Kapoor V, et al. Evaluation of an attenuated vesicular stomatitis virus vector expressing interferon-beta for use in malignant pleural mesothelioma: heterogeneity in interferon responsiveness defines potential efficacy. *Hum Gene Ther* January 2010;**21**(1):51–64.

119. Peng KW, Myers R, Greenslade A, Mader E, Greiner S, Federspiel MJ, et al. Using clinically approved cyclophosphamide regimens to control the humoral immune response to oncolytic viruses. *Gene Ther* March 2013;**20**(3):255–61.

120. Qiao J, Wang H, Kottke T, White C, Twigger K, Diaz RM, et al. Cyclophosphamide facilitates antitumor efficacy against subcutaneous tumors following intravenous delivery of reovirus. *Clin Cancer Res* January 1, 2008;**14**(1):259–69.

121. Lun XQ, Jang JH, Tang N, Deng H, Head R, Bell JC, et al. Efficacy of systemically administered oncolytic vaccinia virotherapy for malignant gliomas is enhanced by combination therapy with rapamycin or cyclophosphamide. *Clin Cancer Res* April 15, 2009;**15**(8):2777–88.

122. Cerullo V, Diaconu I, Kangasniemi L, Rajecki M, Escutenaire S, Koski A, et al. Immunological effects of low-dose cyclophosphamide in cancer patients treated with oncolytic adenovirus. *Mol Ther* September 2011;**19**(9):1737–46.

123. Guo ZS, Parimi V, O'Malley ME, Thirunavukarasu P, Sathaiah M, Austin F, et al. The combination of immunosuppression and carrier cells significantly enhances the efficacy of oncolytic poxvirus in the pre-immunized host. *Gene Ther* December 2010;**17**(12):1465–75.

124. Mingozzi F, Chen Y, Edmonson SC, Zhou S, Thurlings RM, Tak PP, et al. Prevalence and pharmacological modulation of humoral immunity to AAV vectors in gene transfer to synovial tissue. *Gene Ther* April 2013;**20**(4):417–24.

125. Mingozzi F, Chen Y, Murphy SL, Edmonson SC, Tai A, Price SD, et al. Pharmacological modulation of humoral immunity in a nonhuman primate model of AAV gene transfer for hemophilia B. *Mol Ther* July 2012;**20**(7):1410–6.

126. Power AT, Wang J, Falls TJ, Paterson JM, Parato KA, Lichty BD, et al. Carrier cell-based delivery of an oncolytic virus circumvents antiviral immunity. *Mol Ther* January 2007;**15**(1):123–30.

127. Munguia A, Ota T, Miest T, Russell SJ. Cell carriers to deliver oncolytic viruses to sites of myeloma tumor growth. *Gene Ther* May 2008;**15**(10):797–806.

128. Coukos G, Makrigiannakis A, Kang EH, Caparelli D, Benjamin I, Kaiser LR, et al. Use of carrier cells to deliver a replication-selective herpes simplex virus-1 mutant for the intraperitoneal therapy of epithelial ovarian cancer. *Clin Cancer Res* June 1999;**5**(6):1523–37.

129. Ilett EJ, Prestwich RJ, Kottke T, Errington F, Thompson JM, Harrington KJ, et al. Dendritic cells and T cells deliver oncolytic reovirus for tumour killing despite pre-existing anti-viral immunity. *Gene Ther* May 2009;**16**(5):689–99.

130. Thorne SH, Liang W, Sampath P, Schmidt T, Sikorski R, Beilhack A, et al. Targeting localized immune suppression within the tumor through repeat cycles of immune cell-oncolytic virus combination therapy. *Mol Ther* September 2010;**18**(9):1698–705.

131. Mader EK, Maeyama Y, Lin Y, Butler GW, Russell HM, Galanis E, et al. Mesenchymal stem cell carriers protect oncolytic measles viruses from antibody neutralization in an orthotopic ovarian cancer therapy model. *Clin Cancer Res* December 1, 2009;**15**(23):7246–55.

132. Wei J, Wahl J, Nakamura T, Stiller D, Mertens T, Debatin KM, et al. Targeted release of oncolytic measles virus by blood outgrowth endothelial cells in situ inhibits orthotopic gliomas. *Gene Ther* November 2007;**14**(22):1573–86.

133. Kottke T, Diaz RM, Kaluza K, Pulido J, Galivo F, Wongthida P, et al. Use of biological therapy to enhance both virotherapy and adoptive T-cell therapy for cancer. *Mol Ther* December 2008;**16**(12):1910–8.

134. Dudek AZ. Endothelial lineage cell as a vehicle for systemic delivery of cancer gene therapy. *Transl Res* September 2010;**156**(3):136–46.

135. Bodempudi V, Ohlfest JR, Terai K, Zamora EA, Vogel RI, Gupta K, et al. Blood outgrowth endothelial cell-based systemic delivery of antiangiogenic gene therapy for solid tumors. *Cancer Gene Ther* December 2010;**17**(12):855–63.

136. Ahmed AU, Rolle CE, Tyler MA, Han Y, Sengupta S, Wainwright DA, et al. Bone marrow mesenchymal stem cells loaded with an oncolytic adenovirus suppress the anti-adenoviral immune response in the cotton rat model. *Mol Ther* October 2010;**18**(10):1846–56.

137. Garcia-Castro J, Alemany R, Cascallo M, Martinez-Quintanilla J, Arriero Mdel M, Lassaletta A, et al. Treatment of metastatic neuroblastoma with systemic oncolytic virotherapy delivered by autologous mesenchymal stem cells: an exploratory study. *Cancer Gene Ther* July 2010;**17**(7):476–83.

138. Yong RL, Shinojima N, Fueyo J, Gumin J, Vecil GG, Marini FC, et al. Human bone marrow-derived mesenchymal stem cells for intravascular delivery of oncolytic adenovirus Delta24-RGD to human gliomas. *Cancer Res* December 1, 2009;**69**(23):8932–40.

139. Mader EK, Butler G, Dowdy SC, Mariani A, Knutson KL, Federspiel MJ, et al. Optimizing patient derived mesenchymal stem cells as virus carriers for a phase I clinical trial in ovarian cancer. *J Transl Med* 2013;**11**(20).

140. Diaz RM, Galivo F, Kottke T, Wongthida P, Qiao J, Thompson J, et al. Oncolytic immunovirotherapy for melanoma using vesicular stomatitis virus. *Cancer Res* March 15, 2007;**67**(6):2840–8.

141. Hodi FS, O'Day SJ, McDermott DF, Weber RW, Sosman JA, Haanen JB, et al. Improved survival with ipilimumab in patients with metastatic melanoma. *N Engl J Med* August 19, 2010;**363**(8):711–23.

Chapter 18

T Cell-Based Gene Therapy of Cancer

Saar Gill[1] and Michael Kalos[2]

[1]*Division of Hematology-Oncology, Department of Medicine, University of Pennsylvania, Perelman School of Medicine, Philadelphia, PA, USA;*
[2]*Department of Pathology and Laboratory Medicine, University of Pennsylvania, Perelman School of Medicine, Philadelphia, PA, USA*

LIST OF ABBREVIATIONS

aAPC Artificial antigen presenting cells
APC Antigen presenting cell
CAR Chimeric antigen receptor
CDRs Complementarity determining regions
CRS Cytokine release syndrome
scFv Single chain variable fragment
GM-CSF Granulocyte macrophage–colony-stimulating factor
GVHD Graft versus host disease
HAMA Human–anti-mouse antibodies
IFN Interferon
IL-2 Interleukin-2
MART1 Melanoma antigen recognized by T cells-1
PBMC Peripheral blood mononuclear cells
TALEN Transcription activator-like effector nuclease
TCR T cell receptor
TIL Tumor-infiltrating lymphocytes

18.1 INTRODUCTION: T CELL-BASED IMMUNOTHERAPY

T lymphocytes are multifunctional effector cells with exquisite antigen specificity and a crucial role in protection from infection and malignancy. The importance of T cells in the control of malignancy is inferred from studies showing increased susceptibility to cancer in immunodeficient patients and experimental animals, and from decades of experience with *allogeneic stem cell transplantation* that have established the central role of T cells in the graft-versus-tumor response.[1] Many cases of progressive human tumor are heavily infiltrated with T cells, and the implication that such infiltrating T cells fail to recognize the tumor effectively or are functionally impaired in some manner has been experimentally validated.[2–5] Over the years, considerable effort has been invested in attempts to (re)activate or otherwise trigger such T cells to mediate an antitumor effector activity via a plethora of vaccine-based approaches.[6–8] With few exceptions, these approaches have failed to fulfill the promise of activating potent antitumor T cell immunity, at least in part because they fail to effectively address the issue of immune tolerance, whether at the level of the available repertoire (central tolerance) or at the level of tumor-induced immune suppression (peripheral tolerance). More recently, approaches to reverse the functional impairment of T cells have been developed, with dramatic clinical outcomes in a subset of patients with selected malignancies, further highlighting the ability of an unleashed T cell response to effectively mediate potent antitumor immunity.[9]

The adoptive transfer of T lymphocytes after ex vivo expansion is an alternative approach long thought to hold promise as a strategy to overcome the fundamental issue of immune tolerance. Initial approaches to apply adoptive T cell transfer to target malignancy involved *leukapheresis* of peripheral blood mononuclear cells (PBMC) from patients followed by ex vivo expansion and reinfusion along with exogenous cytokine stimulation using interleukin-2 (IL-2). However, randomized clinical trials showed this approach to be no more effective than IL-2 infusion alone and this approach has been essentially abandoned.[10,11]

The Rosenberg group at the National Cancer Institute pioneered the harvest, expansion, and adoptive transfer of tumor-infiltrating lymphocytes (TIL). The premise was that most T cells infiltrate a tumor because of recognition of

FIGURE 18.1 Requirements for successful adoptive cellular therapy. After infusion, effect or cells must proliferate, traffic, recognize target, recruit effector functions, and avoid rejection, exhaustion, and senescence to persist until tumor elimination.

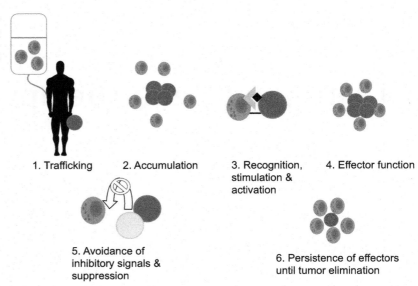

1. Trafficking   2. Accumulation   3. Recognition, stimulation & activation   4. Effector function

5. Avoidance of inhibitory signals & suppression   6. Persistence of effectors until tumor elimination

tumor-associated peptides via their T cell receptor (TCR), and hence expansion of TIL rather than PBMC should select for tumor-reactive T cells without a priori knowledge of their cognate tumor antigen. Although infusion of TIL has induced remissions in a substantial percentage of patients with melanoma, this approach comes with a caveat: It appears that *myeloablative chemoradiotherapy* is necessary for maximal clinical activity, which makes it difficult to ascribe the clinical responses to the TIL infusion alone.[12] Nonetheless, this appears to be a promising approach for the treatment of selected patients with melanoma, with up to 28% of patients achieving a complete response. These studies highlighted an important concept in adoptive cellular immunotherapy, specifically the correlation of response with prolonged T cell persistence.[12]

Recent technological advances have facilitated efficient expression of transgenes in T cells, enabling the redirection of T cells to target antigens of choice by introduction of a predetermined immunoreceptor. Whereas the ability to redirect T cell specificity to target antigens of choice has perhaps been the most obvious development in the field of T cell-based gene therapy, clearly other aspects of T cell biology are amenable to genetic manipulation and could have an important role in the success of adoptive immunotherapy. These elements include long-term functional persistence, trafficking of effector cells and accumulation at the tumor site, and engagement of co-stimulatory receptors to mediate a robust effector response. In addition, infused T cells must be able to resist mechanisms of exhaustion, senescence, and immunosuppression by the tumor microenvironment, avoid the host humoral and cellular immune responses, and be amenable to deletion on demand to mitigate potential toxicity issues (Figure 18.1).

This chapter will focus on a discussion of results from the published literature that provide insights about these aspects of T cell biology, with the objective of identifying specific biological parameters that can be manipulated to engineer biologically enhanced and more potent T cell products.

18.2 EX VIVO T CELL EXPANSION

Like any pharmaceutical product, genetic modification of T cells for clinical use must occur ex vivo in a *good manufacturing practice* (GMP)-grade laboratory. Viral-based gene transfer occurs more efficiently in proliferating cells, particularly when using gamma retroviral vectors.[13] Hence, the requirement for stimulation of T cells to induce proliferation and enhance gene transfer provides a dual opportunity to infuse into the patient T cells that are not only gene-modified but augmented in number. Methods for transfer of genetic material to redirect T cells will be discussed in a subsequent section. Methods for T cell expansion are discussed here.

The optimal dose for T cell infusion is unknown. Unlike small molecules or antibodies, T cells are living drugs that have the capacity to replicate and expand in vivo, and hence it may not be possible to demonstrate a strict dose–response relationship. Nonetheless, early trials in which unmanipulated, nonexpanded, nonredirected T cells from allogeneic stem cell donors were infused into recipients (so-called "donor lymphocyte infusions") showed that as few as 1×10^7/kg T cells could induce complete remission in some patients with leukemia.[14] As will be discussed later in this chapter, clinical trials of genetically engineered T cells have employed a wide range of doses for modified cells, with minimal and maximal doses specified by individual protocols and often dependent on the employed manufacturing process.

Methods for ex vivo T cell expansion are summarized in Table 18.1. The results of a typical T cell expansion using the anti-CD3/CD28 bead system is shown in Figure 18.2. Schemata for ex vivo expansion of T cells vary but rely on the

TABLE 18.1 Methods for Ex vivo Expansion Used in Clinical Trials of Genetically Engineered T Cells

Method	Typical Culture Time	Comments	References
Anti-CD3/CD28 beads	12 days	Typical expansion of T cells 26- to 525-fold	15,16
Anti-CD3/CD28 beads + IL-2	11–17 days	Typical expansion of T cells 49- to 385-fold	17,18
Virally infected APC + IL-2	5–6 weeks	EBV-transformed LCL pulsed with CMV and adenoviral peptides 21 ± 10-fold	19
Anti-CD3 and IL-2 with irradiated feeder cells	6 weeks		20,21
Anti-CD3 and IL-2 ± irradiated feeder cells	8–26 days		22,23
Artificial APC	10 days to 2 months as required	K562 cell line coated with anti-CD3, anti-CD28 antibodies and expressing CD137 ligand; Typical expansion 150-fold	24,25
		K562 cell line expressing CD19, CD137 ligand, CD86, CD64, and membrane-bound IL-15 and IL-21; Up to 40,000-fold expansion	26

IL, interleukin; EBV, Epstein–Barr virus; APC, antigen presenting cell; LCL, lymphoblastoid cell lines; CMV, cytomegalovirus.

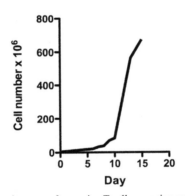

FIGURE 18.2   Typical growth expansion curve for ex vivo T cell expansion using the anti-CD3/CD28 bead system.

provision of a combination of signal 1 (TCR–mediated signaling using agonistic anti-CD3 antibodies such as clone OKT3), signal 2 (co-stimulation provided by agonistic anti-CD28 antibodies), and cytokine signaling (IL-2, IL-15, and IL-21).

The anti-CD3 and anti-CD28 signals can be delivered by soluble, plate-bound, or bead-based approaches, with some suggestion that immobilization of the antibodies is superior.[27] These approaches do not faithfully recapitulate the physiologic immune synapse between a T cell and its antigen presenting cell (APC). Therefore, methods using artificial antigen presenting cells (aAPC) have been developed. Here, the K562 tumor cell line is transduced to express a variety of stimulatory molecules such as CD137 ligand, CD86, or membrane-bound IL-15. T cell expansion after aAPC stimulation may yield higher numbers and more favorable T cell phenotypes than expansion after bead-based stimulation, but it is more complicated from a regulatory viewpoint because of the requirement for a tumor cell line as the APC. This is reflected in the fact that no trials of adoptive T cellular therapy clinical trials using aAPC-based expansion have been approved to date. The production of a master cell bank for K562 aAPC may be helpful in this regard.[24,26]

18.3 MODIFICATION STRATEGIES FOR T CELL REDIRECTION

In nature, antigen recognition by T cells occurs via the TCR, a complex composed of the highly variable α and β chain (or γ and δ chain) recognition domains that interact with the dimeric signal transduction machinery of the CD3 complex, including the CD3 γ, δ, ε, and ζ chains. Antigen recognition specificity is provided by the hypervariable complementary determining regions

(CDRs) of the α and β chains, although TCR–target interaction is degenerate; that is to say, a given target may be recognized by several different TCRs and a given TCR could recognize several different targets. The TCR recognizes peptide fragments that are embedded within the groove of a *major histocompatibility complex* (MHC) molecule. In contrast, antibody-based recognition relies on interactions between some or all six CDRs in the heavy and light chains of the antibody and their cognate antigen; unlike the TCR–peptide–MHC interaction, antibodies bind to the surface of antigens, recognizing amino acid sequences that may be discontinuous in the primary structure but contiguous in the folded, tertiary structure of the protein. This means that antibody recognition, unlike TCR recognition, does not depend on intracellular processing and MHC-based presentation of antigen. Conversely, only TCR can recognize intracellular antigens (after these are processed and presented on the appropriate MHC molecule). The affinity of antibody binding tends to be higher than that of TCR binding, although the strength of T cell binding to its target cell is enhanced by the summation of multiple TCR–antigen forces resulting in enhanced avidity. These considerations are important in the selection of antibody-based or TCR-based tumor antigen targeting approaches.

18.3.1 Approaches to Impart Antigen Specificity

Imparting antigen specificity to T cells requires the transfer of genetic material encoding a functional receptor into primary human T cells. Successful transfer of genetic material encoding α and β chains of a particular TCR into a heterogeneous population of T cells means that all of the T cells are now redirected to recognize the new target, although these T cells still express their endogenous TCR, which may compete or mis-pair with the introduced TCR with deleterious consequences.[28,29] T cells can also be redirected by the introduction of genetic material encoding a chimeric molecule composed of the antigen-recognition domains of an antibody and the signal transduction domains of a TCR. These chimeric molecules are termed chimeric antigen receptor (CAR), chimeric immunoreceptor, or T body. Studies have been published using both these approaches to generate a novel immunoreceptor.[30,31] TCR redirection requires the transduction of genetic material encoding TCR α and β chains of the desired specificity. CAR-based redirection involves the transfer of genetic material encoding a single chain variable fragment (scFv) based on an antibody of desired specificity.[32,33] The introduced antigen-binding domains (scFv) are fused to T cell signaling domains and associate with the native TCR signal transduction machinery leading to T cell activation, induction of effector functions, and antigen-driven proliferation. The differences between transgenic TCR and CAR approaches are summarized in Table 18.2 and Figure 18.3.

Advances in gene transfer (outlined in a subsequent section) have allowed the field to develop rapidly, and clinical results using both TCR-redirected[22,35–37] and CAR-redirected[38–40] T cells have been published (Table 18.3).

TABLE 18.2 Comparisons of CAR- and TCR-Based Targeting Strategies

	CAR	TCR
Antigens targeted	Surface only	Surface or intracellular
Requirement for antigen processing and presentation	No	Yes
Generation of new specificities	Relatively easy, dependent on the availability of antibodies	More challenging, depends on the identification of relevant tumor-specific TCR-a/b sequences
Ligand density	Correlates with target antigen expression	Dependent on target antigen expression and processing
Signal potency	High and can be further enhanced with co-stimulatory molecules	Relates to the affinity of the native TCR complex but can be enhanced by mutagenesis
Surface expression	Stable	Depends on efficient a/b chain pairing and potentially interacts with the native TCR chains to induce mis-pairing
Potential immunogenicity	High if sequences are derived from heterologous species; sequences can be easily humanized	Low if sequences are of human origin
Off-target effects	Can target normal tissues with low levels of expression; there are few surface markers with truly tumor-restricted expression	High if mis-pairing occurs or there is degeneration of specificity during affinity enhancement Can potentially target normal tissues with low levels of expression even without degeneration of specificity

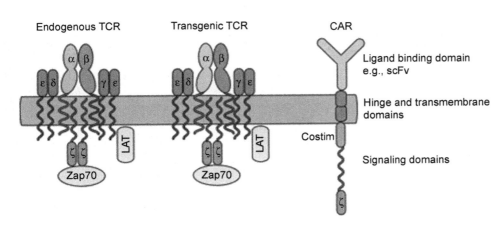

FIGURE 18.3 Diagrammatic representation of transgenic TCR and CAR structures. TCR, T cell receptor; CAR, chimeric antigen receptor; scFv, single chain variable fragment. *From Ref. 34, with permission.*

TABLE 18.3 Some Potential Methods for Gene Transfer into T Cells

Gene Transfer Method	Principle	Payload Size	Target	Integration into Genome	Comments	Used in T Cell Immuno-therapy
Viral Vector	**Mammalian Viruses have Evolved to Transfer RNA or DNA into Mammalian Cells**					
Gamma retroviral	Use of the VSV G glycoprotein allows viral entry into host cells via membrane fusion rather than via specific co-receptors	~6.5 kb	Dividing cells only (pre-integration complex cannot enter the nucleus of non-dividing cells)	Yes; stable integration of reverse-transcribed viral RNA Random integration sites	Oncogenic transformation observed when viral genomes integrate near cellular oncogenes Silencing has been observed	Yes
Lentiviral	Use of VSV G glycoprotein allows viral entry into host cells via membrane fusion rather than via specific co-receptors	~8 kb	Dividing as well as nondividing cells	Yes; stable integration of reverse-transcribed viral RNA	No genotoxicity observed, despite preferential integration into transcriptionally active sites	Yes
Adenoviral	Infection of target cell with some tissue tropism based on cell surface receptors	~36 kb	Dividing as well as nondividing cells	No; episomal	Host immune response is stimulated, leading to rejection	No
Adeno-associated virus	Infection of target cell with some tissue tropism based on cell surface receptors	~4.8 kb	Dividing as well as nondividing cells	No; episomal, incapable of prolonged high-level expression in dividing cells	Require a helper virus Nonpathogenic	No
Nonviral						
DNA EP	Electroporation of linearized plasmid DNA leads to chromosomally integrated vector in some T cell clones	No practical limitation	T cells must be carefully cloned and expanded under drug selection	Random		Yes

Continued

TABLE 18.3 Some Potential Methods for Gene Transfer into T Cells—cont'd

Gene Transfer Method	Principle	Payload Size	Target	Integration into Genome	Comments	Used in T Cell Immuno-therapy
mRNA EP	Electroporation of in vitro-transcribed mRNA encoding desired sequence	No practical limitation	Any cell	No; translated protein expression is lost as cells divide and mRNA levels decline owing to normal cellular processes	Allows a transient biodegradable expression of desired sequence	Yes
Transposon	DNA plasmids encoding transposon/transposase system (e.g., Sleeping Beauty) are electroporated into cells	No practical limitation	Any cell	Yes; random sites	Lower efficiency than lentiviral techniques	Yes

Because of the impact of central tolerance on the T cell repertoire to self-antigens, tumor antigen-specific T cells isolated from cancer patients or from healthy volunteers typically are of low affinity. Attempts to overcome this issue have included the isolation of T cells in a xenogeneic setting (e.g., vaccination of mice transgenic for human leukocyte antigen-A2)[4] or allogeneic setting (during graft versus host disease (GVHD)),[41] or by generating T cells to gender-restricted antigens (such as prostate or ovary) from peripheral blood cells of the opposite gender.[42] An alternative approach involves the engineering of high-affinity TCR by directed evolution of isolated tumor antigen-specific TCR-α/β chains using any one of a number of elegant genetic approaches, such as phage display libraries, or alternatively by strategies to enhance avidity by improving the steric behavior of the TCR.[42–46] Naturally, such efforts must be carefully controlled to ensure retained specificity for the target antigen,[47] and could theoretically introduce *neoepitopes* that could be targeted by the recipient's immune system, leading to rejection of the infused product. An additional drawback to TCR-based approaches is the potential for mis-pairing between the introduced TCR-α/β chains and the endogenous α/β chains, leading to decreased expression of the introduced TCR, or the generation of novel TCR complexes with undesired (autoreactive) specificities.

In the first clinical trial using the TCR transduction approach, melanoma patients were treated with gp100-specific T cells, and although there was no toxicity, the surface expression of the transgenic TCR was surprisingly low.[37] These complications could be averted by knockdown of the endogenous TCR chains or by introducing cysteine residues to facilitate preferential pairing between the transferred TCR chains.[48] Subsequent trials using melanoma antigen recognized by T cells-1 (MART1)-specific TCR-transgenic T cells for melanoma led to a 30% objective response in patients, along with off-tumor on-target toxicity manifesting as destruction of normal melanocytes in the skin, eye, and ear.[22]

A particularly instructive but unfortunate result of TCR engineering to enhance affinity is off-target toxicity, as has been demonstrated in two patients treated with T cells transduced with a high-affinity TCR to the cancer-testis antigen MAGE-A3. Both patients had lethal cardiac complications, retrospectively ascribed to cross-recognition by the transgenic TCR of a peptide derived from an unrelated muscle protein, titin,[49] accompanied by significant infiltration of CD3+ lymphocytes and diffuse myocyte necrosis (Figure 18.4).

CARs are essentially a chimeric TCR-antibody molecule, also known as a T body.[50] CARs are composed of an extracellular target-binding module, a transmembrane domain anchoring the CAR into the cell membrane, and an intracellular signaling domain. The target-binding domain is usually derived from scFv determinants isolated from antibodies linked in a single chain through a linker polypeptide sequence. The transmembrane domain is typically derived from the CD8 hinge or from the immunoglobulin-4 hinge regions. The intracellular region is usually composed of the ζ chain of the TCR complex and is responsible for transmitting the TCR engagement signal to the intracellular machinery. Recent studies established

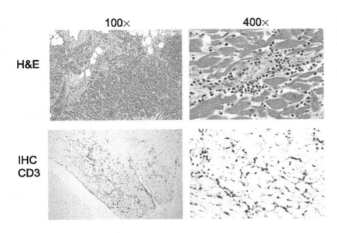

FIGURE 18.4 **Infiltration of CD3+ lymphocytes in myocardium biopsy tissue from two patients who received TCR-engineered lymphocytes.** Top panels: Hematoxylin–eosin. Lower panels: Immunohistochemical (IHC) staining with anti-CD3. This research was originally published in *Blood*. Linette et al.[49] © The American Society of Hematology.

that the incorporation of co-stimulatory or activating molecules such as CD28, CD134, or CD137 can augment ζ chain signaling with important effects on cell proliferation and response.[51]

Construction of a CAR relies on identification of a suitable antibody to a cell surface molecule of interest, as described below. Because CAR recognition does not rely on peptide processing or presentation by MHC, the number of target epitopes is stoichiometrically equal to the number of target antigen molecules on the cell surface, and every target molecule expressed on the membrane represents a potential CAR-triggering epitope, contributing to the potency of this approach. Conversely, a major limitation of CAR-based approaches is that only cell surface molecules can be targeted. Furthermore, because there are nonhuman sequences in the scFv domain of mouse antibodies, there is a potential for CAR T cells to be immunogenic.

18.3.2 Gene Transfer into T Cells

T cells can be genetically engineered at the DNA or RNA level. Scientific considerations include the size and number of transgenes to be transferred, and the requirement for permanent or transient expression. Practical considerations include the cost of manufacturing and whether clinical-grade vector systems are available. These considerations inform the ideal characteristics of the cellular product, which depend on the required setting. For example, parameters for the treatment of a life-threatening malignancy may differ from those for correction of a genetic defect or for the treatment of an incurable but manageable chronic disease. These approaches are summarized in Table 18.3.

18.3.3 Viral Approaches

Gamma retroviral or lentiviral vectors can integrate into the host genome leading to permanent transgene expression and have low intrinsic immunogenicity. These are therefore the most commonly used approaches. Gamma retrovirus-based transduction requires replicating cells for viral integration and retroviral transduction of hematopoietic stem cells has been associated with insertional oncogenesis[52]; in contrast, lentiviral vectors can integrate into nondividing cells, are less susceptible to silencing by host restriction factors, and can deliver larger DNA sequences than retroviruses.[13,53–55] Despite the concern about insertional mutagenesis when using integrating viral vectors, long-term follow-up of clinical studies supports the safety of using these vectors in T cells,[56] at least in part because lentiviral integration sites are not random and do not favor proto-oncogenes.[57] A new system that has not yet entered clinical trials is based on foamy virus vectors, which possess favorable integration properties and are not pathogenic in humans.[58] Where transient gene transfer is sufficient, nonintegrating viruses such as the adenovirus-based Ad5–35 vectors can be used and achieve high efficiency; an example of this application is in the transient expression of zinc-finger nucleases to permanently delete genes of interest.[59]

18.3.4 Nonviral Approaches

Nonviral approaches benefit from a lower manufacturing cost and are in principle less immunogenic than viral approaches. Although such approaches are perceived to be safer because they are not dependent on stable integration of transgene sequences into host DNA, their safety record is shorter than that of retro- or lentiviral vectors. DNA electroporation achieves efficient transient transgene expression but long-term expression is rare. A clinical trial with an anti-CD20 CAR demonstrated feasibility of production but poor engraftment and antitumor efficacy in the engineered cells.[39] Using transposon/hyperactive transposase technology such as the Sleeping Beauty system, it is possible to deliver a large payload with persistent high-level transgene expression and clinical testing is under way.[60–63]

18.3.5 RNA Electroporation

Electroporation of lymphocytes using in vitro transcribed RNA can lead to transient protein expression for up to 1 week and is associated with efficiencies approaching 100% with much lower toxicity than electroporation with DNA plasmids. RNA electroporation has been used to deliver messages for TCR or CAR, chemokine receptors, or cytokines.[64–68]

18.3.6 Novel Approaches

A number of novel technologies to disrupt specific target genomic sequences have been developed. Such technologies, which include innovative platforms based on zinc-finger nuclease and transcription activator-like effector nuclease (TALEN) technology, allow for the ability to engineer T cell populations that are bi-allelically disrupted for specific genes.[69,70] Although appropriate implementation of such technologies will require more complete understanding of gene activity in a systems biology-based context, such technologies offer a tremendous potential to interrupt T cell expression of particular proteins that may be deleterious to therapeutic function.

Efforts to optimize gene delivery must go hand in hand with optimized expansion techniques using GMP.[17,24,71–73] The parallel development by several different groups of multiple gene transfer methods provides flexibility in terms of our future ability to tailor different gene transfer modalities to the various treatment settings. It makes most sense to investigate novel therapeutics with untested toxicity profiles using T cells with transient expression, such as after RNA electroporation.[74] In contrast, when investigators have gained experience with the toxicity profile and behavior of a novel genetically engineered T cell product, a permanent gene transfer system such as lenti- or retroviral transduction may be employed.[75]

18.3.7 Definition of Appropriate Target Antigens

A major focus of immunotherapy-based strategies to target cancer has been the identification of antigens uniquely expressed or overexpressed by tumor cells and to use those antigens as immunogens to trigger antigen-specific T cell responses in patients.[7,76] A multitude of candidate antigens and immunologically relevant epitopes have been identified over the years.[77] It would be overly simplistic to expect that one or even a few candidate antigens might be expressed uniquely on tumors compared with normal tissue yet broadly expressed across tumor types; however, several candidate antigens have been identified that are either overexpressed or aberrantly expressed by tumors: in particular, antigens that are expressed developmentally (e.g., the cancer-testis antigens such as MAGE or NY-ESO-1) or dramatically overexpressed in tumors compared with normal tissues (such as Her-2/neu or EpCAM).[78,79]

Most results published to date have targeted cancer-testis antigens such as MAGE-A3 (myeloma and melanoma), NY-ESO1 (melanoma and synovial sarcoma),[35] CEA (colorectal carcinoma),[80] MART1, and gp100 (melanoma).[22] As mentioned, despite the perceived advantage that these targets are tumor-specific, results have clearly illustrated the complexity of target selection, with both on-target off-tumor and off-target effects reported.[22,49] These issues merit significant consideration and concern as the breadth of affinity-enhanced TCR in clinical evaluation increases.

As mentioned, the use of an antibody-derived, scFv-based CAR for T cell redirection is limited to the targeting of cell surface antigens. There are few instances of truly tumor-specific expression of a cell surface antigen, and one of the challenges of CAR-based T cell immunotherapy is to find a suitable antigen. Most clinical trials to date have focused on antigens such as CD19 or CD20 in which the antigen is specific to B cells and in which extensive experience with antibody-based B cell depleting therapy (for example, Rituximab) has shown that prolonged iatrogenic B cell depletion is well tolerated. Results from clinical studies that have targeted B cell lineage antigens have highlighted the considerable promise of CAR-based approaches, with strong clinical responses including long-lasting and ongoing remissions in a number of cases.[15,18,23,75] The use of differing CAR delivery systems (retrovirus, lentivirus, plasmid electroporation, or others), co-stimulatory domains (none, CD28, CD137, or a combination thereof), ex vivo expansion approaches (aAPCs or anti-CD3 antibody based) may in part account for the differing efficacy and toxicities observed. Beyond targeting B cell lineage surface antigens, there has been considerable interest in targeting solid tumors using CAR technology and clinical efforts to this end are under way (see below). Collectively, these studies indicate that the targets of CAR-based therapy must be carefully selected to avoid unwanted adverse effects.[81] An additional observation from collective experience with CART-19 therapy in B cell acute and chronic lymphoid leukemias is that serious toxicity can arise owing to tumor lysis or a cytokine release syndrome (CRS) as the result of the robust proliferation and activation of T cells.[15,75]

Based on the recent success of CAR trials, the next few years are likely to see a dramatic expansion of clinical trials targeting a broad range of malignancies. Each trial is likely to have to contend with potential toxicities specific to the target antigen and/or tissue. For example, trials of T cells redirected to acute myeloid leukemia antigens are likely to be associated with significant myeloid toxicity or granulocytopenia, attempts to target epithelial cancers are likely to be associated with tissue-specific epithelial toxicity, and a unique and creative solution will have to be found to circumvent T cell fratricide in any approach to the treatment of T cell malignancies with CAR T cells.

18.4 APPROACHES TO ENHANCE T CELL ACTIVITY

18.4.1 Selecting the Right T Cell Population

T cells for use in cellular therapy have been derived from peripheral blood[16,82,83] or the tumor bed[84–86] and have been infused fresh or more commonly after ex vivo stimulation or enrichment for virus specificity.[87,88] T cells exhibit many of the features required for adoptive cellular therapy, as defined above (Figure 18.1). After infection or immunization, antigen-specific T cells typically undergo expansion, plateau, and contraction phases. A long-standing objective of adoptive T cell therapy has been to recapitulate this vaccine effect with robust T cell expansion, antitumor activity, contraction, and long-term functional persistence of a memory T cell subset. During the effector phase, T cells employ multiple effector mechanisms that may be contact-dependent (granule based or death-receptor mediated cytotoxicity) or noncontact-dependent (cytokines). Polyfunctionality—that is, T cells exhibiting multiple different effector functions simultaneously—appears to correlate with enhanced in vivo function in infective models[89] and can also be demonstrated for engineered T cells.[90] After exerting their function, T cells acquire a memory phenotype with the capacity for recall response[91,92] and this persistence allows them to participate in the control of future infections or malignant relapse. Adoptively transferred T cells have now been empirically shown to persist long term as memory cells after a single infusion.[56,75]

GMP laboratory-based manipulation of the T cell product before infusion introduces the potential for further engineering of the product by selecting specific T cell subsets to infuse. Preclinical nonhuman studies have shown that adoptive transfer of younger, less differentiated CD8 T cells derived from the T_{CM} compartment leads to superior persistence compared with the transfer of more differentiated memory cells derived from the T_{EM} compartment.[93] However, whether this manipulation is superior to the transfer of T cells that have been expanded from a bulk population of cells has not been established. Notably, the phenotype of cells that have been expanded ex vivo tends to be homogeneous and consistent with activated, effector memory T cells.

Segregation of T cells into CD4 helper or CD8 cytotoxic, naive, central memory, effector memory, and other subtypes has allowed the assignment of a division of labor to the different subsets. Although it is tempting to try to select a particular subtype (for example, CD8 cytotoxic lymphocytes) for adoptive cellular therapy, there is a risk inherent in applying a reductionist approach to the immune system. For example, there is compelling evidence that CD4 T cells are required for CD8 memory formation, clearly an important quality in the control of tumor.[94,95] Nonetheless, animal studies have clearly shown that some T cell subtypes possess desirable qualities in vivo such as enhanced engraftment, antitumor effect, or survival, and have led to promising reports on the adoptive transfer of central memory T cells,[96–99] Th17 cells,[100] or so-called memory stem cells.[101] Cell culture conditions can be optimized to promote the expansion of T central memory cells.[102] Despite these interesting preclinical observations, there are as yet no published clinical comparisons of the infusions of T cell subsets alone or as a bulk population.

18.4.2 Enhancing T Cell Proliferation and Survival

Elimination of viral pathogens by T cells requires proliferation, expansion, and contraction, and is usually followed by long-term persistence of T cells in the form of memory cells for ongoing surveillance. Although most clinical trials using engineered T cells reported to date have been beset by poor persistence of the infused product and thus the failure of some trials could be the result of inherently poor proliferative capacity or rejection by the host of the transferred product,[99,103,105] a report documented long-term persistence of T cells engineered to target CD19, fundamentally recapitulating what occurs after a viral infection.[75] It is now well recognized that stimulation of T cells via their TCR without a second co-stimulatory signal induces tolerance. Thus, whereas first-generation CARs relied solely on intracellular transduction of the recognition signal via the CD3ξ-chain, second- and third-generation CAR constructs have incorporated a second signal in the form of CD28, CD134, or CD137 signaling domains. Such domains have been shown to enhance T cell persistence and/or function.[90,106–108] These co-stimulatory molecules are usually transduced in cis as part of the CAR complex; expression of co-stimulatory molecules in trans on the same cell but not in series with the receptor (whether chimeric or native) has also been performed successfully using CD28.[109] Ongoing efforts have focused on evaluation of signaling domains that have the potential to generate qualitatively unique T cell populations, such as signaling domains from inducible T cell costimulator (ICOS) that have the potential to drive T cells to differentiate into unique subsets. With few exceptions, the optimal co-stimulatory domain(s) have not been determined and, indeed, the optimal co-stimulatory signals to be provided may depend on the T cell expansion strategy, use of exogenous cytokines, and characteristics of the tumor model.[51] Figure 18.5 illustrates the stimulatory and inhibitory molecules involved in T cell responses. To date, only a minority of these molecules have been explored in terms of their potential role in enhancing genetically engineered T cell immunotherapy.

Since the discovery of IL-2 as T cell growth factor,[111] other gamma-chain cytokines have been shown to be important for T cell expansion and activation. IL-7 and IL-15 may have differential roles on T cell effector mechanisms and, indeed, on T cell polarization; immunosuppressive regulatory T cells are particularly sensitive to IL-2 stimulation owing to their high-level expression of CD25, the IL-2 receptor α chain.[112] There are few reports of exogenous administration of cytokines after transfer of redirected T cells,[37] but advanced molecular techniques allow the co-delivery of cytokines or cytokine receptor on T cells such that they are supercharged to proliferate. For example, transduction of the T cells with chimeric granulocyte macrophage–colony-stimulating factor (GM-CSF)-IL-2 receptors that deliver an IL-2 signal upon binding GM-CSF[113] leads to an autocrine IL-2-mediated proliferative response. Transduction of IL-15 into T cells appears to induce particularly robust proliferation that is autonomous of exogenous cytokines and is associated with constitutive telomerase activity and with clonal growth, hence with a potential for malignant transformation.[114] Therefore, despite the elegance of a genetic approach in which a bicistronic vector encodes both the transgenic immunoreceptor and autocrine cytokine provision, it

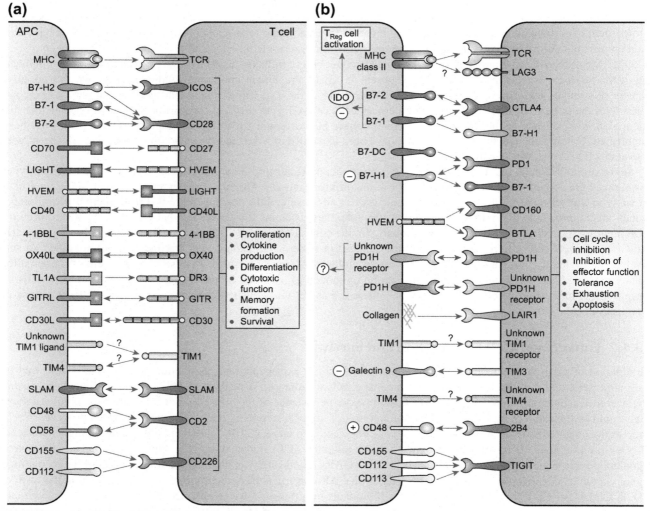

FIGURE 18.5   Co-signaling interactions at the lymphocyte-antigen presenting cell (APC)/target interphase. (a) Co-stimulation of T cells following interaction with counter-receptors on APCs. (b) Co-inhibition of T cells following interaction with counter-receptor on APCs. MHC, major histocompatibility complex; TCR, T cell receptor; *From Ref. 110, with permission.*

may be that exogenous, at-will administration of selected cytokines may be a safer option. ILs-7, -15 or -21 have been shown to have particularly attractive properties in stimulation of T cell proliferation in the non-transgenic setting, with preferential expansion of naive cells (IL-7),[115] CD8+ central memory cells (IL-15),[116] or an early memory phenotype (IL-21).[26]

Enhanced survival can be mediated by overexpression of pro-survival signals such as telomerase,[117] antiapoptotic genes such as Bcl2 or Bcl-x$_L$,[118,119] or the downregulation of proapoptotic molecules such as Fas.[120] Resistance to suppression via inhibitory receptors can be imparted via expression of a dominant negative transforming growth factor-β receptor,[121,122] or potentially a dominant negative PD1 receptor.

It is important to note however that any technique that leads to a significant uncoupling of T cell survival or proliferation from an antigen-specific signal may induce a risk of uncontrolled lympho-proliferation and thus parallel approaches must be developed to both monitor and deal with this potential unintended complication.

18.4.3 Trafficking and Homing to Tumor Tissue

Unless they encounter tumor during their initial journey through the bloodstream after intravenous injection, T cells must have a mechanism to traffic to the site of disease, arrest there, and accumulate by proliferation and persistence. The receptors and mechanisms for endogenous T cell trafficking have been identified and are probably similar for genetically engineered T cells.[123] Innovative work to transduce in chemokine receptors has allowed increased local accumulation and enhanced tumor rejection by T cells in published studies.[66,124] The identification of specific barriers to T cell trafficking into solid tumors will likely lead to further progress in this area.[125,126]

18.4.4 Overcoming Exhaustion and Host Immunosuppression

Early attempts to generate T cells for use in adoptive transfer depended on the use of prolonged ex vivo expansion protocols that resulted in the generation of terminal effector cells with poor ability to survive in vivo. More recent expansion approaches have resulted in the generation of T cells with the ability to functionally persist long-term in patients with phenotypic features of memory T cells and residual capacity to recognize and respond to tumor.[75] In contrast, antigen-specific T cells infused into melanoma patients showed high expression of inhibitory receptors and low proliferative capacity, which likely explain the inability of the cells in this study to mediate regression of melanoma.[98] Careful study of the mechanisms of failure as well as success of T cell-based adoptive therapy trials will lead to rational problem-solving; for example, antibody-based or genetic blockade of inhibitory signals may enhance the function of infused T cells. This remains an important area of active investigation with the potential to lead to strategies that overcome peripheral tolerance and immune suppression.

18.4.5 Avoiding Rejection by the Native Immune System

Ex vivo manipulation of the T cell product to enhance its potency has the potential to make the product more immunogenic after transfer. Use of serum-free cell culture techniques has reduced the potential immunogenicity associated with the use of xenobiotic sera, but there is still the potential for nonself-translated open reading frames present in vector sequences leading to rejection of infused cells. This phenomenon has been demonstrated in a number of cases.[20,127] Similarly, because most CARs are derived from murine antihuman scFV, there is the potential for development of both humoral and cellular responses against CAR sequences or epitopes from the retroviral vector backbone and these responses correlate with rejection and disappearance of the infused cells from the circulation.[128,129]

18.5 MITIGATION OF ADVERSE EVENTS AND SAFETY CONSIDERATIONS

Clinical adverse events related to the infusion of T cells may be caused by nonspecific infusion reactions, unanticipated cross-specificity or recognition of low levels of target antigen on normal tissues, or genomic insertional events as a result of vector integration. Even the transfer of autologous non-modified lymphocytes can lead to infusion reactions related to robust cytokine release or the presence of trace contaminant carryover from the manufacturing process, with occasional tissue-specific immune reactions noted, such as vitiligo, thyroiditis, or colitis.[83] It is therefore not surprising that the transfer of tumor-specific T cells could lead to infusion reactions and a CRS. Patients across several clinical studies have experienced this complication,[15,18,23,130] manifesting high fevers, hypotension, and hypoxemia and laboratory evidence of macrophage activation. The severity of the reaction may be associated with the magnitude of subsequent response. Transgene-related, on-target toxicity occurs as a result of recognition of the target molecule by modified T cells. Well-documented instances of this phenomenon include the recognition of MART1-expressing skin and eye tissues in patients receiving MART1-specific TCR-transgenic T cells for the treatment of melanoma as well as severe liver toxicity in patients receiving carbonic anhydrase IX-redirected T cells for the treatment of kidney cancer.[37,131]

These instances of on-target off-tissue toxicity parallel the ablation of normal B cells that is seen after the infusion of anti-CD19 gene-modified T cells.[75] Engineering mechanisms to eliminate T cells or suppress T cell function on demand is a critically important and active area of investigation, and is of particular importance because of the observation that high-dose steroids do not adequately reverse a florid CRS.[130] Inhibition of IL-6 signaling appears to be particularly useful for ameliorating CRS.[130] Elimination of introduced haploidentical T cells after hematopoietic cell transplantation can be achieved in the setting of GVHD[132] using an inducible caspase-9 system that allows the conditional delivery of apoptotic signals to the transduced cells, representing an elegant improvement on the previous suicide gene methodologies.[133,134] Selected methods to eliminate introduced T cells are outlined in Table 18.4.

TABLE 18.4 Selected Methods to Eliminate Introduced T Cells

System	Principle	Advantage	Disadvantage	Used in Transgenic T Cell Trials?
Routine Clinical Practice for T Cell Depletion				
Steroids	Induction of T cell apoptosis, inhibition of proliferation and reduction of IL-2 production	Well studied and readily availably	Insufficiently active against highly activated T cells	15
ATG	Polyclonal anti-T cell antibody leads to T cell apoptosis and elimination by immune system	Well studied and readily available	Nonspecific depletion of all T cells; prolonged half-life; infusion reactions; immuno-suppressive	No
Alemtu-zumab	Monoclonal anti-CD52 antibody depletes T cells and other immune effector cells	Well-studied and readily available	Nonspecific depletion of all T cells, B cells, NK cells; prolonged half-life; infusion reactions; immunosuppressive	No
Suicide Gene Systems				
HSV TK	Gene for herpes simplex virus thymidine kinase is transduced into T cell product, making T cells susceptible to subsequent administration of antiviral drug ganciclovir	Well studied system; ganciclovir is readily available	Subject to anti-transgene immune response and subsequent elimination of infused T cell product; less active in nonproliferating cells	20,127,135
iCasp9	Gene for caspase-9 is fused to modified human FK-binding protein that is activated upon exposure to synthetic dimerizing drug. This initiates a caspase cascade downstream of initial mitochondrial apoptosis pathway	Rapid, profound T cell elimination within minutes to hours	Incomplete eradication of T cells may occur, requiring repeated administration	132,133
Fas	Gene for intracellular death signaling domain of Fas is fused to modified human FK-binding protein that is activated upon exposure to a synthetic dimerizing drug	Rapid elimination within hours	Insufficient transgene expression in some resting cells leads to failure of elimination of small fraction of T cells	136
Antibody-Based				
CD20	Extracellular portion of CD20 is incorporated into gene-modified T cells, rendering them susceptible to elimination by clinically used anti-CD20 antibodies	Requires addition of small sequence encoding extracellular portion of CD20 receptor; can be used as selection marker	Relies on presence of intact immune system for elimination; kinetics may be slower than that of suicide genes	137,138
EGFR	Extracellular portion of EGF receptor is incorporated into gene-modified T cells, rendering them susceptible to elimination by clinically used anti-EGFR antibodies (Erbitux)	Requires addition of small sequence encoding extracellular portion of EGF receptor; can be used as selection marker	Relies on presence of intact immune system for elimination; kinetics may be slower than that of suicide genes	139

IL-2, interleukin-2.

18.6 TRANSLATION OF ENGINEERED T CELL THERAPY TO THE CLINIC

A summary of selected clinical studies of CAR and TCR-based T cell therapy for cancer is presented in Table 18.5.

18.6.1 Optimizing Preconditioning and Infusion Schedule

Both clinical and preclinical data suggest that pre-infusion lympho-depletion enhances the proliferation and survival of adoptively transferred T cells.[18,145] Lympho-depletion is thought to create homeostatic space, deplete cytokine sinks (in the form of other immune cells competing for the cytokines), and increase homeostatic cytokines produced by the inflammatory reaction to chemo- or radiotherapy; these changes lead to the induction of a memory phenotype after homeostatic proliferation of naive T cells and to enhanced effector function.[85,146,147] No clinical trials compare outcomes after genetically engineered T cell infusion following lympho-depletion or no lympho-depletion, or whether particular chemotherapeutic or radiation regimens should be used. Some data suggest that some regimens can preferentially induce immunogenic cell death,[148,149] and the inclusion of radiation has been found to induce higher levels of the homeostatic cytokine IL-15.[150] Whether T cells should be infused in one dose or in a fractionated schedule is currently unknown, although data from the bone marrow transplantation literature suggest that the latter is safer; for practical reasons, fractionation provides the opportunity to withhold doses in case of severe infusion-related toxicity.[151,152]

18.6.2 Response Assessment and Biomarkers of Activity

The principal questions to answer in cancer therapy are ultimately disease response and toxicity of treatment. Beyond those measures, critically important information about treatment can be derived by employing broad, quality supported, and integrateable platforms to evaluate patient samples.[153] Unlike traditional pharmaceuticals, T cell products are viable biological entities that can expand in vivo, and over time manifest variable phenotypes and functional activities and potentially unpredictable pharmacokinetics. Therefore, T cell immunotherapy trials require the development and evaluation of biomarker platforms and assays that describe the biological properties of the cell product. These classes can be broadly divided into (1) quantity of infused cells, (2) phenotype and functional competence, (3) impact on patient biology, and (4) recipient's immune response to the infused product.[154] Peripheral blood is clearly the most accessible tissue for these assays, but it may not reflect the true T cell number or activity at the level of the targeted tissue; ideally the involved site should be sampled.[155]

Quantitative assessment of the presence of infused T cells relies on flow cytometric antibody-mediated detection of a unique surface molecule. This molecule can be the idiotype of the scFv used to create a CAR, MHC class I multimers to detect TCR transgenic T cells of known specificity, TCR-β chain family antibody panels, or a reporter molecule introduced as part of the transgenic receptor.[75,154,156] Reliable flow cytometric detection requires that the frequency of the cells of interest be at least 0.2–0.5% of total T cells. An important caveat of this methodology is that activation-induced internalization of the TCR or CAR complex can lead to underestimation of the number of genetically modified T cells. Genetic engineering of the T cell product introduces transgenes or selection markers that can be detected by qualitative polymerase chain reaction (such as the TCR, CAR, or HSV thymidine kinase) and has been used in a number of clinical trials.[20,75] Although polymerase chain reaction is considerably more sensitive than flow cytometry, this approach lacks the power of flow cytometry to interrogate single cells for additional phenotypic or functional markers. The infused product may represent a small fraction of the total patient T cell load (perhaps 0.1% or less), but homeostatic or antigen-driven proliferation can lead to substantial in vivo expansion.

Detection of cell surface expression of CAR or TCR allows a direct evaluation by multicolor flow cytometry of markers of T cell differentiation, activation, and functional status. For example, memory T cells demonstrate enhanced persistence and have been shown to be associated with efficacy.[93,157,158] Other markers of differentiation such as CD45RA, CD45RO, CD62L, CCR7, CD27, and CD28, as well as activation markers such as CD25, CD127, CD57, and CD137 and inhibitory markers such as PD-1 and CTLA4, can be used to gain insight into the functional capacity of the infused T cells.[159,160] The importance of persistence and the establishment of memory are highlighted by relevant clinical observations and may be associated with telomere length and with particular T cell memory subsets.[161,162] More direct measures of T cell functional capacity can be detected by analyzing production of cytokines such as interferon (IFN)-γ, TNF-α, IL-2, degranulation by CD107a mobilization, and the levels of perforin and granzyme.[163–165] Efforts to validate and optimize strategies to evaluate the phenotype and functionality of antigen-specific T cells are ongoing.[166] It may be that upstream detection of the phosphorylation of key signaling molecules is equally important in detecting T cell activation.[167] These and other assays may be performed on serial blood samples from the patient infused with the T cell product and will likely inform our future ability

TABLE 18.5 A Summary of Selected Clinical Studies of CAR and TCR-Based T Cell Therapy for Cancer

Target Antigen	Patients (n)	CAR or TCR/ Delivery System/ Co-stimulatory Molecules/ Other	Lympho-depletion	Efficacy	Toxicity	Comments
Carbonic Anhydrase (CAIX)						
Lamers[131]/ Lamers[129]	Renal cell carcinoma (11)	CAR/retroviral/ none	No	No objective responses Limited persistence	Hepatotoxicity owing to expression of CAIX on biliary epithelium	Humoral and/or cellular responses against CAIX CAR and/or vector epitopes in most patients
CD19						
Jensen[19] (CD19 or CD20)	Diffuse large B cell lymphoma or follicular lymphoma (4)	CAR/retroviral/ none/thymidine kinase suicide gene	Yes (treated 3–4 weeks after autologous transplant or chemotherapy)	No objective responses; progression in all patients Limited persistence	Grade 3 lymphopenia in 2 patients	Cellular anti-transgene responses in 2 patients
Porter[15]/ Kalos[31]	CLL (3)	CAR/ lentirviral/4-1BB	Yes	CR (2), PR (1)	Cytokine storm (fever, hypotension)	Complete response persists at >2 years (MK, unpublished)
Brentjens[18]	CLL (8) ALL (2)	CAR/retroviral/ CD28	4 CLL with lympho-depletion	Response correlated with lympho-depleting chemotherapy Best response was marked LN reduction for 3 months	Fever, chills, hypotension	Most responses were stable disease; persistence enhanced in patients receiving lympho-depleting chemotherapy
Savoldo[140]	B-NHL (6)	CAR/retroviral[a]	No	No objective responses	None reported	Persistence was limited to T cells transduced with CD28-containing construct
Kochenderfer[23]	B-NHL (8)	CAR/retroviral/ CD28	Yes	PR (5), CR (1), SD (1)	Fever, hypotension, renal failure	Long-term persistence (2)
Grupp[130]	ALL (2)	CAR/lentiviral/ 4-1BB	Yes	CR (2)	Severe cytokine storm	CR 2 years (1) Relapse with CD19-leukemia (1)
Brentjens[141]	ALL (5)	CAR/retroviral/ CD28	Yes	MRD	Hypoxia, hypotension	NA (went on to allo-hematopoietic cell transplanta-tion) (4) Relapse at 1.5 months (1)
CD20						
Till[39]	B-NHL (7)	CAR/plasmid electroporation/ none	No	PR (1), SD (4)	Grade 1–2 toxicity associated with IL-2 administration	Persistence up to 9 weeks

TABLE 18.5 A Summary of Selected Clinical Studies of CAR and TCR-Based T Cell Therapy for Cancer—cont'd

Target Antigen	Patients (n)	CAR or TCR/ Delivery System/ Co-stimulatory Molecules/ Other	Lympho-depletion	Efficacy	Toxicity	Comments
CEA						
Parkhurst[80]	Colorectal cancer (3)	TCR/retroviral/ NA	Yes	PR lasting 6 months (1)	Severe transient colitis in all patients	Development of HAMA
ErbB2						
Morgan[142]	Colorectal cancer (1)	CAR/retroviral/ CD28 and 4-1BB	Yes	NA	Fatal, early lung inflammation	Recognition of normal lung cells expressing target
Folate Receptor						
Kershaw[103]	Ovarian cancer (14); 6/14 received allogeneic cells	CAR/retroviral/ none	No	No objective responses	Toxicity mostly associated with IL-2 injection	Limited persistence and lack of trafficking of CART cells to tumor
GD2						
Pule[38,143]	Neuroblas-toma (26)	CAR/retroviral/ none	No	CR in 3/11 patients with active disease	Pain at site of tumor deposits	Long-term persistence noted; both activated T cells and EBV-specific T cells were used
Lewis Y						
Ritchie[144]	Acute myeloid leukemia (4)	CAR/retroviral/ CD28	Yes	Stable disease (2) Transient response (2)	Fevers and rigors (1)	Relapse in 3 of 4 within 1–5 months
L1 Adhesion Molecule						
Park[104]	Neuroblastoma (6)	CAR/retroviral/ none/thymidine kinase suicide gene	No	PR (1/6)	Cytopenias, pneumonitis	Limited persistence in 5 of 6 patients
MART1, gp100						
Morgan[37]/ Johnson[22]	Melanoma (43)	TCR/retroviral/ NA	Yes	PR (8), CR (1)	Destruction of melanocytes in skin, eye, and ear	Long-term persistence in all
NY-ESO1						
Robbins[35]	Synovial cell sarcoma; Melanoma (22)	TCR/retroviral/ NA	Yes	Objective responses, (9) including CR (2)	None	

CAR, chimeric antigen receptor; CR, complete response; EBV, Epstein–Barr virus; HAMA, human anti-mouse antibodies; LN, lymph node; NA, not applicable; PR, partial response; SD, stable disease.
[a]Two patients received two different T cell products, one of which had been transduced with an anti-CD19 CAR alone, and the other with an anti-CD19 CAR along with the CD28 co-stimulatory molecule.

to prospectively define the salient features on characterization and release testing of manufactured product before infusion into patients.

Insights about product bioactivity can be obtained by evaluating the impact of the treatment on patient biology. A classic example is the vitiligo and ocular toxicity associated with the destruction of normal melanocytes after melanoma T cell therapy.[168,169] In the case of genetically engineered T cells, serious or fatal adverse effects have been noted.[130,142,170] These highlight the challenges of potent T cell-based therapies and underscore the need to identify and develop early biomarker signatures to predict and track pathologic consequences of T cell reactivity before they become clinically relevant. For instance, the development of severe cytokine storm (elevation of IFN-γ, TNF-α, GM-CSF, IL-6, and IL-10) after T cell infusion was detectable by multiplex serum cytokine assessment and associated with severe clinical illness[142] and delayed tumor lysis correlated with robust in vivo expansion of CAR T cells with dramatic and transient increases in serum proinflammatory cytokines along with rapid clinical response.[15,130,154] Even broader analyses of T cell bioactivity can be performed using molecular array and proteomics-based approaches[171,172]; evaluation of the role of these methodologies in T cell immunotherapy biomarker discovery is a nascent and exciting field.

The observation that a patient with B cell acute lymphoblastic leukemia relapsed with CD19 disease after anti-CD19-redirected T cell therapy simultaneously highlights both the power of this technology and a potential drawback: Antigen-loss escape mutants can occur in rapidly proliferating malignant cells under the specific selective immune pressure of the engineered T cell.[130] This concept is familiar to treating oncologists who rarely, if ever, use a single drug to treat cancer, and presages the development of combination therapies employing multiple targets for T cell therapy.

18.7 CONCLUSIONS AND FUTURE DIRECTIONS

The significant potential of T cell-based gene therapy as an effective approach to target cancer is beginning to be realized. However, a number of important challenges should be addressed before the broader application of these approaches.

For CAR-based approaches, an important issue still to be resolved involves the nature and complexity of a co-stimulatory molecule or molecules to be included in the construct. Most clinical studies have employed the TCR-ξ chain alone or with CD28 or CD137, because these co-stimulatory molecules have shown increased potency in animal models.[51,90,173] However, it is not yet known which co-stimulatory domain (or combination thereof) would be expected to deliver the optimal activating and pro-proliferation signal. A study goes some way toward this by directly comparing two constructs with different signaling domains, demonstrating that CD28 co-stimulation induces superior expansion and persistence compared with no co-stimulatory domain.[140] This approach provides a powerful and compelling mechanism to evaluate CAR with related but unique structural determinants to identify modules that mediate superior homing or other functionally relevant properties to impart to genetically engineered T cells. Although CAR constructs with 4-1BB and TCR-ξ signaling domains have demonstrated remarkable efficacy to target leukemias, it may be that the targeting of solid tumors will require additional yet-undefined signaling functions.

The observation that a relatively small dose of CART-19 cells could eradicate pounds of tumor suggests that genetically engineered T cells can function as serial killers.[15] The mechanism by which T cells first bind with high affinity to their target, then disengage and move on to a second and third target is at present unknown but is important to understand to facilitate engineering of these properties into T cell products. Other important biologic insights include the resistance or susceptibility of genetically engineered T cells to the induction of anergy, apoptosis, or activation-induced cell death and their relative sensitivity to inhibition by the immunosuppressive tumor microenvironment.[174,175]

The use of murine antihuman scFv to generate CAR constructs may lead to clinically significant human–anti-mouse antibodies (HAMA) with subsequent rejection of the infused engineered T cells. There are reports of HAMA formation after CAR-based T cell therapy[176] and patients may need to be specifically immunosuppressed (as, for example, is the case with many patients receiving anti-CD19 T cell therapy, whose humoral immune systems are impaired at baseline) or that in future greater effort should be made to employ humanized antibodies.

With regard to TCR-based approaches, reports on rapid cloning and functional testing of TCR receptor chains from T cells with biologically relevant antitumor activity should enable rapid progress, and the application of these technologies to tumor-infiltrating T cells may allow us to identify relevant and useful TCR specificities for novel tumor antigens.[177,178]

To date, the optimal methodology for delivering genetic material into T cells has not been determined. Both retroviral and lentiviral technologies result in high transduction efficiency, but clinical-grade production is costly and time-consuming and they are limited by the total size of DNA that can be included in the viral vectors, as well as by the theoretical potential for insertional oncogenesis. An exciting alternative to virus-based approaches is RNA electroporation, which is cost-effective and efficient, and allows for the co-transfer of genes that promote co-stimulation, homing, accessory effector function, or combinatorial targeting of more than one antigen.[179] In addition, because the transferred RNA does not integrate stably but rather remains episomal, the transferred specificity is temporary, limited to the biological half-life of RNA, and thus represents an important safety feature in the case of unanticipated toxicity.

Finally, one of the most important issues to resolve is that of off-tumor on-target and off-tumor off-target specificity. This issue is practically impossible to address in preclinical studies, because animal models are generally uninformative for species-specific toxicity and furthermore toxicity has the potential to be patient-specific. Comprehensive repositories of normal human tissue could be used in functional screening for off-target toxicities by genetically engineered T cells. The design of innovative molecular switches to modulate T cell function on demand will likely be instrumental in effectively dealing with this ongoing issue.

The past decade of progress has provided fascinating observations and a tantalizing view of the future of T cell-based therapy. Results from CAR-based clinical trials at the University of Pennsylvania unequivocally demonstrated that the potential for engineered T cells to mediate dramatic antitumor activity is a reality; successfully treated patients remain in sustained complete responses for more than 3 years after treatment. CAR-based clinical trials at other institutions provide independent evidence to further support the promise of CAR-based approaches.[180] In parallel, engineered T cell strategies are also being developed to target viral disease, most notably HIV, with promising early clinical results reported using T cells engineered to express HIV-inhibitory elements[181] or T cells engineered to be genetically edited for the CCR5 locus.[182] The field is now poised for larger trials with sufficient power to rigorously test these observations, as well as for the development of more industrialized strategies to support the commercialization of these approaches.[183] We expect that the coming decade will provide much excitement, fresh insights, and new hope in our quest to successfully apply adoptive T cell therapy to treat disease.

DISCLOSURE OF POTENTIAL CONFLICTS OF INTEREST

The authors have read the journal's policy on disclosure of potential conflicts of interest.

MK is named on patent applications and patents pertaining to CAR-based technologies, including CART-19. SG is named on patent applications pertaining to CART-123 technology.

This chapter conforms to the relevant ethical guidelines for human and animal research.

GLOSSARY

Allogeneic stem cell transplantation Transfer of healthy hematopoietic stem cells from a normal donor into a recipient by intravenous injection to reconstitute a healthy immune and blood-forming system. Also known as bone marrow transplantation, this is a clinical procedure that is used to treat some patients with leukemia.

CD3 A pan-T cell antigen that is part of the normal T cell receptor complex. Ligation of CD3 using activating antibodies leads to nonspecific T cell activation.

CD28 An activating molecule on the surface of T cells. Its natural ligands are CD80 and CD86 on the surface of antigen presenting cells.

Good manufacturing practices (GMP) A set of principles or guidelines to ensure that pharmaceutical products are safe and of high quality, and include hygiene, controlled environmental conditions, standard operating procedures, consistency, clear labeling, record taking and annotation, and the ability to track specific batches.

Leukapheresis Removal of large numbers of white blood cells from the circulation by placement of a large-bore intravenous cannula in one vein, allowing the diversion of blood flow through a machine that separates the blood components using centrifugal forces, followed by a return to the patient of the remaining constituents of the blood.

Major histocompatibility molecule (MHC) A set of cell surface molecules on the surface of all cells (MHC class I) or on antigen presenting cells (MHC class II). T cells recognize antigenic peptides that are presented on the MHC molecules.

Myeloablative chemoradiotherapy Treatment of the patient with high doses of chemotherapy and/or radiotherapy, leading to irreversible suppression of the blood-forming ability of the marrow.

Neoepitope A new epitope: that is, a region of an antigen that is generated by modification of the original antigen. This can lead to recognition of a self-antigen as foreign.

REFERENCES

1. Gill S, Porter DL. Reduced-intensity hematopoietic stem cell transplants for malignancies: harnessing the graft-versus-tumor effect. *Annu Rev Med* 2013;**64**:101–17.
2. Baitsch L, Baumgaertner P, Devêvre E, et al. Exhaustion of tumor-specific CD8+ T cells in metastases from melanoma patients. *J Clin Invest* 2011;**3**(20):23–5. http://dx.doi.org/10.1172/JCI46102DS1.
3. Galon J, Costes A, Sanchez-Cabo F, et al. Type, density, and location of immune cells within human colorectal tumors predict clinical outcome. *Science* 2006;**313**(5795):1960–4. http://dx.doi.org/10.1126/science.1129139.

4. Lustgarten J, Dominguez AL, Cuadros C. The CD8+ T cell repertoire against Her-2/neu antigens in neu transgenic mice is of low avidity with anti-tumor activity. *Eur J Immunol* 2004;**34**(3):752–61. http://dx.doi.org/10.1002/eji.200324427.

5. Zhang L, Conejo-Garcia JR, Katsaros D, et al. Intratumoral T cells, recurrence, and survival in epithelial ovarian cancer. *N Engl J Med* 2003;**348**(3):203–13. http://dx.doi.org/10.1056/NEJMoa020177.

6. Bendandi M. Idiotype vaccines for lymphoma: proof-of-principles and clinical trial failures. *Nat Rev Cancer* 2009;**9**(9):675–81. http://dx.doi.org/10.1038/nrc2717.

7. Kalos M. Tumor antigen-specific T cells and cancer immunotherapy: current issues and future prospects. *Vaccine* 2003;**21**(7–8):781–6. Available at: http://www.ncbi.nlm.nih.gov/pubmed/12531359.

8. Tanimoto T, Hori A, Kami M. Sipuleucel-T immunotherapy for castration-resistant prostate cancer. *N Engl J Med* 2010;**363**(20):1966. http://dx.doi.org/10.1056/NEJMc1009982#SA1. author reply 1967–8.

9. Pardoll DM. The blockade of immune checkpoints in cancer immunotherapy. *Nat Rev Cancer* 2012;**12**(4):252–64. http://dx.doi.org/10.1038/nrc3239.

10. Law T, Motzer R, Mazumdar M. Phase III randomized trial of interleukin-2 with or without lymphokine-activated killer cells in the treatment of patients with advanced renal cell carcinoma. *Cancer* 1995;**76**(5):824–32.

11. Rosenberg S, Lotze M, Muul L, et al. Observations on the systemic administration of autologous lymphokine-activated killer cells and recombinant interleukin-2 to patients with metastatic cancer. *N Engl J Med* 1985;**313**(23):1485–92. http://dx.doi.org/10.1056/NEJM198512053132327.

12. Rosenberg SA, Dudley ME. Adoptive cell therapy for the treatment of patients with metastatic melanoma. *Curr Opin Immunol* 2009;**21**(2):233–40. http://dx.doi.org/10.1016/j.coi.2009.03.002.

13. Amado RG, Chen ISY. Lentiviral vectors – the promise of gene therapy within reach? *Science* 1999;**285**(5428):674–6. http://dx.doi.org/10.1126/science.285.5428.674.

14. Mackinnon BS, Papadopoulos EB, Carabasi MH, et al. Adoptive immunotherapy evaluating escalating doses of donor leukocytes. *Blood* 1995;**86**:1261–8.

15. Porter DL, Levine BL, Kalos M, Bagg A, June CH. Chimeric antigen receptor-modified T cells in chronic lymphoid leukemia. *N Engl J Med* 2011;**365**(8):725–33. http://dx.doi.org/10.1056/NEJMoa1103849.

16. Porter DL, Levine BL, Bunin N, et al. A phase 1 trial of donor lymphocyte infusions expanded and activated ex vivo via CD3/CD28 costimulation. *Blood* 2006;**107**(4):1325–31. http://dx.doi.org/10.1182/blood-2005-08-3373.

17. Hollyman D, Stefanski J, Przybylowski M, et al. Manufacturing validation of biologically functional T cells targeted to CD19 antigen for autologous adoptive cell therapy. *J Immunother* 2009;**32**(2):169–80. http://dx.doi.org/10.1097/CJI.0b013e318194a6e8.

18. Brentjens RJ, Rivière I, Park JH, et al. Safety and persistence of adoptively transferred autologous CD19-targeted T cells in patients with relapsed or chemotherapy refractory B-cell leukemias. *Blood* 2011:4817–28. http://dx.doi.org/10.1182/blood-2011-04-348540.

19. Micklethwaite KP, Savoldo B, Hanley PJ, et al. Derivation of human T lymphocytes from cord blood and peripheral blood with antiviral and anti-leukemic specificity from a single culture as protection against infection and relapse after stem cell transplantation. *Blood* 2010;**115**(13):2695–703. http://dx.doi.org/10.1182/blood-2009-09-242263.

20. Jensen MC, Popplewell L, Cooper LJ, et al. Antitransgene rejection responses contribute to attenuated persistence of adoptively transferred CD20/CD19-specific chimeric antigen receptor redirected T cells in humans. *Biol Blood Marrow Transplant* 2010;**16**(9):1245–56. http://dx.doi.org/10.1016/j.bbmt.2010.03.014.

21. Jensen MC, Clarke P, Tan G, et al. Human T lymphocyte genetic modification with naked DNA. *Molr Ther* 2000;**1**(1):49–55. http://dx.doi.org/10.1006/mthe.1999.0012.

22. Johnson LA, Morgan RA, Dudley ME, et al. Gene therapy with human and mouse T-cell receptors mediates cancer regression and targets normal tissues expressing cognate antigen. *Blood* 2009;**114**(3):535–46. http://dx.doi.org/10.1182/blood-2009-03-211714.

23. Kochenderfer JN, Dudley ME, Feldman SA, et al. B-cell depletion and remissions of malignancy along with cytokine-associated toxicity in a clinical trial of anti-CD19 chimeric-antigen-receptor-transduced T cells. *Blood* 2012;**119**:2709–20. http://dx.doi.org/10.1182/blood-2011-10-384388.

24. Maus MV, Thomas AK, Leonard DGB, et al. Ex vivo expansion of polyclonal and antigen-specific cytotoxic T lymphocytes by artificial APCs expressing ligands for the T-cell receptor, CD28 and 4-1BB. *Nat Biotechnol* 2002;**20**(2):143–8. http://dx.doi.org/10.1038/nbt0202-143.

25. Ye Q, Loisiou M, Levine BL, et al. Engineered artificial antigen presenting cells facilitate direct and efficient expansion of tumor infiltrating lymphocytes. *J Transl Med* 2011;**9**(1):131. http://dx.doi.org/10.1186/1479-5876-9-131.

26. Singh H, Figliola MJ, Dawson MJ, et al. Reprogramming CD19-specific T cells with IL-21 signaling can improve adoptive immunotherapy of B-lineage malignancies. *Cancer Res* 2011;**71**(10):3516–27. http://dx.doi.org/10.1158/0008-5472.CAN-10-3843.

27. Levine B, Mosca J, Riley J, et al. Antiviral effect and ex vivo CD4+ T cell proliferation in HIV-positive patients as a result of CD28 costimulation. *Science* 1996;**272**:1939–43.

28. Bendle GM, Linnemann C, Hooijkaas AI, et al. Lethal graft-versus-host disease in mouse models of T cell receptor gene therapy. *Nat Med* 2010;**16**(5):565–70. http://dx.doi.org/10.1038/nm.2128.

29. Ahmadi M, King JW, Xue S, et al. CD3 limits the efficacy of TCR gene therapy in vivo. *Blood* 2011;**118**:3528–37. http://dx.doi.org/10.1182/blood-2011-04-346338.

30. June CH. Science in medicine adoptive T cell therapy for cancer in the clinic. *J Clin Invest* 2007;**117**(6):1466–76. http://dx.doi.org/10.1172/JCI32446.1466.

31. Kalos M. Muscle CARs and TcRs: turbo-charged technologies for the (T cell) masses. *Cancer Immunol Immunother* 2012;**61**(1):127–35. http://dx.doi.org/10.1007/s00262-011-1173-5.

32. Clay TM, Custer MC, Sachs J, Hwu P, Rosenberg SA, Nishimura MI. Efficient transfer of a tumor antigen-reactive TCR to human peripheral blood lymphocytes confers anti-tumor reactivity. *J Immunol* 1999;**163**(1):507–13. Available at: http://www.ncbi.nlm.nih.gov/pubmed/10384155.

33. Cooper LJN, Kalos M, Deborah A, Riddell SR, Greenberg PD, Lewinsohn DA. Transfer of specificity for human immunodeficiency virus type 1 into primary human T lymphocytes by introduction of T-cell receptor genes. *J Virol* 2000;**74**(17):8207–12. http://dx.doi.org/10.1128/JVI.74.17.8207-8212.2000.

34. Kalos M, June CH. Adoptive T cell transfer for cancer immunotherapy in the era of synthetic biology. *Immunity* 2013;**39**:49–60.

35. Robbins PF, Morgan RA, Feldman SA, et al. Tumor regression in patients with metastatic synovial cell sarcoma and melanoma using genetically engineered lymphocytes reactive with NY-ESO-1. *J Clin Oncol* 2011;**29**(7):917–24. http://dx.doi.org/10.1200/JCO.2010.32.2537.

36. Varela-Rohena A, Molloy PE, Dunn SM, et al. Control of HIV-1 immune escape by CD8 T cells expressing enhanced T-cell receptor. *Nat Med* 2008;**14**(12):1390–5. http://dx.doi.org/10.1038/nm.1779.

37. Morgan RA, Dudley ME, Wunderlich JR, et al. Cancer regression in patients after transfer of genetically engineered lymphocytes. *Science* 2006;**314**(5796):126–9. http://dx.doi.org/10.1126/science.1129003.

38. Pule MA, Savoldo B, Myers GD, et al. Virus-specific T cells engineered to coexpress tumor-specific receptors: persistence and antitumor activity in individuals with neuroblastoma. *Nat Med* 2008;**14**(11):1264–70. http://dx.doi.org/10.1038/nm.1882.

39. Till BG, Jensen MC, Wang J, et al. Adoptive immunotherapy for indolent non-Hodgkin lymphoma and mantle cell lymphoma using genetically modified autologous CD20-specific T cells. *Blood* 2008;**112**(6):2261–71. http://dx.doi.org/10.1182/blood-2007-12-128843.

40. Jena B, Dotti G, Cooper LJN. Review article redirecting T-cell specificity by introducing a tumor-specific chimeric antigen receptor. *Blood* 2010;**116**(7):1035–44. http://dx.doi.org/10.1182/blood-2010-01-043737.

41. Amir AL, van der Steen DM, van Loenen MM, et al. PRAME-specific Allo-HLA-restricted T cells with potent antitumor reactivity useful for therapeutic T-cell receptor gene transfer. *Clin Cancer Res* 2011;**17**(17):5615–25. http://dx.doi.org/10.1158/1078-0432.CCR-11-1066.

42. Friedman RS, Spies AG, Kalos M. Identification of naturally processed CD8 T cell epitopes from prostein, a prostate tissue-specific vaccine candidate. *Eur J Immunol* 2004;**34**(4):1091–101. http://dx.doi.org/10.1002/eji.200324768.

43. Chervin AS, Aggen DH, Raseman JM, Kranz DM. Engineering higher affinity T cell receptors using a T cell display system. *J Immunol Methods* 2008;**339**(2):175–84. http://dx.doi.org/10.1016/j.jim.2008.09.016.

44. Li Y, Moysey R, Molloy PE, et al. Directed evolution of human T-cell receptors with picomolar affinities by phage display. *Nat Biotechnol* 2005;**23**(3):349–54. http://dx.doi.org/10.1038/nbt1070.

45. Robbins PF, Li YF, El-gamil M, et al. Single and dual amino acid substitutions in TCR CDRs can enhance antigen-specific T cell functions. *J Immunol* 2008;**180**:6116–31.

46. Kuball J, Hauptrock B, Malina V, et al. Increasing functional avidity of TCR-redirected T cells by removing defined *N*-glycosylation sites in the TCR constant domain. *J Exp Med* 2009;**206**(2):463–75. http://dx.doi.org/10.1084/jem.20082487.

47. Zhao Y, Bennett AD, Zheng Z, et al. CD4+ T cells with the ability to recognize and kill tumor. *J Immunol* 2007;**179**:5845–54.

48. Kuball J, Dossett ML, Wolfl M, et al. Facilitating matched pairing and expression of TCR chains introduced into human T cells. *Blood* 2007;**109**(6):2331–8. http://dx.doi.org/10.1182/blood-2006-05-023069.

49. Linette GP, Stadtmauer EA, Maus MV, et al. Cardiovascular toxicity and titin cross-reactivity of affinity-enhanced T cells in myeloma and melanoma. *Blood* 2013;**122**(6):863–71. http://dx.doi.org/10.1182/blood-2013-03-490565.

50. Gross G, Waks T, Eshhar Z. Expression of immunoglobulin-T-cell receptor chimeric molecules as functional receptors with antibody-type specificity. *Proc Natl Acad Sci USA* 1989;**86**(24):10024–8. Available at: http://www.pubmedcentral.nih.gov/articlerender.fcgi?artid=298636&tool=pmcentrez&rendertype=abstract.

51. Milone MC, Fish JD, Carpenito C, et al. Chimeric receptors containing CD137 signal transduction domains mediate enhanced survival of T cells and increased antileukemic efficacy in vivo. *Mol Ther* 2009;**17**(8):1453–64. http://dx.doi.org/10.1038/mt.2009.83.

52. Hacein-Bey-Abina S, Von Kalle C, Schmidt M, et al. LMO2-associated clonal T cell proliferation in two patients after gene therapy for SCID-X1. *Science* 2003;**302**(5644):415–9. http://dx.doi.org/10.1126/science.1088547.

53. Naldini L, Blömer U, Gallay P, et al. In vivo gene delivery and stable transduction of nondividing cells by a lentiviral vector. *Science* 1996;**272**(5259):263–7. Available at: http://www.ncbi.nlm.nih.gov/pubmed/8602510.

54. Ellis J. Silencing and variegation of gammaretrovirus and lentivirus vectors. *Hum Gene Ther* 2005;**16**(11):1241–6. http://dx.doi.org/10.1089/hum.2005.16.1241.

55. Wolf D, Goff SP. Host restriction factors blocking retroviral replication. *Annu Rev Genet* 2008;**42**:143–63. http://dx.doi.org/10.1146/annurev.genet.42.110807.091704.

56. Scholler J, Brady TL, Binder-Scholl G, et al. Decade-long safety and function of retroviral-modified chimeric antigen receptor T cells. *Sci Transl Med* 2012;**4**(132):132ra53. http://dx.doi.org/10.1126/scitranslmed.3003761.

57. Bushman FD. Retroviral integration and human gene therapy. *J Clin Invest* 2007;**117**(8):2083–6. http://dx.doi.org/10.1172/JCI32949.based.

58. Lindemann D, Rethwilm A. Foamy virus biology and its application for vector development. *Viruses* 2011;**3**(5):561–85. http://dx.doi.org/10.3390/v3050561.

59. Perez EE, Wang J, Miller JC, et al. Establishment of HIV-1 resistance in CD4+ T cells by genome editing using zinc-finger nucleases. *Nat Biotechnol* 2008;**26**(7):808–16. http://dx.doi.org/10.1038/nbt1410.

60. Dupuy AJ, Akagi K, Largaespada DA, Copeland NG, Jenkins NA. Mammalian mutagenesis using a highly mobile somatic Sleeping Beauty transposon system. *Nature* 2005;**436**(7048):221–6. http://dx.doi.org/10.1038/nature03691.

61. Hackett PB, Largaespada DA, Cooper LJN. A transposon and transposase system for human application. *Mol Ther* 2010;**18**(4):674–83. http://dx.doi.org/10.1038/mt.2010.2.

62. Kebriaei P, Huls H, Jena B, et al. Infusing CD19-directed T cells to augment disease control in patients undergoing autologous hematopoietic stem-cell transplantation for advanced B-lymphoid malignancies. *Hum Gene Ther* 2012;**23**(5):444–50. http://dx.doi.org/10.1089/hum.2011.167.

63. Hackett PB, Largaespada DA, Switzer KC, Cooper LJN. Evaluating risks of insertional mutagenesis by DNA transposons in gene therapy. *Transl Res* 2013;**161**(4):265–83. http://dx.doi.org/10.1016/j.trsl.2012.12.005.

64. Zhao Y, Zheng Z, Cohen CJ, et al. High-efficiency transfection of primary human and mouse T lymphocytes using RNA electroporation. *Mol Ther* 2006;**13**(1):151–9. http://dx.doi.org/10.1016/j.ymthe.2005.07.688.

65. Yoon SH, Lee JM, Cho HI, et al. Adoptive immunotherapy using human peripheral blood lymphocytes transferred with RNA encoding Her-2/neu-specific chimeric immune receptor in ovarian cancer xenograft model. *Cancer Gene Ther* 2009;**16**(6):489–97. http://dx.doi.org/10.1038/cgt.2008.98.

66. Mitchell DA, Karikari I, Cui X, Xie W, Schmittling R, Sampson JH. Selective modification of antigen-specific T cells by RNA electroporation. *Hum Gene Ther* 2008;**19**(5):511–21. http://dx.doi.org/10.1089/hum.2007.115.

67. Rowley J, Monie A, Hung C-F, Wu T-C. Expression of IL-15RA or an IL-15/IL-15RA fusion on CD8+ T cells modifies adoptively transferred T-cell function in cis. *Eur J Immunol* 2009;**39**(2):491–506. http://dx.doi.org/10.1002/eji.200838594.

68. Zhao Y, Moon E, Carpenito C, Paulos CM, Liu X. Multiple injections of electroporated autologous T cells expressing a chimeric antigen receptor mediate regression of human disseminated tumor. *Cancer Res* 2010;**70**:9053–61. http://dx.doi.org/10.1158/0008-5472.CAN-10-2880.

69. Cannon P, June C. Chemokine receptor 5 knockout strategies. *Curr Opin HIV AIDS* 2011;**6**(1):74–9. http://dx.doi.org/10.1097/COH.0b013e32834122d7.

70. Reyon D, Tsai SQ, Khayter C, Foden JA, Sander JD, Joung JK. FLASH assembly of TALENs for high-throughput genome editing. *Nat Biotechnol* 2012;**30**(5):460–5. http://dx.doi.org/10.1038/nbt.2170.

71. Varela-Rohena A, Carpenito C, Perez EE, et al. Genetic engineering of T cells for adoptive immunotherapy. *Immunol Res* 2008;**42**:166–81. http://dx.doi.org/10.1007/s12026-008-8057-6.Genetic.

72. Kalamasz D, Long SA, Taniguchi R, Buckner JH, Berenson RJ, Bonyhadi M. Optimization of human T-cell expansion ex vivo using magnetic beads conjugated with anti-CD3 and Anti-CD28 antibodies. *J Immunother* 2004;**27**(5):405–18. Available at: http://www.ncbi.nlm.nih.gov/pubmed/15314550.

73. Yang S, Dudley ME, Rosenberg SA, Morgan RA. A simplified method for the clinical-scale generation of central memory-like CD8+ T cells after transduction with lentiviral vectors encoding antitumor antigen T-cell receptors. *J Immunother* 2010;**33**(6):648–58. http://dx.doi.org/10.1097/CJI.0b013e3181e311cb.

74. Beatty GL, Haas AR, Maus MV, et al. Mesothelin-specific chimeric antigen receptor mRNA-engineered T cells induce anti-tumor activity in solid malignancies. *Cancer Immunol Res* 2014;**2**(2):112–20. http://dx.doi.org/10.1158/2326-6066.CIR-13-0170.

75. Kalos M, Levine BL, Porter DL, et al. T cells with chimeric antigen receptors have potent antitumor effects and can establish memory in patients with advanced leukemia. *Sci Transl Med* 2011;**3**(95):95ra73. http://dx.doi.org/10.1126/scitranslmed.3002842.

76. Sadelain M, Brentjens R, Rivière I. The promise and potential pitfalls of chimeric antigen receptors. *Curr Opin Immunol* 2009;**21**:215–23. http://dx.doi.org/10.1016/j.coi.2009.02.009.

77. Novellino L, Castelli C, Parmiani G. A listing of human tumor antigens recognized by T cells: March 2004 update. *Cancer Immunol Immunother* 2005;**54**(3):187–207. http://dx.doi.org/10.1007/s00262-004-0560-6.

78. Caballero OL, Chen Y-T. Cancer/testis (CT) antigens: potential targets for immunotherapy. *Cancer Sci* 2009;**100**(11):2014–21. http://dx.doi.org/10.1111/j.1349-7006.2009.01303.x.

79. Old LJ. Cancer vaccines: an overview. *Cancer Immun* 2008;**8**:1–8.

80. Parkhurst MR, Yang JC, Langan RC, et al. T cells targeting carcinoembryonic antigen can mediate regression of metastatic colorectal cancer but induce severe transient colitis. *Mol Ther* 2011;**19**(3):620–6. http://dx.doi.org/10.1038/mt.2010.272.

81. Amos SM, Duong CPM, Westwood JA, et al. Autoimmunity associated with immunotherapy of cancer. *Blood* 2011;**118**(3):499–509. http://dx.doi.org/10.1182/blood-2011-01-325266.

82. O'Reilly RJ, Dao T, Koehne G, Scheinberg D, Doubrovina E. Adoptive transfer of unselected or leukemia-reactive T-cells in the treatment of relapse following allogeneic hematopoietic cell transplantation. *Semin Immunol* 2010;**22**(3):162–72. http://dx.doi.org/10.1016/j.smim.2010.02.003.

83. Rapoport AP, Stadtmauer EA, Aqui N, et al. Rapid immune recovery and graft-versus-host disease-like engraftment syndrome following adoptive transfer of costimulated autologous T cells. *Clin Cancer Res* 2009;**15**(13):4499–507. http://dx.doi.org/10.1158/1078-0432.CCR-09-0418.

84. Besser MJ, Shapira-Frommer R, Treves AJ, et al. Clinical responses in a phase II study using adoptive transfer of short-term cultured tumor infiltration lymphocytes in metastatic melanoma patients. *Clin Cancer Res* 2010;**16**(9):2646–55. http://dx.doi.org/10.1158/1078-0432.CCR-10-0041.

85. Dudley ME, Wunderlich JR, Robbins PF, et al. Cancer regression and autoimmunity in patients after clonal repopulation with antitumor lymphocytes. *Science* 2002;**298**(5594):850–4. http://dx.doi.org/10.1126/science.1076514.

86. Tran KQ, Zhou J, Durflinger KH, et al. Minimally cultured tumor-infiltrating lymphocytes display optimal characteristics for adoptive cell therapy. *J Immunother* 2008;**31**(8):742–51. http://dx.doi.org/10.1097/CJI.0b013e31818403d5.

87. Leen AM, Christin A, Myers GD, et al. Cytotoxic T lymphocyte therapy with donor T cells prevents and treats adenovirus and Epstein-Barr virus infections after haploidentical and matched unrelated stem cell transplantation. *Blood* 2009;**114**(19):4283–92. http://dx.doi.org/10.1182/blood-2009-07-232454.

88. Louis CU, Straathof K, Bollard CM, et al. Adoptive transfer of EBV-specific T cells results in sustained clinical responses in patients with locoregional nasopharyngeal carcinoma. *J Immunother* 2010;**33**(9):983–90. http://dx.doi.org/10.1097/CJI.0b013e3181f3cbf4.

89. Appay V, Douek DC, Price DA. CD8+ T cell efficacy in vaccination and disease. *Nat Med* 2008;**14**(6):623–8. http://dx.doi.org/10.1038/nm.f.1774.

90. Carpenito C, Milone MC, Hassan R, et al. Control of large, established tumor xenografts with genetically retargeted human T cells containing CD28 and CD137 domains. *Proc Natl Acad Sci USA* 2009;**106**(9):3360–5. http://dx.doi.org/10.1073/pnas.0813101106.

91. Wong P, Pamer EG. CD8 T cell responses to infectious pathogens. *Annu Rev Immunol* 2003;**21**(2):29–70. http://dx.doi.org/10.1146/annurev.immunol.21.120601.141114.

92. Zhang N, Bevan MJ. CD8(+) T cells: foot soldiers of the immune system. *Immunity* 2011;**35**(2):161–8. http://dx.doi.org/10.1016/j.immuni.2011.07.010.

93. Berger C, Jensen MC, Lansdorp PM, Gough M, Elliott C, Riddell SR. Adoptive transfer of effector CD8 + T cells derived from central memory cells establishes persistent T cell memory in primates. *J Clin Invest* 2008;**118**:294–305. http://dx.doi.org/10.1172/295.

94. Sun JC, Bevan MJ. Defective CD8 T cell memory following acute infection without CD4 T cell help. *Science* 2003;**300**(5617):339–42. http://dx.doi.org/10.1126/science.1083317.

95. Sun JC, Williams MA, Bevan MJ. CD4+ T cells are required for the maintenance, not programming, of memory CD8+ T cells after acute infection. *Nat Immunol* 2004;**5**(9):927–33. http://dx.doi.org/10.1038/ni1105.

96. Seder RA, Darrah PA, Roederer M. T-cell quality in memory and protection: implications for vaccine design. *Nat Rev Immunol* 2008;**8**(4):247–58. http://dx.doi.org/10.1038/nri2274.

97. Maus MV, Kovacs B, Kwok WW, et al. Extensive replicative capacity of human central memory T cells. *J Immunol* 2004;**172**(11):6675–83. Available at: http://www.ncbi.nlm.nih.gov/pubmed/15153483.

98. Wang A, Chandran S, Shah SA, et al. The stoichiometric production of IL-2 and IFN-mRNA defines memory T cells that can self-renew after adoptive transfer in humans. *Sci Transl Med* 2012;**4**(149):149ra120. http://dx.doi.org/10.1126/scitranslmed.3004306.

99. Klebanoff CA, Gattinoni L, Torabi-Parizi P, et al. Central memory self/tumor-reactive CD8+ T cells confer superior antitumor immunity compared with effector memory T cells. *Proc Natl Acad Sci USA* 2005;**102**(27):9571–6. http://dx.doi.org/10.1073/pnas.0503726102.

100. Muranski P, Boni A, Antony P a, et al. Tumor-specific Th17-polarized cells eradicate large established melanoma. *Blood* 2008;**112**(2):362–73. http://dx.doi.org/10.1182/blood-2007-11-120998.

101. Gattinoni L, Lugli E, Ji Y, et al. A human memory T cell subset with stem cell-like properties. *Nat Med* 2011;**17**(10):1290–7. http://dx.doi.org/10.1038/nm.2446.

102. Zhang H, Snyder KM, Suhoski MM, et al. 4-1BB is superior to CD28 costimulation for generating CD8+ cytotoxic lymphocytes for adoptive immunotherapy. *J Immunol* 2007;**179**(7):4910–8. Available at: http://www.ncbi.nlm.nih.gov/pubmed/17878391.

103. Kershaw MH, Westwood JA, Parker LL, et al. A phase I study on adoptive immunotherapy using gene-modified T cells for ovarian cancer. *Clin Cancer Res* 2006;**12**:6106–15. http://dx.doi.org/10.1158/1078-0432.CCR-06-1183.

104. Park JR, Digiusto DL, Slovak M, et al. Adoptive transfer of chimeric antigen receptor re-directed cytolytic T lymphocyte clones in patients with neuroblastoma. *Mol Ther* 2007;**15**(4):825–33. http://dx.doi.org/10.1038/mt.

105. Zhou J, Shen X, Huang J, et al. Telomere length of transferred lymphocytes correlates with in vivo persistence and tumor regression in melanoma patients receiving cell transfer therapy. *J Immunol* 2012;**175**:7046–52.

106. Finney HM, Akbar AN, Alastair DG. Activation of resting human primary T cells with chimeric receptors: costimulation from CD28, inducible costimulator, CD134 and CD137 in series with signals from the TCR zeta chain. *J Immunol* 2004;**172**:104–13.

107. Imai C, Mihara K, Andreansky M, et al. Chimeric receptors with 4-1BB signaling capacity provoke potent cytotoxicity against acute lymphoblastic leukemia. *Leukemia* 2004;**18**(4):676–84. http://dx.doi.org/10.1038/sj.leu.2403302.

108. Pulè MA, Straathof KC, Dotti G, Heslop HE, Rooney CM, Brenner MK. A chimeric T cell antigen receptor that augments cytokine release and supports clonal expansion of primary human T cells. *Mol Ther* 2005;**12**(5):933–41. http://dx.doi.org/10.1016/j.ymthe.2005.04.016.

109. Topp MS, Riddell SR, Akatsuka Y, Jensen MC, Blattman JN, Greenberg PD. Restoration of CD28 expression in CD28⁻ CD8+ memory effector T cells reconstitutes antigen-induced IL-2 production. *J Exp Med* 2003;**198**(6):947–55. http://dx.doi.org/10.1084/jem.20021288.

110. Chen L, Flies D. Molecular mechanisms of T cell co-stimulation and co-inhibition. *Nat Rev Immunol* 2013;**13**:227.

111. Lotze MT, Strausser JL, Rosenberg SA. In vitro growth of cytotoxic human lymphocytes. II. Use of T cell growth factor (TCGF) to clone human T cells. *J Immunol* 1980;**124**(6):2972–8. Available at: http://www.ncbi.nlm.nih.gov/pubmed/6989911.

112. Ma A, Koka R, Burkett P. Diverse functions of IL-2, IL-15, and IL-7 in lymphoid homeostasis. *Annu Rev Immunol* 2006;**24**:657–79. http://dx.doi.org/10.1146/annurev.immunol.24.021605.090727.

113. Evans LS, Witte PR, Feldhaus AL, et al. Expression of chimeric granulocyte-macrophage colony-stimulating factor/interleukin 2 receptors in human macrophage colony-stimulating factor-dependent growth. *Hum Gene Ther* 1999;**10**:1941–51.

114. Hsu C, Jones SA, Cohen CJ, et al. Cytokine-independent growth and clonal expansion of a primary human CD8+ T-cell clone following retroviral transduction with the IL-15 gene. *Blood* 2007;**109**(12):5168–77. http://dx.doi.org/10.1182/blood-2006-06-029173.

115. Sportès C, Hakim FT, Memon SA, et al. Administration of rhIL-7 in humans increases in vivo TCR repertoire diversity by preferential expansion of naive T cell subsets. *J Exp Med* 2008;**205**(7):1701–14. http://dx.doi.org/10.1084/jem.20071681.

116. Lugli E, Goldman CK, Perera LP, et al. Transient and persistent effects of IL-15 on lymphocyte homeostasis in nonhuman primates. *Blood* 2010;**116**(17):3238–48. http://dx.doi.org/10.1182/blood-2010-03-275438.

117. Rufer N. Transfer of the human telomerase reverse transcriptase (TERT) gene into T lymphocytes results in extension of replicative potential. *Blood* 2001;**98**(3):597–603. http://dx.doi.org/10.1182/blood.V98.3.597.

118. Charo J, Finkelstein SE, Grewal N, Restifo NP, Robbins PF, Rosenberg SA. Bcl-2 overexpression enhances tumor-specific T-cell survival. *Cancer Res* 2005;**65**(5):2001–8. http://dx.doi.org/10.1158/0008-5472.CAN-04-2006.

119. Eaton D, Gilham S, O A, H RE. Retroviral transduction of human peripheral blood lymphocytes with bcl-x L promotes in vitro lymphocyte survival in pro-apoptotic conditions. *Gene Ther* 2002;**9**:527–35. http://dx.doi.org/10.1038/sj/gt/3301685.

120. Dotti G, Savoldo B, Pule M, et al. Human cytotoxic T lymphocytes with reduced sensitivity to Fas-induced apoptosis. *Blood* 2005;**105**(12):4677–84. http://dx.doi.org/10.1182/blood-2004-08-3337.

121. Zhang Q, Yang X, Pins M, et al. Adoptive transfer of tumor-reactive transforming growth factor-beta-insensitive CD8+ T cells: eradication of autologous mouse prostate cancer. *Cancer Res* 2005;**65**(5):1761–9. http://dx.doi.org/10.1158/0008-5472.CAN-04-3169.

122. Lacuesta K, Buza E, Hauser H, et al. Assessing the safety of cytotoxic T lymphocytes transduced with a dominant negative transforming growth factor-beta receptor. *J Immunother* 2006;**29**(3):250–60. http://dx.doi.org/10.1097/01.cji.0000192104.24583.ca.

123. Nolz JC, Starbeck-Miller GR, Harty JT. Naive, effector and memory CD8 T cell trafficking: parallels and distinctions. *Immunotherapy* 2011;**3**(10):1223–33. http://dx.doi.org/10.2217/imt.11.100.

124. Moon EK, Carpenito C, Sun J, et al. Expression of a functional CCR2 receptor enhances tumor localization and tumor eradication by retargeted human t cells expressing a mesothelin-specific chimeric antibody receptor. *Clin Cancer Res* 2011;**17**(14):4719–30. http://dx.doi.org/10.1158/1078-0432.CCR-11-0351.

125. Harlin H, Meng Y, Peterson A, et al. Chemokine expression in melanoma metastases associated with. *Cancer Res* 2009;**69**:3077–85.

126. Fisher DT, Chen Q, Appenheimer MM, et al. Hurdles to lymphocyte trafficking in the tumor microenvironment: implications for effective immunotherapy. *Immunol Invest* 2006;**35**(3–4):251–77. http://dx.doi.org/10.1080/08820130600745430.

127. Berger C, Flowers ME, Warren EH, Riddell SR. Analysis of transgene-specific immune responses that limit the in vivo persistence of adoptively transferred HSV-TK-modified donor T cells after allogeneic hematopoietic cell transplantation. *Blood* 2006;**107**(6):2294–302. http://dx.doi.org/10.1182/blood-2005-08-3503.

128. Davis JL, Theoret MR, Zheng Z, Lamers CHJ, Rosenberg SA, Morgan RA. Development of human anti-murine T-cell receptor antibodies in both responding and nonresponding patients enrolled in TCR gene therapy trials. *Clin Cancer Res* 2010;**16**(23):5852–61. http://dx.doi.org/10.1158/1078-0432.CCR-10-1280.

129. Lamers CHJ, Willemsen R, van Elzakker P, et al. Immune responses to transgene and retroviral vector in patients treated with ex vivo-engineered T cells. *Blood* 2011;**117**(1):72–82. http://dx.doi.org/10.1182/blood-2010-07-294520.

130. Grupp SA, Kalos M, Barrett D, et al. Chimeric antigen receptor-modified T cells for acute lymphoid leukemia. *N Engl J Med* 2013;**368**(16):1509–18. http://dx.doi.org/10.1056/NEJMoa1215134.

131. Lamers CHJ, Langeveld SCL, Groot-van Ruijven CM, Debets R, Sleijfer S, Gratama JW. Gene-modified T cells for adoptive immunotherapy of renal cell cancer maintain transgene-specific immune functions in vivo. *Cancer Immunol Immunother* 2007;**56**(12):1875–83. http://dx.doi.org/10.1007/s00262-007-0330-3.

132. Di Stasi A, Tey S-K, Dotti G, et al. Inducible apoptosis as a safety switch for adoptive cell therapy. *N Engl J Med* 2011;**365**(18):1673–83. http://dx.doi.org/10.1056/NEJMoa1106152.

133. Quintarelli C, Vera JF, Savoldo B, et al. Co-expression of cytokine and suicide genes to enhance the activity and safety of tumor-specific cytotoxic T lymphocytes. *Blood* 2007;**110**(8):2793–802. http://dx.doi.org/10.1182/blood-2007-02-072843.

134. Griffioen M, van Egmond EHM, Kester MGD, Willemze R, Falkenburg JHF, Heemskerk MHM. Retroviral transfer of human CD20 as a suicide gene for adoptive T-cell therapy. *Haematologica* 2009;**94**(9):1316–20. http://dx.doi.org/10.3324/haematol.2008.001677.

135. Bonini C, Ferrari G, Verzeletti S, et al. HSV-TK gene transfer into donor lymphocytes for control of allogeneic graft-versus-leukemia. *Science* 1997;**276**(5319):1719–24. http://dx.doi.org/10.1126/science.276.5319.1719.

136. Berger C, Blau CA, Huang M-L, et al. Pharmacologically regulated Fas-mediated death of adoptively transferred T cells in a nonhuman primate model. *Blood* 2004;**103**(4):1261–9. http://dx.doi.org/10.1182/blood-2003-08-2908.

137. Introna M, Barbui AM, Bambacioni F, et al. Genetic modification of human T cells with CD20: A strategy to purify and lyse transduced cells with Anti-CD20 antibodies. *Hum Gene Ther* 2000;**11**:611–20.

138. Griffioen M, van Egmond EHM, Kester MGD, Willemze R, Falkenburg JHF, Heemskerk MHM. Retroviral transfer of human CD20 as a suicide gene for adoptive T-cell therapy. *Haematologica* 2009;**94**(9):1316–20.

139. Terakura S, Yamamoto TN, Gardner RA, Turtle CJ, Jensen MC, Riddell SR. Generation of CD19-chimeric antigen receptor modified CD8+ T cells derived from virus-specific central memory T cells. *Blood* 2012;**119**(1):72–82. http://dx.doi.org/10.1182/blood-2011-07-366419.

140. Savoldo B, Ramos CA, Liu E, et al. Brief report CD28 costimulation improves expansion and persistence of chimeric antigen receptor – modified T cells in lymphoma patients. *J Clin Invest* 2011;**121**(5):1822–6. http://dx.doi.org/10.1172/JCI46110DS1.

141. Brentjens RJ, Davila ML, Riviere I, et al. CD19-targeted T cells rapidly induce molecular remissions in adults with chemotherapy-refractory acute lymphoblastic leukemia. *Sci Transl Med* 2013;**5**(177):177ra38. http://dx.doi.org/10.1126/scitranslmed.3005930.

142. Morgan RA, Yang JC, Kitano M, Dudley ME, Laurencot CM, Rosenberg SA. Case report of a serious adverse event following the administration of T cells transduced with a chimeric antigen receptor recognizing ERBB2. *Mol Ther* 2010;**18**(4):843–51. http://dx.doi.org/10.1038/mt.2010.24.

143. Louis CU, Savoldo B, Dotti G, et al. Antitumor activity and long-term fate of chimeric antigen receptor-positive T cells in patients with neuroblastoma. *Blood* 2011;**118**(23):6050–6.

144. Ritchie DS, Neeson PJ, Khot A, et al. Persistence and efficacy of second generation CAR T Cell against the LeY antigen in acute myeloid leukemia. *Mol Ther* 2013;**21**(11):2122–9. http://dx.doi.org/10.1038/mt.2013.154.

145. Pegram HJ, Lee JC, Hayman EG, et al. Tumor-targeted T cells modified to secrete IL-12 eradicate systemic tumors without need for prior conditioning. *Blood* 2012;**119**(18):4133–41. http://dx.doi.org/10.1182/blood-2011-12-400044.

146. Muranski P, Boni A, Wrzesinski C, et al. Increased intensity lymphodepletion and adoptive immunotherapy – how far can we go? *Nat Clin Pract Oncol* 2006;**3**(12):668–81.

147. Wrzesinski C, Restifo NP. Less is more: lymphodepletion followed by hematopoietic stem cell transplant augments adoptive T-cell-based anti-tumor immunotherapy. *Curr Opin Immunol* 2005;**17**(2):195–201. http://dx.doi.org/10.1016/j.coi.2005.02.002.

148. Machiels JH, Reilly RT, Emens LA, et al. Cyclophosphamide, doxorubicin, and paclitaxel enhance the antitumor immune response of granulocyte/macrophage-colony stimulating factor-secreting whole-cell vaccines in HER-2/neu tolerized mice. *Cancer Res* 2001;**61**:3689–97.

149. Obeid M, Tesniere A, Ghiringhelli F, et al. Calreticulin exposure dictates the immunogenicity of cancer cell death. *Nat Med* 2007;**13**(1):54–61. http://dx.doi.org/10.1038/nm1523.

150. Miller JS, Soignier Y, Panoskaltsis-Mortari A, et al. Successful adoptive transfer and in vivo expansion of human haploidentical NK cells in patients with cancer. *Blood* 2005;**105**(8):3051–7. http://dx.doi.org/10.1182/blood-2004-07-2974.

151. Porter DL, Collins RH, Hardy C, et al. Treatment of relapsed leukemia after unrelated donor marrow transplantation with unrelated donor leukocyte infusions. *Blood* 2000;**95**:1214–21.

152. Peggs KS, Thomson K, Hart DP, et al. Dose-escalated donor lymphocyte infusions following reduced intensity transplantation: toxicity, chimerism, and disease responses. *Blood* 2004;**103**(4):1548–56. http://dx.doi.org/10.1182/blood-2003-05-1513.

153. Kalos M. An integrative paradigm to impart quality to correlative science. *J Transl Med* 2010;**8**:26. http://dx.doi.org/10.1186/1479-5876-8-26.

154. Kalos M. Biomarkers in T cell therapy clinical trials. *J Transl Med* 2011;**9**(1):138. http://dx.doi.org/10.1186/1479-5876-9-138.

155. Melenhorst JJ, Scheinberg P, Chattopadhyay PK, et al. High avidity myeloid leukemia-associated antigen-specific CD8+ T cells preferentially reside in the bone marrow. *Blood* 2009;**113**(10):2238–44. http://dx.doi.org/10.1182/blood-2008-04-151969.

156. Jena B, Maiti S, Huls H, et al. Chimeric antigen receptor (CAR)-specific monoclonal antibody to detect CD19-specific T cells in clinical trials. *PLoS One* 2013;**8**(3):e57838. http://dx.doi.org/10.1371/journal.pone.0057838.

157. Robbins PF, Dudley ME, Wunderlich J, et al. Cutting edge: persistence of transferred lymphocyte clonotypes correlates with cancer regression in patients receiving cell transfer therapy. *J Immunol* 2004;**173**(12):7125–30. Available at: http://www.pubmedcentral.nih.gov/articlerender.fcgi?artid=2175171&tool=pmcentrez&rendertype=abstract.

158. Powell DJ, Dudley ME, Robbins PF, Rosenberg SA. Transition of late-stage effector T cells to CD27+ CD28+ tumor-reactive effector memory T cells in humans after adoptive cell transfer therapy. *Blood* 2005;**105**(1):241–50. http://dx.doi.org/10.1182/blood-2004-06-2482.

159. Gratama JW, Kern F. Flow cytometric enumeration of antigen-specific T lymphocytes. *Cytometry A* 2004;**58**(1):79–86. http://dx.doi.org/10.1002/cyto.a.90005.

160. Chattopadhyay PK, Betts MR, Price DA, et al. The cytolytic enzymes granzyme A, granzyme B, and perforin: expression patterns, cell distribution, and their relationship to cell maturity and bright CD57 expression. *J Leukoc Biol* 2009;**85**(1):88–97. http://dx.doi.org/10.1189/jlb.0208107.

161. Turtle CJ, Swanson HM, Fujii N, Estey EH, Riddell SR. A distinct subset of self-renewing human memory CD8+ T cells survives cytotoxic chemotherapy. *Immunity* 2009;**31**(5):834–44. http://dx.doi.org/10.1016/j.immuni.2009.09.015.

162. Till BG, Jensen MC, Wang J, et al. CD20-specific adoptive immunotherapy for lymphoma using a chimeric antigen receptor with both CD28 and 4-1BB domains: pilot clinical trial results. *Blood* 2012;**119**(17):3940–50. http://dx.doi.org/10.1182/blood-2011-10-387969.

163. Betts MR, Harari A. Phenotype and function of protective T cell immune responses in HIV. *Curr Opin HIV AIDS* 2008;**3**(3):349–55. http://dx.doi.org/10.1097/COH.0b013e3282fbaa81.

164. Gaucher D, Therrien R, Kettaf N, et al. Yellow fever vaccine induces integrated multilineage and polyfunctional immune responses. *J Exp Med* 2008;**205**(13):3119–31. http://dx.doi.org/10.1084/jem.20082292.

165. Makedonas G, Hutnick N, Haney D, et al. Perforin and IL-2 upregulation define qualitative differences among highly functional virus-specific human CD8 T cells. *PLoS Pathog* 2010;**6**(3):e1000798. http://dx.doi.org/10.1371/journal.ppat.1000798.

166. Lin Y, Gallardo HF, Ku GY, et al. Optimization and validation of a robust human T cell culture method for monitoring phenotypic and polyfunctional antigen-specific CD4 and CD8 T-cell responses. *Cytotherapy* 2009;**11**(7):912–22. http://dx.doi.org/10.3109/14653240903136987.

167. Suni MA, Maino VC. Flow cytometric analysis of cell signaling proteins. In: Kalyuzhny AE, editor. *Methods in molecular biology. Methods in molecular biology*, ;**vol. 717**. Clifton (NJ): Springer Science; 2011. pp. 155–69. http://dx.doi.org/10.1007/978-1-61779-024-9.

168. Yee C, Thompson JA, Roche P, et al. Melanocyte destruction after antigen-specific immunotherapy of melanoma: direct evidence of T cell-mediated vitiligo. *J Exp Med* 2000;**192**(11):1637–44. Available at: http://www.pubmedcentral.nih.gov/articlerender.fcgi?artid=2193107&tool=pmcentrez&rendertype=abstract.

169. Palmer DC, Chan C-C, Gattinoni L, et al. Effective tumor treatment targeting a melanoma/melanocyte-associated antigen triggers severe ocular autoimmunity. *Proc Natl Acad Sci USA* 2008;**105**(23):8061–6. http://dx.doi.org/10.1073/pnas.0710929105.

170. Brentjens R, Yeh R, Bernal Y, Riviere I, Sadelain M. Treatment of chronic lymphocytic leukemia with genetically targeted autologous T cells: case report of an unforeseen adverse event in a phase I clinical trial. *Mol Ther* 2010;**18**(4):666–8. http://dx.doi.org/10.1038/mt.2010.31.

171. Gajewski TF, Louahed J, Brichard VG. Gene signature in melanoma associated with clinical activity: a potential clue to unlock cancer immunotherapy. *Cancer J* 2010;**16**(4):399–403. http://dx.doi.org/10.1097/PPO.0b013e3181eacbd8.

172. Stroncek DF, Jin P, Wang E, Ren J, Sabatino M, Francesco M. Global transcriptional analysis for biomarker discovery and validation in cancer and hematological malignancy biologic therapies. *Mol Diagn Ther* 2009;**13**(3):181–93. http://dx.doi.org/10.2165/01250444-200913030-00003.

173. Brentjens RJ, Santos E, Nikhamin Y, et al. Genetically targeted T cells eradicate systemic acute lymphoblastic leukemia xenografts. *Clin Cancer Res* 2007;**13**(18 Pt 1):5426–35. http://dx.doi.org/10.1158/1078-0432.CCR-07-0674.

174. Riley JL, June CH. The CD28 family: a T-cell rheostat for therapeutic control of T-cell activation. *Blood* 2005;**105**(1):13–21. http://dx.doi.org/10.1182/blood-2004-04-1596.

175. Loskog A, Giandomenico V, Rossig C, Pule M, Dotti G, Brenner MK. Addition of the CD28 signaling domain to chimeric T-cell receptors enhances chimeric T-cell resistance to T regulatory cells. *Leukemia* 2006;**20**(10):1819–28. http://dx.doi.org/10.1038/sj.leu.2404366.

176. Lamers CHJ, Sleijfer S, Vulto AG, et al. Treatment of metastatic renal cell carcinoma with autologous T-lymphocytes genetically retargeted against carbonic anhydrase IX: first clinical experience. *J Clin Oncol* 2006;**24**(13):e20–2. http://dx.doi.org/10.1200/JCO.2006.05.9964.

177. Birkholz K, Hofmann C, Hoyer S, et al. A fast and robust method to clone and functionally validate T-cell receptors. *J Immunol Methods* 2009;**346**(1–2):45–54. http://dx.doi.org/10.1016/j.jim.2009.05.001.

178. Seitz S, Schneider CK, Malotka J, et al. Reconstitution of paired T cell receptor alpha- and beta-chains from microdissected single cells of human inflammatory tissues. *Proc Natl Acad Sci USA* 2006;**103**(32):12057–62. http://dx.doi.org/10.1073/pnas.0604247103.

179. Barrett DM, Zhao Y, Liu X, et al. Treatment of advanced leukemia in mice with mRNA engineered T Cells. *Hum Gene Ther* 2011;**22**(12):1575–86. http://dx.doi.org/10.1089/hum.2011.070.

180. Davila ML, Riviere I, Wang X, et al. Efficacy and toxicity management of 19-28z CAR T cell therapy in B cell acute lymphoblastic leukemia. *Sci Transl Med* 2014;**6**(224):224ra25. http://dx.doi.org/10.1126/scitranslmed.3008226.

181. Tebas P, Stein D, Binder-scholl G, et al. Antiviral effects of autologous CD4 T cells genetically modified with a conditionally replicating lentiviral vector expressing long antisense to HIV. *Blood* 2013;**121**(9):1524–33. http://dx.doi.org/10.1182/blood-2012-07-447250.

182. Tebas P, Stein D, Ww T, et al. Gene editing of CCR5 in autologous CD4 T cells of persons infected with HIV. *N Engl J Med* 2014;**370**(10):901–10. http://dx.doi.org/10.1056/NEJMoa1300662.Gene.

183. Levine BL, June CH. Perspective: assembly line immunotherapy. *Nature* 2013;**498**:S17. http://dx.doi.org/10.1038/498S17a.Perspective.

Chapter 19

Current Status of Gene Therapy for Brain Tumors*

Jianfang Ning and Samuel D. Rabkin
Department of Neurosurgery, Molecular Neurosurgery Laboratory, Brain Tumor Research Center, Massachusetts General Hospital and Harvard Medical School, Boston, MA, USA

LIST OF ABBREVIATIONS

Ad Adenovirus
BBB Blood–brain barrier
CAR Chimeric antigen receptor
CD Cytosine deaminase
CEA Carcinoembryonic antigen
CED Convection-enhanced delivery
CPA Cyclophosphamide
CRAd Conditionally replicative adenovirus
5-FC 5-Fluorocytosine
Flt3L Fms-like tyrosine kinase 3 ligand
GBM Glioblastoma
GCV Ganciclovir
GM-CSF Granulocyte macrophage colony stimulating factor
GSCs Glioblastoma stem cells
HSV Herpes simplex virus
i.c. Intracerebral
IFN Interferon
MLV Murine leukemia virus
MSCs Mesenchymal stromal (stem) cells
MV Measles virus
MYXV Myxoma virus
NIS Sodium iodide symporter
NDV Newcastle disease virus
NSCs Neural stem cells
OV Oncolytic virus
PNP Purine nucleoside phosphorylase
PV Poliovirus
RCR Replication-competent gamma-retrovirus
s.c. Subcutaneous
sCE Secreted carboxylesterase
TK Thymidine kinase
TRAIL Tumor necrosis factor-related apoptosis-inducing ligand
VSV Vesicular stomatitis virus
VV Vaccinia virus

*This chapter is based on a published review.[1]

Translating Gene Therapy to the Clinic. http://dx.doi.org/10.1016/B978-0-12-800563-7.00019-1

19.1 INTRODUCTION

Brain tumors encompass a range of neoplasms that occur in the central nervous system, with primary tumors arising in the brain and metastatic tumors infiltrating from distal organs. Brain metastases occur about 10 times more frequently than primary tumors, with approximately 10% of cancer patients developing brain metastases, which significantly decrease survival.[2] Primary tumors range from benign low-grade tumors, which in adults are the majority, such as meningioma, pituitary, and nerve sheath tumors, to malignant, high-grade tumors, such as glioblastoma (GBM), other gliomas, and lymphoma. GBM is the most common and deadliest primary brain tumor in adults. While the incidence of GBM is relatively low, about 3.2/100,000 in the United States compared to 124, 49, and 21/100,000 for breast cancer, colon cancer, and melanoma, respectively, the survival rate is exceedingly poor, 14% at 2 years.[3] This low survival rate persists despite multimodal therapy (maximal surgical resection, radiation, and temozolomide chemotherapy), which has had limited impact on disease progression.[4]

There are a number of factors that contribute to this grim outcome for GBM patients. (1) Surgery and radiation are often limited by proximity to critical functional structures in the brain. (2) The tumor cells tend to be highly invasive, migrating into the brain parenchyma, and along blood vessels and white matter tracks, so the tumor margin is rarely defined and recurrences typically occur within a few cm. (3) The presence of glioblastoma stem cells (GSCs) that are tumor initiating and propagating, and in some cases radio- and chemotherapy resistant, provide seeds for recurrence. (4) The heterogeneous histopathology and genotype complicate targeting specific oncogenic pathways. (5) The blood–brain barrier (BBB), even though diminished in tumors, limits therapeutic delivery. (6) The immune-privileged status of the brain and immune-suppressive features of the tumor limit systemic immune responses. Thus, there is an urgent need for new therapeutic strategies that can overcome these barriers to treatment, and gene therapy may hold promise for the treatment of GBM.

Gene therapy originated with the concept of replacing a mutated or deleted gene with a normal functional version that could correct the clinical disorder arising from the loss of the gene product. This strategy is beginning to bear fruit for a number of monogenic diseases such as severe immunodeficiencies and hereditary blindness. Although cancer is a genetic disease, it is almost never a monogenic disease. The genetic alterations involve not only loss-of-function but also gain-of-function, and the genetic-epigenetic landscape is constantly evolving. Thus, the strategy for gene therapy of cancer, and brain tumors, has shifted from gene correction to cancer cell destruction, targeting not just the cancer cells but also "normal" cells within or recruited to the tumor. GBM is a good candidate for gene therapy because: tumors remain local within the brain and very rarely metastasize to other tissues, so local administration is appropriate; the brain is physiologically isolated from the rest of the body, which limits potential systemic toxicities; tumors can often be neurosurgically accessed for vector administration; sophisticated imaging paradigms are available in the clinic; and the majority of cells in the brain are postmitotic, which allows for specific targeting of dividing tumor cells.

In considering a gene therapy strategy for brain tumors, two basic questions need to be addressed: Which gene or sequence should be delivered, and how should it be delivered? The types of genes and their mechanisms of action can be grouped into: (1) oncolytic viruses (OVs) that selectively replicate in and kill tumor cells while sparing normal cells; (2) cytotoxic and suicide genes whose gene products convert nontoxic prodrugs to cytotoxic metabolites; (3) immunomodulatory genes for activating or enhancing antitumor immune responses, i.e., inhibiting innate and regulatory immune cells, and/or promoting tumor migration and proliferation of antigen-presenting cells and adaptive immune cells; (4) genes that disrupt the tumor microenvironment, such as inhibiting angiogenesis; and (5) interfering RNAs (siRNA, shRNA, RNAi, miR, miRi) to silence protumorigenic genes. Like most cancer therapies, combinations of these approaches should improve potency, and delivering multiple genes is relatively easy compared to developing new drugs. This chapter will describe these strategies in detail for the treatment of brain tumors, focusing on GBM, and current preclinical and clinical studies.

19.2 GENE DELIVERY VEHICLES FOR BRAIN TUMORS

Efficient delivery of genetic materials to the tumor is vital for successful gene therapy. There are a variety of delivery vehicles available for gene therapy in the brain; nonreplicating viral vectors, replication-competent OVs, cell-based (stem, progenitor, or immune cells), and synthetic vectors (liposomes, nanoparticles), with pros and cons for each.

19.2.1 Replication-Defective Viruses

Viruses have evolved as effective vehicles for horizontal gene transfer and therefore have been a preferred approach for gene therapy. Replication-defective or nonreplicating virus vectors are deleted for genes essential for viral replication or, in some cases, all open reading frames. This limits toxicity due to lack of virus replication, expression of viral genes, and antivector immune responses, and provides space for therapeutic transgenes. Two viruses have been predominantly used for glioma gene therapy; adenovirus (Ad) and retroviruses.[5] Specificity for tumor cells can be improved by additional modifications: retargeting virus receptor-binding proteins to cancer selective cell-surface molecules (transductional targeting) through physical mechanisms (i.e., bispecific adapters, antibodies, avidin-biotin) or genetic mechanisms (i.e., envelope pseudotyping, fiber modification, artificial ligands); and targeting therapeutic gene expression (i.e., transcriptionally regulated gene expression).[6] The genetic payload for each virus is limited by the inherent packaging constraints of the virus capsid and the extent of deletion of the virus genome. The administered dose of a replication-defective vector corresponds to the maximum number of gene-transduced cells possible. Unfortunately, virus infectivity and spread in the tumor tends to be poor due to low efficiency of infection, physical barriers that limit virus dispersion, and rapid clearance by innate immune cells, so the proportion of transduced tumor cells is small compared to the input virus. Thus, any effects of the delivered and expressed gene(s) must be amplified so they affect nontransduced cells, so-called bystander effects (Figure 19.1).

19.2.2 Oncolytic Viruses

OVs are selectively replication-competent in cancer cells and, thus, able to amplify themselves after initial administration and potentially spread throughout the tumor (Figure 19.1).[6,7] The genetic engineering of herpes simplex virus (HSV) for selective replication in glioma cells opened the door for the modern development of OVs.[7] The genetic payload in this case is the viral genome, driving replication and cytotoxicity. The selectivity of OVs for cancer cells is premised on the altered cellular physiology arising from defects commonly found across many cancer types, such as lack of antiviral responses, activation of oncogenic pathways, loss of tumor suppressors, and defective apoptosis.

Some viruses—Newcastle disease virus (NDV), reovirus (RV), myxoma virus (MYXV), and measles virus (MV)—have an inherent proclivity to specifically target cancer cells, and upon virus replication cause significant cell death and tumor regression. Other viruses (HSV, Ad, vaccinia virus (VV), vesicular stomatitis virus (VSV), and poliovirus (PV)) need to be genetically engineered to engender oncolytic activity and safety.[7] It is especially important in the brain that OVs be highly attenuated for neurovirulence and nonpathogenic. In most cases, this involves additional mutations or the use of vaccine strains (i.e., MV, PV, VV). In addition to their inherent cytotoxicity, OVs often induce antitumor immune responses.[8] Strategies to improve virus spread and persistence include: combinations with drugs that reduce innate immune responses, "cloaking" viruses inside carrier cells, infusion via convection-enhanced delivery (CED) or ultrasound; and engineering OVs to express genes that facilitate their physical dispersal. OVs tested for the treatment of GBM include both DNA (HSV, Ad, VV) and RNA viruses (NDV, MV, RV, VSV, PV) (Table 19.1).[19,40,41] In addition to direct cytotoxic and inflammatory effects, OVs can also act as gene therapy vectors, delivering therapeutic transgenes to the tumor.[8] So-called armed OVs contain therapeutic transgenes inserted into their genome, which are then expressed in infected tumor cells[42] (Table 19.2).

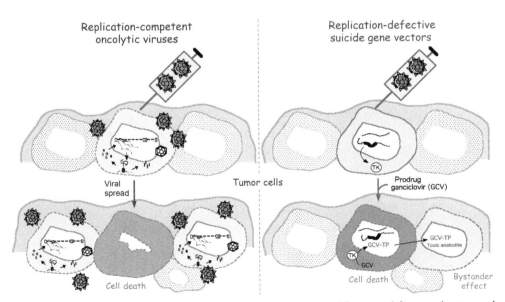

FIGURE 19.1 Gene therapy strategies for brain tumors; oncolytic versus suicide gene. Viruses used for gene therapy can be divided into two general categories; replication-competent oncolytic viruses (OVs) that selectively replicate in tumor cells and are typically cytotoxic (left) and replication-defective virus vectors that deliver therapeutic transgenes to infected cells for cytotoxicity (right). Because OVs replicate in tumor cells, they can be amplified in situ and can potentially spread throughout the tumor. Replication-defective viruses are dependent upon the activity of the delivered gene(s), often a "suicide gene," which encodes a drug-activating enzyme (i.e., thymidine kinase (TK)), that converts a nontoxic prodrug (ganciclovir (GCV)) into a cytotoxic drug or anabolite (GCV-TP).

TABLE 19.1 Oncolytic Viruses Used for Brain Tumors

Oncolytic Virus Name	Virus	Mutations (for Tumor Selectivity and Safety)	Model System (Mouse or Rat/Cell Type)	References
G207	HSV	$\gamma34.5\Delta$, ICP6⁻, LacZ⁺	Nude, i.c. U87	9,10
G47Δ	HSV	$\gamma34.5\Delta$, ICP6⁻, ICP47Δ, LacZ⁺	Nude, i.c. U87, GSC	9,11
MG18L	HSV	U_S3^-, ICP6⁻	Nude, i.c. GSC	12
Δ68H-6	HSV	$\gamma34.5$ BDDΔ, ICP6⁻	Nude, i.c. U87, GSC	13
rQnestin34.5	HSV	$\gamma34.5\Delta$, ICP6⁻, GFP⁺, nestin promoter-34.5	Nude, i.c. U87ΔEGFR	14
R-LM113	HSV	gDΔ-anti HER2 scFv	SCID, i.c. mGBM-HER2	15
KNE	HSV	gB^mut, gDΔ-anti EGFR scFv	Nude, i.c. GBM30 GSC	16
Ad-Δ24RGD	Ad	$E1A_{24bp\Delta}$, RGD-4C	Nude, i.c. U87	17
ICOVIR-5	Ad	$E1A_{24bp\Delta}$, RGD-4C, E2F promoter-E1A	Nude, i.c. U87MG	18
CRAd-S-pk7	Ad	Polylysine modified fiber knob, survivin promoter-E1A	Nude, i.c. U87	19
Ad5/35.GΔ-Ki	Ad	GFAP promoter-E1A, Ki67 promoter-E4, Ad 35 fiber knob	Nude, i.c. U251	20
vvDD	VV	TKΔ, VGFΔ	Rat, i.c. F98 & RG2	21
GLV-1h68	VV	F14.5L⁻, TK⁻, HA⁻	Nude, s.c. C6 Nude, i.c. U87	22 23
LIVP 1.1.1	VV	TK⁻	Nude, i.c. U87	23
vMyxgfp	MYXV	VV promoter-GPF	SCID, i.c., human GSCs	24
H-1PV	Parvo		Nude rat, i.c. U87 Wistar rats, i.c. RG2	25
LuIII	Parvo		SCID, s.c. U373/U87	26
PVS-RIPO	PV	HRV2 IRES Sabin vaccine strain	Nude rat, intrathecal U87MGΔEGFR; Nude, i.c. U-118	27
V4UPM	NDV	V4 vaccine variant	Nude, s.c. U87	28
Hitchner B1	NDV		SCID, s.c. U87	29
MV F^miR7	MV	Edmonston-B vaccine strain, miR7⁺	NOD/SCID, s.c. U87	30
MV-GFP-H_AA-IL-13	MV	Edmonston-B vaccine strain, hIL13⁺, H^blind (CD46⁻, SLAM⁻), EGFP⁺	Nude, i.c. GSC	31
MV-141.7 MV-AC133	MV	H^blind (CD46⁻ & SLAM⁻), anti-CD133 scFv, EGFP⁺	SCID, i.c. infected GSC NCH644	32
MV-CEA	MV	Edmonston-B vaccine strain, CEA⁺	Nude, i.c. GSC GBM6, 12	33
VSV^ΔM51	VSV	ΔM51	Nude, i.c. U87	34
VSV-M51R	VSV	M51R	Nude, s.c. U87	35
VSV-rp30	VSV	P_{mut}, L_{mut} (VSVwt serial passage)	SCID, i.c. U87	36
srVSV	VSV	VSV-ΔL +VSVΔG	SCID, s.c. G62	37
Reovirus serotype 3	RV	Dearing strain	Fischer rats, i.c. RG2 and 9L Nude, i.c. U87 and U251	38 39

The mutations listed endow tumor selectivity that is not necessarily limited to brain tumors. The model systems involve the implantation of glioma cells (human, except for mouse mGBM-HER2 and rat C6, F98, and RG2) into mice (nude or SCID) or rats. Abbreviations: Ad, adenovirus; BDD, $\gamma34.5$ beclin-1 binding domain; CEA, carcinoembryonic antigen; EGFR, epidermal growth factor receptor; GBM, glioblastoma; GFP, green fluorescent protein; GSC, glioblastoma stem cell; HA, hemagglutinin; HRV2, human rhinovirus type 2; HSV, herpes simplex virus; i.c., intracerebral; IRES, internal ribosomal entry site; MV, measles virus; MYXV, myxoma virus; NDV, Newcastle disease virus; PV, poliovirus; RV, reovirus; s.c., subcutaneous; scFv, single-chain variable fragment; TK, thymidine kinase; VGF, vaccinia growth factor; VSV, vesicular stomatitis virus; VV, vaccinia virus.

TABLE 19.2 Gene Therapy Vectors "Armed" with Therapeutic Genes

Vector Name	Vector Type	Transgene(s)	Model System (Mouse or Rat/Cell Type)	References
ACE-CD	MLV	CD/5-FC	Nude, i.c. U87 Rat, i.c. RG2	43
Toca 511	MLV	Yeast optimized CD/5-FC	Balb/c, i.c, CT26; and B6C3F1, i.c. Tu-2449	44
ACE-PNP	MLV	PNP/F-araAMP	Nude, i.c. U87	45
MGH2.1	HSV	CYP2B1/CPA shiCE/CPT-11	Nude, i.c. Gli36ΔEGFR	46
CRAdRGDflt-IL24	Ad	mda7/IL-24	Nude, i.c. D54MG	47
Ad5/35.IR-E1A/TRAIL	Ad	Ad 35 fiber knob, E1BΔ, TRAIL	SCID, s.c. U87	48
CRAd5/11-D24.TRAIL/arresten	Ad	Ad 5/11 chimeric fiber E1A$_{24bpΔ}$, TRAIL, arresten	Nude, s.c. U87	49
oHSV-TRAIL	HSV (G47Δ)	HSV IE4/5 promoter-secreted TRAIL	SCID, i.c. LN229	50
JX-594	VV (TK⁻)	Murine GM-CSF	Rat, i.c. RG2; C57BL/6, i.c. GL261	51
G47Δ-Flt3L	HSV	Flt3L	C57BL/6, i.c. CT-2A	52
M002	HSV	Murine IL-12	B6D2F1, i.c. 4C8; SCID, i.c. D54MG	53
G47Δ-mIL12	HSV	Murine IL-12	Nude, i.c. human GSC MGG4 C57BL/6, i.c. murine GSC 005	54,55
UCB-MSC-IL12M	MSC	IL-12p40 N-glycosylation mutant	C57BL/6, i.c. GL26	56
AF-MSC-endostatin-sCE2	MSC	Endostatin, sCE2	Nude, i.c. U87MG+CPT11 prodrug	57
hNSC-aaTSP-1	NSC	Secretable TSP-1 (aa412-499), GFP	SCID, i.c. Gli36vIII	58
G47Δ-mAngio	HSV	Angiostatin	Nude, i.c. U87	59
34.5ENVE	HSV (rQnestin)	Vstat120	Nude, i.c. U87ΔEGFR, Gli36Δ5, X12-V2	58,60
G47Δ-PF4	HSV	PF4 (CXCL4)	Nude, s.c. U87	58
Ad-isthmin	RD-Ad	Xenopus isthmin	Nude, i.c. U251	61
MV-NIS	MV	NIS/¹³¹I	Nude, s.c. U251 Nude, i.c. primary GBM	62

The model systems involve the implantation of cancer cells (human except for mouse 005, 4C8, B6C3F1, CT-2A, CT26, GL261, and Tu-2449, and rat RG2) into mice (nude, SCID, or inbred) or rats. Abbreviations: 5-FC, 5-fluorocytosine; Ad, adenovirus; CPA, cyclophosphamide; CD, cytosine deaminase; GBM, glioblastoma; GFP, green fluorescent protein; GM-CSF, granulocyte macrophage colony stimulating factor; GSC, glioblastoma stem cell; HSV, herpes simplex virus; i.c., intracerebral; MLV, murine leukemia virus; MSC, mesenchymal stromal (stem) cell; MV, measles virus; NIS, sodium iodide symporter; NSC, neural stem cell; PNP, purine nucleoside phosphorylase; RD-Ad, replication-deficient Ad; shiCE, secreted human intestinal carboxylesterase; sCE2, carboxylesterase 2; TK, thymidine kinase; TRAIL, tumor necrosis factor-related apoptosis-inducing ligand; TSP-1, thrombospondin-1; UCB-MSC, umbilical cord blood-derived MSC; Vstat, vasculostatin; VV, vaccinia virus.

19.2.3 Cell-Based Delivery Vehicles

Many adult stem or progenitor cells (neural stem cells (NSCs)) and mesenchymal stromal (stem) cells (MSCs) have an inherent ability to migrate to tumors in the brain, likely due to "injury" signals, thus they provide an attractive vehicle for gene therapy.[63] NSCs are multipotent progenitors of neural cells (neurons, astrocytes, oligodendrocytes), thus endogenous to the brain. They were the first stem/progenitor cells tested for homing to brain tumors and can migrate toward the tumor and cross the BBB even when injected peripherally. From a safety standpoint, stem cells must be nontumorigenic and non-immunogenic, and should not differentiate into functional cells that could interfere with normal brain activity. MSCs are nonhematopoeitic progenitor cells that play an important role in tissue repair.

Tumor homing of MSCs relies on the expression of soluble inflammatory mediators that often accompany tumor progression. One of the limitations of the xenograft model in mice, as opposed to the clinical situation, is that the tumor implantation process itself is a tissue injury, which may facilitate homing of administered stem/progenitor cells. MSCs have some advantages over other stem cells because they are easily acquired from patients (from bone marrow or adipose tissue) and expanded ex vivo.[63] However, they carry the risk of adversely contributing to the tumor microenvironment by promoting angiogenesis, stroma formation, and immunosuppressive effects. To act as gene therapy vectors, NSCs and MSCs are transduced with therapeutic transgenes, typically using retroviral vectors, delivering cytotoxic, antiangiogenic, and immunomodulatory genes.[5,63] Because the transgenes are usually expressed constitutively, they are active during cell migration and in "normal" tissue environments. For cytotoxic therapies, there is a challenge for the transduced stem/progenitor cells to remain viable long enough to impact the tumor. An NSC expressing cytosine deaminase (CD), HB1-F3. CD, is currently in clinical trial for GBM (Table 19.3).

In addition to acting as classical gene delivery vectors, stem/progenitor cells are also being used as cell carriers for OVs.[63] In this case, the OV-infected stem/progenitor cells can either be delivered intravascularly, evading antivirus immunity (innate and adaptive), to extravasate into the brain, or delivered intracranially at a distance from the tumor. Eventually, the OV will replicate and eliminate the carrier cells. To be useful as OV cell carriers, the stem/progenitor cells must be

TABLE 19.3 Recent and Ongoing Gene Therapy Clinical Trials in Patients with Glioblastoma

Vector Name	Vector Type	Phase	Protocol	Results	References (ClinicalTrials.gov)
1716 (Seprehvir®; Virttu Biologics)	HSV	I/II	10^3 to 10^5 and 10^5 pfu intratumoral (3 trials; $n=45$)	Replication in tumor	41
G207	HSV	Ib	1.5×10^8 pfu at time of biopsy; 1×10^9 pfu into tissue surrounding resected tumor 2–5 days after biopsy	Median survival 6.6 mo. from G207. No toxicity	64
G47Δ	HSV	I	Dose escalation, 3 doses		WHO JPRN-UMIN000002661
M032 (hIL-12)	HSV	I	Dose escalation: 1×10^5–3×10^9 pfu ($n=3$–6). 15% of MTD through catheters implanted in tumor; then 85% of MTD into resected tumor site		Gene transfer protocol 0801-899
ONYX-015	Ad	I	10^7–10^{10} pfu into resection cavity ($n=24$)	No efficacy observed	19
Delta-24-RGD-4C (DNX-2401; DNAtrix)	Ad	I	Dose escalation—8 doses ($n=3$)		NCT00805376
H-1PV	Parvo	I/IIa	Dose escalation—intratumoral and intravenous followed by into tumor margin ($n=18$)		NCT01301430
Reolysin® (Oncolytics Biotech)	RV	I-II	Dose escalation up to 1×10^9 $TCID_{50}$	Safe, well tolerated	NCT00528684
NDV-HUJ	NDV	I-II	1. Dose escalation of 0.1, 0.32, 0.93, 5.9, and 11 BIU IV followed by 3 cycles of 55 BIU 2. 3 cycles of 11 BIU and then 2 doses of 11 BIU weekly	No MTD, no toxicity, 1/11 complete response	41
MV-CEA	MV	I	1. MV administered to resected cavity 2. MV administered i.t. and then resected cavity		NCT00390299

Continued

TABLE 19.3 Recent and Ongoing Gene Therapy Clinical Trials in Patients with Glioblastoma—cont'd

Vector Name	Vector Type	Phase	Protocol	Results	References (ClinicalTrials. gov)
PVS-RIPO	PV	I	IT injection through catheters (CED) implanted at time of biopsy of 1×10^8–1×10^{10} TCID$_{50}$		NCT01491893
Toca 511 (CD; Tocagen)	MLV	I–II	Dose escalation: Single IT or resection cavity injection, 3–4 weeks later 6-day cycles with oral 5-FC (130 mg/kg) repeated monthly		NCT01156584 NCT01470794
Toca 511 (CD; Tocagen)	MLV	I	Injection (4 dose levels) into resection cavity, 7 weeks later 8-day cycle of oral 5-FC repeated 3×		NCT01470794
AdV-tk (GliAtak™; Advantagene)	RD-Ad	Ib/II	Dose escalation: 3×10^{10}–3×10^{11} vector into tumor bed at resection followed by valacyclovir, radiation, TMZ (Ib; $n = 13$)	Safe, signs of immune cell infiltrate	65 NCT00589875
AdV-tk	RD-Ad	II	Survival rate and recurrence-free survival rate		NCT00870181
AdvHSV-tk (sitimagene ceradenovec; Ark Ther)	RD-Ad	III	Resection, perilesional vector (1×10^{12} vp) followed by GCV (days 5–19)	Improved time to death or reintervention, but not overall survival	66
HB1.F3 (CD)	NSC	I	HB1.F3 into tumor bed following resection. Oral 5-FC every 6 h for 4–10 days		NCT01172964

Abbreviations: 5-FC, 5-fluorocytosine; Ad, adenovirus; BIU, billion (10^9) infectious units; CD, cytosine deaminase; CEA, carcinoembryonic antigen; CED, convection-enhanced delivery; GCV, ganciclovir; HSV, herpes simplex virus; IT, intratumoral; MLV, murine leukemia virus; MTD, maximum tolerated dose; MV, measles virus; NCT#, from clinicaltrials.gov; NDV, Newcastle disease virus; NSC, neural stem cell; pfu, plaque forming unit; PV, poliovirus; RD-Ad, replication-defective Ad; TCID$_{50}$, 50% tissue culture infectious dose; TK, thymidine kinase; TMZ, temozolomide.

susceptible to infection by the OV, virus replication must be arrestable or slow enough for tumor homing, and infection should not alter the physiology of the cells during transit. NSCs were first shown to act as carrier cells for oncolytic HSV, following mimosine-induced growth arrest.[67] However, the bulk of the studies have utilized oncolytic Ad-infected NSCs and MSCs, including Ad5Delta24-RGD, which has been in clinical trial (Table 19.3).[5,63] When compared directly, human NSCs loaded with conditionally replicative adenovirus (CRAd)-S-pk7 (Table 19.1) were significantly better than loaded MSCs in extending survival of mice with intracranial gliomas.[63] Oncolytic Ad-infected HB1-F3.CD NSCs, in clinical trial (Table 19.3), retained tumor homing, supported Ad replication, and delivered infectious virus to tumor cells, including GSCs, in vivo.[68] To decrease virus toxicity in vitro prior to administration, N-acetylcysteine amide, an antioxidant, was added and shown to increase the viability of oncolytic Ad-infected NSCs, virus production, and efficacy in vivo.[69]

T cells can be genetically reprogrammed with chimeric antigen receptors (CARs) so they recognize specific tumor cell antigens and enhance T cell effector functions. CARs consist of an extracellular antigen-recognition domain coupled to an intracellular signaling domain, often CD3ζ and CD28. Patient isolated T cells are often transduced with a retrovirus construct containing the genetically engineered CAR.[70] In early clinical trials, this strategy has elicited impressive results in patients with leukemia and neuroblastoma. For glioma, T cells have been targeted to EphA2,[71] IL13Rα2,[72] and EGFRvIII[70] receptors.

19.2.4 Synthetic Vectors

Nanotechnology, the use of synthetic vectors or nanoparticles for gene therapy, can overcome a major barrier to drug delivery in the brain, the BBB, as well as protect genetic material or viruses from degradation or neutralization. The formulations for nanotechnology include liposomes (spherical vesicles made up of a lipid bilayer), polymers (natural or synthetic and biodegradable, often with hydrophobic core and hydrophilic shell), dendrimers (repetitively branched polymers), nanoparticles (inorganic; iron, gold, or silica), and combinations thereof.[73] A liposome carrying the interferon β (IFNβ) gene has been tested in glioma patients, and IFNβ was found in the resection cavity of some patients.[5,74] Cationic polymers form nanocomplexes with DNA and RNA, making them attractive for gene delivery. A PEI nanoparticle modified with myristic acid for crossing the BBB, efficiently delivered tumor necrosis factor-related apoptosis-inducing ligand (TRAIL) cDNA to intracranial gliomas and improved survival.[5,75] Delivery is determined by size, shape, charge, and steric stabilization (i.e., polyethylene glycol (PEG)), which affects stability, biodistribution, cell penetration, crossing the BBB, and toxicity.[73] In addition, their surfaces can be modified with ligands for tumor-specific targeting. Recently, Jensen and colleagues reported a new spherical nucleic acid (SNA) nanoparticle conjugate, consisting of densely packed siRNA surrounding an inorganic gold nanoparticle core that can be labeled with gadolinium for magnetic resonance imaging (MRI) or fluorochromes for fluorescent imaging.[76] When delivered systemically in mice, the SNA nanoparticles crossed the BBB/blood–tumor barrier and were taken up by glioma cells.[76] Nanotechnology has many advantages for gene therapy in the brain: (1) the size of DNA is not limited as in viruses; (2) any nucleic acid can be packaged in the same carrier; (3) they are biocompatible and do not elicit immune responses; (4) they cross the BBB; and (5) they are amenable to a huge variety of modifications and engineering. Some of these strategies (i.e., cationic polymer, PEG, dendrimer coating) can also be applied to virus vectors to improve their biodistribution.[77]

19.3 GENE THERAPY STRATEGIES FOR BRAIN TUMORS

We have grouped the gene therapy strategies according to mechanism of action into five broad categories: (1) OVs; (2) cytotoxic; (3) immune-stimulatory; (4) tumor microenvironment disrupting; and (5) RNA interference (RNAi). While OVs are inherently cytotoxic, they are described separately, except where armed OVs express cytotoxic transgenes. Cytotoxic gene therapy includes gene products that are directly cytotoxic and prodrug activating "suicide" genes. Immune-stimulatory genes should boost adaptive antitumor immune responses, stimulating tumor antigen recognition and activity of effector cells. The tumor microenvironment is composed of "normal" cells and extracellular matrix, and the predominant target is angiogenesis. Blocking gene expression of proteins contributing to tumorigenesis can be achieved using RNAi. Combinations of multiple strategies are likely to enhance overall activity, and gene therapy vectors/vehicles easily accommodate multiple therapeutic transgenes.

19.3.1 Oncolytic Virus Therapy for Brain Tumors

A large number of OVs have been developed for cancer therapy over the past few decades, many of which have been evaluated in brain tumor models and translated into clinical trials for glioma (Table 19.1). The brain has features that raise unique safety concerns, often restricting the range of applicable mutant viruses that can be used. Because OVs are replication-competent, these concerns are greater than for other vectors that are not able to replicate (i.e., replication-deficient viruses, cell-based vehicles, synthetic vectors). Thus, it is important to consider the biology and genetic manipulations of each virus to ensure an appropriate balance between safety and efficacy in the brain. There are a number of recent reviews of OV virus therapy for glioma that provide different perspectives on this topic.[8,19,40]

HSV is a human DNA virus with a large genome (~150 kb) that has great potential for GBM therapy because of its natural neurotropism.[9] To convert HSV from a human pathogen to an OV, the HSV genome is genetically engineered to introduce mutations/deletions in nonessential genes that engender cancer selectivity, such as ICP6, the large subunit of ribonucleotide reductase, Us3, apoptosis inhibitor, and γ34.5, the major neurovirulence gene and inhibitor of protein shutoff and autophagy.[9,10] Deletion of nonessential genes also provides space for insertion of large and multiple transgenes. The ability to eliminate neurovirulence genes and the availability of antiviral drugs means that this lethal pathogen can be used safely in the brain.[9] Three oncolytic HSVs have been or are in clinical trial for GBM: 1716, G207, and G47Δ[10] (Table 19.3). They all contain deletions of both copies of γ34.5. In addition, both G207 and G47Δ have a LacZ insertion disrupting the UL39 gene, encoding ICP6, which further enhances specificity and safety by limiting replication to dividing tumor cells. G207 and 1716 were very efficacious in glioma cell line models and demonstrated safety in clinical trials (Table 19.3).

While the deletion of γ34.5 renders oncolytic HSV tumor-selective and nonpathogenic, it also attenuates replication in tumor cells. A more significant problem is that HSVs with deleted γ34.5 have minimal or no replication in GSCs, thought to be critical for tumor eradication.[11] To overcome these limitations, a number of strategies have been used: (1) a second-site suppressor of γ34.5Δ that places the late Us11 gene, a PKR inhibitor, under control of the immediate-early ICP47 promoter, in G47Δ[11]; (2) expressing γ34.5 under control of the glioma-specific nestin promoter, in rQNestin34.5[14]; (3) deleting only the beclin1 binding domain of γ34.5 that blocks autophagy but not other functions of γ34.5, in Δ68H-6[13]; and (4) deletion of Us3, an antiapoptotic gene, with γ34.5 in MG18L.[12] All these oncolytic HSVs also lacked ICP6, which greatly improves safety in the brain (Table 19.1).[9] One hallmark of GBM is the presence of hypoxic microenvironments. In contrast to other therapies, G207 replicated better in glioma cell lines in vitro under hypoxic compared with normoxic conditions,[78] while G47Δ replicated similarly in hypoxic and nonhypoxic regions of GSC-derived tumors.[79] Targeting virus entry to glioma selective receptors is an alternate strategy for endowing oncolytic activity to HSV.[9] This requires mutating viral glycoprotein gD to block HVEM and nectin1 receptor binding and gB heparan sulfate binding, while maintaining membrane fusion activity.[9] Retargeting polypeptides like anti-HER2 scFv for HER2 receptor (in R-LM113), anti-EGFR scFV for EGFR (in KNE) (Table 19.1), and IL-13 for IL-13Rα2 (in R5111) can be inserted into gD.[15,16]

Ads are human DNA viruses that have been extensively studied as gene therapy vectors and oncolytic agents. CRAds typically have deletions in early genes, E1A or E1B, that inactivate cellular Rb and p53 respectively, to target tumor cells which typically have mutations in these tumor suppressor genes or pathways.[80] ONYX-015, an E1B-55kD gene deleted CRAd was one of the earliest OVs tested in GBM clinical trials, with no toxicity observed; however, there was no significant efficacy[81] (Table 19.3). Ads with deletions in E1A, especially a 24 base pair deletion (Delta-24), have been developed as base vectors to accommodate additional genetic modifications to enhance anti-GBM activity.[80] Transcriptional targeting can be used to selectively drive expression of early genes (i.e., E1A) in CRAds for tumor selective replication, or therapeutic transgene expression for gene delivery. A number of promoters/enhancers that are active in glioma cells (GFAP, nestin, midkine) or cancer cells generally (TERT, survivin, VEGFR-1, E2F, Ki67) have been used to drive E1A expression in CRAds and consequently to selective replication in glioma cells[80] (Table 19.1).

One of the major limitations with using Ad5 (oncolytic and replication defective) for gene therapy is the limited expression of coxsackie-adenovirus receptor (CAR) on cancer cells, including GBM. Therefore, a major avenue for improving oncolytic Ad is through altering its tropism to more selectively bind to glioma cells. One way is to delete the CAR-binding domain in the fiber and replace it with a receptor selective polypeptide, for example: an Arg-Gly-Asp (RGD) peptide that binds to integrins αvβ3 and αvβ5 (Delta-24-RGD), an EGFRvIII binding peptide (Delta24-EGFR), or a polylysine motif (Ad5.pK7) to bind heparan sulfate proteoglycans[80] (Table 19.1). Different Ad serotypes have different cellular receptors, so using different serotype or xenotype (species) fiber knobs or chimeric fibers can also alter Ad tropism as well as avoid neutralizing antibodies. Using an Ad3 fiber knob, which binds to CD80, CD86, and an unknown receptor, or an Ad5 fiber chimera (Ad5/3), generated viruses with greatly increased glioma infectivity and cytotoxicity in vitro. Other Ad5 chimeras, with canine Ad1 (CK1) or porcine Ad (PK) fiber, were more efficient than Ad5/3.[80] CRAds with Ad16p (binds to CD46, overexpressed in GBM) or chimpanzee CV23 efficiently infected GSCs.[82] A screen of 16 Ad5 fiber chimeras on primary glioma cell cultures identified B-group viruses (Ad11, Ad35, Ad50 (binds to CD46)) as having greatly increased infectivity compared with Ad5.[83] These same strategies can be used for targeting replication-defective Ad vectors.

Poxviruses. VV, a large DNA virus, is the vaccine agent used in the eradication of smallpox. With a rapid lytic replication cycle that occurs in the cytoplasm and high immunogenicity, it has demonstrated success as an OV. Some vaccine strains that do not replicate in mammalian cells (i.e., MVA) have been used as replication-defective vectors for cancer vaccines.[84] The ease of genetic manipulation and large genome size allow the incorporation of therapeutic transgenes in armed OVs or gene delivery vectors.[84] Although the vaccine strains are attenuated, there is still concern about replication-competent VV in the brain. Therefore, additional mutations, such as in thymidine kinase (TK) and vaccinia growth factor, have been introduced to target cancer cells. A double-deleted VV, vvDD (Table 19.1), was cytotoxic in human and rat glioma cell lines in vitro and prolonged the survival of immunocompetent rats bearing intracerebral rat gliomas.[21] GLV-1h68 (Table 19.1), with reporter genes inserted into VV genes F14.5L, TK, and A56R, was quite attenuated for neurovirulence,[22] however, systemic delivery to mice with intracerebral U87 xenografts was ineffective.[23] The cytotoxicity of JX-594 (Table 19.2) was significantly better than RV or VSVΔM51 in mouse GL261 glioma cells, and it extended survival in rat and mouse immunocompetent glioma models.[51] As with other OVs, the combination with radiotherapy significantly improved survival compared with radiation alone, and radiation increased intratumoral replication of single mutant VV LIVP1.1.1 (Table 19.1).[23]

MYXV is a rabbit-specific poxvirus that has oncolytic activity in mammalian cells deficient in IFN responses, which comprises most cancer cells, including human glioma cells. MYXV (Table 19.1) replicated in and killed human glioma cell lines and GSCs in vitro, and extended survival of SCID mice bearing three different GSC lines.[24] Combination with rapamycin increased infection of half of the GSC lines, which correlated with enhanced efficacy in tumor-bearing mice.[24] While rodent glioma cell lines were somewhat sensitive to MYXV in vitro, MYXV was minimally effective in syngeneic rodent glioma tumors in vivo, except in combination with rapamycin.[85]

Autonomous parvoviruses are small single-stranded DNA viruses that preferentially replicate in transformed cells.[86] The viral NS1 protein alone is able to kill oncogene-transformed cells. Rodent parvovirus H-1PV replicates in human cancer cells, including GBM, yet is nonpathogenic after experimental infection of humans.[86] Intratumoral injection of H-1PV was efficacious in human gliomas in immunodeficient mice and rat gliomas in immunocompetent rats[25] (Table 19.1).

Comparing therapy of RG2 gliomas in immunodeficient and immunocompetent rodents indicated that adaptive immunity plays a significant role in antitumor efficacy.[86] Unlike other OVs, H-1PV was found to cross the BBB, and cured rats of intracerebral gliomas after intravenous administration.[25] LuIII (Table 19.1), an autonomous parvovirus of unknown host, was shown to have broad glioma specificity in a screen of 12 parvoviruses for oncolytic activity against five glioma cell lines, with H-1PV only killing two.[26]

Retroviruses are positive-strand RNA viruses with an RNA genome that is reverse-transcribed into DNA that is integrated into the host genome. Retroviruses only integrate and replicate in mitotic cells, which provides specificity for dividing tumor cells. Typically, retroviruses have been used as replication-deficient gene delivery vectors; however, replication-competent gamma-retroviruses (RCRs), derived from murine leukemia virus (MLV), provide for amplification of the gene delivery vector and spread throughout the tumor.[43] RCRs are nonlytic and noncytotoxic, so to be oncolytic, they must express therapeutic transgenes, usually cytotoxic or suicide genes (Table 19.2, discussed below). RCRs can incorporate up to 8 kb of foreign DNA, and transgenes are stably expressed because of integration.[43]

PV is a human positive-strand RNA picornavirus that is neurotropic. Its receptor, CD155, is highly expressed on GBM cells, but also on normal cells such as motor neurons.[27] To eliminate neurotoxicity, its internal ribosomal entry site was replaced with a nonneurotoxic internal ribosomal entry site from human rhinovirus type 2. To further improve safety, PVS-RIPO (Table 19.1), is derived from the Sabin polio vaccine strain.[27] Genetic instability of RNA viruses is a serious safety concern; however, in vivo serial passage in gliomas did not alter PVS-RIPO specificity. PVS-RIPO replication in glioma cells is promoted by activation of Mnk1 and stimulation of cap-independent translation.[27]

NDV is an avian paramyxovirus with a negative-strand RNA genome and is nonpathogenic in humans. Defects in antiviral immunity, resistance to apoptosis, and induction of autophagy, found in many cancers including GBM, endow cancer specificity for NDV.[87] NDV replicates rapidly in cancer cells and has pleiotropic immunostimulatory effects in vivo, beneficial properties for an OV. It is only recently that reverse genetics has been developed for NDV, so a variety of naturally occurring strains have been used as oncolytic NDV against GBM. These range from pathogenic velogenic and mesogenic (MTVH68) to nonpathogenic lentogenic (NDV-HUJ, Hitchner B1) poultry strains[87] (Table 19.1). Another vaccine strain, V4UPM, inhibited tumor growth and induced apoptosis in U87MG subcutaneous tumor-bearing mice.[87]

MV is a human paramyxovirus with similar oncolytic activities as NDV; however, MV is known to be neurotropic and in rare human cases causes encephalitis.[62] The attenuated Edmonston vaccine strain has been used as the backbone for most recombinant oncolytic MV, and can accommodate up to 6 kb of inserted RNA.[62] The MV hemagglutinin (H) protein binds to CD46 receptors, often highly expressed in GBM. Unlike other RNA viruses, MV exhibits robust genetic stability after passaging in human cells. Oncolytic MV has been shown to infect, replicate in, and kill human medulloblastoma and glioma cell lines,[62,88] as well as GSCs, and prolong survival of GSC tumor-bearing mice.[33] The MV fusion (F) protein causes membrane fusion and syncytia formation, which leads to apoptosis. There are a number of ways to target oncolytic MV to GBM. MV replication can be restricted in non-GBM cells by insertion of a brain-specific microRNA (miRNA-7) target sequence that is downregulated in glioma[62] (Table 19.1). MV can also be retargeted to bind to GBM-specific receptors after ablating H binding to CD46 and SLAM, and inserting IL13 for IL13Rα2 binding or anti-CD133 scFv for binding to CD133 on GSCs into H[31,32] (Table 19.1). Two transgenes (carcinoembryonic antigen (CEA) and sodium iodide symporter (NIS)) have been inserted into MV to permit noninvasive monitoring of OV replication in the clinic. MV-expressing CEA (MV-CEA) has demonstrated success in several GBM animal models, including intracranial GSC xenografts in nude mice, where serum levels of CEA were a measure of virus replication[62] (Table 19.1). The safety of MV-CEA was demonstrated after intracerebral injection in rhesus macaques,[62] in advance of a clinical trial (Table 19.3). An MV expressing the NIS (MV-NIS) (Table 19.2) allows for in vivo monitoring of ^{99m}Tc or ^{123}I uptake or can be used in combination with ^{131}I for radiotherapy.[62] The combination with cytotoxic ^{131}I only modestly improved survival in both GBM and medulloblastoma models.[62,88]

VSV is a negative-strand RNA rhabdovirus, with ubiquitous receptor binding and broad species and tissue tropism. VSV is highly sensitive to innate type I IFN responses, which are often lacking in tumor cells, including GBM, allowing VSV to specifically target tumor cells. Although VSV has not been associated with any human disease, it is neurotoxic in animal models so efforts have been made to attenuate its neuropathogenicity.[40] Mutations in the VSV-M protein, in particular at Met-51, renders the virus unable to block antiviral innate responses, which improves safety but does not affect replication in cancer cells. VSVΔM51 is efficacious in killing human glioma xenografts even after systemic delivery,[34] and M51R VSV was even effective in U87 glioma cells overexpressing antiapoptotic Bcl-X[35] (Table 19.1). To further improve glioma specificity, screens have been undertaken for VSV mutants with enhanced glioma specificity. VSV-rp30 (Table 19.1), with single mutations in the P and L genes, was isolated by serial passage on glioma cells in vitro followed by lack of adsorption to fibroblasts. Unfortunately, this virus was still quite cytotoxic to normal human glia.[89] VSV-p1-GFP was identified in a screen for selective cytotoxicity in glioma cells and not normal glial cells, even in the presence of IFN-α, and found to be nonneurovirulent in mice.[89] Recently, a semireplication-competent VSV vector system was created (srVSV; Table 19.1), where viruses lacking the viral polymerase (DL) were combined with viruses lacking the glycoprotein (DG) so that only cells infected with both viruses could replicate. This virus combination was as efficacious as wild-type VSV in subcutaneous human G62 xenografts and, as opposed to wild-type virus, did not cause any neurotoxicity.[37]

RV is a double-stranded RNA virus that is nonpathogenic to humans. Serotype 3 was shown to have oncolytic activity due to a requirement for activated Ras and/or inactive PKR signaling, which are often present in cancer cells including GBM.[39] All four serotypes of RV have oncolytic activity against primary GBM cells in vitro.[90] RV serotype 3 was very effective in vitro in 20 of 24 glioma cell lines and in vivo in intracranial glioma models in immunodeficient mice and immunocompetent rats.[38,39] Interestingly, in rats with bilateral tumors where a single glioma was injected, only the ipsilateral and not contralateral tumor was reduced in size, suggesting no immune-mediated effects.[38] After high-dose intracerebral injection of RV into cynomolgus monkeys, some clinical signs were observed and virus RNA was detected in the urine.[38] RV is one of the only genetically unmodified OVs to enter clinical trial (Reolysin®; Table 19.3).[41]

19.3.2 Cytotoxic or Suicide Gene/Prodrug Therapy

Cytotoxic chemotherapy and radiotherapy have been the standards of care for GBM patients, with limited success because of their negative effects on surrounding healthy tissue, systemic toxicity, and small therapeutic indices. Cytotoxic gene therapy is tumor selective and offers a way to overcome some of the limitations of systemic chemotherapy. The earliest gene therapy clinical trial for GBM utilized a "suicide" gene therapy strategy, which continues to be one of the most frequently employed strategies. Suicide gene therapy involves cancer-selective delivery of a drug-activating enzyme (the suicide gene) that converts a systemically administered nontoxic prodrug to cytotoxic metabolites.[91] Suicide genes are usually absent or expressed at very low levels in mammalian cells, thus the cytotoxic metabolites would be generated locally in the tumor.

The most common and best-studied suicide gene/prodrug combination is HSV TK/ganciclovir (GCV). TK catalyzes the phosphorylation of GCV, converting it into GCV-monophosphate, which is subsequently converted by cellular kinases to GCV-triphosphate, a cytotoxic nucleoside analog that can be incorporated into the DNA of actively proliferating cells, disrupting DNA replication and halting cell division.[91] Thus, phosphorylated GCV is only toxic to dividing cells. GCV-triphosphate was found to kill surrounding tumor cells that had not been transduced with the suicide gene, the so-called bystander effect (Figure 19.1), which is an important element of the therapy.[91] Unfortunately, GCV-triphosphate cannot cross the cell membrane and the bystander effect is dependent upon gap junctions or release from dead cells.

Most of the TK/GCV clinical trials have used replication-defective retroviruses or Ad for gene delivery, and they were hampered by poor transduction of cancer cells and low activity of the activated prodrug.[4] To overcome these limitations, a number of approaches have been described: TK mutants with increased activity, alternate delivery methods, other suicide genes/prodrugs with diffusible and more toxic prodrug metabolites, and combinations thereof.[91] Because TK/GCV therapy was shown to induce an antitumor immune response, combinations with immunostimulatory genes (more details in next section) are an obvious strategy. A promising combination for GBM is replication-defective Ad expressing TK with the immune-stimulatory cytokine fms-like tyrosine kinase 3 ligand (Flt3L) to recruit dendritic cells. A bicistronic gutless adenovirus vector with inducible Flt3L, HC-Ad-TK/TetOn-Flt3L (Table 19.4), has been generated for clinical use.[93] Helper-dependent, gutless, or high-capacity Ad vectors have all their viral genes eliminated, so they induce only minimal anti-Ad immune responses and provide cloning space for up to 35kb. Intracranial injection of up to 10^9 viral particles of HC-Ad-TK/TetOn-Flt3L followed by doxycycline for Flt3L expression and GCV was safe, whereas 10^{10} viral particles had toxic side effects.[93] In contrast, 10^8 viral particles significantly extended survival of glioma-bearing rats, which was associated with the development of immune memory against GBM antigens.[93]

The bacterial or yeast enzyme CD/5-fluorocytosine (5-FC) prodrug system is similar to the TK system, where CD converts 5-FC into the cytotoxic metabolite 5-FU that is phosphorylated and inhibits thymidine synthetase.[91] In contrast to GCV, 5-FU diffuses out of cells. ACE-CD and Toca 511 (Table 19.2), RCRs that express CD, were shown to prolong survival in human and mouse immunocompetent glioma models.[43,44] Delivery of viral vectors and broad distribution within the tumor is a problem with all gene therapy strategies, especially in the brain. One promising technique is CED, using a fluid pressure gradient and bulk flow to infuse vectors within the extracellular fluid space. CED of Toca 511 resulted in broader distribution within the tumor than direct injection, and enhanced survival after 5-FC administration.[106] CD/5-FC cytotoxicity can be enhanced by coexpressing uracil phosphoribosyl transferase (UPRT), which phosphorylates 5-FU to 5-FU-monophosphate.[91] Purine nucleoside phosphorylase (PNP), which converts F-araAMP to diffusible toxic 2-FA, was inserted into an RCR (ACE-PNP; Table 19.2). ACE-PNP alone had no effect on intracranial glioma growth, while administration of F-araAMP significantly extended survival, although all mice eventually succumbed to disease.[45] Oncolytic HSV MGH2.1, in a G207-like backbone, expresses two prodrug activating enzymes: CYP2B1, converting cyclophosphamide

TABLE 19.4 Nonreplicating Cytotoxic Gene Therapy Vectors and Cells

Vector Name	Vector Type	Transgene	Model System (Mouse or Rat/Cell Type)	Additional Treatment	References
LCMV-GP pseudotyped	Lentivirus	HSV-TK	Rat, i.c. GSC	GCV	92
VSV-G pseudotyped	Lentivirus	HSV-TK	Rat, i.c. GSC	GCV	92
HC-Ad-TK/TetOn-Flt3L	Ad	HSV-TK, Tet-On Flt3L	Rat, i.c. CNS-1	Valacyclovir	93
Ad-stTRAIL	Ad	TRAIL	Nude, i.c. U87, U251		94
Ad.5/3-mda7	Ad	mda7/IL-24	Nude, i.c. primary GBM		95
Ad-mhiL-4.TRE.mhIL-13-PE	Ad	Mutated IL-13 fused to PE	Nude, i.c. U251 Rag1$^{-/-}$, i.c. human primary xenograft C57/B6, i.c. GL26-H2		96
HB1-F3.CD (F3-CD)	NSC	CD	Rat, i.c. U373MG Nude, i.c. U251, U87, Nude, i.c. medulloblastoma Daoy Nude, i.c. breast MDA-MB435	5-FC	97 98
F3.rCE	NSC	rCE	Nude, subdural medulloblastoma D283	CPT-11	99
MSCtk	MSC	HSV-TK	Rat, i.c. C6	GCV	100
TK-MSC	MSC	HSV-TK	Nude, i.c. U87	GCV	101
CDy-AT-MSC	MSC	CD::UPRT	Rat, i.c. C6	5-FC	102
UCB-TRAIL	MSC	stTRAIL	Nude, i.c. U87		103
MSC-S-TRAIL	MSC	Secreted, extracellular TRAIL fused to hFlt3L	SCID, i.c. GSC		104
hAT-MSC.TRAIL	MSC	TRAIL	Rat, i.c. F98		105

The model systems involve the implantation of cancer cells (human except for mouse G26-H2 and rat C6, CNS-1, and F98) into mice (nude, Rag1$^{-/-}$, or SCID) or rats. Abbreviations: 5-FC, 5-fluorocytosine; Ad, adenovirus; CD, cytosine deaminase; GBM, glioblastoma; GCV, ganciclovir; GSC, glioblastoma stem cell; HSV, herpes simplex virus; i.c., intracerebral; LCMV, lymphocytic choriomeningitis virus; MSC, mesenchymal stromal (stem) cell; NSC, neural stem cell; PE, *Pseudomonas* exotoxin; rCE, rabbit carboxylesterase; stTRAIL, secretable trimeric TRAIL; TK, thymidine kinase; TRAIL, tumor necrosis factor-related apoptosis-inducing ligand; UPRT, uracil phosphoribosyl transferase; VSV, vesicular stomatitis virus.

(CPA) into the alkylating agent phosphoramide mustard; and secreted carboxylesterase (sCE), converting irinotecan (CPT-11) into topoisomerase I inhibitor SN-38.[46] A toxicology study, in preparation for a clinical trial, found that MGH2.1 was safe after intracranial injection in nontumor-bearing mice, but the combination with CPA and CPT-11 caused some lethality at 5×10^6 pfu.[46] CPA also inhibits innate immune responses that block virus replication and spread, so its administration improves HSV oncolytic activity.[46] Other cytotoxic strategies include secreted proapoptotic proteins, such as TRAIL or mda-7/IL-24, that are selectively active in tumor cells[95,107] and cytotoxins, such as *Pseudomonas* exotoxin fused to a ligand (IL-13) for a glioma-specific receptor[96] (Table 19.4). Oncolytic HSV expressing sTRAIL (Table 19.2) is able to overcome TRAIL resistance of glioma cells by inducing caspase-mediated apoptosis in vitro and in vivo, extending the survival of glioma-bearing mice compared to oHSV alone.[50]

To overcome the limited number and distribution of suicide gene transduced glioma cells after replication-deficient retrovirus injection, RCRs, migratory stem or progenitor cells (NSCs and MSCs), or other vectors have been evaluated. Replication-deficient pseudotyped lentiviral vectors had much higher transduction efficiency in vivo, including with GSCs, than retroviral vectors, and GCV treatment greatly prolonged survival.[92] A human immortalized NSC line, HB1-F3, transduced with *Escherichia coli* CD, has been developed for treating brain tumors.[97] It was efficacious in a number of brain tumor models (glioma, medulloblastoma), and metastatic breast cancer[97] (Table 19.4). HB1-F3.CD cells displayed glioma tropism after injection contralateral to the tumor, even in the presence of dexamethasone or after radiation.[98] The HB1-F3.

CD NSCs are currently in clinical trial for GBM (Table 19.3). NSCs have also been generated expressing sCE and demonstrated efficacy, with long-term survivors, in a medulloblastoma model that was sensitive to CPT-11 alone.[99]

MSCs (human and mouse) can migrate in the brain or after carotid artery injection in a range of GBM mouse models, including to human GSC-derived and immunocompetent genetically induced tumors.[104,108] Some MSCs alone have been shown to suppress glioma growth through inhibition of angiogenesis and downregulation of PDGF signaling.[109] Different viral vectors have been used to transduce MSCs, including retrovirus, lentivirus, Ad, and baculovirus.[101,104] A number of suicide genes (TK, CD, CD::UPRT, sTRAIL) have been introduced into MSCs and shown to be effective with prodrug in treating intracranial glioma models (Table 19.4).[101,102] Unfortunately, some of these studies coimplanted the MSCs with the glioma cell lines, which obviates the migratory advantage of this strategy, or used human MSCs to treat immunogenic gliomas in rats. TRAIL is another popular class of transgenes to introduce into stem/progenitor cells, and several studies have shown therapeutic effects and increased apoptosis in glioma xenograft models (Table 19.4). As normal cells, MSCs are resistant to TRAIL; however, many glioma cells are also resistant to TRAIL, despite the expression of TRAIL receptors.[104] To overcome this, MSC-TRAIL was combined with a lipoxygenase inhibitor MK886, which increased DR5 (TRAIL receptor) and decreased antiapoptotic protein expression, leading to prolonged survival in vivo.[110] Differentiated embryonic stem cells or induced pluripotent stem cells have some potential advantages over adult stem cells; they can be stably genetically modified, proliferate indefinitely, and could be established with a whole range of HLA types for immunologic matching. Mouse embryonic stem cells expressing TRAIL or mda-7 have been generated by site-specific recombination and then terminally differentiated into astrocytes, which migrate in vivo and induce apoptosis.[63] However, treatment activity in orthotopic glioma models has not been demonstrated.

19.3.3 Immune-Stimulatory Gene Therapy

The immune-privileged state of the brain is a major obstacle to immunotherapy against brain tumors. The brain is somewhat isolated from peripheral immune responses because it lacks lymphatics that impede immune cells from exiting the brain parenchyma, and has only a limited supply of antigen-presenting cells. To further confound this, the GBM microenvironment is very immune suppressed, with elevated regulatory T cells and myeloid-derived suppressor cells.[111] Despite these challenges, significant progress with immune-mediated gene therapy strategies has been achieved against glioma.[111] Recruitment of antigen-presenting cells to the tumor and presentation of tumor-associated antigens are critical to the generation of an adaptive immune response. That is the basis for the TK-Flt3L replication-defective Ad combination described above. The addition of replication-defective Ad vectors expressing IL-2 or IFNγ further enhanced Ad-Flt3L/TK efficacy.[112] A similar strategy has been described using oncolytic HSV-expressing Flt3L. G47Δ-Flt3L significantly extended survival in the mouse CT2A glioma model, with about 40% long-term survivors compared with G47Δ-Empty (no transgene), which had minimal effect (Table 19.2).[52] It is important to bear in mind that OV infection alone often induces inflammatory responses and immunogenic cancer cell death, which induces innate and adaptive responses that can be beneficial and/or detrimental.

Another immunotherapy strategy is to express cytokines to enhance adaptive immune responses. JX-594, a TK-deleted VV expressing granulocyte macrophage colony stimulating factor (GM-CSF; Table 19.2), is currently in clinical trials for peripheral tumors.[113] JX-594 inhibited tumor growth and increased survival in two immunocompetent GBM models, which was associated with increased GM-CSF-dependent inflammation.[51] Most human GSCs tested in vitro were susceptible to JX-594 replication and cytotoxicity, although considerably less so than U87 glioma cells.[51] IL-12 is one of the more potent antitumor cytokines, driving a Th1 response. Several groups have delivered IL-12 using a variety of gene therapy vectors. An oncolytic HSV expressing murine IL-12 (MOO2; Table 19.2) was more efficacious in an immunocompetent glioma model than its parental, no transgene, oncolytic HSV.[53] Importantly, it elicited no toxicity after intracerebral inoculation in HSV-susceptible nonhuman primates, but did increase activation of lymphocytes.[53] For clinical trials, a similar oncolytic HSV was constructed expressing human IL-12 (M032; Table 19.3).

One issue with preclinical studies of immunotherapy for brain tumors is the dearth of representative animal models. A new mouse GSC-derived intracranial tumor model was recently described using 005 GSCs isolated from GBM arising in lentivirus-transduced transgenic mice.[54] Tumors arising after 005 GSC implantation have hallmarks of human GBM: morphologically heterogeneous with multinucleated giant cells, cells expressing stem cell markers, invasive growth, angiogenic with aberrant vasculature, and poorly immunogenic with an immunosuppressive microenvironment.[54] This model has been used to evaluate G47Δ-mIL12 (Table 19.2), which had multifaceted effects on the tumors and their microenvironment, killing GSCs and bulk tumor cells, inhibiting tumor growth in a T-cell dependent fashion, inducing IFNγ synthesis, reducing immunosuppressive tumor Tregs, and inhibiting angiogenesis.[54] IL-12 has also been expressed from MSCs, after transduction with replication-defective Ad (UCB-MSC-IL12M; Table 19.2), which prolonged survival of C57Bl/6 mice bearing intracranial GL26 tumors after intratumoral injection.[56] The surviving mice developed long-term immunity, which protected them from rechallenge with GL26 cells in the ipsilateral or contralateral hemispheres.[56]

19.3.4 Tumor Microenvironment Disrupting

Targeting the tumor microenvironment is an attractive approach because it consists of normal cells that should not develop resistance to the therapy. Normally, neovascularization is a tightly regulated balance between naturally occurring angiogenesis activators and inhibitors. In tumors, however, the dividing cancer cells outgrow the normal vasculature, increasing hypoxia and the expression of proangiogenic factors.[58] GBM is a highly vascularized tumor, but there are limitations with current antiangiogenic drugs such as bevacizumab (Avastin®), an anti-VEGF monoclonal antibody, which does not significantly improve overall survival and has the detrimental effect of increasing glioma invasiveness.[58] Developing alternative strategies, such as local expression in the tumor or combination strategies (i.e., targeting multiple angiogenic pathways) might be better approaches, as angiogenesis inhibition is cytostatic and not cytotoxic.[58] A number of antiangiogenic factors have been expressed from oncolytic HSV. Angiostatin, an endogenous inhibitor of angiogenesis, was inserted into G47Δ.[59] A single treatment of G47Δ-mAngio (Table 19.2) significantly extended survival of glioma-bearing mice over nontransgene containing G47Δ, and this was associated with decreased microvascular density and VEGF expression.[59] This could be further improved by combining a lower dose of G47Δ-mAngio with a low (noninvasive) dose of bevacizumab.[59] IL-12, in addition to its immunostimulatory activity, is also antiangiogenic, so that G47Δ-mIL12 was efficacious in human glioma models.[55] When combined with G47Δ-mAngio, it synergized in inhibiting glioma growth, curing mice with U87 gliomas and extending survival in a human GSC (MGG4) model.[55] A number of other naturally occurring angiogenesis inhibitors have been expressed in gliomas.[58] Vasculostatin (Vstat120), a proteolytic fragment of BAI1 cloned into transcriptionally targeted oHSV 34.5ENVE, and CXCL4 (PF4) in G47Δ (Table 19.2) improved the efficacy of oncolytic HSV in human glioma models, and this was associated with decreased neovascularization.[58] Isthmin, an angiogenesis inhibitor derived from the brain of *Xenopus*, delivered with a replication-defective Ad (Table 19.2), inhibited angiogenesis and intracranial tumor growth.[61] Secreted thrombospondin-1, expressed from NSCs after lentivirus transduction, reduced tumor angiogenesis and growth.[58] Combining antiangiogenic gene therapy with suicide gene therapy can also improve efficacy, for example, oncolytic Ad expressing arresten (collagen IV fragment) and TRAIL,[49] and MSCs transduced with endostatin (an endogenous angiogenesis inhibitor) and sCE (Table 19.2).[57]

19.3.5 Interfering RNA

Some proteins are essential for the survival and/or proliferation of tumor cells, thus are promising targets for inhibition or inactivation. A gene therapy strategy to target such proteins is through RNAi. Major targets for RNAi therapy include "undruggable" oncogenes and genes that are involved in angiogenesis, metastasis, survival, antiapoptosis, and resistance to chemotherapy.[114] RNAi as a tumor therapeutic has advantages over traditional pharmaceutical drugs: they are easily designed, requiring only knowledge of the target mRNA sequence; the identification of selective and active sequences is fast; and production/synthesis is rapid and scalable.[114] Conversely, they can have off-target effects that are specific or nonspecific, induce innate immune responses, or saturate the endogenous RNAi machinery. A number of mRNAs have been targeted with RNAi in glioma models, either as RNA packaged in synthetic delivery vehicles (Bcl2L12, EGFR, pleiotrophin, hTERT) or expressed from virus vectors (Hec1, EGFR, Rad51, Bcl2, hTERT).[76,114] MicroRNAs are endogenous small RNAs that regulate almost a third of mRNAs. Because they are not 100% sequence complementary to their targets, a single miRNA can affect a large number of different transcripts. MiR-124a, which inhibits STAT3 signaling, is downregulated in GBM and microglia/macrophages. Intravenous administration of pre-miR-124a in liposomes was efficacious in treating mouse gliomas (GL261 and genetically engineered RCAS tumors) in immunocompetent mice but not in nude mice or after depletion of CD4+ or CD8+ T cells, demonstrating a role for T cell immunity.[115] A dsRNA targeting tenascin-C (ATN-RNA) is the only RNAi strategy evaluated in the clinic. ATN-RNA (80 µg) was injected into the tumor border in glioma patients (WHO grades II–IV) after subtotal resection.[114,116] Because of the study design, it is difficult to draw conclusions about efficacy; however, patients receiving ATN-RNA had improved median survival and KPS changes compared to patients receiving brachytherapy.[116]

19.4 STATUS OF CLINICAL TRIALS FOR GBM

Gene therapy clinical trials for GBM have been promising as far as safety is concerned, with no maximum tolerated dose reached in any trial; however, overall benefits have been mixed and phase III clinical trials have only been completed with HSV-TK gene therapy.[41] The earliest gene therapy trials for GBM, starting in 1977, used vector-producing cells with replication-defective MLV expressing TK. This culminated in an international phase III trial, where 248 patients with newly diagnosed GBM were randomized to standard therapy (surgical resection and radiotherapy at that time) or standard therapy plus adjuvant gene therapy with MLV-TK vector producer cells during surgery followed by GCV on days 14–27. While this trial convincingly demonstrated safety, there was no significant difference between the two arms in time to progression or overall survival.[41] Recent gene therapy clinical trials for GBM are described in Table 19.3. To increase transduction efficiency, replication-defective Ad vectors expressing TK have been tested in the clinic. An early phase I trial, comparing replication-defective retrovirus with Ad, found that overall survival was twice as long with Ad-TK.[4] A phase Ib trial combining AdV-tk and valacyclovir, in place of GCV, with early radiation in newly diagnosed GBM patients, reported no safety issues, with encouraging survival rates and significant T cell infiltration in four of four patient posttreatment specimens.[65] A European randomized phase III trial with AdvHSV-tk (Sitimagene ceradenovec® from Ark Therapeutics, London, UK) in 250 newly diagnosed GBM patients amenable to complete resection was recently completed.[66] Unfortunately, there was no difference in overall survival between the treatment and control groups, although the time to reintervention or death (composite primary endpoint) was improved. The number of patients with serious adverse events or adverse events related to treatment was higher in the AdvHSV-tk group.[66] Interestingly, from an immunotherapy standpoint, treated patients with high baseline antibody titer had a significantly better composite primary endpoint. A confounding issue was the emergence of temozolomide as standard of care after the trial began, with a post-hoc analysis suggesting that patients with unmethylated MGMT promoter benefited most from treatment.[66]

Greater optimism has recently been generated with preclinical success of newer gene therapy strategies, especially armed OVs, that have just begun to enter clinical trials for GBM.[8] OVs make up the majority of currently active clinical trials for GBM. Second-generation OVs have demonstrated safety in humans in early clinical trials, and these viruses and third-generation viruses are being further pursued. Oncolytic HSV G207, which exhibited safety and anecdotal efficacy in an early phase 1 clinical trial, was examined in a phase 1b trial. This trial demonstrated only a marginal increase in survival of patients; however, it was the first study to demonstrate oncolytic HSV replication in the tumor.[64] It has been followed in Japan by a clinical trial using G47Δ, a third-generation derivative of G207. A retargeted and tumor-specific CRAd, AdV-delta24-RGD, is the only oncolytic Ad currently in phase I clinical trials for GBM patients. Several RNA viruses are being assessed in the clinic. Reolysin, a nonengineered RV, was well tolerated and safe in a phase I trial and has completed enrollment of a phase I/II trial.[41] Intravenous NDV-HUJ was well tolerated in GBM patients in a phase 1 study; no toxicity was observed, no maximum tolerated dose achieved, and 1 out of 11 patients had a complete response to treatment.[41] Poliovirus PVS-RIPO is in phase I trial where it is being delivered by CED, and has resulted in large clinical and radiographic responses. MV-CEA, expressing CEA for noninvasive, real-time monitoring of viral gene expression, is currently in phase I trial for recurrent GBM.[4] There are currently two phase I clinical trials of Toca 511, expressing an optimized CD, either intratumoral or into the resection cavity using real-time MRI guidance for injection.

19.5 CURRENT CHALLENGES AND FUTURE DIRECTIONS

Treating malignant tumors in the brain remains a daunting therapeutic challenge; however, new gene therapy strategies provide numerous opportunities to attack the tumors via multiple mechanisms, and preclinical studies are showing great promise. The distinctive environment in the brain, where the BBB restricts access, coupled to the highly invasive nature of GBM, presents unique obstacles compared with tumors in other organs in the body. Thus, delivery often requires direct injection into the tumor bed at the time of biopsy or into the resection cavity or border at surgery, and remains a major impediment to effective therapy, especially for replication-defective vectors. Replication-competent OVs and tumor-homing stem or progenitor cells have the advantage of spread throughout the tumor and motility. Developing new and effective delivery techniques, routes of administration, and vehicles for genetic material remains a goal for brain tumor gene therapy. The immune-privileged status of the brain, with a lack of lymphatics, and the overall immunosuppressive character of GBM, make these tumors a difficult challenge for immunotherapy, but one important to overcome. The ability to easily combine gene therapy strategies (cytotoxic, antiangiogenic, and immune-stimulatory) is a strength of this approach and an area ripe for expanded study. Successful cancer therapy typically involves multimodal approaches, and gene therapy is no different, so that combining gene therapy with other therapeutic modalities, including standards-of-care, is likely to improve outcomes for brain tumor patients. While there are currently no gene therapy products approved to treat brain tumors, which has limited investor and corporate interest, it is critical that there be continued efforts to translate new vectors/vehicles and strategies to the clinic. The diversity of approaches and targets of brain tumor gene therapy provide a plethora of opportunities to improve the outcomes for patients.

REFERENCES

1. Murphy AM, Rabkin SD. Current status of gene therapy for brain tumors. *Transl Res* 2013;**161**(4):339–54.
2. Kaye AH, Laws ER. *Brain tumors.* 3rd ed. Saunders; 2011.
3. Ostrom QT, Gittleman H, Farah P, Ondracek A, Chen Y, Wolinsky Y, et al. CBTRUS statistical report: primary brain and central nervous system tumors diagnosed in the United States in 2006-2010. *Neuro Oncol* 2013;**15**(Suppl. 2):ii1–56.
4. Tobias A, Ahmed A, Moon KS, Lesniak MS. The art of gene therapy for glioma: a review of the challenging road to the bedside. *J Neurol Neurosurg Psychiatry* 2013;**84**(2):213–22.
5. Kwiatkowska A, Nandhu MS, Behera P, Chiocca EA, Viapiano MS. Strategies in gene therapy for glioblastoma. *Cancers (Basel)* 2013;**5**(4):1271–305.
6. Miest TS, Cattaneo R. New viruses for cancer therapy: meeting clinical needs. *Nat Rev Microbiol* 2014;**12**(1):23–34.
7. Russell SJ, Peng KW, Bell JC. Oncolytic virotherapy. *Nat Biotechnol* 2012;**30**(7):658–70.
8. Zemp FJ, Corredor JC, Lun X, Muruve DA, Forsyth PA. Oncolytic viruses as experimental treatments for malignant gliomas: using a scourge to treat a devil. *Cytokine Growth Factor Rev* 2010;**21**(2–3):103–17.
9. Rabkin S. Oncolytic HSV vectors for cancer therapy. In: Weller S, editor. *Alphaherpesviruses: molecular virology.* Norfolk (UK): Caister Academic Press; 2011. pp. 401–43.
10. Parker JN, Bauer DF, Cody JJ, Markert JM. Oncolytic viral therapy of malignant glioma. *Neurotherapeutics* 2009;**6**(3):558–69.
11. Wakimoto H, Kesari S, Farrell CJ, Curry Jr WT, Zaupa C, Aghi M, et al. Human glioblastoma-derived cancer stem cells: establishment of invasive glioma models and treatment with oncolytic herpes simplex virus vectors. *Cancer Res* 2009;**69**(8):3472–81.
12. Kanai R, Wakimoto H, Martuza RL, Rabkin SD. A novel oncolytic herpes simplex virus that synergizes with phosphoinositide 3-kinase/Akt pathway inhibitors to target glioblastoma stem cells. *Clin Cancer Res* 2011;**17**(11):3686–96.
13. Kanai R, Zaupa C, Sgubin D, Antoszczyk SJ, Martuza RL, Wakimoto H, et al. Effect of gamma34.5 deletions on oncolytic herpes simplex virus activity in brain tumors. *J Virol* 2012;**86**(8):4420–31.
14. Kambara H, Okano H, Chiocca EA, Saeki Y. An oncolytic HSV-1 mutant expressing ICP34.5 under control of a nestin promoter increases survival of animals even when symptomatic from a brain tumor. *Cancer Res* 2005;**65**(7):2832–9.
15. Gambini E, Reisoli E, Appolloni I, Gatta V, Campadelli-Fiume G, Menotti L, et al. Replication-competent herpes simplex virus retargeted to HER2 as therapy for high-grade glioma. *Mol Ther* 2012;**20**(5):994–1001.
16. Uchida H, Marzulli M, Nakano K, Goins WF, Chan J, Hong CS, et al. Effective treatment of an orthotopic xenograft model of human glioblastoma using an EGFR-retargeted oncolytic herpes simplex virus. *Mol Ther* 2013;**21**(3):561–9.
17. Fueyo J, Alemany R, Gomez-Manzano C, Fuller GN, Khan A, Conrad CA, et al. Preclinical characterization of the antiglioma activity of a tropism-enhanced adenovirus targeted to the retinoblastoma pathway. *J Natl Cancer Inst* 2003;**95**(9):652–60.
18. Jiang H, Gomez-Manzano C, Lang FF, Alemany R, Fueyo J. Oncolytic adenovirus: preclinical and clinical studies in patients with human malignant gliomas. *Curr Gene Ther* 2009;**9**(5):422–7.
19. Auffinger B, Ahmed AU, Lesniak MS. Oncolytic virotherapy for malignant glioma: translating laboratory insights into clinical practice. *Front Oncol* 2013;**3**(32).
20. Hoffmann D, Meyer B, Wildner O. Improved glioblastoma treatment with Ad5/35 fiber chimeric conditionally replicating adenoviruses. *J Gene Med* 2007;**9**(9):764–78.
21. Lun XQ, Jang JH, Tang N, Deng H, Head R, Bell JC, et al. Efficacy of systemically administered oncolytic vaccinia virotherapy for malignant gliomas is enhanced by combination therapy with rapamycin or cyclophosphamide. *Clin Cancer Res* 2009;**15**(8):2777–88.
22. Zhang Q, Liang C, Yu YA, Chen N, Dandekar T, Szalay AA. The highly attenuated oncolytic recombinant vaccinia virus GLV-1h68: comparative genomic features and the contribution of F14.5L inactivation. *Mol Genet Genomics* 2009;**282**(4):417–35.
23. Advani SJ, Buckel L, Chen NG, Scanderbeg DJ, Geissinger U, Zhang Q, et al. Preferential replication of systemically delivered oncolytic vaccinia virus in focally irradiated glioma xenografts. *Clin Cancer Res* 2012;**18**(9):2579–90.
24. Zemp FJ, Lun X, McKenzie BA, Zhou H, Maxwell L, Sun B, et al. Treating brain tumor-initiating cells using a combination of myxoma virus and rapamycin. *Neuro Oncol* 2013;**15**(7):904–20.
25. Geletneky K, Kiprianova I, Ayache A, Koch R, Herrero YCM, Deleu L, et al. Regression of advanced rat and human gliomas by local or systemic treatment with oncolytic parvovirus H-1 in rat models. *Neuro Oncol* 2010;**12**(8):804–14.
26. Paglino JC, Ozduman K, van den Pol AN. LuIII parvovirus selectively and efficiently targets, replicates in, and kills human glioma cells. *J Virol* 2012;**86**(13):7280–91.
27. Goetz C, Dobrikova E, Shveygert M, Dobrikov M, Gromeier M. Oncolytic poliovirus against malignant glioma. *Future Virol* 2011;**6**(9):1045–58.
28. Zulkifli MM, Ibrahim R, Ali AM, Aini I, Jaafar H, Hilda SS, et al. Newcastle diseases virus strain V4UPM displayed oncolytic ability against experimental human malignant glioma. *Neurol Res* 2009;**31**(1):3–10.
29. Alkassar M, Gartner B, Roemer K, Graesser F, Rommelaere J, Kaestner L, et al. The combined effects of oncolytic reovirus plus Newcastle disease virus and reovirus plus parvovirus on U87 and U373 cells in vitro and in vivo. *J Neurooncol* 2011;**104**(3):715–27.
30. Leber MF, Bossow S, Leonard VH, Zaoui K, Grossardt C, Frenzke M, et al. MicroRNA-sensitive oncolytic measles viruses for cancer-specific vector tropism. *Mol Ther* 2011;**19**(6):1097–106.
31. Allen C, Paraskevakou G, Iankov I, Giannini C, Schroeder M, Sarkaria J, et al. Interleukin-13 displaying retargeted oncolytic measles virus strains have significant activity against gliomas with improved specificity. *Mol Ther* 2008;**16**(9):1556–64.
32. Bach P, Abel T, Hoffmann C, Gal Z, Braun G, Voelker I, et al. Specific elimination of CD133+ tumor cells with targeted oncolytic measles virus. *Cancer Res* 2013;**73**(2):865–74.

33. Allen C, Opyrchal M, Aderca I, Schroeder MA, Sarkaria JN, Domingo E, et al. Oncolytic measles virus strains have significant antitumor activity against glioma stem cells. *Gene Ther* 2013;**20**(4):444–9.

34. Lun X, Senger DL, Alain T, Oprea A, Parato K, Stojdl D, et al. Effects of intravenously administered recombinant vesicular stomatitis virus (VSV(deltaM51)) on multifocal and invasive gliomas. *J Natl Cancer Inst* 2006;**98**(21):1546–57.

35. Cary ZD, Willingham MC, Lyles DS. Oncolytic vesicular stomatitis virus induces apoptosis in U87 glioblastoma cells by a type II death receptor mechanism and induces cell death and tumor clearance in vivo. *J Virol* 2011;**85**(12):5708–17.

36. Wollmann G, Rogulin V, Simon I, Rose JK, van den Pol AN. Some attenuated variants of vesicular stomatitis virus show enhanced oncolytic activity against human glioblastoma cells relative to normal brain cells. *J Virol* 2010;**84**(3):1563–73.

37. Muik A, Dold C, Geiss Y, Volk A, Werbizki M, Dietrich U, et al. Semireplication-competent vesicular stomatitis virus as a novel platform for oncolytic virotherapy. *J Mol Med (Berl)* 2012;**90**(8):959–70.

38. Yang WQ, Lun X, Palmer CA, Wilcox ME, Muzik H, Shi ZQ, et al. Efficacy and safety evaluation of human reovirus type 3 in immunocompetent animals: racine and nonhuman primates. *Clin Cancer Res* 2004;**10**(24):8561–76.

39. Maitra R, Ghalib MH, Goel S. Reovirus: a targeted therapeutic–progress and potential. *Mol Cancer Res* 2012;**10**(12):1514–25.

40. Wollmann G, Ozduman K, van den Pol AN. Oncolytic virus therapy for glioblastoma multiforme: concepts and candidates. *Cancer J* 2012;**18**(1):69–81.

41. Mohyeldin A, Chiocca EA. Gene and viral therapy for glioblastoma: a review of clinical trials and future directions. *Cancer J* 2012;**18**(1):82–8.

42. Kaur B, Cripe TP, Chiocca EA. "Buy one get one free": armed viruses for the treatment of cancer cells and their microenvironment. *Curr Gene Ther* 2009;**9**(5):341–55.

43. Tai CK, Kasahara N. Replication-competent retrovirus vectors for cancer gene therapy. *Front Biosci* 2008;**13**:3083–95.

44. Ostertag D, Amundson KK, Lopez Espinoza F, Martin B, Buckley T, Galvao da Silva AP, et al. Brain tumor eradication and prolonged survival from intratumoral conversion of 5-fluorocytosine to 5-fluorouracil using a nonlytic retroviral replicating vector. *Neuro Oncol* 2012;**14**(2):145–59.

45. Tai CK, Wang W, Lai YH, Logg CR, Parker WB, Li YF, et al. Enhanced efficiency of prodrug activation therapy by tumor-selective replicating retrovirus vectors armed with the *Escherichia coli* purine nucleoside phosphorylase gene. *Cancer Gene Ther* 2010;**17**(9):614–23.

46. Kasai K, Nakashima H, Liu F, Kerr S, Wang J, Phelps M, et al. Toxicology and biodistribution studies for MGH2.1, an oncolytic virus that expresses two prodrug-activating genes, in combination with prodrugs. *Mol Ther Nucleic Acids* 2013;**2**:e113.

47. Kaliberova LN, Krendelchtchikova V, Harmon DK, Stockard CR, Petersen AS, Markert JM, et al. CRAdRGDflt-IL24 virotherapy in combination with chemotherapy of experimental glioma. *Cancer Gene Ther* 2009;**16**(10):794–805.

48. Wohlfahrt ME, Beard BC, Lieber A, Kiem HP. A capsid-modified, conditionally replicating oncolytic adenovirus vector expressing TRAIL leads to enhanced cancer cell killing in human glioblastoma models. *Cancer Res* 2007;**67**(18):8783–90.

49. Li X, Mao Q, Wang D, Zhang W, Xia H. A fiber chimeric CRAd vector Ad5/11-D24 double-armed with TRAIL and arresten for enhanced glioblastoma therapy. *Hum Gene Ther* 2012;**23**(6):589–96.

50. Tamura K, Wakimoto H, Agarwal AS, Rabkin SD, Bhere D, Martuza RL, et al. Multimechanistic tumor targeted oncolytic virus overcomes resistance in brain tumors. *Mol Ther* 2013;**21**(1):68–77.

51. Lun X, Chan J, Zhou H, Sun B, Kelly JJ, Stechishin OO, et al. Efficacy and safety/toxicity study of recombinant vaccinia virus JX-594 in two immunocompetent animal models of glioma. *Mol Ther* 2010;**18**(11):1927–36.

52. Barnard Z, Wakimoto H, Zaupa C, Patel AP, Klehm J, Martuza RL, et al. Expression of fms-like tyrosine kinase 3 ligand by oncolytic herpes simplex virus type I prolongs survival in mice bearing established syngeneic intracranial malignant glioma. *Neurosurgery* 2012;**71**(3):741–8.

53. Markert JM, Cody JJ, Parker JN, Coleman JM, Price KH, Kern ER, et al. Preclinical evaluation of a genetically engineered herpes simplex virus expressing interleukin-12. *J Virol* 2012;**86**(9):5304–13.

54. Cheema TA, Wakimoto H, Fecci PE, Ning J, Kuroda T, Jeyaretna DS, et al. Multifaceted oncolytic virus therapy for glioblastoma in an immunocompetent cancer stem cell model. *Proc Natl Acad Sci USA* 2013;**110**(29):12006–11.

55. Zhang W, Fulci G, Wakimoto H, Cheema TA, Buhrman JS, Jeyaretna DS, et al. Combination of oncolytic herpes simplex viruses armed with angiostatin and IL-12 enhances antitumor efficacy in human glioblastoma models. *Neoplasia* 2013;**15**(6):591–9.

56. Ryu CH, Park SH, Park SA, Kim SM, Lim JY, Jeong CH, et al. Gene therapy of intracranial glioma using interleukin 12-secreting human umbilical cord blood-derived mesenchymal stem cells. *Hum Gene Ther* 2011;**22**(6):733–43.

57. Yin J, Kim JK, Moon JH, Beck S, Piao D, Jin X, et al. hMSC-mediated concurrent delivery of endostatin and carboxylesterase to mouse xenografts suppresses glioma initiation and recurrence. *Mol Ther* 2011;**19**(6):1161–9.

58. Gatson NN, Chiocca EA, Kaur B. Anti-angiogenic gene therapy in the treatment of malignant gliomas. *Neurosci Lett* 2012;**527**(2):62–70.

59. Zhang W, Fulci G, Buhrman JS, Stemmer-Rachamimov AO, Chen JW, Wojtkiewicz GR, et al. Bevacizumab with angiostatin-armed oHSV increases antiangiogenesis and decreases bevacizumab-induced invasion in U87 glioma. *Mol Ther* 2012;**20**(1):37–45.

60. Yoo JY, Haseley A, Bratasz A, Chiocca EA, Zhang J, Powell K, et al. Antitumor efficacy of 34.5ENVE: a transcriptionally retargeted and "Vstat120"-expressing oncolytic virus. *Mol Ther* 2012;**20**(2):287–97.

61. Yuan B, Xian R, Ma J, Chen Y, Lin C, Song Y. Isthmin inhibits glioma growth through antiangiogenesis in vivo. *J Neurooncol* 2012;**109**(2):245–52.

62. Msaouel P, Opyrchal M, Domingo Musibay E, Galanis E. Oncolytic measles virus strains as novel anticancer agents. *Expert Opin Biol Ther* 2013;**13**(4):483–502.

63. Binello E, Germano IM. Stem cells as therapeutic vehicles for the treatment of high-grade gliomas. *Neuro Oncol* 2012;**14**(3):256–65.

64. Markert JM, Liechty PG, Wang W, Gaston S, Braz E, Karrasch M, et al. Phase Ib trial of mutant herpes simplex virus G207 inoculated pre-and post-tumor resection for recurrent GBM. *Mol Ther* 2009;**17**(1):199–207.

65. Chiocca EA, Aguilar LK, Bell SD, Kaur B, Hardcastle J, Cavaliere R, et al. Phase IB study of gene-mediated cytotoxic immunotherapy adjuvant to up-front surgery and intensive timing radiation for malignant glioma. *J Clin Oncol* 2011;**29**(27):3611–9.

66. Westphal M, Yla-Herttuala S, Martin J, Warnke P, Menei P, Eckland D, et al. Adenovirus-mediated gene therapy with sitimagene ceradenovec followed by intravenous ganciclovir for patients with operable high-grade glioma (ASPECT): a randomised, open-label, phase 3 trial. *Lancet Oncol* 2013;**14**(9):823–33.

67. Herrlinger U, Woiciechowski C, Sena-Esteves M, Aboody KS, Jacobs AH, Rainov NG, et al. Neural precursor cells for delivery of replication-conditional HSV-1 vectors to intracerebral gliomas. *Mol Ther* 2000;**1**(4):347–57.

68. Ahmed AU, Thaci B, Tobias AL, Auffinger B, Zhang L, Cheng Y, et al. A preclinical evaluation of neural stem cell-based cell carrier for targeted antiglioma oncolytic virotherapy. *J Natl Cancer Inst* 2013;**105**(13):968–77.

69. Kim CK, Ahmed AU, Auffinger B, Ulasov IV, Tobias AL, Moon KS, et al. *N*-acetylcysteine amide augments the therapeutic effect of neural stem cell-based antiglioma oncolytic virotherapy. *Mol Ther* 2013;**21**(11):2063–73.

70. Morgan RA, Johnson LA, Davis JL, Zheng Z, Woolard KD, Reap EA, et al. Recognition of glioma stem cells by genetically modified T cells targeting EGFRvIII and development of adoptive cell therapy for glioma. *Hum Gene Ther* 2012;**23**(10):1043–53.

71. Chow KK, Naik S, Kakarla S, Brawley VS, Shaffer DR, Yi Z, et al. T cells redirected to EphA2 for the immunotherapy of glioblastoma. *Mol Ther* 2013;**21**(3):629–37.

72. Kong S, Sengupta S, Tyler B, Bais AJ, Ma Q, Doucette S, et al. Suppression of human glioma xenografts with second-generation IL13R-specific chimeric antigen receptor-modified T cells. *Clin Cancer Res* 2012;**18**(21):5949–60.

73. Tzeng SY, Green JJ. Therapeutic nanomedicine for brain cancer. *Ther Deliv* 2013;**4**(6):687–704.

74. Yoshida J, Mizuno M, Fujii M, Kajita Y, Nakahara N, Hatano M, et al. Human gene therapy for malignant gliomas (glioblastoma multiforme and anaplastic astrocytoma) by in vivo transduction with human interferon beta gene using cationic liposomes. *Hum Gene Ther* 2004;**15**(1):77–86.

75. Li J, Gu B, Meng Q, Yan Z, Gao H, Chen X, et al. The use of myristic acid as a ligand of polyethylenimine/DNA nanoparticles for targeted gene therapy of glioblastoma. *Nanotechnology* 2011;**22**(43):435101.

76. Jensen SA, Day ES, Ko CH, Hurley LA, Luciano JP, Kouri FM, et al. Spherical nucleic acid nanoparticle conjugates as an RNAi-based therapy for glioblastoma. *Sci Transl Med* 2013;**5**(209):209ra152.

77. Kwon OJ, Kang E, Choi JW, Kim SW, Yun CO. Therapeutic targeting of chitosan-PEG-folate-complexed oncolytic adenovirus for active and systemic cancer gene therapy. *J Control Release* 2013;**169**(3):257–65.

78. Aghi MK, Liu TC, Rabkin S, Martuza RL. Hypoxia enhances the replication of oncolytic herpes simplex virus. *Mol Ther* 2009;**17**(1):51–6.

79. Sgubin D, Wakimoto H, Kanai R, Rabkin SD, Martuza RL. Oncolytic herpes simplex virus counteracts the hypoxia-induced modulation of glioblastoma stem-like cells. *Stem Cells Transl Med* 2012;**1**(4):322–32.

80. Nandi S, Lesniak MS. Adenoviral virotherapy for malignant brain tumors. *Expert Opin Biol Ther* 2009;**9**(6):737–47.

81. Chiocca EA, Abbed KM, Tatter S, Louis DN, Hochberg FH, Barker F, et al. A phase I open-label, dose-escalation, multi-institutional trial of injection with an E1B-Attenuated adenovirus, ONYX-015, into the peritumoral region of recurrent malignant gliomas, in the adjuvant setting. *Mol Ther* 2004;**10**(5):958–66.

82. Ranki T, Hemminki A. Serotype chimeric human adenoviruses for cancer gene therapy. *Viruses* 2010;**2**(10):2196–212.

83. Brouwer E, Havenga MJ, Ophorst O, de Leeuw B, Gijsbers L, Gillissen G, et al. Human adenovirus type 35 vector for gene therapy of brain cancer: improved transduction and bypass of pre-existing anti-vector immunity in cancer patients. *Cancer Gene Ther* 2007;**14**(2):211–9.

84. Guse K, Cerullo V, Hemminki A. Oncolytic vaccinia virus for the treatment of cancer. *Expert Opin Biol Ther* 2011;**11**(5):595–608.

85. Lun X, Alain T, Zemp FJ, Zhou H, Rahman MM, Hamilton MG, et al. Myxoma virus virotherapy for glioma in immunocompetent animal models: optimizing administration routes and synergy with rapamycin. *Cancer Res* 2010;**70**(2):598–608.

86. Rommelaere J, Geletneky K, Angelova AL, Daeffler L, Dinsart C, Kiprianova I, et al. Oncolytic parvoviruses as cancer therapeutics. *Cytokine Growth Factor Rev* 2010;**21**(2–3):185–95.

87. Lam HY, Yeap SK, Rasoli M, Omar AR, Yusoff K, Suraini AA, et al. Safety and clinical usage of newcastle disease virus in cancer therapy. *J Biomed Biotechnol* 2011;**2011**:718710.

88. Hutzen B, Pierson CR, Russell SJ, Galanis E, Raffel C, Studebaker AW. Treatment of medulloblastoma using an oncolytic measles virus encoding the thyroidal sodium iodide symporter shows enhanced efficacy with radioiodine. *BMC Cancer* 2012;**12**:508.

89. Wollmann G, Tattersall P, van den Pol AN. Targeting human glioblastoma cells: comparison of nine viruses with oncolytic potential. *J Virol* 2005;**79**(10):6005–22.

90. Alloussi SH, Alkassar M, Urbschat S, Graf N, Gartner B. All reovirus subtypes show oncolytic potential in primary cells of human high-grade glioma. *Oncol Rep* 2011;**26**(3):645–9.

91. Duarte S, Carle G, Faneca H, Lima MC, Pierrefite-Carle V. Suicide gene therapy in cancer: where do we stand now? *Cancer Lett* 2012;**324**(2):160–70.

92. Huszthy PC, Giroglou T, Tsinkalovsky O, Euskirchen P, Skaftnesmo KO, Bjerkvig R, et al. Remission of invasive, cancer stem-like glioblastoma xenografts using lentiviral vector-mediated suicide gene therapy. *PLoS One* 2009;**4**(7):e6314.

93. Puntel M, AKM, GM, Farrokhi C, Vanderveen N, Paran C, Appelhans A, et al. Safety profile, efficacy, and biodistribution of a bicistronic high-capacity adenovirus vector encoding a combined immunostimulation and cytotoxic gene therapy as a prelude to a phase I clinical trial for glioblastoma. *Toxicol Appl Pharmacol* 2013;**268**(3):318–30.

94. Liu Y, Lang F, Xie X, Prabhu S, Xu J, Sampath D, et al. Efficacy of adenovirally expressed soluble TRAIL in human glioma organotypic slice culture and glioma xenografts. *Cell Death Dis* 2011;**2**:e121.

95. Hamed HA, Yacoub A, Park MA, Eulitt PJ, Dash R, Sarkar D, et al. Inhibition of multiple protective signaling pathways and Ad.5/3 delivery enhances mda-7/IL-24 therapy of malignant glioma. *Mol Ther* 2010;**18**(6):1130–42.

96. Candolfi M, Xiong W, Yagiz K, Liu C, Muhammad AK, Puntel M, et al. Gene therapy-mediated delivery of targeted cytotoxins for glioma therapeutics. *Proc Natl Acad Sci USA* 2010;**107**(46):20021–6.

97. Kim SU. Neural stem cell-based gene therapy for brain tumors. *Stem Cell Rev* 2011;**7**(1):130–40.

98. Aboody KS, Najbauer J, Metz MZ, D'Apuzzo M, Gutova M, Annala AJ, et al. Neural stem cell-mediated enzyme/prodrug therapy for glioma: preclinical studies. *Sci Transl Med* 2013;**5**(184):184ra59.

99. Lim SH, Choi SA, Lee JY, Wang KC, Phi JH, Lee DH, et al. Therapeutic targeting of subdural medulloblastomas using human neural stem cells expressing carboxylesterase. *Cancer Gene Ther* 2011;**18**(11):817–24.

100. Amano S, Li S, Gu C, Gao Y, Koizumi S, Yamamoto S, et al. Use of genetically engineered bone marrow-derived mesenchymal stem cells for glioma gene therapy. *Int J Oncol* 2009;**35**(6):1265–70.

101. Bak XY, Lam DH, Yang J, Ye K, Wei EL, Lim SK, et al. Human embryonic stem cell-derived mesenchymal stem cells as cellular delivery vehicles for prodrug gene therapy of glioblastoma. *Hum Gene Ther* 2011;**22**(11):1365–77.

102. Altaner C, Altanerova V, Cihova M, Ondicova K, Rychly B, Baciak L, et al. Complete regression of glioblastoma by mesenchymal stem cells mediated prodrug gene therapy simulating clinical therapeutic scenario. *Int J Cancer* 2014;**134**(6):1458–65.

103. Kim SM, Lim JY, Park SI, Jeong CH, Oh JH, Jeong M, et al. Gene therapy using TRAIL-secreting human umbilical cord blood-derived mesenchymal stem cells against intracranial glioma. *Cancer Res* 2008;**68**(23):9614–23.

104. Sasportas LS, Kasmieh R, Wakimoto H, Hingtgen S, van de Water JA, Mohapatra G, et al. Assessment of therapeutic efficacy and fate of engineered human mesenchymal stem cells for cancer therapy. *Proc Natl Acad Sci USA* 2009;**106**(12):4822–7.

105. Choi SA, Hwang SK, Wang KC, Cho BK, Phi JH, Lee JY, et al. Therapeutic efficacy and safety of TRAIL-producing human adipose tissue-derived mesenchymal stem cells against experimental brainstem glioma. *Neuro Oncol* 2011;**13**(1):61–9.

106. Yin D, Zhai Y, Gruber HE, Ibanez CE, Robbins JM, Kells AP, et al. Convection-enhanced delivery improves distribution and efficacy of tumor-selective retroviral replicating vectors in a rodent brain tumor model. *Cancer Gene Ther* 2013;**20**(6):336–41.

107. Yerbes R, Palacios C, Lopez-Rivas A. The therapeutic potential of TRAIL receptor signalling in cancer cells. *Clin Transl Oncol* 2011;**13**(12):839–47.

108. Doucette T, Rao G, Yang Y, Gumin J, Shinojima N, Bekele BN, et al. Mesenchymal stem cells display tumor-specific tropism in an RCAS/Ntv-a glioma model. *Neoplasia* 2011;**13**(8):716–25.

109. Ho IA, Toh HC, Ng WH, Teo YL, Guo CM, Hui KM, et al. Human bone marrow-derived mesenchymal stem cells suppress human glioma growth through inhibition of angiogenesis. *Stem Cells* 2013;**31**(1):146–55.

110. Kim SM, Woo JS, Jeong CH, Ryu CH, Lim JY, Jeun SS. Effective combination therapy for malignant glioma with TRAIL-secreting mesenchymal stem cells and lipoxygenase inhibitor MK886. *Cancer Res* 2012;**72**(18):4807–17.

111. Agarwalla PK, Barnard ZR, Curry Jr WT. Virally mediated immunotherapy for brain tumors. *Neurosurg Clin N Am* 2010;**21**(1):167–79.

112. Mineharu Y, Muhammad AK, Yagiz K, Candolfi M, Kroeger KM, Xiong W, et al. Gene therapy-mediated reprogramming tumor infiltrating T cells using IL-2 and inhibiting NF-κB signaling improves the efficacy of immunotherapy in a brain cancer model. *Neurotherapeutics* 2012;**9**(4):827–43.

113. Breitbach CJ, Thorne SH, Bell JC, Kirn DH. Targeted and armed oncolytic poxviruses for cancer: the lead example of JX-594. *Curr Pharm Biotechnol* 2012;**13**(9):1768–72.

114. Guo D, Wang B, Han F, Lei T. RNA interference therapy for glioblastoma. *Expert Opin Biol Ther* 2010;**10**(6):927–36.

115. Wei J, Wang F, Kong LY, Xu S, Doucette T, Ferguson SD, et al. miR-124 inhibits STAT3 signaling to enhance T cell-mediated immune clearance of glioma. *Cancer Res* 2013;**73**(13):3913–26.

116. Rolle K, Nowak S, Wyszko E, Nowak M, Zukiel R, Piestrzeniewicz R, et al. Promising human brain tumors therapy with interference RNA intervention (iRNAi). *Cancer Biol Ther* 2010;**9**(5):396–406.

Index

Note: Page numbers followed by "f" denotes figures; "t", tables; "b",boxes.

Printed in the United States
By Bookmasters